Lecture Notes in Artificial Intelligence 1932

Subseries of Lecture Notes in Computer Science
Edited by J. G. Carbonell and J. Siekmann

Lecture Notes in Computer Science
Edited by G. Goos, J. Hartmanis and J. van Leeuwen

Springer
*Berlin
Heidelberg
New York
Barcelona
Hong Kong
London
Milan
Paris
Singapore
Tokyo*

Zbigniew W. Raś Setsuo Ohsuga (Eds.)

Foundations of Intelligent Systems

12th International Symposium, ISMIS 2000
Charlotte, NC, USA, October 11-14, 2000
Proceedings

 Springer

Series Editors

Jaime G. Carbonell, Carnegie Mellon University, Pittsburgh, PA, USA
Jörg Siekmann, University of Saarland, Saarbrücken, Germany

Volume Editors

Zbigniew W. Raś
University of North Carolina, Department of Computer Science
Charlotte, NC 28223, USA
E-mail: ras@uncc.edu
and Polish Academy of Sciences, Institute of Computer Science
01-237 Warsaw, Poland

Setsuo Ohsuga
Waseda University, Department of Science and Technology
61-414, 3-4-1, Ohkubo, Shinjuku-ku, 169-8555 Tokyo, Japan
E-mail: ohsuga@fd.catv.ne.jp or ohsuga@ohsuga.info.waseda.ac.jp

Cataloging-in-Publication Data applied for

Die Deutsche Bibliothek - CIP-Einheitsaufnahme

Foundations of intelligent systems : 12th international symposium ;
proceedings / ISMIS 2000, Charlotte, NC, USA, October 11 - 14, 2000.
Zbigniew W. Ra´s ; Setsuo Ohsuga (ed.). - Berlin ; Heidelberg ; New
York ; Barcelona ; Hong Kong ; London ; Milan ; Paris ; Singapore ;
Tokyo : Springer, 2000
  (Lecture notes in computer science ; Vol. 1932 : Lecture notes in
  artificial intelligence)
  ISBN 3-540-41094-5

CR Subject Classification (1998): I.2, H.3, H.2

ISBN 3-540-41094-5 Springer-Verlag Berlin Heidelberg New York

This work is subject to copyright. All rights are reserved, whether the whole or part of the material is concerned, specifically the rights of translation, reprinting, re-use of illustrations, recitation, broadcasting, reproduction on microfilms or in any other way, and storage in data banks. Duplication of this publication or parts thereof is permitted only under the provisions of the German Copyright Law of September 9, 1965, in its current version, and permission for use must always be obtained from Springer-Verlag. Violations are liable for prosecution under the German Copyright Law.

Springer-Verlag Berlin Heidelberg New York
a member of BertelsmannSpringer Science+Business Media GmbH
© Springer-Verlag Berlin Heidelberg 2000
Printed in Germany

Typesetting: Camera-ready by author, data conversion by PTP-Berlin, Stefan Sossna
Printed on acid-free paper   SPIN: 10781187   06/3142   5 4 3 2 1 0

Preface

This volume contains the papers selected for presentation at the Twelfth International Symposium on Methodologies for Intelligent Systems - ISMIS 2000, held in Charlotte, N.C., 11–14 October, 2000. The symposium was co-organized by the College of Information Technology at UNC-Charlotte and the Polish-Japanese Institute of Information Technology. It was sponsored by the US Army Research Office, NCR Data Mining Laboratory, College of IT at UNC-Charlotte, and others.

ISMIS is a conference series that was started in 1986 in Knoxville, Tennessee. Since then it has been held in Charlotte (North Carolina), Knoxville (Tennessee), Torino (Italy), Trondheim (Norway), Warsaw (Poland), and Zakopane (Poland).

The program committee selected the following major areas for ISMIS 2000: Evolutionary Computation, Intelligent Information Retrieval, Intelligent Information Systems, Knowledge Representation and Integration, Knowledge Discovery and Learning, Logic for Artificial Intelligence, and Methodologies.

The contributed papers were selected from 112 full draft papers by the following program committee: A. Biermann, P. Bosc, J. Calmet, S. Carberry, N. Cercone, J. Chen, W. Chu, B. Croft, J. Debenham, S.M. Deen, K. DeJong, R. Demolombe, B. Desai, T. Elomaa, F. Esposito, A. Giordana, J. Grzymala-Busse, M. Hadzikadic, H. Hamilton, D. Hislop, K. Hori, W. Kloesgen, Y. Kodratoff, J. Komorowski, J. Koronacki, W. Kosinski, R. Kostoff, B.G.T. Lowden, D. Maluf, D. Malerba, R.L. de Mantaras, S. Matwin, R. Meersman, Z. Michalewicz, R. Michalski, R. Mizoguchi, M. Mukaidono, L. De Raedt, V. Raghavan, E. Rosenthal, L. Saitta, A. Skowron, V.S. Subrahmanian, S. Tsumoto, T. Yamaguchi, G.P. Zarri, M. Zemankova, N. Zhong, J.M. Zytkow. Additionally, we acknowledge the help in reviewing papers from: E. El-Kwae, S. Ferilli, M. Kryszkiewicz, and A. Wieczorkowska.

We wish to express our thanks to Jaime Carbonell, Bruce Croft, Philip Emmerman, Bill Harris, and Ryszard Michalski who presented invited talks at the symposium. Also, we are thankful to Zbigniew Michalewicz for organizing the Special Session on Evolutionary Computation. We express our appreciation to the sponsors of the symposium and to all who submitted papers for presentation and publication in the proceedings. Our sincere thanks go to the Organizing Committee of ISMIS 2000. Also, our thanks are due to Alfred Hofmann of Springer-Verlag for his continuous help and support.

August 2000   Zbigniew W. Raś
                                                                                                 Setsuo Ohsuga

Table of Contents

Invited Papers

Information Retrieval Based on Statistical Language Models............. 1
   W. Bruce Croft

Intelligent Agent Battlespace Augmentation 12
   Philip J. Emmerman and Uma Y. Movva

Learning and Evolution: An Introduction to Non-darwinian Evolutionary
Computation 21
   Ryszard S. Michalski

Regular Papers

1A Knowledge Discovery and Learning

Can Relational Learning Scale Up?................................... 31
   Attilio Giordana, Lorenza Saitta, Michele Sebag, and Marco Botta

Discovering Geographic Knowledge: The INGENS System................ 40
   Donato Malerba, Floriana Esposito, Antonietta Lanza, and
   Francesca A. Lisi

Temporal Data Mining Using Hidden Periodicity Analysis................ 49
   Weiqiang Lin and Mehmet A. Orgun

Mining N-most Interesting Itemsets 59
   Ada W.-c. Fu, Renfrew W.-w. Kwong, and Jian Tang

1B Intelligent Information Retrieval

Repository Management in an Intelligent Indexing Approach for
Multimedia Digital Libraries 68
   B. Armani, E. Bertino, B. Catania, D. Laradi, B. Marin,
   and G.P. Zarri

Logic-Based Approach to Semistructured Data Retrieval 77
   Mohand-Saïd Hacid and Farouk Toumani

High Quality Information Retrieval for Improving the Conduct and
Management of Research and Development........................... 86
   Ronald N. Kostoff

Signature-Based Indexing for Retrieval by Spatial Content in Large
2D-String Image Databases 97
   Essam A. El-Kwae

2A Knowledge Discovery and Learning

Refining Logic Theories under OI-Implication......................... 109
   Floriana Esposito, N. Fanizzi, S. Ferilli, and G. Semeraro

Rule Quality Measures Improve the Accuracy of Rule Induction:
An Experimental Approach 119
   Aijun An and Nick Cercone

A Dynamic Approach for Knowledge Discovery of Web Access Patterns... 130
   Alaaeldin Hafez

Data Reduction via Conflicting Data Analysis 139
   M. Boussouf and M. Quafafou

A Comparison of Rule Matching Methods Used in AQ15 and LERS 148
   Jerzy W. Grzymala-Busse and Pankaj Shah

2B Evolutionary Computation

Evolving Behaviors for Cooperating Agents 157
   Jeffrey K. Bassett and Kenneth A. De Jong

Evolving Finite-State Machine Strategies for Protecting Resources 166
   William M. Spears and Diana F. Gordon

A Method of Generating Program Specification from Description of Human
Activities 176
   Shuhei Kawasaki and Setsuo Ohsuga

PLAtestGA: A CNF-Satisfiability Problem for the Generation of Test
Vectors for Missing Faults in VLSI Circuits 186
   Alfredo Cruz

Evaluating Migration Strategies for an Evolutionary Algorithm Based on
the Constraint-Graph that Solves CSP................................ 196
   Arturo Núñez and María-Cristina Riff

3A Methodologies

Relative Robustness: An Empirical Investigation of Behaviour Based and
Plan Based Paradigms as Environmental Conditions Change 205
   Jennifer Kashmirian and Lin Padgham

A Heuristic for Domain Independent Planning and Its Use in an Enforced
Hill-Climbing Algorithm 216
　　Jörg Hoffmann

Planning while Executing: A Constraint-Based Approach 228
　　R. Barruffi, M. Milano, and P. Torroni

3B Intelligent Information Systems

Problem Decomposition and Multi-agent System Creation for Distributed
Problem Solving 237
　　Katsuaki Tanaka, Michiko Higashiyama, and Setsuo Ohsuga

A Comparative Study of Noncontextual and Contextual Dependencies 247
　　S.K.M. Wong and C.J. Butz

Extended Query Answering Using Integrity Rules 256
　　Barry G.T. Lowden and Jerome Robinson

4A Knowledge Discovery and Learning

Finding Temporal Relations: Causal Bayesian Networks vs. C4.5 266
　　Kamran Karimi and Howard J. Hamilton

Learning Relational Clichés with Contextual LGG 274
　　Johanne Morin and Stan Matwin

Design of Rough Neurons: Rough Set Foundation and Petri Net Model ... 283
　　J.F. Peters, A. Skowron, Z. Suraj, L. Han, and S. Ramanna

Towards Musical Data Classification via Wavelet Analysis 292
　　Alicja Wieczorkowska

4B Logic for AI

Annotated Hyperresolution for Non-horn Regular Multiple-Valued Logics . 301
　　James J. Lu, Neil V. Murray, and Erik Rosenthal

Fundamental Properties on Axioms of Kleene Algebra 311
　　Tomoko Ninomiya and Masao Mukaidono

Extending Entity-Relationship Models with Higher-Order Operators 321
　　Antonio Badia

Combining Description Logics with Stratified Logic Programs in
Knowledge Representation 331
　　Jianhua Chen

5A Knowledge Discovery and Learning

Emergence Measurement and Analyzes of Conceptual Abstractions During
Evolution Simulation in OOD 340
    Mourad Oussalah and Dalila Tamzalit

Using Intelligent Systems in Predictions of the Bacterial Causative Agent
of an Infection 349
    *Diana R. Cundell, Randy S. Silibovsky, Robyn Sanders, and
Les M. Sztandera*

An Intelligent Lessons Learned Process 358
    *Rosina Weber, David W. Aha, Hector Muñoz-Ávila, and
Leonard A. Breslow*

5B Intelligent Information Systems

What the Logs Can Tell You: Mediation to Implement Feedback in
Training 368
    David A. Maluf and Gio Wiederhold

Top-Down Query Processing in First Order Deductive Databases under
the DWFS 377
    C.A. Johnson

Discovering and Resolving User Intent in Heterogeneous Databases 389
    Chris Fernandes and Lawrence Henschen

6A Learning and Knowledge Discovery

Discovering and Matching Elastic Rules from Sequence Databases 400
    Sanghyun Park and Wesley W. Chu

Perception-Based Granularity Levels in Concept Representation 409
    Lorenza Saitta and Jean-Daniel Zucker

Local Feature Selection with Dynamic Integration of Classifiers 417
    Alexey Tsymbal and Seppo Puuronen

Prediction of Ordinal Classes Using Regression Trees 426
    *Stefan Kramer, Gerhard Widmer, Bernhard Pfahringer, and
Michael de Groeve*

6B Intelligent Information Retrieval

Optimal Queries in Information Filtering 435
    Ali H. Alsaffar, Jitender S. Deogun, and Hayri Sever

Automatic Semantic Header Generator 444
    Bipin C. Desai, Sami S. Haddad, and Abdelbaset Ali

On Modeling of Concept Based Retrieval in Generalized Vector Spaces ... 453
  *Minkoo Kim, Ali H. Alsaffar, Jitender S. Deogun, and
  Vijay V. Raghavan*

Template Generation for Identifying Text Patterns 463
  Cécile Boisson and Nahid Shahmehri

7A Knowledge Discovery and Learning

Qualitative Discovery in Medical Databases 474
  David A. Maluf and Jiming Liu

Finding Association Rules Using Fast Bit Computation: Machine-Oriented
Modeling 486
  Eric Louie and Tsau Y. Lin

Using Closed Itemsets for Discovering Representative Association Rules ... 495
  Jamil Saquer and Jitender S. Deogun

Legitimate Approach to Association Rules under Incompleteness 505
  Marzena Kryszkiewicz and Henryk Rybinski

7B Logic for AI

A Simple and Tractable Extension of Situation Calculus to Epistemic
Logic 515
  Robert Demolombe and Maria del Pilar Pozos Parra

Rule Based Abduction 525
  Sai K. Lakkaraju and Yan Zhang

An Efficient Proof Method for Non-clausal Reasoning 534
  E. Altamirano and G. Escalada-Imaz

An Intelligent System Dealing with Complex Nuanced Information within
a Statistical Context 543
  D. Pacholczyk and F. Dupin de Saint Cyr

8A Learning and Knowledge Discovery

On the Complexity of Optimal Multisplitting 552
  Tapio Elomaa and Juho Rousu

Parametric Algorithms for Mining Share-Frequent Itemsets 562
  Brock Barber and Howard J. Hamilton

Discovery of Clinical Knowledge in Hospital Information Systems:
Two Case Studies 573
  Shusako Tsumoto

Foundations and Discovery of Operational Definitions 582
  Jan M. Żytkow and Zbigniew W. Raś

A Multi-agent Based Architecture for Distributed KDD Process 591
  Chunnian Liu, Ning Zhong, and Setsuo Ohsuga

8B Knowledge Representation

Towards a Software Architecture for Case-Based Reasoning Systems 601
  Enric Plaza and Josep-Lluís Arcos

Knowledge Representation in Planning: A PDDL to OCL_h Translation ... 610
  R.M. Simpson, T.L. McCluskey, D. Liu, and D.E. Kitchin

A Method and Language for Constructing Multiagent Systems........... 619
  Hiroyuki Yamauchi and Setsuo Ohsuga

A Formalism for Building Causal Polytree Structures Using Data
Distributions 629
  M. Ouerd, B.J. Oommen, and Stan Matwin

Abstraction in Cartographic Generalization 638
  Sébastien Mustière, Lorenza Saitta, and Jean-Daniel Zucker

Author Index **645**

Information Retrieval Based on Statistical Language Models

W. Bruce Croft

Computer Science Department
University of Massachusetts, Amherst, MA 01003-4610
croft@cs.umass.edu
http://ciir.cs.umass.edu

Abstract. The amount of on-line information is growing exponentially. Much of this information is unstructured and language-based. To deal with this flood of information, a number of tools and language technologies have been developed. Progress has been made in areas such as information retrieval, information extraction, filtering, speech recognition, machine translation, and data mining. Other more specific areas such as cross-lingual retrieval, summarization, categorization, distributed retrieval, and topic detection and tracking are also contributing to the proliferation of technologies for managing information. Currently these tools are based on many different approaches, both formal and ad hoc. Integrating them is very difficult, yet this will be a critical part of building effective information systems in the future. In this paper, we discuss an approach to providing a framework for integration based on language models.

1 Introduction

Tools for managing language-based information have become essential components of modern information systems. This type of information, in the form of unstructured or semi-structured text (e.g. HTML or XML), is found throughout the applications that are driving our economy. In addition, the increase in the use of speech input and data, OCR, and metadata descriptions of images and video, has resulted in text becoming a *lingua franca* for information systems. Although considerable progress has been made with language-based tools such as information retrieval, filtering, categorization, extraction, summarization, and mining, their performance is unreliable and the effects of integrating them are unpredictable. One of the major reasons for this is the lack of a unifying formal framework for developing and combining language-based technologies. Instead, the tools are based on many different approaches and theories, often implicit and sometimes ad hoc. If a single unifying framework and architecture for information management could be created, it would enable the development of significantly more effective tools, support integration, and substantially advance our understanding of the processes underlying information access and organization. A growing number of researchers believe that such a framework can be based on statistical language models.

The language modeling approach has been applied, with considerable success, to speech recognition and machine translation [10, 4, 3]. More recently, there have been breakthroughs in applying this approach to information retrieval and extraction [16, 15, 1, 12, 22, 2].

The use of language models is attractive for several reasons. Building an information system using language models allows us to reason about the design and empirical performance of the system in a principled way, using the tools of probability theory. In addition, we can leverage the work that has been carried out in the speech recognition community in the past thirty years on such problems as smoothing and combining language models for multiple topics and collections. The language modeling approach applies naturally to a wide range of information system technologies, such as distributed retrieval, cross-language IR, summarization and filtering.

Much remains to be done to establish language modeling as a unifying framework. We need to show how language models can represent documents, topics, databases, languages, queries, and even people. We need to develop efficient algorithms for acquiring, comparing, and summarizing language models of different types and granularities. We need to show how statistical language models can describe the crucial functions in a language-based information system, such as information retrieval, filtering, and summarization. Finally, we need to demonstrate that the performance of the language-based functions improves as a result of using a language modeling framework. Figure 1 gives an overview of the representation and function aspects of the language model framework.

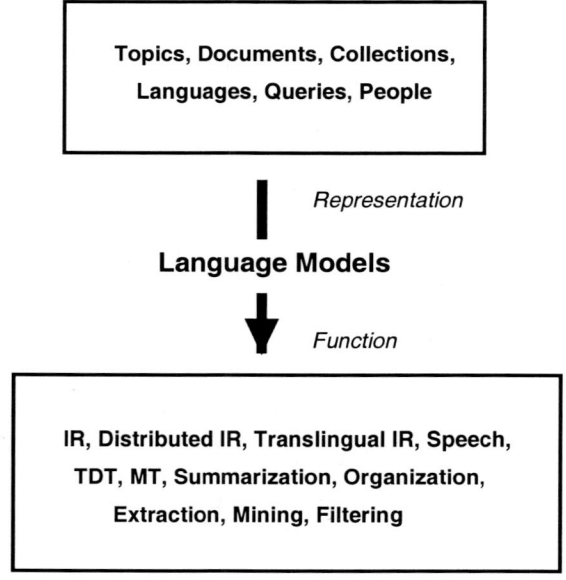

Fig. 1. Overview of Language Model Framework

A project addressing these issues has begun as a collaboration between researchers at the University of Massachusetts and Carnegie Mellon University.

In this paper, I will focus on a more detailed description of the issues involved in applying language models to information retrieval. The next section describes how language models can transform the view of document representations and indexing models. Section 3 discusses how language modeling approaches to IR approach relevance. In other words, how is the concept of relevance incorporated in the overall retrieval model. Section 4 shows how combination of evidence or data fusion can be implemented using language models.

2 Language Models and Indexing

Over the past three decades, probabilistic models of document retrieval have been studied extensively. In general, these approaches can be characterized as methods of estimating the probability of relevance of documents to user queries. One component of a probabilistic retrieval model is the indexing model, i.e., a model of the assignment of indexing terms to documents.

A well-known example of an indexing model is the 2-Poisson model [8]. The success of the 2-Poisson model has been somewhat limited but it should be noted that Robertson's *tf* weight, which has been quite successful, was intended to behave similarly to the 2-Poisson model [18]. Other probabilistic indexing models have also been proposed (e.g. [7]).

Estimating the probability that an index term is "correct" for that document is difficult. As a result, heuristic *tf.idf* weights are used in the retrieval algorithms based on these models. In order to avoid these weights and the awkwardness of modeling the correctness of indexing, Ponte and Croft [16] proposed a *language modeling* approach to information retrieval. The phrase "language model" is used by the speech recognition community to refer to a probability distribution that captures the statistical regularities of the generation of language [10]. Generally speaking, language models for speech attempt to predict the probability of the next word in an ordered sequence. For the purposes of document retrieval, Ponte and Croft modeled occurrences at the document level without regard to sequential effects, although they showed that it is possible to model local predictive effects for features such as phrases. Mittendorf and Schauble [13] used a similar approach to construct a generative model for retrieval based on document passages.

The approach to retrieval described in Ponte and Croft [16] is to infer a language model for each document and to estimate the probability of generating the query according to each of these models. Documents are then ranked according to these probabilities. In this approach, collection statistics such as term frequency, document length and document frequency are integral parts of the language model and do not have to be included in an ad hoc manner. The score for a document in the simple unigram model used in Ponte and Croft is given by:

$$P(Q|D) = \prod_{w \in Q} P(w|D) \prod_{w \notin Q} (1 - P(w|D))$$

where $P(Q|D)$ is the estimate of the probability that a query can be generated for a particular document, and $P(w|D)$ is the probability of generating a word given a particular document (the language model).

Much of the power of this simple model comes from the estimation techniques used for these probabilities, which combine both maximum likelihood estimates and background models. This part of the model benefits directly from the extensive research done on estimation of language models in fields such as speech recognition and machine translation. More sophisticated models that make use of bigram and even trigram probabilities are currently being investigated [12, 19].

The idea of a language model representing the text written in specific documents leads directly to the possibility of using language models to represent topics in domains and users' views of domains. Establishing a context for the query is a crucial part of achieving effective retrieval. The query "star wars" can be interpreted very differently in the context of missile defense systems rather than Hollywood films. Many approaches have been tried to identify and use context, mostly in the form of query expansion techniques. For example, the Local Context Analysis technique [23] identifies words and phrases associated with the query context by analyzing retrieved documents. This technique, although one of the most successful in terms of improving retrieval effectiveness, is ad-hoc and cannot distinguish multiple contexts for a given query. The language model approach appears to provide a more principled way of describing and using context that will lead to substantially more effective retrieval.

Language models for important topics could be based on groups of similar documents. We call these topic models to distinguish them from models based on individual documents. To generate topic models for a set of documents, the documents would first need to be clustered or grouped, and then a model could be estimated for each group. Note that this represents a different form of the clustering hypothesis [17], which states that closely associated documents tend to be relevant to the same requests. Instead, we are assuming that closely associated documents will have the same underlying language model. Xu and Croft [22] used this technique to represent databases using multiple language models for distributed search.

A variation of this approach would be to cluster document passages and allow multiple topic models to be associated with a given document. In a very similar approach, Hoffman [9] describes how mixture models based on *latent classes* can represent documents and queries. The latent classes are generated using clustering based on the EM (Expectation-Maximization) algorithm [11], and Hoffman shows how this approach is related to Latent Semantic Indexing [6]. The use of mixture models to represent queries and documents makes it clear that many of the previous uses of expansion and clustering techniques in IR can be described as smoothing techniques in the language model framework.

3 Language Models and Relevance

The Ponte and Croft model uses a relatively simple definition of relevance that is based on the probability of generating a query text. This definition does not easily describe some of the more complex phenomena involved with information retrieval. The language model approach can, however, be extended to incorporate more general notions of relevance. For example, Berger and Lafferty [1] show how a language modeling approach based on machine translation provides a basis for handling synonymy and polysemy. Related tasks, such as question answering and summarization, also provide a challenge for a model of relevance.

Early probabilistic models of retrieval, such as the binary independence model, viewed retrieval as a classification problem [17]. Documents were treated as belonging to either the relevant (R) or non-relevant classes for a particular query. In this model, documents are ranked according to the probability $P(R|D)$, or the probability that a particular document D belongs to the relevant class. Fuhr [7] extended this model to non-binary document representations and made the conditioning on the query explicit. In his model, documents are ranked by $P(R|D,Q)$. Turtle and Croft [20] proposed a Bayesian net model that calculates $P(I|D)$, which is the probability that a particular information need I is satisfied for a given document. This is a different way of describing relevance, but is otherwise quite similar to the previous models. In this model (described in more detail in the next section), the query is represented as intermediate propositions that describe the information need.

Ponte [15] views a query as the user's description of an ideal or relevant document. More specifically, the description is treated as a text sample. The task of the retrieval system, in his view, is to rank documents by $P(Q|M_D)$, which is the probability that the query text can be generated by the language model M associated with a given document D. This view of retrieval, however, does not easily describe the question-answering task and is regarded by some researchers to be an inadequate model of relevance.

Miller et al [12] described a simple probabilistic model for ranking documents by $P(D\ is\ R|Q)$, which is described as the probability that D is relevant given the query Q. Using Bayes Rule, this can be transformed into

$$\frac{P(Q|D\ is\ R)P(D\ is\ R)}{P(Q)}.$$

This model is somewhat awkward, and although the term $P(Q|D\ is\ R)$ is treated in Miller et al [12] as being the same as the Ponte probability, it is not because of the constraint that the document is relevant. In the absence of relevance information, it is difficult to apply this model.

Berger and Lafferty's model [1] has similarities to the Ponte model in that they view the user generating a query as a sample of an ideal document. The task of the system is then to find the *a posteriori* most likely documents given the query and the specific user U. In other words, documents are ranked by

$$P(D|Q,U) = \frac{P(Q|D,U)P(D|U)}{P(Q|U)}.$$

The denominator $P(Q|U)$ is fixed for a given query and user. The term $P(D|U)$ is a "document prior" that can be used, for example, to discount short documents. If we assume a uniform prior, this is the same as the Ponte model.

Another formulation of the retrieval process, which we are currently investigating, views the query as a sample or a description of an underlying language model M_Q. This language model describes the information need. In other words, instead of the user having a "perfect document" in mind, this approach assumes that the user has some idea of the characteristics of good documents and can describe these characteristics in terms of relative frequencies, co-occurrences, and other phenomena that can be captured in a language model. The task of the retrieval system is viewed as first to estimate M_Q and then use this model to retrieve documents (or answers).

In this approach, we estimate M_Q using a mixture of document models. A "pooled" mixture could be found using the document models that optimize

$$\arg\max_{\{T_1...T_k\}} P(Q|M_{\{T_1...T_k\}}).$$

This part of the retrieval process is very similar to the Ponte model, but here the document models are being used to smooth the model of the information need. Then, for each document in the collection, we compute the posterior likelihood that the smoothed model M_Q is the source from which D was generated: $P(M_Q|D)$. Applying Bayes Rule, we rank documents by the equivalent $\log P(D|M_Q)/P(D)$ (the prior $P(M_Q)$ does not affect the ranking). This model is also similar to the Berger and Lafferty approach, but there are important differences. In particular, the process of forming the language model of the information need allows the query to be something other than a text sample. Queries formulated using query languages (such as Boolean operators) or as questions can be accommodated. More development of this model of relevance and experimental validation remain to be done.

4 Language Models and Combination of Evidence

Combining multiple sources of evidence about relevance has been shown many times to be an effective approach to IR.

The inference network framework, developed by Turtle and Croft [20] and implemented as the INQUERY system [5], was explicitly designed for combining multiple representations and retrieval algorithms into an overall estimate of the probability of relevance. This framework uses a Bayesian network [14] to represent the propositions and dependencies in the probabilistic model (Figure 2). The network is divided into two parts: the document network and the query network. The nodes in the document network represent propositions about the observation of documents (D nodes), the contents of documents (T nodes), and representations of the contents (K nodes). Nodes in the query network represent propositions about the representations of queries (K nodes and Q nodes) and satisfaction of the information need (I node). This network model corresponds closely to a framework for combining classifiers as indicated by the labels on the boxes in Figure 2.

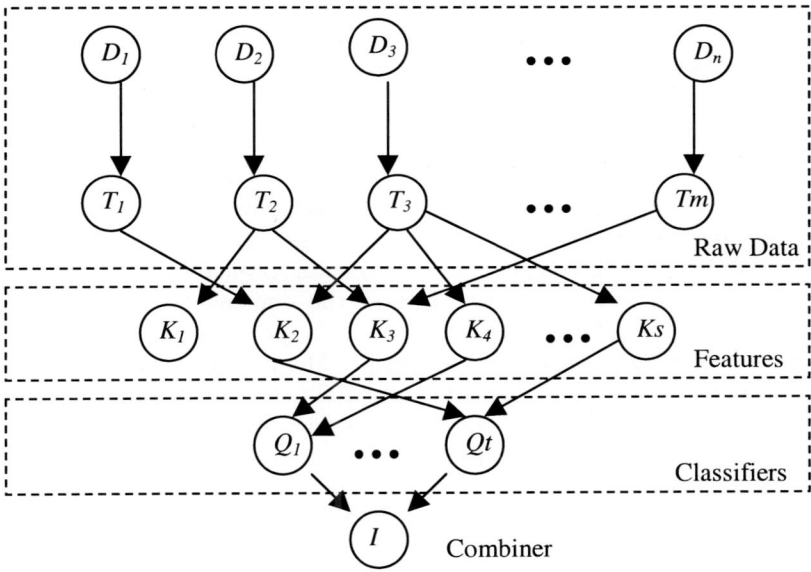

Fig. 2. Bayesian net model of information retrieval

In this model, all nodes represent propositions that are binary variables with the values *true* or *false*, and the probability of these states for a node is determined by the states of the parent nodes. For node A, the probability that A is *true* is given by:

$$p(A) = \sum_{S \subseteq \{1,\ldots,n\}} \alpha_S \prod_{i \in S} p_i \prod_{i \notin S} (1 - p_i)$$

where α_S is a coefficient associated with a particular subset S of the n parent nodes having the state *true*, and p_i is the probability of parent i having the state *true*. Some coefficient settings result in very simple but effective combinations of the evidence from parent nodes. For example, if $\alpha_S = 0$ unless all parents have the state *true*, this corresponds to a Boolean *and*. In this case, $p(A) = \prod_{i=1}^{n} p_i$.

The most commonly used combination formulas in this framework are the average and the weighted average of the parent probabilities. These formulas are the same as those shown in other research to be the best combination strategies for classifiers and discussed earlier in the paper. The combination formula based on the average of the parent probabilities comes from a coefficient setting where the probability of A being *true* depends only on the number of parent nodes having the state *true*. The weighted average comes from a setting where the probability of A depends on the specific parents that are *true*. Parents with higher weight have more influence on the state of A. The INQUERY search system provides a number of these "canonical" combination formulas as query operators. The three described above are #*and*, #*sum*, and #*wsum*.

In the INQUERY system, different document representations are combined by constructing nodes corresponding to propositions about each representation (i.e. is this document represented by a particular term from a representation vocabulary) and constructing queries using those representation nodes. The queries for each representation are combined using operators such as #*wsum*.

In the inference net model, the probabilities associated with the query node propositions are computed from the probabilities associated with representation nodes. The probabilities associated with representation nodes, however, can be computed from evidence in the raw data of the documents. For example, a *tf.idf* formula is used in INQUERY to compute the probability of a word-based representation node for a particular document. The use of heuristic estimation formulas, the lack of knowledge of prior probabilities, and the lack of training data means that the outputs of the inference net (document scores) do not correspond closely to real probabilities.

The language model framework described previously can readily incorporate new representations, can produce accurate probability estimates, and can be incorporated into the general Bayesian net framework. Miller et al [12], point out that estimating the probability of query generation involves a mixture model that combines a variety of word generation mechanisms. They describe this combination using a Hidden Markov Model with states that represent a unigram language model ($P(w|D)$), a bigram language model ($P(w_n|w_{n-1}, D)$), and a model of general English ($P(w|English)$), and mentions other generation processes such as a synonym model and a topic model. Hoffman [9], and Berger and Lafferty [1] also describe the generation process using mixture models, but with different approaches to representation. Put simply, incorporating a new representation into the language model approach to retrieval involves estimating the language model (probability distribution) for the features of that representation and incorporating that new model into the overall mixture model. The standard technique for calculating the parameters of the mixture model is the EM algorithm. This algorithm can be applied to training data that is pooled across queries and this, together with techniques for smoothing the maximum likelihood estimates, results in more accurate probability estimates than a system using *tf.idf* weights without training, such as INQUERY.

There is a strong relationship between the Ponte approach to language modeling for IR and the inference net. Figure 3 shows the unigram language model approach represented using a simplified part of the network from Figure 2. The W nodes that represent the generation of words by the document language model replace the K nodes representing index terms describing the content of a document. The Q node represents the satisfaction of a particular query. In other words, the inference net computes the value of $P(Q \text{ is true})$. In the Ponte and Croft model, the query is simply a list of words. In that model, Q is *true* when the parent nodes representing words present in the query are *true* and the words not in the query are *false*. The document language model gives the probabilities of the *true* and *false* states for the W nodes.

As we mentioned in the last section, however, we can regard the query as having an underlying language model, similar to documents. This language model is associated with the information need of the searcher and can be described by $P(W_1, \ldots, W_n | Q)$. This probability is directly related (by Bayes rule) to the probability $P(Q | W_1, \ldots, W_n)$ that is computed by the inference network.

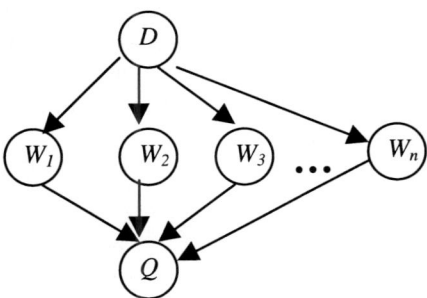

Fig. 3. The language model approach represented in a Bayesian net

The inference network, therefore, provides a mechanism for comparing the document language model to the initial specification of the searcher's language model, which is the first part of the retrieval process described in the last section.

5 Conclusion

Research on language model-based information retrieval systems is beginning to bear fruit. Experiments indicate that these systems will be more flexible and effective than systems based on ad-hoc approaches or other probabilistic models. There have also been some early promising results in related areas such as summarization [21]. This, combined with the established track record of the language model approach for tasks such as speech recognition, provides substantial encouragement for further study of a language model framework for integrating language technologies. The increasing interest in applications such as question answering, cross-language retrieval, and information mining gives additional impetus to the development of this framework.

Acknowledgments

The recent approach to modeling relevance was developed with Victor Lavrenko. This material is based on work supported in part by the National Science Foundation, Library of Congress and Department of Commerce under cooperative agreement number EEC-9209623, and SPAWARSYSCEN-SD under grant number N66001-99-1-8912. Any opinions, findings and conclusions or recommendations expressed in this material are the authors' and do not necessarily reflect those of the sponsor.

References

1. A. Berger and J. Lafferty, "Information retrieval as statistical translation," Proceedings of the 22nd International Conference on Research and Development in Information Retrieval (SIGIR'99), pp. 222–229, 1999.
2. D. Bikel, R. Schwartz, and R. Weischedel, "An Algorithm that Learns What's in a Name," *Machine Learning* 34(1-3), 1999.
3. P. Brown, V. Della Pietra, S. Della Pietra, and R. Mercer. "The mathematics of statistical machine translation," *Computational Linguistics*, 19(2), 1993.
4. P. F. Brown, J. Cocke, S. A. Della Pietra, V. J. Della Pietra, F. Jelinek, J. D. Lafferty, R. L. Mercer, and P. S. Roossin, "A statistical approach to machine translation," Computational Linguistics, vol. 16, no. 2, pp. 79-85, Jun. 1990.
5. J. Callan, W.B. Croft, and J. Broglio, "TREC and TIPSTER experiments with INQUERY," *Information Processing and Management*, 31(3), pp. 327–343, 1995.
6. S. Deerwester, S. Dumais, G. Furnas, T. Landauer, and R. Harshman, "Indexing by latent semantic analysis," *Journal of the American Society for Information Science*, 41, pp. 391–407, 1990.
7. N. Fuhr, "Probabilistic models in information retrieval," *Computer Journal*, 35, pp. 243–255, 1992.
8. S.P. Harter, "A probabilistic approach to automatic keyword indexing," *Journal of the American Society for Information Science*, 26, pp. 197-206, 1975.
9. T. Hofmann, "Probabilistic latent semantic indexing," In *Proceedings of the 22nd ACM SIGIR Conference on Research and Development in Information Retrieval*, pp. 50–57, 1999.
10. F. Jelinek, *Statistical Methods for Speech Recognition*, MIT Press, 1997.
11. C. Manning and H. Schutze, *Foundations of statistical natural language processing*, MIT Press, Cambridge, 1999.

12. D. Miller, T. Leek, and R. Schwartz, "A Hidden Markov Model information retrieval system," In *Proceedings of the 22nd ACM SIGIR Conference on Research and Development in Information Retrieval*, pp. 214–221, 1999.
13. E. Mittendorf and P. Schauble, "Document and passage retrieval based on Hidden Markov Models," In *Proceedings of the 17th ACM SIGIR Conference on Research and Development in Information Retrieval*, pp. 318–327, 1994.
14. J. Pearl, *Probabilistic reasoning in intelligent systems: Networks of plausible inference*, Morgan Kaufmann, San Mateo, 1988.
15. J. Ponte, *A Language Modeling Approach to Information Retrieval*. Ph.D. thesis, University of Massachusetts at Amherst, 1998.
16. J. Ponte and W. B. Croft, "A language modeling approach to information retrieval," Proceedings of the 21st International Conference on Research and Development in Information Retrieval (SIGIR'98), pp. 275–281, 1998.
17. C.J. Van Rijsbergen, *Information Retrieval*. Butterworths, London, 1979.
18. S. Robertson, S. Walker, M. Hancock-Beaulieu, A. Gull, and M. Lau (1992). "Okapi at TREC," In *Proceedings of the Text REtrieval Conference (TREC-1)*, Gaithersburg, Maryland.
19. F. Song, and W.B. Croft, "A general language model for information retrieval," in Proceedings of the Conference on Information and Knowledge Management (CIKM), pp. 316–321, 1999.
20. H. Turtle and W.B. Croft, "Evaluation of an inference network-based retrieval model," *ACM Transactions on Information Systems*, 9(3), pp. 187–222, 1991.
21. M. Witbrock and V. Mittal, "Ultra-summarization: A statistical approach to generating highly condensed non-extractive summaries," Proceedings of the 22nd International Conference on Research and Development in Information Retrieval (SIGIR'99), 1999.
22. J. Xu and W.B. Croft, "Cluster-based language models for distributed retrieval," in Proceedings of ACM SIGIR 99, pp. 254-261, 1999.
23. J. Xu and W.B. Croft, "Improving the effectiveness of information retrieval with local context analysis," *ACM Transactions on Information Systems*, **18**, pp. 79-112, 2000.

Intelligent Agent Battlespace Augmentation

Philip J. Emmerman and Uma Y. Movva

U.S. Army Research Laboratory, Adelphi, MD 20783

Abstract. The anticipated dynamics of the future battlefield will require greatly increased mobility, information flow, information assimilation, and responsiveness from a tactical operation center (TOC) and platforms (tanks, armored personnel carriers, etc.). Three significant and related trends in the evolution of the tactical battlefield address these requirements. The first is the increased automation of the brigade nerve center or TOC. Much of this automation will be provided by software agent technology. The second is the digitization of current battlefield platforms. This digitization greatly reduces the uncertainty concerning these platforms and enables automated information exchange between these platforms and their TOC. The third is the rapid development of robotics or physical agents for numerous battlefield tasks such as clearing buildings of hazards (such as snipers) or performing wingman functions for a future combat vehicle. This paper illustrates the potential synergy between these seemingly disparate developments, particularly related to battlefield visualization, multi-resolution analysis, software agents, and physical agents. Battlefield visualization programs are currently focussed on representing the physical environment. This greatly contributes to situation awareness at the TOC and platform levels. As intelligent agents, both software and physical agents, are developed, battlefield visualization must be enhanced to include the state, behavior, and results of the actions of these agents. Multi-resolution data and analysis will enhance visualization, software agent and physical agent performance.

Introduction

There is widespread dissatisfaction with the design and functionality of current Army tactical operation centers (TOCs) [4]., due primarily to their lack of mobility, inefficiency, and high complexity. The extensive hardware, software, and manpower resources needed to operate a current TOC severely limit the required mobility needed for a future nonlinear, dynamic battlefield. A greatly increased level of automation is needed both to significantly lower the human resources required and to improve information flow. Figure 1 depicts the size and mobility envisioned for a future TOC.

The TOC exists to support the tactical commander in understanding the current state of the battlefield and in predicting its future state. It also provides planning, monitoring, and reaction functions to the commander. The situation awareness that results enables rapid and effective decision making and leadership. Although the TOC is the information and control center of the tactical battlefield, it must also be able to project its critical information to a commander on a remote platform such as a tank or helicopter, observing or interacting with vital positions on the battlefield. Because the TOC is an information integration and fusion node, it is an essential part of a highly distributed and mobile force. A scalable, extensible, and adaptable visualization and software agent architecture and rich application set are required to achieve the increased efficiency envisioned. Most low-level information retrieval, dissemination, and analysis will be performed or controlled by these agents.

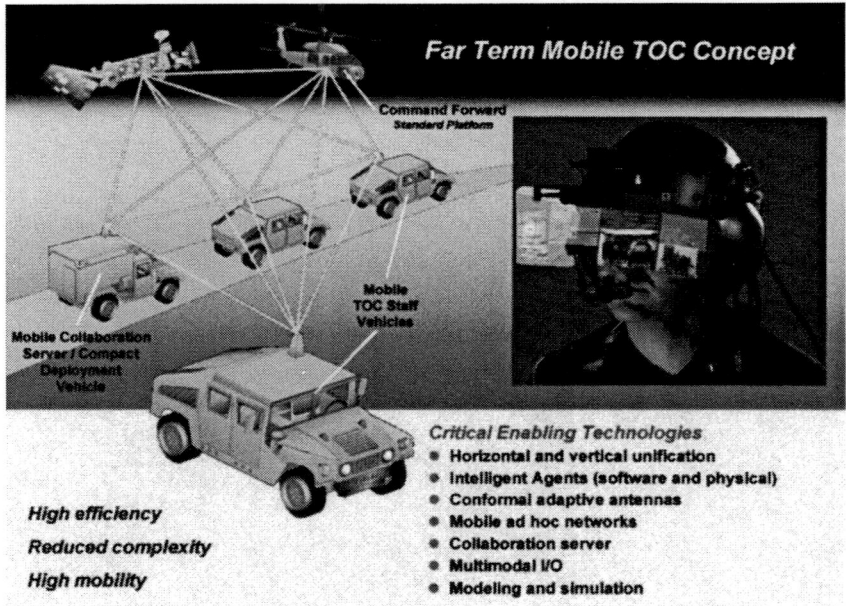

Fig. 1. Mobile future TOC concept.

Battlefield visualization technology and software agent technology are closely linked because of the need to visualize and interact with both the agents and the results of their analysis. Automated communications between the TOC and its associated platforms (human or robotic) will be agent based. The digitization of the lower echelons of the army strongly enhances the coupling of the TOC and the tactical platforms, enabling the automated exchange of data and information, as well as access to more advanced applications by means of an agent environment. This automated information exchange will greatly reduce the latency of information, reduce uncertainty, and enable a more real-time control system approach in the battlefield. Figure 2 illustrates this exchange, where agents are classified according to their battlefield functional area.

Physical agents are expected to be ubiquitous on the future battlefield, significantly lowering the risk to our soldiers. They will be present in a myriad of shapes, sizes, and capabilities. Because these physical agents are to complement future manned systems, they must be able to collaborate not only amongst themselves but also with their manned partners. Their missions will range from scout missions (reconnaissance, surveillance, and target acquisition) to urban rescue. Robotic sentinels and remote communication systems would reduce the soldier workload of a future TOC. Teams of small robots deployed by manned or unmanned mother ships will explore (for hazards) and define buildings before manned occupation. Figure 3 depicts an urban scenario [1]. The Army has both cross-country and urban mission robot programs in development. Robust mobility, collaborative military behavior, and effective soldier robot interaction are major development areas. These robots must be able to operate in these battlefield environments approximately at the same tempo as the manned forces.

Fig. 2. TOC-platform agent interaction.

The information gathered by these agents will be sent to a mother ship or TOC and be visualized by human controllers. The high-level control and interaction between the mother ship and its agents will be based on software agent technology, analogous to the TOC/platform interaction. Software agents will be monitoring the robot disposition and communicating with the robot controller. A future combat system could be augmented by these small robots, thereby increasing its urban effectiveness.

Software Agent Applications

Figure 4 illustrates the relationship between software agent applications and visualization. Software agents provide much of the analysis of battlefield data. Both the results of this analysis and the state and behavior of these agents need to be visualized. Of the myriad possible battlefield agent applications, this paper focuses on several that require scalability and extensibility of the agent approach.

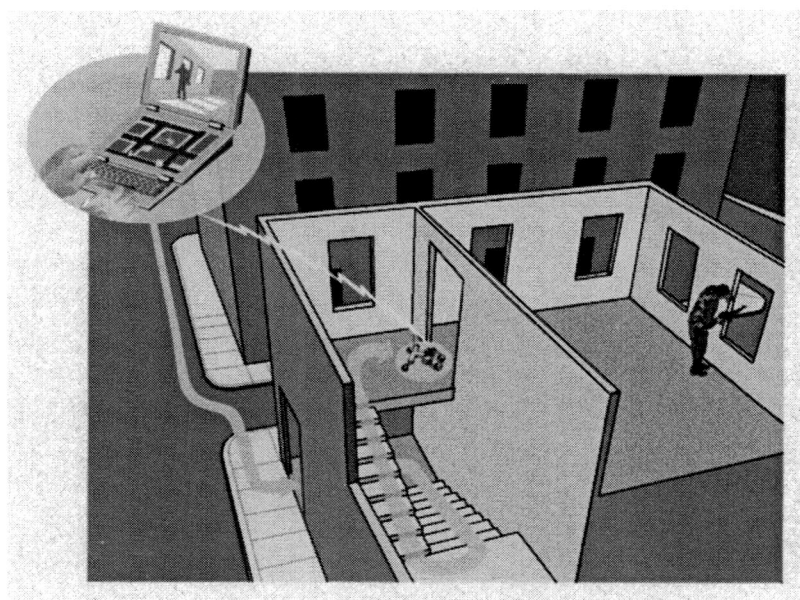

Fig. 3. Small robot urban scenario

Consider initially the basic sentinel application, where agents must be able to dynamically monitor and analyze battlefield activity and perform alert functions. These agents are assigned to monitor either fixed areas on the battlefield or areas associated with entities (fixed or moving). The following are two examples of monitor agents scenarios:
1. Assign an agent to monitor a specific area of interest where if enemy armor is detected in force before the blue force occupies the nearby hills, the blue
2. commander and the maneuvering units must be alerted. This agent, although fixed spatially, must have spatial and temporal reasoning.
3. Assign an agent to monitor a maneuvering blue force battalion, and alert it if any enemy radar is capable of detecting it as it performs its planned maneuver path. This agent has mobility (not fixed to a geographic area) in addition to spatial reasoning.

Although these sentinel agent applications seem simple, significant temporal and spatial reasoning is required to minimize unnecessary alerts.

Now consider a broader agent application scenario. The TOC brigade commander has selected a maneuver course of action plan that calls for the synchronized movement, enemy engagement,and logistics resupply of the brigade. The plan has been disseminated and the maneuver platforms have begun executing this course of action. This plan implementation stimulates significant agent activity both in the TOC as well as in the maneuver platforms. A global maneuver monitor agent in the TOC interacts with the maneuver monitor agents in the platforms. The platform synchronization monitor agents have the task of alerting the human platform commander if the maneuver entity cannot execute its maneuver plan. This agent would also alert the TOC maneuver monitor agent of any execution problems. A

TOC intelligence agent continuously monitors and retrieves any pertinent enemy information that would affect this operation. For example, suppose a radar is detected near the planned path of one of the maneuver battalions. This intelligence agent alerts both the TOC maneuver plan agent as well as the affected platform agents (maneuver and intelligence). At the TOC, a fire support agent generates an attack plan to disable this enemy sensor asset. This plan is presented to the TOC commander and is refused because the commander considers the available fire support assets insufficient. At the affected platforms, the platform maneuver agent generates a reactive maneuver plan and if acceptable to the local commander, the plan is executed. A platform logistics monitor agent keeps track of local resources (fuel, ammunition, spare parts, etc.) and disseminates this information to the TOC logistics agent. The TOC logistics agent continuously monitors the resupply plan that supports this engagement. If the planned resupply points become inadequate because of excessive engagement times or maneuver, the TOC logistics agent redefines the resupply points.

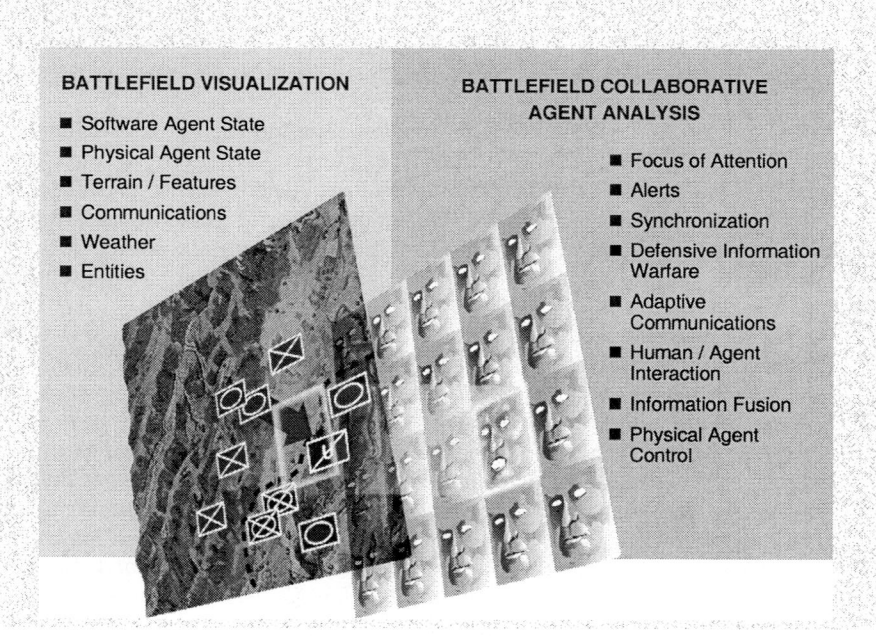

Fig. 4. Intelligent agent battlefield applications and visualization.

This example application indicates that monitoring, alerting, dissemination and retrieval agents are needed for each of the major battlefield functions (such as maneuver, intelligence, and logistics) at both the TOC and the lead platforms. Many applications are possible within each of the functional areas. Some of which may differ, within each functional area such as maneuver, at the TOC and the platform. Because of the complexities inherent in creating and interacting with a large set of agents, it is essential that the human/agent interaction be intuitive and not cumbersome. Since many agent applications will be oriented toward entities or areas in the battlefield, an effective battlefield visualization approach representing the agents and their behaviors is essential.

Battlefield Visualization
We introduce here a multi-resolution approach to visualization as well as analysis. Most of the current emphasis of the Army battlefield visualization program is on providing a global infrastructure with the ability to visualize the battlefield environment (terrain, weather, entities, features, communications, etc.) at whatever resolution is required and available. This enables the commander to have a custom global view of the battlefield as well as a high-resolution local view to support critical decisions. This same infrastructure supports high-fidelity local views for the platform commanders as well as the ability to jump to any other local view in the world (as long as data is available) to support training or preparation for deployment. This scalability provides a single visualization approach suitable for both TOC and platform applications, including robotic platforms. Figure 5 illustrates a coupled 2D/3D visualization approach.

A 2D/3D approach in necessary since soldiers are very familiar with two-dimensional maps and can maintain their global situation awareness. However the 2D representation is not as effective for visualization of high-resolution, complex terrain. 3D representation is excellent for high-resolution, complex terrain, but it is very easy to lose a global perspective (get lost) in all the detail presented. Presenting both views simultaneously eliminates many of the problems inherent in a single-view approach.

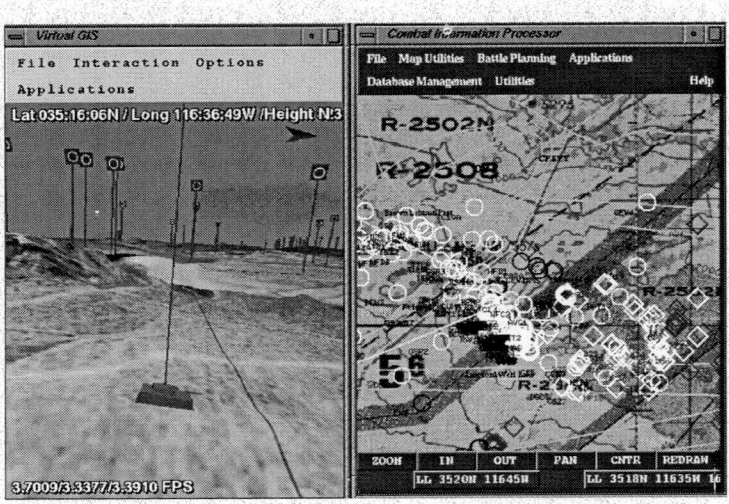

Fig. 5. Coupled 2D/3D visualization.

Many sources of environmental data are available, albeit with widely varying resolution and coverage. It is therefore necessary for any visualization system to work with multresolution data (elevation and imagery). Software agents will use this multiresolution data for responsive planning and mission execution. While robots do not visualize, they must reason about their environment. Although the robotic platforms will have effective local perception, this multiresolution environmental data will enable them to create reactive plans(implemented by software agents), similar to the agent activity in human platforms.

Military planners currently use digital terrain and elevation data along with digital feature data to plan. Because the currently available elevation data are so coarsely sampled (100 m or 30 m post spacing), these planned routes may contain numerous, significant obstacles. In order to traverse these routes, the manned or unmanned vehicles must sense and react to these obstacles. As the number of reactions increases, the time to complete the mission also increases. Fortunately, under the battlefield visualization umbrella, there are programs, that are developing the technology to both rapidly generate and visualize much higher resolution data (1 m). This would enable an operator to visualize the planned routes and manually detect obstacles. If the planning and execution analysis could use the high-resolution data, then many of the obstacles that fall within the 1 to 100 m range could be detected and avoided in the plan. However, the cost for this high-resolution analysis is increased processing time, since the route-planning algorithms would be using much more data. A multiresolution analysis would use high-resolution data only when the environmental complexity required it. This would greatly decrease the processing cost for most areas. Because the cost for reactive planning is high, particularly in robotic platforms, significant mission savings (time) are expected. Figure 6 illustrates the need for high-resolution data.

The original plan developed with 100 m elevation post spacing does not recognize a significant obstacle to the planned maneuver. With 1 m data, the resultant plan does not require reactive planning.

Agent/Visualization Implementation

The Army Research Laboratory (ARL) and the University of Maryland (UMD) have recently integrated a software agent architecture with a 2D/3D multiresolution visualization research testbed [3]. The University of Maryland has developed a software agent architecture called Interactive Maryland Platform for Agents Collaborating Together (IMPACT) [2,5], and ARL has developed a large-scale battlefield visualization testbed, the Combat Information Processor (CIP). IMPACT was used to agentize the legacy client/server-based CIP and provide the initial sentinel agent functionality described in this paper. This functionality was added by agentizing the CIP control measure and entity servers. Figure 7 represents the human computer interface of this agent application.

Conclusions

The Army must take advantage of the synergy between its visualization, software agent, and physical agent technology developments. Without a holistic approach, multiple competing visualization and software agent designs will proliferate. Even with a single optimal design approach for human/agent interaction, this research and development must address the ability of the human controller to assimilate and act on the state of the battlefield and direct his agents rapidly enough to satisfy future battlefield dynamics. An effective physical and software agent interaction would be perceived to be non-intrusive and would provide all the necessary focussed information for rapid decision making. A software agent application architecture may be sufficient to perform many of the manpower intensive tasks at both the TOC and the individual platforms. These tasks have been categorized similarly to the battlefield functional areas. Although myriad applications are possible, spanning a widely dispersed level of complexity, a number of low-level applications can also be very effective in TOC automation. It is critical that the agent approach be scalable, extensible, and adaptable to address the broad application area of the tactical battlefield. Many of these tasks can be implemented with generic low- level monitor, alert, retrieve, and disseminate functions.

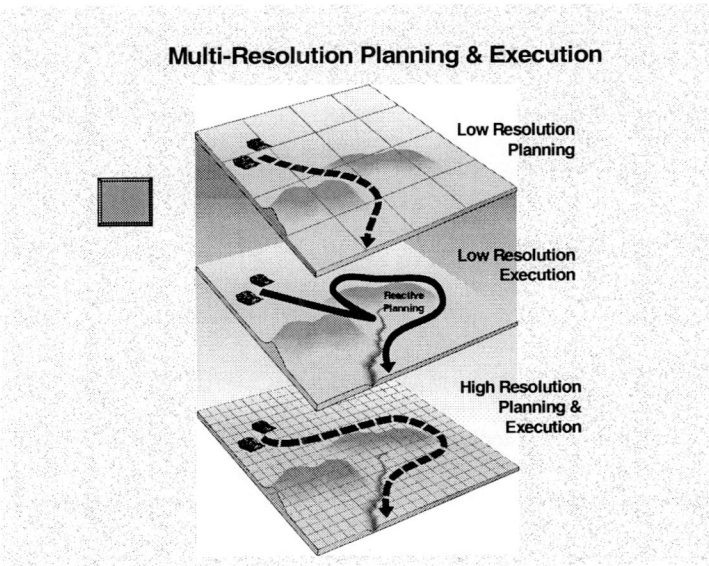

Fig. 6. Multi-resolution planning

There still is concern that the human/agent interaction may be too encumbering for the commanders and staff involved. Closely coupling the agent interaction with battlefield visualization should make the interaction more intuitive. Also, an embedded training application for decision making that uses an this agent approach will accelerate the acceptance of this approach. This embedded training would include the ability to rapidly construct scenarios to continuously improve the commander's and staff's decision making. If this training capability is embedded, the operators will automatically train on the use of this agent approach and develop a trust in these agents.

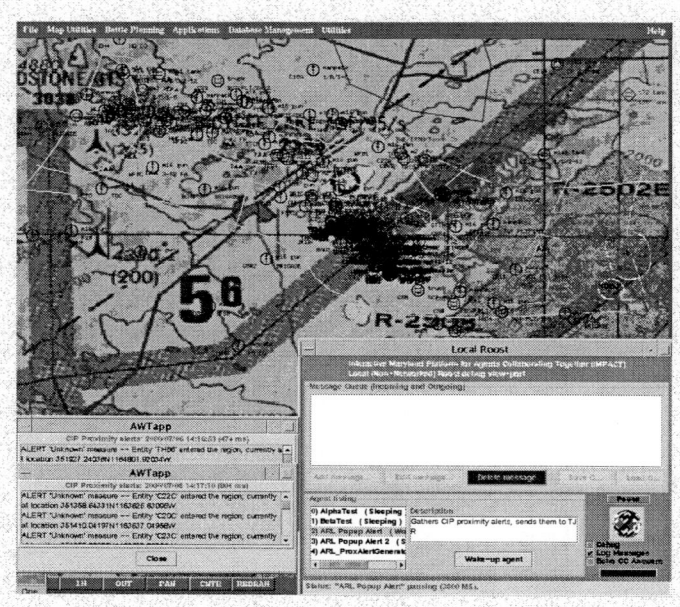

Fig. 7. Sentinel visualization interface.

References

1. Budulas, P.P., Young, S. H., and Emmerman, P. J., *Mother Ship and Physical Agents Collaboration*, Proceeding SPIE 1999
2. Eiter,T., Subramanian, V.S.,and Rogers, T.J., *Heterogeneous Active Agents, III: Polynomially Implementable Agents*, Artificial Intelligence Journal, Vol. 117, Nr. 1, pps 107-167, Feb. 2000.
3. Emmerman, P.J., Gasarch, C., Movva, U.Y., Rogers, T. J., Subrahmanian, V.S., and Tokarcik, L., *An Agent Based Combat Information Processor System*, Proceedings Fusion 2000 Conference
4. Emmerman, P.J., Grills, J. P., Johnson, J. E., and Rodriguez, A., *Future Army*
5. *Tactical Operation Center Concept*, Proceedings 1999 CCRT Conference.
6. Subrahmanian, V.S., Bonatti, P., Dix, J., Eiter, T., Kraus, S., Ozcan, F., and Ross, R., *Heterogeneous Agent Systems: Theory and Implementation*, MIT Press, 2000.

Learning and Evolution:
An Introduction to Non-darwinian Evolutionary Computation

Ryszard S. Michalski

Machine Learning and Inference Laboratory
School of Computational Sciences
George Mason University, Fairfax, VA, USA
and
Institute of Computer Science
Polish Academy of Sciences, Warsaw, Poland

michalski@gmu.edu

Abstract. The field of evolutionary computation has drawn inspiration from Darwinian evolution in which species adapt to the environment through random variations and selection of the fittest. This type of evolutionary computation has found wide applications, but suffers from low efficiency. A recently proposed non-Darwinian form, called *Learnable Evolution Model* or *LEM*, applies a learning process to guide evolutionary processes. Instead of random mutations and re-combinations, LEM performs hypothesis formation and instantiation. Experiments have shown that LEM may speed-up an evolution process by two or more orders of magnitude over Darwinian-type algorithms in terms of the number of births (or fitness evaluations). The price is a higher complexity of hypothesis formation and instantiation over mutation and recombination operators. LEM appears to be particularly advantageous in problem domains in which fitness evaluation is costly or time-consuming, such as evolutionary design, complex optimization problems, fluid dynamics, evolvable hardware, drug design, and others.

1 Introduction

In his prodigious treatise "On the Origin of Species by Means of Natural Selection," Darwin conceived the idea that the evolution of species is governed by "one general law, leading to the advancement of all organic beings, namely, multiply, vary, let the strongest live and the weakest die" (Darwin, 1859). In such biological or natural evolution, new organisms are created via asexual reproduction with variation (mutation) or via sexual reproduction (recombination). The underlying assumption is that the evolution process is not guided by some "external mind," but proceeds through semi-random modifications of genotypes through mutation and recombination, and progresses to more advanced forms due to the principle of the "survival of the fittest."

In Darwinian evolution, individuals thus serve as holders and transmitters of their genetic material. Their life experiences play no role in shaping their offspring's properties. Jean-Baptiste Lamarck's[1] idea that traits learned during the lifetime of an individual could be directly transmitted to progeny has been rejected as biologically viable because it is difficult to construe a mechanism through which this could occur.[2] Many scientists believe, however, that there is another mechanism through which learned traits might influence evolution, namely, the so-called *Baldwin effect* (Baldwin, 1896). This effect stems from the fact that, due to learning, certain individuals can survive even though their genetic material may be suboptimal. In this way, some traits that otherwise would not survive are passed on to the next generations. Some researchers argue that under certain conditions such learning may actually slow genetic change and thus slow the progress of evolution (Anderson, 1997).

More than a century after Darwin introduced his theory of evolution, computer scientists adopted it as a model for implementing evolutionary computation (e.g., Holland, 1975; Goldberg, 1989; Michalewicz, 1996; Koza et al. 1999). Their efforts have led to the development of several major approaches, such as genetic algorithms, evolutionary strategy, genetic programming, and evolutionary programs. These and related approaches, viewed jointly, constitute the rapidly growing field of evolutionary computation (see, e.g., Baeck, Fogel, M. Mitchell; 1996, Banzhaf et al., 1999; Zalzala, 2000).

Methods of evolutionary computation based on principles of Darwinian evolution use various forms of mutation and/or recombination as variation operators. These operators are easy to implement and can be applied without any knowledge of the problem area. Therefore, Darwinian-type evolutionary computation has found a very wide range of applications, including many kinds of optimization and search problems, automatic programming, engineering design, game playing, machine learning, evolvable hardware, and many others.

The Darwinian-type evolutionary computation is, however, semi-blind: the mutation is a random, typically small, modification of a current solution; the crossover is a semi-random recombination of two or more solutions; and selection is a sort of parallel hill climbing. In this type of evolution, the generation of new individuals is not guided by principles learned from past generations, but is a form of the trial and error process executed in parallel. Consequently, computational processes based on Darwinian evolution tend to be not very efficient. Low efficiency has been the major obstacle in applying Darwinian-type evolutionary computation to highly complex

---

[1] Jean-Baptiste Lamarck, a French naturalist (1744-1829), who proposed a theory that the experience of an individual can be encoded in some way and passed to the genome of the offspring.
[2] Recent studies show that Lamarckian evolution appears to apply in the case of antibody genes (Steel and Blanden, 2000).

problems. The objective of many research efforts in this area has been thus to increase the efficiency of the evolutionary process.

In modeling computational processes after principles of biological evolution, the field of evolutionary computation has followed a long-practiced tradition of looking to nature when seeking technological solutions. The imitation of bird flying by mythological Icarus and Daedalus is an early example of such efforts. In seeking technological solutions, the "imitate-the-nature" approach, however, frequently does not lead to the best engineering results. Modern examples of successful solutions that are not imitations of nature include balloons, automobiles, airplanes, television, electronic calculators, computers, etc.

This paper discusses a recently proposed, non-Darwinian form of evolutionary computation, called *Learnable Evolution Model* or *LEM*. In LEM, new individuals are created by hypothesis formation and instantiation, rather through mutation or recombination. This form of evolutionary computation attempts to model "intellectual evolution"---the evolution of ideas, technical solutions, human organizations, artifacts, etc.---rather than biological evolution. In contrast to Darwinian evolution, an intellectual evolution is guided by an "intelligent mind," that is, by humans who analyze advantages and disadvantages of previous generation of solutions and use the developed understanding in creating next generation of solutions. It is due to the intellectual evolution that the process of evolving the automobile, airplane or computer from primitive prototypes to modern forms was astonishingly rapid, taking just few human generations.

The idea and the first version of the LEM methodology were introduced in (Michalski, 1998). A more advanced and comprehensive version is in (Michalski, 2000). Its early implementation, LEM1, produced very encouraging results on selected function optimization problems (Michalski and Zhang, 1999). Subsequent experiments with a more advanced implementation, LEM2, confirmed earlier results and added new highly encouraging ones (e.g., Cervone et al., 2000, Cervone, Kaufman and Michalski, 2000).

The following sections briefly describe LEM and its relationship to Darwinian-type evolutionary computation, and then summarize results of testing experiments.

2 LEM vs. Darwinian Evolutionary Computation

Darwinian-type evolutionary algorithms can be generally viewed as stochastic techniques for performing parallel searches in a space of possible solutions. They simulate natural evolution by creating and evolving a population of individuals until a termination condition is met. Each individual in the population represents a potential solution to a problem. Such a solution can be represented as a vector of parameters, an instantiation of function arguments, an engineering design, a concept description, a control strategy, a pattern, a computer program, etc. A precondition for applying an evolutionary algorithm is the availability of a method for evaluating the quality (fitness) of individuals from the viewpoint of the given goal.

A general schema of an evolutionary computation consists of the following steps:

1. **Initialization**
   t := 0
   Create an initial population P(t) and evaluate fitness of its individuals.
2. **Selection**
   t := t+1
   Select a new population from the current one based on their fitness: P(t) := Select(P(t-1))
3. **Modification**
   Apply *change operators* to generate new individuals: P(t) := Modify(P(t))
4. **Evaluation**
   Evaluate fitness of individuals in P(t)
5. **Termination**
   If P(t) satisfies the *termination condition*, then END, otherwise go to step **2**.

Different evolutionary algorithms differ in the way individuals are represented, created, evaluated, selected and modified. They may also use different orders of steps in the above schema, employ single or multiple criteria in fitness evaluation, assume different termination conditions, and simultaneously evolve more that one population. Some algorithms (specifically, genetic algorithms) make a distinction between the *search space* and the *solution space*. The search space is a space of encoded solutions ("genotypes"), and the solution space is the space of actual solutions ("phenotypes"). Encoded solutions have to be mapped onto the actual solutions before the solution quality or fitness is evaluated.

As mentioned earlier, in Darwinian-type (henceforth, also called conventional) evolutionary algorithms, change operators are typically some forms of mutation and/or recombination. Mutation is a unary transformation operator that creates new individuals by modifying previous individuals. Recombination is an n-ary operator (where n is typically 2) that creates new individuals by combining parts of n individuals. Both operators are typically semi-random, in the sense that they make random modifications within certain constraints.

The selection operator selects individuals for the next population. Typical selection methods include proportional selection (the probability of selecting an individual is proportional to its fitness), tournament selection (two or more individuals compete for being selected on the basis of their fitness), and ranking selection (individuals are sorted according to their fitness and selected according to probabilities associated with different ranks on the sorted list). The termination condition evaluates the progress of the evolutionary process and decides whether to continue it or not.

Learnable Evolution Model, briefly, LEM, also follows this general schema. Its fundamental difference from Darwinian-type algorithms lies in step 4, as it generates new individuals in very different way. In contrast to semi-random change operators employed in Darwinian-type algorithms, LEM conducts a reasoning process in generating new individuals. Specifically, it applies operators of *hypothesis formation* and *hypothesis instantiation*.

The operator of hypothesis formation selects from a population a group of high-performing individuals, called the H-group, and a group of low-performing individuals, called the L-group, according to their fitness. The H-group and L-group may be selected from the current population or from a sequence of past populations. These groups can be selected using a *population-based* method, a fitness-based method, or a combination of the two. The population-based method applies High and Low Population Thresholds (HPT and LPT) in selecting individuals, and fitness-based method applies High and Low Fitness Thresholds (HFT and LFT). The thresholds can be fixed or may change in the process of evolution. For details, see (Michalski, 2000).

The H-group and L-group are then supplied to a machine learning program that generates a general hypothesis distinguishing between high performing from low performing individuals. Such a hypothesis can be viewed as a theory explaining the differences between the two groups. Alternatively, it can be viewed as a characterization of the sub-areas of the search space that are likely to contain the top performing individuals (the best solutions). Once such a hypothesis has been generated, the algorithm generates new individuals that satisfy the hypothesis.

In principle, any inductive learning method can be used for hypothesis formation. LEM1 and LEM2 implementations of the LEM methodology has used the AQ-type learning method (specifically, AQ15 and AQ18, respectively; see Wnek et al., 1995; Kaufman and Michalski, 2000b). This method appears to be particularly advantageous for LEM, because it employs *attributional calculus* as the representation language (Michalski, 2000b). Attributional calculus adds to the conventional logic operators new operators, such as *internal disjunction, internal conjunction, attribution relation,* and the *range operator*, which are particularly useful for characterizing groups of similar individuals. Attributional calculus stands between propositional calculus and predicate calculus in terms of its representational power.

New individuals are generated by a hypothesis instantiation operator that instantiates the given hypothesis in various ways. To very simply illustrate, suppose that a hypothesis was generated by an AQ-type learning program and expressed in the form of two *attributional rules* (these rules are in a simplified form to facilitate explanation):

Rule 1: $[x = a \vee c]$ & $[y = 2.3 .. 4]$ & $[z > 5]$ (sup=80)
Rule 2: $[x = b \vee d \vee e]$ & $[z = 3.5 .. 6.4]$ (sup=15)   (1)

where the domains of attributes x, y, and z are: $D(x) = \{a,b,c,d,e,f\}$, and $D(y)$ and $D(z)$ range over real numbers between 0 and 10.

Rules in (1) characterize two subareas of the search space that contain high performing individuals. The first rule states that high performing individuals appear in the area in which the variable x has value a or c, the variable y takes value between 2.3 and 4, and the variable z takes value greater than 5. The parameter sup (support) indicates that this rules covers 80 individuals in the H-group. The second rule describes an alternative set of conditions, namely, that high performing individuals appear also in the area in which differ x takes value b or d or e, and z takes values from the real interval between 3.5 and 6.4. The second rule covers sup=15 individuals in the H-group. Note that Rule 2 does not include variable y. This means that this variable was found irrelevant for differentiating between high and low performing individuals.

The hypothesis (1) is a generalization of the set of individuals in the H-group. Thus, it may potentially cover many other, unobserved individuals. The instantiation operator instantiates the hypothesis in different ways, that is, generates different individuals that satisfy conditions of the rules. For example, using hypothesis (1), the operator may generate such individuals as:

<a, 2, 6>, <c, 3.5, 9.1>, <a, 2.1, 6.4>   (based on Rule 1)
<d, 2, 6>, <e, 5.5, 4.3>, <b, 2.2, 4.5>   (based on Rule 2)   (2)

Since variable y is not present in Rule 2, any values of y could be selected from $D(y)$ to instantiate this rule. In our experiments, variables not present in the rule were instantiated to values selected randomly from among those that appeared in individuals of the training set (H-group and L-group).

The newly generated individuals are combined with the previous ones, and a new population is selected using some selection method. Again, an H-group and L-group are generated and operations of hypothesis generation and instantiation are repeated. The process continues until a *LEM termination condition* is met, e.g., the (presumably) global or a satisfactory solution has been found.

The above-described process of creating new individuals by operators of hypothesis formation (through inductive generalization) and hypothesis instantiation (by generating individuals satisfying the hypothesis) constitutes the *Machine Learning Mode* of LEM. A general form of LEM includes two versions: *uniLEM*, which repetitively applies the Machine Learning mode until a termination condition is satisfied, and *duoLEM*, which toggles between Machine Learning and Darwinian Evolution mode, switching from one mode to another when the termination condition for the given mode is satisfied (when there is little progress in executing the mode).

The Darwinian Evolution Mode executes one of the existing conventional evolutionary algorithms.

A comprehensive explanation of various details of the LEM methodology and its variants is in (Michalski, 1999).

3 A Simple Illustration of LEM

To illustrate LEM, let us consider a very simple search problem in a discrete space. The search space is spanned over four discrete variables: x, y, w, and z, with domains {0,1}, {0,1}, {0,1}, and {0,1,2}, respectively. Figure 1A, presents this space using the General Logic Diagram or GLD (Michalski, 1978; Zhang, 1997). Each cell of the diagram represents one individual. For example, the uppermost cell marked by 7 represents the vector: <0, 0, 0, 2>. The initial population is visualized by cells marked by dark dots (Figure 1A). The numbers next to the dots indicate the fitness value of the individual. The search goal is to determine individuals with the highest fitness, represented by the cell marked by an x (with the fitness value of 9).

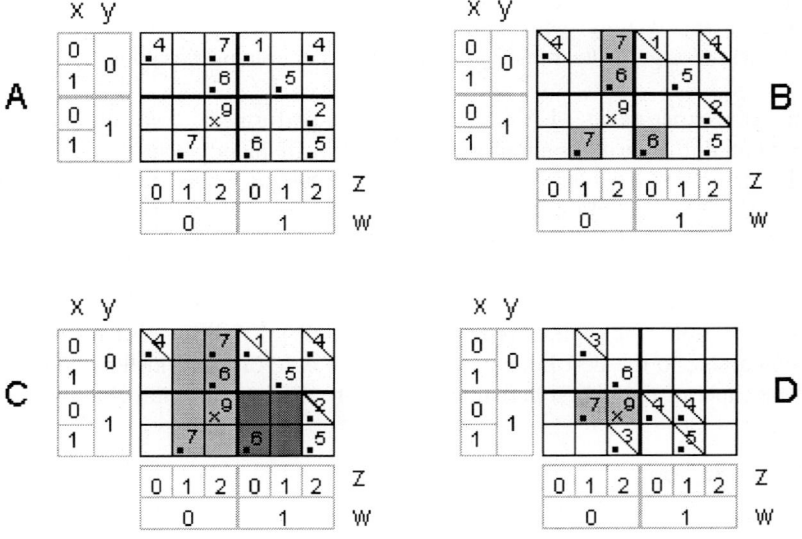

Figure 1. The search space and four states of the LEM search process.

We assume that descriptions discriminating between an H-group and an L-group are in the form of attributional rules learned AQ-type learning programs. Figure 1B presents the H-group individuals (the gray-shaded cells) and L-group individuals (crossed cells) determined from the initial population. The shaded areas in Figure 1C represent two attributional rules discriminating between the H-group and the L-group: [w = 0] & [z = 1 v 2] and [y = 1] & [w = 1] & [z = 0 v 1].

Figure 1D shows individuals in the H-group (shaded cells) and the L-group (crossed cells) generated by instantiating rules in Figure 1C. The shaded area in Figure 1D represents a rule that discriminates between these groups: [x=0] & [y=1] & [w=0] & [z=1v2]. This rule was obtained through incremental specialization of the parent rule, and covers two individuals. The global solution will be located in the next iteration.

4 Summary of Testing Experiments

To test the LEM methodology, it has been implemented in a general-purpose form in programs LEM1 (Michalski and Zhang, 1999) and LEM2 (Cervone, 1999). It was also employed in program ISHED1, specifically tailored to problems of optimizing heat exchangers (Kaufman and Michalski, 2000a). Both LEM1 and LEM2 were applied to a range of function optimization problems. LEM1 was also successsulfy applied to a problem in filter design (Colleti et. al, 1999). LEM2 was tested in a wide range of experiments dealing with optimizing different types of functions with different numbers of arguments, ranging from 4 to180 continuous variables.

In all experiments LEM2 strongly outperformed conventional evolutionary computation algorithms employed in the study, frequently achieving two or more order of magnitude speedups in terms of the number of births (or function evaluations). Results from LEM2 were also significantly better than the best results from conventional evolutionary algorithms published on a website. These and other recent results have been described in (Cervone et. al, 2000a; Cervone et al., 2000b). Results from experiments with ISHED1 were presented in (Kaufman and Michalski, 2000a). According to the collaborating expert, ISHED1's heat exchanger designs were comparable to the best human designs in the case of uniform flow of refrigerant, and were superior to the best human designs in the case of non-uniform flow.

5 Conclusion

Experimental studies conducted so far have strongly demonstrated that the proposed Learnable Evolution Model can significantly speed up evolutionary computation processes in terms of the number of births (or fitness evaluations). These speed-ups have been achieved at the cost of higher complexity of operators generating new individuals (hypothesis formation and instantiation). An open problem is thus to study trade-offs associated with the LEM application to different problem domains. It is safe to say, however, that LEM is likely to be highly advantageous in problem areas in which computation of the evaluation function is costly or time-consuming. Such areas include engineering design, complex optimization problems, fluid dynamics, evolvable hardware, drug design and automatic programming.

Another limiting aspect of LEM is that in order to apply it, the machine learning system must be able to work with the given representation of individuals. For example, if individuals are represented as attribute-value vectors, rule and decision tree learning systems can be applied. If they are represented as relational structures, a structural learning system must be employed.

Concluding, among the open problems for further research on LEM are to understand the benefits, trade-offs, advantages and disadvantages of LEM versus Darwinian-type evolutionary algorithms in different problem domains.

Acknowledgments

The author thanks Guido Cervone, Ken Kaufman and Liviu Panait for an excellent collaboration on the LEM project and the experimental validation of the LEM methodology. This research has been conducted in Machine Learning and Inference Laboratory at George Mason University. The Laboratory's research on this project has been supported in part by the National Science Foundation under Grants No. IIS-9904078 and IRI-9510644.

References

Anderson R.W. (1997), The Baldwin Effect, in *Handbook of Evolutionary Computation*, Section C3.4.1, pp. C3.:4:1-C3.4:15., IOP Publishing Ltd and Oxford: Oxford University Press.

Baeck, T., Fogel, D.B. and Michalewicz, Z. (eds.) (1997), *Handbook of Evolutionary Computation*. IOP Publishing Ltd and Oxford: Oxford University Press.

Baldwin, J.M. (1896), A New Factor in Evolution, *American Naturalist*, vol. 30, pp.441-51.

Banzhaf, W., Nordin P., Keller R.E., and Francone F.D (1998), *Genetic Programming: An Introduction*, Morgan Kaufman Publishers, Inc., San Francisco, CA, 1998.

Cervone, G., Michalski, R.S., Kaufman K., Panait, L. (2000), Combining Machine Learning with Evolutionary Computation: Recent Results on LEM, *Proceedings of the Fifth International Workshop on Multistrategy Learning* (MSL2000), Michalski, R.S. and Brazdil, P. B. (eds.), Guimaraes, Portugal, June 5-7.

Cervone, G., Kaufman, K.A., and Michalski, R.S. (2000), Experimental Validations of the Learnable Evolution Model, *Proceedings of the 2000 Congress on Evolutionary Computation*, La Jolla, California.

Coletti, M., Lash, T., Mandsager, C., Michalski, R.S., and Moustafa, R. (1999), Comparing Performance of the Learnable Evolution Model and Genetic Algorithms on Problems in Digital Signal Filter Design. *Proceedings of the 1999 Genetic and Evolutionary Computation Conference (GECCO)*.

Darwin, C. (1859), On the Origin of Species by Means of Natural Selection, or the Preservation of Favoured Races in the Struggle for Life, John Murray, London.

Goldberg, D.E. (1989), *Genetic Algorithms in Search, Optimization and Machine Learning*. Addison-Wesley.

Holland, J. (1975), *Adaptation in Artificial and Natural Systems*. Ann Arbor: The University of Michigan Press.

Kaufman, K. A. and Michalski, R.S. (2000a), Applying Learnable Evolution Model to Heat Exchanger Design, *Proceedings of the Seventeenth National Conference on Artificial Intelligence and Twelfth Conference on Innovative Applications of Artificial Intelligence* (AAAI-2000/IAAI-2000), Austin, Texas.

Kaufman, K.A. and Michalski, R.S. (2000b), The AQ18 Machine Learning and Data Mining System: An Implementation and User's Guide. *Reports of the Machine Learning Laboratory*, George Mason University, Fairfax, VA (to appear).

Koza, J.R., Bennett, F. H. III, Andre D., Keane M. A. (1999), *Genetic Programming III: Darwinian Invention and Problem Solving*, Morgan Kaufmann Publishers, San Francisco, CA.

Michalewicz, Z. (1996), *Genetic Algorithms + Data Structures = Evolutionary Programs*. Springer Verlag, Third edition.

Michalski, R.S. (1998), Learnable Evolution: Combining Symbolic and Evolutionary Learning. *Proceedings of the Fourth International Workshop on Multistrategy Learning (MSL'98)*, 14-20.

Michalski, R.S. (2000a), LEARNABLE EVOLUTION MODEL: Evolutionary Processes Guided by Machine Learning. *Machine Learning* 38(1-2).

Michalski, R.S. (2000b), Natural Induction: A Theory and Methodology of the AQ Approach to Machine Learning and Data Mining. *Reports of the Machine Learning Laboratory*, George Mason University, Fairfax, VA (to appear).

Michalski. R.S. and Zhang, Q. (1999), Initial Experiments with the LEM1 Learnable Evolution Model: An Application to Function Optimization and Evolvable Hardware. *Reports of the Machine Learning and Inference Laboratory*, MLI 99-4, George Mason University, Fairfax, VA.

Mitchell, M. (1996). *An Introduction to Genetic Algorithms*. Cambridge, MA: MIT Press.Back, Thomas, *Optimization by Means of Genetic Algorithms*, ENCORE.

Steele E. J. and Blanden R. V. (2000), Lamarck and Antibody Genes, *Science*, Vol. 288, No. 5475, pp. 2318.

Wnek, J., Kaufman, K., Bloedorn, E. and Michalski, R.S. (1995), Inductive Learning System AQ15c: The Method and User's Guide, *Reports of the Machine Learning and Inference Laboratory*, MLI 95-4, George Mason University, Fairfax, VA.

Can Relational Learning Scale Up?

Attilio Giordana[1], Lorenza Saitta[1], Michele Sebag[2], and Marco Botta[3]

[1] DISTA, Università del Piemonte Orientale, Alessandria, Italy
attilio@unipmn.it
[2] LMS, École Polytechnique, Palaiseau, France
sebag@cmapx.polytechnique.fr
[3] Dipartimento di Informatica, Università di Torino, Torino, Italy
botta@di.unito.it

Abstract. A key step of supervised learning is testing whether a candidate hypothesis covers a given example. When learning in first order logic languages, the covering test is equivalent to a Constraint Satisfaction Problem (CSP). For critical values of some order parameters, CSPs present a phase pransition, that is, the probability of finding a solution abruptly drops from almost 1 to almost 0, and the complexity dramatically increases. This paper analyzes the complexity and feasibility of learning in first order logic languages with respect to the phase transition of the covering test.

1 Introduction

This paper is concerned with supervised learning from structured examples, termed *relational learning* [Qui90] or *Inductive Logic Programming* (ILP) [MDR94]. Relational learning involves an additional difficulty, compared to learning in attribute-value languages: in first order logic, the covering test — testing whether a candidate hypothesis covers a given example — can be formulated as a Constraint Satisfaction problem (CSP), which is a NP-hard task. The phase transition manifests itself as an abrupt change, with respect to some *order parameters* of the problem class, of the probability for a problem instance to be satisfiable. This change is usually coupled with a peak in computational complexity [HHW96]; the "hardest-on-average" instances lie in the phase transition, or *mushy* region.

Previous work has provided strong evidence of the existence of a phase transition for the relational covering test [GS00]. The mushy region was empirically localized and found to be relevant to relational learning; this region is very likely to be visited when learning non-toy relational concepts. Even worse, the phase transition might have deceptive effects on the learning search; according to preliminary experiments, the mushy region seems to *attract* the learning search, no matter what the region of the "true" target concept is.

In this paper, we focus on the actual effects of the phase transition on relational learning, through systematic experiments on artificial problems. We define some hundreds of target concepts, located in the under-constrained, over-constrained and mushy regions. For any given target concept, we construct a learning and a test set. The well-known top-down relational learner FOIL [Qui90] is run on these training sets, and the theories learnt by FOIL are compared to the true target concepts. A *Failure Region* appears, where FOIL fails to either identify or accurately approximate the target concept. The reasons for FOIL's behavior are analyzed and some explanations are given.

2 Phase Transition in Hypothesis Testing

We restrict ourselves to the simplest case of concept learning in first order logic, i.e., learning a 0-ary conjunctive relation. The target concept φ is thus described as a conjunctive formula implicitly existentially quantified. Let E be a universe where φ is evaluated. E is said a positive example of φ if it contains a model of φ, and a negative example otherwise.

Let us briefly recall our previous results. Let any example E be given as a conjunction of ground literals $\alpha_i(v_{i_1},..v_{i_K})$, where α_i is a predicate symbol and v_i denotes a constant of the application domain [MDR94]. The set of literals built on a same predicate symbol α_i is termed a relation. The complexity of example E is characterized from two parameters: the number L of distinct constants and the average size N of the relations occurring in E. Similarly, the complexity of hypothesis φ_t is characterized from its number n of variables and its total number m of literals. The CSP defined as "test whether φ_t covers E" is finally characterized from the 4-tuple (n, N, m, L).

Extending the work by [Pro96], [BGS99] have shown the occurrence of a phase transition in the covering test with respect to the order parameters (m, L). Fig. 1(a) plots the probability P_{cov} for a hypothesis to cover an example as a function of m and L; n and N are set to 10 and 100, respectively. The contour plots of

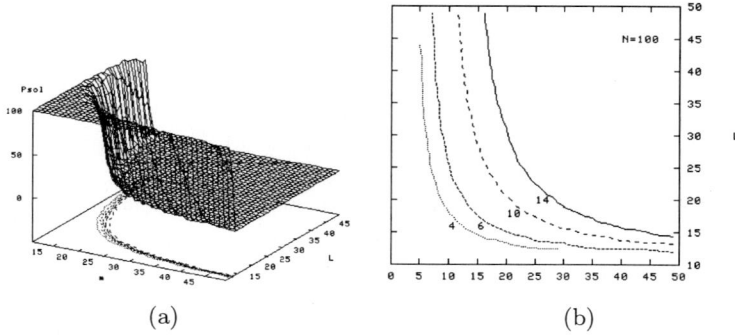

(a) (b)

Fig. 1. (a) P_{cov} in the plane (m, L); $n = 10$, $N = 100$; (b) $P_{cov} = 0.5$ for $n = 4, 6, 10, 14$; $N = 100$.

the crossover point ($P_{cov} = .5$) for different numbers n of variables is given in Fig. 1(b); the phase transition shifts toward the upper right as n increases, and the computational cost (not shown here) increases exponentially. Let φ be a concept and let m_φ denote its number of literals; let, moreover, $L_{\varphi,cr}$ be the critical number of constants for which the pair $(m_\varphi, L_{\varphi,cr})$ falls within the phase transition region. Any problem of deciding whether φ covers an aexample E lies on the vertical line $m = m_\varphi$ in the above landscape (Fig. **??**). Depending on the number L of constants in E, three possibilities are distinguished. For $L > L_{\varphi,cr}$, the covering test lies in the over-constrained region (*NO* region); assuming that E corresponds to a uniformly generated universe, E is almost surely a negative example of φ. Symmetrically, when $L < L_{\varphi,cr}$, the covering test lies in the under-constrained region (*YES* region), and E would be a positive example of φ. Last, if $L \approx L_{\varphi,cr}$, E might be a positive or negative example with about equal probability (again, assuming a uniform random generation of E). This implies that two different concepts having m_φ literals *cannot be distinguished* with respect to their coverage of uniformly generated examples, *except if* those examples involve about $L_{\varphi,cr}$ constants.

On the contrary, assuming a non-random example distribution, we may expect any rate of successful and unsuccessful covering tests in any region of the plane (m, L). In the following, both random and non-random distributions will be considerd.

3 Experiment Goal and Setting

Our study is based on artificial learning problems. Each learning problem is characterized as a triplet $(\varphi, \mathcal{E}_L, \mathcal{E}_T)$, where φ denotes the "true" target concept, and $\mathcal{E}_L$ and $\mathcal{E}_T$ respectively denote the learning and the test sets. Two restrictions have been done: φ only contains binary predicates and all predicates in the examples are relevant, i.e. they appear in φ.

A total number of 451 problems have been constructed, each characterized by a 4-tuple (n, N, m, L). The number m of literals in φ varies in the interval $[5 \div 30]$; the number L of constants in the examples varies in the interval $[11 \div 40]$. The number n of distinct variables in φ is set to 4; the relation size is set to $N = 100$ in all examples. In this way, a wide region of the plane (m, L), including the phase transition, is covered.

The problems have been generated using the random generator described by [BGS99], which guarantees a uniform distribution of both the target concepts and the examples. All training and test sets contain 100 positive and 100 negative examples each. As noted earlier, such an even distribution is quite unlikely when the pair (m, L) falls outside the mushy region and the examples follow a uniform distribution. We thus repair the training and test sets, by turning some negative examples into positive ones, by adding a model of φ when φ belongs to the *NO* region; symmetrically, some positive examples are turned into negative ones by removing all models of φ, when φ belongs to the *YES* region.

The relational learning goal is to discover either the very description of the target concept φ, or some accurate approximation $\hat{\varphi}$ of it. Besides the computational cost, two issues have been specifically considered:

Predictive accuracy. As usual, the accuracy is given by the percentage of test examples correctly classified by the hypothesis $\hat{\varphi}$ produced by the learner. The accuracy is considered satisfactory iff it is greater than 80% (the point of this threshold value will be discussed later on).
Concept identification. It must be emphasized that a high predictive accuracy does *not* imply that the learner has discovered the actual target concept φ. The two issues must therefore be distinguished. The identification is considered satisfactory iff the structure of $\hat{\varphi}$ is close to that of the true target concept φ, i.e., if $\hat{\varphi}$ is conjunctive.

Most experiments have been done using FOIL [Qui90], which basically performs a top-down exploration. It starts with the most general hypothesis, and iteratively specializes its current hypothesis φ_t by taking its conjunction with the "best" literal $\alpha_i(x_j, x_k)$ according to some statistical criterion (Information Gain [Qui90] or Minimum Description Length (MDL) [Ris78]). When specializing further the current hypothesis does not improve the criterion, φ_T is retained, all positive examples covered by φ_T are removed from the training set, and the search is restarted, unless the training set is empty. The final hypothesis $\hat{\varphi}$ returned by FOIL is the disjunction of all retained partial hypotheses φ_T.

4 Results

Predictive Accuracy. Each relational problem $(\varphi, \mathcal{E}_L, \mathcal{L}_T)$ is defined from the size m of the target concept and the number L of constants in the examples, and represented as a point in the plane (m, L). Fig 2 reports the predictive accuracy of the hypotheses $\hat{\varphi}$ learned with FOIL A successful case (resp. a failure case) corresponds to a "+" (resp. ".").

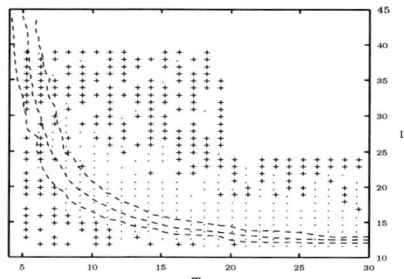

Fig. 2. Relational learning with FOIL. The *Failure region* (.) and the *Success region* (+). The upper (resp. lower) curve indicates the (m, L) points such that any random concept φ with m literals subsumes with probability .1 (resp. .9) a random example generated from L constants.

There are marked differences between successful and failure cases: the predictive accuracy usually is either very high ($\geq 95\%$) or comparable to that of random guessing ($\leq 58\%$) (Table 1). Other experiment made with other learners suggest that the failure region seems almost independent from the success criterion and the learning strategy. Overall, the experiments suggest that relational learning succeeds iff either the target concept is sufficiently small ($m \leq 6$), or the relational problem is sufficiently far away from the phase transition. The latter condition was unexpected, as it states that longer concepts (extreme right region) might be easier to learn than shorter ones (close to the phase transition). This point will be discussed further in Section 5.

Concept Identification. Table 1 reports the characteristics of the learnt theories *versus* the target concept: the first two columns recalls the coordinates m and L of the relational learning problem; columns 3 and 4 give the number of conjunctive hypotheses in $\hat{\varphi}$ and their average number of literals, respectively. Columns 5, 6 and 7 give the predictive accuracy of $\hat{\varphi}$ on the training and test sets, and the computational cost of learning (in seconds on a Sparc Enterprise 450). Last column gives the problem category, explained below. Table 1 shows three categories of relational problems.

Table 1. Target concept φ and learnt hypothesis $\hat{\varphi}$

φ		$\hat{\varphi} = \varphi_{T_1} \vee \ldots \vee \varphi_{T_K}$		Performances			
m	L	#φ_{T_i}	#lit (φ_{T_i})	ε_L	ε_T	CPU Time	
8	16	1	8	100	100	106.2	E
10	13	1	14	100	99	144.2	E
10	16	8	11.75	88	48.5	783.5	H
11	13	1	11	100	100	92.2	E
11	15	6	13.5	85	53.5	986.2	H
12	13	3	14	98.5	83	516.4	A
φ belongs to the YES region (lower left)							
15	29	1	6	100	100	185.3	A
15	35	2	6	97.5	84.5	894.6	A
18	35	1	6	100	100	201.0	A
21	18	8	4.13	81.5	58	1394.9	H
25	24	1	6	100	99	135.9	A
29	17	1	12	100	99.5	144.9	A
φ belongs to the NO region (upper right)							
6	28	12	8.08333	91.5	50.5	815.4	H
7	28	11	7.63636	91.5	60.5	1034.2	H
8	27	1	7	100	100	58.8	E
13	26	1	9	100	99	476.8	A
17	14	8	15	93	46	294.6	H
18	16	8	8.875	91	58.5	404.0	H
26	12	3	24.3333	80	58	361.4	H
φ belongs to the phase transition region							

E. *Easy* problems. FOIL finds a conjunctive hypothesis $\hat{\varphi}$ which equals φ or differs from φ by at most one literal, and correctly classifies (almost) all training and test examples. Easy problems lie in the YES or in the mushy regions, for low values of m.

A. *Approximable* problems. FOIL finds a conjunctive hypothesis $\hat{\varphi}$, which correctly classifies (almost) all training and test examples, but largely over-generalizes φ (e.g. $\hat{\varphi}$ has 6 literals instead of 18).
Approximable problems are mostly in the NO region, far away from the phase transition.

H. *Hard* problems. FOIL learns a disjunctive hypothesis $\hat{\varphi}$, involving many conjunctive hypotheses φ_T (between 6 and 15) of various sizes, and each φ_T only covers a few training examples. The predictive accuracy of $\hat{\varphi}$ is not much better than random guess on the test set. In other words, such cases involve the emergence of the true concept is a conjunctive one. Incidentally, the computational cost reaches its maximum for hard problems; this results from the number of hypotheses learned and from the fact that they lie close to the phase transition (see next paragraph).

Hard problems lie on or close to the phase transition, for high values of m.

These results confirm the fact that a high predictive accuracy does not imply that the true concept φ has been discovered. It is true that FOIL succeeds whenever it correctly discovers a single conjunctive concept $\hat{\varphi}$; but $\hat{\varphi}$ might be a wild generalization of φ. Obviously, there is no way one can distinguish between easy and approximable problems in real-world applications.

Location of the hypotheses. Let us examine the hypotheses learnt by FOIL. Except for those easy problems located in the *YES* region, conjunctive hypotheses φ_T lie in the mushy region (Fig. 3). More precisely, in the easy problems located in the mushy region, FOIL discovers the true concept; in approximable problems, FOIL discovers a generalization of the true concept, lying in the mushy region; in hard problems, FOIL retains seemingly random disjuncts, most of them lying in the mushy region. As previously noted [GS00], the phase transition behaves as an *attractor* of the learning search.

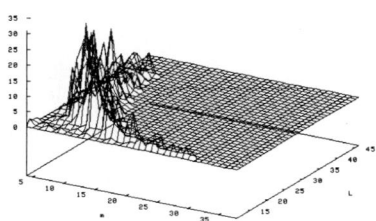

Fig. 3. Histogram of the conjunctive hypotheses φ_t.

5 Interpretation

The above results raise at least three questions. Why does the learning search end up in the mushy region ? When and why is the target concept correctly identified ? When and why should a relational learner fail to approximate the target concept ? Some tentative answers are proposed in this section.

5.1 The Phase Transition Is an Attractor

FOIL constructs a series of candidate hypotheses. It starts with a single literal φ_1, and specializes φ_t to obtain φ_{t+1}. The series of hypotheses thus forcedly starts in the *YES* region, then it might come to visit the mushy region, and possibly thereafter the *NO* region. Each φ_t is required to be representative, covering sufficiently many positive examples; the last hypothesis φ_T is such that it is sufficiently correct, covering no or few negative examples. We examine the implications of this search strategy, depending on the location of the target concept φ.

Case 1: φ belongs to the phase transition region.
By construction, φ would cover a random example with probability around .5; examples need little repairing (Section 3) in order to get evenly distributed training and test sets. Hence:
  • No hypothesis in the *YES* region can be correct as it likely covers all training examples. The search must go on until reaching the mushy region.
  • Symmetrically, any hypothesis in the *NO* region would hardly cover any training example, hence it is not representative. The search thus should stop at the very beginning of the *NO* region, and preferably before, that is, in the mushy region.
Therefore in this case, a top-down learner is bound to produce hypotheses φ_T lying in the mushy region.

Case 2: φ belongs to the *NO* region.
Here, negative examples do not need to be repaired; hence, any hypothesis in the *YES* region will cover them; thus the search must go on at least until reaching the mushy region. On the other hand, any hypothesis in the *NO* region should be correct, and there is no need to continue the search. Top-down learning is thus bound to produce hypotheses φ_T lying in the mushy region, or on the verge of the *NO* region.

Case 3: φ belongs to the *YES* region.
The situation is different here, since there exist correct hypotheses in the *YES* region, namely the target concept itself, and possibly many specializations thereof. Should these hypotheses be discovered (the chances for such a discovery are discussed in the next subsection), it would not be necessary to continue the search. In any case, the search should stop before reaching the *NO* region, for the following reason: positive examples do not need to be repaired; any hypothesis in the *NO* region would cover none of them. Then, top-down learning is bound to produce hypotheses φ_T in the *YES* or in the mushy region.

The above remarks explain why the phase transition constitutes an attractor for top-down learning.

5.2 Correct Identification of the Target Concept

As the information gain relies on the number of models of a candidate hypothesis. But any hypothesis in the *YES* region admits many models in any random example. The number of models associated to any literal is thus hardly meaningful, except when the current hypothesis is close to the target concept. Further, the variance in the number of models blinds the selection of the literals. Complementary experiments [BGSS00] show that the variance reaches its maximum as hypotheses reach the phase transition.

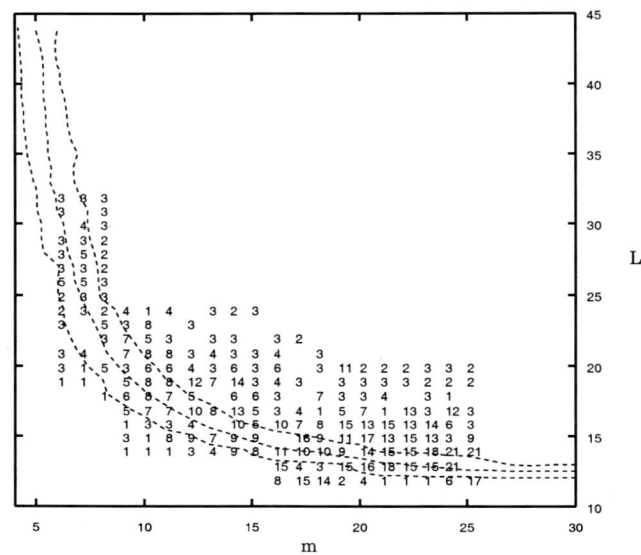

Fig. 4. Minimum size of φ_t before the information gain becomes reliable.

Fig. 4 reports, at coordinates (m, L), the minimal level t_m of the specialization process where the information gain becomes reliable. Fig. 4 could be thus interpreted as a *reliability map* of the information gain.

Note that, for most problems in the mushy region or on the borderline between the mushy and the *NO* regions, t_m takes high values, denoting a poor ability to find any correct path; moving farther away from the phase transition, t_c gradually decreases.

5.3 Good Approximation of the Target Concept

According to the above discussion, relational learning is doomed to fail when either the size m of the target concept and/or the number L of constants in the application domain, are high. Still, when both m and L are high (upper right region in Fig. 2), FOIL succeeds, and finds highly accurate hypotheses.

This can be explained as follows. Let us assume that the target concept φ belongs to the *NO* region.

We show that any generalization ξ of the target concept φ will almost surely correctly classify any training or test examples, provided that ξ belongs to the *NO* region: negative examples are randomly constructed; hence, any hypothesis in the *NO* region will be correct; in particular, ξ is correct. On the other hand, any example covered by φ is also covered by ξ; this implies that ξ covers all positive examples. Finally, any generalization of φ that belongs to the *NO* region is complete and (almost surely) correct.

It follows that, if the learning search happens to examine a generalization ξ of φ which is close to the *NO* region, ξ will be considered an optimal hypothesis, which will stop the search. The success of relational learning, with respect to predictive accuracy, thus depends on the probability of finding a generalization ξ of φ on the edge of the phase transition.

6 Conclusion

The present study shed some light on the limitations of several up-to-date relational learners. One major result of the paper is the fact that the learning search, be it based on top-down or genetic-like exploration, is trapped in the mushy region. This result is supported by the systematic experiments reported here and also by complementary experiments [GS00] on real-world applications. A second result is the fact that there is a large "blind spot" in the concept landscape. Any concept φ in this area could not be learned from examples; all relational learners considered in the present study failed to learn anything better than random guess from the available examples. This blind spot reflects the criteria used to guide the search, which actually mislead it.

References

[BGS99] M. Botta, A. Giordana, and L. Saitta. Relational learning: Hard problems and phase transitions. In *Proceedings of th 16th International Joint Conference on Artificial Intelligence*, pages 1198–1203, Stockholm, Sweden, 1999.

[BGSS00] M. Botta, A. Giordana, L. Saitta, and M. Sebag. Relational learning: Hard problems and phase transition. In *Selected papers from AIIA'99*, volume to appear. Springer-Verlag, 2000.

[GS00] A. Giordana and L. Saitta. Phase transitions in relational learning. *Machine Learning*, x:to appear, 2000.

[HHW96] T. Hogg, B.A. Huberman, and C.P. Williams, editors. *Artificial Intelligence: Special Issue on Frontiers in Problem Solving: Phase Transitions and Complexity*, volume 81(1-2). Elsevier, 1996.

[MDR94] S. Muggleton and L. De Raedt. Inductive logic programming: Theory and methods. *Journal of Logic Programming*, 19:629–679, 1994.

[Pro96] P. Prosser. An empirical study of phase transitions in binary constraint satisfaction problems. *Artificial Intelligence*, 81:81–110, 1996.

[Qui90] R. Quinlan. Learning logical definitions from relations. *Machine Learning*, 5:239–266, 1990.

[Ris78] J. Rissanen. Modeling by shortest data description. *Automatica*, 14:465–471, 1978.

Discovering Geographic Knowledge: The INGENS System

Donato Malerba   Floriana Esposito   Antonietta Lanza   Francesca A. Lisi

Dipartimento di Informatica, Università degli Studi di Bari
via Orabona 4, 70125 Bari, Italy
{malerba | esposito | lanza | lisi}@di.uniba.it

Abstract. INGENS is a prototypical GIS which integrates machine learning tools in order to discover geographic knowledge useful for the task of topographic map interpretation. It embeds ATRE, a novel learning system that can induce recursive logic theories from a set of training examples. An application to the problem of recognizing four morphological elements in topographic maps of the Apulia region is also illustrated.

1   Introduction

Data stored in many geographical information systems (GIS) concern *topographic maps*, which show relief, vegetation, hydrography and man-made features of a land portion [4]. Some map management functions implemented in current GIS are storage, retrieval and visualization on different scales. Nevertheless, the interpretation of topographic maps is an equally important facility which is rarely supported in a GIS. Indeed, information given in topographic map legends or in GIS models is often insufficient to recognize geographic objects of interest for a given application. For example, a study of the drawing instruction of Bavarian cadastral maps pointed out that symbols for road, pavement, roadside, garden and so on were defined neither in the legend nor in the GIS model of the map [8]. These objects require a process of map interpretation, which can be quite complex in some cases. The detection of morphologies characterizing the territory described in a topographic map, the selection of important environmental elements, both natural and artificial, and the recognition of forms of territorial organization require abstraction processes and deep domain knowledge that only human experts have. Although these are the patterns which geographers, geologists and town planners are interested in, they are never explicitly represented in topographic maps or in GIS.

In order to acquire the necessary knowledge for map interpretation, we propose to extend a GIS with a training facility and a learning capability, so that each time a user wants to query its database on some geographic objects not explicitly modeled, he/she can prospectively train the system to recognize such objects and to create a special user view. Both examples and counter-examples are provided by the expert user by means of the GIS interface. The symbolic representation of the training examples is automatically extracted from the maps, although it is still controlled by the user who can select a suitable level of abstraction and/or aggregation of data. The learning module of the information system implements one or more inductive learning algorithms that can generate models of geographic objects from the chosen representations of training examples.

INGENS (INductive GEographic iNformation System) is a prototypical GIS devoted to manage topographic maps of the Apulia region (Italy) to support land planning. Its logical architecture is described in the next section. The distinguishing feature of INGENS is its inductive learning capability, which is used to discover geographic knowledge of interest to town planners. In Section 3, the main characteristics and the high level algorithm of a novel learning system currently embedded in INGENS is described. This system, named ATRE, has been applied to map interpretation tasks to locate important environmental and morphological concepts on topographic maps. Section 4 is devoted to the explanation of some preliminary results. The paper concludes with a brief discussion on future work.

2 INGENS software architecture and object data model

The software architecture of INGENS is reported in Figure 1. The *Map Repository* is the database instance that contains the actual collection of maps stored in INGENS. Geographic data are organized according to a hybrid *tessellation – topological* object-oriented model.[1] The tessellation model follows the usual topographic practice of superimposing a regular grid on a map to simplify the localization process. Indeed each map in the repository is divided into square cells of same size. The raster image of a cell is stored together with its coordinates and component objects. In the topological model of each cell it is possible to distinguish two different hierarchies:

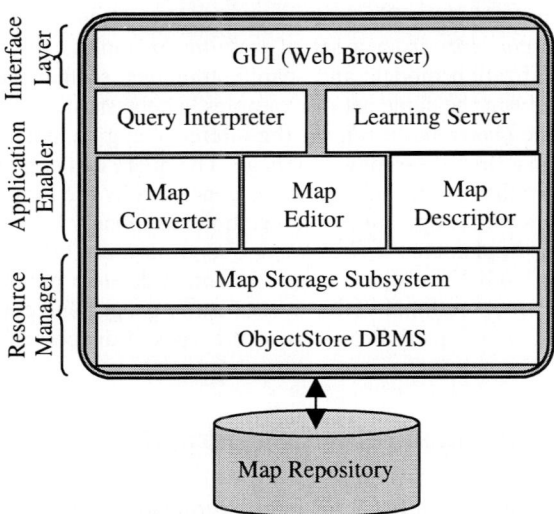

Fig. 1. INGENS three-layered software architecture.

---

[1] The object-oriented database management system (OODBMS) used to store data is ObjectStore 5.0 by Object Design, Inc.

physical and *logical*. The former describes the geographical objects by means of the most appropriate physical entity, that is point, line or region, while the latter expresses the semantics of geographic objects (hydrography, orography, administrative or political boundary, and so on), independently of their physical representation.

The *Map Storage Subsystem* is involved in storing, updating and retrieving items to and from the map repository. As *resource manager*, it represents the only access path to the data contained in the repository by multiple, concurrent clients.

The layer of the *application enablers* makes several functionalities available to the different users of the system. Users are classified in four categories:

- Administrators, who are responsible for GIS management.
- Map maintenance users, whose main task is updating the Repository.
- Sophisticated end users, who can train the system to learn operational definitions of geographic objects not explicitly modeled in the database.
- Casual end users, who occasionally access the database and may need different information each time. Casual users cannot train INGENS.

The *Map Converter* is a suite of tools which support the acquisition of maps from external sources, namely raster images from scanners and geographic objects from files of maps in a proprietary vector format. Currently, INGENS can automatically acquire information from vector maps in the MAP87 format defined by the Italian Military Geographic Institute (IGMI) (http://www.nettuno.it/fiera/igmi/igmit.htm). Since these maps contain static information on orographic, hydrographic and administrative boundaries alone, a *Map Editor* is required in order to integrate and/or modify this information. The *Map Descriptor* is the application enabler responsible for the automated generation of first-order logic descriptions of geographic objects. The *Learning Server* provides a suite of learning systems that can be run by multiple users to train INGENS. Currently, two inductive learning systems are available in the suite: INDUBI/CSL [6] and ATRE [7]. Both systems can induce first-order logic descriptions of some *concepts* from a set of training examples. Nevertheless, they adopt different generalization models and search strategies, so that they can induce different descriptions from the same set of examples. The system ATRE is described in the next section. The *Query Interpreter* is the inference engine that allows any user to formulate a query in a first-order logic language. The query can contain both spatial and aspatial descriptors that can be automatically generated by the Map Descriptor, as well as new descriptors whose operational description has already been learned.

The interface layer implements a *Graphical User Interface* (GUI), which allows the four categories of INGENS users to create/ maintain/delete a repository of maps, train the system to learn operational definitions of some geographic concepts, choose a specific map repository and query/browse it on the basis of the content of its maps.

3 Learning classification rules for geographical objects

Sophisticated end users may train INGENS in order to learn operational definitions of geographical objects that are not explicitly modeled in the database. The system ATRE, which is presented in this section, can induce recursive logical theories and can autonomously discover concept dependencies, the latter being an important issue for many map interpretation problems.

Here the term *logical theory* (or simply *theory*) denotes a set of first-order definite clauses. An example of a logical theory is the following:

downtown(X) ← *high_business_activity(X), onthesea(X).*
residential(X) ← *close_to(X,Y), downtown(Y), low_business_activity(X).*
residential(X) ← *close_to(X,Y), residential(Y), low_business_activity(X).*

It expresses sufficient conditions for the two concepts of "main business center of a city" and "residential zone," which are represented by the unary predicates *downtown* and *residential*, respectively.

The learning problem solved by ATRE can be formulated as follows:
Given
- a set of concepts $C_1, C_2, ..., C_r$ to be learned,
- a set of observations O described in a language L_O,
- a background knowledge BK described in a language L_{BK},
- a language of hypotheses L_H,
- a generalization model Γ over the space of hypotheses,
- a user's preference criterion PC,

Find
a (possibly recursive) logical theory T for the concepts $C_1, C_2, ..., C_r$, such that T is complete and consistent with respect to O and satisfies the preference criterion PC.

The *completeness* property holds when the theory T explains all observations in O of the r concepts C_i, while the *consistency* property holds when the theory T explains no counter-example in O of any concept C_i. The satisfaction of these properties guarantees the correctness of the induced theory with respect to O.

As to the representation languages L_O, L_{BK}, L_H, the basic component is the *literal*, which takes two distinct forms:

$f(t_1, ..., t_n) = Value$ (simple literal) $f(t_1, ..., t_n) \in [a..b]$ (set literal),

where *f* and *g* are function symbols called *descriptors*, t_i's and s_i's are terms, and $[a..b]$ is a closed interval. Descriptors can be either *nominal* or *linear*, according to the ordering relation defined on its domain values. Some examples of literals are: $color(X)=blue$, $distance(X,Y)=63.9$, $width(X) \in [82.2 .. 83.1]$, and $close\_to(X,Y)=true$.

The last example shows the lack of predicate symbols in the representation languages adopted by ATRE. Indeed, ATRE can deal with *classical negation*, ¬, but not with *negation by failure*, *not* [5]. Thus, the first-order literals $p(X,Y)$ and $\neg p(X,Y)$ are represented as $f_p(X,Y)=true$ and $f_p(X,Y)=false$, respectively, where f_p is the function symbol associated to the predicate p. Henceforth, for the sake of simplicity, we will adopt the usual notation $p(X,Y)$ and $\neg p(X,Y)$.

The *language of observations* L_O is *object-centered*, meaning that observations are represented as ground multiple-head clauses, called *objects*, with a conjunction of simple literals in the head. An instance of an object is the following:

downtown(zone₁) ∧ *residential(zone₂)* ← *close_to(zone₁, zone₂), onthesea(zone₁),*
         high_business_activity(zone₁), low_business_activity(zone₂).

which is semantically equivalent to the set of definite clauses:

downtown(zone₁) ←   *close_to(zone₁, zone₂), onthesea(zone₁),*
         high_business_activity(zone₁), low_business_activity(zone₂).
residential(zone₂) ←   *close_to(zone₁, zone₂), onthesea(zone₁),*
         high_business_activity(zone₁), low_business_activity(zone₂).

Multiple-head clauses are peculiar to ATRE and present two main advantages with respect to definite clauses: higher comprehensibility and efficiency. The former is basically due to the fact that multiple-head clauses provide us with a compact description of multiple properties to be predicted in a complex object like those we may have in map interpretation. The second advantage is the possibility to have a unique representation of known properties shared by a subset of observations. In fact, ATRE distinguishes *objects* from *examples*, which are described as pairs <H, OID> where *H* is a literal in the head of the object indicated by the object identifier *OID*. Examples can be considered as *positive* or *negative*, according to the concept to be learned. For instance, $\langle downtown(zone_1)=true, O_1\rangle$ is a positive example of the concept *downtown(X)=true*, a negative example of the concept *downtown(X)=false*, and it is neither a positive nor a negative example of the concept *residential(X)=true*.

The *language of hypotheses* L_H is that of *linked, range-restricted* definite clauses [2] with simple and set literals in the body and one simple literal in the head. The interval *[a..b]* in a set literal $f(X_1, ..., X_n) \in [a..b]$ is computed according to the same criterion used in INDUBI/CSL [6]. Some examples of clauses induced by ATRE are given in the next section.

The *background knowledge* defines any relevant domain knowledge. It is expressed in a language L_{BK} with the same constraints as the language of hypotheses. The following is an example of spatial background knowledge:

close_to(X,Y) ← *adjacent(X,Y)*

which states that two adjacent zones are also close.

Theories generated by ATRE can be easily translated into sets of Datalog definite clauses with built-in predicates [1], thus allowing to extend notions and properties of standard first-order logics (e.g., resolution) to ATRE definite clauses.

Regardless of the representation language adopted, a key part of the induction process is the search through a space of hypotheses. A generalization model provides a basis for organizing this search space, since it establishes when a hypothesis explains a positive/negative example and when a hypothesis is more general/specific than another. A novel generalization model, named *generalized implication* [7], is adopted by ATRE.

The main learning procedure is shown in Figure 2. To illustrate the algorithm, let us consider the following input data:

Objects O_1 $downtown(zone_1) \land \neg residential(zone_1) \land residential(zone_2) \land$
$\neg downtown(zone_2) \land \neg downtown(zone_3) \land residential(zone_4) \land$
$\neg downtown(zone_4) \land \neg downtown(zone_5) \land \neg residential(zone_5) \land$
$\neg residential(zone_6) \land downtown(zone_7) \land \neg residential(zone_7) \leftarrow$
$onthesea(zone_1), high\_business\_activity(zone_1), close\_to(zone_1, zone_2),$
$low\_business\_activity(zone_2), close\_to(zone_2, zone_4), adjacent(zone_1, zone_3),$
$onthesea(zone_3), low\_business\_activity(zone_3), low\_business\_activity(zone_4),$
$close\_to(zone_4, zone_5), high\_business\_activity(zone_5), adjacent(zone_5, zone_6),$
$low\_business\_activity(zone_6), close\_to(zone_6, zone_8), low\_business\_activity(zone_8),$
$close\_to(zone_1, zone_7), onthesea(zone_7), high\_business\_activity(zone_7).$

BK $close\_to(X,Y) \leftarrow adjacent(X,Y)$
$close\_to(X,Y) \leftarrow close\_to(Y,X)$
Concepts C_1 $downtown(X)=true$
C_2 $residential\_zone(X)=true$
PC   Minimize/maximize negative/positive examples explained by the theory.

```
procedure learn_recursive_theory(Objects, BK, {C₁,...,Cₙ}, PC)
SatObjects := saturate_objects(Objects, BK)
Examples := generate_pos_and_neg_examples(Objects, {C₁,...,Cᵣ})
LearnedTheory := ∅
Concepts := {C₁,...,Cᵣ}
repeat
    ConsistentClauses := parallel_conquer(Concepts, Examples, PC)
    Clause := find_best_clause (Consistent_clauses, PC )
    ConsistentTheory:=verify_global_consistence(Clause,Learned_theory,Objects,Examples)
    LearnedTheory := ConsistentTheory ∪ {Clause}
    Objects := saturate_objects(SatObjects, LearnedTheory)
    Examples := update_examples(LearnedTheory,Examples)
    foreach Cᵢ in Concepts do
        if pos_examples(Cᵢ)= ∅ then Concepts := Concepts / {Cᵢ}
        endif
    endforeach
until Concepts = ∅
return LearnedTheory
```

Fig. 2. ATRE: Main procedure.

The first step towards the generation of inductive hypotheses is the *saturation* of all objects with respect to the given *BK* [9]. In this way, information that was implicit in the object, given the background knowledge, was made explicit. In the above example, the saturation of O_1 involves the addition of the nine literals logically entailed by BK, that is: *close_to(zone₂, zone₁), close_to(zone₁, zone₃), close_to(zone₃, zone₁), close_to(zone₇, zone₁), close_to(zone₄, zone₂), close_to(zone₅, zone₄), close_to(zone₅, zone₆), close_to(zone₆, zone₅),* and *close_to(zone₈, zone₆).*

Initially, all positive and negative examples (pairs ⟨L,OID⟩) are generated for every concept to be learned, the learned theory is empty, while the set of concepts to be learned contains all C_i. With reference to the above input data, the system generates two positive examples for C₁ (*downtown(zone₁)* and *downtown(zone₇)*), two positive examples for C₂ (*residential(zone₂)* and *residential(zone₄)*), and eight negative examples equally distributed between C₁ and C₂ (¬*downtown(zone₂),* ¬*downtown(zone₃),* ¬*downtown(zone₄)* ∧ ¬*downtown(zone₅),* ¬*residential(zone₁),* ¬*residential(zone₅),* ¬*residential(zone₆),* ¬*residential(zone₇)*).

The procedure *parallel_conquer* generates a set of consistent clauses, whose minimum number is defined by the user. For instance, by requiring the generation of at least one consistent clause with respect to the above examples, this procedure returns the following set of clauses:

downtown(X) ← *onthesea(X), high_business_activity(X).*
downtown(X) ← *onthesea(X), adjacent(X,Y).*
downtown(X) ← *adjacent(X,Y), onthesea(Y).*

The first of these is selected according to the preference criterion (procedure *find_best_clause*). In fact, the hypothesis space of the concept *residential* has been simultaneously explored, just when the three consistent clauses for the concept *downtown* have been found, no consistent clause for *residential* has been discovered yet. Thus, the *parallel_conquer* procedure stops, since the number of consistent clauses is greater than one.

Since the addition of a consistent clause to the partially learned theory may lead to an augmented, inconsistent theory, the procedure *verify_global_consistence* makes necessary checks and possibly reformulates the theory in order to recover the consistency property without repeating the learning process from scratch. The reformulation is based on the *layering* technique, which is peculiar to ATRE. The learned clause is used to resaturate the object. Continuing the previous example, the two literals added to O_1 are *downtown($zone_1$)* and *downtown($zone_7$)*. This operation enables ATRE to generate also definitions of the concept *residential* that depend on the concept *downtown*. Indeed, at the second iteration the procedure *parallel_conquer* returns the clause:

residential(X) ← *close_to(X,Y), downtown(Y), low_business_activity(X).*

and by resaturating the object with both learned clauses, it becomes possible to generate a recursive clause at the third iteration, namely:

residential(X) ← *close_to(X,Y), residential(Y), low_business_activity(X).*

At the end of each iteration, the procedure *update_examples* tags positive examples explained by the current learned theory, so that they are no longer considered for the generation of new clauses. The loop terminates when all positive examples are tagged, meaning that the learned theory is complete and consistent.

4. Application to Apulian map interpretation

INGENS has been applied to the recognition of four morphological elements in topographic maps of the Apulia region (Italy), namely *regular grid system of farms*, *fluvial landscape*, *system of cliffs* and *royal cattle track*. Such elements are deemed relevant for the environmental protection, and are of interest to town planners. A regular grid system of farms is a particular model of rural space organization that originated from the process of rural transformation. The fluvial landscape is characterized by the presence of waterways, fluvial islands and embankments. The system of cliffs presents a number of terrace slopes with the emergence of blocks of limestone. A royal cattle track is a road for the transhumance that can be found exclusively in the South-Eastern part of Italy.

The considered territory covers 131 km^2 in the surroundings of the Ofanto River, spanning from the zone of Canosa until the Ofanto mouth. The examined area is covered by five map sheets on a scale of 1:25000 produced by the IGMI. The gridding step chosen for the segmentation of the territory delimits, for each map, square observation units of 1 Km^2 each. This is the same gridding system superimposed over IGMI topographic chart on a scale of 1:25000. Thus there is a one-to-one mapping between observation units in the chart and single cells in the database. Each cell has to be described in the logic formalism of ATRE objects.

First-order logic descriptions of the maps are generated by applying algorithms derived from geometrical, topological, and topographical reasoning. Since descriptors are quite general they can also be used to describe maps on different scales. A partial description of a cell containing fifty-two distinct objects is given in Figure 3. The whole description is a clause with three hundred and forty literals in the body.

```
class(x1)=other ←
    contain(x1,x2)=true, contain(x1,x3)=true, ..., contain(x1,x53)=true,
    type_of(x2)=canal_line, type_of(x3)=vegetation, ..., type_of(x53)=vegetation,
    color(x2)=blue, color(x3)=black,..., color(x53)=black, trend(x2)=straight,
    trend(x8)=straight, ..., trend(x51)=curvilinear, extension(x2)=184.057,
    extension(x8)=170.074, ..., extension(x51)=982.207,
    geographic_direction(x2)=north_west, geographic_direction(x8)=north_est, ...,
    geographic_direction(x49)=north_est, shape(x16)=cuspidal, shape(x18)=cuspidal, ...,
    shape(x44)=cuspidal, density(x4)=low, density(x6)=low, ..., density(x52)=low,
    relation(x8,x10)=almost_parallel, relation(x8,x14)=almost_parallel, ...,
    relation(x46,x49)=almost_parallel, distance(x8,x10)=463.09,
    distance(x8,x14)=423.111, ..., distance(x46,x49)=477.322
```

Fig. 3. Partial logical description of a cell. Constant $x1$ represents the whole cell, while all other constants denote the fifty-two enclosed geographic objects. Distances and extensions are expressed in meters.

Then the problem of recognizing the four morphological elements can be reformulated as the problem of labeling each cell with at most one of four labels. Unlabelled cells are considered uninteresting with respect to the goal of environmental protection. Globally 131 cells were selected, each of which was assigned to one of the following five classes: system of farms, fluvial landscape, system of cliffs, royal cattle track and other. The last class simply represents "the rest of the world," and no classification rule is generated for it. Indeed, the cells assigned to it are not interesting with respect to the problem of environmental protection under study, and they are always used as negative examples when ATRE learns classification rules for the remaining classes. Forty-five cells from the map of Canosa were selected to train the system, while the remaining eighty-six cells were randomly selected from the other four maps. The preference criterion *PC* maximizes both the number of explained positive examples and the number of clause literals.

A fragment of the logical theory induced by ATRE is reported in the following:

class(X1) = *fluvial_landscape* ← *contain(X1,X2)*, *color(X2)=blue*,
    type_of(X2)=river,trend(X2)=curvilinear, extension(X2) $\in$ *[325.00..818.00]*.
class(X1) = *fluvial_landscape* ← *contain(X1,X2)*, *type_of(X2)=river*, *color(X2)=blue*,
    relation(X3,X2)=almost_perpendicular, extension(X2) $\in$ *[615.16..712.37]*,
    trend(X3)=straight.
class(X1)=system_of_farms ← *contain(X1,X2)*, *color(X2)=black*,
    relation(X2,X3)=almost_perpendicular, relation(X3,X4)=almost_parallel,
    type_of(X4)=interfarm_road, geographic_direction(X4)=north_est,
    extension(X2) $\in$ *[362.34 .. 712.25]*, *color(X3)=black*,
    type_of(X3)=farm_road, color(X4)=black, trend(X2)=straight.

The first two clauses explain all training observations of a fluvial landscape, while the third clause is a partial description of the concept *system_of_farms*.

In order to test the accuracy of the induced theory, the *Query Interpreter* was provided with both the eighty-six observations, reserved for the test phase, and the theory itself. Test cells were recognized with a predictive accuracy of over 95%. These results are promising, although they are affected by the careful selection of both a suitable representation of observations and a training set. Results of a previous experiment on a smaller scale map of the same region (1:50000) are reported in [3].

5. Conclusions and future work

Knowledge of the meaning of symbols listed in map legends is not generally sufficient to recognize interesting geographic patterns on a topographic map, so that GIS users are asked to formulate quite complex queries to describe such patterns. In fact, these user queries are operational definitions of abstract concepts often reported in specialist texts and handbooks. To support GIS users in their activity a new approach has been proposed in this paper. The idea is asking users a set of classified instances of the patterns of their interest, and then applying machine learning tools and techniques to generate the operational definitions for such patterns. These definitions can be subsequently used to search for new instances not in the training set, or to facilitate the formulation of a query. INGENS is a prototypical GIS with learning capabilities that has been designed and implemented to provide users with a training facility. An application of the system to the problem of Apulian map interpretation has been briefly described, and preliminary experimental results are presented. The learning system used in this application is ATRE, whose innovative features have been briefly explained.

INGENS can be extended in various directions. Currently, a set of generalization and abstraction operators has been implemented to provide the user with some tools that simplify the complex descriptions produced by the *Map Descriptor*. These operators are similar to those commonly used in on-line analytical processing (OLAP) tools. For the future, we plan to embed a system for the discovery of spatial association rules in the Learning Server.

References

1. Ceri, S., Gottlob, G., and Tanca, L. 1989. What you always wanted to know about Datalog (and never dared to ask), IEEE Transactions on Knowledge and Data Engineering, 1(1), pp. 146-166.
2. De Raedt, L. 1992, Interactive theory revision: An inductive logic programming approach. London: Academic Press.
3. Esposito, F., Lanza, A., Malerba, D., and Semeraro, G. 1997. Machine learning for map interpretation: an intelligent tool for environmental planning. Applied Artificial Intelligence. 11(7-8), October-December:673-695.
4. Laurini, R., and Thompson, D. 1992. Fundamentals of Spatial Information Systems. Academic Press.
5. Lloyd, J.W. 1987, Foundations of logic programming. 2nd ed. Berlin: Springer-Verlag.
6. Malerba, D., Esposito, F., Semeraro, G. , and Caggese, S. 1997. Handling Continuous Data in Top-down Induction of First-order Rules. In AI*IA 97: Advances in Artificial Intelligence, ed. M. Lenzerini, Lecture Notes in Artificial Intelligence 1321, pp. 24-35. Springer: Berlin.
7. Malerba, D., Esposito, F. and Lisi, F.A. 1998. Learning Recursive Theories with ATRE. In Proc. of the 13th European Conf. on Artificial Intelligence, ed. H. Prade, pp. 435-439, John Wiley & Sons: Chichester (UK).
8. Mayer, H. 1994. Is the knowledge in map-legends and GIS-models suitable for image understanding? In International Archives of Photogrammetry and Remote Sensing 30, 4, pp. 52-59.
9. Rouveirol, C. 1994. Flattening and saturation: Two representation changes for generalization. Machine Learning. 14(2), February: 219-232.

Temporal Data Mining Using Hidden Periodicity Analysis

Weiqiang Lin* and Mehmet A. Orgun

Department of Computing, Macquarie University
Sydney, NSW 2109, Australia
E-mail: {wlin,mehmet}@comp.mq.edu.au

Abstract. Data mining, often called knowledge discovery in databases (KDD), aims at semiautomatic tools for the analysis of large data sets. This report is first intended to serve as a timely overview of a rapidly emerging area of research, called temporal data mining (that is, data mining from temporal databases and/or discrete time series). We in particular provide a general overview of temporal data mining, motivating the importance of problems in this area, which include formulations of the basic categories of temporal data mining methods, models, techniques and some other related areas. This report also outlines a general framework for analysing discrete time series databases, based on hidden periodicity analysis, and presents the preliminary results of our experiments on the exchange rate data between US dollar and Canadian dollar.

Keywords: data mining, temporal databases, temporal data analysis, time series, statistical theory, hidden periodicity analysis.

1 Introduction

Data Mining, also known as Knowledge Discovery in Databases(KDD), aims at semiautomatic tools for the analysis of large, realistic data sets. It is a rapidly evolving area of research, that is at intersection of several disciplines, including statistics, pattern recognition, databases, optimization, visualization and high-performance computing. There are some very important and challenging research problems in data mining, for instance, the application of data mining techniques and tools to different types of databases such as temporal databases, spatial databases and so on. This paper focusses on issues and challenges for mining temporal databases.

Temporal data mining is concerned with discovering qualitative and quantitative temporal patterns in a temporal database or in a discrete-valued time series (DTS) dataset. Recently, there has been special attention to two kinds of major problems in the literature: similarity problem and periodicity problem. Although there are various results to date on discovering periodic patterns and similarity patterns in discrete-valued time series (DTS) datasets (e.g. [2]), a general theory of discovering patterns for DTS data analysis is not well known.

---

* Communicating author: W. Lin, Department of Computing, Macquarie University, Sydney, NSW 2109, Australia, Phone: +61 2 9850-9514, Fax: +61 2 9850-9551, Email: wlin@comp.mq.edu.au

In this article, we first provide an overview of temporal data mining (TDM), define some of the key ideas, identify a variety of challenging problems, both in the theory and the systems, and motivate their importance. We then propose a general framework for analysing a DTS and then focus on the special problem of discovering patterns using hidden periodicity analysis.

The rest of the paper is organized as follows. Section 2 starts discussion with temporal databases (in particular time-series databases) and then moves onto current issues in temporal data mining. Section 3 provides a few definitions for temporal patterns, and the theory and methods of hidden periodicites analysis. Section 4 moves onto experimental analysis for pattern discovery on US dollar versus Canadian dollar exchange rates. The paper concludes with a brief summary.

2 An Overview of Temporal Data Mining

2.1 What Is a Temporal Database?

Time is an important aspect of real world phenomena. Conventional databases model an enterprise as it changes dynamically by a snapshot at particular points in time. Traditional databases store only the current state of characteristic of the data, so when new data become valid, old ones are overwritten (or lost). But in many situations, this kind of databases is inadequate. They can not easily handle historical queries, because they are not designed to model the way in which the entities represented in the database change over time.

Due to the importance of time-varying data, efforts have been made to design Temporal Databases (TDB) which support some aspect of time such as valid time, logical time and transaction time [3]. TDBs are able to overcome this limitation of traditional databases by not overwriting attribute data information, but instead storing valid time ranges with them, which can be used to determine their validity at particular times, including the present. There are numerous time concepts proposed to date for storing information in temporal databases such as: valid time, denoting the time a fact was true in reality and transaction time, representing the time the information was entered into the database. In addition to these two concepts which are of general interest, there are also user-defined time (time fields in a traditional database), decision time, absolute time and relative time.

These kinds of time induce different types of databases. A traditional database supporting neither vaild nor transaction time is termed a snapshot database, since it contains only a snapshot of the real world. A valid-time database contains the entire history of the enterprise, as is best known now. A transaction-time database supports transaction-time and hence allows rolling back the database to a previous state. We adopt the definition of Temporal Database provided by Tansel *et al* [4] as follows:

Definition 1 *A Temporal Database (TDB) is real world database that maintains past, present, and future data.*

A TDB model may support one or more of the time concepts. There has been a great deal of interest in temporal databases over the last decade with the number of papers published in the area rising steadily. There are numerous models for temporal databases which have been designed using both object oriented and relational database as the underlying database models. The basic understanding of temporal databases has progressed to the point where a standard temporal language and infrastructure have been proposed [5]. Most of the work done to date has identified the basic properties of temporal information and various data models and associated algebras and query languages have been produced in order to manipulate data with a temporal component.

Although research in temporal databases is now quite mature, the development of a general purpose temporal data mining system still remains in its infancy. According to temporal characteristics, objects in temporal databases can be classified into three categories [6]: (1)Time-invariant objects, (2) Time-varying objects, and (3) Time-series objects.

In rest of the section, we focus only on temporal databases for time-series objects, which are often called time-related databases. Time-related databases are of growing importance in many modern database applications, such as data mining, data warehousing and so on.

2.2 Temporal Data Mining

Temporal data mining is to perform time series analysis on the information held in a temporal database. Statistical methods provide a natural way of analysing time-related information in a temporal database.

Definition 2 *Temporal Data Mining deals with problems of knowledge discovery from large Temporal Databases.*

A relevant and important question is how to apply data mining techniques on a temporal database and how to interpret the results. For instance, sequential/temporal patterns are mined to analyse a collection (subset) of records over periods of different variables/or time as whole records (set) of variables/or time. Few sequential/temporal techniques have been developed, based on Discrete Fourier Transformation, to map a time sequence to frequency domain. Other techniques used in the discovery of sequential/temporal patterns include dynamic time wrapping, neural networks and rough sets.

According to techniques of data mining and theory of statistical time series analysis, the theory of temporal data mining may involve following areas of investigation:

1. Temporal data mining tasks include:
    - Temporal data characterization and comparison,
    - Temporal clustering analysis,
    - Temporal classification,
    - Temporal association rules,
    - Temporal pattern analysis,
    - Temporal prediction and trend analysis.
2. A new temporal data model may need to be developed based on:
    - Temporal data structures,
    - Temporal semantics, or

3. A new temporal data mining concept may need to be developed based on the following:
   - the task of temporal data mining can be seen as a problem of extracting an interesting part of the logical theory of a model, and
   - the theory of a model may be formulated in a logical formalism able to express quantitative knowledge and approximate truth.

There are two kinds of problems that have been studied in temporal data mining area in recent years: (1) the similarity problem, that is, finding a time sequence (or TDB) or subtime sequence similar to a given sequence (or query) or finding all pairs of similar subtime sequences in the time sequence; and (2) the periodical problem, that is, finding periodic patterns in TDB.

Similarity Problems. In data mining applications, it is often necessary to search within a time series database (e.g,TDB) for those series that are similar to a given query series. This kind of a question is part of the general problem often called the similarity search problem. The Similarity Search Problem in time-series objects TDBs is to find out how many of them are similar to one another (or to compare with a given series) within the same or between different time-series set(s) which may be one-dimensional or multi-dimensional.

There are two main categories for similarity problems in time-series objects TDBs:

- All-Occurrences Sub-Sequence Matching (AOSM): given a query series Q of length n and a TDB with length $N(N \gg n)$, find all occurrences of a contiguous subsequence within the TDB that matches Q approximately. The matching is under the condition that the query series Q is small, and we look for a subset of the TDB that best matches the query Q.
- All-Occurrences Whole-Sequence Matching (AOWSM): given a query series Q of length n and a set of number N of data sequences (TDBs) with the same length n, find all occurrences of the TDBs that match Q approximately. The matching is under the condition that the sequences to be compared have the same length and we look for the TDB that match the query sequence Q.

Many data mining techniques have been applied in similarity problems such as classification, regression and clustering/segmentation. The main steps for solving the similarity problem are as follows:

- define similarity: this step allows us to find similarities between sequences with different scaling factors and baseline values.
- choose a query sequence: this step allows us to find what we want to know from large sequences (e.g, characteristic, classification)
- processing algorithms for TDB: this step allows us to use some statistical methods (e.g, transformation, wavelet analysis) on a TDB to remove noisy data, and interpolate missing data.
- processing an approximate algorithm: this step allows us to build up the classification scheme for a time-series TBD according to the definition of similarity by using some data mining techniques (e.g, visualisation).

In search for similarity, a given query series may have some different types of matching, such as full match, match with shift, match with scaling, match with combination of scaling and shifting, approximate match and so on. Also lots of techniques in these areas have made use of statistical analysis theory such as wavelet analysis, multi-fourier analysis and various statistical transformations. Some results of similarity problems and case studies have been published in the literature(e.g, [7]). The results from a similarity search in a time-series TDB can be used for association, prediction, and so on.

Periodical Problems. The periodic problem involves finding periodic patterns or, cyclicity occurring in a time-series TDB. The problem is related to two concepts: *pattern* and *interval*. In any selected sequence of a TDB, we are interested in finding the patterns that repeat over time and their recurring intervals (period), or finding the repeating patterns of a sequence (or TDB) as well as the interval which corresponds to the pattern period. The basic variations of preiodical problems include: *value-based, trend-base, partial pattern* and *complete pattern* problems. There are two main categories for periodical problems:

1. Fixed Period Periodicity Search: This kind of a periodicity search algorithm is based on data cubes and OLAP operations combined with some sequential pattern search strategies to discover large periodic patterns in a time series (or TDB).
2. Arbitrary Periodicity Search: This kind of periodicity search algorithms are based on mathematical techniques: sequential algorithms, forward optimization algorithms and backward optimization algorithms.

A general theory for Searching Periodical Problems in a time-series TDB is still lacking, but we can consider some main steps for searching periodical problems:

– determine some definitions of the concept of a period under some assumptions so that we know what kind of a periodicity search we want to perform on the TDB.
– build up a set of algorithms that allow us to use properties of periodic time series for finding out periodic patterns from a subset of the TDB.
– apply simulation algorithms to find patterns from the whole TDB.

If a time-series TDB contains unhealthy data such as noisy data and missing data, then the results of a periodicity search will not be useful.

Note: A lot of techniques have been involved in this kind of problems by using pure mathematical analysis such as function data distribution analysis and so on (e.g, [1]).

Discussion. In a time-series TDB, sometimes similarity and periodical search problems are difficult even when there are many existing methods, but most of the methods are either inapplicable or prohibitively expensive. In fact, similarity and periodical search problems can be combined into the problem of finding interesting sequential patterns in TDBs. Since sequential patterns are essentially associations over temporal data, they utilize some of the ideas initially proposed for the discovery of association rules (e.g, [8,9]). In recent years, some new algorithms have been developed such as:

- generalized sequential pattern (GSP) algorithm: it essentially performs a level-wise or breadth-first search of the sequence lattice spanned by the subsequence relation,
- sequential pattern discovery using equivalence classes (SPADE) algorithm: it decomposes the original problem into smaller sub-problems using equivalence classes on frequent sequences.

2.3 Some Research Challenges

We list below some of the challenges that are of particular relevance to data mining.

- Develop a general theory for a foundation of temporal data warehousing, supporting multiple granularities and multiple lines of evolution, for data mining purposes.
- Develop a general asymptotic theory (parameter estimating) for temporal database models by statistical tools(using multivariate functional analysis).
- Develop a general modelling theory on different temporal databases for forecasting of the processes with or without parameters.
- Develop data reduction methods for removing redundant or irrelevant data.
- Build up a general technique for mapping a local multivariate time series, using part of temporal data, to k-dimensional space such that the dissimilarties (or interesting properties) are preserved.
- Develop new temporal data mining methodologies such as statistical tools, neural nets and ad-hoc query-based mining etc.

In the rest of the paper, we return our attention to a particular temporal data mining method based on a well-known statistical tool, called hidden periodicity analysis[10].

3 Hidden Periodicity Analysis

This section first provides a few definitions and results to formalize what we mean by periodic and similar patterns, and then discusses hidden periodicity analysis in more detail.

3.1 Temporal Patterns

Without loss of generality, we consider the bivariate data $(X_1, Y_1), \ldots, (X_n, Y_n)$, which form an independent and identically distributed sample from a population (X, Y). Then the data as being generated from the model is

$$\mathbf{Y} = m(\mathbf{X}) + \sigma(\mathbf{X})\varepsilon$$

where $E(\varepsilon) = 0$, $Var(\varepsilon) = 1$, and X and ε are independent.

We assume that for every successive pair of two time points in DTS $t_{i+1} - t_i = f(t)$ is a function (in most cases, $f(t)$ = constant). For every succession of three time points: X_j, X_{j+1} and X_{j+2}, the triple value of (Y_j, Y_{j+1}, Y_{j+2}) has only different " 9-states" (or, called 9 local features). If we let states: S_s is the same state as prior one, S_u is the go-up state compare with prior one and S_d is the go-down state compare with prior one, then we have the state-space $S = \{s1, s2, s3, s4, s5, s6, s7, s8, s9\} = \{(Y_j, S_u, S_u), (Y_j, S_u, S_s), (Y_j, S_u, S_d), (Y_j, S_s, S_u), (Y_j, S_s, S_s), (Y_j, S_s, S_d), (Y_j, S_d, S_u), (Y_j, S_d, S_s), (Y_j, S_d, S_d) \}$.

Definition 3 Let $h = \{h_1, h_2, \ldots, \}$ be a sequence. If $h_j \in S$ for every $h_j \in h$, then the sequence h is called a Structural Base sequence and the subsequence h_{sub} of h is called a sub-Structural Base sequence.

If h_{sub} is a periodic sequence, then h_{sub} may be called sub-structural periodic sequence(e.g, full or partial periodic pattern in B). Also h is a structural periodic sequence (existence periodic pattern(s)).

Note that a sequence is called a full periodic sequence if its every point in time contributes (precisely or approximately) to the cyclic behavior of the overall time series (that is, there are cyclic patterns with the same or different periods of repetition).

A sequence is called a partial periodic sequence if the behavior of the sequence is periodic at some but not all points in the time series.

Definition 4 Let $y = \{y_1, y_2, \ldots.\}$ be a real value sequence. If $y_{sub} = \{y_1, y_2, \ldots .y_N\}$ ($y_j \in Y, j = 1, 2, \ldots, N$) and N is the size of a subset of y, then it may be called a value-point process. If y_j with $0 \le y_k < 1$ (mod 1) for all N,then we say that y is uniformly distributed if every subinterval of [0, 1] gets its fair share of the terms of the sequence in the long run. More precisely, if

$$\lim_{n \to \infty} \frac{number\ of\ \{j \le n : y_k \in J\}}{n} = length\ of\ J$$

for all subintervals J of [0, 1).

Definition 5 Let $y = \{y_1, y_2, \ldots.\}$ be a sequence of real numbers with $I - \delta < y_k < I + \delta$ for all k. We say that y has an approximate constant sequence distribution. More generally, if $h(t) - \delta < y_k < h(t) + \delta$ for all k, we say that y has an approximate distribution function $h(t)$.

We have the following results([11]):

Lemma 1. *A discrete-valued dataset contains periodic patterns if and only if there exist structural periodic patterns and periodic value-point processes with or without an independently identical distribution (i.i.d.).*

Lemma 2. *In a discrete-valued dataset, there exist similarity patterns if and only if there exist structural base periodic patterns and similarity value-point distribution with or without an independently identical distribution.*

3.2 Methods of Hidden Periodicity Analysis

We briefly introduce the hypothesis testing method of Grenander [10] for detecting hidden periodicities in noisy data. Suppose that the model of general observations of a sub-set is that of

$$x(t) = \sum_{n=1}^{P} \xi_n e^{i\lambda_n t} + \eta(t), \qquad t \in \mathbb{Z}$$

where P is known, ξ_n and λ_n are unkown parameters, $\eta(t)$ is independently identical distributed (i.i.d. $N(0, \sigma^2)$) and σ is unkown parameter [1]. Then Grenander suggested the following testing procedure:

$$H_0 : x(t) = \gamma(t), \quad t = 0, \pm 1, \pm 2, \ldots$$

$$H_1 : M = r \ (\xi(t) \text{ possesses } r \text{ frequency components}, \ 1 \leq r \leq m)$$

In the hypothesis testing, the parameter r is assumed to be known a priori. Since in usual cases r is unknown, we apply the testing step by step, i.e. first put $r = 1$ and apply the testing. If H is rejected then put $r = 2$ and so on, until, say $r = p + 1$, when H is accepted then we estimate the order as p. If $r = 1$ is rejected, then it means $\xi(t)$ is not a white noise series, so the distribution of $P\{g(r)\ z\}$ has to be changed.

4 Discovery of Temporal Patterns

4.1 Structural Pattern Discovery

From the point of view of our new method in data analysis, we use squared distance functions which are provided by a class of positive semidefinite quadratic forms. Specifically, if $\mathbf{u} = (u_1, u_2, \cdots, u_p)$ denotes the p-dimensional observation of each different distance of patterns in a state on an object that is to be assigned to one of the g prespecified groups, then, for measuring the squared distance between $\mathbf{u}$ and the centroid of the ith group, we can consider the function

$$D^2(i) = (\mathbf{u} - \bar{\mathbf{y}})' \mathbf{M} (\mathbf{u} - \bar{\mathbf{y}})$$

where $\mathbf{M}$ is a positive semidefinite matrix to ensure the $D^2(i) \geq 0$. Different choices of the matrix $\mathbf{M}$ lead to different metrics, and the class of squared distance functions represented by above equation is not unduly narrow.

4.2 Point-Value Pattern Discovery

Here, we introduce an enhancement of an approach for modelling discrete-valued time series through hidden periodicity analysis. On the value-point pattern discovery, suppose that the model of observations is of subsection 3.2

The first stage of method for detecting the characteristics of those records is to use the linear regression analysis. We may assume linear model is $\mathbf{Y} = \mathbf{X}\beta + \varepsilon$. The linear model based upon least square estimation (LSE) is $\hat{\beta} = (\mathbf{X}^T\mathbf{X})^{-1}\mathbf{X}^T\mathbf{Y}$. Then we have: $\hat{\beta} \sim N(\beta, Cov(\hat{\beta}))$. Particularly, for $\hat{\beta}_i$ we have $\hat{\beta}_i \sim N(\beta_i, \sigma_i^2)$, where $\sigma_i^2 = \sigma^2 a_{ii}$, and a_{ii} is the ith diagonal element of $(\mathbf{X}^T\mathbf{X})^{-1}$.

Now, for each value-point, we may fit a linear model as above and parameters can be estimated under LSE. Therefore, we first remove the trend effect of each curve from the original record by subtracting the above regression function at x from the corresponding value to obtain a comparatively stationary series. Then the problem can be formulated as the hidden periodicity analysis of discrete-valued time series.

---

[1] In fact, Grenander considered the model $\xi(t) = \sum_{k=1}^{P} A_k cos(\omega_k t) + \gamma(t)$ where P is known, A_k, ω_k are unkown parameters, $\gamma(t)$ is i.i.d. $N(0, \sigma^2)$ and σ is unkown parameter

4.3 Experimental Results

For brevity, we only present a few experimental results for both structural and value-point pattern discovery, on the exchange rate between the US and Canadian dollars.

Structural pattern discovery experiments. We are investigating the sample of the structural base to test the naturalness of the similarity and periodicity on Structural Base distribution. We consider 9 states in the state-space of structural distribution: $S = \{s1, s2, s3, s4, s6, s7, s8, s9\}$. In summary, some results for the structural base experiments are as follows (e.g., see right Figure of 1).

- Structural distribution in a practical transition of states is a hidden periodic distribution with a periodic length function $f(t)$.
- There exist some partial periodic patterns in a practicular transition of states based on a distance shifting function $d(t)$.
- There also exist some similarity patterns with a small distance shifting in a practicular transition of states.

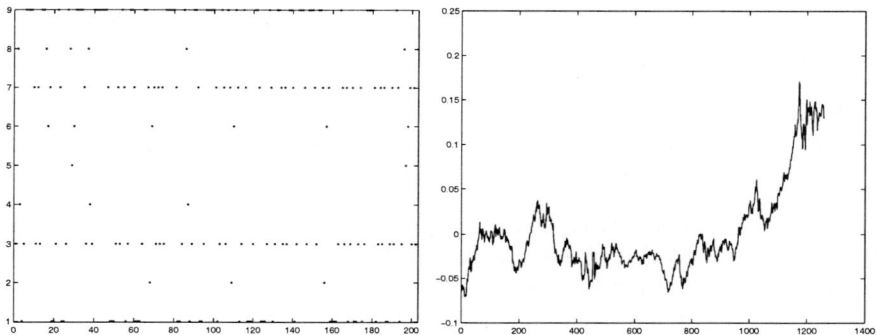

Fig. 1. Left: 200 business days of daily U.S. dollar exchange rate against Canadian dollar in states base. Right: After removal the trend effect represented by linear regression function w(t) for the 1257 business days.

Value-point pattern discovery experiments. Suppose the exchange rate value can be modelled as

$$Y_i = m(t)Y_{i+k} + \varepsilon_i, \qquad (k > 0 \text{ and fixed integer})$$

We then adjust off the trend by the linear regression function Y_i and a new series

$$w(t) = v(t) - Y(t), \qquad t = 1, 2, \ldots, N$$

may be obtained, where $v(t)$ is the original record.

Then we may use the linear regression by hidden perodicity analysis for the new value base series $w(t)$. Some results for the value-point of experiments are given below:

- there does not exist any full periodic pattern, but there exist some partial periodic patterns with a distance shifting function $d_j(t)$,
- there exist similarity patterns, etc.

5 Concluding Remarks

This paper has reviewed current research problems and challenges in temporal data mining. It has also presented a new method based on Hidden Periodicity Analysis for finding patterns in discrete-valued time series databases. The method described in this paper is still in its preliminary stages. But it guarantees finding different patterns with structural and valued probability distribution of a real-dataset. The method can be implemented using a straightforward algorithm, and the results of preliminary experiments are promising.

Acknowledgements. This research has been supported in part by an Australian Research Council (ARC) grant and a Macquarie University Research Grant (MURG). Thanks are also due to for Chit Swe many fruitful discussions.

References

1. R. Agrawal and R. Srikant. Mining sequential patterns. In *International Conference on Database Engineering*, pages 3–14. IEEE Computer Society, 1995.
2. C. Bettini. Mining temportal relationships with multiple granularities in time sequences. *IEEE Transactions on Data & Knowledge Engineering*, 1998.
3. R. Snodgrass. Temporal databases. *IEEE Computer*, 19, 1982.
4. S.Gadia S.Jajodia A.Segev A.U.Tansel, J.Clifford and R. Snodgrass, editors. *Temporal databases theory, design and implementation*. Benjamin Publishing Compang, 1993.
5. S.Jajodia andS.Sripada O.Etzion, editor. *Temporal databases: Research and Practice*. Springer-Verlag,LNCS1399, 1998.
6. J. Y. Lee, R.Elmasri, and J.Won. An integrated temporal data model incorporating time series concept. 1997.
7. Michael K. Ng and Zhexue Huang. Temporal data mining with a case study of astronomical data analysis. In G. Golub, S. H. Lui, F. Luk, and R. Plemmons, editors, *Proceedings of the Workshop on Scientific Computing 97*, pages 258–264. Springer-Verlag, Hong Kong, March 1997.
8. M.J.Zaki. Fast mining of sequential patterns in very large databases. *Uni. of Rochester Technical report*, 1997.
9. R. Agrawal, K.I.Lin, H.S.Sawhney, and K.shim. Fast similarity search in the presence of noise, scaling, and translation in time-series databases. In *21th International Conference on Very Large Data Bases proceedings*, 1995.
10. Grenander U and Rosenblatt M, editors. *Statistical analysis of stationary time series*. Wiley, 1957.
11. W. Lin and M. A. Orgun; Applied hidden periodicity analysis for mining discrete-valued time-series databases. In M. Gergatsoulis and P. Rondogiannis (editors), *Intensional Programming II*, World-Scientific Publishing Company, Singapore, ISBN 981-02-4095-3, Scheduled Spring 2000.

Mining N-most Interesting Itemsets

Ada Wai-chee Fu   Renfrew Wang-wai Kwong   Jian Tang

Department of Computer Science and Engineering
The Chinese University of Hong Kong, Hong Kong
{adafu, wwkwong}@cse.cuhk.edu.hk

Abstract. Previous methods on mining association rules require users to input a minimum support threshold. However, there can be too many or too few resulting rules if the threshold is set inappropriately. It is difficult for end-users to find the suitable threshold. In this paper, we propose a different setting in which the user does not provide a support threshold, but instead indicates the amount of results that is required.

1 Introduction

In recent years, there have been a lot of studies in association rule mining. An example of such a rule is :

$$\forall x \in persons, buys(x, "biscuit") \Rightarrow buys(x, "orangejuice")$$

where x is a variable and $buy(x,y)$ is a predicate that represents the fact that the item y is purchased by person x. This rule indicates that a high percentage of people that buy biscuits also buy orange juice at the same time, and there are quite many people buying both biscuits and orange juice.

Typically, this method requires the users to specify the minimum support threshold, which in the above example is the minimum percentage of transactions buying both biscuits and orange juice in order for the rule to be generated. However, it is difficult for the users to set this threshold to obtain the result they want. If the threshold is too small, a very large amount of results are mined. It is difficult to select the useful information. If the threshold is set too large, there may not be any result. Users would not have much idea about how large the threshold should be. Here we study an approach where the user can set a threshold on the amount of results instead of the threshold.

We observe that solutions to multiple data mining problems including mining association rules [2,4], mining correlation [3], and subspace clustering [5], are based on the discovery of large itemsets, i.e. itemsets with support greater than a user specified threshold. Also, the mining of large itemsets is the most difficult part in the above methods. Therefore, we would like to mine the interesting itemsets instead of interesting association rules with the constraint on the number of large itemsets instead of the minimum support threshold value. The resulting interesting itemsets are the *N-most interesting itemsets* of size k for each $k \geq 1$.

2 Definitions

Similar to [4], we consider a database D with a set of transactions T, and a set of items $I = i_1, i_2, ..., i_n$. Each transaction is a subset of I, and is assigned a transaction identifier $<TID>$.

Definition 1. *A k-itemset is a set of items containing k items.*

Definition 2. *The* **support** *of a k-itemset (X) is the ratio of number of transactions containing X to the total number of transactions in D.*

Definition 3. *The N-most interesting k-itemsets : Let us sort the k-itemsets by descending support values, let S be the support of the N-th k-itemset in the sorted list. The N-most interesting k-itemsets are the set of k-itemsets having support $\geq S$.*

Given a bound m on the itemset size, we mine the N-most interesting k-itemsets from the transaction database D for each $k, 1 \leq k \leq m$.

Definition 4. *The N-most interesting itemsets is the union of the N-most interesting k-itemsets for each $1 \leq k \leq m$. That is, N-most interesting itemset = N-most interesting 1-itemset $\cup$ N-most interesting 2-itemset $\cup$ ... $\cup$ N-most interesting m-itemset. We say that an itemset in the N-most interesting itemsets is* **interesting**.

Definition 5. *A* **potential** *k-itemset is a k-itemset that can potentially form part of an interesting $(k+1)$-itemset.*

Definition 6. *A* **candidate** *k-itemset is a k-itemset that potentially has sufficient support to be interesting and is generated by joining two potential $(k-1)$-itemsets.*

A potential k-itemset is typically generated by grouping itemsets with support greater than a certain value. A candidate k-itemset is generated as in the apriori-gen function.

3 Algorithms

In this section, we propose two new algorithms, which are *Itemset-Loop* and *Itemset-iLoop*, for mining N-most interesting itemsets. Both of the algorithms have a flavor of the Apriori algorithm [4] but involve backtracking for avoiding any missing itemset. The basic idea is that we automatically adjust the support thresholds at each iteration according to the required number of itemsets. The notations used for the algorithm are listed below.

P_k	Set of potential k-itemsets, sorted in descending order of the support values.
$support_k$	The minimum support value of the N-th k-itemset in P_k.
$lastsupport_k$	The support value of the last k-itemset in P_k.
C_k	Set of candidate k-itemsets.
I_k	Set of interesting k-itemsets.
I	Set of all interesting itemsets. (N-most interesting itemsets)

3.1 Mining N-most Interesting Itemsets with *Itemset-Loop*

This algorithm has the following inputs and outputs.

Inputs : A database D with the transaction T, the number of interesting itemsets required (N), the bound on the size of itemsets (m).

Outputs : N-most Interesting k itemsets for $1 \leq k \leq m$

Method : In this algorithm, we would find some k-itemsets that we call the *potential* k-itemsets. The potential k-itemsets include all the N-most interesting k-itemsets and also extra k-itemsets such that two potential k-itemsets may be joined to form interesting $(k+1)$-itemsets as in the Apriori algorithm.

First, we find the set P_1 of potential *1*-itemsets. Suppose we sort all 1-itemset in descending order of support. Let S be the support of the N-th 1-itemset in this ordered list. Then P_1 is the set of 1-itemsets with support greater than or equal to S. At this point P_1 is the N-most interesting 1-itemsets. The candidate *2*-itemsets (C_2) are then generated from the potential 1-itemsets.

The potential *2*-itemsets P_2 are generated from candidate *2*-itemsets. P_2 is the N-most interesting 2-itemsets among the itemsets in C_2. If $support_2$ is greater than $lastsupport_1$, it is unnecessary for looping back. This is the pruning effect. If $support_2$ is less than or equal to $lastsupport_1$, it means that we have not uncovered all 1-itemsets of sufficient support that may generate a 2-itemset with support greater than $support_2$. The system will loop back to find new potential 1-itemsets whose supports are not less than $support_2$. P_1 is augmented with these 1-itemsets, and the value of $lastsupport_1$ is also updated. C_2 is generated again from P_1. The new potential 1-itemsets may produce candidate potential *2*-itemsets having support $\geq$ the value of $support_2$ in the above. P_2 is generated again from C_2, it now contains the N-most interesting 2-itemsets from C_2. The values of $support_2$ and $lastsupport_2$ are updated.

For mining potential *3*-itemsets, the system will find the candidate *3*-itemsets from P_2 with the Apriori-gen algorithm. After finding *3*-itemsets, $support_3$ and $lastsupport_3$, it will compare $support_3$ and $lastsupport_1$.

Algorithm 1 : Itemset-Loop
var: $1 < k \leq m$, $support_k$, $lastsupport_k$, N, C_k, P_k, D

$(P_1, support_1, lastsupport_1) = $ find_potential_1_itemset(D, N);
$C_2 = $ gen_candidate(P_1);
for ($k=2; k < m; k++$){
  $(P_k, support_k, lastsupport_k) = $ find_N_potential_k_itemset(C_k, N, k);
  if $k < m$ then $C_{k+1} = $ gen_candidate(P_k); }
$I_k = N$-most interesting k-itemsets in P_k;
$I = \cup_k I_k$;
return (I);

find_N_potential_k_itemset(C_k, N, k)
{
  $(P_k, support_k, lastsupport_k)=$find_potential_k_itemset$(C_k, N)$;
  $newsupport = support_k$;

  for($i=2; i <= k; i++$) $updated_i = $ FALSE;
  for($i=1; i < k; i++$) {
    if ($i = 1$) {
      if ($newsupport \leq lastsupport_i$) {
        $(P_i, support_i, lastsupport_i) = $ find_potential_1_itemsets_with_support$(D, newsupport)$;
        if $i < k$ then $C_{i+1} = $ gen_candidate(P_i);
        if C_{i+1} is updated then $updated_{i+1} = $ TRUE; } }
    else {
      if ($newsupport \leq lastsupport_i$ or $updated_i = $ TRUE) {
        $(P_i, support_i, lastsupport_i) = $ find_potential_k_itemsets_with_support$(C_i, newsupport)$;
        if $i < k$ then $C_{i+1} = $ gen_candidate(P_i);
        if C_{i+1} is updated then $updated_{i+1} = $ TRUE; } }
    if (no. of k-itemsets $< N$ and $i = k$ and $k = m$) {
      $newsupport = $ reduce$(newsupport)$;
      for($j=2; j <= k; j++$) $updated_j = $ FALSE;
      $i = 1$; } }
  return$(P_k, support_k, lastsupport_k)$;
}

Fig. 1. Itemset-Loop

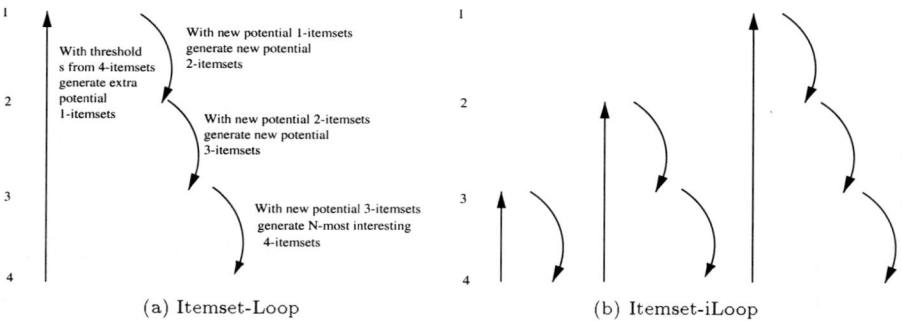

Fig. 2. Sketch of the iterations in the step for mining N-most interesting 4-itemsets

- If $lastsupport_1$ is greater than $support_3$, it means that there may be some relevant 1-itemsets missing. P_1 will be augmented by including 1-itemsets whose supports are $\geq support_3$. The value of $lastsupport_1$ is updated accordingly. The set C_2 candidate 2-itemsets will be generated from P_1 again. After that P_2 is generated from C_2 including all itemsets with support $\geq support_3$. $lastsupport_2$ is updated accordingly.
- If $lastsupport_1$ is not greater than $support_3$, $support_3$ will be compared with $lastsupport_2$ of P_2. similar processing is applied to update P_2, C_3 and P_3.

This process is iterated with larger and larger itemsets and stops at the user specified bound m on the itemset size. Figure 2 (a) illustrates the idea. Next we describe the functions used.

find_potential_1_itemset(D,N) : This function finds the N-most interesting 1-itemsets and returns these itemsets as the potential 1-itemsets together with their supports. The itemsets are sorted in descending order of the supports and are placed in P_1. In order to obtain the support values, this function scans all the transaction records in the database.

The minimum support among the return itemsets is recorded as $support_1$ and also $lastsupport_1$.

gen_candidate(P_k) : This function generates the candidate $(k+1)$-itemsets from potential k-itemsets using the Apriori-gen function [4]. It will also scan the database to count the support for the newly generated candidate itemsets. A hash tree is used in this process as in [4].

find_N_potential_k_itemset(C_k,N,k) : This function finds the N-most interesting k-itemsets. The system will first compare $support_k$ with $lastsupport_1$. If $support_k \leq lastsupport_1$, the potential 1-itemset is updated by adding all 1-itemsets with support $\geq support_k$. Then candidate 2-itemsets C_2 will be updated if necessary. The process is repeated with l-itemsets for $2 \leq l \leq k$.

find_potential_k_itemset(C_k,N) : This function finds potential k-itemsets from the candidate k-itemsets in C_k. The N-most interesting k-itemsets in C_k is returned. The values of $support_k$ and $lastsupport_k$ are also returned.

find_potential_1_itemset_with_support$(D,newsupport)$: This function finds all potential 1-itemsets with the support $\geq newsupport$. All itemsets with sufficient support are stored into the potential 1-itemset (P_1). These itemsets are returned together with $lastsupport_1$ and $support_1$.

find_potential_k_itemsets_with_support$(C_i,newsupport)$: This function finds the potential k-itemsets with the $newsupport$ value and the candidate k-itemsets. The candidates in C_i are scanned and those having support $\geq newsupport$ are returned. These are returned as P_k, the values of $lastsupport_k$, and $support_k$ are also updated and returned.

reduce$(newsupport)$: This function reduces the $newsupport$ value for mining N potential k-itemsets if there are no enough N potential k-itemsets.

Correctness: The correctness of the algorithm is based on the downward closure of large itemsets : If a k-itemset $X = \{X_1, ..., X_k\}$ is large, then a $(k-1)$-itemset $Y \subset X$ must also be large. When we compute the N largest k-itemsets, and discovers the smallest support of the itemsets is S, then for a $(k-1)$-itemset, if the support is less than S, it cannot form part of an interesting k-itemset. Hence if we have considered all the $(k-1)$-itemsets with support $\geq S$ in the generation of candidate k-itemsets, we have not missed any interesting k-itemsets. Otherwise, the algorithm loops back to uncover all the smaller itemsets to uncover all l-itemsets $l < k$ which have support $\geq S$.

3.2 Second Algorithm : *Itemset-iLoop*

The first approach requires loop back in the k-th iteration to generating itemsets of size 1, 2, ..., $k-1$ in that order, using a support bound S generated at the k-itemsets. One alternative is the following : we loop back first to generate extra $(k-1)$-itemsets using S, then using these extra $(k-1)$-itemsets, we may generate more k-itemsets. With the newly generated k-itemsets, if any, we may be able to to come up with a support bound S' greater than S. With S', we may require the generation of less itemsets of size less than $k-1$. This process can be repeated with itemsets of size $k-2$, $k-3$, ... 1. Hence we propose a second algorithm based on this technique. The second proposed algorithm is similar to the first algorithm except that at the k-th iteration, instead of loop backing to the generation of potential 1-itemsets, we loop back first to examine $(k-1)$-itemsets. The algorithm is called *Itemset-iLoop*. This algorithm has the same inputs and outputs as Algorithm *itemset-Loop*.

Method : The functions in the algorithm are the same as the corresponding functions in *Itemset-Loop* algorithm except for the following:

find_N_potential_k_itemset(C_k,N,k) : This function finds n potential k-itemsets given the candidate k-itemsets C_k and a new support, $support_k$. If $support_k \geq lastsupport_{k-1}$, it is not necessary to update P_{k-1}. If $support_k < lastsupport_{k-1}$, the potential $(k-1)$-itemsets (P_{k-1}) will be updated. The missing $(k-1)$-itemsets, which have support greater than or equal to $support_k$, will be inserted into (P_{k-1}). Then candidates C_k and P_k with $support_k$, and $lastsupport_k$ will be updated. After this, the system will compare $support_k$ with $lastsupport_{k-2}$, the potential $(k-2)$-itemsets (P_{k-2}) may be updated in a similar manner. Then the potential $(k-1)$-itemsets, $support_{k-1}$, $lastsupport_{k-1}$, the potential k-itemsets, $support_k$, and $lastsupport_k$ will be updated accordingly. This is repeated with $lastsupport$ for indices $k-3$, $k-4$, ... 1. In each case, we compare $support_k$ with all $lastsupport_i$ where $i < k$, and update P_i if necessary. P_j may be updated at every pass, where $j > i$, if P_i is updated.

Note that the first two iterations are the same as that in Algorithm *Itemset-Loop*. Figure 2 (b) is a sketch of the iterations for mining potential *4*-itemset.

Algorithm 2 : Itemset-iLoop
find_N_potential_k_itemset(C_k,N,k)
{
  (P_k,$support_k$,$lastsupport_k$)=find_potential_k_itemset(C_k,N);
  $newsupport = support_k$;
  for($i=k-1;i \geq 1;i=i-1$) {
    if($newsupport \leq lastsupport_i$) {
      for($j=i;j \leq k;j$++) {
        if($j = 1$) {
          P_j = find_potential_1_itemset_with_support(D,$newsupport$); }
        else {
          P_j = find_potential_k_itemset_with_support(C_j,$newsupport$); }
        if($j = k$) {
          $newsupport = support_k$; }
        if($j \neq k$) {
          C_{j+1} = gen_candidate(P_j); }
  } } }
  if (no. of k-itemsets $< N$ and $i = 1$ and $k = m$) {
    $newsupport$ = reduce($newsupport$);
    $i = k - 1$; } }
  return(P_k,$support_k$,$lastsupport_k$);
}

Fig. 3. Itemset-iLoop

4 Experimental Results

In this section, we present the performance analysis of the algorithms *Itemset-Loop* and *Itemset-iLoop* and comparison with the Apriori algorithm [4]. All experiments were carried out on a SUN ULTRA 5_10 machine running SunOS 5.6. The workstation has 128MB memory. The hash-tree data structure [4] is used for keeping candidate itemsets. Both synthetic datasets and real datasets were used.

The real data comes from census of United States 1990. The US census database is available at the web site of IPUMS-98[1]. The experiments are based on two sets of real data: a small database with 5577 tuples and 77 different items, and a large database with 57972 tuples and 77 different items. For each database, we investigate the performance under different values of N in the N-most interesting itemsets. The different values of N are 5, 10, 15, 20, 25, and 30. We mine itemsets up to size 4, hence k-itemsets are mined for $1 \leq k \leq 4$. For the function *reduce(newsupport)* in our proposed algorithms, we choose a factor of 0.8, meaning that when the function is called, the value of newsupport is reduced to be 0.8 times its original value.

In Figure 4(a) and 4(b), we show the performance of the *Itemset-Loop* algorithm, the *Itemset-iLoop* algorithm, and the Apriori algorithm with different support thresholds for the small and the large databases respectively. We perform the algorithms *Itemset-Loop* and *Itemset-iLoop* first and take the minimum support thresholds under every N, where N are 5, 10, 15, 20, 25, and 30 after mining *4-itemsets*. And we use the notations $minsup$ to represent these thresholds.

---

[1] The URL of IPUMS-98 is http://www.ipums.umn.edu/.

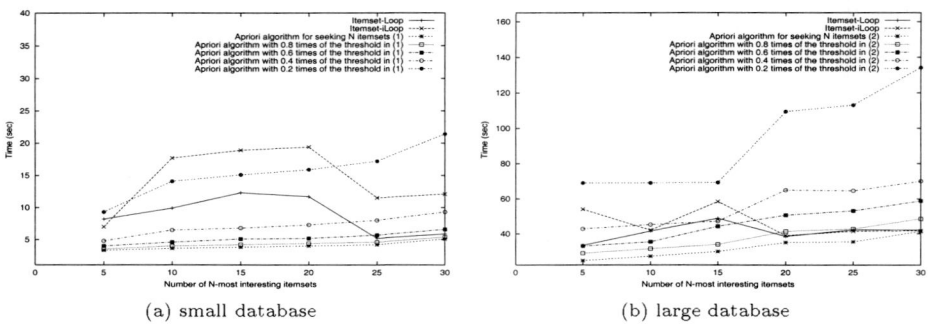

(a) small database (b) large database

Fig. 4. Performance with the growth of the number of N-most interesting itemsets

For the tiny database, the thresholds are found to be $\{0.097, 0.069, 0.062, 0.06, 0.058, 0.054\}$ for $N = 5, 10, 15, 20, 25, 30$, respectively. For the large database, the thresholds are found to be $\{0.22, 0.22, 0.22, 0.14, 0.13, 0.11\}$.[2] We apply the Apriori algorithm with these thresholds to measure the execution time. We also apply the Apriori algorithm with 0.8, 0.6, 0.4, and 0.2 of these thresholds, which we call $minsup_{0.8}$, $minsip_{0.6}$, $minsup_{0.4}$, and $minsup_{0.2}$, respectively.

In general, the performance of *Itemset-Loop* algorithm is better than that of *Itemset-iLoop* algorithm. This is because the *Itemset-Loop* algorithm loops back to the *1*-itemset first every time and updates k-itemset for $k > 1$ if necessary. On the other hand, the *Itemset-iLoop* algorithm loops back to check $(k\text{-}1)$-itemsets first and does comparisons. Then it loops back to check $(k\text{-}2)$-itemsets and updates $(k\text{-}1)$-itemsets and k-itemset if necessary, and so on so for. It may involve more back-tracking than the *Itemset-Loop* algorithm. The Apriori algorithm can provide the optimum results if the user knows the exact maximum support thresholds that can generate the N-most interesting results. We refer to this threshold as the optimal threshold. Otherwise, the proposed algorithms perform better.

We have studied the execution time for every pass using the *Itemset-Loop* and the *Itemset-iLoop* algorithms. Since we only record N or a little bit more for each itemsets for every k-itemset at the first step, it may be necessary to loop back for updating the result in both algorithms proposed. In general the increase of N leads to the increase of execution time. However, sometimes less looping back is necessary for a greater value of N and a decrease in execution time is recorded.

Table 1 shows the total number of unwanted itemsets generated by the Apriori algorithm in the large database when the guess of the thresholds is not optimal. The thresholds of $minsup_i$, where i=0.8, 0.6, 0.4 and 0.2, are used,

---

[2] Notice that the optimal thresholds can vary by orders of magnitude from case to case, and it is very difficult to guess the optimal thresholds.

N	$minsup_{0.8}$	$minsup_{0.6}$	$minsup_{0.4}$	$minsup_{0.2}$
5	251	395	583	1236
10	255	379	567	1220
15	105	250	437	947
20	372	541	921	1624
25	374	587	957	1854
30	467	800	1016	2322

Table 1. Number of unwanted itemsets generated by Apriori (large database)

$minsup_i$ is i times the optimal minimum support thresholds. We can see that the unwanted information can increase very dramatically with the deviation from the optimal thresholds.

We have also carried out another set of experiments on synthetic data. The results are similar in that the proposed method is highly effective and can outperform the original method by a large margin if the guess of the minimum support threshold is not good. For the interest of space the details are not shown here.

5 Conclusion

We proposed two algorithms for the problem of mining *N-most interesting k*-itemsets. We carried out a number of experiments to illustrate the performance of the proposed techniques. We show that the proposed methods do not introduce much overhead compared to the original method even with the optimal guess of the support threshold. For thresholds that are not optimal by a small factor, the proposed methods have much superior performance in both efficiency and the generation of useful results.

References

1. N. Megiddo, R. Srikant : Discovering Predictive Association Rules. Proc. of the 4th Int'l Conf. on Knowledge Discovery in Databases and Data Mining (1998)
2. J. Han, Y. Fu : Discovery of Multiple-Level Association Rules from Large Databases. Proc. of the 21st Int'l Conf. on Very Large Data Bases (1995) 420-431
3. S. Brin, R. Motwani, C. Silverstein : Beyond Market Baskets: Generalizing Association Rules to Correlations. Proc. of the 1997 ACM SIGMOD International Conference on Management of Data (1997) 265-276
4. R. Agrawal, R. Srikant : Fast Algorithms for Mining Association Rules. Proc. of the 20th Int'l Conf. on Very Large Data Bases (1994) 487-499
5. R. Agrawal, J. Gehrke, D. Gunopulos, P. Raghavan : Automatic Subspace Clustering of High Dimensional Data for Data Mining Application. Proc. of the 1996 ACM SIGMOD Int'l Conf. on Management of Data (1998) 94-105

Repository Management in an Intelligent Indexing Approach for Multimedia Digital Libraries

B. Armani[1], E. Bertino[2], B. Catania[1], D. Laradi[3], B. Marin[3], and G.P. Zarri[3]

[1]Dipartimento di Informatica e Scienze dell'Informazione, University of Genova, Italy
{armani,catania}@disi.unige.it
[2]Dipartimento di Scienze dell'Informazione, University of Milano, Italy
bertino@dsi.unimi.it
[3] Centre National de la Recherche Scientifique (CNRS), Paris, France
{laradi,marin,zarri}@ivry.cnrs.fr

Abstract. Metadata represent the vehicle by which digital documents can be efficiently indexed and retrieved. The need for such kind of information is particularly evident in multimedia digital libraries, which store documents dealing with different types of media (text, images, sound, video). In this context, a relevant metadata function consists in superimposing some sort of conceptual organization over the unstructured information space proper to these digital repositories, in order to facilitate the intelligent retrieval of the original documents. To this purpose, the usage of conceptual annotations seems quite promising. In this paper, we propose a two-steps annotation approach by which conceptual annotations, represented in NKRL [7], [8], are associated with multimedia documents and used during retrieval operations. We then discuss how documents and metadata can be stored and managed on persistent storage.

Introduction

It is today well recognized that an effective retrieval of information, from large bodies of multimedia documents contained in current digital libraries, requires, among other things, a characterization of such documents in terms of some *metadata*. A relevant metadata function consists in superimposing some sort of conceptual organization over the unstructured information space often typical of digital libraries, in order to facilitate the intelligent retrieval of the original documents. Querying or retrieving various types of digital media is executed directly at the metadata level.

Among the classes of metadata proposed by the scientific literature, only *content-specific metadata* „reflect the semantics of the media object in a given context" and provide a sufficient degree of generality [1]. Unfortunately, as well known, a veritable access by semantic content is particularly difficult to achieve, especially for non-textual material (images, video, audio). In those cases, content-based access is often supported by the use of simple keywords, or of features mainly related with the physical structure of multimedia documents (such as colour, shape, texture, etc.) [4]. In order to overcome the limitations of such approaches, *conceptual annotations* have been introduced for describing in some depth the context of digital objects [2], [3], [6]. However, the current approaches, often based on the use of simple ontologies in a description logic style, have several limitations in terms of description of complex semantic contents (e.g., of complex events) events.

To alleviate these problems, we propose a different approach for building up conceptual annotations to be used for indexing documents stored in a thematic multimedia library. With thematic multimedia library we mean a library storing documents concerning a given application domain. Our approach is based on a *two steps* annotation process:

- During the first step, any interesting multimedia document is annotated with a simple Natural Language (NL) caption in the form of a short text, representing a general, neutral description of the content of the document. In the case of textual documents, the interesting parts of the text, or the text itself, could represent the NL caption. This approach corresponds to the typical process of annotating a paper document, by underlying the interesting parts or writing down remarks and personal opinions. In the case of other media documents, the NL caption may represent the semantic content of the document and additional observations associated with it.
- During the second phase, annotations represented by NL captions are (semi-automatically) converted into the final conceptual annotations. We propose to represent the final conceptual annotations in NKRL (*Narrative Knowledge Representation Language*) [7], [8]. In NKRL, the metaknowledge associated with a document consists not only in a set of concepts and instances of concepts (individuals) but also in a structured set of more complex structures (occurrences) obtained through the instantiations of general classes of events called *templates*. This approach is actually tested in the context of an European project, CONCERTO [9].

Note that the use of a two-steps annotation process guarantees a high level of flexibility in querying. First of all, this approach provides a general solution for the mixed media access. This means that a single metadata query can retrieve information from data that pertain to different media since the same mechanism is used to represent their content. Moreover, the first annotation step is quite useful in supporting a similarity-based indexing. Indeed, by associating similar captions to different documents we make them „similar" from the point of view of the content and therefore of the retrieval.

In designing an architecture supporting the approach described above, the component dealing with the storage and the management of all the types of knowledge (documents, templates, concepts, and conceptual annotations) on secondary storage plays a fundamental role, since its implementation strongly influences the performance of the overall system. The aim of this paper is that of presenting a proposal for designing and implementing such component, that we call *Knowledge Manager*. For this task, we have followed a Web-Based approach. In particular, the Knowledge Manager has been implemented as a true server manager that can be hosted on a generic machine connected over Intranet/Internet networks to the clients requiring such services. The advantage of this approach is that the software component we have designed can be easily used by other architectures, based on the use of NKRL or similar languages for encoding conceptual annotations.

The paper is organized as follows. Section 2 introduced NKRL whereas Section 3 introduces an approach for the internal representation of such language. The Knowledge Manager architecture is then presented in Section 4. Finally, Section 5 presents some concluding remarks.

NKRL as a Metalanguage for Document Annotations

In the following, we briefly review the basic characteristics of NKRL (*Narrative Knowledge Representation Language*) (we refer the reader to [7], [8], [10] for additional details).

The core of NKRL consists of a set of general representation tools that are structured into four integrated components, described in the following.

Definitional and enumerative components. The *definitional component* of NKRL supplies the tools for representing the important notions (*concepts*) of a given domain. In NKRL, a concept is, therefore, a definitional data structure associated with a symbolic label like `human_being`, `city_`, etc. Concepts (*definitional component*) and individuals (*enumerative component*) are represented essentially as frame-based structures. All NKRL concepts are inserted into a generalization/specialization hierarchy that corresponds to the usual ontology of terms and is called H_CLASS(es).

The *enumerative component* of NKRL concerns the formal representation of the instances (individuals) (`lucy_`, `wardrobe_23`) of the concepts of H_CLASS. Throughout this paper, we will use the italic type style to represent a `concept_` and the roman style to represent an `individual_`.

Descriptive and factual components. The dynamic processes describing the interactions among the concepts and individuals in a given domain are represented by making use of the *descriptive* and *factual* components. The *descriptive component* concerns the tools used to produce the formal representations (*predicative templates* or simply *templates*) of general classes of narrative events, like 'moving a generic object', 'formulate a need', 'be present somewhere'. In contrast to the binary structure used for concepts and individuals, templates are characterized by a threefold format where the central piece is a predicate, i.e., a named relation that exists among one or more arguments introduced by means of roles. The general format of a predicative template is therefore the following:

$$(P_i\ (R_1\ a_1)\ (R_2\ a_2)...\ (R_n\ a_n))$$

In the previous expression, P_i denotes the symbolic label identifying the predicative template, R_k, $k = 1,...,n$, denote generic roles, and a_k, $k = 1,...,n$, denote the role arguments. The predicates pertain to the set BEHAVE, EXIST, EXPERIENCE, MOVE, OWN, PRODUCE, RECEIVE, and the roles to the set SUBJ(ect), OBJ(ect), SOURCE, BEN(e)F(iciary), MODAL(ity), TOPIC, CONTEXT. Templates are structured into an inheritance hierarchy, H_TEMP(lates), which corresponds to a taxonomy (ontology) of events. The instances (*predicative occurrences*) of the predicative templates, i.e., the representation of single, specific events like

„Tomorrow, I will move the wardrobe" or „Lucy was looking for a taxi" are in the domain of the *factual component*.

Example 1. The NKRL sentence presented in Figure 1 codes an information like: „On April 5th, 1982, Gordon Pym is appointed Foreign Secretary by Margaret Thatcher", that can be directly found in a textual document, contained in an historical digital library. The subject of this event is Gordon Pym, represented as a particular instance (gordon_pym) of the concept individual_person. The object of this event is the position Gordon Pym is appointed to, represented by the concept foreign_secretary_pos. Finally, the source of this event is Margaret Thatcher (represented by the instance margaret_thatcher) since she is responsible for the event. In the predicative occurrence, temporal information is represented through two temporal attributes, date-1 and date-2. They define the time interval in which the meaning represented by the predicative occurrence holds. In c1, this interval is reduced to a point on the time axis, as indicated by the single value, the timestamp 5-april-82, associated with the temporal attribute date-1; this point represents the beginning of an event because of the presence of begin (a *temporal modulator*).

```
c1)  OWN  SUBJ gordon_pym
          OBJ  foreign_secretary_pos
          SOURCE margaret_thatcher
          [begin]
          date-1: (5-april-82)
          date-2:
```

Fig. 1. Annotation of a WWW textual document

In the previous example, the arguments associated with roles are simple. However, NKRL also provides a specialized sublanguage, AECS, supporting the construction of *structured arguments* by using four operators: the disjunctive (ALTERNative) operator, the distributive (ENUMerative) operator, the collective (COORDination) operator, and the attributive (SPECIFication) operator.

Predicative occurrences can also be combined together, through the use of specific second order structures, called *binding occurrences*. Each binding occurrence is composed of a binding operator and a list of predicative or binding occurrences, representing its arguments. Each document (NL caption, in the considered framework) is then associated with a single *conceptual annotation*, corresponding to the binding occurrence representing its semantic content.

In order to query NKRL occurrences, *search patterns* have to be used. Search patterns are NKRL data structures representing the general framework of information to be searched for, within the overall set of conceptual annotations. A search pattern is a data structure including, at least, a predicate, a predicative role with its associated argument, where it is possible to make use of explicit variables, and, possibly, the indication of the temporal interval where the unification holds. As an example, the conceptual annotation in Figure 1 can be successfully unified with a search pattern like: „When was Gordon Pym appointed Foreign Secretary?", presented in Figure 2. The variable ?x means that we want to know the instant when the event happened.

We refer the reader to [7], [8] for additional details on these topics.

```
(?w IS-PRED-OCCURRENCE
 :predicate OWN
 :SUBJ gordon_pym
 :date-1 ?x
```

Fig. 2. A simple example of an NKRL search pattern

A Representation Language for NKRL

The usual way of implementing NKRL has been, until recently, that of making use of a three-layered approach: Common Lisp + a frame/object oriented environment (e.g., CRL, Carnegie Representation Language, in the NOMOS project) + NKRL. In order to ensure a high level of standardization, we are now realizing, in the context of the CONCERTO project [9], a new version of NKRL, implemented in Java and RDF-compliant (RDF = Resource Description Format) [5].

RDF is a proposal for defining and processing WWW metadata that is developed by a specific W3C Working Group (W3C = World Wide Web Consortium). The model, implemented in XML (eXtensible Markup Language), makes use of Directed Labeled Graphs (DLGs) where the nodes, that represent any possible Web resource (documents, parts of documents, collections of documents, etc.) are described basically by using attributes that give the named properties of the resources. No predefined 'vocabulary' (ontologies, keywords, etc.) is in itself a part of the proposal. The values of the attributes may be text strings, numbers, or other resources. In the last versions of the RDF Model and Syntax Specifications new, very interesting, constructs have been added [5]. Among them, of particular interest are the 'containers', i.e., tools for describing collections of resources. In an NKRL context the containers are used to represent the structured arguments created by making use of the operators of the AECS sublanguage (see Section 2).

A first, general problem to be solved to set up an RDF-compliant version of NKRL has concerned the very different nature of the RDF and NKRL data structures. The first are 'binary' ones, i.e., based on the usual organization into 'attribute – value' pairs. The second are 'tripartite', i.e., are organized around a 'predicate', whose 'arguments' are introduced through a third, functional element, the 'role'. To provide the conversion into RDF format, the NKRL data structures have been represented as intertwined binary 'tables' that describes the RDF-compliant, general structure of an NKRL template.

Example 3. Consider the predicative occurrence presented in Figure 1. The RDF/XML description of c1 is presented in Figure 3. In general, the RDF text associated with each predicative occurrence is composed of several tags, all nested inside the <CONCEPTUAL_ANNOTATION> tag and belonging to two different namespaces: rdf and ca. The first namespace describes the standard environment under which RDF tags are interpreted. The second namespace describes specific tags defined in the context of our specific application. More precisely, the tag <ca:Template_i> is used to specify that the predicative occurrence is an instance of the template identified by Template_i. The identifier of the occurrence is an attribute of such tag (occ11824 in our example). The other tags specify the various roles of the predicative occurrence, together with the associated arguments. Additional tags are used to represent temporal information and modulators.

The Knowledge Manager Architecture

Four main modules compose the architecture supporting our approach:

```
<?xml version="1.0" ?>
<!DOCTYPE DOCUMENTS SYSTEM "CA_RDF.dtd">
<CONCEPTUAL_ANNOTATION>
   <rdf:RDF xmlns:rdf="http://www.w3.org/1999/02/22-rdf-syntax-ns#"
            xmlns:ca="http://projects.pira.co.uk/concerto#">
     <rdf:Description about="occ11824">
     <rdf:type resource="ca:Occurrence"/>
        <ca:instanceOf>Template43</ca:instanceOf>
        <ca:predicateName>Own</ca:predicateName>
        <ca:subject rdf:ID="Subj43" rdf:parseType="Resource">
           <ca:filler>gordon_pym</ca:filler>
        </ca:subject>
        <ca:object rdf:ID="Obj43" rdf:parseType="Resource">
           <ca:filler>foreign_secretary_pos<ca:filler>
        </ca:object>
        <ca:source rdf:ID="Source43" rdf:parseType="Resource">
           <ca:filler>margaret_thatcher</ca:filler>
        </ca:source>
        <ca:listOfModulators>
           <rdf:Seq><rdf:li>begin</rdf:li></rdf:Seq>
        </ca:listOfModulators>
        <ca:date1>05/04/1982</ca:date1>
     </rdf:Description>
   </rdf:RDF>
</CONCEPTUAL_ANNOTATION>
```

Fig. 3. The RDF format of a predicative occurrence

- *Acquisition module*, providing a user-friendly interface by which the user can insert documents and associate with them some short NL captions.
- *Annotation module*, that is in charge of the translation of the NL captions into the NKRL format.
- *Knowledge Manager module*, implementing the basic features for storing and managing NKRL concepts, templates, original documents, and the associated conceptual annotations on persistent storage.
- *Query module*, applying sophisticated mechanisms to retrieve all documents satisfying certain user criteria, by using conceptual annotations.

In the context of the proposed architecture, the Knowledge Manager plays a fundamental role. Indeed, since it manages the repositories on secondary storage, its implementation strongly influences the performance of the overall system. In the current architecture, the Knowledge Manager has been implemented as a server, following a Web-based approach, by using Internet derived technologies for the communication protocol and metadata representation. In particular, the Knowledge Manager is organized according to a three-tier architecture, represented in Figure 4. The first level corresponds to the repository management on persistent storage, through the use of a specific database management system (IBM DB2 in our case); the second level is an application level, providing an easy programming interface (through a Java API) to the repository. Finally, the third level consists of a specific interface language (called KMIL) to provide access to the Knowledge Manager through a Web-Based approach. In the following, the repositories and their management as well as the communication protocol are described in more details.

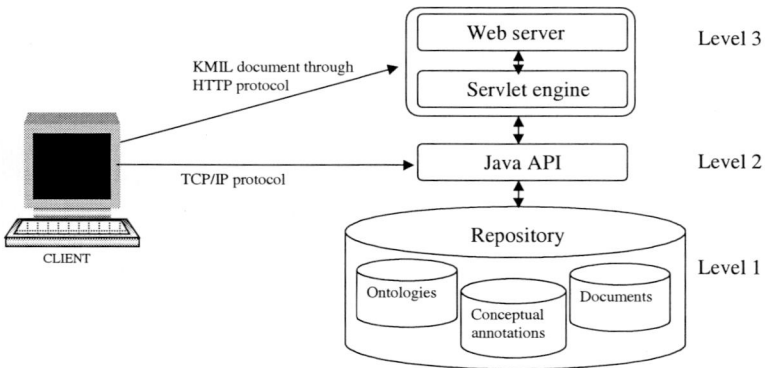

Fig. 4. General architecture of the Knowledge Manager

The Repositories and Their Management

In order to deal with NKRL data structures, we designed three distinct but interrelated repositories. The first repository is the *Document Repository*, storing the original documents, together with the corresponding NL captions. In order to deal with conceptual annotations, the H_TEMP and H_CLASS ontologies are stored in the *Ontology Repository*. The concrete conceptual annotations, generated by the Annotation Module, are then stored in the *Conceptual Annotation Repository*.

The Conceptual Annotation Repository is certainly the most critical one since user queries are executed against it. It contains two main types of data: predicative occurrences and binding occurrences. Each predicative occurrence is characterized, among the others, by its XML/RDF text and the identifier of the template it is an instance of. For each template, we also maintain the set of predicative occurrences representing the leaves of the subtree rooted by it in the H_TEMP. The use of this information optimizes query processing since a search pattern always selects a set of predicative occurrences that are instances of a single template. Each binding occurrence is internally characterized, among the others, by the binding operator and the identifiers of its arguments (i.e., binding or predicative occurrences).

Each document is then associated with a single conceptual annotation, arbitrarily complex, describing atomic information, through the use of predicative occurrences, and combined information, through the use of binding occurrences. The repository maintains the relationship between documents and the associated conceptual annotations. It is important to note that, to guarantee a high level of flexibility, we assume that each occurrence can be associated with different documents. This corresponds to the situation in which different documents refer similar or equal events or contain similar or equal images or sound.

Since RDF can be implemented by using XML, in order to store conceptual annotations and templates, we choose *IBM DB2 Universal Database* together with the XML extender, recently released by IBM. The repositories are then managed through the use of a Java API, implementing specific operation to be executed against the repositories. Each operation, before execution, is translated into some SQL commands to be executed by DB2. The use of a Java API provides a high level of portability for the system we have developed. Moreover, since several packages for implementing an

XML parser in Java are currently available, this choice fits well in the overall system architecture. Among the supported operations, queries against the Conceptual Annotation Repository intensively use the functionalities supported by IBM DB2 and IBM DB2 XML Extender to retrieve predicative occurrences starting from given selection conditions.

The Communication Protocol and the Interface Language

The Knowledge Manager services can be executed under two different modalities (see Figure 4). In a local environment, the Java API operations are directly called and executed. In a remote environment, communication is performed through the HTTP protocol. The use of HTTP guarantees an efficient access to the Knowledge Manager from any software module located at any site on the Internet. In order to guarantee a standard communication between modules, services have to be expressed by means of an XML document. Such document has to be constructed according to a specific XML language, called *Knowledge Manager Interface Language* (KMIL). KMIL requests can be sent by using an HTTP post action to a Knowledge Manager front-end Servlet running under a specific HTTP servlet engine. This solution has the advantage that the Knowledge Manager can be hosted on a generic machine, becoming strongly independent from other modules of the architecture. All requests sent to the Knowledge Manager are then captured by a Web Server that activates a specific Java Servlet for the execution of the requested services, through the use of the Java API, on the underlying DBMS. As a result, an XML document containing the result of the computation is returned to the calling module.

Example 4. Suppose that the conceptual annotation of Figure 1 has to be inserted into the Conceptual Annotation Repository. This can be specified by using the KMIL document presented in Figure 5. Such document contains a <KMIL-ACTION> tag for the document and the predicative occurrence that have to be inserted, respectively, together with all the required information. This information is then used to consistently update the content of the Conceptual Annotation and Document repositories.

```
<?xml version="1.0"?>
    <!DOCTYPE KMIL-SESSION SYSTEM "KmilIn.dtd">
      <KMIL-SESSION>
        <KMIL-ACTION serial_number="1">
          <KMIL-INSERT-Document IdDoc="doc132" >
            <TEXT>
              On April 5th, 1982, Gordon Pym is appointed Foreign
              Secretary by Margaret Thatcher
            </TEXT>
          </KMIL-INSERT-Document>
        </KMIL-ACTION>
        <KMIL-ACTION serial_number="2">
          <KMIL-INSERT-PredOcc IdPO="occ11824" Doc="doc132">
            <TEXT> RDF Text </TEXT>
          </KMIL-INSERT-PredOcc>
        </KMIL-ACTION>
      </KMIL-SESSION>
```

Fig. 5. Example of a KMIL request

Concluding Remarks

In this paper we have presented an approach for indexing and retrieving multimedia digital documents through the use of conceptual annotations, describing in details the component entrusted with the management of documents and conceptual annotations in secondary storage. The techniques presented in this paper are now being exploited in the framework of the Esprit project CONCERTO (*CONCEptual indexing, querying and ReTrieval Of digital documents*, Esprit 29159) [9]. The aim of such project is to improve current techniques for indexing, querying and retrieving textual documents, mainly concerning the socio-economical and the biotechnology contexts. Future work includes the definition of specialized techniques for storing and indexing conceptual annotations. In particular, disk placement and caching techniques for conceptual annotations are currently under investigation in order to improve the performance of the system.

Acknowledgements. We would like to thank Pietro Leo, from IBM, for several useful comments and suggestions about the design of the proposed architecture.

References

1. S. Boll, W. Klas, and A. Sheth. Overview on Using Metadata to Manage Multimedia Data. In *Multimedia Data Management*, pages 1-24, 1998. McGraw Hill, New York.
2. D. Fensel, S. Decker, M. Erdmann, M., and R. Studer. Ontobrocker: Or How to Enable Intelligent Access to the WWW. In *Proc. of the 11th Banff Knowledge Acquisition for KBSs Workshop, KAW'98*. University of Calgary, 1998.
3. J. Heflin, J. Hendler, and S. Luke. SHOE: A Knowledge Representation Language for Internet Applications. Technical Report CS-TR-4078, Univ. of Maryland, College Park (MA), 1999.
4. IEEE Computer - Special Issue on Content-Based Image Retrieval Systems. *IEEE Computer*, 28(9), 1995.
5. O. Lassila and R. Swick. Resource Description Framework (RDF) Model and Syntax Specification. Technical report, W3C, 1999.
6. A. Levy, A. Rajaraman, and J. Ordille. Querying Heterogeneous Information Sources Using Sources Descriptions. In *Proc. of the 22nd Int. Conf. on Very Large Databases* (VLDB-96), pages 251-262, 1996.
7. G. Zarri. NKRL, a Knowledge Representation Tool for Encoding the 'Meaning' of Complex Narrative Texts. *Natural Language Engineering* - Special Issue on Knowledge Representation for Natural Language Processing in Implemented Systems, 3:231-253, 1997.
8. G. Zarri. Representation of Temporal Knowledge in Events: The Formalism, and Its Potential for Legal Narratives. *Information & Communications Technology Law* - Special Issue on Models of Time, Action, and Situations, 7:213-241, 1998.
9. G. Zarri et al. CONCERTO, An Environment for the 'Intelligent' Indexing, Querying and Retrieval of Digital Documents. In *LNCS 1609: Proc. of the 11th Int. Symp. on Methodologies for Intelligent Systems*, pages 226-234, Varsavia, Poland, 1999. Springer Verlag.
10. G. Zarri and L. Gilardoni. Structuring and Retrieval of the Complex Predicate Arguments Proper to the NKRL Conceptual Language. In *LNCS 1079: Proc. of the 9th Int. Symp. on Methodologies for Intelligent Systems*, pages 398-417, Zakopane, Poland, 1996. Springer Verlag.

Logic-Based Approach to Semistructured Data Retrieval

Mohand-Saïd Hacid[1] and Farouk Toumani[2]

[1] Department of Computer Sciences, Purdue University
West Lafayette, IN 47907, USA
E-mail: mshacid@cs.purdue.edu
[2] LIMOS-ISIMA
Campus des Cezeaux - B.P. 125
63173 AUBIERE - France
E-mail: ftoumani@sp.isima.fr

Abstract. We investigate logic-based query language for semistructured data, that is data having irregular, partial or only implicit structure. A typical example is the data found on the Web. We present the syntax and semantics of *SemLog*, a logic for querying and restructuring semistructured data, and show how this language can be used to query video data.

Keywords: Intelligent information retrieval, Logic for databases, Integrating navigation and search, Video retrieval.

1 Introduction

Semistructured data models are intended to capture data that are not intentionally structured, that are structured heterogeneously, or that evolve so quickly that the changes cannot be reflected in the structure. A typical example is the World-Wide Web with its HTML pages, text files, bibliographies, biological databases, etc. A semistructured database essentially consists of objects, which are linked to each other by attributes.

Semistructured Data represent a particularly interesting domain for query languages. Computations over semistructured data can easily become infinite, even when the underlying alphabet is finite. This is because the use of path expressions (i.e., compositions of labels) is allowed, so that the number of possible paths over any finite alphabet is infinite. Query languages for semistructured data have been recently investigated mainly in the context of algebraic programming [2,4].

In this paper, we explore a different approach to the problem, an approach based on logic programming, instead of algebraic programming. In particular, we develop an extension of *Datalog*[1] for manipulating semistructured data. It has both a clear declarative semantics and an operational semantics. The semantics are based on fixpoint theory, as in classical Logic programming [9]. The language of terms uses five countable, disjoint sets: a set of atomic values ($\mathcal{D}_1$), a set of objects ($\mathcal{D}_2$), a set of labels ($\mathcal{D}_3$), a set of object variables ($\mathcal{V}$), and a set of path variables ($\tilde{\mathcal{V}}$). A path variable is

---
[1] *Data*base *log*ic

a variable ranging over paths. The universe of paths over $\mathcal{D}_3$ is infinite. Thus, to keep the semantics of programs finite, we do not evaluate rules over the entire universe, $\mathcal{D}_3^*$, but on a specific *active domain*. We define the *active domain* of a database to be the set of constants (objects and labels) occurring in the database. We then define the *extended active domain* to include both the constants in the *active domain* and all path expressions resulting from the composition of labels in the *active domain*. The semantics of our language is defined with respect to the *extended active domain*. In particular, substitutions range over this domain when rules are evaluated.

The *extended active domain* is not fixed during query evaluation. Instead, whenever a new path expression is created, the new path and all paths resulting from its concatenation with already existing paths are added to the *extended active domain*.

Paper outline: In Section 2, we introduce the data model. In Section 3, we develop our language and give its syntax and semantics. Section 4 provides an application example. We conclude in Section 5 by anticipating on the necessary extensions.

2 Data Model

Recent research works propose to model semistructured data using "*lightweight*" data models based on labeled directed graphs [4,1]. Informally, the vertices in such graphs represent objects and the labels on the edges convey semantic information about the relationship between objects. The vertices without outgoing edges (sink nodes) in the graph represent atomic objects and have values associated with them. The other vertices represent complex objects. An example of a semistructured database in the style of OEM [13] is given figure 1.

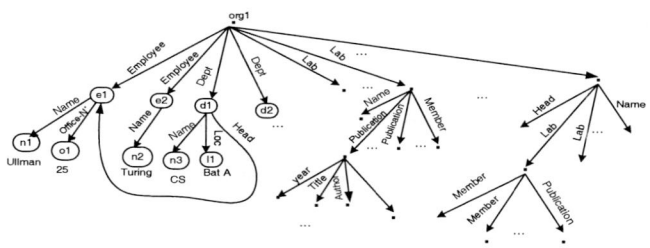

Fig. 1. Example Semistructured Data

Path expressions describe path along the graph, and can be viewed as compositions of labels. For example, the expression

Residence.City

describes a path that starts in an object, continues to the residence of that object, and ends in the city of that residence.

In this paper, we assume that the usual base types String, Integer, Real, etc., are available. In addition, we shall use a new type *Feature* for labels that would correspond to attribute names. We write numbers and features literally (the letter usually capitalized) and use quotation marks for strings, e.g., "car". In what follows we make the simplifying assumption that labels can be symbols, strings, integers, etc; in fact, the type of labels is just the discriminated union of these base types. In addition, we consider only semistructured data whose graph is acyclic.

Like [11] we represent the graph using two base relations:

- $link(FromObj, ToObj, Label)$: This relation contains all the edge information. For instance, $link(org_1, e_2, Employee)$ refers to an edge labeled *Employee* from the object org_1 to the object e_2. There may be more than one edge from org_1 to e_2.
- $atomic(Obj, Value)$: It contains all the value information. For instance, the fact $atomic(n_1, "Ullman")$ states that object n_1 is atomic and has value *Ullman*.

We assume that each atomic object has exactly one value, and each atomic object has no outgoing edges. We consider that the data comes in as an instance over link and atomic satisfying these two conditions. We use the term database in the following for such a data set.

Let p be a path expression of the form $a_1.a_2\ldots a_n$, where each a_i is a label.

Then, $link(o_1, o_2, p)$ holds if there are objects $o'_1, \ldots, o'_{n-1}$ such that $link(o_1, o'_1, a_1)$, ..., $link(o'_{n-1}, o_2, a_n)$ hold.

3 The Language

This section develops the syntax and semantics of the language for specifying programs over semistructured data.

3.1 Syntax

The language of terms uses three countable, pair-wise disjoint sets:

1. A set $\mathcal{D}$ of constant symbols. This set is the union of three pair-wise disjoint sets:
   - $\mathcal{D}_1$: a set of atomic values
   - $\mathcal{D}_2$: a set of entities, also called object entities
   - $\mathcal{D}_3$: a set of labels
2. A set $\mathcal{V}$ of variables called object variables and value variables, and denoted $X, Y, \ldots$
3. A set $\check{V}$ of variables called path variables, and denoted $\alpha, \beta, \ldots$

Definition 1. (Predicate Symbol) *we define the following predicate symbols:*

- *The predicate symbol* **link** *with arity 3*
- *The binary predicate symbol* **atomic**
- *The user specified intentional predicates (ordinary[2] predicates)*

We model semistructured data by a program P which contains, besides the set of facts built from *link* and *atomic*, the following rule :

$$link(X, Y, \alpha.\beta) :- link(X, Z, \alpha), link(Z, Y, \beta)$$

---
[2] Ordinary predicates can be of any arity.

This rule says that if an object Z can be reached from an object X by a path α and from Z we can reach an object Y by a path β, then there is a path $\alpha.\beta$ from X to Y.

Other ordinary facts can be specified by rules of the form:

$$H(\bar{X}) :- L_1(\bar{Y}_1), \ldots, L_n(\bar{Y}_n)$$

for some $n \geq 0$, where $\bar{X}, \bar{Y}_1, \ldots, \bar{Y}_n$ are tuples of variables or constants. We require that the rules are safe, i.e., a variable that appears in $\bar{X}$ must also appear in $\bar{Y}_1 \cup \ldots \cup \bar{Y}_n$. The predicates $L_1, \ldots, L_n$ may be either *link* or *atomic*, or ordinary predicates. In the following, we use the term (positive) atoms to make reference to $L_1, \ldots, L_n$.

Example 1. Figure 2 shows a fragment of semistructured data of figure 1.

$atomic(o_1, 25)$ $link(org_1, e_1, Employee)$
$atomic(n_2, "Turing")$ $link(org_1, d_1, Dept)$
$atomic(l_1, "BatA")$ $link(e_1, n_1, Name)$
     $link(d_1, n_3, Name)$
     $link(d_1, e_1, Head)$

Fig. 2. A set of facts

The intensional part of the program P contains the single rule :

$$link(X, Y, \alpha.\beta) :- link(X, Z, \alpha), link(Z, Y, \beta)$$

Given the semistructured data of figure 1, the query:

$$Answer(X) :- link(org_1, Y, \alpha.Author.Name), atom(Y, X)$$

returns all the author names, reachable from the root object org_1 by any path having as prefix any path (here given by the (possible) value of the path variable α), and as suffix the path *Author.Name*.

3.2 Semantics

Our language has a declarative model-theoretic and fixpoint semantics.

Model-theoretic semantics. Recall that $\mathcal{V}$ denotes a set of variables called object and value variables, and $\tilde{\mathcal{V}}$ denotes a set of variables called path variables. Let $\check{\mathcal{V}} = \mathcal{V} \cup \tilde{\mathcal{V}}$.

Let var_1 be a countable function that assigns to each syntactical expression a subset of $\mathcal{V}$ corresponding to the set of object and value variables occurring in the expression, and var_2 be a countable function that assigns to each expression a subset of $\tilde{\mathcal{V}}$ corresponding to the set of path variables occurring in the expression.

Let $var = var_1 \cup var_2$. If $E_1, \ldots, E_n$ are syntactic expressions, then $var(E_1, \ldots, E_n)$ is an abbreviation for $var(E_1) \cup \ldots \cup var(E_n)$.

A ground atom A is an atom for which $var(A) = \emptyset$. A ground rule is a rule r for which $var(r) = \emptyset$.

Definition 2. (Extension) *Given the set $\mathcal{D}_3$ of labels, the extension of $\mathcal{D}_3$, written $\mathcal{D}_3^{ext}$, is the set of path expressions containing the following elements:*

- *each element in $\mathcal{D}_3$*
- *for each ordered pair p_1, p_2 of elements of $\mathcal{D}_3^{ext}$, the element $p_1.p_2$*

Definition 3. (Extended Active Domain) *The active domain of an interpretation $\mathcal{I}$, noted $\mathcal{D}_\mathcal{I}$ is the set of elements appearing in $\mathcal{I}$, that is, a subset of $\mathcal{D}_1 \cup \mathcal{D}_2 \cup \mathcal{D}_3$. The extended active domain of $\mathcal{I}$, denoted $\mathcal{D}_\mathcal{I}^{ext}$, is the extension of $\mathcal{D}_\mathcal{I}$, that is, a subset of $\mathcal{D}_1 \cup \mathcal{D}_2 \cup \mathcal{D}_3^{ext}$.*

Definition 4. (Interpretation) *Given a program P, an interpretation $\mathcal{I}$ of P consists of:*

- *A domain $\mathcal{D}$*
- *A mapping from each constant symbol in P to an element of domain $\mathcal{D}$*
- *A mapping from each n-ary predicate symbol in P to a relation in $(\mathcal{D}^{ext})^n$*

Definition 5. (Valuation) *A valuation v_1 is a total function from $\mathcal{V}$ to the set of elements $\mathcal{D}_1 \cup \mathcal{D}_2$. A valuation v_2 is a total function from $\tilde{\mathcal{V}}$ to the set of elements $\mathcal{D}_3^{ext}$. Let $v = v_1 \cup v_2$. v is extended to be identity on $\mathcal{D}$ and then extended to map free tuples to tuples in a natural fashion.*

Definition 6. (Atom Satisfaction) *Let $\mathcal{I}$ be an interpretation. A ground atom L is satisfiable in $\mathcal{I}$ if L is present in $\mathcal{I}$.*

Definition 7. (Rule Satisfaction) *Let r be a rule of the form :*

$$r : A \leftarrow L_1, \ldots, L_n$$

where $L_1, \ldots, L_n$ are (positive) atoms. Let $\mathcal{I}$ be an interpretation, and v be a valuation that maps all variables of r to elements of $\mathcal{D}_\mathcal{I}^{ext}$. The rule r is said to be true (or satisfied) in interpretation $\mathcal{I}$ for valuation v if $v[A]$ is present in $\mathcal{I}$ whenever each $v[L_i], i \in [1, n]$ is satisfiable in $\mathcal{I}$.

Fixpoint Semantics. The fixpoint semantics is defined in terms of an immediate consequence operator, T_P, that maps interpretations to interpretations. An interpretation of a program is any subset of all ground atomic formulas built from predicate symbols in the language and elements in $\mathcal{D}^{ext}$. Each application of the operator T_P may create new atoms. We show below that T_P is **monotonic** and **continuous**. Hence, it has a **least fixpoint** that can be computed in a **bottom-up** iterative fashion.

Recall that the language of terms has three countable disjoint sets: a set of atomic values ($\mathcal{D}_1$), a set of entities ($\mathcal{D}_2$), and a set of labels ($\mathcal{D}_3$). A path expression is an element of $\mathcal{D}_3^{ext}$. We define $\mathcal{D}^{ext} = \mathcal{D}_1 \cup \mathcal{D}_2 \cup \mathcal{D}_3^{ext}$.

Lemma 1. *If $\mathcal{I}_1$ and $\mathcal{I}_2$ are two interpretations such that $\mathcal{I}_1 \subseteq \mathcal{I}_2$, then $\mathcal{D}_{\mathcal{I}_1}^{ext} \subseteq \mathcal{D}_{\mathcal{I}_2}^{ext}$.*

Definition 8. (Immediate Consequence Operator) Let P be a program and $\mathcal{I}$ an interpretation. A ground atom A is an immediate consequence for $\mathcal{I}$ and P if either $A \in \mathcal{I}$, or there exists a rule $r : H \leftarrow L_1, \ldots, L_n$ in P, and there exists a valuation v, based on $\mathcal{D}_\mathcal{I}^{ext}$, such that:

- $A = v(H)$, and
- $\forall i \in [1, n]$, $v(L_i)$ is satisfiable.

Definition 9. (T-Operator) The operator T_P associated with program P maps interpretations to interpretations. If $\mathcal{I}$ is an interpretation, then $T_P(\mathcal{I})$ is the following interpretation:

$$T_P(\mathcal{I}) = \mathcal{I} \cup \{A \mid A \text{ is an immediate consequence for } \mathcal{I} \text{ and } P\}$$

Theorem 1. (Continuity & monotonicity) The operator T_P is continuous and monotonic.

In the following, we illustrate the use of our query language for video data retrieval.

4 An Example: Video Databases

Digital video is content-rich information carrying media of massive proportion. In fact, the data volume of video is about seven orders of magnitude larger than a structured data record [6]. Video data also carries temporal and spatial information. Moreover, the structure of video data and the relationships among them is very complex and ill-defined. These unique characteristics pose great challenges for the management of video data in order to provide efficient and content-based user access. One of the main problems is that defining schema information for some video data in advance turns out to be very difficult [5] and thus a semistructured approach has to be considered.

We consider two layers for representing video content (figure 3):

(1) *Feature & Content Layer*. It contains video visual features (e.g., color, shape, motion). This layer is characterized by a set of techniques and algorithms allowing to retrieve video sequences based on the similarity of visual features.
(2) *Semantic Layer*. This layer contains objects of interest, their descriptions, and relationships among objects based on extracted features. Objects in a video sequence are represented in the semantic layer as visual entities. Instances of visual objects consist of conventional attributes (e.g., name, actorID, date, etc.).

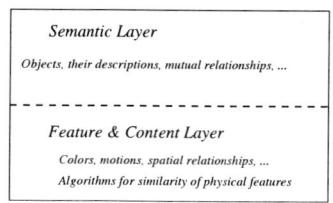

Fig. 3. Two layers for video content

Figure 4 shows a fragment of the semantic content of a video database. Although a real-world video database would of course be much, much larger, this example concisely captures the sort of structure (or lack thereof) needed to illustrate the features of our language. As illustrated by figure 4, the structure of the content describing a video differs from a category to another, and even within the same category. Here, the attribute name **Frames** links an abstract object to a concrete object name which denotes the name of the image sequences stored in the Feature & Content Layer.

It is easy to see that the proposed query language can be used to navigate the Semantic Layer of this database. In the following, we extend the language to accommodate the Feature & Content Layer.

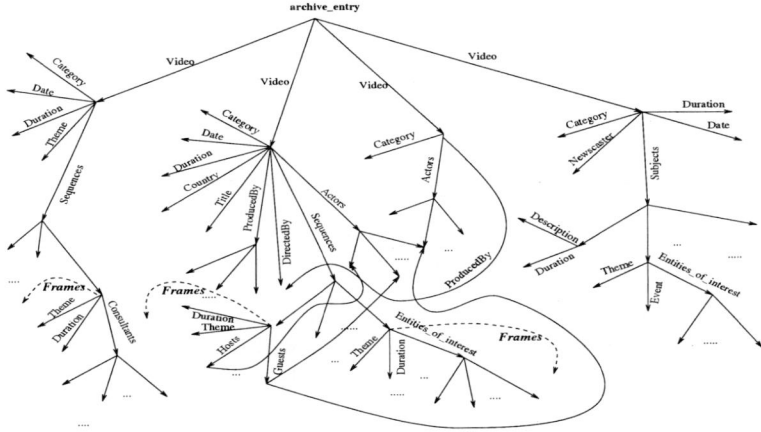

Fig. 4. A fragment of a video database content

Extension to Feature & Content Layer

Video data in the Feature & Content Layer can be organized in a hierarchy of units with individual frames at the base level, and higher level segments such as shots, scenes, and episodes. This model facilitates querying and composition at different levels and thus enables a very rich set of temporal and spatial operations. Examples of temporal operations are "follows", "contains", and "transition". Examples of spatial operations are "parallel to" and "below".

If we can regard a piece of video data as a set of images, then *Query-by-Example* methods developed for images (see, for example, [8]) can be used to retrieve video data by audiovisual content. For example, [12] implemented a system which makes retrieval of video data possible by specifying the motion of an object observed in video data by giving an example. An example of an object motion is specified by making a mouse move, and then a trajectory and velocity are sampled in accordance with the movement.

In the following, by exploiting this notion of procedural attachment [10], we provide an extension of our language, leading to a rule-based, constraint query language for video data retrieval.

Definition 10. (Rule) A rule *in our extended language has the form:*

$$r : H \leftarrow L_1, \ldots, L_n \& c_1, \ldots, c_m$$

where H is an atom, $n, m \geq 0$, $L_1, \ldots, L_n$ are (positive) literals, and $c_1, \ldots, c_m$ are constraints.

Definition 11. (Rule Satisfaction) *Let r be a rule of the form:*

$$r : A \leftarrow L_1, \ldots, L_n \& c_1, \ldots, c_m$$

$L_1, \ldots, L_n$ are (positive) atoms, and $c_1, \ldots, c_m$ are constraints. Let $\mathcal{I}$ be an interpretation, and v be a valuation that maps all variables of r to elements of $\mathcal{D}^{ext}$. The rule r is said to be true (or satisfied) in interpretation $\mathcal{I}$ for valuation v if $v[A]$ is present in $\mathcal{I}$ whenever:

- each $v(c_i)$, $i \in [1, m]$ is satisfiable, and
- each $v[L_i]$, $i \in [1, n]$ is satisfiable in $\mathcal{I}$.

Given the previous database fragment, the query:

q(Y) :- link(X, Y, α.sequences), link(Y, Z, duration),
     atomic(Z, Z'), link(Z, Z",frames), atomic(Z", F) &
     Z' < 20, **similar-color**(F, fd)

Would be "Find a set of sequences, with a duration below 20, and the video clip (i.e., the filler of the attribute frames, here F) of each sequence in the answer set is similar to the video clip fd regarding color". Here fd is the name of a video clip stored in the **Feature & Content Layer**. **similar-color** is a symbol with an attachment. The attached program will be executed in the **Feature & Content Layer**.

5 Conclusion

There is a growing interest in semistructured data, and this field offers many new challenges to IR research. As semistructured data proliferate, aids to browsing and filtering become increasingly important tools for managing such exponentially growing information resources and for dealing with access problems. We believe that formal settings will help understanding related modeling and querying problems. This will lead to the development of robust systems in order to effectively integrate, retrieve and correlate semistructured data.

We have presented a logic-based language for querying semistructured data, given its formal semantics, and applied it to video data. Several interesting directions to pursue:

- In the language we presented, navigational queries are expressed using variables ranging over path expressions in the graph representing the data. An important aspect to be considered is the use of path constraints [3] to take advantage of local knowledge about the data graph.
- An important and critical problem is the discovery of the structure implicit in our video data. This is especially important, since video data are often accessed in an explorative or browse mode. For that, it may be useful to build a layer of classes on top of our data model. The classes can be defined by rules and populated by computing a greatest fixpoint [11].
- Due to the visual nature of video data, a user may be interested in results that are similar to the query. Thus, the query system should be able to perform exact as well as partial or fuzzy matching. The first investigations reported in [7] constitute a nice basis.

References

1. Serge Abiteboul. Querying Semi-Structured Data. In *Proceedings of the International Conference on Database Theory (ICDT'97), Delphi, Greece*, pages 1–18, Janvier 1997.
2. Serge Abiteboul, Dallan Quass, Jason McHugh, Jennifer Widom, and Janet L. Wiener. The Lorel Query Language for Semistructured Data. *International Journal on Digital Libraries*, 1(1):68–88, 1997.
3. Serge Abiteboul and Victor Vianu. Regular Path Queries with Constraints. In *Proceedings of the Sixteenth ACM SIGACT-SIGMOD-SIGART Symposium on Principles of Databasests (PODS'97), Tucson, Arizonaem Sy*, pages 122–133. ACM Press, May 1997.
4. Peter Buneman, Susan Davidson, Gerd Hillebrand, and Dan Suciu. A Query Language and Optimization Techniques for Unstructured Data. In *Proceedings of the ACM SIGMOD International Conference (SIGMOD'96), Montreal, Canada*, pages 505–516, June 1996.
5. Cyril Decleir, Mohand-Said Hacid, and Jacques Kouloumdjian. A Database Approach for Modeling and Querying Video Data. In *In Proceedings of the 15th International Conference on Data Engineering (ICDE'99), Sydney, Australia*, March 1999.
6. Ahmed K. Elmagarmid and Haitao Jiang. *Video Database System: Issues, Products and Applications*. Kluwer, 1997.
7. Ronald Fagin. Fuzzy Queries in Multimedia Database Systems. In Jan Paredaens, editor, *Proceedings of the 1998 ACM SIGACT-SIGMOD-SIGART Symposium on Principles of Database Systems (PODS'98)*, pages 1–10, Seattle, Washington, USA, 1998. Invited Paper.
8. Myron Flickner, Harpreet Sawhney, Wayne Niblack, Jonathan Ashley, Qian Huang, Byron Dom, Monika Gorkani, Jim Hafner, Denis Lee, Dragutin Petkovic, D. Steele, and P. Yanker. Query by Image and Video Content: The QBIC System. In Marc T. Maybury, editor, *Intelligent Multimedia Information Retrieval*, chapter 1, pages 7–22. 1996.
9. John W. Lioyd. *Foundations of Logic Programming*. Springer-Verlag, 1987. Second edition.
10. Karen L. Myers. Hybrid Reasoning Using Universal Attachment. *Artificial Intelligence*, (67):329–375, 1994.
11. Svetlozar Nestorov, Serge Abiteboul, and Rajeev Motwani. Extracting Schema from Semistructured Data. In Laura M. Haas and Ashutosh Tiwary, editors, *Proceedings of the ACM SIGMOD International Conference on Management of Data (SIGMOD'98)*, pages 295–306, Seattle, Washington, USA, June 1998. ACM Press.
12. A. Yoshitaka, Y. Hosoda, M. Yoshimitsu, M. Hirakawa, and T. Ichikawa. VIOLONE : Video Retrieval by Motion Example. *Journal of Visual languages and Computing*, 7:423–443, 1996.
13. J.Widom Y.Papakonstantinou, H.Garcia Molina. Object Exchange Across Heterogeneous Information Sources. In *Proceedings of the 11th International Conference on Data Engineering (ICDE'95), Taipei, Taiwan*, pages 251–260, Mars 1995.

High Quality Information Retrieval for Improving the Conduct and Management of Research and Development

Ronald N. Kostoff [1]

Office of Naval Research, 800 N. Quincy St., Arlington, VA 22217
Internet: kostofr@onr.navy.mil

Abstract. The purpose of the present paper is to convey the importance of high quality information retrieval for maximizing progress in R&D, and to present generic protocols for constructing high quality literature queries. The paper begins with an example of the information retrieval limitations characteristic of present R&D practices, describes requirements for conducting high quality information retrieval, and presents a proposal for expanding dissemination and widening access to high quality information retrieval methods. Retrieval of medical R&D information was selected as an illustrative example.

1. Introduction

For the past decade, the author has been developing methods for extracting useful information from large S&T text databases [1, 2]. These methods have been based upon the latest information technology concepts and algorithms, and can offer literature searches that are extremely comprehensive with high signal-to-noise ratios. As part of a recent assessment of information retrieval techniques [3], the author examined many biomedical studies that included literature searches. The Science Citation Index (SCI) Abstracts of these studies contained the queries used for the literature surveys. These queries had the following characteristics:
1) The source data came almost exclusively from Medline alone, except for those studies whose objective was to survey the Web resources available for the target medical issue;
2) The focus of most of the studies seemed to concentrate around narrowly defined medical problems, with little indication offered that supporting or related medical/ technical areas were of any interest;
3) The reported queries contained 3-6 phrases on average;
4) The phrases were either searcher-generated, or were the indexed terms from the Medline Mesh taxonomy. No evidence was presented that an exhaustive search of author-generated terms was performed.

---

[1] THE VIEWS PRESENTED IN THIS PAPER ARE SOLELY THOSE OF THE AUTHOR AND DO NOT REPRESENT THE VIEWS OF THE DEPARTMENT OF THE NAVY.

Queries with the above characteristics result in a deficient retrieved information base. These deficiencies translate into limitations on the credibility and quality of study results and subsequent research and development (R&D), for the following reasons.

1) Searches that do not access the myriad databases available, and queries that do not result in comprehensive retrievals of the information available in the databases actually searched, result in only a fraction of the existing knowledge being available for study and R&D exploitation.

2) Searches and queries not designed to a) access literatures directly supportive of the target literature and b) access literatures related to the target literature by some common or intermediary thread, will not provide the insights and discoveries from these other disciplines that often result in innovations in the target discipline of primary interest [4]

3) Queries that are severely restricted in length, that rely in large measure on generic indexer-supplied terms, and have not been extensively iterated with the author-supplied language in the source database, will be inadequate in capturing the myriad ways in which different authors describe the same concept, and will also yield many records that are non-relevant to the main technical themes of the study.

In summary, these types of simple limited queries can result in two serious problems: a substantial amount of relevant literature is not retrieved, and a substantial amount of non-relevant literature is retrieved. As a result, the potential user is either overwhelmed with extraneous data, or is uninformed about existing valuable information, leading to potential duplication of effort and/ or R&D based on incomplete use of existing data. All the subsequent data processing, both human and computerized, cannot compensate for these deficiencies in the base data quality. In contrast to these typical biomedical study Medline queries reported in the SCI Abstracts, the author's group has been developing information retrieval techniques [1, 2] using an iterative relevance feedback approach. The source database queries result in retrieval of very comprehensive source database records that encompass direct and supporting literatures with very high ratios of desired/ undesired records. Some of the queries consist of hundreds of terms [5, 6]. In those specific cases, queries of this magnitude are necessary to achieve the retrieval comprehensiveness and 'signal-to-noise' ratio required. Queries of a specific size are not a query development target; rather, the query development process produces a query of sufficient magnitude to achieve the target objectives of comprehensiveness and high relevance ratio.

The reader interested in more details about the query development protocols discussed above, as well as the larger text mining context in which they are imbedded, is encouraged to contact the author. An excellent overview of information retrieval techniques is contained in [7]. Many detailed information retrieval technique descriptions can be found in the TREC Conferences' Proceedings on the NIST Web site, and the SIGIR Conferences' Proceedings on the ACM Web site.

2. Importance of High Quality Information Retrieval to Support S&T

Information retrieval is one component of a larger information extraction and integration process. To extract useful information from large volumes of semi-structured and unstructured S&T text, sophisticated text mining (TM) techniques have been generated [5, 8, 9, 10, 11, 12]. TM could address the following specific issues that arise repeatedly in the conduct and management of R&D:

What R&D is being done globally; Who is doing it; What is the level of effort;
Where is it being done; What are the major thrust areas; What are the relationships among major thrust areas; What are the relationships between major thrust areas and supporting areas, including the performing and archiving infrastructure; What is not being done; What are the promising directions for new research; What are the innovations and discoveries?

These issues can be divided into two categories: infrastructure (who, where, when) and technical (what, why). To address these issues comprehensively, TM techniques typically have four major generic components:
(1) Information retrieval to select raw textual data on which the information processing will be performed;
(2) Bibliometrics to identify the people, archival, institutional, and regional infrastructure of the topical domain being analyzed;
(3) Computational linguistics to extract topical themes of interest, and relationships among these themes and the infrastructure components:
(4) Visualization and/or other types of information display that summarize the TM analyses and results for the users/ customers.

While all four data mining components are important for a high quality useful product, good information retrieval is fundamental to the quality of the results and the latter three components. All the sophisticated bibliometrics and computational linguistics processing cannot compensate for insufficient or unfocused base data.
In order to maintain awareness of global R&D, effectively exploit its results, and remain at the cutting edge of R&D, the medical researchers, clinicians, and sponsors need to understand:
1) R&D done in the past, to both exploit it presently and not repeat mistakes that were made in the previous development;
2) R&D being conducted presently, to both leverage existing programs for optimal resource use and avoid duplication;
3) R&D planned to be conducted, to allow a) strategic budgetary planning for future R&D transitions; b) planning strategic cost-sharing in areas of common interest; and c) withdrawal of planned budgets from areas of peripheral interest that will be addressed elsewhere.

Any technology specialty community requires this information both in R&D areas directly related to its technologies of interest, and in allied and disparate technical fields as well. These supporting technical areas can serve as sources of innovation and discovery for advancing the prime technical areas (4), and can help remove the underlying critical path barriers that serve as roadblocks to progress along the primary technical paths. Some of the most revolutionary discoveries from TM/ information retrieval have occurred in the medical field, resulting from linking disparate literatures to the primary target literature [11-16].

Because of this interlocking nature of R&D, results from many different types of R&D efforts are required to produce advances in any specific area. For example, advances in biomedical instrumentation require underlying advances in materials, electronics, signal processing, mathematical analysis, physics, chemistry, energy conversion, radiation sciences, solid and fluid mechanics, robotics and micro-technology, and other technologies depending on specific applications. Maximum advances in non-invasive medical diagnostics require access to the latest science and engineering literature in remote sensing, non-destructive evaluation, signal and image processing, pattern recognition, multi-source data fusion, fluid dynamics, acoustics, robotics, materials, electronics, and many other disciplines.

R&D sponsors with broad mission areas have an additional problem; their R&D needs are very eclectic. Results from many different types of R&D are required in order to accomplish the overall objectives of the sponsoring organizations. However, any organization can afford to sponsor only a small fraction of the R&D necessary to provide the technical foundations for accomplishing its broader mission objectives. It is imperative for any organization (that requires significant technological advances to accomplish its broad mission objectives) to maintain awareness of all the R&D being performed globally. This continual awareness will allow the agency or company to leverage and exploit the results of externally-sponsored R&D in a timely manner for its own and national benefit.

The technical community needs access to a variety of sources for this global R&D information, to gain the full spectrum of perspectives on available R&D. These sources include human contacts, literature, multi-media, and physical sources. Advanced information retrieval techniques that can address the literature in particular are becoming available. These advanced information retrieval methods could be used as the cornerstone of a process that would both extract information directly from the text sources as well as use the preliminary extracted information as a gateway to the other data sources. For example, simple processing of the very comprehensive information retrieved by these advanced methods will identify R&D performers, journals, organizations, and sponsors [5, 6, 8]. These sources can then be contacted to provide a more personal type of information retrieval, and supplement the literature-based approach extensively.

3. Problems with Present Information Retrieval Approaches

The information retrieval/literature surveys performed by the R&D community have not kept pace with the breadth and expansion of literature available. Present information retrieval approaches have four major intrinsic limitations:

1) They access only a fraction of available source databases, due to a combination of lack of knowledge of the existing databases, lack of interest in making the effort required to identify the complete scope of existing databases, and lack of appropriate tools and techniques to readily access the full spectrum of available data sources.

2) They are typically limited to narrowly focused literatures, either due to the surveyor's lack of interest in going beyond the directly focused target area, or the surveyor's lack of knowledge about techniques and tools available to readily access allied and disparate literatures from which insights and discovery could be extrapolated.

3) They devote insufficient effort to query development, due to lack of time and other resources and/ or lack of understanding of the consequences of severely deficient queries on the quality of their subsequent R&D.

4) They are typically based on user-supplied terminology rather than database author-supplied terminology, due to lack of understanding the value of using author generated terms and/ or lack of knowledge of the tools and techniques available to extract query terms efficiently from the authors' own writings.

4. Requirements and Mechanics of High Quality Information Retrieval

4.1. Requirements

A high quality query should have the following operational characteristics:
1) Retrieve the maximum number of records in the technical discipline of interest
2) Retrieve substantial numbers of records in closely allied disciplines
3) Retrieve substantial numbers of records in disparate disciplines that have some connection to the technical discipline of interest
4) Retrieve records in aggregate with high signal-to-noise ratio (number of desirable records large compared to number of undesirable records)
5) Retrieve records with high marginal utility (each additional query term will retrieve large ratio of desirable to undesirable records)
6) Minimize query size to conform to limit requirements of search engine(s) used

Development of a high quality query requires:
1) Incorporation of technical experts;

2) In-depth understanding by the study performers of the contents and structure of the potential databases to be queried;
3) Sufficient technical breadth of the study performers in aggregate to understand the potentially different meanings and contexts that specific technical phrases could have when used in different technical areas and by different technical cultures (e.g., SPACE SATELLITES, SATELLITE CLINICS, SATELLITE TUMORS);
4) Understanding of the relation of these database contents to the problem of interest; and
5) Substantial time and effort on the part of the technical expert(s) and supporting information technologist(s).

Development of a high quality query is complex and time consuming, with attendant non-negligible costs. The stringent and complex development requirements run counter to the unfounded assertions being promulgated by information technology algorithm developers and vendors: sophisticated tools exist that will allow low-cost non-experts to perform comprehensive and useful data retrieval and analysis with minimal expenditures of time and resources.

4.2. Mechanics

In order to meet the requirements for a high quality information retrieval process described in the previous section, the query development process generically needs to be full-text based and iterative, with relevance feedback and associated query expansion occurring during each iteration. A small core group of documents relevant to the topic of interest is identified using a test query. Unique characteristics of these core documents are identified from bibliometrics (authors, journals, institutions, sponsors, citations) and computational linguistics (phrase frequency and phrase proximity) analysis. Patterns of bibliometrics and phrase relationships in existing fields are identified, the test query is modified (by some combination of human experts and intelligent agents) with new search term combinations that follow the newly identified patterns, and the process is repeated. In addition, patterns of bibliometrics and phrase relationships that reflect extraneous non-relevant material are identified, and search terms that have the ability to remove non-relevant documents from the database are added to the modified query. This iterative procedure continues until convergence is obtained, where relatively few new documents are found or few non-relevant documents are identified, even though new search terms are added.

The specific steps used in these generic relevance feedback approaches are summarized as follows:
1) Definition of study scope;
2) Generation of query development strategy;
3) Generation of test query;
4) Retrieval of records from database; selection of sample;
5) Division of sample records into relevant and non-relevant categories, or gradations of relevance;
6) Identification of bibliometric and linguistic patterns characteristic of each category.

In addition to using computational linguistics for characteristic pattern matching in the semi-structured databases' text fields, the author has used bibliometrics for pattern matching in the following other fields to retrieve more relevant records:
6a) Author Field; 6b) Journal Field; 6c) Institution Field; 6d) Sponsor Field;
6e) Citation Field;

There are at least three ways in which the citation field can be used to help identify additional relevant papers.
6ei) Papers that Cite Relevant Documents;
6eii) Papers Cited by Relevant Documents;
6eiii) Other Papers Cited by Paper that Cites Relevant Documents;

7) Identify marginal value of adding bibliometric and linguistic patterns to the query;
8) Construct modified query;
9) Repeat process until convergence obtained.

While the generic query development process is systematic as presented, it is neither mechanistic nor automated easily. Judgement must be used at each detailed step, especially when using the linguistic patterns from the text fields to assist in the generation of new query terms. Some of the complexities in the linguistics pattern identification will be summarized.

Linguistic patterns uniquely characteristic of each category (relevant and non-relevant records) are selected to modify the query. The underlying assumption is that records in the source database that have the same linguistic patterns as the relevant records from the sample will have a high probability of being relevant, and records in the source database having the same linguistic patterns as the non-relevant records from the sample will also be non-relevant. Linguistic patterns characteristic of the relevant records modify the query such that additional relevant records are retrieved from the source database. Linguistic patterns characteristic of the non-relevant records modify the query such that existing and additional non-relevant records are not retrieved.
To expand the relevant records retrieved, a phrase from the sample records should be added to the query if it:
1) appears predominately in the relevant record category;
2) has a high marginal utility based on the sample;
3) has reasons for its appearance in the relevant records that are understood well; and
4) IS PROJECTED TO RETRIEVE ADDITIONAL RECORDS FROM THE SOURCE DATABASE (E.G., SCI) MAINLY RELEVANT TO THE SCOPE OF THE STUDY.

If the candidate query phrase extracted from the sample was part of the test query, the source database occurrence projection is straight-forward. If the candidate query phrase extracted from the sample was not part of the test query, the actual source database occurrence ratio in relevant and non-relevant records may be far different from the projection based on the ratio of frequency of occurrence in each sample category. The IR example discussed in the next paragraph is an excellent demonstration of the mis-estimate of total source database occurrence possible with use of a phrase derived from the linguistic patterns of the sample but not part of the initial test query.

As an example from the query development in a recent TM study on the discipline of TM (3), the phrase IR (an abbreviation for information retrieval used in many SCI Abstracts) was characteristic of predominantly relevant sample records, had a very high absolute frequency of occurrence in the sample, and had a high marginal utility based on the sample. However, it was not 'projected to retrieve additional records from the source database mainly relevant to the scope of the study'. A test query of IR in the Science Citation Index source database showed that it occurred in 65740 records dating back to 1973. Examination of only the first thirty of these records showed that IR is used in science and technology as an abbreviation for InfraRed (physics), Immuno-Reactivity (biology), Ischemia-Reperfusion (medicine), current(I) x resistance(R) (electronics), and Isovolume Relaxation (medical imaging). IR occurs as an abbreviation for information retrieval in probably one percent of the total records retrieved containing IR, or less. As a result, the phrase IR was not selected as a stand-alone query modification candidate.

Consider the implications of this real-world example. Assume a query consists of 200 terms. Assume 199 of these terms are selected correctly, according to the guidelines above. If the 200th term were like IR above, then the query developer would have been swamped with an overwhelming deluge of unrelated records. ONE MISTAKE IN QUERY SELECTION JUDGEMENT can be fatal for a high signal-to-noise product.

Careful judgement must be exercised when selecting each candidate phrase. When potentially dominant relevant query modification terms extracted from the sample are being evaluated, one has to consider whether substantial amounts of non-relevant records will also be retrieved from use of the query term in the source database. When potentially dominant non-relevant query modification terms extracted from the sample are being evaluated, one has to consider whether substantial amounts of relevant records will not be retrieved.

Thus, the relation of the candidate query term to the objectives of the study, and to the contents and scope of the total records in the full source database (i.e., all the records in the Science Citation Index, not just those retrieved by the test query), must be considered in query term selection. The quality of this selection procedure will depend upon the expert(s)' understanding of both the scope of the study and the different possible meanings of the candidate query term across many different areas of R&D. *This strong dependence of the query term selection process on the overall study context and scope makes the 'automatic' query term selection processes reported in the published literature very suspect.*

5. Improving the Dissemination of High Quality Information Retrieval Processes and Queries

This final section proposes an option for increasing the dissemination of technical discipline queries to relevant communities. The background, need, and proposed alternatives are outlined. Finally, one specific additional application is addressed briefly.

5.1. Background

The previous sections of this paper have shown the importance and complexity of, and effort required to develop, high quality database queries. Yet, the dissemination of these queries to other potential users by their developers leaves much to be desired. Other than inclusion in published papers, the queries are essentially not distributed.

A fundamental axiom in the R&D community is that a comprehensive literature survey should be performed before R&D is proposed and initiated. Some if not most, Federal agencies require that such surveys be performed before R&D is started. The degrees to which these requirements are enforced and the survey quality and comprehensiveness are assessed, are unknown. Thus, there may be a lot of 're-inventing of the wheel', as each research group conducts surveys in topical areas similar to those surveyed previously. In addition, if the 're-invented' literature search is not of the same caliber as the original (due to poorer queries), the prospective researcher will not have the comprehensive global data to exploit, and the possibility for duplication of effort increases.

5.2. Need

If there were some type of query repository, with stringent query quality requirements, much of this redundancy could be avoided. Even if the objectives of the prospective literature survey were somewhat different from the objectives used to develop a previous query, the completed query could be used as a credible starting point for the desired query. Many researchers will have neither the time nor tools nor specialized information technology capability to perform comprehensive queries. Especially for resource intensive queries of the type described previously, widespread availability of these substantive queries could be of high value for a wide variety of researchers. The question arises as to how best to make these substantive resource-intensive queries widely available to the potential user community.

One feasible method would be to establish a Web site at one of the existing data repositories (e.g., NTIS, DTIC). Queries would be submitted to the site manager, subjected to some review, then posted on the site. Sample guidance for query submission and content is shown below. Both the broad-based technical journals (e.g., Science, Nature) and the specialty technical journals (e.g., JAMA, NEJM, Journal of Aircraft) would be used to inform readers of the new query titles that have been added to the repository. This option does not over-burden the expensive journal real estate, but does inform interested readers of the full query's location.

5.3. Guidance for Submitting Queries to Repository

One component of the overall repository maintenance protocol would be guidance to the query developers for submitting their completed query to the repository. The query developers would supply information describing the values of each of the parameters on which the query depends. The required information follows.
1) Identify Contents of Specific Source Databases Used;
2) Specify Fields of Source Database Used to Develop Query;
3) Specify Goals and Objectives of the Study whose Literature will be Retrieved by the Query;
4) Specify the Philosophy and Strategy used to Develop the Query;
5) Specify the Technical Backgrounds and Perspectives of Query Developers;
6) Describe the Features of the Search Engine Used, and any Limitations that Impacted the Final Query;
7) Describe any Other Events or Phenomena Pertinent to the Final Query Form;
8) Describe the Query Metrics (Records Retrieved, Relevant Fraction).

5.4. Specific Additional Application

One of the functions of the repository could be to serve as an enforcement mechanism for credible literature surveys for Federal grant recipients, if the repository operation is designed properly. Each prospective researcher would perform the requisite literature search, and submit the query and associated documentation to the repository gate-keepers. The query and supportive documentation would be reviewed by topical domain experts, and any deficiencies due to poor technique, game-playing, or other reasons, would result in rejection of the query. Not until a credible high quality query were accepted would the R&D be allowed to proceed. This would insure that duplications of effort are minimized, and the latest documented findings of the global R&D community are available to the prospective R&D performer(s) for exploitation.

6. Summary and Conclusions

Information retrieval plays a central role in modern day R&D. Present-day information retrieval techniques in wide use have limited capabilities compared to what the state of information technology can provide. Use of these inadequate retrieval techniques can result in an excess of non-relevant records, and retrieval of a fraction of the relevant records available, all of which translates into waste of limited R&D resources. *State-of-the-art information retrieval capabilities require time for high quality query development, and the costs of this development are not negligible.* The potential for net cost savings, due to the elimination of duplication and use of complete data possible with use of advanced queries, is high. Once these high quality queries have been developed, they should be disseminated to the broadest segment of the technical community, and archived. One mechanism for accomplishing this dissemination and archiving is through establishment of a query repository, with attendant advertising of the repository's contents through the technical journals, bulletin boards, professional society home pages, and other dissemination forums.

References

1. Kostoff, R.N., Eberhart, H.J., and Toothman, D.R., "Database Tomography for Information Retrieval", Journal of Information Science, 23:4, 1997
2. Kostoff, R.N., and Toothman, D.R., "Simulated Nucleation for Information Retrieval" to be Submitted for Publication, 2000
3. Kostoff, R.N., Toothman, D.R., and Humenik, J.A., " A Text Mining Study of Text Mining", to be Submitted for Publication, 2000
4. Kostoff, R.N., "Science and Technology Innovation". Technovation. 19 October 1999. Also, www.dtic.mil.dtic/kostoff/index.html
5. Kostoff, R.N., Green, K.A., Toothman, D.R., and Humenik, J.A., "Database Tomography Applied to an Aircraft Science and Technology Investment Strategy". Journal of Aircraft. 23:4, July-August 2000
6. Kostoff, R.N., Coder, D., Wells, S., Toothman, D.R., and Humenik, J., "Surface Hydrodynamics Roadmaps Using Bibliometrics and Database Tomography". Submitted for Publication, 2000
7. Greengrass, E., "Information Retrieval: An Overview", National Security Agency, TR-R52-02-96, 28 February 1997
8. Kostoff, R.N., Braun, T., Schubert, A., Toothman, D.R., and Humenik, J., "Fullerene Roadmaps Using Bibliometrics and Database Tomography". JCICS, Jan-Feb 2000
9. Kostoff, R.N., Eberhart, H.J., and Toothman, D.R.,"Hypersonic and Supersonic Flow Roadmaps Using Bibliometrics and Database Tomography". Journal of American Society for Information Science. 15 April 1999
10. Watts, R.J., Porter, A.L., Cunningham, S., and Zhu, D.H., "TOAS Intelligence Mining; Analysis of Natural Language Processing and Computational Linguistics" Principles of Data Mining and Knowledge Discovery, 1263:323-334-1997
11. Smalheiser, N.R., Swanson, D.R., "Using ARROWSMITH: A Computer Assisted Approach to Formulating and Assessing Scientific Hypotheses" Comput Meth Prog Bio 57: (3), 1998
12. Smalheiser, N.R., Swanson, D.R., " Assessing a Gap in the Biomedical Literature – Magnesium – Deficiency and Neurologic Disease", Neurosci Res Commun 15: (1), 1994
13. Swanson, D.R., "Fish Oil, Raynauds Syndrome, and Undiscovered Public Knowledge", Perspect Biol Med .30: (1), 1986
14. Smalheiser, N.R., Swanson, D.R., "Calcium-Independent Phospholipase A (2) and Schizophrenia". Arch Gen Psychiat 55 (8), 1998
15. Swanson, D.R., Smalheiser, N.R., "An Interactive System for Finding Complementary Literatures: A Stimulus to Scientific Discovery". Artif Intell 91 (2), 1997
16. Swanson, D.R., "Computer – Assisted Search for Novel Implicit Connections in Text Databases". Abstr Pap Am Chem S 217, 1999

Signature-Based Indexing for Retrieval by Spatial Content in Large 2D-String Image Databases

Essam A. El-Kwae
University of North Carolina at Charlotte
Department of Computer Science
9201 University City Boulevard
Charlotte, NC 28223
eelkwae@uncc.edu
http://www.coe.uncc.edu/~eelkwae

Abstract. Image matching and content-based spatial similarity assessment based on the 2D-String image representation has been extensively studied. However, for large image databases, matching a query against every 2D-String has prohibitive cost. Indexing techniques are used to filter irrelevant images so that image matching algorithms can only focus on relevant ones. Current 2D-String indexing techniques are not efficient for handling large image databases. In this paper, the Two Signature Multi-Level Signature File (*2SMLSF*) is used as an efficient tree structure that encodes image information into two types of binary signatures. The *2SMLSF* significantly reduces the storage requirements, responds to more types of queries, and its performance significantly improves over current techniques. For a simulated image databases of 131,072 images, a storage reduction of up to 35% and a querying performance improvement of up to 93% were achieved.

1. Introduction

Several logical image representation techniques for spatial similarity retrieval have been previously proposed such as ΘR-Strings [1], Spatial Orientation Graphs (*SOG*) [2,3] and the symbolic image [4]. Efforts are underway within the MPEG-7 (formally called the Multimedia Content Description Interface) standard to standardize multimedia content description [5]. Symbolic projection methods for image representation based on the 2D-String were introduced in [6]. Among those techniques, the 2D-String is the most studied. Various extensions of 2D-Strings have been proposed as the 2D-G String [7], the 2D-C String [8] and the 2D-C^+ String [9] to deal with situations of overlapping objects with complex shapes. The 2D-String representation changes the problem of pictorial information retrieval into a problem of 2D sub-sequence matching. 2D-Strings allow for matching images based on the perception of the objects and the spatial relations that exist between them, thus providing high-level object-oriented search rather than search based on the low-level image primitives of objects such as color, texture, and shape.

To extract the *2D*-String of a grayscale image, image understanding and pattern recognition techniques [e.g. 10] are used to extract the pictorial objects included in the image. In spite of the amount of image segmentation research in the past years, a general algorithm for segmenting general images has not yet been developed. However, segmentation has been somewhat successful on specific applications such as medical imaging, e.g. in brain MR images [11]. Although computationally expensive, the process of image understanding is performed only once when the image is inserted into the database.

In a large image database, matching a query image sequentially with every database image is not feasible. Indexing techniques are used to filter out irrelevant images to improve the search speed. A successful indexing mechanism should not drop relevant images (*true disposals*), and should try to minimize the number of irrelevant images (*false alarms*). The index should be dynamic, capable of answering different types of queries, and should be efficient in terms of performance and storage space required. Several techniques for indexing *2D*-String databases have been proposed [12-14]. However, those techniques may only be used for answering specific queries and their performance degrades considerably in large image databases.

Signature files have been widely employed in information retrieval of both formatted and unformatted data [14-18] and recently to image databases [19,20]. Signatures are commonly calculated using superimposed coding in which each object (or object pair) in an image is hashed into a word signature. An image signature is generated by superimposing (*OR*ing) all its individual signatures. To resolve a query, the query signature is generated and matched (*AND*ed) against image signatures. Fig.1 shows an example for generating an image signature from object signatures and the results of matching different query signatures to the image signature.

The main contribution of this paper is introducing an efficient indexing technique for large *2D*-String image databases based on the *two-signature multi-level signature file* (*2SMLSF*) [19] and providing comparisons of the *2SMLSF* to existing indexing techniques using a large simulated image database. Comparisons of the *2SMLSF* to existing *2D*-String indexing techniques revealed that the *2SMLSF* tree significantly reduces the storage requirements, responds to more types of queries and significantly improves the search performance. The rest of this paper is organized as follows: In section 2, an overview of *2D*-Strings is given. In section 3, several *2D*-String indexing techniques are discussed. The *2SMLSF* technique is introduced in section 4. Comparisons of the *2SMLSF* to existing techniques are given in section 5 followed by conclusions in section 6.

2. *2D*-String Overview

The *2D*-String of a symbolic picture [6] transforms an image into a *two dimensional* string by projecting the objects of that picture along the x- and y- coordinates. Thus, the *2D*-String is a pair of *1D* strings *(u, v)*, *u* represents the spatial relationships between the pictorial objects along the *X*-axis while *v*, represents those along the *Y*-axis. In *u* and *v*, "<" denote is-west-of and is-south-of relationships, respectively. For example, consider the symbolic picture f shown in Fig.2.(a) The symbols *a, b, c* and *d* represent pictorial objects. The *2D-String* representation of *f* is *S=(d<ab<c, a<bc<d)*.

A spatial query is also represented by a *2D-String*. Thus, the problem of image retrieval becomes that of *2D* subsequence matching. In [6], three types of *2D* subsequences were defined, namely, type-*0*, type-*1* and type-*2*. A string *u* is a type-*i* subsequence of string *v*, if *u* is contained in *v* and if $a_1 w_1 b_1$ is a substring of *u*, a_1 matches a_2 in *v* and b_1 matches b_2 in *v*, then:

(type-*0*) $r(b_2) - r(a_2) \geq r(b_1) - r(a_1)$ or $r(b_1) - r(a_1) = 0$
(type-*1*) $r(b_2) - r(a_2) \geq r(b_1) - r(a_1) > 0$ or $r(b_2) - r(a_2) = r(b_1) - r(a_1) = 0$
(type-*2*) $r(b_2) - r(a_2) = r(b_1) - r(a_1)$

where *r(x)*, the rank of a symbol *x*, is defined to be one plus the number of "<" symbols preceding *x*. Let *(u, v)* and *(u', v')* be the respective *2D-String* representations of *f* and *f'*. Then, *(u', v')* is a type-*i* *2D* subsequence of *(u, v)*, if *u'* is type-*i* *1D* subsequence of *u* and *v'* is type-*i* *1D* subsequence of *v*. The picture *f'* is then said to be type-*i* subpicture of *f*. In this paper, type-*2* spatial relationships are used in all the experiments performed. For example, consider the four images *f, f_1, f_2* and *f_3* shown in Fig.2 [6]. The *2D-String* representation of *f, f_1, f_2* and *f_3* are: f: (d < ab < c, a < bc < d), f_1: (a < c, a < c), f_2: (d < a, a < d) and f_3: (d < ac, a < dc), then: the type-*0* subpictures of *f* are *f_1, f_2* and *f_3*, the type-*1* subpictures of *f* are *f_1* and *f_2*, and the type-*2* subpicture of *f* is *f_1*

Object Sigs.: (D): 001 000 110 010
(E): 010 001 100 010
(H): 001 000 110 010
(J) : 001 010 110 000

Image Signature: 011 011 110 010

(c) Object and image signatures

(a) Example Image with 4 objects: Dog(D), Deer (E), Jockey (J) and Horse (H)

Queries	Signature	Result
1) Deer	010 001 100 010	Match
2) Desk	000 010 100 101	No Match
3) Deer & Jockey	011 011 110 010	Match
4) Car	010 010 100 000	False Drop

{E<JH<D, H<D<E}
(b) *2D*-String rep.

(d) Sample queries and matching results

Fig.1. Signature generation and comparison based on superimposed coding

d		
	b	c
	a	

(a) f

	c	
a		

(b) f_1

	d	
	a	

(c) f_2

	d	c
	a	

(d) f_3

Fig.2. *2D*-String example

3. Previous Work on Indexing *2D*-String Image Databases

In [21], the *2D* longest common subsequence method for *2D*-String matching was proposed. The problem of string matching is transformed into a maximal common subgraph (clique) which has exponential complexity. In addition, each query must be matched against all images in the *2D*-String database.

In [12], a *2D*-String was indexed based on all object pairs included in the image. For each pair o_i and o_j, an ordered triplet is created (o_i, o_j, r_{ij}) and entered into a hash table, where r_{ij} is the spatial relationship between the two objects. Each pair of query objects acts as a separate query used to retrieve the set of images stored at the corresponding hash table address. The intersection of the retrieved sets constitutes the candidate set of images. This addressing scheme requires that all images are known in advance. A preprocessing step is needed to derive a perfect hash function, which ceases to be perfect when new images are inserted into the database.

Another approach is based on groups of two or more objects, called "image subsets", was introduced [13]. All image subsets from *2* up to a specified size K_{max} are produced. The number of image subsets becomes very large especially when $K_{max} > 5$ and n (the number of objects per image) > 10 which renders this method unsuitable for large image databases. A separate hash table is created for image subsets with the same number of objects up to a K_{max} objects. Queries with a number of objects $>K_{max}$ have to be decomposed into multiple smaller queries. In a simulation test on a 1000 image database, the retrieval response time was slower, in some cases, than that of a sequential search. In addition, this technique has a significant storage overhead.

The two level signature file (*2LSF*) [20] uses a two level signature file to represent the *2D*-Strings in the image database. In *2LSF*, several *2D*-Strings are grouped as a block. Each *2D*-String is associated with a record (leaf) signature and a block (root) signature. The bit-sliced two-level signature file (*BS2LSF*) introduced in [22] uses bit-transposed files to improve the performance of image retrieval of the *2LSF* at the expense of insertion cost. The S-tree [23] is a multilevel signature file that creates higher level signatures by superimposing signatures at lower levels. As more signatures are included, the bit density of the signatures will increase rendering the method useless due to a large number of false alarms. The multilevel signature file (*MSLF*) [14] is a multi-level extension to the *2LSF* for text retrieval. The bit density problem of the S-tree does not exist in the *MLSF* since signatures at higher levels have longer lengths and are generated independently from those at lower levels.

4. The Two Signature Multi-level Signature File Technique (*2SMLSF*)

The *Two Signature Multi-Level Signature File (2SMLSF)* (Fig.3.) [19] creates a tree structure where images or groups of images are represented by binary signatures, namely Type_S at the leaf level and Type_O at all other levels. The equations used to calculate w, the signature weight or the number or ones, and m, the signature width at different levels of the tree, are obtained so that the global false drop probability is minimized [17,18]. The one bits are randomly chosen and each of the possible signatures is equally likely to be chosen, then m and w may be calculated as follows:

$$w = \left(\frac{1}{\ln 2}\right) \ln\left(\frac{1}{P^f}\right) \quad (1)$$

$$m = \left(\frac{1}{\ln 2}\right)^2 s \ln\left(\frac{1}{P^f}\right) \quad (2)$$

where P^f is the false drop probability and s is the number of distinct items to be encoded to create a signature.

A multilevel signature file is a forest of b-ary trees with every node, except leaf nodes, in the structure having b child nodes. The number of levels in the structure is h. The trees are assumed to be complete b-ary trees ($n = b^h$). Local parameters representing the value of some global parameter p at level i are denoted p_i. To further simplify the analysis, it is assumed that the local false drop probability is the same at every level. The relationship between the global and local false drop probabilities is:

$$p_i^f = p_j^f \quad \forall i,j \quad \text{then} \quad p^f = \frac{\prod_{i=1}^{h}(bp_i^f)}{n} = \prod_{i=1}^{h} p_i^f \quad \text{and} \quad p_i^f = (p^f)^{1/h} \quad (3)$$

The *2SMLSF* uses two types of signatures for each image. *Type_O* signatures used at all levels except the leaf level and are based only on the objects included in the image while *Type_S* signatures are used only at the leaf level and are based on the included objects in addition to their spatial relationships. For an image I with x objects, there exists $x(x-1)/2$ object pairs. From equation (2), the storage requirement of the two types of signatures is given by the following equations:

$$m_o = \left(\frac{1}{\ln 2}\right)^2 x \ln\left(\frac{1}{p_f}\right) \qquad m_s = \left(\frac{1}{\ln 2}\right)^2 \frac{x(x-1)}{2} \ln\left(\frac{1}{p_f}\right)$$

The ratio of storage requirement of both types of signatures is then calculated as:

$$\frac{m_S}{m_O} = \frac{(x-1)}{2} \quad (4)$$

A *Type_S* signature requires more storage than a *Type_O* signature whenever $x \geq 3$. For large image databases, a substantial reduction in storage and improvement in query performance will be achieved when *Type_O* signatures are used. However, *Type_S* signatures may answer exact queries about the included objects and their spatial relationships while *Type_O* signatures may only answer existential queries about the objects included in an image. Current signature based methods for 2D-String indexing [20,22] use only *Type_S* signatures for indexing. In the *2SMLSF*, both types of signatures are used for image encoding, which allows the *2SMLSF* to respond to both types of queries.

4.1. Index Creation in the 2SMLSF

The algorithm used to create the *2SMLSF* signature tree (Fig.4.) creates h independent signatures for each image, one for each level in the tree. At the leaf level, each pairwise spatial relationship contained in a *2D-String* is represented by a spatial string. For any two objects A and B where A is less than B in alphabetical order, let $r(x)$ be the rank of object x in a *1D* string. The type-2 spatial character, $V^2(A, b)$, denoting the type-2 spatial relationship between A and B is defined as follows [22]:

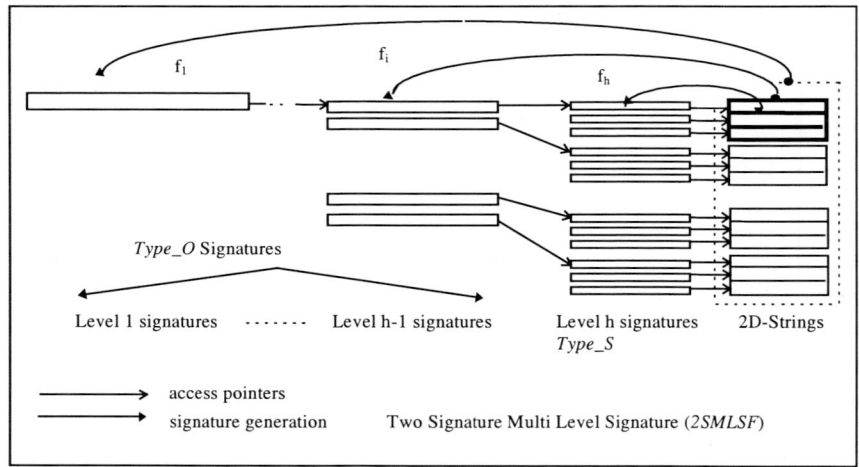

Fig.3. The Two Signature Multi Level Signature File (*2SMLSF*)

(type-2) $V^2(A, b) = $ "00"   if $r(A) = r(B)$
               $V^2(A, b) = $ "1"   + $Str(r(B) - r(A))$ if $r(A) < r(B)$
               $V^2(A, b) = $ "2"   + $Str(r(B) - r(A))$ if $r(A) > r(B)$

where "+" denotes string concatenation and *Str(x)* is a transformation function from integer x into string "x". The type-2 spatial string, $S^2(A, B)$, is the concatenation of the two symbols A, B and the two type-2 spatial characters $V^2_u(A,B)$ and $V^2_v(A,B)$, where $V^2_u(A,B)$ and $V^2_v(A,B)$ are the type-2 spatial characters of A, B in *1D* strings u and v, respectively. For example, in $S=(u, v) = (ad < b < c, ac < b < d)$, $r_u(a) = 1 < r_u(c) = 3$ and $r_v(a) = 1 = r_v(c) = 1$. The type-2 spatial string $S^2(a, c)$ representing the type-2 spatial relationships between a and c is: $S^2(a, c) = $ "ac1200"

The leaf signature for an object pair of width m_h and weight w_h is calculated using equations (1) and (2) using w_h hash functions. The image signature is created by superimposing (*ORing*) all object pair signatures in the image. For any other level in the tree, an image signature of width m_i and weight w_i is based only on the objects included in the image. Object signatures are then superimposed to generate the image signature at level i. All the signatures of images in the same block are then superimposed to generate the block signature. In general, 2^i image signatures are superimposed to generate the block signature at level $h-i$. A pointer to the signatures at the next lower level is associated with each non-leaf block signature and a pointer from the leaf signature to the corresponding *2D*-String is created. The *2D*-String then points to the corresponding physical image.

An example image database of *8* images is shown in Fig.5.(a). Each image has between *4* and *6* objects selected from *10* distinct objects. The corresponding *MLSF* has *3* levels (log_2 8). The total number of signature bits in the tree (excluding pointers) is *278* bits. The corresponding *2SMLSF* (Fig.5.(b)) also has *3* levels (log_2 8). The total number of signature bits in the tree (excluding pointers) is *192* bits, a saving of about *30%* over the *MLSF*. Note that the leaf level signatures in both techniques are the same since both techniques use Type-*S* signatures at the leaf.

4.2. Query Processing in the *2SMLSF*

Two types of queries may be submitted to the *2SMLSF*, the first, *Type_S* queries, "Find all images which include the set of objects and satisfy all spatial relations in the query image", and the second, *Type_O* queries, "Find all images which include the set of objects in the query image". Only *Type_S* queries may be submitted to the *2LSF*, the *MLSF*, and most other indexing techniques discussed above.

When a query image q is submitted to the *2SMLSF*, a 2D-String representation is created for q, then h signatures labeled $q_1, q_2, ..., q_h$ are generated for q using the algorithm in Fig.4. Starting at the root, the query signature q_1 is ANDed to all root signatures. If the result is not exactly q_1, it is certain that there are no images underneath the root that satisfy this query (unsuccessful search). If for a certain root signature, the result is exactly q_1, there is a chance that some images underneath this signature satisfy the query. The search then resumes with the signatures in the next level underneath qualifying root signatures.

For a *Type_S* query, the search is repeated until the leaf level is reached. All leaf signatures that match the q_h signature are query candidates. For a *Type_O* query, the search is repeated until the level right above the leaf level is reached since this is the last level that uses *Type_O* signatures. All matching signatures at this level are query candidates. The false drop probability of the *Type_S* queries is p^f while that for the *Type_O* queries is slightly higher since the search stops above the leaf level.

$$p_o^f = \prod_{i=1}^{h-1} p_i^f = (p^f)^{(h-1)/h} \qquad (5)$$

since $p_i^f < 1$ and $h > 1$, then $(p_i^f)^{(h-1)/h} > p_i^f$. The difference between the two probabilities decreases by increasing the number of levels in the tree. Due to the information loss in representing images by signatures, a false drop may occur. The *2D*-String pointed to by the query candidates are passed to a spatial similarity algorithm [e.g.3,12], which performs a detailed checking in order to exclude false drops and rank the other candidates based on their degree of similarity to the query image.

5. Evaluation of the 2SMLSF

In [19], analytical comparisons were performed between the *2SMLSF*, the *2LSF* [14,20], and the *MLSF* [14]. In this paper, simulations of a large image database were carried out. Two criteria were used for comparison, the amount of storage required (M) and the total number of bits (B) compared during query processing. Two parameters were used to quantify the comparisons, the storage reduction ratio (*SRR*) and the computation reduction ratio (*CRR*) calculated as follows:

$$SRR = \frac{M_{2LSF} - M_{2SMLSF}}{M_{2LSF}} \qquad CRR = \frac{B_{2LSF} - B_{2SMLSF}}{B_{2LSF}}$$

===

Input Given the values of: n : number of images in the database
                                p^f : global false drop probability
                                x : average number of objects per image

Procedure
   Step 1:   // initialization
                $h = \log_b n$ and $b=2$. // h is the number of levels
                $\forall i, i = 1, 2, ..., h$

$$p_i^f = \left(p^f\right)^{1/h} \quad \text{// local false drop probability at level i}$$

$$w_i = \left(\frac{1}{\ln 2}\right) \ln\left(\frac{1}{p_i^f}\right) \quad \text{// signature weight}$$

$$m_i = \left(\frac{1}{\ln 2}\right)^2 s_i \ln\left(\frac{1}{p_i^f}\right) \quad \text{// signature width}$$

   Step 2:   // For every image I, create a leaf Type_S signature (SIG$_{Ih}$)
                $\forall$ Image I, I = 1, 2, ..., n
                   SIG$_{Ih}$=0
                   $\forall$ Object $i \in I$, i = 1, 2, ..., x_I
                       $\forall$ Object $j \in I$, j = 2, 3, ..., x_I
                          SP$_{ijh}$ = f(ID$_i$, ID$_j$, Spatial_Relation$_{ij}$)
                          SIG$_{ijh}$ = create_signature(w_h, m_h, SP$_{ijh}$)   // as in section 4.1.
                          SIG$_{Ih}$= OR(SIG$_{Ih}$, SIG$_{ijh}$)
                Add a pointer from the leaf signature to the *2D-String* of this image.
   Step 3:   // create *h-1* signatures at other levels as follows
                $\forall$ Level L, L = h-1, h-2, ..., 1
                   $n_b = 2^L$ // number of images per block
                   $\forall$ Block B, B = 1, 2, ..., n/n_b
                       SIG$_{BL}$=0
                       $\forall$ Image I$\in$B, I = 1, 2, ..., n_b
                          SIG$_{IL}$ = 0
                          $\forall$ Object $i \in I$, i = 1, 2, ..., x_I
                              SP$_{iL}$ = f(ID$_i$)
                              SIG$_{iL}$ = create_signature(w_h, m_h, SP$_{iL}$)   // as in section 4.1.
                              SIG$_{IL}$ = OR(SIG$_{IL}$, SIG$_{iL}$)
                       SIG$_{BL}$ = OR(SIG$_{BL}$, SIG$_{IL}$)
                Adjust pointers to next level.

===

Fig.4. The 2SMLSF tree creation algorithm

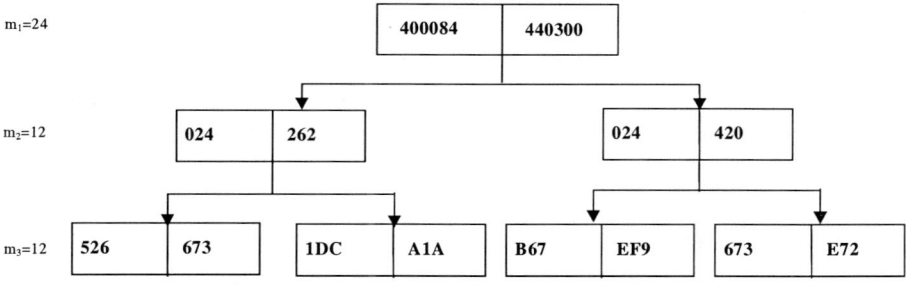

(a) The example image database and *2D*-String representation

(b) The corresponding *2SMLSF* representation
Fig.5. Image database example

In [14], it was shown that the storage requirement for the *SLSF* (Single Level Signature File), *2LSF*, and *MLSF* is the same, if they all have the same global false drop probability while the *MLSF* significantly reduces the number of bits compared during query processing. It was also shown that the optimum blocking factor is $b=2$, i.e. when each node in the *MLSF* includes exactly 2 signatures.

Several simulations were carried out between the *2SMLSF* and the *MLSF* [14]. In the first simulation, the number of objects per image was chosen at random between 5 and 9 objects, images were divided into 3x3 blocks while the false drop probability was chosen to be 0.1. The simulation studied the effect of increasing the number of images on the storage requirement and the query performance. The number of images was changed from *8K* (8192) images to *128K* (131,072) (Fig.6.). The average *SRR* improvement for this experiment was 35%. For performance comparisons, two types of queries were considered, random queries in which query images are created randomly and selected queries, in which query images are selected from images in the image database. Thus, selected queries are guaranteed to be in the database, while random queries may lead to unsuccessful searches. The *CRR* was *89.19%* on the average for random queries (Fig.7) and *50.12%* on the average for selected queries (Fig.8).

In the second simulation, the number of images in the database was kept constant at *32K* (32768 images) while the number of objects per image was varied from 3 to 7 objects. Again, images were divided into 3x3 blocks while the false drop probability was chosen to be 0.1. The average *SRR* for this experiment was 24% (Fig.9.). The *CRR* was *79.39%* on the average for random queries (Fig.10) and *38.32%* on the average for selected queries (Fig.10). Those results confirm the significant performance improvement of the *2SMLSF* for indexing large *2D*-String image databases.

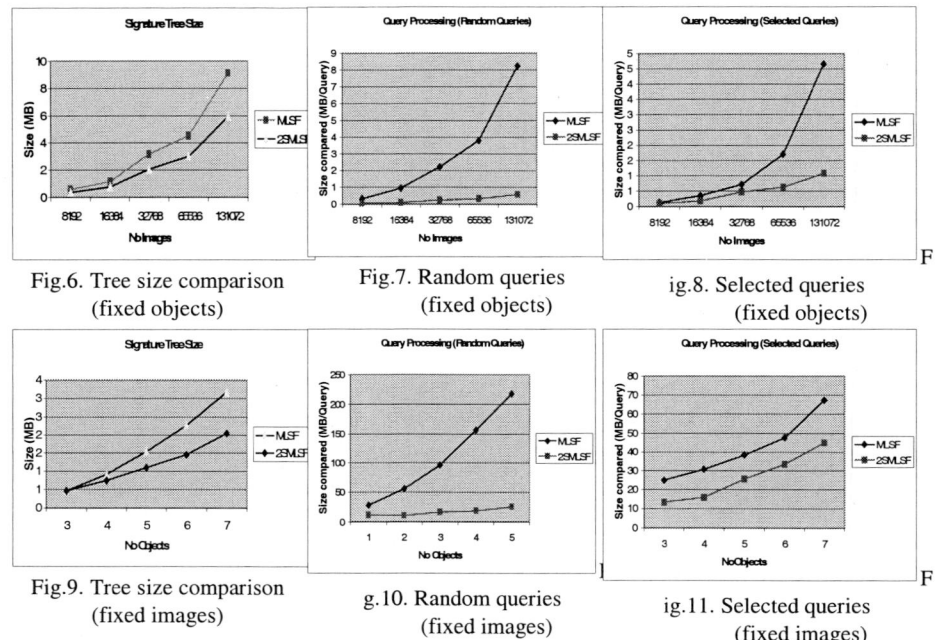

Fig.6. Tree size comparison (fixed objects)

Fig.7. Random queries (fixed objects)

ig.8. Selected queries (fixed objects)

Fig.9. Tree size comparison (fixed images)

g.10. Random queries (fixed images)

ig.11. Selected queries (fixed images)

6. Conclusions

The Two Signature Multi-Level Signature File (*2SMLSF*) was used for indexing large *2D*-String image databases. The *2SMLSF* is based on signature representation, which does not allow any true dismissals. However, false alarms may occur with a certain controlled probability p^f. The value of p^f may be reduced on the expense of additional storage. Two types of signatures are generated for each *2D*-String. *Type_S* signatures stored at the leaf and are based on the included domain objects and their spatial relationships and *Type_O* signatures used at all the other levels of the signature tree and are based only on the domain objects included in the image. Simulation comparisons of the *2SMLSF* to the *MLSF* in terms of storage requirements (*SRR*) and search performance (*CRR*) were performed. Simulations of image databases with a variable number of images up to *128K* images and a variable number of objects per image have confirmed that the proposed indexing technique significantly improves both the *SRR* (up to *35%*) and the *CRR* (up to *93%*) of existing techniques. In addition, the *2SMLSF* can answer both general and exact queries while the *MLSF* can only answer exact queries. Future extension of this work includes testing the indexing technique in a real environment such as medical image databases and extending the *2SMLSF* to other *2D*-String types such as *2D-G*, *2D-C*, and *2D-C+* strings.

References

[1] GUDIVADA, V. N., "θR-String: A Geometry-Based Representation for Efficient and Effective Retrieval of Images by Spatial Similarity," Technical Report CS-95-02, School of Electrical Engineering and Computer Science, Ohio University, 1995.
[2] GUDIVADA, V. N. and RAGHAVAN, V., "Design and Evaluation of Algorithms for Image Retrieval by Image Similarity," ACM Trans. on Info. Sys. 13, April 1995, 115-144.
[3] EL-KWAE, E. and KABUKA, M., "A Robust Framework for Content-Based Retrieval by Spatial Similarity in Image Databases," ACM Trans. on Info. Sys. 17(2), April 1999, 174-198.
[4] GUDIVADA, V. N. and JUNG, G. S., "An Algorithm for Content-Based Retrieval in Multimedia Databases," Proc. of the Int'l Conf. on MM Comp. and Sys., Japan, June 17-23, 1995, 90-97.
[5] Jane Hunter, "MPEG-7 Behind the Scenes," D-Lib Magazine, 5(9), Sep. 1999.
[6] CHANG, S. K., SHI, Q. Y., and YAN, C. W., "Iconic Indexing by 2-D Strings," IEEE Transactions on Pattern Analysis and Machine Intelligence 9(3), May 1987, 413-428.
[7] CHANG, S. K. and JUNGERT, E.,"Pictorial Data Management Based Upon the Theory of Symbolic Projections," Journal of visual Languages and Computing 2(3), Sep. 1991, 195-215.
[8] LEE, S. Y. and HSU, S. Y., "Spatial Reasoning and Similarity Retrieval of Images Using *2D-C* String Knowledge Representation," Pattern Recognition 25(3), 1992, 305-318.
[9] HUANG, P. W. and JEAN, Y. R., "Using $2D\ C^+$-Strings As Spatial Knowledge Representation For Image Database Systems," Pattern Recognition 27(9), 1994, 1249-1257.
[10] EL-KWAE, E. and KABUKA, M., "A Boolean Neural Network Approach for Image Understanding," Proc. of Artificial Neural Network in Engineering Con. (ANNIE'96), St Louis, Missouri, Nov. 10-13, 1996, 437-442.
[11] C. Tsai, B. S. Manjunath, and R. Jagadeesan, "Automated Segmentation of Brain MR Images," Pattern Recognition 28(12), 1995, 1825-1837.
[12] CHANG, C. C. and LEE, S., "Retrieval of Similar Pictures on Pictorial Databases," Pattern Recognition 24(7), 1991, 675-680.
[13] PETRAKIS, E. and ORPHANOUDAKIS, s., "A Generalized Approach for Image Indexing and Retrieval Based on 2-D Strings," First Workshop on Spatial Reasoning, Norway, Aug. 1993.
[14] LEE, D. L., KIM, Y. M. and PATEL, G., "Efficient Signature File Methods for Text Retrieval," IEEE Transactions on Knowledge and Data Engineering 7(3), Jun. 1995, 423-435.
[15] Faloutsos, C., "Access Methods for Text," ACM Computing Surveys 17, 1985, 49-74.
[16] FALOUTSOS, C. and CHRISTODOULAKIS, S., "Signature Files: An Access Method for Documents and Its Analytical Performance Evaluation," ACM Trans. on Ofc. Info. sys. 4, Oct. 1984, 267-288.
[17] ROBERTS, C. S., "Partial-Match Retrieval via Method of Superimposed Coding," Proceedings of IEEE 67(12), Dec. 1979, 1624-1642.
[18] DAVIS, R. S. and K. RAMAMOHANARAO, K,.A, "Two-Level Superimposed Coding Scheme for Partial Match Retrieval," Information Systems 8(4), 1983, 273-280.

[19] EL-KWAE, E. and KABUKA, M., "Efficient Content-Based Indexing of Large Image Databases," ACM Transactions on Information Systems, (to appear).
[20] LEE S. Y. and SHAN, M. K., "Access Methods of Image Databases," International Journal of Pattern Recognition and Artificial Intelligence 4, 1990, 27-44.
[21] LEE, S.Y., SHAN, M. K. YANG, W., "Similarity Retrieval of Iconic Image Databases," Pattern Recognition 22(6), 1989, 675-682.
[22] TSENG, J., HWANG, T. and YANG, W., "Efficient Image Retrieval Algorithms for Large Spatial Databases," Int'l Journal of Pattern Recognition and Art. Int. 8(4), 1994, 919-944.
[23] DEPPISCH, U., "S-Tree: A dynamic balanced signature index for office retrieval," Proc. of the ACM Conf. on Research and Development in Info. Ret., Pisa, Italy, Sep. 1986, 77-87.

Refining Logic Theories under OI-implication

F. Esposito, N. Fanizzi, S. Ferilli, G. Semeraro

Dipartimento di Informatica - Università di Bari
{esposito, fanizzi, ferilli, semeraro}@di.uniba.it

Abstract. We present a framework for theory refinement operators fulfilling properties that ensure the efficiency and effectiveness of the learning process. A refinement operator satisfying these requirements is defined *ideal*. Past results have demonstrated the impossibility of defining ideal operators in search spaces ordered by the logical implication or the θ-subsumption relationships. By assuming the *object identity* bias over a space defined by a clausal language ordered by logical implication, we obtain OI-implication, a novel ordering relationship, and show that ideal operators can be defined for the resulting search space.

1 Introduction

In this paper we continue our work presented in [4, 14, 3] on the definition of a framework fulfilling properties that are deemed as desirable for the incremental inductive synthesis of logic-based knowledge. Such properties are ensured by the notion of *ideality* of the definable refinement operators, that provides efficiency and effectiveness of the learning process.

Ideal operators have been proven not to exist in the spaces ordered by the classical notions of implication or θ-subsumption [15]. Our framework relies on the *Object Identity* assumption, that, when applied to the standard ordering relationships, induces changes upon the corresponding search spaces that allow for the existence of ideal operators.

After introducing the assumption for the search space induced by θ-subsumption, yielding the θ_{OI}-subsumption relationship [4, 14], we now weaken the implication ordering, obtaining a OI-implication [5, 3]. It has been shown how to define ideal operators in clausal spaces ordered by θ_{OI}-subsumption [4, 14]. Extending our framework to spaces ordered by OI-implication, we intend to investigate whether even in these search spaces ideal operators can be specified.

The remainder of the paper is organized as follows. Section 2 recalls the basic notions of the representation language and introduces the new ordering relationships that we propose, while Section 3 deals with the operators for searching in the resulting spaces. In Section 4, a novel framework, based on Object Identity, that overcomes negative results on standard search spaces is presented. Lastly, Section 5 draws some conclusions.

2 Preliminaries

In our framework, we adopt a representation language $\mathcal{L}$ expressing theories as *logic programs* made up of *clauses*[1]. It is based essentially on the following assumption.

Assumption 2.1 (Object Identity). *In a clause, terms denoted with different symbols must be distinct, i.e. they represent different entities of the domain.*

This notion constitutes the basis of the novel generality orderings proposed in the paper.

2.1 Generality Orderings

Essentially, generalization can be cast as a search problem [9]. Hence, a major issue is the algebraic organization underlying the search space.

Definition 2.2. *Given a set $\mathcal{S}$, a binary relation $\preceq$ on $\mathcal{S}$ is a quasi-ordering on $\mathcal{S}$ iff it is reflexive and transitive; a quasi-ordering $\preceq$ induces an equivalence relationship, denoted with $\sim$, such that: $\forall C, D \in \mathcal{S} : C \sim D$ iff $C \preceq D \wedge D \preceq C$. Given two clauses C and D, we say that such a relationship holds properly, denoted with $C \prec D$, when $C \preceq D \wedge D \npreceq C$.*

Implication and θ-subsumption are the standard ordering relationships investigated in inductive logic programming. We weaken them in order to obtain more manageable relationships leading to the definition of a form of implication that complies with the object identity assumption.

θ_{OI}-subsumption. In order to cope with the *object identity* principle, we have derived a new ordering relationship from the classic θ-subsumption, that induces a quasi-ordering upon the (Datalog [2]) clausal spaces [14, 3].

We discuss further properties required to substitutions in order to fulfill object identity. In fact, a substitution can be regarded as a function mapping variables to terms. In particular, we are interested here in a specific type of injective mappings.

Definition 2.3. *Given a set of terms T, let σ be a substitution. We say σ is an OI-substitution w.r.t. T iff $\forall t_1, t_2 \in T : t_1 \neq t_2 \Rightarrow t_1\sigma \neq t_2\sigma$.*

Hence, we introduce a new relationship, based on θ-subsumption, which complies with Assumption 2.1:

Definition 2.4. *Given two clauses C and D, C θ-subsumes D under object identity (C θ_{OI}-subsumes D) iff $\exists \sigma$ OI-substitution w.r.t. terms(C) such that $C\sigma \subseteq D$. Then, we say that C is more general or equivalent to D (resp. D is more specific or equivalent to C) under object identity and we write $D \preceq_{OI} C$.*

---
[1] Basic notions about clausal representation can be found in [8, 11].

θ_{OI}-subsumption is strictly a weaker relationship than standard implication and θ-subsumption [14].

Since θ_{OI}-subsumption maps each literal of the subsuming clause onto a single literal in the subsumed one, equivalent clauses under $\leq_{OI}$ must have the same number of literals. Thus, a search space ordered by θ_{OI}-subsumption is made up of non-redundant clauses[2]. As a consequence, it is possible to prove the following result:

Proposition 2.1. *Let C and D be two clauses. Then $C \sim_{OI} D$ iff they are alphabetic variants.*

Implication and θ-subsumption. A characterization of implication with respect to θ-subsumption was given by Bain and Muggleton [1]. This result bridges the gap between these two relationships. Indeed, it states that logical implication between clauses can be divided in two separate steps: a derivation by resolution [8] and then a subsumption step.

We recall a special case of the subsumption theorem, recently re-proven with respect to various resolution mechanisms (general, linear, SLD) [11]. In our framework, we deal with linear resolution, hence the following definition is needed:

Definition 2.5. *Let T be a set of clauses. Then, the n-th linear resolution of T, denoted by $\mathcal{L}^n(T)$, is defined inductively as follows:*
- $\mathcal{L}^1(T) = T$
- $\mathcal{L}^n(T) = \{R \mid C \in \mathcal{L}^{n-1}(T), D \in T \cup \mathcal{L}^k(T), k < n, \text{ and } R \text{ is a resolvent of } C \text{ and } D\}$ $(n > 1)$

Now we can state the corresponding subsumption theorem as follows:

Theorem 2.1 (Subsumption Theorem). *Let C, D be clauses (D non-tautological). Then $C \Rightarrow D$ iff $\exists E \in \mathcal{L}^n(\{C\})$ ($n \geq 0$) such that E θ-subsumes D.*

C can be resolved with itself only, or with one of its resolvents. *Self resolution* is possible when C is a *recursive*[3] clause. Otherwise, implication for non recursive (or even non ambivalent) clauses is equivalent to θ-subsumption [6].

2.2 OI-implication

Derived forms of implication are studied here in order to comply with the object identity assumption. Given a notion of θ_{OI}-subsumption, and using Theorem 2.1, we can define a novel generalization ordering. The goal here is to define constructively implication under object identity.

First, we have to define a form of resolution coping with the object identity assumption. From the notion of OI-substitution, we can specify a notion of a unifier fulfilling Assumption 2.1:

---

[2] A clause is called *redundant* when it is equivalent, w.r.t. a given ordering, to one of its subsets.
[3] A clause is *recursive* iff there exist literals $A, \neg B$ such that A is unifiable with a variant of B.

Definition 2.6. *Given a finite set of simple expressions S, we say that θ is an OI-unifier iff $\exists E \ \forall E_i \in S : E_i \theta = E$ and θ is an OI-substitution w.r.t. $terms(E_i)$. An OI-unifier θ for S is called a* most general OI-unifier *(mgu$_{OI}$) for S iff, for each OI-unifier σ of S, there exists an OI-substitution τ such that $\sigma = \theta \tau$.*

Differently from [3], we derive our definition of OI-resolution from [13]:

Definition 2.7. *Given the clauses C and D, standardized apart, the clause R is an OI-resolvent of C and D iff, given two subsets $M \subseteq C$ and $N \subseteq D$ such that $\{M, \overline{N}\}$ [4] is unifiable via the mgu$_{OI}$ θ, it holds that:*

$$R = ((C \setminus M) \cup (D \setminus N))\theta.$$

As mentioned before, we will consider only the case of linear resolution [11]. We will denote with $\mathcal{L}_{OI}$ the linear OI-resolution operator, and with $\mathcal{L}^*_{OI}$ its closure. If C can be derived by means of zero or more (linear) resolution steps from the set of clauses T, this will be denoted with $T \vdash_{OI} C$ or $C \in \mathcal{L}^n_{OI}(T)$, $n > 0$. We can now define the form of implication, that copes with object identity.

Definition 2.8. *Let C and D be any two clauses. C implies D under object identity (equivalently, C OI-implies D), denoted $C \Rightarrow_{OI} D$ iff either D is a tautological clause or there exists a clause $E \in \mathcal{L}^*_{OI}(\{C\})$ such that E θ_{OI}-subsumes D. In this case we say that C is* more general or equivalent to D *(resp. D is* more specific or equivalent to C*) under OI-implication. Equivalence under OI-implication is denoted by $\Leftrightarrow_{OI}$.*

It is easy to see that OI-implication is strictly a stronger ordering relationship than θ_{OI}-subsumption.

3 Refinement Operators

Our learning problem is cast as a search problem. In this section we focus on the properties of the operators that perform this search.

Theory refinement is triggered by new evidence made available to be assimilated in a knowledge base. Generally speaking, the canonical inductive paradigm requires the fulfillment of the properties of *completeness* and *consistency* for the synthesized theory with respect to a set of input examples. When an inconsistent (respectively, incomplete) hypothesis is detected, a specialization (respectively, generalization) of the hypothesis is required in order to restore this property of the theory. Roughly speaking, in the former case weaker clauses must be searched; in the latter, stronger clauses are needed or new ones are to be introduced. Formally, in terms of the adopted ordering:

Definition 3.1. *Given a quasi-ordered set of clauses $(\mathcal{L}, \preceq)$, a* refinement operator *is a mapping from $\mathcal{L}$ to $2^\mathcal{L}$ such that:*
- *$\forall C \in \mathcal{L} : \rho(C) \subseteq \{D \in \mathcal{L} | \ D \preceq C\}$ (downward refinement operator)*
- *$\forall C \in \mathcal{L} : \delta(C) \subseteq \{D \in \mathcal{L} | \ C \preceq D\}$ (upward refinement operator)*

---
[4] We indicate with $\overline{L}$ the complement of a (set of) literal(s).

A notion of closure upon refinement operators will be useful when proving the completeness property for the operators.

Definition 3.2. *Given a quasi-ordered set $(\mathcal{L}, \preceq)$, let τ be a refinement operator and $C \in \mathcal{L}$. The closure of τ (in symbols τ^*) for C is such that:*
$\tau^*(C) = \bigcup_{n \geq 0} \tau^n(C) = \tau^0(C) \cup \tau^1(C) \cup \ldots \cup \tau^n(C) \cup \ldots$
where $\tau^n(C)$ is inductively defined as:
- $\tau^0(C) = \{C\}$
- $\tau^n(C) = \{D| \exists E \in \tau^{n-1}(C) : D \in \tau(E)\}$

Ultimately, refinement operators should construct chains of refinements from the initial hypotheses to target ones. The next definition introduces this notion.

Definition 3.3. *In a quasi-ordered set $(\mathcal{L}, \preceq)$, given a refinement operator τ, a sequence of clause $C_0, C_1, \ldots, C_n$ in $\mathcal{L}$ is a τ-chain iff $C_i \in \tau(C_{i-1})$, $1 \leq i \leq n$.*

Properties of the Refinement Operators. We specify the properties that confer *ideality* to a refinement operator by recalling the definitions in [15]. First, we define the property that is fundamental to construct refinement operators that are actually mechanizable.

A major source of inefficiency in computing refinements may come from clauses that turn out to be equivalent to the starting ones. Indeed, it is desirable that the chain of refinements leads directly to target elements. Depending on the search algorithm adopted, refinements that are equivalent to some element already discarded introduce a lot of useless computation. As to the effectiveness of the search, a refinement operator should be able to build chains between two any comparable elements of the search space (or their equivalent representatives). This means that a complete refinement operator can derive any comparable element in a finite number of steps.

The following definitions formally specify these concepts:

Definition 3.4. *In a quasi-ordered set $(\mathcal{L}, \preceq)$, a refinement operator τ is locally finite iff $\forall C \in \mathcal{L} : \tau(C)$ is finite and computable.*

A downward (resp. upward) refinement operator ρ (δ) is proper iff $\forall C \in \mathcal{L} : D \in \rho(C)$ implies $D \prec C$ (resp. $\forall C \in \mathcal{L} : D \in \delta(C)$ implies $C \prec D$).

A downward (resp. upward) refinement operator ρ (δ) is complete iff $\forall C, D \in \mathcal{L}$, $D \prec C$ implies $\exists E \in \mathcal{L} : E \in \rho^*(C)$ and $E \sim D$ (resp. $C \prec D$ implies $\exists E \in \mathcal{L} : E \in \delta^*(C)$ and $E \sim D$).

The combination of these three properties confers effectiveness and efficiency to an operator. Indeed, local finiteness and completeness ensure the presence of a computable refinement chain to a target element. Besides, properness makes the refinement process more efficient, by avoiding the search of equivalent clauses. The following definition accounts for all of them.

Definition 3.5. *In a quasi-ordered set $(\mathcal{L}, \preceq)$, a downward (resp. upward) refinement operator ρ (δ) is ideal iff it is locally finite, proper and complete.*

Nonexistence conditions for ideal refinement operators in unrestricted set of clauses ordered by θ-subsumption are given in [15]:

Theorem 3.1. *In an unrestricted search space $(\mathcal{C}, \leq_\theta)$, with at least one predicate symbol of arity > 1, an ideal upward refinement operator does not exist.*

Similar results apply for downward refinement in the same search space or also when a stronger ordering relationship like implication is adopted.

4 Ideal Operators for OI-implication

In this section we propose refinement operators for spaces ordered by θ_{OI}-subsumption and OI-implication for an unrestricted search space. We also present the results on their ideality, though for brevity the proofs are omitted, but they can be found in [5].

We show here how it is possible to define ideal refinement operators in function-free clausal spaces under the weaker, but more mechanizable, ordering induced by θ_{OI}-subsumption. Ideal refinement operators in this space have been defined in [14, 3].

Given the notion of OI-substitution, we extend the definition of the relationship $\leq_{OI}$ (and then the refinement operators) to the case of unrestricted search spaces ordered by θ_{OI}-subsumption. Indeed, an easy characterization that can be made is that D θ_{OI}-subsumes C whenever an OI-substitution σ exists, such that $D\sigma \subseteq C$. We extend the definition of the refinement operators to include also the case of functions.

Definition 4.1. *Let C be a clause. Then $D \in \rho_{oI}(C)$ when one of these conditions holds:*

1. $D = C\theta$, where $\theta = \{X/a\}$, $X \in vars(C)$, $a \notin consts(C)$;
2. $D = C\theta$, where $\theta = \{X/f(Y_1,\ldots,Y_n)\}$, f is an n-ary function symbol $(n > 0)$ and $X \in vars(C)$;
3. $D = C \cup \{L\}$, where L is a literal, such that: $L \notin C$.

$D \in \delta_{oI}(C)$ *when one of these conditions holds:*

1. $D = C\sigma$, where $\sigma = \{a/X\}$, $a \in consts(C), X \notin vars(C)$;
2. $D = C\sigma$, where $\theta = \{f(Y_1,\ldots,Y_n)/X\}$, f is an n-ary function symbol $(n > 0)$ such that $f(Y_1,\ldots,Y_n) \in terms(C)$ and $X \notin vars(C)$;
3. $D = C \setminus \{L\}$, where L is a literal, such that: $L \in C$.

Even in this case, we obtain ideal refinement operators for this search space.

Theorem 4.1. *In an unrestricted clausal space, the operators ρ_{oI} and δ_{oI} are ideal refinement operators.*

When dealing with non recursive clauses, generalization and specialization under implication (respectively OI-implication) correspond to the cases considered for the θ-subsumption (θ_{OI}-subsumption), because of Gottlob's theorem [6].

Now, we want an operator for computing the resolution and inverse resolution steps, in the case of recursive clauses (for the subsumption theorem). [10] introduced the notions of *powers* and *roots* as operations where a clause is resolved with itself. They can be considered as refinement operators.

Definition 4.2. *Let C be a clause. A clause D is an n-th power of C iff D is a variant of a clause in $\mathcal{L}^n(C)$ ($n \geq 1$). We also say that C is an n-th root of D.*

By exploiting the subsumption theorem, a way to obtain downward refinements of a clause in a search space ordered by implication is to self-resolve a clause n times to obtain an n-th power or to apply the downward refinement operator used for θ_{OI}-subsumption. Conversely, upward refinements require to compute an n-th root of a clause, and again, to employ an upward refinement operator for θ_{OI}-subsumption. Both cases are not practically feasible, since we do not know *a priori* the n to stop at in the process of self-resolution. Moreover, while it is clear how to compute n-th powers by using linear resolution, in order to find downward refinements of a clause, the dual is a more complex task since it yields inversion steps.

Inverting OI-Resolution. We deal with the problem of inverting resolution by adapting the technique presented in [7] to our framework. Specifically, we start by defining a way to construct parent clauses given the resolvent, then we generalize this process with the aim of constructing OI-ancestors of the starting clause.

Given a clause R, it can be considered as the resolvent of two clauses C and D according to the definition of resolution, such that:
$$R = ((C \setminus M) \cup (D \setminus N))\theta.$$
where C and D are standardized apart and θ is an mgu_{OI} both for $\{M, \overline{N}\}$.

Besides, the most specific parent clauses resolve upon just one literal — M and N are singletons L_C and L_D, respectively — and inherit all literals from the OI-resolvent, hence θ has to be an mgu_{OI} also for $\{C \setminus \{L_C\}, D \setminus \{L_D\}\}$ [5]:
$$(C \setminus \{L_C\})\theta = (D \setminus \{L_D\})\theta = R$$
Thus:
$$C = R \cup \{L\} \text{ and } D = R \cup \{\overline{L}\}.$$
where $L = L_C = \overline{L_D}$.

Hence, by introducing a new literal, we have obtained two parent clauses. This applies to the cases of ambivalent clauses. It holds:

Proposition 4.1. *Let R be a clause and L be a literal. Then: $\{R\} \Leftrightarrow_{OI} (R \cup \{L\}, R \cup \{\overline{L}\})$.*

It is also provable [5] that:

Proposition 4.2. *Let C and D be two clauses and R an OI-resolvent of C and D. Then there exists a literal L such that $C \geq_{OI} R \cup \{L\}$ and $D \geq_{OI} R \cup \{\overline{L}\}$.*

---
[5] The extension of unifiers to from simple expressions formulæ is straightforward.

By using the same technique iteratively, applied this time to invert more than just one resolution step, we compute clauses from which R follows in two steps and so on, by introducing other literals. Namely, given: $\{R \cup \{L\}, R \cup \{\overline{L}\}\}$ we can decide to invert either of the two parent clauses. Then we can apply the or-introduction technique to the clause chosen (say, the first), obtaining:
$$\{R \cup \{L\} \cup \{L'\}, R \cup \{L\} \cup \{\overline{L'}\}, R \cup \{\overline{L}\}\}.$$
This technique can be extended by defining the following notion:

Definition 4.3. *Let C be a clause and Ω be a sequence of literals. Then, a set of clauses S is or-introduced from C by Ω iff:*

1. $S = \{C\}$ and $\Omega = []$ or
2. $S = (S' \setminus D) \cup \{D \cup \{L\}, D \cup \{\overline{L}\}\}$ and $\Omega = [L_1, \ldots, L_n, L]$, where S' is a set of clauses or-introduced from C by $[L_1, \ldots, L_n]$ and $D \in S'$.

Logical equivalence holds after this step of inversion [5]:

Theorem 4.2. *Let S be a set of clauses or-introduced from clause C. Then $S \equiv_{oI} \{C\}$.*

A sequence of resolutions can be inverted by applying or-introduction of a sequence of literals [5]:

Theorem 4.3. *Let T be a set of clauses, D a clause in $\mathcal{L}_{oI}^n(T)$. Then there exists a set of clauses S or-introduced from D, $\exists C \in T$ such that $\forall E \in S : C$ θ_{oI}-subsumes E.*

OI-Expansions. We have seen how starting from a clause it is possible to obtain a set of generalizations that is logically equivalent to an n-th power of the clause while this θ_{oI}-subsumes the clauses in the set. The goal is to reduce resolution to subsumption mechanisms. Thus, we come to the actual computation of the upward refinements of clauses by using the notion of expansions.

Definition 4.4. *Let C be a clause and Ω a sequence of literals. Then a clause E is an OI-expansion of C by Ω iff E is a least general generalization under θ_{oI}-subsumption of a set of clauses or-introduced from C by the sequence Ω.*

The notion of *least general generalization under θ_{oI}-subsumption* (lgg_{oI}) [14] used in this definition is a transposition of Plotkin's lgg [12].

Idestam-Almquist shows that his technique is practically infeasible, since it leads to an exponential growth of the computed expansion. Indeed, he proves that if n is the number of literals or-introduced to compute an expansion E of a clause C such that $|C| = m$, then the maximal cardinality of E is $(m+n)^{n+1}$. Instead, in our framework [5]:

Theorem 4.4. *Let C be a clause ($|C| = m$), S a set of clauses or-introduced from C by $[L_1, \ldots, L_n]$, and E an lgg_{oI} of S. Then $|E| \leq (m+n)$.*

Another important property about OI-expansions of a clause is that they are logically equivalent to it.

Theorem 4.5. *Let C be a clause and E its OI-expansion by some sequence Ω. Then $C \Leftrightarrow_{OI} E$.*

The main result is the following [5].

Theorem 4.6. *Given two clauses C and D, D non tautological, if C OI-implies D then there exists an expansion E of D such that C θ_{OI}-subsumes E.*

Hence, we can define refinement operators δ'_{OI} and ρ'_{OI} for spaces ordered by OI-implication:

Definition 4.5. *Let C be a clause, then:*
- $D \in \delta'_{OI}(C)$ iff $\exists E$, E expansion of C, and $D \in \delta_{OI}(E)$;
- $D \in \rho'_{OI}(C)$ iff $E \in \mathcal{L}^n_{OI}(\{C\})$, for some n, and $D \in \rho_{OI}(E)$.

As regards the properties of these operators, we have already remarked as computing the n-th powers of a clause in the definition of ρ'_{OI} is a merely theoretical issue, for the algorithm cannot know *a priori* which n to stop at. This yields a non locally finite operator. It is, instead, surely a proper and complete operator for the properness and completeness of ρ_{OI}. Conversely,

Theorem 4.7. *In a space ordered by OI-implication, δ'_{OI} is an ideal upward refinement operator.*

5 Conclusions and Future Work

Many problems encountered in ILP are theoretically or practically infeasible. Therefore, biasing them can help to find solutions in significant, yet restricted cases. This work is an effort in this direction: in our framework, the language was not deprived of representation power, however the complexity of the refinement operators was reduced because of the bias on the search space.

The work presented regarded the definition of a framework fulfilling the property of ideality of refinement operators, that guarantees for the efficiency and effectiveness of the learning process. While such operators have been proven not to exist in the spaces ordered by the notions of implication or θ-subsumption, in our framework, relying on the Object Identity assumption, we have weakened the implication ordering, obtaining OI-implication, that allows for the existence of ideal operators in the corresponding search spaces.

Future work will concern a deeper investigation of the properties of OI-implication. OI-implication seems to be promising since it appears more mechanizable than implication, yet the relationships holding between this ordering and the others presented in this work deserve further study. For the moment, we have stated that OI-implication is strictly weaker than unconstrained implication and stronger than θ_{OI}-subsumption. In addition, we have given an ideal upward refinement operator for search spaces ordered by OI-implication. A model-theoretic definition of this notion ought to be given together with the proof of its decidability. Hence, it should be easy to define ideal downward operators.

References

[1] M. Bain and S.H. Muggleton. Non-monotonic learning. In S.H. Muggleton, editor, *Inductive Logic Programming*. Academic Press, London, U.K., 1992.

[2] S. Ceri, G. Gottlob, and L. Tanca. *Logic Programming and Databases*. Springer, 1990.

[3] F. Esposito, N. Fanizzi, S. Ferilli, and G. Semeraro. Ideal theory refinement under object identity. In *Proceedings of the 17th International Conference on Machine Learning - ICML2000*. Morgan Kaufmann, 2000. (forthcoming).

[4] F. Esposito, A. Laterza, D. Malerba, and G. Semeraro. Locally finite, proper and complete operators for refining datalog programs. In Z.W. Raś and M. Michalewicz, editors, *Proceedings of the 9th International Symposium on Methodologies for Intelligent Systems - ISMIS96*, volume 1079 of *LNAI*, pages 468–478. Springer, 1996.

[5] N. Fanizzi. *Refinement Operators in Multistrategy Incremental Learning*. Ph.D. thesis, Dipartimento di Informatica, Università di Bari, Italy, 1999.

[6] G. Gottlob. Subsumption and implication. *Information Processing Letters*, 24(2):109–111, 1987.

[7] P. Idestam-Almquist. *Generalization of Clauses*. Ph.D. thesis, Stockholm University and Royal Institute of Technology, Kiesta, Sweden, 1993.

[8] J.W. Lloyd. *Foundations of Logic Programming*. Springer, 2nd edition, 1987.

[9] T.M. Mitchell. Generalization as search. *Artificial Intelligence*, 18:203–226, 1982.

[10] S.H. Muggleton. Inverting implication. In S. Muggleton and K. Furukawa, editors, *Proceedings of the 2nd International Workshop on Inductive Logic Programming*, ICOT Technical Memorandum TM-1182, 1992.

[11] S.-H. Nienhuys-Cheng and R. de Wolf. *Foundations of Inductive Logic Programming*, volume 1228 of *LNAI*. Springer, 1997.

[12] G.D. Plotkin. A note on inductive generalization. *Machine Intelligence*, 5:153–163, 1970.

[13] J.A. Robinson. A machine-oriented logic based on the resolution principle. *Journal of the ACM*, 12(1):23–41, January 1965.

[14] G. Semeraro, F. Esposito, D. Malerba, N. Fanizzi, and S. Ferilli. A logic framework for the incremental inductive synthesis of datalog theories. In N.E. Fuchs, editor, *Proceedings of the 7th International Workshop LOPSTR97*, volume 1463 of *LNCS*, pages 300–321. Springer, 1998.

[15] P.R.J. van der Laag. *An Analysis of Refinement Operators in Inductive Logic Programming*. Ph.D. thesis, Erasmus University, Rotterdam, NL, 1995.

Rule Quality Measures Improve the Accuracy of Rule Induction: An Experimental Approach

Aijun An and Nick Cercone

Department of Computer Science, University of Waterloo
Waterloo, Ontario N2L 3G1 Canada
Email: {aan, ncercone}@uwaterloo.ca

Abstract. Rule quality measures can help to determine when to stop generalization or specification of rules in a rule induction system. Rule quality measures can also help to resolve conflicts among rules in a rule classification system. We enlarge our previous set of statistical and empirical rule quality formulas which we tested earlier on a number of standard machine learning data sets. We describe this new set of formulas, performing extensive tests which also go beyond our earlier tests, to compare these formulas. We also specify how to generate formula-behavior rules from our experimental results, which show the relationships between a formula's performance and the characteristics of a dataset. Formula-behavior rules can be combined into formula-selection rules which can select a rule quality formula before rule induction takes place. We report the experimental results showing the effects of formula-selection on the predictive performance of a rule induction system.

1 Introduction

A rule induction system generates decision rules from a set of training data. The set of decision rules determines the performance of a classifier that exploits the rules to classify unseen objects. It is therefore important for a rule induction system to generate decision rules that have high predictability or reliability. These properties are commonly measured by a function called rule quality. A rule quality measure is needed in both the rule induction and classification processes. In rule induction, a rule quality measure can be used as a criterion in the rule specification andor generalization process. In classification, a rule quality value can be associated with each rule to resolve conflicts when multiple rules are satisfied by the example to be classified.

We survey a number of statistical and empirical rule quality measures, some of which have been discussed by Bruha [7,8] and An and Cercone [3]. In our earlier work [3], we evaluated some of these formulas on a smaller collection of data sets. One contribution of this paper is to include more formulas in our experiments and the tests also go beyond our earlier tests by including data sets in the experiments. In our evaluation, ELEM2 [2] is used as the basic learning and classification algorithms. We report the experimental results from using these formulas in ELEM2 and compare the results by indicating the significance level of the difference between each pair of the formulas. In addition, the relationship between the performance of a formula and a dataset is obtained by automatically generating formula-behavior rules from a dataset that describes the experimental results for the formulas and the characteristics of the datasets. The formula-behavior rules are further combined into formula-selection rules which can be employed by ELEM2 to select a rule quality formula before inducing rules from a dataset. We report the experimental results showing the effects of formula-selection on ELEM2's predictive performance.

2 Rule Quality Measures

Many rule quality measures are derived by analyzing the relationship between a decision rule R and a class C. The relationship can be depicted by a 2×2 contingency table [4,7]:

Table 1. Contingency Table with Absolute Frequencies

	Class C	Not class C	
Covered by rule R	n_{rc}	$n_{r\bar{c}}$	n_r
Not covered by R	$n_{\bar{r}c}$	$n_{\bar{r}\bar{c}}$	$n_{\bar{r}}$
	n_c	$n_{\bar{c}}$	N

where n_{rc} is the number of training examples covered by rule R and belonging to class C; $n_{r\bar{c}}$ is the number of training examples covered by R but not belonging to C, etc; N is the total number of training examples; n_r, $n_{\bar{r}}$, n_c and $n_{\bar{c}}$ are marginal totals, e.g., $n_r = n_{rc} + n_{r\bar{c}}$, which is the number of examples covered by R. The contingency table can also be presented using relative rather than absolute frequencies as follows:

Table 2. Contingency Table with Relative Frequencies

	Class C	Not class C	
Covered by rule R	f_{rc}	$f_{r\bar{c}}$	f_r
Not covered by R	$f_{\bar{r}c}$	$f_{\bar{r}\bar{c}}$	$f_{\bar{r}}$
	f_c	$f_{\bar{c}}$	1

where $f_{rc} = \frac{n_{rc}}{N}$, $f_{r\bar{c}} = \frac{n_{r\bar{c}}}{N}$, and so on.

2.1 Empirical Formulas

Empirical rule quality formulas are based on intuitive logic. We describe two empirical formulas that combine two basic characteristics of a rule: consistency and coverage. Using the elements of the contingency table, the consistency of a rule R can be defined as $cons(R) = \frac{n_{rc}}{n_r}$ and its coverage as $cover(R) = \frac{n_{rc}}{n_c}$.

Weighted Sum of Consistency and Coverage. Michalski [13] proposes to use the weighted sum of the consistency and coverage as a measure of rule quality as follows:

$$Q_{WS} = w_1 \times cons(R) + w_2 \times cover(R)$$

where w_1 and w_2 are user-defined weights with their values belonging to $(0,1)$ and summed to 1. This formula is applied in an incremental learning system YAILS [15]. The weights in YAILS are specified automatically as: $w_1 = 0.5 + \frac{1}{4}cons(R)$ and $w_2 = 0.5 - \frac{1}{4}cons(R)$. These weights depend on consistency. The larger the consistency, the more influence consistency has on rule quality.

Product of Consistency and Coverage. Brazdil and Torgo [6] propose to use a product of consistency and coverage as rule quality:

$$Q_{Prod} = consR \times f(cover(R))$$

where f is an increasing function. The authors conducted a large number of experiments and chose to use the following form of f: $f(x) = e^{x-1}$. This setting of f makes the difference in coverage have smaller influence on rule quality, which results in the rule quality formula to prefer consistency.

2.2 Measures of Association

A measure of association indicates a relationship between the classification for the columns and the classification for the rows in the 2×2 contingency table.

Pearson χ^2 Statistic. The χ^2 statistic is based on the assumption: if the classification for the columns is independent of that for the rows, the frequencies in the cells of the contingency table should be proportional to the marginal totals. The χ^2 value is given by

$$\chi^2 = \sum \frac{(n_o - n_e)^2}{n_e}$$

where n_o is the observed absolute frequency of examples in a cell, and n_e is the expected absolute frequency of examples for the cell. For example, for the upper-left cell, $n_o = n_{rc}$ and $n_e = \frac{n_r n_c}{N}$. The value $\frac{(n_o - n_e)^2}{n_e}$ is computed for each cell of the table individually and the values for all cells are added to yield the value of χ^2. This value measures whether the classification of examples by rule R and one by class C are related. The lower the χ^2 value, the more likely it is that the correlation between R and C is due to chance.

G2 Likelihood Ratio Statistic. The G2 likelihood ratio measures the distance between two distributions: the observed frequency distribution of examples among classes satisfying the rule R and the expected frequency distribution of the same number of examples under the assumption that the rule R selects examples randomly. The value of this statistic can be obtained using the absolute frequencies in the contingency table as follows:

$$G2 = 2(\frac{n_{rc}}{n_r}log\frac{n_{rc}N}{n_r n_c} + \frac{n_{r\bar{c}}}{n_r}log\frac{n_{r\bar{c}}N}{n_r n_{\bar{c}}})$$

where the logarithm is of base e. The lower the G2 value, the more likely it is that the apparent association between the two distributions is due to chance. Both the χ^2 and the likelihood ratio statistics are distributed asymptotically as χ^2 with one degree of freedom.

2.3 Measures of Agreement

A measure of agreement concerns the association of the elements of a contingency table on its main diagonal only [7].

Cohen's Formula. We can measure the *actual* agreement by simply summing up the main diagonal using the relative frequencies: $f_{rc}+f_{\bar{r}\bar{c}}$. A *chance* agreement occurs if the row variable is independent of the column variable, which can measured by $f_r f_c + f_{\bar{r}} f_{\bar{c}}$. Cohen [9] suggests to compare the actual agreement with the chance agreement by using the normalized difference of the two which we can use as a rule quality measure:

$$Q_{Cohen} = \frac{f_{rc} + f_{\bar{r}\bar{c}} - (f_r f_c + f_{\bar{r}} f_{\bar{c}})}{1 - (f_r f_c + f_{\bar{r}} f_{\bar{c}})}$$

When both elements f_{rc} and $f_{\bar{r}\bar{c}}$ are reasonably large, Cohen's statistic gives a higher value which indicates the agreement on the main diagonal.

Coleman's Formula. Coleman [5,7] defines a measure of agreement that indicates an association between the first column and any particular row in the contingency table. Bruha [7] suggests using a modified version of Coleman's measure for the purpose of rule quality definition, which actually responds to the agreement on the upper-left element of the contingency table. The formula is also derived by normalizing the difference between the actual and chance agreement as follows:

$$Q_{Coleman} = \frac{f_{rc} - f_r f_c}{f_r - f_r f_c}.$$

C1 and C2 Formulas. Further analysis indicates that Coleman's formula does not properly comprise the coverage (i.e. completeness) of a rule. On the other hand, Cohen's statistic is more completeness-based. Therefore, Bruha [8] modified Coleman's formula in two ways, which yields formulas C1 and C2:

$$Q_{C1} = Q_{Coleman} \times \frac{2 + Q_{Cohen}}{3}$$

$$Q_{C2} = Q_{Coleman} \times \frac{1 + cover(R)}{2}$$

where the coefficients 2, 3 and 1, 2 are used for the normalization purpose.

2.4 Measure of Information

The measure of information is another statistical measurement that can be used to define rule quality. Given a class C, the amount of information necessary to correctly classify an instance into class C whose prior probability is $P(C)$ is defined as [12] $-log P(C)$ [bit], where the log function is of base 2. Now given a rule R, the amount of information we need to correctly classify an instance into class C is $-logP(C|R)$ [bit], where $P(C|R)$ is the posterior probability of C given R. Therefore, the amount of information obtained by the rule R is $-logP(C) + logP(C|R)$ [bit]. Kononenko and Bratko [12] call the value of this formula the *information score*, which measures the amount of information the rule R contributes. Using frequencies to estimate the probabilities, the formula can be written as

$$Q_{IS} = -log\frac{n_c}{N} + log\frac{n_{rc}}{n_r}.$$

2.5 Measure of Logical Sufficiency

The logical sufficiency measure is a standard likelihood ratio statistic, which have been applied to measure rule quality [10,1]. Given a rule R and a class C, the degree of logical sufficiency of R with respect to C is defined by

$$Q_{LS} = \frac{P(R|C)}{P(R|\bar{C})}$$

where P denote probability. A rule for which Q_{LS} is large means that the observation of R is encouraging for the class C - in the extreme case of Q_{LS} approaching infinity, R is sufficient to establish C in a strict logical sense. On the other hand, if Q_{LS} is much less than unity, then the observation of R is discouraging for C. Using frequencies to estimate the probabilities, the formula can be expressed as $Q_{LS} = \frac{\frac{n_{rc}}{n_c}}{\frac{n_{r\bar{c}}}{n_{\bar{c}}}}$.

2.6 Measure of Discrimination

Another statistical rule quality formula is the measure of discrimination, which is applied in ELEM2 [2]. The formula was inspired by a query term weighting formula used in the probability-based information retrieval. The formula measures the extent to which a query term can discriminate between relevant and non-relevant documents [14]. If we consider a rule R as a query term in an information retrieval setting, positive examples of a class C as relevant documents, and negative examples as non-relevant documents, then the following formula can be used to measure the extent to which the rule R can discriminate between the positive and negative examples of the class C:

$$Q_{MD} = log\frac{P(R|C)(1 - P(R|\bar{C}))}{P(R|\bar{C})(1 - P(R|C))}$$

where P denotes probability. The formula can be estimated using the frequencies as $Q_{MD} = log\frac{\frac{n_{rc}}{n_{\bar{r}c}}}{\frac{n_{r\bar{c}}}{n_{\bar{r}\bar{c}}}}$.

3 Experiments with Rule Quality Measures

3.1 Experimental Design

We evaluate the rule quality formulas described in Section 2 by determining how different rule quality formulas affect the predictive performance of a rule induction system, ELEM2. In ELEM2, a rule quality formula is used in both post-pruning and classification processes. In post-pruning, removal of an attribute-value pair depends on whether it will decrease the quality value of the rule. In classification, the rule quality formula is used to help resolve conflicts among rules. In our experiments, we run versions of ELEM2, each of which uses a different rule quality formula. The χ^2 statistic is used in two ways, in both of which the χ^2 formula is used as the ELEM2 rule quality measure. They differ in the method to post-prune a generated rule.

1. $Q_{\chi^2_{.05}}$ In post-pruning, the removal of an attribute-value pair depends on whether the rule quality value after removing an attribute-value pair is greater than $\chi^2_{.05}$, i.e., the tabular χ^2 value for the significance level of 0.05 with one degree of freedom. If the calculated value is greater than tabular $\chi^2_{.05}$, then remove the attribute-value pair; otherwise check other pairs or stop post-pruning if all pairs have been checked.
2. $Q_{\chi^2_{.05+}}$ In post-pruning, an attribute-value pair is removed if and only if the rule quality value Q_{after} after removing an attribute-value pair is greater than $\chi^2_{.05}$ and Q_{after} is no less than the rule quality value before removing the attribute-value pair.

The G2 statistic, denoted as $Q_{G2.05+}$, is used in the same way as $Q_{\chi^2_{.05+}}$, i.e., a pair is removed in post-pruning if and only if the value of $Q_{G2.05+}$ is greater than $\chi^2_{.05}$ and the removal does not cause the rule quality value to decrease.

Our experiments are conducted using 27 benchmark datasets obtained from the UCI Repository of Machine Learning database. The datasets represent a mixture of characteristics ranging from 2 to 10 classes, from 4 to 64 condition attributes, and from 24 to 7491 examples.

3.2 Results

On each dataset, we conduct the ten-fold evaluation of a rule quality measure using ELEM2. The results in terms of predictive accuracy mean on each dataset for each formula are shown in Figure 1. The average of the accuracy means for each formula over the 27 datasets is shown in Table 3, where the rule quality formulas are listed in decreasing order of average accuracy means. Whether a

Table 3. Average of accuracy means for each formula over the datasets.

	Q_{C2}	Q_{WS}	Q_{C1}	Q_{LS}	Q_{MD}	$Q_{Coleman}$	$Q_{G2.05+}$	Q_{IS}	Q_{Prod}	$Q_{\chi^2_{.05+}}$	Q_{Cohen}	$Q_{\chi^2_{.05}}$
Average	81.89	81.71	81.61	81.38	80.95	80.65	79.94	79.87	79.59	78.44	78.08	72.42

formula with a higher average is significantly better than a formula with a lower average is determined by paired t-tests. The t-test results in terms of p-values

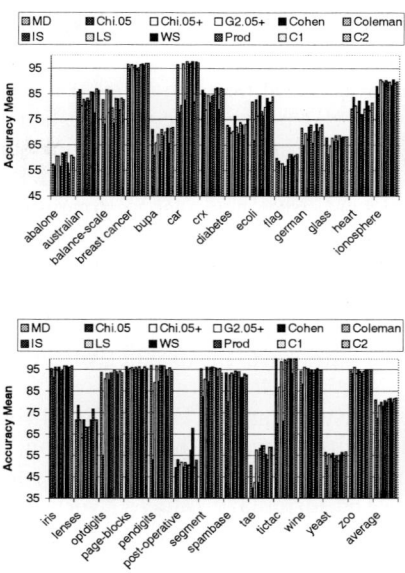

Fig. 1. Results on the 27 datasets

are reported in Table 4. A small p-value indicates that the null hypothesis (the difference between the two formulas is due to chance) should be rejected in favor of the alternative at any significance level above the calculated value. In Table 4, the p-values that are smaller than 0.05 are shown in bold-type to indicate that the formula with higher average is significantly better than the formula with the lower average at the 5% significance level.

Table 4. Significance levels (p-values from paired t-test) of improvement.

	Q_{C2}	Q_{WS}	Q_{C1}	Q_{LS}	Q_{MD}	$Q_{Coleman}$	$Q_{G2.05+}$	Q_{IS}	Q_{Prod}	$Q_{\chi^2_{.05+}}$	Q_{Cohen}	$Q_{\chi^2_{.05}}$
Q_{C2}	NA	0.5483	0.2779	0.1645	**0.0261**	**0.0078**	**0.0011**	**0.0016**	0.033	**0.0017**	**0.0072**	**0.0003**
Q_{WS}	-	NA	0.8074	0.4866	0.072	**0.0465**	**0.0052**	**0.0086**	**0.0286**	**0.0028**	**0.0107**	**0.0004**
Q_{C1}	-	-	NA	0.3328	0.0721	**0.0094**	**0.0012**	**0.0014**	0.0871	**0.0052**	**0.0161**	**0.0006**
Q_{LS}	-	-	-	NA	0.3665	**0.0278**	**0.0006**	**0.0022**	0.1343	**0.0122**	**0.028**	**0.0009**
Q_{MD}	-	-	-	-	NA	0.5435	0.0911	0.1037	0.2231	**0.0115**	**0.0336**	**0.0012**
$Q_{Coleman}$	-	-	-	-	-	NA	**0.0378**	0.0626	0.3573	0.0609	0.0908	**0.0024**
$Q_{G2.05+}$	-	-	-	-	-	-	NA	0.8234	0.7648	0.2104	0.2295	**0.0056**
Q_{IS}	-	-	-	-	-	-	-	NA	0.8059	0.2532	0.2632	**0.0058**
Q_{Prod}	-	-	-	-	-	-	-	-	NA	0.2213	0.2428	**0.0031**
$Q_{\chi^2_{.05+}}$	-	-	-	-	-	-	-	-	-	NA	0.6246	**0.0067**
Q_{Cohen}	-	-	-	-	-	-	-	-	-	-	NA	**0.0083**
$Q_{\chi^2_{.05}}$	-	-	-	-	-	-	-	-	-	-	-	NA

Generally speaking, we can say that, in terms of predictive performance, Q_{C2}, Q_{WS}, Q_{C1}, Q_{LS} and Q_{MD} are comparable even if their performance may not agree on a particular dataset. The same for $Q_{Coleman}$, $Q_{G2.05+}$, Q_{IS} and Q_{Prod}, and $Q_{\chi^2_{.05+}}$ and Q_{Cohen}. The performance of $Q_{G2.05+}$ and Q_{IS} are not only comparable, but also similar on each particular dataset (seen from Figure 1), which indicates that the two formulas have similar trends with regard to n_{rc}, n_r, n_c and N in the contingency table.

4 Learning from the Experimental Results

From the experimental results, we posit that, even if on some datasets (such as the *breast cancer* dataset) the performance of the learning system is not very sensitive to the rule quality formula used, the performance greatly depends on the formula on most of the other datasets. It would be desirable that we can apply a "right" formula that gives the best performance among other formulas on a particular dataset. For example, even though the formula $Q_{\chi^2_{.05}}$ is not a good formula in general, it performs better than other formulas on some datasets such as *heart* and *lenses*. If we can find the conditions under which each formula leads to a good performance of the learning system, we can select "right formulas" for different datasets and can improve the predictive performance of the learning system further.

To find out this regularity, we use ELEM2 to learn the formula selection rules from the experimental results shown in the last section. The learning problem is divided into (1) learning the rules for each rule quality formula that describe the conditions under which the formula produces "very good", "good", "medium" or "bad" results, and (2) combining the rules for all the formulas that describe the conditions under which the formulas give the "very good" results. The resulting set of rules is the formula-selection rules that can be used by the ELEM2 classification procedure to perform formula selection.

4.1 Data Representation

For the purpose of learning formula-behavior rules, i.e., the rules that describe the conditions under which a formula leads to "very good", "good", "medium", or "bad" performance, we construct training examples from the above results and the dataset characteristics. First, on each dataset, we decide the relative performance of each formula as "very good", "good", "medium", or "bad". For example, on the *balance-scale* dataset, we say that the formulas whose accuracy mean is above 85% produce "very good" results; the formulas whose accuracy mean is between 80% and 85% produce "good" results; the ones with the mean between 75% and 80% are "medium" and other formulas give "bad" results. Then, for each formula, we construct a training data set in which an training example describes the characteristics of a dataset and also a description in term of whether the formula produces "very good", "good", "medium", or "bad" result on this dataset. Thus, to learn the rules for each formula, we have 27 training examples. The characteristics of a data set is described in terms of number of examples, number of attributes, number of classes and the class distribution. A sample of training examples for learning the behavior rules of the formula Q_{IS} is shown in Table 5.

4.2 The Learning Results

ELEM2 with its default rule quality formula (Q_{MD}) is used to learn the "behavior" rules from the training dataset constructed for each formula. Table 6 lists some of these behavior rules for each formula, where N stands for the number

Table 5. Sample of training examples for learning the behavior of a formula

Number of			Class	
Examples	Attributes	Classes	Distribution	Performance
4177	8	3	Even	Very Good
690	14	2	Even	Medium
625	4	3	Uneven	Bad
683	9	2	Uneven	Medium
1728	6	4	Uneven	Good

of examples, NofA is the number of attributes, NofC is the number of classes, and "No. of Support Datasets" means the number of the datasets that support the corresponding rule. These rules summarize the predictive performance of each formula in terms of characteristics of datasets. We further build a set of

Table 6. Formula Behavior Rules

Formula	Condition	Decision	Rule Quality	No. of Support Datasets
Q_{C2}	(768<N≤1728)	Very good	1.30	4
	(N≤653)and(NofA>10)and(NofC≤7)	Good	1.36	5
Q_{WS}	(625<N≤1728)and(NofA>8)and(ClassDistr!=Even)	Very good	1.48	4
	(N>336)and(NofC>5)	Good	1.38	4
Q_{C1}	(N>270)and(8<NofA≤15)	Very good	1.66	5
	(15<NofA≤57	Good	1.43	7
Q_{LS}	(N>2310)	Very good	1.45	5
	(N≤87)	Bad	2.41	2
Q_{MD}	(N>768)and(8<NofA≤16)	Very good	2.04	3
	(351<N≤4601)and(NofA>13)	Good	1.23	6
$Q_{Coleman}$	(N>958)and(NofC≤5)	Very good	1.79	5
	(N≤87)	Bad	1.51	2
$Q_{G2.05+}$	(N>101)and(10<NofA≤18)and(NofC>2)	Very good	2.04	3
	(270<N≤690)and(NofA≤15)	Medium	2.25	6
Q_{IS}	(N>150)and(NofC>2)and(ClassDistr=Even)	Very good	2.13	4
	(N≤101)	Bad	1.50	3
Q_{Prod}	(N≤214)and(NofA>7)and(NofC≤6)	Very good	1.80	3
	(N>768)and(8<NofA≤57)	Medium	1.70	6
$Q_{\chi^2_{.05+}}$	(N≤178)and(NofA>9)	Very good	1.67	2
	(N≤214)and(4<NofA≤9)	Bad	1.39	3
Q_{Cohen}	(345<N≤1484)and(NofA≤8)	Very good	1.80	3
	(4<NofA≤6)	Bad	1.63	3
$Q_{\chi^2_{.05}}$	(9<NofA≤14)and(NofC≤2))	Very good	1.91	2
	(N>24)	Bad	0.98	20

formula-selection rules by combining all the "very good" rules, i.e., the rules that predicts "very good" performance for each formula, and use them to select a "right" formula for a (new) dataset. For formula selection, we can use the ELEM2 classification procedure that takes formula-selection rules to classify a data set into a class of using a particular formula.

4.3 ELEM2 with Multiple Rule Quality Formulas

With formula-selection rules, ELEM2 has the flexibility of using different formulas on different datasets. To see how this strategy works, we conduct ten-fold evaluation of the "flexible" ELEM2 on the 27 datasets we used before. The result is shown in Figure 2, in which the average accuracy mean from the "flexible" ELEM2 (labeled as "Combine" in the graph) is compared with the ones from using individual formulas. We also conduct paired t-tests to see how much the

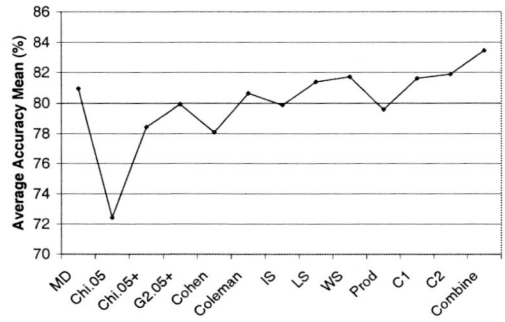

Fig. 2. Average of accuracy means of each formula on the 22 datasets

flexible ELEM2 improves over the ELEM2 with a single rule quality formula. The p-values from the t-test are shown in Table 7. We can see that "Combine" improves all the single formulas significantly.

Table 7. Significance levels of the improvement of "Combine" over individual formulas

	Q_{C2}	Q_{WS}	Q_{C1}	Q_{LS}	Q_{MD}	$Q_{Coleman}$	$Q_{G2_{.05+}}$	Q_{IS}	Q_{Prod}	$Q_{\chi^2_{.05+}}$	Q_{Cohen}	$Q_{\chi^2_{.05}}$
p-value	0.0139	0.0009	0.0154	0.008	0.0025	0.0006	0.0002	0.0002	0	0.0001	0.0005	0

5 Conclusions

We have described and experimented with various statistical and empirical formulas for defining rule quality measures. All formulas are applicable to a rule induction system for the purpose of post-pruning and classification, but their performance varies among the datasets. The empirical formulas, especially Q_{WS}, work very well even if they are not backed by statistical theories. Among statistical formulas, Q_{C2}, Q_{C1}, Q_{LS} and Q_{MD} work the best on the tested dataset and are comparable with Q_{WS}.

To determine the regularity of the rule quality formula's performance in terms of dataset characteristics, we used our learning system to induce formula-behaviour rules from a dataset constructed from the experimental results for different formulas. These rules provided ideas about the situations in which a formula leads to very good, good, medium or bad performance. These rules were also combined and used to automatically select a rule quality formula before rule induction begins. Our experiment showed that this selection of rule quality formula can lead to significant improvement over the rule induction system using a single rule quality formula. Future work includes testing our conclusions on more datasets to obtain more reliable formula-behavior rules. With more datasets available, we will test the formula-selection rules on the datasets that are different from the datasets used for generating the rules.

Acknowledgment. The authors are members of the Institute for Robotics and Intelligent Systems (IRIS) and wish to acknowledge the support of the Networks of Centres of Excellence of the Government of Canada, the Natural Sciences and Engineering Research Council, and the participation of PRECARN Associates Inc.

References

1. Ali, K. and Pazzani, M. 1993. "HYDRA: A noise-tolerant relational concept learning algorithm". *Proceedings of the Thirteenth International Joint Conference on Artificial Intelligence (IJCAI'93)*, Chambery, France. Morgan Kaufmann.
2. An, A. and Cercone, N. 1998. "ELEM2: A Learning System for More Accurate Classifications." *Lecture Notes in Artificial Intelligence 1418*.
3. An, A. and Cercone, N. 1999. "An Empirical Study on Rule Quality Measures", *Proceedings of the Seventh International Workshop on Rough Sets, Fuzzy Sets, Data Mining, and Granular-Soft Computing*, Yamaguchi, Japan.
4. Arkin, H. and Colton, R. R. 1970. *Statistical Methods*. Barnes & Noble Inc., New York.
5. Bishop, Y.M.M, Fienberg, S.E. and Holand, P.W. 1991. *Discrete Multivariate Analysis: Theory and Practice*. The MIT Press.
6. Brazdil, P. and Torgo, L. 1990. "Knowledge Acquisition via Knowledge Integration". In: *Current Trends in Knowledge Acquisition*, IOS Press.
7. Bruha, I. 1993. "Quality of Decision Rules: Empirical and Statistical Approaches". *Informatica*, 17, pp.233-243.
8. Bruha, I. 1996. "Quality of Decision Rules: Definitions and Classification Schemes for Multiple Rules", in Nakhaeizadeh, G. and Taylor, C. C. (eds.): Machine Learning and Statistics, The Interface. Jone Wiley & Sons Inc.
9. Cohen, J. 1960. "A Coefficient of Agreement for Nominal Scales". *Educational and Psych. Meas.* 22, pp.37-46.
10. Duda, R., Gaschnig, J. and Hart, P. 1979. "Model Design in the Prospector Consultant System for Mineral Exploration". In D. Michie (ed.), *Expert Systems in the Micro-electronic Age*. Edinburgh University Press, Edinburgh, UK.
11. Holte, R., Acker, L. and Porter, B. 1989. "Concept Learning and the Problem of Small Disjuncts". *Proceedings of the Eleventh International Joint Conference on Artificial Intelligence*, Detroit, Michigan.
12. Kononenko, I. and Bratko, I. 1991. "Information-Based Evaluation Criterion for Classifier's Performance". *Machine Learning*, 6, 67-80.
13. Michalski, R.S. 1990. "Pattern Recognition as Rule-Guided Inductive Inference". *IEEE Transactions on Pattern Analysis and Machine Intelligence*, PAMI-2, 4.
14. Robertson, S.E. and Sparck Jones, K. 1976. "Relevance Weighting of Search Terms". *Journal of the American Society for Information Science*. Vol.27. pp.129-146.
15. Torgo, L. 1993. "Controlled Redundancy in Incremental Rule Learning". *ECML-93*, pp.185-195.

A Dynamic Approach for Knowledge Discovery of Web Access Patterns[*]

Alaaeldin Hafez[1]
ahafez@cacs.louisiana.edu
The Center for Advanced Computer Studies
University of Louisiana at Lafayette
Lafayette, LA 70504, USA

Abstract. The emergence of the World Wide Web (Web) technology and the advance of data capturing techniques have lead to exponential growth in amounts of data being stored in Web server logs. This growth in turn has motivated researchers to seek new techniques for the extraction of knowledge implicit or hidden in such data. Designing a web site is a complex problem. Web Server logs provide an opportunity to observe users interacting with the site and make improvements to that site's structure and presentation. In this paper, we motivate the need for a Dynamic data mining approach for mining user access patterns that uses previous mining results during previous time periods. We present an efficient approach that uses latest results of data mining and new changes in Web server logs to generate new mining rules. The proposed approach is shown to be effective for solving problems related to efficiency of handling data updates and accuracy of data mining results. The proposed approach does not depend on the technique used to generate new frequent user access patterns during the current episode (time period). In our analysis, we have used an *Apriori-Like* algorithm as a local algorithm to generate frequent user access patterns. The experimental results show that, comparing to *Apriori-like* techniques, our dynamic approach improves the efficiency of the mining process.

Keywords: Knowledge Discovery, Data Mining, Web Mining, User Access Patterns, Association Mining, Web Structure.

1 Introduction

With the growing popularity of the World-Wide Web (Web) and the rapid progress of the Web technology, hundreds of millions of transactions are processed every day through the Web. Web servers keep log entries (files) for all transactions that are accessing their sites, and the sizes of those log files are increasing by tens of megabytes every day. Server logs reveal an enormous amount of information about users, server behavior, changes in sites, and potential benefits of new technical developments. Most institutions have not been able to perform an effective use of Web server log files for enhancing and improving server performance and design

---

[*] This research was supported in part by the U.S. Department of Energy, Grant No. DE-FG02-97ER1220.
[1] on leave from The Department of Computer Science and Automatic Control, Faculty of Engineering, Alexandria University, Alexandria, Egypt

improvement. Mining information and knowledge from the Web transaction data has become a prominent and important research and application area.

The behavior of user access patterns can be detected by using the history contained in Web server log files [10,11,12,13]. Analyzing and capturing similarities in this behavior can enhance system performance and identify user interests. Many studies have been conducted to understand user motivation and reaction, analyze system performance, and improve system design [5,13]. Applying data mining techniques on Web logs discovers interesting access patterns that can be used to restructure server sites in an efficient way. Unfortunately most of the existing data mining techniques are iterative and require many disk scans over transaction (log) files [1,3,8].

Web applications require up to date mining of information from data that changes on a regular basis [7]. Thousands of remote sites (URLs) are daily created and removed. In such an environment, frequent or occasional updates may change the status of some interesting patterns discovered earlier [13]. Discovering knowledge is an expensive operation [5,6]. It requires extensive access of secondary storage that can become a bottleneck for efficient processing. Running of data mining algorithms from scratch, each time there is a change in data, is not an efficient strategy. Updating previously discovered knowledge could solve many problems that data mining techniques have faced for years; that is, inability to handle data updates, lack of accuracy of data mining results, and poor performance.

Association mining that discovers dependencies among values of an attribute was introduced by Agrawal et al.[1] and has emerged as an important research area. The problem of association mining, also referred to as the *market basket* problem, is formally defined as follows. Let $I = \{i_1, i_2, \ldots, i_n\}$ be a set of items and $S = \{s_1, s_2, \ldots, s_m\}$ be a set of transactions, where each transaction $s_i \in S$ is a set of items that is $s_i \subseteq I$. An *association rule* denoted by $X \Rightarrow Y, X, Y \subset I,$ and $X \cap Y = \Phi$, describes the existence of a relationship between the two itemsets X and Y.

Several measures have been introduced to define the *strength* of the relationship between itemsets X and Y such as *support, confidence*, and *interest*. The definitions of these measures, from a probabilistic model are given below.

I. *Support* $(X \Rightarrow Y) = P(X,Y)$, or the percentage of transactions in the database that contain both X and Y.
II. *Confidence* $(X \Rightarrow Y) = P(X,Y)/P(X)$, or the percentage of transactions containing Y in those transactions containing X.
III. *Interest*$(X \Rightarrow Y) = P(X,Y)/P(X)P(Y)$ represents a test of statistical independence.

Agrawal et al [2], introduced the problem of mining sequential patterns over such databases. Two algorithms, *AprioriSome* and *Apriori-Like* [2], have been presented to solve this problem, and their performances have been evaluated using synthetic data. The two algorithms have comparable performances. *AprioriSome* has performed better when the minimum number of users required to deem a sequential pattern to be interesting is low.

In this paper, we propose an approach that dynamically updates knowledge obtained from the data mining process during previous time periods. Transactions over a long duration are divided into a set of consecutive episodes. We propose a modified structure for keeping updated log transactions. The proposed structure facilities the use of different association mining techniques. Our approach discovers current frequent user access patterns by using updates that have occurred during the current time period along with the frequent user access patterns that have been discovered in the previous time period.

In section 2, we give the formal definition of the problem of discovering frequent user access patterns. The proposed structure of Web transaction log and the dynamic approach are described in section 3. Our experimental results are presented in section 4. The experimental results are discussed and the paper is concluded in section 5.

2   Problem Definition

In the *original* Web log file *OF*, each request received by the Web server creates a Web log entry e that contains three components: $User(e)$ denotes the user-id of that user who originated the request, $Time(e)$ is the time-stamp of that request, and $url(e)$ is the set of requested URLs [2,4,10]. Examples 2.1 and 2.2 demonstrate, for a given Web server, the original log file *OF* and the current log file *CF*.

Example 2.1 The *original* Web log file *OF*

e	User(e)	Time(e)	url(e)
1	1	4	{a,b,c,d}
2	2	6	{a,c}
3	1	8	{a,b,d}
4	2	10	{a,c,f}
5	3	14	{c}
6	2	16	{a,f}
7	3	18	{c,f}
8	4	20	{a}

Example 2.2 The *current* Web log file *CF*

e	User(e)	Time(e)	url(e)	e	User(e)	Time(e)	url(e)
1	1	4	{a,b,c,d}	9	3	24	{c,d}
2	2	6	{a,c}	10	5	24	{a,d}
3	1	8	{a,b,d}	11	1	26	{a,c}
4	2	10	{a,c,f}	12	2	30	{a,c,f}
5	3	14	{c}	13	5	32	{a,c}
6	2	16	{a,f}	14	3	36	{a,b,c}
7	3	18	{c,f}	15	6	36	{b,c}
8	4	20	{a}	16	5	40	{b,c}

We adopt the same definitions used in [2] to define the terms *sequential pattern, support, confidence, and frequent k-sequence.*

sequential pattern   is defined as a set of one or more URLs that are accessed sequentially.
support(X)   is defined as the ratio of users who have requested sequential pattern *X*.

confidence(X⇒Y) is defined as the ratio of users who have requested sequential pattern *X* and *Y* among users who have requested sequential pattern *X*.

frequent k-sequence is defined as a set of *k urls* that are accessed sequentially, and has support greater than or equal a support threshold *minsup*

Example 2.3 Let *minsup* =0.5. The *frequent k-sequences* derived from Web log file *OF*, defined in Example 2.1, are

Frequent 1-sequence	a	{(1,4),(2,6),(1,8),(2,10), (2,16),(4,20)}	support(a)=0.75
	c	{(1,4),(2,6),(2,10), (3,14),(3,18)}	support(b)=0.75
	f	{(2,10),(2,16),(3,18)}	support(f)=0.5
Frequent 2-sequence	ac	{(1,4),(2,6),(2,10)}	support(ac)=0.5
	cf	{(2,10),(3,18)}	support(cf)=0.5

Example 2.4 Let *minsup* =0.5. The *frequent k-sequences* derived from Web log file *CF*, defined in Example 2.2, are

Frequent 1-sequence	A	{(1,4),(2,6),(1,8),(2,10),(2,16),(4,20),(5,24), (1,26),(2,30),(5,32),(3,36)}	support(a)=0.833
	B	{(1,4),(1,8),{3,36),(6,36),(5,40)}	support (b)=0.75
	C	{(1,4),(2,6),(2,10)(3,14),(3,18),(3,24),(1,26), (2,30),(5,32),(3,36),(6,36),(5,40)}	support (c)=0.833
	D	{(1,4),(1,8),(3,24),(5,24)}	support (d)=0.5
Frequent 2-sequence	Ac	{(1,4),(2,6),(2,10),(1,26),(2,30),(5,32),(3,36)}	support (ac)=0.75
	Bc	{(1,4),(3,36),(6,36),(5,40)}	support (bc)=0.75

3 The Dynamic Approach

Knowledge discovery of patterns is defined as locating those patterns in which accesses to different resources consistently occurring together, or accesses from a particular place occurring at regular times [4,11,12]. In our approach, we define a structure for keeping log transactions. Rather than describing log entries with respect to their entry order, we map the original structure of Web log files into an equivalent structure where, for each *URL*, there exists a set *ID(URL)* such that each element in *ID(URL)* is a pair <user-id, time-stamp>. Formally speaking, for a given log file *F* and a Web page *URL*, *ID(URL)*= {(User(e),Time(e))| ∀e∈ F, URL=Url(e)}. In examples 3.1 and 3.2, we demonstrate, for a given Web server, the proposed mappings of original log file *OF* and the current log file *CF*, respectively.

Example 3.1 The mapping of the *original* Web log file *OF* defined in example 2.1 is

URL	ID(URL)
A	{(1,4),(2,6),(1,8),(2,10),(2,16),(4,20)}
B	{(1,4),(1,8)}
C	{(1,4),(2,6),(2,10),(3,14),(3,18)}
D	{(1,4),(1,8)}
F	{(2,10),(2,16),(3,18)}

Example 3.2 The mapping of the *current* Web log file *CF* defined in example 2.2 is

URL	ID(URL)
a	{(1,4),(2,6),(1,8),(2,10),(2,16),(4,20),(5,24),(1,26),(2,30),(5,32),(3,36)}
b	{(1,4),(1,8),{3,36),(6,36),(5,40)}
c	{(1,4),(2,6),(2,10)(3,14),(3,18),(3,24),(1,26),(2,30),(5,32),(3,36),(6,36),(5,40)}
d	{(1,4),(1,8),(3,24),(5,24)}
f	{(2,10),(2,16),(3,18),(2,30)}

Web log files keep information of all accesses including those accesses to those canceled Web pages, that could be canceled along time ago. Mining algorithms should keep a list of those canceled pages in order to not counting those deleted Web pages, but still scanning the whole log file is considered. In the context of Web mining [9,10,11,12], It is a better strategy to use those mining results collected in the last mining session and only apply the mining procedure only on those transactions added to the Web log. In this paper, we propose a dynamic algorithm for mining user access patterns, that treats Web log transactions as sequences over periods of time and uses the latest discovered (in a previous session) association rules to improve the efficiency of the mining process.

The storage requirement for keeping the structure of transaction updates is $\alpha \tilde{N}$ where $\tilde{N}$ is the number of disk blocks needed to store the transaction updates, and α is the reduction factor caused by grouping user ID's of the same Web page.

In this section, we introduce the notions of *continuous pattern*, *non-continuous pattern*, *uprising pattern*, and *non-uprising pattern*.

Definition 3.1 A sequence X is a *continuous pattern* through two time periods T_1 and T_2 if
$$\text{support}_{T_1}(X) \geq minsup \quad \text{and} \quad \text{support}_{T_2}(X) \geq minsup$$

Definition 3.2 A sequence X is a *non-continuous pattern* through two time periods T_1 and T_2 if
$$\text{support}_{T_1}(X) \geq minsup \quad \text{and} \quad \text{support}_{T_2}(X) < minsup$$

Definition 3.3 A sequence X is an *uprising pattern* through two time periods T_1 and T_2 if
$$\text{support}_{T_1}(X) < minsup \quad \text{and} \quad \text{support}_{T_2}(X) \geq minsup$$

Definition 3.4 A sequence X is a *non-uprising pattern* through two time periods T_1 and T_2 if
$$\text{support}_{T_1}(X) < minsup \quad \text{and} \quad \text{support}_{T_2}(X) < minsup$$

In order to minimize the number of disk scans and keep only the necessary information, we only consider Web log changes in time period T_i along with the results obtained during time period T_{i-1}, for $i=2,3,\ldots$. Time values and repeated user-id's are omitted from *k-sequences*. A new parameter $disp_{T_i}(X)$ is defined to reflect the displacement of the *k-sequence* X in the time period T_i.

Definition 3.5 Let X be a *k-sequence* of URLs and $ID(X)$ be a set of pairs (u_j, t_j), $j=1,2,\ldots, D$, and $\tau_{i-1} < t_j \leq \tau_i$; $\tau_i = \tau_{i-1} + T_i$, where u_j and t_j are a user-id and its time-stamp, respectively. *Displacement* of X in time period T_i is defined as

$$disp_{T_i}(X) = \frac{\sum_{\substack{j=1 \\ \tau_{i-1} < t_j(X) \leq \tau_i}}^{D} t_j(X)}{D}$$

Example 3.3 Let *minsup* =0.5. The *frequent k-sequences* derived from transaction file *OF* during period T_1 and those updated transactions in transaction file *CF* during period T_2, defined in Examples 3.1 and 3.2, respectively, are

	T_1		$disp_{T_1}$		T_2		$disp_{T_2}$
a	{1,2,4}	support(a)=0.75	10.66	a	{1,2,3,5}	Support(a)=0.75	29.6
c	{1,2,3}	support(b)=0.75	10.4	b	{3,5,6}	Support(b)=0.5	37.33
f	{2,3}	support(f)=0.5	14.66	c	{1,2,3,5,6}	Support (c)=0.833	32
ac	{1,2}	support(ac)=0.5	6.66	ac	{1,2,3,5}	Support (ac)=0.75	31
cf	{2,3}	support(cf)=0.5	14	bc	{3,5,6}	Support (bc)=0.5	37.33

We consider only those *patterns* defined in definitions 3.1 to 3.3, which are divided into two categories,

Category 1 *continuous patterns* and *uprising patterns*.
Category 2 *non-continuous patterns*.

patterns in *category 1* are automatically included as *frequent k-sequences*. *non-continuous patterns* (in *category 2*) are considered *frequent k-sequences* if

$$\delta_{T_i}(X)\, support_{T_{i-1}}(X) + (1 - \delta_{T_i}(X))\, support_{T_i}(X) \geq minsup, \text{ where } 0 \leq \delta_{T_i}(X) \leq 1 \quad (1)$$

In our experimental work, we choose the value of $\delta_{T_i}(X)$ to be dependent on the behavior of *k-sequence* X through time periods T_{i-1} and T_i, where

$$\delta_{T_i}(X) = \frac{(\tau_i - disp_{T_{i-1}}(X))}{(disp_{T_i}(X) - disp_{T_{i-1}}(X))}$$

Algorithm *DynamicApriori*

$f_1^+(T_2) = \{$ *frequent 1 - sequences*$\}$; //continuous frequent 1-sequences and uprising frequent 1-sequences,
$f_1^-(T_2) = \{$ *non - continues 1 - sequences*$\}$;
$f_1(T_2) = f_1^+(T_2) \cup \{x \mid x \in f_1^+(T_1) \cap f_1^-(T_2)$ and $\delta_{T_2}(x)\, support_{T_1}(x) + (1 - \delta_{T_2}(x))\, support_{T_2}(x) \geq minsup\}$
for $(k=2; f_{k-1}(T_2) \neq \emptyset; k++)$ do
    begin
        $C_k = AprioriGen(f_{k-1}(T_2))$;
        forall transactions $t \in F_{T_2}$ do
            forall candidates $c \in C_k$ do
                if $c \subseteq t$ then c.count++;
        $f_k^+(T_2) = \{c \in C_k \mid c.count \geq minsup\}$;
        $f_k^-(T_2) = \{c \in C_k \mid \exists\ c \in f_{k-1}(T_1)$ and $c.count < minsup\}$;
        $f_k(T_2) = f_k^+(T_2) \cup \{x \mid x \in f_k^+(T_1) \cap f_k^-(T_2)$ and
                $\delta_{T_2}(x)\, support_{T_1}(x) + (1 - \delta_{T_2}(x))\, support_{T_2}(x) \geq minsup\}$
    end;
return $\cup f_k(T_2)$;

function $AprioriGen(f_{k-1})$
    insert into C_k
    select $l_1, l_2, \ldots, l_{k-1}, c_{k-1}$
    from $f_{k-1}\ l, f_{k-1}\ c$
        where $l_1 = c_1 \wedge l_2 = c_2 \wedge \ldots \wedge l_{k-2} = c_{k-2} \wedge l_{k-1} < c_{k-1}$;
    delete all items $c \in C_k$ such that (k-1)-subset of c is not in F_{k-1};
return C_k;

Figure 3.1 The *DynamicApriori* Algorithm

As we mentioned before, the Dynamic approach can use any data mining technique, as a local technique, to generate frequent user access patterns. In this paper, we demonstrate our dynamic approach using the *Apriori-Like* algorithm. The *Apriori-Like* algorithm is slightly modified to reflect those new factors needed to perform the dynamic mining. As it is shown in Figure 3.1, The *DynamicApriori* algorithm mainly follows the main outlines of the *Apriori-Like* algorithm. The *DynamicApriori* algorithm is decomposed into two modules,

- Using previous frequent sequences, all sequences that satisfy inequality (1) are generated.
- From those itemsets generated in the first module, generate all association rules that satisfy certain *minconf* value.

In the *DynamicApriori* algorithm, the number of disk accesses required is

$$m \, \alpha \, \bar{N}$$

where $\bar{N}$ is the number of disk blocks needed to store the transaction updates, α is the reduction factor caused by grouping user ID's of the same Web page, and m is the size of maximal frequent user pattern.

4 Performance Results

The *DynamicApriori* algorithm has been tested and compared to the performance of the *Apriori-Like* algorithm, using the following assumptions:

- Total Time length is one year
- *minsup* values are uniformly distributed over the range 0.05 and 0.2
- Users *inter-arrival* time is exponentially distributed with means 1 minute and 5 minutes.
- Users and URLs (Web pages) are normally distributed (generated from uniform distributions with means 10000 users and 100 URLs, and 50000 users and 250 URLs).
- Number of URLs per user is uniformly distributed with mean 20.

In the *DynamicApriori* algorithm, the one year time interval is equally divided into equal time periods. Five different period sizes; 1,2,3,4 and 6 months, have been considered.

In our experimental results, we compare the number of disk accesses of the *DynamicApriori* algorithm and the *Apriori-Like* algorithm. Our experiments use the same time interval in both algorithms. As an example, for 1 month time period, both algorithms have been executed 12 times, and the accumulated results are compared. The frequent sequences generated by the two algorithms are compared, and the ratios between the number of same frequent sequences generated by both algorithms and the number of frequent sequences generated by each of the two algorithms are calculated. In figures 4.1 and 4.2, we give the results of our experimental results. In figure 4.1, the number of disk accesses needed for the *Apriori-Like* algorithm are compared to those of the *DynamicApriori* algorithm. We have found that, for small time periods, the difference between the two algorithms is large (for 1 month period, almost 120 times), which is acceptable due to the following two factors:

- The dynamic approach uses only those transaction updates, not the whole transaction file.

- The size of transaction (mapped) file is reduced after eliminating time values and duplicate user-ids.

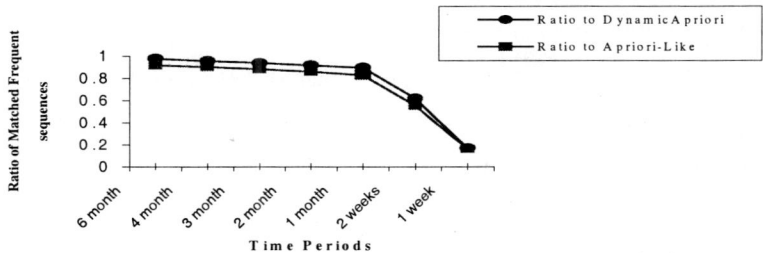

Figure 4.2 Matched Frequent Sequences

In figure 4.2, we compare the ratio of matched frequent sequences, i.e., those found by both algorithms. For time periods greater than or equal 1 month, the results have shown that the *DynamicApriori* algorithm has generated most of the frequent

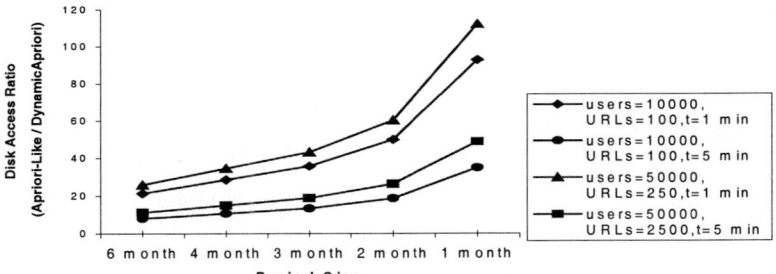

Figure 4.1 Performance Evaluation of PeriodicApriori Algorithm

sequences that have been generated by the *Apriori-Like* algorithm. For those frequent sequences that have been generated by the *DynamicApriori* algorithm and not by the *Apriori-Like* algorithm, and vice versa, we have carefully studied their behavior, and found out that our approach missed only those frequent sequences with low time span.

5 Discussion and Conclusions

In this paper, we have introduced a dynamic approach for Knowledge Discovery of Web Access Patterns. In this approach, the time space is divided into time periods, and the association mining procedure is applied only on one time interval and uses those association rules discovered in the previous time interval. To demonstrate our dynamic approach, we have used the *Apriori-Like* algorithm as a local algorithm to generate frequent user patterns during time periods. A set of experiments has been performed, and the results of the *DynamicApriori* algorithm and the *Apriori-Like* algorithm are compared. Although we have used synthetic data to run our experiments, we have carefully chosen the distribution functions that reflect the behavior of users and Web pages.

In our experimental work, we choose the value of $\delta(X)$ to be dependent on the behavior of *k-sequence* X through the different time periods. We believe that by applying different techniques to choose the values of $\delta(X)$, we may much further improve the performance of our approach.

The experimental results have shown that the *DynamicApriori* algorithm has efficiently generated frequent sequences, which are used to generate association rules. Depending on the time interval length, the ratio between the number of disk blocks accessed by the two algorithms ranged between 8.0357 and 112.069, in favor of the *DynamicApriori* algorithm. For a reasonable time interval (greater than or equal 1 month), *DynamicApriori* algorithm has generated most of the frequent sequences that have been generated by the *Apriori-Like* algorithm. For those frequent sequences that have been generated by the *DynamicApriori* algorithm and not by the *Apriori-Like* algorithm, and vice versa, we have carefully studied their behavior, and found out that our approach missed only those frequent sequences with low time span. These results favor our approach and prove that our algorithm not only produces frequent sequences but also implicitly performs time analysis on the discovered frequent sequences.

References

[1] R. Agrawal, T. Imilienski, and A. Swami, "Mining Association Rules between Sets of Items in Large Databases," Proc. of the ACM SIGMOD Int'l Conf. On Management of data, May 1993.2
[2] R. Agrawal and R. Srikant, "Mining Sequential Patterns", In Proc. 11th Intl. Conf. On Data Engineering, Taipi, Taiwan, March 1995.3
[3] C. Brunk, J. Kelly, and R. Kohavi, "MineSet: An Inegrated System for Data Mining", In Proc. 3rd Int. Conf. Knowledge Discovery and Data Mining (KDD'97), NewPort Beach, California , Aug. 1997.9
[4] M. Chen, J. Park, and P. YU, "Data Mining for Path Traversal Patterns in a Web Environment", Proc. 16th Untl. Conf. Distributed Computing Systems, May 1996.12
[5] J. Cumming, "Hits and Misses: A Year Watching the Web", In Proc. 6th Int. World Wide Web Conf., santa Clara, California, April 1997.14
[6] A. Hafez, J. Deogun, and V. Raghavan ,"The Item-Set Tree: A Data Structure for Data Mining", DaWaK' 99 Conference, Florence, Italy, Aug. 1999.17
[7] C. Kurzke, M. Galle, and M. Bathelt, "WebAssist: a user profile specific information retrieval assistant," Seventh International World Wide Web Conference, Brisbone, Australia, April 1998.19
[8] H. Mannila, H. Toivonen, and A. Verkamo, "Efficient Algorithms for Discovering Association Rules," AAAI Workshop on Knowledge Discovery in databases (KDD-94) , July 1994. 21
[9] M. Perkowitz and O. Etzioni, "Adaptive Sites: Automatically Learning from User Access Patterns", In Proc. 6th Int. World Wide Web Conf., santa Clara, California, April 1997.22
[10] G. Rossi, D. Schwabe, and F. Lyardet, "Improving Web Information Systems with Navigational Patterns," The Eighth International World Wide Web Conference, Toronto, Canada, May 1999.24
[11] T. Sullivan, "Reading Reader Reaction: A Proposal for Inferential Analysis of Web Server Log Files", In Proc. 3rd Conf. Human Factors & The Web, Denver, Colorado, June 1997.28
[12] L. Tauscher and S. Greenberg, "How People Revisit Web Pages: Empirical Findings and Implications for the Design of History Systems", International. Journal of Human Computer Studies, Special Issue on World Wide Web Usability, 47, 1997.29
[13] C. Wills, and M. Mikhailov, "Towards a Better Understanding of Web Resources and Server Responses for Improved Caching," The Eighth International World Wide Web Conference, Toronto, Canada, May 1999.31

Data Reduction via Conflicting Data Analysis

M. Boussouf and M. Quafafou

IRIN, Université de Nantes, 2 rue de la Houssinière,
BP 92208 - 44322, Nantes Cedex 03, France.
{boussouf,quafafou}@irin.univ-nantes.fr

Abstract. This paper introduces a new method for instances selection. The conceptual framework and the basic notions used by this method are those of an extended rough set theory, called α-rough set theory. In this context we formalize a notion of conflicting data, which is at the basis of a conflict normalization method used for instances selection. Extensive experiments are performed to show the efficiency and the accuracy of models built from the reduced datasets. The selection methodology and its results are discussed.

1 Introduction

One way to achieve an efficient processing of very large data is to reduce the number of input data without losing the main information and without decreasing the quality of the extracted knowledge. To deal with the problem of data reduction, many methods have been proposed. Generally, two main approaches are distinguished: statistical sampling techniques and clustering or prototyping approaches. For instance, Quinlan [10] used windowing approach in ID3 to learn on subsets of tuples, Catlett [3] considered windowing in C4.5, who uses stratification according to the decision attribute, John and Langley [7] discuss static versus dynamic sampling, Toivenen [12] and Zaki et al. [14] examine applications of random sampling for finding association rules, Reinartz [11] reuses a variant of leader clustering algorithm of Hatingan[6]. He proposed a similarity-driven sampling approach, which is based on two steps: sorting and stratification. Dubes et al. [4] discuss clustering methodologies. Whereas, Zhang[15] proposes a data summarization algorithm using a single scan incremental process to create a hierarchical tree of sub-clusters summarizing the original dataset.

2 Data Analysis and Rough Sets

2.1 Rough Sets Overview

Rough Sets Theory (RST) is an extension of set theory. It was introduced by Z. Pawlak [9] in 1982 to offer a framework for handling imperfect data. It is a mathematical tool, which deals with vagueness and uncertainty. There has been a fast-growing interest in the rough set theory, which has proved to be very useful in practice. Successful applications have been developed in medicine, decision analysis, banking, market research, knowledge discovery, and so on. Before presenting our investigations, we will first review the main concepts of rough set theory.

Information system: In rough sets theory, an information system has a data table form. Formally, an information system S is a 4-tuple $S = (U, Q, V, g)$, where: U is a finite set of objects; Q is a finite set of attributes; $V = \cup V_q$, where V_q is a domain of attribute q; g is an information function assigning a value of attribute for every object and every attribute, i.e., $g : U \times Q \mapsto V$, such that for every $x \in U$ and for every $q \in Q$, $g(x, q) \in V_q$.

Indiscernibility relation: Let K be a subset of attributes, the indiscernibility relation, denoted I_K ($\subseteq U \times U$), is assumed to be an equivalence relation, which is defined as follows:

$$x \ I_K \ y \ \Leftrightarrow \ \forall \ p \in K \ x_p = y_p$$

Consequently, x is related to y if, and only if, they have the same value for all attributes in K. The pair (U, I_K) is called a Pawlak approximation space. The relation I_K is an equivalence relation, which partitions the space U into disjoint subsets. The quotient set U/I_K consists of equivalence classes of I_K, also called elementary sets.

Approximations of sets: A key idea in rough set theory is the approximation of concepts using two operators, which assign to any subset of the universe, $X \subseteq U$, two approximations called *lower* and *upper* approximations denoted respectively I_{lower} and I_{upper}.

$$I_{lower}(X) = \{x \in U \mid I_K(x) \subseteq X\} \ , \ I_{upper}(X) = \{x \in U \mid I_K(x) \cap X \neq \phi\}$$

The I_{lower} approximation of X is the set of elements, which certainly belong to X, whereas the I_{upper} approximation of X is the set of elements, which possibly belong to X. Elements which are probably in X but do not certainly belong to X define a doubtful region called the boundary region, i.e., $Bound(X) = I_{upper}(X) - I_{lower}(X)$. We say that a set X is rough (inexact) when its boundary is a non empty set. In this paper, we introduce a new method for data reduction based on conflicting data analysis. The notion of conflicting data is based on the relationship between boundaries of concepts.

2.2 Conflicting Data Analysis

The notion of conflict plays an important role in different domains like business and military operations. Different formal models of conflict have been proposed [5][8]. We use the notion of boundary to express conflictual relations between concepts. The normalization of this conflict leads to the selection of a subset of instances. Let us first introduce a binary relation between two instances. According to a subset of attributes K, two instances x and y are said to be allied, denoted $\mathcal{X}_P(x, y)$, if they have the same value of all attributes: $\mathcal{X}_P(x, y) = 1$ if $x_p = y_p \ \forall \ p \in P$ and 0 otherwise.

Definition 1 (Conflicting instances) *Let D be the set of condition attributes, C the decision attribute, $Q = D \cup C$. We say that $x, y \in U$ are conflicting if, and only if, they are redundant or inconsistent:*

- **Redundant instances** *have the same value for both condition and decision attributes, i.e. $\mathcal{X}_Q(x,y) = 1$.*
- **Inconsistent instances** *have the same value for condition attributes, but different values for decision attribute, i.e., $\mathcal{X}_D(x,y) = 1 \wedge \mathcal{X}_C(x,y) = 0$.*

2.3 Extending Rough Set Theory

Rough set theory is formulated using a basic notion of indiscernibility between objects of the universe, which is based on binary relations. Many different studies have been developed to extend rough set theory by replacing the classical equivalence relation by different kind of binary relations. The choice of a given indiscernibility relation directly alters the interpretation of rough sets.

Definition 2 (α–Indiscernibility) *Let K be subset of attributes, $\alpha \in [0,1]$, and I_K the Pawlak indiscernibility relation. Two instances x, y of the universe U are said to be α-indiscernible, denoted $x\ I_K^\alpha\ y$, if, and only if:*

$$\exists\ K' \subseteq K\ |\ xI_{K'}\ y\ \text{and}\ f(K,K') \geq \alpha$$

Consequently, the semantic of the indiscernibility relation is rich as the function f can be defined according to the prior domain knowledge. In what follows we consider the following domain-independent function f: $f(K,K') = \frac{|K'|}{|K|}$, where $|K|$ denotes the cardinality of K.

The extension of the well known definition of indiscernibility I_K is important, especially for high dimensional spaces. In fact, the relation I_K tends to break down in high dimensional spaces. The main reason is that the resulted partitioning of the universe is probably very fine when the cardinality of the set of attributes K is very high: for any pair of objects of the universe, it likely exists few dimensions for which this objects are indiscernible. Different algorithms have proposed to deal with this problem especially in the context of clustering of high dimensional spaces. Our formalization is a new way to consider this problem in the context of rough set theory and met the approach developed for fast algorithms for projected clustering developed by Aggarwal et al.[1].

The use of this parameterized indiscernibility relation leads to a weak definition of conflict. In fact, we can easily express the two main notions of conflicting data, introduced before, i.e., redundancy and inconsistency, using I_K^α. Two instances x,y are said to be:

- *Redundant* iff $x\ I_Q^1\ y$, (i.e., $\mathcal{X}_Q(x,y) = 1$).
- *Inconsistent* iff $x\ I_D^1\ y\ \wedge\ \neg(\ x\ I_C^1\ y)$ (i.e., $\mathcal{X}_D(x,y) = 1$ and $\mathcal{X}_C(x,y) = 0$

Consequently, the binary relation I_K^α allows us to express a weak notion of conflict.

Definition 3 (α–conflicting instances) *Two instances x, y of the universe U are said to be conflicting at the level α if, and only if:*

$$\exists\ K \subseteq D \text{ such that } \mathcal{X}_K(x,y) = 1 \text{ and } f(D, K) \geq \alpha.$$

Thus, two instances x, y weakly conflicting are said to be weak redundant, respectively weak inconsistent, if they are weakly conflict and $\mathcal{X}_C(x, y) = 1$, respectively $\mathcal{X}_C(x, y) = 0$. Data reduction can be viewed as normalizing conflicts. We resolve the conflict between instances by selecting only one instance of dominant concept in each conflicting group. The strongly conflicting instances (corresponding to classical framework of rough sets) are obtained when $\alpha=1$. The parameter α influences and controls the number of conflicting instances, i.e., when α decreases, the size of (weakly) conflicting group increases. Consequently, the number of selected instances decreases. The process of selection will be detailed in the next section.

3 Conflicting Data Normalization

3.1 Foundations

The goal of this section is the introduction of the concept of conflicting data normalization and the description of a method supporting the normalization process. We have underlined two types of conflict, i.e., redundancy, inconsistency and weak conflict. For this reason the normalization process is divided into two steps: (1) redundancy reduction and (2) inconsistency normalization. All inconsistent instances belong to a set, which is equal to the union of boundaries of all concepts C_i defined by the following constraint $"C = C_i''"$ where C is the decision attribute. The set of all inconsistent instances is called Global Boundary and denoted $GB = \bigcup Bound(C_i)$. Whereas, the redundant instances belong to the complement of the set GB, i.e., $U - GB$, which is equal to the union of the lower approximations of all concepts C_i. The normalization process reduces the redundancy and normalizes the inconsistency as follows:

- **Redundancy:** let's consider that a set of instances $T = \{x_1, x_2, ..., x_n\}$ are redundant, which means that $\mathcal{X}_D(x_i, x_j) = 1$ and $\mathcal{X}_C(x_i, x_j) = 1$ for all i, j in $\{1, 2, ..., n\}$. These instances are identical, we keep only one among them all the other instances are deleted.
- **Inconsistency:** as we have seen before the GB contains all inconsistent instances. Let $T \in GB/I_D$ and θ be an operator such that:

$$\theta(T) = \{C_i \in GB/I_D \mid T \cap C_i \neq \phi\}$$

The result of the operator θ is the set of conflicting concepts given the set T of conflicting instances. Only dominant concepts are kept. We define the operator ψ as follows:

$$\psi(T) = \{C_k\ :\ |C_k \cap T| = Max\{|C_i \cap T|\ :\ C_i \in \theta(T)\}\}$$

The cardinality of a set X is denoted $|X|$. The result of the operator ψ is the set of dominant concepts given a set of concepts $\theta(T)$. The operator ψ carries out a voting operation between conflicting concepts. The inconsistency normalization means the replacement of each set T of GB/I_D by only one instance representing the dominant concepts. If there are m dominant concepts, m instances from T are randomly selected, each one represents a dominant concept. Thus, instances of T that belong to non dominant concepts are deleted and a pruning operation is realized on dominant concepts.

The previous normalization depends on the parameter α. Considering only strict conflicting data, i.e., $\alpha = 1$, which means that all attributes are used to distinguish instances. However, taking into account all attributes in the situation where we consider high dimensional spaces is a real obstacle for conflict analysis. In fact, the more the cardinality of GB is low the more the conflict is reduced. For this reason, we can vary the value of α from 1 to 0 to evaluate the weakness of conflict between concepts.

In order to evaluate the cost of this approach, let us assume that there are N instances. In the worst case, the number of comparisons to compute the approximations, i.e., to find all conflicts, is equal to $\frac{N(N-1)}{2}$ and the conflict normalization needs N comparisons. Consequently, the total comparisons needed to select instances is $\frac{N(N-1)}{2} + N$. So, the complexity is $O(N^2)$.

3.2 Conflict Normalization by Hand

This simple example considered here shows how the normalization method works step by step. Let us consider the information system drawn in Table 1. The universe U contains 16 instances and equals to $\{1, 2, ..., 16\}$. The set of attributes $Q = \{q_1, q_2, q_3, q_4\}$ is divided into condition attributes $D = \{q_1, q_2, q_3\}$ and decision attribute $C = \{q_4\}$. The partitioning produced when we consider condition attributes

Table 1.Example

#	q_1	q_2	q_3	q_4	#	q_1	q_2	q_3	q_4
1	x	y	z	c1	9	y	x	y	c3
2	x	y	z	c3	10	x	x	y	c2
3	x	y	z	c1	11	x	y	x	c1
4	x	y	z	c2	12	z	y	y	c3
5	x	x	x	c3	13	z	y	y	c3
6	y	y	y	c2	14	z	z	x	c3
7	z	z	z	c1	15	y	x	y	c2
8	y	x	y	c2	16	y	x	y	c3

and decision attribute are respectively $U/I_D = \{\{1,2,3,4\},\{5\},\{6\},\{7\},\{8,9, 15,16\},\{10\},\{11\},\{12,13\},\{14\}\}$ and $U/I_C = \{\{1,3,7,11\},\{2,5,9,12,13,14, 16\},\{4,6,8,10,15\}\}$. According to the latter partitioning, we obtain three concepts $C_1 = \{1,3,7,11\}$, $C_2 = \{4,6,8,10,15\}$ and $C_3 = \{2,5,9,12,13,14,16\}$.

The approximations and boundaries of the three concepts are: $I_{lower}(C_1) = \{7,11\}$, $I_{upper}(C_1) = \{1,2,3,4,7,11\}$, $Bound(C_1) = \{1,2,3,4\}$, $I_{lower}(C_2) = \{6,10\}$, $I_{upper}(C_2) = \{1,2,3,4,6,8,9,10,15,16\}$, $Bound(C_2) = \{1,2,3,4,8,9, 15,16\}$, $I_{lower}(C_3) = \{5,12,13,14\}$, $I_{upper}(C_3) = \{1,2,3,4,5,8,9,12,13,14,15, 16\}$, $Bound(C_3) = \{1,2,3,4,8,9,15,16\}$.

Only one indiscernible subset of instances, i.e., $\{12, 13\}$, which belongs to the lower approximation of the concept C_3 is considered during the redundancy reduction phase. It is replaced by a randomly selected instance from $\{12, 13\}$. The global boundary set, i.e., GB, is equal to the set $\{1, 2, 3, 4, 8, 9, 15, 16\}$. Consequently, the set of indiscernible instances subset in terms of condition attributes is $GB/I_D = \{\{1, 2, 3, 4\}, \{8, 9, 15, 16\}\}$. Let's consider a subset T_1 of indiscernible instances belonging to GB such that $T_1 = \{1, 2, 3, 4\}$. According to the definition of the operator θ we obtain $\theta(T_1) = \{C_1, C_2, C_3\}$. The voting procedure produces the dominant conflicting concept $\psi(T_1) = \{C_1\}$. Consequently, we replace the subset T_1 by a randomly selected instance from $C_1 = \{1, 3\}$. Similarly, we apply the same procedure to the subset $T_2 = \{8, 9, 15, 16\}$, which will be replaced by two instances randomly selected to represent respectively $\{9, 16\}$ and $\{8, 15\}$.

4 Optimization for Large Datasets

We cannot directly apply the conflict normalization method described before to large datasets because its time complexity is quadratic. To deal with this problem we have reduced the number of comparisons necessary to compute approximations and to determine conflicting data. To achieve this goal we propose an incremental clustering algorithm. Given a indiscernibility threshold α and a maximal number of clusters, this algorithm is described in two steps: (1) Building clusters: the content of each cluster is summarized by a vector of values, noted D^*. Each entry D_q^* of this vector represents the most frequented value of the attribute q ($q \in 1...|D|$) in the current cluster. This step is achieved by comparing incrementally all objects with the representative vector of each built cluster. If they are α-indiscernible then the current object is inserted in the cluster and the vector D is updated, otherwise, a new cluster is created; (2) after building the clusters, we apply the normalization process on each clusters, which is achieved by choosing the nearest object of D^* of dominant concepts, i.e., by applying the ψ operator on each cluster.

CCDN-Algorithm: Our algorithm, called CCDN-Algorithm (Clustering based Conflicting Data Normalization Algorithm) can be summarized as follows:

Input   N : Training set size; M : Maximal number of clusters;
         D : Predictive attributes; C : Class; α : indiscernibility threshold;
Output   ReducedInstancesSet;
   for i:=1 **to** N Insert(i,Cluster,α);
      /* Insert the ith object in a cluster among already created clusters.*/
      /* Let ClusterNumber be the number of created clusters (ClusterNumber $\leq$ M)*/
      for j:=1 **to** ClusterNumber Add(ReducedInstancesSet,BestObjects(Cluster[i]));
         /* Select the best object(s) from each cluster */
   Return(ReducedInstancesSet);

In order to evaluate the cost of this algorithm let us assume that there are N objects and the number of allowable clusters is M ($M << N$). For the insert function, in the worst case, the number of comparisons to insert an object in a cluster among M ones is equal to M. So, to process all objects, the maximal number of comparisons is NM. For the second function, in order to select the best objects of each clusters, we need N comparisons. Consequently, the complexity of CCDN algorithm is $O(NM)$.

5 Experimental Results

In order to evaluate the proposed instance selection method we run experiments on 12 real-world datasets taken from the UCI Irvine repository [2], their characteristics are summarized in Table 2.

Table 2. Datasets considered: the size of training set and test set, the number attributes, class cardinality and the percentage of numeric attributes.

Base	Train	Test	#Att	#class	%N.	Base	Train	Test	#Att	#class	%N.
Australian	552	138	14	2	43	Mushroom	6499	1625	22	2	0
Pima	615	153	8	2	100	Pendegit	7494	3498	16	10	100
Vehicle	677	169	18	4	100	Letter	16000	4000	16	26	100
Segment	1848	462	20	7	95	Adult	32561	16281	14	2	43
Abalone	3133	1044	8	3	88	Shuttle	43500	14500	8	7	100
Annthyroid	3772	3428	21	3	29	Covertype	387342	193670	54	7	18

Original data are transformed using discretization method proposed by Van de Merckt in [13]. We have used C4.5 system [10] to construct a decision tree from both the original data and the selected instances. We separate randomly the original dataset, which do not contains specific test set into training set (80%) and test set (20%); for Covertype dataset, we use 66% for training and 33% for testing. Firstly, our method is used to select a subset of instances. Its results are compared with random and stratified sampling methods. For these latter methods, we repeat sampling and data mining 10 times and we present average results. The size of considered samples is determined by our method.

In order to evaluate the cost of our proposed algorithm, we draw the time evolution of conflicting data identification for the largest dataset, i.e., Covertype dataset. We have shown that the complexity of the algorithm CCDN is linear (Section 4). Fig. 1 underlines this linearity feature with the Covertype dataset considering 48 attributes, i.e., $\alpha = 48/54 = 0.89$.

The results drawn in Table 3 show that the size of returned sample is lower than 25% for original datasets for 11 datasets among 12, it is lower than 10% for 7 datasets.

The instances selected using the conflicting data normalization based method lead to a model, which is more accurate than the ones extracted from a sample using random or stratified sampling technique, the differences equal 5.36% and 4.28% respectively. However, this is not the only contribution of our work. In fact, the size of the sample is generally given by the user before the selection

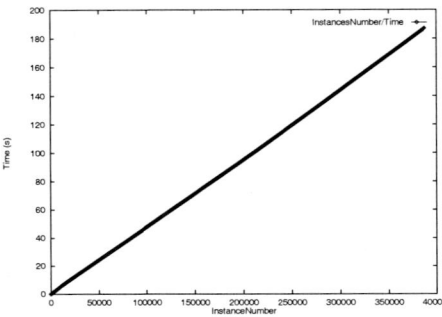

Fig. 1. Time evolution for Covertype dataset

of instances. The problem is, which size to choose? Our method does not need this information, it is part of its result. Besides, we have introduced a formal notion based on conflicting data, which can play an important role for data understanding.

Table 3. The accuracy of C4.5, the percentage of selected instances, the accuracy of C4.5 using selected instances and C4.5 accuracy (with the same percentage) using random and stratified sampling.

Dataset	All-C4.5	Select.%	CCDN-C4.5	Rand-C4.5	Strat-C4.5
Australian	85.5	7.25	85.5	80.53±3.0	82.19±1.5
Pima	76.5	10.57	77.1	71.62±2.1	72.35±1.6
Vehicle	71.0	12.85	72.2	64.15±2.8	63.53±2.3
Segment	92.4	16.72	92.4	83.31±10.7	87.37±1.1
Abalone	63.6	5.20	63.4	58.21±2.0	58.51±1.4
Annthyroid	94.0	2.82	93.2	93.57±0.4	92.94±0.2
Mushroom	100	0.46	99.0	91.78±4.20	89.85±5.0
Pendegit	91.7	23.87	89.9	84.77±7.3	87.44±0.5
Letter	79.4	46.60	76.7	67.91±8.0	74.25±0.3
Adult	85.4	2.50	84.1	83.76±1.2	83.37±0.7
Shuttle	99.8	0.16	99.7	96.65±2.1	95.33±2.3
Covertype	62.2	1.66	62.5	54.94±1.21	57.03±0.9
MEAN	83.46		82.98	77.60	78.70

The size of the sample produced by CCDN-Algorithm can be very low without decreasing the quality of the knowledge induced from the selected instances. For instance, only 68 instances (0.16%) are selected from Shuttle dataset, which contains 43500 instances. The quality of classification is the same for both the selected instances and all data. Also, only 0.46% are selected from mushroom original data and the accuracy is decreased with only 1%. The most important result is obtained with the largest dataset, i.e., Covertype dataset. Among 387342 instances, our algorithm chooses only 1.66% (6445 instances) and the accuracy of the reduced model is slightly improved comparing with original model.

6 Conclusion

This paper tackles the problem of mining efficiently a very large data and proposes two solutions based on instances selection via conflict normalization. The proposed method is developed in a conceptual framework, which is an extension of Rough Set Theory called α-Rough Sets. The contribution of this paper is threefold (1) formalization of a notion of conflicting concept, (2) proposition of a method for conflict normalization to select a subset of instances, (3) proposition of a heuristic algorithm to avoid the quadratic complexity, (4) presentation of results of extensive experiments on different datasets.

References

1. Aggarwal, C.C., Procopiuc, C., Wolf, J.L., Yu, P.S., Park, J.S.: Fast Algorithms for Projected Clustering. SIGMod'99, **28(2)** (1999) 61–72
2. Blake, C.L., Merz, C.J.: UCI Repository of machine learning databases. "http://www.ics.uci.edu/~mlearn/MLRepository.html" (1998)
3. Catlett, J.: MegaInduction : Machine Learning on Very Large Data-bases. Sydney, Australia, ICML (1988) 87–99
4. Dubes, R., Jain, A.K.: Clustering methodologies in Exploratory Data Analysis. Advances in Computers"New York, Academic Press , **19** (1980)
5. Hart, H.: Structures of Influences and Cooperation-Conflict. International Interaction **1** (1974) 141–162
6. Hartingan, J. A.: Clustring Algorithms. John Willy & Sons Inc.; New York (1975)
7. John, G.H., Langley, P.: Static versus Dynamic Sampling for Data Mining. KDD'96, (1996) 367–370
8. Pawlak, Z.: On Conflicts. Int. J. of Man-Machine Studies, **21(2)** (1984) 127–134
9. Pawlak, Z.: Rough sets: theoretical aspects of reasoning about data. Theory and decision library. Series D: System theory, knowledge engineering, and problem solving. London, Kluwer Academic, **9**, (1991)
10. Quinlan, J.R.: C4.5: Programs for Machine Learning. Kaufmann M., California, (1993)
11. Reinartz, T.: Similarity-Driven Sampling for Data Mining. PKDD (1998) 423–431
12. Toivonen, H.: Sampling Large Databases for Finding Association Rules. VLDB (1996) 134–145
13. Van de Merckt, T.: Decision Trees in Numerical Attributes Spaces, IJCAI (1993)
14. Zaki, M.J., Parthasarathy, S., Li, W., Ogihara, M.: Evaluation of Samling for Data Mining of Association Rules, in Proceedings of the 7th Workshop on Research Issues in Data Engineering, Scheuermann, P. (eds.), Birmingham, England, (1997)
15. Zhang, T.: Data Clustering for Very Large Datasets Plus Applications. Technical Report, University of Wisconsin, Computer Sciences Department (1997) TR1355

A Comparison of Rule Matching Methods Used in AQ15 and LERS

Jerzy W. Grzymala-Busse[1] and Pankaj Shah[2]

[1] Department of Electrical Engineering and Computer Science
University of Kansas, Lawrence, KS 66045, USA
[2] Systems Material Handling Co., Olathe, KS 66062, USA

Abstract. This paper focuses on a performance comparison of two rule matching (classification) methods, used in data mining systems AQ15 and LERS. All rule sets used in our experiments were induced by the LERS (Learning from Examples using Rough Sets) system from ten typical input data sets. Then these rule sets were truncated using three different criteria: *t-weight*, *u-weight* and the *strongest rule*. The truncation process was performed using six different cut-off values for *t*-weight, six different cut-off values for *u*-weight and using the strongest rule option. Hence for each of the input rule files thirteen truncated rule sets were created. Performance was measured by a classification error rate. The objective of this study was to determine the best overall method of classification and the best truncation option.

Keywords. Knowledge discovery and data mining, rule induction, classification systems, rule truncation, AQ15, LERS, ten-fold cross validation, Wilcoxon matched-pairs signed rank test.

1 Introduction

In this paper input data were presented in the form of a table, called a *decision table*. The columns are labeled by *variables*. One of the variables is called a *decision* and the remaining variables are called *attributes*. The rows represent *examples*. A *concept* is defined as a set of all examples having the same decision value. All examples belonging to the concept *C* are called positive examples for *C* and all remaining examples are called negative examples for *C*. The concepts are described in the form of rule sets [3, 9].

Like many other data mining systems, AQ15 and LERS were primarily designed to induce rules from training examples. Both systems are equipped with modules for rule matching used for classification of new, unseen examples.

In our research all rule sets used for experiments were induced by the LEM2 algorithm [4] of LERS, since the rule matching module of LERS would not recognize rules in the format of AQ15 and its successors, AQ17 and AQ18. Thus we compare only rule matching (classification) performance of AQ15 and LERS, restricted to the rule sets induced by LERS.

The rule set, induced by the inductive learning process, may be used for interpreting regularities hidden in the input data, for visualization of these regularities,

or for classification of unseen examples [5, 10, 14]. During the classification process of unseen data, an attempt is made by all rules to match each example. For each example, the following are different possible outcomes of the classification process:
- The example is exclusively and correctly classified as a member of the correct concept,
- The example is exclusively and incorrectly classified as a member of a wrong concept,
- The example is correctly classified and incorrectly classified at the same time, i.e., some rules classify it as a member of the correct concept, while other rules classify it as a member of a wrong concept,
- The example is not classified by any rule.

An example is completely classified by a rule if all the attribute-value pairs of the rule match the attribute values of the example. The example is partially classified if only some of the attribute-value pairs of the rule match the attribute values of the example. The example is not classified at all if none of the attribute values of the example match any of the attribute-value pairs for all rules.

Truncation [10] is a method of reducing the rule set by deleting weak rules, describing a few training examples. Concepts can match different examples with varying degrees of precision and have context-dependent meaning. Instead of seeking a strict match, the system determines the degree of similarity between the concept description and the given example, and compares it with the results from matching the example with other concept descriptions. The concept that gives the best match is assigned to the example. Michalski's truncation algorithm was originally designed for AQ15, but the algorithm works for rule sets generated by LERS as well.

This paper presents a performance comparison of AQ15 and LERS classification systems with rule set truncation. In our experiments we used the following assumptions:
- All rule sets were induced by the LERS system,
- *Michalski's truncation algorithm* was used to prune the rule sets,
- 10-fold cross validation process [15] was used to validate results,
- AQ15 and LERS were used to classify the unseen data with the truncated rule sets,
- Wilcoxon matched-pairs signed rank test [7] was used to compare the AQ15 and LERS classification systems.

2 Michalski's Rule Truncation Algorithm

In AQ15 [10], each rule is associated with a pair of weights: t and u, representing the total number of training examples correctly classified by the rule, and the number of training examples uniquely and correctly classified by the rule, respectively. The t-weight may be interpreted as a strength of a rule, an idea used also in the LERS classification method, while u-weight is a measure how much the rules differ from each other. The rule with the highest t-weight may be interpreted as describing the most typical examples of the concept, while rules with the lowest u-weights can be viewed as describing exceptional examples [10].

In AQ15 there are two methods of recognizing the concept membership of an example: the *strict match* and the *flexible match*. In the strict match, an example must satisfy all conditions of the rule. In the flexible match, a degree of similarity between the example and the rule is determined.

During the truncation process [10], we remove weak rules, with the value of *t*-weight or *u*-weight not exceeding some cut-off and applying flexible matching to classify an example. By removing the weak rules the total number of rules describing the concept is reduced. This may result in rules that may not match the examples completely as they would have before the truncation process. Thus, a truncated rule set is simpler but it requires a more sophisticated classification method: flexible matching is used to classify the example. By applying a flexible match the example may still be very closely related to the correct concept and thus may be correctly recognized. An interesting problem is to test how the rule set truncation method affects the accuracy of classification. Results of similar research were reported in [1].

3 AQ15 Classification System

The original classification module of AQ15 was called ATEST [13]. AQ15 provided three methods of rule testing. In our paper we re-implemented and tested one of them, the method described in [10]. In this method, during the process of recognizing the example against a set of rules, there are three possible outcomes [10]:

- Only one rule may classify the example (SINGLE_MATCH case),
- More rules than one rule classify the example (MULTIPLE_MATCH case),
- No rule recognizes the example (NO_MATCH case).

When recognizing the examples, each of the above categories requires a different evaluation procedure.

In the SINGLE_MATCH case the classification is straightforward. If the rule decision is equal to the known decision for the example, the example is counted as correctly classified. If not it is considered as wrongly classified. In the case of MULTIPLE_MATCH and NO_MATCH cases, the classification procedures are more complicated.

MULTIPLE_MATCH case: In this case there are more rules than one that classify the example. The system selects the most probable decision. Let us consider n concepts, $C_1, C_2, ..., C_n$, that classify the example e. Each concept C_i is described by a rule set. In AQ15, a rule is called a *complex*, and it is said that the rule set is a disjunction of complexes (Cpx), each complex (rule) in turn is a conjunction of *selectors* (*Sel*). The estimate of probability, *EP* of a concept C_i is defined as the probabilistic sum of *EP*s of its complexes. If the rule set for C_i consists of a disjunction of two complexes Cpx_1 and Cpx_2, then the corresponding estimate of probability is computed in the following way:

$$EP(C_i, e) = EP(Cpx_1, e) + EP(Cpx_2, e) - EP(Cpx_1, e) * EP(Cpx_2, e),$$

where *EP* of a complex Cpx_j in the context of the example e is the ratio of the total number of positive examples classified by the complex Cpx_j (i.e., the *weight* of Cpx_j)

to the total number of training examples, if the complex recognizes the example e, and it is equal to 0 otherwise:

$$EP(Cpx_j, e) = \begin{cases} \dfrac{Weight\ (Cpx_j)}{\#\ examples} & \text{if complex } Cpx_j \text{ recognizes example } e, \\ 0 & \text{otherwise.} \end{cases}$$

The most probable concept is the one with the largest EP.

NO_MATCH case: In this case there are no complexes that classify the example e. The system uses flexible matching to determine the best complex that suggests the most probable decision. One way to perform such flexible matching is to measure the fit between attribute values of the example and the concepts. A measure of fit (MF) is defined as follows: MF of a concept C_i to an example e is computed as a probabilistic sum for a disjunction of all complexes. Let us say that the concept C_i consists of a disjunction of two complexes Cpx_1 and Cpx_2, the measure of fit for C_i is defined as [10]:

$$MF(C_i, e) = MF(Cpx_1, e) + MF(Cpx_2, e) - MF(Cpx_1, e) * MF(Cpx_2, e),$$

where MF of a complex Cpx_j to an example e is defined as the product of MFs for a all selectors of Cpx_j, weighted by the proportion of training examples covered by Cpx_j :

$$MF(Cpx_j) = \prod_k MF(Sel_k, e) * \dfrac{Weight\ (Cpx_j)}{\#\ examples},$$

where $Weight\ (Cpx_j)$ is the number of training examples covered by Cpx_j. MF of a selector Sel_k and an example e is 1 if the selector is satisfied by the example, i.e., if one of the example's attribute values is equal to the selector values. If no selector value is equal to the attribute value of the example, its MF is proportional to the amount of the decision space covered by the selector, i.e., it is the ratio of the number of attribute values in the selector to the total number of all possible values of the attribute:

$$MF(Sel_k, e) = \begin{cases} 1 & \text{if selector } Sel_k \text{ is satisfied by } e, \\ \dfrac{\#\ values}{Domain\ size} & \text{otherwise.} \end{cases}$$

Note that the measure of fit, MF, is a generalization of the estimate of probability EP. When all selectors in a complex are satisfied, the measure of fit is equal to the estimate of probability [10]. There exists another possibility of defining $MF(Cpx_j)$. In AQ18 this feature is extended to selectors [8].

4 LERS Classification System

Rule sets were induced using the LEM2 algorithm of the LERS system [4]. In LERS, inconsistencies in training data are handled using rough set theory [11, 12]. Data are consistent if for any two different examples with all attribute values the same, the decision values are also the same.

When the data are completely classified, LERS uses the following factors: *strength*, *specificity* and *support* to classify an example [5]. The original approach was introduced in [2, 6]. When data are partially classified, LERS uses an additional factor called a *matching factor* to determine the best concept to classify the example. Strength is a measure of how well the rule has performed during classification of training data. It is computed as the number of examples correctly classified by the rule. Obviously, rules that correctly classified more examples are stronger. Specificity is a measure of complexity of the rule. Rules with larger numbers of attribute-value pairs are more specific. Matching factor is a measure of how well the rule matched the attribute values of an example. It is computed as the ratio of the number of matched attribute-value pairs of a rule to an example to the total number of attribute-value pairs of the rule.

Support is computed as the sum of scores for all matching rules from one concept. It is defined as follows:

$$\sum_{\text{partially matching rules } R \text{ describing } C} \text{Strength}(R) * \text{Specificity}(R) * \text{Matching factor}(R)$$

The concept C for which support is the largest is a winner and the example is classified as being a member of C.

5 Experiments

There were ten typical, well-known data files used to compare the performance of the two classification methods. The basic facts of the data files are described in Table 1.

Table 1

Data file	Number of examples	Number of attributes	Number of concepts	Missing attribute values
lymphography	148	18	4	no
breast-cancer	286	9	2	yes
iris	150	4	3	no
hepatitis	155	19	2	yes
soybean	307	35	19	yes
primary-tumor	339	17	21	yes
house	435	16	2	yes
wisconsin	625	9	9	no
mammography	1284	12	2	no
bupa	345	6	2	no

Table 2

Method	lymphography		breast-cancer		iris	
	LERS	AQ15	LERS	AQ15	LERS	AQ15
t = 1	20.53	16.89	32.87	32.87	4.67	4.67
t = 2	18.24	17.57	30.42	30.42	4.67	5.33
t = 3	19.59	18.92	28.67	29.02	4.67	6.00
t = 4	17.57	16,89	28.32	28.67	3.33	5.33
t = 5	17.57	16.22	27.97	28.32	4.00	4.67
t = 10	21.62	18.92	27.97	27.62	4.00	5.33
u = 1	18.24	16.89	32.87	32.87	4.67	4.67
u = 2	18.92	15.54	30.77	31.12	4.00	8.67
u = 3	24.32	22.30	28.67	28.32	11.33	23.33
u = 4	27.03	23.65	28.67	27.62	26.67	29.33
u = 5	26.35	26.35	29.37	27.27	28.67	31.33
u = 10	52.70	47.30	–	–	30.00	32.67
Strongest	25.00	25.00	35.92	27.88	6.00	10.00

This research is focused on the classification performance of two methods AQ15 and LERS. For each of the ten data files, the following options were used to compare the performance of AQ15 and LERS:
- Truncate option = t-weight with cut-off weight = 1, 2, 3, 4, 5, and 10,
- Truncate option = u-weight with cut-off weight = 1, 2, 3, 4, 5, and 10,
- Truncate option = strongest rule (each concept is described by the rule with largest t-weight).

Table 3

Method	hepatitis		soybean		primary-tumor	
	LERS	AQ15	LERS	AQ15	LERS	AQ15
t = 1	19.35	18.71	18.57	18.89	64.60	64.60
t = 2	19.35	18.71	18.24	18.89	64.01	64.31
t = 3	19.35	18.71	18.57	19.22	65.19	65.19
t = 4	20.65	19.35	19.54	20.52	65.78	65.72
t = 5	18.71	19.35	22.15	21.50	68.14	68.44
t = 10	20.00	20.65	49.19	50.49	70.50	71.09
u = 1	20.65	18.71	18.57	18.89	64.60	64.60
u = 2	22.58	20.00	19.87	18.89	64.01	64.90
u = 3	23.87	19.35	23.13	21.82	68.73	68.14
u = 4	23.87	20.65	25.08	24.76	69.03	69.32
u = 5	31.61	20.65	24.76	25.08	69.91	70.21
u = 10	–	–	53.09	58.31	82.89	85.25
Strongest	44.92	20.86	24.10	29.97	66.96	73.4

For each of the data files and the thirteen truncation options, the average error rates for AQ15 and LERS classification systems were computed using the Wilcoxon matched-pairs signed rank test. Results of our experiments are presented in Tables 2–4.

In Tables 2–4, "–" indicates that truncation resulted in elimination of all rules (rules were too short, i.e., not specific enough).

6 Conclusions

The following conclusions can be derived from our experiments, using the Wilcoxon matched-pairs signed rank two-tailed test with a 5% significance level. Overall, for every individual input data file and for all thirteen results of different truncation options, no method performed significantly better than any other.

Using the same Wilcoxon matched-pairs signed rank two-tailed test with a 5% significance level; for five input data files, one of the classification methods performed better than the other; for the remaining five input data files a significant difference in performance did not occur. In the five cases where one of the classification methods performed better than the other, for two input data files AQ15 performed better with the files lymphography and iris, for the remaining three input data files (iris, primary-tumor and mammography) LERS performed better. Thus these two classification systems do not differ significantly.

Similarly, we tested if any of the classification methods were better in conjunction with any of the thirteen specific truncation methods used in our experiments. None of these thirteen truncation methods resulted in any significant difference in performance of the two classification systems.

However, for a specific data set we may observe a difference in performance between the two classification methods. Also, for a specific rule set some truncation may result in better performance. Therefore, we may conclude that for any specific

Table 4

Method	house		wisconsin		mammography		bupa	
	LERS	AQ15	LERS	AQ15	LERS	AQ15	LERS	AQ15
t = 1	6.44	6.44	22.24	22.56	31.54	32.71	36.81	37.68
t = 2	6.44	6.44	19.84	20.00	31.78	32.40	37.39	38.26
t = 3	6.44	6.44	18.88	18.88	31.78	32.17	39.13	37.68
t = 4	6.44	6.44	17.92	17.92	32.63	32.71	40.58	39.71
t = 5	6.21	6.44	17.92	17.92	33.96	33.80	42.61	40.58
t = 10	6.67	6.44	22.24	17.92	38.47	38.63	42.03	42.03
u = 1	6.44	6.44	22.24	22.56	31.54	32.71	36.81	37.68
u = 2	5.75	5.75	19.84	20.00	31.85	33.02	37.1	36.81
u = 3	5.29	7.13	18.56	18.40	29.91	31.85	38.55	37.39
u = 4	5.75	8.97	19.20	17.92	31.00	32.87	40.58	42.32
u = 5	7.36	6.90	21.92	17.92	30.67	33.96	44.35	42.32
u = 10	12.87	18.62	27.20	17.92	34.35	39.80	–	–
Strongest	7.36	14.94	31.36	19.84	34.81	34.03	41.16	39.71

data set a classification system and a truncation method should be selected individually. Hence, there is no best universal approach to classification of unseen cases and truncation of rule sets.

Acknowledgment

The authors would like to thank reviewers for their invaluable suggestions.

References

1. Bergadano, F., Matwin, S., Michalski, R. S., and Zhang, J.: Learning Two-Tiered Descriptions of Flexible Concepts: The POSEIDON System, *Machine Learning* **8** (1992), 5–43.
2. Booker, L. B., Goldberg, D. E., and Holland, J. F.: Classifier Systems and Genetic Algorithms. In Carbonell, J. G. (ed.): *Machine Learning. Paradigms and Methods*. The MIT Press (1990) 235–282.
3. Grzymala-Busse, J. W.: *Managing Uncertainty in Expert Systems*. Kluwer Academic Publishers, Boston, MA (1991).
4. Grzymala-Busse, J. W.: LERS—A System for Learning from Examples Based on Rough Sets. In: Slowinski, R. (ed.): *Intelligent Decision Support. Handbook of Applications and Advances of the Rough Sets Theory*. Kluwer Academic Publishers, Boston, MA (1992) 3–18.
5. Grzymala-Busse, J. W.: Managing Uncertainty in Machine Learning from Examples. Proc. of the Third Intelligent Information Systems Workshop, Wigry, Poland, June 6–11, 1994, 70–84.
6. Holland, J. H., Holyoak K. J., and Nisbett, R. E.: *Induction. Processes of Inference, Learning, and Discovery*. The MIT Press (1986).
7. Hamburg, M.: *Statistical Analysis for Decision Making*. Harcourt Brace Jovanovich Inc. (1983) Third Edition.
8. Kaufman, K. A. and Michalski, R. S.: The AQ18 System for Machine Learning: User's Guide, Reports of Machine Learning and Inference Laboratory, MLI 00-3, George Mason University, Fairfax, VA, 2000.
9. Michalski, R. S.: A Theory and Methodology of Inductive Learning. In: Michalski, R. S., Carbonell, J. G., Mitchell T. M. (eds.): *Machine Learning. An Artificial Intelligence Approach*, Morgan Kauffman (1983) 83–134.
10. Michalski, R. S., Mozetic, I., Hong, J. and Lavrac, N.: The AQ15 Inductive Learning system: An Overview and Experiments. Department of Computer Science, University of Illinois, Rep. UIUCDCD-R-86-1260 (1986).
11. Pawlak, Z.: Rough sets. *International Journal Computer and Information Sciences* **11** (1982) 341–356.
12. Pawlak, Z.: *Rough Sets. Theoretical Aspects of Reasoning about Data*. Kluwer Academic Publishers, Boston, MA (1991).
13. Reinke, R. E.: Knowledge Acquisition and Refinement Tools for the ADVISE META_EXPERT System, M. S. Thesis, Reports of the Intelligent Systems Group, ISG 84-4, UIUCDCS-F-84-921, Department of Computer Science, University of Illinois, Urbana, July 1984.

14. Slowinski, R. and Stefanowski, J.: 'RoughDAS' and 'RoughClass' Software Implementations of the Rough Set Approach. In: Slowinski, R. (ed.): *Intelligent Decision Support. Handbook of Applications and Advances of the Rough Sets Theory.* Kluwer Academic Publishers, Boston, MA (1992) 445–456.
15. Weiss, S. M. & Kulikowski, C. A.: *Computer Systems That Learn: Classification and Prediction Methods from Statistics, Neural Nets, Machine Learning, and Expert Systems.* Morgan Kaufmann Publishers (1991).

Evolving Behaviors for Cooperating Agents

Jeffrey K. Bassett and Kenneth A. De Jong

George Mason University
Computer Science Department
Fairfax, VA 22030
jbassett@cs.gmu.edu, kdejong@gmu.edu

Abstract. A good deal of progress has been made in the past few years in the design and implementation of control programs for autonomous agents. A natural extension of this work is to consider solving difficult tasks with teams of cooperating agents. Our interest in this area is motivated in part by our involvement in a Navy-sponsored micro air vehicle (MAV) project in which the goal is to solve difficult surveillance tasks using a large team of small inexpensive autonomous air vehicles rather than a few expensive piloted vehicles. Our approach to developing control programs for these MAVs is to use evolutionary computation techniques to evolve behavioral rule sets. In this paper we describe our architecture for achieving this, and we present some of our initial results.

1 Introduction

One of the most challenging aspects of building intelligent systems is the design and implementation of control programs for intelligent autonomous agents. Manually designing and implementing control programs that are sufficiently robust to handle dynamically changing environments and uncertainty has proved to be extremely difficult. As a consequence, there has been considerable interest in the use of machine learning techniques to help automate this process.

A good deal of progress has been made in this area in the past few years using a variety of representations (rules, neural nets, fuzzy logic, etc.) and a variety of learning techniques (symbolic, reinforcement, evolutionary, etc.). A natural extension of this work is to consider solving difficult tasks with teams of cooperating agents.

Our interest in this area is motivated in part by our involvement in a Navy-sponsored micro air vehicle (MAV) project in which the goal is to solve difficult surveillance tasks using a large team of small inexpensive autonomous air vehicles rather than a few expensive piloted vehicles. Our approach to developing control programs for these MAVs is to leverage off the successes in using evolutionary computation techniques to evolve behavioral rule sets for single-agent systems. In this paper we summarize related work, we describe our architecture, and we present some of our initial results. We conclude with a discussion of future work.

2 Background

Our approach to developing teams of cooperating agents is to represent agent behaviors as sets of rules and evolve these rule sets using evolutionary computation techniques. There has been a good deal of work done in this area for single agents, but not cooperating teams of agents. At the same time, there has been work done on "collective robotics" using other techniques. In this section we summarize relevant work in these two areas.

2.1 Rule Learning Using Evolutionary Algorithms

One of the earliest rule evolving approaches is Holland's classifier system [4]. In this system a population of rules is maintained. These rules both compete for space and priority, while also cooperating to produce an appropriate classification for the given input.

An alternative approach is to maintain a population of *rule-sets* which can vary in length. Examples of these are Smith's LS-1 system [9], and the GABIL system which uses a GA for concept learning [5]. Typically these types of systems build rules which have more of a stimulus-response quality.

The SAMUEL system [3] [8], arguably one of the more successful rule evolving systems, uses an interesting hybrid of these two approaches. Individuals are implemented as rule-sets, but SAMUEL also uses a rule bidding system and credit assignment mechanism similar to those found in a classifier system.

Wu, Schultz and Agah implemented a rule learning system for MAVs using a GA [11]. Their GA implementation was very much a canonical GA, with a binary representation, and proportional selection. Fitness was measured using a simulated environment, and each individual defined a variable length rule-set.

2.2 Collective Robotics

Collective robotics involves the use of robot teams which cooperate to perform a task or set of tasks [1]. Teams have several inherent advantages including the ability to distribute themselves, do problem decomposition, and perform parallel processing.

Robot soccer is one of the most popular domains for studying collective robotics. Tucker Balch implemented a soccer simulation to study task differentiation and specialization [2]. The robots were trained using Q-learning, and they would often specialize to playing either a defensive or offensive position.

Other problem domains include multi-robot box pushing [6], and foraging [7] tasks. These problems are often solved by implementing low level swarming behaviors such as *avoid* or *follow*. A learning algorithm is then used to teach the robots to select behaviors and coordinate with other robots.

A common problem in all these experiments was getting the robots to cooperate, particularly when learning algorithms were used. In each case the researchers found that evaluating individuals solely on their own performance wasn't enough. Only when the team was evaluated as a whole did cooperation occur.

3 Our EA Architecture

Our ultimate goal is to evolve heterogeneous team of specialized agents that collectively perform specific tasks. Our strategy for accomplishing this is to start simple and incrementally add complexity. Our first simplification is to assume that the teams consist of homogeneous agents, i.e., they are all executing the same task program. This allows us to focus on evolving a single program which, when simultaneously executed by a team of agents, produces collective cooperative behavior, and it allows us to take advantage of existing work on evolving single agent behaviors.

However, there are still a number of important design decisions that need to be made, such as how rule sets are represented internally in our EA, how rule sets are modified over time, etc. We discuss these design decisions in the following subsections.

3.1 Representation

In our architecture an individual in the population represents a complete set of rules, and its representation is a string in which all the rules are concatenated. The ordering of the rules is not important. From generation to generation, the length of individuals in the population will tend to vary in size. The system has parameters which define a minimum and maximum size for an individual, as well as an initial size.

Each rule is a fixed length binary string. Rules are composed of a condition clause and an action clause. The bits in the condition clause are mapped to the agent's sensors, while the bits in the action clause are mapped to the agent's actuators. This allows each agent to perceive its environment and take a corresponding action.

The rule interpretor used by each agent operates as follows. In any given situation, all the rules are compared to the current input from the sensors, and the rule that has the highest match score is executed. There are several possible ways of doing rule matching. For simplicity we have avoided using rule weights and bidding techniques such as in SAMUEL or classifier systems. Rule matching is described in more detail in the description of the agent environment.

3.2 Selection

Our population management scheme is different from a typical GA. We have implemented an ES-like model involving μ parents and λ offspring. Parent selection is deterministic: all individuals produce the same fixed number of offspring.

The selection bias in our architecture is implemented using survival selection. In an ES survivors are chosen in one of two ways: using a "+" strategy involving both the parent and child populations, or using a "," strategy involving only the child population. The former converges more rapidly but is more likely to find a local optimum, while the latter provides a broader but slower search. We have both options implemented, and experimentally choose the one best suited for the particular fitness landscape.

The ES community typically uses truncation selection for determining survivors. We have chosen to use a binary tournament instead because the selection pressure is weaker, allowing for more exploration early in the search.

3.3 Operators

Since the internal representation is binary, we use a standard bit-flip mutation operator. We also implemented both a 1-point and a 2-point crossover operator.

The 1-point crossover is the same operator as the one used by Wu et. al. [11]. Crossover can only occur on rule boundaries. Because individuals can vary in size, this crossover differs from the standard crossover operator used in most GAs. Instead of selecting crossover points at the same location on both parents, different crossover points are selected for each. This means that each child may contain either more or fewer rules than the parents which spawned them. In fact, this is the only mechanism by which rule sets can change in size.

The 1-point crossover operator does relatively little mixing of parental rules and does not produce any new rules. Based on earlier experience, we felt it would be useful to have a more disruptive crossover operator available as well. We chose to implement a 2-point crossover operator that was not restricted to crossing on rule boundaries. Crossover points are chosen by first picking random rule boundaries in both parents, just as with the 1-point crossover. Then a randomly chosen offset is applied to both crossover points to obtain the inter-rule cut point. This is essentially the same crossover used in the GABIL system [5].

3.4 Fitness

The fitness of a particular rule set is obtained via simulation. Agents within the simulation use the rule set to control their behaviors. The agents have a task to perform, and at the end of the simulation they are given a score which indicates how well the task was performed. Since our intention is to have these agents cooperate, they are all evaluated as a team, and all receive the same score. In the current implementation, all agents use the same rule set, and the resulting score from the simulator is used as the fitness of that rule set.

Without any sort of counteracting force, evolving rule sets tend to grow uncontrollably, very much the way Genetic Programs (GPs) do [10]. Parsimony pressure is used to discourage this growth by penalizing the fitness of larger individuals. We have implemented parsimony pressure with the same approach used by Wu, et. al. [11] as described by the formula $f'(i) = f(i) - \alpha l_i f(i)$. The interesting thing to note about this equation is that the penalty gets stronger as the raw fitness increases. This approach allows individuals to grow larger early in the process, perhaps improving the exploration phase of the search. The particular value used for α is experimentally determined.

4 Experimental Methods

4.1 Simulation

Currently all of our experiments involve the use of a simple micro air vehicle MAV simulator. The simulation environment is a 2-D arena surrounded by walls. Nine identical MAVs are placed into the simulator, and are allowed to move and turn on each timestep. The MAVs are like helicopters in that they can hover or move at a slow constant speed. As the MAVs move, they can potentially collide with each other or with the walls surrounding the arena. Any MAV that is involved in a collision is immediately destroyed and removed from the simulation.

Each MAV has 8 sonar sensors placed radially around the vehicle. These sensors have no range information. They return either a 0 or a 1, indicating whether or not there is an object in range in the direction the sensor is pointing. The sensor range can be adjusted as a parameter of the simulation.

The robots also have a surveillance range. They can "look" down and observe objects on the ground. Currently the MAVs pay no attention to what they are observing. Their only goal is to observe as much of the ground as possible at any given time.

The behavior of the MAVs is defined by a set of stimulus-response rules. Each rule is made up of 12 bits, and contains a condition and an action section. The first 8 bits are the condition section, with one bit for each sensor. At each timestep the current sensor readings are compared with all the conditions in the rule set. The rule with the closest match is the winner. If there is a tie, a winner is chosen randomly from among the best matches. There is also a minimum threshold for matches. At least half of the condition bits must match the current sensor configuration. If the winning rule exceeds this threshold, its action is executed.

The action section of the rule consists of two parts, a speed and a turn angle. The speed can have a value of either 0 or 1, where 0 indicates that the plane will not move in the current timestep, and 1 indicates that it will. The second part of the action is the turn angle. This indicates the number of degrees the plane will turn relative to its current heading.

Figure 1 provides a more concrete picture of an MAV simulation via a series of three snapshots from an example run involving a reasonably good set of evolved rules. The goal in this case is for a team of nine MAVS, starting from an initial configuration on the left edge of a surveillance area, to dynamically configure itself (without collisions!) in such a way as to obtain maximal surveillance coverage.

The simulator is stochastic in that the results of a simulation using the same rule set can change from run to run, resulting in a "noisy" fitness evaluation. Consequently, we typically run an individual through several trials. We then assign the average of all the trials as the fitness for the individual.

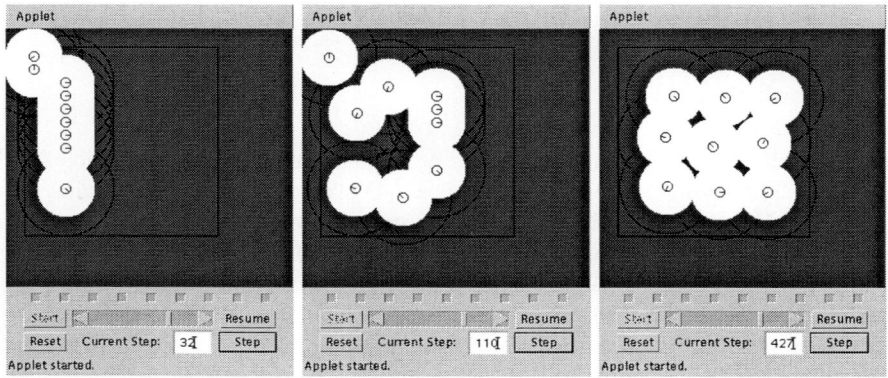

Fig. 1. MAV Simulator

5 Initial Experimental Results

The goal of our initial experiments was to test our design decisions, tune our system, and evaluate its ability to evolve effective rule sets for teams of homogeneous agents. We describe these experiments in the following subsections.

The following parameters were used for our experiments unless stated otherwise. We set both μ and λ to 100. The number of trials was 5, while the crossover and mutation rates were 1.0 and 0.001 respectively. Individuals were limited to between 1 and 200 rules, with an initial starting size of 5 rules.

5.1 Population Management

Recall that we have both "+" and "," population management strategies available for use. The goal of our first set of experiments was to determine how sensitive the results are to this design choice. In general $(\mu + \lambda)$ slightly outperformed (μ, λ), although not by much. As a consequence we adopted the $(\mu + \lambda)$ strategy for the remaining experiments.

5.2 Parsimony Pressure

Another important design choice is the amount of parsimony pressure used. In general, too little parsimony pressure allows the length of individuals to grow indefinitely, and too much parsimony pressure produces compact individuals with suboptimal fitness. What is needed is a pressure point in between these two extremes. The goal of our second set of experiments was to get a rough sense of how parsimony pressure affected our system. We used three different values for parsimony pressure initially: 0, 1/2400 and 1/300. Figures 2 and 3 show that the parsimony pressure does work as expected. Higher parsimony pressures tend to produce smaller individuals, but at the expense of fitness.

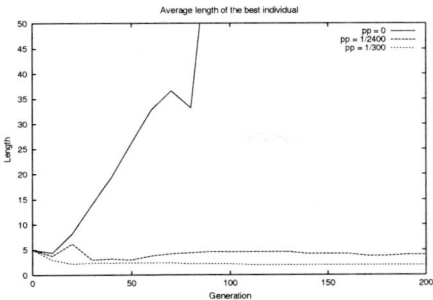 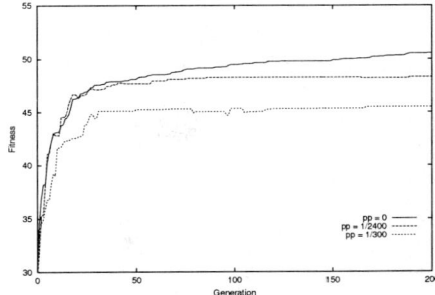

Fig. 2. Length of the best individual averaged over 5 runs for three parsimony pressures.

Fig. 3. Best-so-far curves of raw fitness averaged over 5 runs for three parsimony pressures.

5.3 Crossover

Recall that we have both a one and two point crossover operator implemented. The experiments so far have used the default 1-point crossover operator. Our third set of experiments involved testing the sensitivity of the results to these operators. Since we felt there could be some interaction with parsimony pressure, we tested sensitivity at a variety of pressure points: 0, 1/24000, 1/2400, 1/1200, 1/600 and 1/300. Five runs are performed for 200 generations at each parsimony pressure value.

In figure 4 we plot the raw fitness after 200 generations versus the parsimony pressure used. Again we see that higher parsimony pressures yield individuals with lower raw fitness values. However, we also see that 2-point crossover outperforms the 1-point crossover consistently at all levels of parsimony pressure. Consequently, we made 2-point crossover the default.

5.4 Generalization

Although our system is at this point evolving interesting and effective rule sets, figure 4 is somewhat disconcerting in that fitness declines steadily with increasing parsimony pressure. Ideally, one would hope to see shorter rule sets emerging with more general rules that achieve comparable performance. One possible explanation for why we don't see this is that the rule language itself is not well suited for generalization.

To test this we added classifier-like wildcards to our system by allowing the genes in the condition section of the rules to take on three values: 0, 1 and '*'. We also modified the random initialization and mutation operators so that we could adjust the number of wildcards in our individuals. We added a parameter called "wildcard ratio" which allows us to adjust the bias for the number of wildcards which end up in our rules. It can take a value between 0 and 1. A value of 0.4, for example, would mean that on average 40% of the genes in the condition sections of the rules will be a '*'.

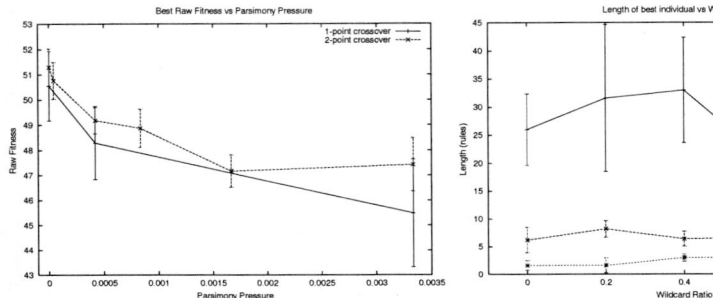

Fig. 4. Raw fitness vs. parsimony pressure is plotted for both the 1-point and 2-point crossover operators.

Fig. 5. Length vs. wildcard ratio bias is plotted using 3 different parsimony pressures: 1/24000, 1/2400, 1/600.

We ran experiments using several values for the wildcard ratio and parsimony pressure. In figure 5 we plot length vs. wildcard ratio for the three different parsimony pressures. Each point on the graph represents the best individual at the end of 200 generations, and is an average of five runs.

As one can see, adding wildcards to the system had little effect on our ability to evolving smaller rule sets. A wild card ratio of zero is equivalent to having no wild cards. As we increase the wild card ratio the evolved rule lengths are basically unchanged regardless of the parsimony pressure. We see two possible explanations for this. First, our rule matching approach differs from the one used in classifier systems. We allow partial matches, and that acts as an alternative, and perhaps competing method of rule generalization. Another more likely explanation is that the problem domain we've chosen is just too simple. We believe that wildcards would be more useful if given a more difficult problem.

6 Conclusions and Future Work

We have completed our initial design and evaluation of an EA designed to evolve behavioral rules for teams of cooperating agents. Building on the work done for single agent systems, we were able to relatively quickly make and test design choices that resulted in the ability to evolve effective rule sets for teams of homogeneous agents. We are now continuing to develop the system further in several ways. First, we believe that the ability to evolve shorter and more general rule sets is important. Our initial experiments with wild cards were not successful. We are working on understanding this better.

Our ultimate goal is to work toward evolving heterogeneous cooperating agents. This initial work involving homogeneous provides a foundation for doing so, but needs further development. Extending our system to include notions of cooperative co-evolution seem quite appropriate here. We will be reporting on this in the near future.

Acknowledgments. This research was funded by a grant from the Naval Research Laboratory.

References

1. R.C. Arkin and T. Balch. Cooperative multiagent robotic systems. In David Kortenkamp, R.P. Bonasso, and R. Murphy, editors, *Artificial Intelligence and Mobil Robots*, Cambridge, MA, 1998. MIT/AAAI Press.
2. T. Balch. Learning roles: Behavioral diversity in robot teams. In *Collected Papers from the 1997 AAAI Workshop on Multiagent Learning*, pages 7–12, Cambridge, MA, July 1997. AAAI Press.
3. J. Grefenstette. Learning rules from simulation models. In *Proceedings of the 1989 International Association of Knowledge Engineers Conference*, pages 117–122, Washington, DC, 1989. IAKE.
4. J. Holland. Escaping brittleness: The possibilities of general-purpose learning algorithms applied to parallel rule-based systems. In J. Carbonell R. Michalski and T. Mitchell, editors, *Machine Learning: An Artificial Intelligence Approach*, Los Altos, 1996. Morgan Kaufman.
5. K. A. De Jong and W. M. Spears. Learning concept classification rules using genetic algorithms. In *IJCAI 91, Proceedings of the 12th International Conference on Artificial Intelligence*, pages 651–656, Sydney, Australia, 1991. Morgan Kaufmann Publishers, Inc.
6. C. Kube and H. Zhang. Collective robotics: From social insects to robots. *Adaptive Behavior*, 2(2):189–219, 1993.
7. M. Mataric. Learning to behave socially. In D. Cliff, P. Husbands, J.A. Meyers, and S. Wilson, editors, *From Animals to Animats 3 (Third International Conference on Simulation of Adaptive Behavior)*, pages 453–462. MIT Press, 1994.
8. A. Schultz and J. Grefenstette. Using a genetic algorithm to learn behaviors for autonomous vehicles. In *Proceedings of the American Institute of Aeronautics and Astronautics Guidance, Navigation and Control Conference*, pages 739–749, Hilton Head, SC, 1992. AIAA.
9. S. Smith. Flexible learning of problem solving heuristics through adaptive search. In William Kaufman, editor, *Proceeding of the Eighth International Joint Conference on Artificial Intelligence*, pages 422–425, Karlsruche, Germany, 1983.
10. T. Soule and J.A. Foster. Effects of code growth and parsimony pressure on populations in genetic programming. *Evolutionary Computation*, 6(4):293–309, 1999.
11. A. Wu, A.C. Schultz, and A. Agah. Evolving control for distributed micro air vehicles. In *IEEE Computational Intelligence in Robotics and Automation Engineers Conference*, 1999.

Evolving Finite-State Machine Strategies for Protecting Resources

William M. Spears and Diana F. Gordon

AI Center, Naval Research Laboratory, Washington DC 20375, USA,
spears@aic.nrl.navy.mil,
WWW home page: http://www.aic.nrl.navy.mil/~spears

Abstract. We are becoming increasingly dependent on large interconnected networks for the control of our resources. One important issue is resource protection strategies in the event of failures and/or attacks. To address this issue we investigated the effectiveness of evolving finite-state machine (FSM) strategies for winning against an adversary in a challenging Competition for Resources simulation. Although preliminary results were promising, unproductive cyclic behavior lowered performance. We then augmented evolution with an algorithm that rapidly detects and removes this cyclic behavior, thereby improving performance dramatically.

1 Introduction

We are becoming increasingly dependent on large interconnected networks for the control of our resources, such as the Internet, communications networks, and power grids. The advantage of these networks is the ability to route resources in a reasonably optimal fashion. However, their interconnectivity, coupled with the lack of global view of what is happening in these networks, can lead to tremendous problems in network reliability. For example, small local failures can easily propagate to entire networks, causing loss of service and corruption of data. Also, deliberate attacks (such as "denial of service" [3] attacks) can easily cause widespread havoc, as poignantly demonstrated recently [11].

Thus one important issue is the development of effective network traversal strategies to protect as many resources as possible from failure and/or attacks, i.e., to maximally restrict the number of resources damaged. To address this issue we have decided to create a "resource protection" simulation that captures the essential aspects of this problem. A "defender" attempts to protect resources before they are damaged by an intentional (or unintentional) "adversary".

Our primary goal then is to create sophisticated reactive strategies for the defender. We use finite-state machines (FSMs) for our strategies, since there are a number of precedents for FSMs being effective strategies for adversarial situations. We use evolutionary algorithms (EAs) to create the FSMs, since there is ample evidence for the effectiveness of this approach [4,5].[1] This paper serves to summarize and highlight the results we have obtained thus far.

---

[1] The evolution of FSMs is often referred to as "evolutionary programming" or "EP".

2 The Competition for Resources Problem

Our current Competition for Resources simulation is a novel two-player game on a toroidal board of squares. Each square corresponds to a resource, and the two players (the "defender" and "adversary") compete for squares on the board. If the board is of size $N \times N$, then the defender will start at square (1,1) and the adversary will start at square (N,N). The remaining squares are initially unoccupied. Since the board grid represents real networks, such as power grids or communication networks, and in the real world networks may be highly interconnected and will have few geophysical boundaries, our board is toroidal (has no edges). In this paper we assume $N = 10$, which is quite challenging.

Each player can only perceive limited information, namely, the status of the north, south, east, and west squares neighboring the current position of the player. The diagonal squares can not be seen. The status of each neighboring square will be one of the following: unoccupied, occupied by that player, or occupied by the opponent. Due to the toroidal nature of the board, the defender and the adversary are close to one another at the beginning of the game. However, because they can not see along diagonals, they can't see one another initially.

Each time step the players alternate taking an action, which consists of moving to a neighboring resource to control/protect that resource. A player can move to an unoccupied square or back to a square that it has previously occupied, but not to a square occupied by the opponent. The player isn't allowed to "stand still" and make no move. However, because each player must follow a path of "owned" resources to its current position, it will always be able to make a move at every time step (it can always back up along the path it has taken). Thus a player can not be "trapped" at a square, i.e., it can not be completely surrounded by the opponent. Once an agent occupies a resource, it controls/protects that resource forever. A game ends when all squares are occupied or time runs out. The agent with the most resources at the end of the game wins.

Throughout this paper the adversary will have a fixed stochastic strategy that the defender must "learn" to defeat. The strategy we have chosen for the adversary is simple, but is surprisingly hard to beat. If the adversary detects any unoccupied neighboring squares, it uniformly randomly moves to one of them. Otherwise it uniformly randomly backtracks to a neighboring square it has previously occupied. Given the game and our adversary, we focus on developing effective strategies for the defender.

3 Overview of Finite State Machines

FSMs can be effective representations of agent plans/strategies, e.g., see [1] or [10]. The type of machine used here allows for indeterminate-length action sequences. Recall from Hopcroft and Ullman [9] that the usual acceptance criterion for finite-length strings is termination in a "final" state. Here we assume that there are no final states, i.e., action sequences of any length are allowed. This provides a good model of embedded agents that are continually responsive to their environment.

Formally, we define the machine M to be a six-tuple $(Q, \Sigma, \Delta, \delta, \lambda, q_1)$. Q is the set of vertices (states) of M, Σ is the alphabet of input symbols (which are agent sensory inputs), and Δ is the alphabet of output symbols (which are agent actions). δ is the transition function from a state and an input to a next state, i.e., $\delta(q_i, x_i) = q_{i+1}$ where $q_i, q_{i+1} \in Q$ and $x_i \in \Sigma$ is a sensory input. λ is the transition function from a state and an input to an output, i.e., $\lambda(q_i, x_i) = a_i$ where $q_i \in Q$, $x_i \in \Sigma$, and $a_i \in \Delta$ is an action. The initial state is q_1.

We assume that the FSMs are *deterministic* and *complete*. The FSMs are deterministic because δ and λ are functions, i.e., for every state and input there is a unique next state and action. The FSMs are complete because there exists a next state and action for every state and input, i.e., δ and λ are fully defined. Deterministic and complete FSMs are strategies that tell the agent precisely what to do in every situation it perceives.

For an FSM strategy for the Competition for Resources simulation, the sensory input x_i shows the status of the neighboring resources immediately to the north, east, south, and west of the agent. The status of each resource can be 0 (unoccupied), 1 (occupied by the defensive agent), or 2 (occupied by the adversary). Thus an input of "2100" specifies that the north resource is owned by the adversary, the east resource is owned by the defensive agent, and that the south and west resources are unoccupied.

4 Evolution of Finite State Machines

In an EA a population of P individual structures is initialized and then evolved from generation t to generation $t+1$ by repeated applications of fitness evaluation, selection, recombination, and mutation. In the context of the Competition for Resources simulation, each individual in the population is an FSM. Each FSM is evaluated by playing the game numerous times, to obtain an estimate of how well that FSM is defending the resources against the adversary. Those FSMs that perform the task better are allowed to have more children, which are created through the processes of mutation and recombination. This process continues generation by generation, until termination.

Representation. For efficiency we chose a simple tabular representation for the FSMs. Rows in a table correspond to states, and columns correspond to inputs. For each state q_i and input x_j, table entry (i,j) has two elements. The first element is the next state, i.e., it is $\delta(q_i, x_j)$. The second element is the action to take given the agent is in this state q_i and sees input x_j, i.e., it is $\lambda(q_i, x_j)$.

The number of states S is user defined. The maximum number of inputs is $3^4 = 81$, since the status of each of the four neighboring squares may have three values (0, 1, and 2). However, the input "2222" will never occur, since that implies that the defender is surrounded by the adversary. This is impossible, since the defender must have been able to get to the square it currently occupies. Thus there are 80 possible inputs and we require a table of size $S \times 80$. Each entry in the table is an "allele" that represents a next-state/action pair. Since each allele is defined uniquely, the FSM is guaranteed to be deterministic and complete. The initial state is always state 1.

Initialization. Throughout this paper a population size of $P = 100$ is assumed, since it produced good results. Each of the P FSMs at generation zero is initialized by using domain-specific knowledge. For any given state (row) and input (column) the next state is chosen uniformly randomly from the set of all S states. However, the choice of action is somewhat more complex. The number of possible actions is maximally four, since the defender may potentially move north, east, south, or west. However, in practice, some of these moves might be impossible, if the adversary owns the neighboring squares.

For example, suppose again that the input is "2100". In this case the north resource is owned by the adversary, and there are only three *legal* moves: east, south, and west. Moving north is *illegal*, since the adversary owns that square. Actions are restricted to those that are legal, and every input has a set of legal moves that are possible. However, since the goal of the game is to capture resources, we also found it useful to define *preferred* moves – those that capture previously unoccupied squares. When the input is "2100", moves to the south or west capture new territory and are thus preferable. During initialization actions are always chosen uniformly randomly from the set of preferred actions, if there are any. If there are no preferred actions, then a legal action is randomly chosen.[2]

Adapting the Number of States. Standard methods for evolving FSMs adapt the number of states [4]. State adaptation raises a number of issues. When a state is deleted, should one really erase the state from the table, or should it simply be made inaccessible? Our prior experience in similar areas has shown that it is often best to make the information inaccessible [2]. Since information has been learned, keeping the information stored serves as a useful memory, which can be re-activated at a later time (if the state is added back to the FSM). Thus we added a "tag bit" to each row of the FSM table. If the tag is 1 the state is accessible. If the tag is 0 the state is inaccessible, but is not destroyed. When a state is added, it is accomplished simply by turning on the tag. Tag bits are subject to an independent mutation operation, that flips the tags with probability 0.001. Since state 1 is always the initial state, it can not be made inaccessible.

Once a state q_j has been made inaccessible, how should the remainder of the FSM (that points to that state) be "repaired"? We investigated two solutions: (1) if state q_i points to q_j, change the pointer so that it points back to q_i and (2) change the pointer to point to any state that is accessible, chosen uniformly randomly. The latter solution performed better.

Adapting FSM Table Entries. Adaptation of the FSM table entries is accomplished with mutation and recombination. Mutation is reasonably straightforward. Each allele (next-state/action pair) in the FSM is chosen with probability p_m. Once an allele is chosen a coin is flipped to see whether the action or the next state is mutated. With probability p the next state is mutated by uniformly randomly choosing a state from the set of all accessible states. With

---

[2] The emphasis on preferred actions enormously helps the initial search of the EA.

probability $1-p$ the action is mutated. If there are any preferred actions the algorithm uniformly randomly chooses one of those. If there are no preferred actions the algorithm uniformly chooses any legal action. Note that this could result in no change (e.g., if there is only one legal action). Experiments indicated that performance was remarkably insensitive to p_m and we use $p_m = 0.001/S$ throughout this paper. Setting p to 0.5 worked well.

We also use P_0 uniform recombination [12]. A proportion p_r of pairs of parents in the population are chosen for recombination. For each pair of parents, a coin is flipped for each of the $S \times 80$ alleles. The allele at the table location (i,j) in the first FSM is swapped with the corresponding allele in the second FSM, with probability P_0. If alleles are swapped, both the next state and the action are swapped. Since only corresponding alleles are swapped, there is no need to worry about possible illegal actions. If an action is legal for one FSM at location (i,j) it must be legal for any other FSM at location (i,j), since the input j is the same. Since parents may have different sets of accessible states, recombination may swap alleles in such a fashion that a next state that was accessible in one child FSM is now inaccessible in the other FSM. In this situation a new next state is chosen uniformly randomly from the set of accessible states (in that FSM). Experiments indicated that performance was very sensitive to p_r and P_0. Using recombination to its fullest extent ($p_r = 1.0$ and $P_0 = 0.5$) worked best.

Fitness Evaluation. Since the adversary in the Competition for Resources game is stochastic, each defender FSM will have to play the game multiple times in order to obtain an estimate of how well it defends the resources. Recall that the player with the most resources at the end of the game wins. In case of a tie, the adversary wins. Given G games, the fitness of a defender FSM is the fraction of games that it wins. This fitness function returns values from 0.0 to 1.0, with 1.0 representing an FSM that won all the games it played. Setting G properly proved to be difficult. Prior work [7] concluded that the overall efficiency of the EA may often be improved by reducing G and by running for more generations. This did not work for us. A low value of G resulted in unacceptable sampling error and a high value was too CPU intensive. We were unable to balance these constraints with an intermediate value.

To solve this difficulty we took a two-phase approach. Initially, we use a low value of G, so that each individual can get a quick evaluation. If that individual is promising (it did better than the best individual seen thus far), it is re-evaluated using a high value of G. If it still beats the best individual thus far, it becomes the new best individual. The idea was to carefully evaluate only those individuals that appeared promising. This approach worked quite well. We used a value of $G = 500$ for the initial evaluation and $G = 10,000$ for the subsequent re-evaluation (if it was performed). Since most individuals were unable to beat the best individual seen thus far, they were not re-evaluated.

Selection and Termination. We use standard fitness-proportional selection [8] with elitism (i.e., the population contains a copy of the best individual that has ever been seen). For a termination criterion we ran the EA for a user-defined number of generations (2500).

5 Experimental Evaluation

We performed two experiments to judge the efficacy of our method. We were interested in answering two questions. First, how many states should be accessible initially? Second, does the adaptive-state EA find the optimal range of accessible states? To address the first question we ran an experiment where each FSM individual is initialized with $S = 10$ states. The experiment consisted of a comparison between the adaptive-state EA in three configurations: one, five, and ten initially accessible states. The only mechanism for adapting the number of accessible states is via the independent mutation operation mentioned above, which flips the accessibility tags. Although we have no "penalty" function per se (that would penalize the FSMs for having more accessible states), the mutation operator provides a slight bias towards having $S/2$ accessible states.

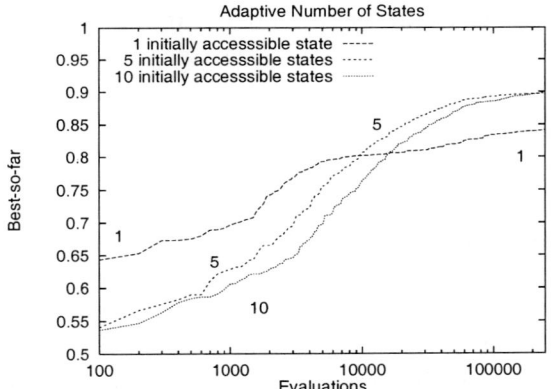

Fig. 1. "Best-so-far" curves for the adaptive-state EA with one, five and ten initially accessible states.

Figure 1 shows the best-so-far curves (the fitness of the best individual seen thus far) for the adaptive-state EA initialized with one, five and ten accessible states. The log plot emphasizes early behavior. Results are averaged over ten independent runs per configuration. On average the adaptive-state EA changed the number of accessible states in the FSMs, from one to five, five to seven, and ten to nine, respectively. One can see that having fewer initially accessible states helps early performance.[3] This is intuitively reasonable, since the adaptive-state EA is initially searching a smaller space in this situation. However, having too few

---

[3] The difference between one and five initially accessible states is statistically significant ($p < 0.04$) everywhere except between 7,000 and 18,000 evaluations. The difference between five and ten initially accessible states is significant ($p < 0.04$) between 2,000 and 19,000 evaluations. The data may not be normally distributed – hence we used an exact Wilcoxon rank-sum test with paired data in this paper.

initially accessible states (e.g., one) hurts later performance. With five and ten initially accessible states performance was quite reasonable, with a final fitness of 0.898 and 0.899 respectively. These results indicate that if the best number of states for solving a problem is not known a priori it may be best to err on the side of having too many, rather than too few.

The second question (whether the adaptive-state EA finds the optimal range of accessible states) is also important, although we have not seen it addressed in the literature. To address it we ran a control (ablation) experiment, where we turned off the adaptation of the number of states. Instead, the EA was run with a fixed number of states S. There were ten configurations (S ranged from one to ten) and ten independent runs per configuration. The results are shown in Table 1. Two points are clear. The first is that for best performance this problem requires FSMs with at least three states (i.e., state information is useful). The second is that performance is fairly comparable in the range of three to ten states.[4] This agrees with the previous experiment (with the adaptive number of states), which always ended in the range of five to nine accessible states.

Table 1. The final fitness of the best individuals for each configuration, averaged over ten runs per configuration. The optimum number of states is between three and ten.

	Fixed Number of States									
	1	2	3	4	5	6	7	8	9	10
Fitness	0.806	0.867	0.893	0.888	0.901	0.903	0.899	0.905	0.896	0.883

In summary, the adaptive-state EA effectively converges to the optimal range of states. Furthermore, its performance is competitive with the best fixed-state results (when started with five or ten initially accessible states). The only cause for concern is that although the adaptive-state EA that started with one initially accessible state ended with roughly five accessible states, the end performance was much poorer (0.841) than the fixed-state results (0.901). This suggests that although states are being made accessible the FSM is not taking full advantage of them. We investigate this possibility further in Section 7.

6 External Behavior of the Evolved FSMs

In order to understand and improve our results, we watched the agents play the simulation. The most noticeable feature was unproductive cycling behavior by the defender. In other words, the defender repeatedly visits a small set of squares on the board, while the adversary continues capturing new squares. Unfortunately, evolution alone can not solve this problem because cycles are inherent to FSMs. Therefore we augmented the FSMs with an auxiliary memory

---

[4] The increase in performance from one to two states and two to three states is significant ($p < 0.003$ and $p < 0.04$, resp.) whereas the other differences are not significant.

and an algorithm to use this memory to detect and eliminate cycles (up to a user-defined maximum length). To detect cycles, we use *behavior checking* (earlier results with *model checking* are described in [13]). Behavior checking examines the dynamic run-time behavior of the agent. Run-time checking of system behavior is a very new topic in the verification community, but some of the results already appear promising (e.g., [6]). Here, we present the first algorithm of which we are aware that does a run-time check for an FSM agent's cyclic behavior.

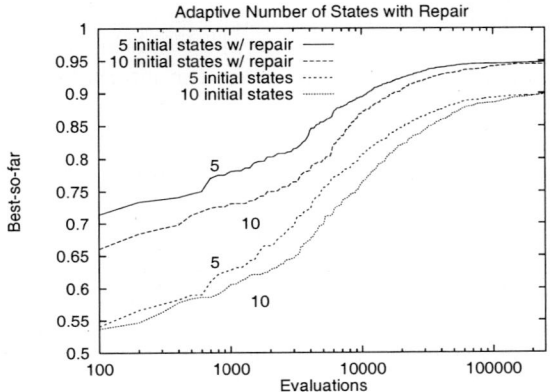

Fig. 2. A comparison of performance with and without cycle checking and repair.

Our behavior checking algorithm is executed while the agents play a game. For a sliding window of t time steps, the defender saves its current state and location on the board (the defender now consists of an FSM and auxiliary memory). The defender uses this auxiliary memory to make a cycle check before every move. If all four immediate neighbors are occupied, then the defender checks whether its current state and location are equal to any other in its window. If yes, a cycle has been identified and a random alternative action is taken to the one recommended by the FSM.[5] We have found that a window size of $2 \times N = 20$ time steps, which identifies cycles up to length 20, works well.

To test the hypothesis that behavior checking and cycle repair will improve performance, we reran the adaptive-state EA experiment with five and ten initially accessible states, but added in cycle detection and repair. This hypothesis is confirmed, as shown in Figure 2.[6] Our best performing defender FSM with repair wins 96% of the games!

---

[5] Of course, this alternative action could also create a cycle, but the behavior checking algorithm will immediately detect that cycle, after that move.
[6] The improvement using detection and repair is statistically significant, $p < 0.01$.

7 Internal Behavior of the Evolved FSMs

Although we achieved excellent performance with the addition of cycle detection and repair, we were still concerned that the adaptive-state EA might not be making good use of newly accessible states. To investigate this further we performed a dynamic internal analysis of the FSMs as they were executed by the defender. While the FSM was executing we counted the number of times that each of the $n \leq 10$ accessible states was actually the next state of a transition. The results for the fixed-state experiment were reassuring – the FSM tended to make reasonably uniform use of all n states. However, this was not true for the adaptive-state experiment. As stated earlier, on average the adaptive-state EA changed the number of accessible states in the FSMs from one to five, five to seven, and ten to nine, respectively. However, the internal analysis shows that only two, four, and six states (respectively) are actually being used to any appreciable degree. This explains the poor performance of the configuration where only one state is initially accessible. Although five states became accessible, only two were actually being used. Clearly the adaptive-state EA is having difficulty making full use of newly accessible states that have never been seen before.

This raises the serious concern that the addition and deletion of states are such disruptive operations that they cause noticeable problems for the evolution of FSMs. Currently we are investigating the application of "gentler" operators that could perform the same role. Simply deleting states (or turning them off) is too disruptive, due to the repair that must be performed afterwards. However, merging two similar states could remove a state in a fashion less deleterious to evolution. This process would be analogous to generalization. Similarly, as opposed to adding states (or turning them on), an alternative operator would clone an existing row of the tabular representation. Accessing this new state would not be deleterious and evolution could proceed to modify it slowly. This provides a process of specialization. We are currently exploring these options.

8 Summary and Future Work

To summarize, this paper has empirically explored issues related to evolving FSMs in the context of the Competition for Resources problem. Our experiments yielded some interesting and useful results. For example, given enough initially accessible states, it was encouraging to find that the adaptive-state EA was able to successfully converge to the optimal range of number of states and was able to provide good performance. However, problems arise when starting with too few initially accessible states, and an analysis indicates that (for the Competition for Resources problem at least) the adaptive-state EA is having some difficulty making good use of newly accessible states. We also found that the ubiquitous presence of cycles hampered the defender's performance significantly. This latter difficulty was greatly diminished by augmenting the FSMs with memory and an algorithm for cycle detection and repair.

Our main focus for the future is to improve the evolution of the FSMs, to make the Competition for Resources game more realistic, and to continue our empirical investigations in the context of newer versions of the game. For example, in the current game resources are all treated equally. In the spirit of game theory, we would like to consider resources having different numeric values, and perhaps have the value of a resource differ for each of the agents. Another possibility is to allow one agent to (with some small probability) "steal" a resource owned by the other agent. Another possibility is to include multiple agents and co-evolution. What is most interesting about this game is how easily it can be changed to represent a wide variety of problems. For example, with minor modifications we have extended the game to represent the epidemiology of virus versus anti-virus spread. In the virus version of the game, each square represents an agent with the virus, anti-virus, or neither. At each time step, an agent having the virus or anti-virus can spread it to one of its neighbors. What one sees on the board when watching this version of the game looks like a "spreading activation". Further pursuit of the virus version both in simulation and in a corresponding mathematical model are currently in progress.

References

1. Carmel, D. and Markovitch, S. (1996) Learning models of intelligent agents. Proceedings of the Thirteenth National Conference on Artificial Intelligence.
2. De Jong, K., Spears, W. and Gordon, D. (1993) Using genetic algorithms for concept learning. Machine Learning Journal, 13 2/3.
3. Denning, D. (1999) Information Warfare and Security. Addison-Wesley, NY.
4. Fogel, L. (1999) Intelligence Through Simulated Evolution: Forty Years of Evolutionary Programming. Wiley Series on Intelligent Systems.
5. Fogel, L., Owens, A., and Walsh, M. (1966) Artificial Intelligence Through Simulated Evolution. John Wiley and Sons, Inc., New York.
6. Gordon, D., Spears, W., Sokolsky, O., and Lee, I. (1999) Distributed spatial control, global monitoring and steering of mobile physical agents. Proceedings of the IEEE International Conference on Information, Intelligence, and Systems.
7. Grefenstette, J. and Fitzpatrick, J. (1985) Genetic search with approximate function evaluations. Proceedings of the Int'l Conference on Genetic Algorithms.
8. Holland, J. (1975) Adaptation in Natural and Artificial Systems. University of Michigan Press.
9. Hopcroft, J. and Ullman, J. (1979) Introduction to Automata Theory, Languages, and Computation. Addison-Wesley, Menlo Park.
10. Jefferson, D., Collins, R., Cooper, C., Dyer, M., Flowers, M., Korf, R., Taylor, C. and Wang, A. (1991) Evolution as a theme in artificial life: the Genesys/Tracker system. Proceedings of Artificial Life II.
11. Levy, S. and Stone, B. (2000) Hunting the hackers. Newsweek, February 21.
12. Spears, W. and De Jong, K. (1991) On the virtues of parameterized uniform crossover. Proceedings of the International Conference on Genetic Algorithms.
13. Spears. W. and Gordon, D. (2000) Evolution of strategies for resource protection problems. Submitted to A. Ghosh and S. Tsutsui, editors, Theory and Application of Evolutionary Computation: Recent Trends. Springer Verlag.

A Method of Generating Program Specification from Description of Human Activities

Shuhei Kawasaki and Setsuo Ohsuga

Waseda University, Department of Inf. and Comp. Science
3-4-1 Okubo Shinjuku-ku, Tokyo 169-8555, JAPAN

Abstract. This paper discusses a method of translating human activities into a program. The final goal of this research is to develop an automatic programming system which can be used easily. A new modeling scheme is introduced to allow human-like representation and to replace the subject of programming from person to computer. A method of translating rules described based on this modeling scheme into program specification is proposed. By using domain which is defined to variables of rules, the optimum program specification can be generated.

1 Introduction

The goal of this paper is to discuss a part of research conducted by the author's group. The final goal of this research is to develop a way of automatic programming. Programming is a special activity by human being. Therefore a concept of activity used in this paper is discussed first. Every activity has its outcome and is executed by some subject. The subject can be a person or the other creatures or a machine. Each subject has its own language and tool for representing and promoting an activity. Accordingly every subject has its own way of promoting activity. Subject's activity is also affected by environment in which the subject is put. Therefore there can be different activities with the same outcome by the different subject and environment. In order for someone to watch and describe the activity by some others correctly, it is necessary to include the subject of the activity and its environment in its representation.

When the subject is a computer, a formal description of its activity is a computer program. Programming is an activity to represent formally an activity of a computer at the higher level. Subject of this activity (programming) has been mostly human being. To automate programming is to replace the subject of the activity from person to computer. In order to achieve this goal, it is necessary to represent formally the activity of programming of which the subject is a computer. Thus the objective of this paper is to describe a way of representing a goal directed activity by human being and translating it into an activity by computer, i.e. program. This goal is divided into two sub-goals; creating a representation of human activity and translating it into computer's activity. The range of representation of human activity is very wide. Among all, such a representation that is very near to computer program so that almost formal transformation into computer program is possible is the ordinary program specification. This form is far from natural representation of human activity and the one is required to take programming technique into account in the representation. The objective of this research therefore is to allow ones human-like representation on human activity and also to discuss a computer technology that can translate it into computer's activity. This paper discusses a method of translating human activity to program code in this vast area. It is assumed that description of human activity has already been given. To make it is another problem.

2 New Modeling Scheme-Model Representation Including Subjects and Objects

Every significant computer activity has some object to which the activity applies. Therefore the scope of consideration is limited to this class. That is, activity without any object is out of consideration. A new modeling scheme is developed. It is to represent every activity with their related objects that concern achieving the goal. It includes two kinds of structures; object model structure and subject model structure. The former is a structure of objects that are included in some activity to be executed by some subject. Objects are organized in a model structure by means of finite structural relations. Typical examples are is-a relation and part-of relation. In order to represent human activity in its natural form for human being, these are not enough but the others are necessary. For example, it is difficult to represent a power set with these structural relations. The power set concept is sometimes important to represent human idea and, accordingly, human activity. Cartesian products, list and graph are also necessary. Thus a least set of structural relations must be defined to represent these structures. Using these basic structural relations an object structure is formed. Usually it is a hierarchy either or both of is-a and pat-of relation and graph in the same level of the hierarchy,

Subjects are also organized in a model structure by means of a structural relation. In reality there is no substantial structural relation as is the case of object structure. Rather subjects exist independently to each other. Every subject has it's own activity. It is represented by a predicate. It includes a subject as a term. There is some relation between activities. An activity A may depend on another activity B in the sense A uses the outcome of B. In this case these activities, and therefore their subjects, are arranged in a hierarchy such that the activity B is put under the activity A. When activities A and B have mutual dependency they form a recursive relation. In this case the activities and their subjects are put on the same level. Thus subjects are organized in a structure via the relation of their activities.

There are two types of activities. The first class activities are those arranged at the top or middle of the subject model hierarchy. Every activity in this class has some dependency relation to the other activity. The second class activities are those arranged at the leaves of the subject model hierarchy and have no dependency relation to the other activity. Every activity can be executed independently to the other activity referring only to the object structure. Thus the global goal of this system can be achieved by executing these activities from the bottom. This is a case of human execution of the activities. Automatic programming is to translate this structure of activities into a program. In this case the order of activity execution in computer must be decided. The first class activities concern the control structure of the produced program. In order to decide this control structure, top-down interpretation of the activity structure is necessary. Each activity at the leaf represents a program unit that is fabricated in the control structure. This unit program is not always very small. In order to make up whole program the automatic programming of the unit program must be assured. Thus automatic programming consists of two stages; automatic programming of unit program and development of control structure. In the following the automatic programming of unit program is discussed.

3 KAUS as Representation Language

A language suited for representing this system is necessary. In order to cope with problem model of the form as discussed in section 2, it must be suited for representing predicate including data-structure as argument and also for describing meta-level operation such as knowledge for describing on other knowledge. KAUS (Knowledge Acquisition and Utilization Language) has been developed for the purpose. In the following, some logical expressions appear as knowledge. In order to keep consistency and integrity of expressions throughout the whole system these must be written in KAUS language. But these are not necessarily written in correct KAUS expressions but locally simplified. It is because KAUS syntax is not included in this volume and also these locally simplified expressions are more comprehensive than correct expressions.

4 Automatic Programming for Unit Program

Problem formulation of automatic programming of unit program is as follow. A unit activity at the lowest level (leaf) of activity hierarchy is presented. It becomes a query to a programming system to be solved and the obtained procedure for solving this problem is translated into a program. The system contains a knowledge base and inference engine to solve problem automatically. The activity as query includes variables with domains to represent input and output variables of program to be generated. The programming system is required to solve the query as problem for all possible cases in the variables of designated domains. This programming process is composed of two stages; generating a specification-tree to represent the program specification and converting this specification-tree to object code. The specification-tree is represented as shown in Fig. 1.

Every activity is represented in the form of logical predicate in this tree. The tree has predicate nodes and rule nodes, and these two kinds of nodes appear alternately. Variable type, either input type or output type, is recorded in every predicate-node. Domain of variables is recorded in rule-node. This domain can

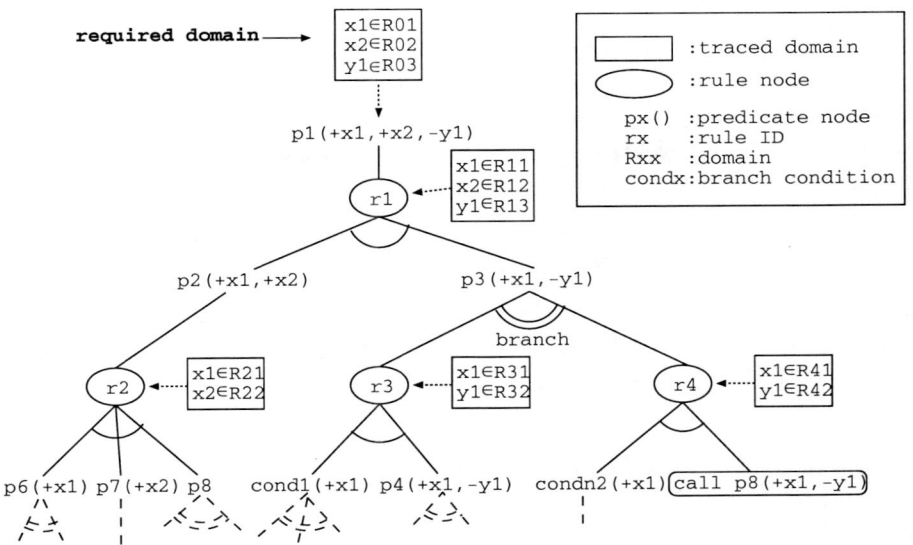

Fig. 1. specification-tree

be narrowed while problem solving as will be shown later on. More than one specification-tree is generated depending on the situation to represent different function and each specification-tree is transformed into program source code as a function call. In each tree, the function name and its process are declared in the top node at beginning, and the top node representing a function is expanded into child nodes to represent its detailed. Nodes inside the tree show a logical structures and input-output relation of processes. Every leaf nodes represents a primitive process, to which a program is prepared. With KAUS, leaf node is either fact predicate or PTA predicate., which process is defined in procedural language. When a specification-tree is generated and is transformed into program, the nodes inside tree are ignored but only leaf nodes are encoded on the basis of depth-first-search because these inner nodes are generated as intermediate products to get the correct set of leaf nodes. A node of which lower nodes are connected by "or" connective is transformed into a branch. Once a specification-trees is generated, it can be transformed into the program of any target language according to the transformation rule. Thus for automatic generation of program, the method of generating specification-tree is important. It is discussed in detail first.

4.1 Specification-Tree Generation

The basic idea of automatic programming in this paper is based on automatic problem solving. Backward reasoning is achieved by the inference mechanism of KAUS. A succeeded path of problem solving in the trace of the process is remained by deleting backtracked paths and is represented as a tree of rules used there. In general it is difficult to solve problems including variables and generate the trees directly because depending on the domains of mutually related variables the different trees can be generated.

Therefore the tree is generated in two steps. First the system chooses a specific value for every variable from its domain, solves this instance problem and generates an instance tree. The tree can be different by the selection of instances for the variables. It shows a possible pattern of the tree.

Then, as the second step, generalize this tree by recovering a variable to each instance. There are two approaches for generalization. One is to apply the instance problem solving repeatedly by selecting a different set of instances each time until no more new pattern of the tree is generated. Then by merging every different pattern into a tree, a general structure of the tree for the query including variables is generated.

With this method selecting sufficient queries is necessary to get the whole patterns of trees to satisfy all values in input domain. The authors adopted first this method. But here is a difficulty in case input domain is continuous or such as the whole integers. Even in this case the number of possible patterns can be finite. But there is no guideline for an efficient section of the instance sets.

The second approach is to use the domain information of variables after an instance problem, or a few instance problems if necessary, was solved and an instance tree was generated. In KAUS expression, domain can be included in every predicate like '(Ax/domain1)(Ay/domain2) predicate1(x, y):- predicate2 (x), predicate3'. By using this domain information, it becomes possible to make the optimized program specification.

The forrowing rule shows how to generate specification-tree.

```
makeSpecificationTree(+Subject, +Query, -SpecificationTree) :-
    makeNewInstanceQuery(+Query, -InitialQuery),
    getInstanceTree(+Subject, +InitialQuery, -InstanceTree),
    analyzeInstanceTree(+Query, +InstanceTree, -AnalyzedTree),
    modifyTree(+InitialTree, -ModifiedTree),
    generalizeTree(+Subject, +ModifiedTree, -GeneralTree),
    OptimizeTree(+GeneralTree, -OptimizedTree),
    identifyConditionNodes(+OptimizedTree, -SpecificationTree).

generalizeTree(+Subject, +CurrentTree, -GeneralTree) :-
    makeNewQuery(+CurrentTree, -Query),
    makeNewInstanceQuery(+Query, -InstanceQuery),
    getInstanceTree(+Subject, +InstanceQuery, -InstanceTree),
    analyzeInstanceTree(+Query, +InstanceTree, -AnalyzedTree),
    mergeTree(+CurrentTree, +AnalyzedTree, -MergedTree),
    modifyTree(+MergedTree, -ModifiedTree),
    generalizeTree(+Subject, +ModifiedTree, -GeneralTree).

generalizeTree(+_, +GeneralTree, -GeneralTree).
```

For readability, the form of this rule is modified from KAUS form. In this expression, "+" mark given ahead to value means that the value is an argument, and "-" means return value.

4.2 An Example of Specification-Tree Generation

The specification-tree generation process is explained here with an example. Whole processes are shown in Fig.2. An operation to generate a specification-tree is started by the predicate *makeSpecificationTree*. *makeNewInstanceQuery* select a specific value from the domain R0 in the query q0. And, new query *NewInstanceQuery* is made with the specific value selected. *getInstanceTree* is called by value *NewInstanceQuery*, and, system actually solve problem. Tracing the problem solving process, *getInstanceTree* make an *InstanceTree*. Then *analyzeInstanceTree* analyzes the input-output type based on *InstanceTree* and traced domain of each variable of all nodes in the *InstanceTree*. *modifyTree* modifies the structure of *AnalyzedTree*. The tree *ModifiedTree* is the initial state of specification-tree.

By generalizing this tree *GeneralTree* to represent a general processing structure is obtained. In *generalizeTree*, *makeNewQuery* and *makeNewInstanceQuery* search such nodes that are possible to select not-yet used rule in the tree being generalized, and generate a query to select not-yet used rules. For example, the node of p1(x) has possibility of being expanded by the rule r2 other than r1. Since $x \in R0$ in p1(x) and $R0 \cap R3 \neq \emptyset$, it is possible to generate a query 'p1(3)?' that expands r2. Using this query, it is possible to obtain an analyzed tree in the same way as that used for obtaining an initial tree. Then *mergeTree* and *modifyTree* merge the tree being generalized and this new analyzed tree. This generalization procedure is repeated until new query is not generated any more by *makeNewQuery*. In this example, generalization by a query 'p2(5)? ' is achieved again and the generalization terminates. By identifying the conditions of optimization and branching in the tree thus obtained, a specification-tree (*SpecificationTree*) is obtained.

Note that the rules that are not used like r3 do not appear in the specification-tree. As is shown, selection of necessary and sufficient rules becomes possible by tracing domain of variables. Accordingly, general rules can be written without taking notice on each case of using to specific problem.

4.3 Domain Tracing

Domain tracing is a most important process in specification-tree generation. To put necessary and sufficient rules into specification-tree becomes possible by tracing and narrowing domain of input-output values. The domain of arguments is traced from top of the tree, and the domain of return value is traced from the bottom of the tree. Fig. 3 shows the example when query '$p1(x,y)$?' is given with x as argument and y as return value. The domain of x is traced from the top and the domain of y is traced from the bottom. In r3, y is function of x, therefore the domain of y ($R7'$) have to be calculated according to the domain

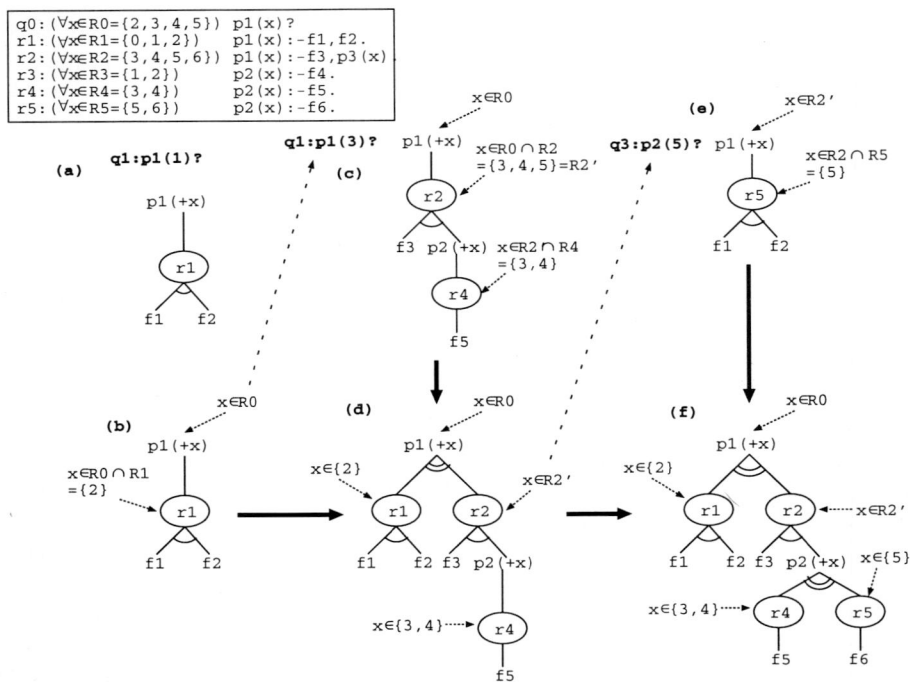

Fig. 2. specification-tree generation

of x. If calculation of all of the primitive predicates is possible, the domain is traced precisely. In many cases, however, this calculation is not possible. In this case, the domain to which calculation is not possible is set to $Univ$ to mean a universe. A larger domain causes selecting unnecessary rules. However in many cases it does not matter even if the domain is set to $Univ$. In case of narrow domain is necessary in the following process in order to select rules exhaustively, however, larger domain causes a problem of rule selection.

There are some cases in which possibility of programming can be revealed during domain tracing. For example, let a predicate $p1$ of which the domain of an attribute x is $R1$ be expanded and it is revealed that $p1$ is to be processed either by $p2$ or $p3$ of which the domain of the argument x is $R2$ or $R3$ respectively. If $R1 \supset R2 \cup R3$, some inputs for $p1$ in $R1$ cannot be processed only by $R2$ and $R3$. Then programming becomes impossible with existing rules. If there are such nodes other than those nodes that are to be used for judging the conditions as will be discussed below, the system has to tell user that the necessary rule for programming is not enough and suuply the lacked knowledge.

When specification-tree has a recursive structure including loop, the domain of the variable that is transmitted inside loop change in every loop operation. Since it is difficult to trace the domain that changes in every cycle, the domain is made $Univ$. Those variables that are not changed in the loop operation are given

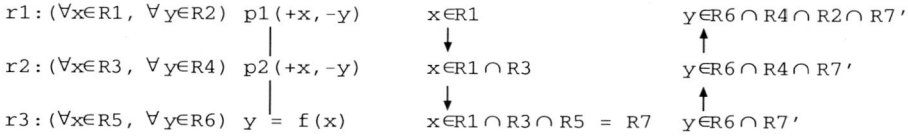

Fig. 3. domain tracing

the domains that are the domains at just before entering the loop. It is necessary here to investigate the method of tracing the domain that is transmitted inside loop as far as possible, but in many case this is not urgent because putting it as $Univ$ does no bring large defect.

4.4 Structure Extraction

Different from procedural program, an activity defined in the form of predicate does not have information on process sequences. In order to get a process sequences, let system solve a problem, and get the specification-tree which is equivalent to the process system performed. Before structure extraction, a specification-tree generated in this way may have many duplicate or overlapping sub-trees. These sub-trees are extracted from the main tree as an independent function. Recursive process and loop process has a sub-tree of the same form, which appears repeatedly increasing the depth in the tree with the number of repetition. A special treaty is necessary in order to prevent the tree becoming of infinite depth.

This sub-tree is extracted and processed separately. This is achieved by cutting this sub-tree as a separate specification-tree. Only a calling method to this extracted part is left to the original tree. After then every specification-tree that has the same top node is merged into a single tree. After then there is no such sub-trees that have the same predicate node in the same tree and let the number of specification-trees becomes at most equal to the number of rules.

As an example of sub-structure extraction, a case including a loop is shown in Fig. 4. This example shows that a structure is to be modified by adding an analyzed instance tree $T3$ to a tree being composed of $T1$ and $T2$ and being generalized. The node surrounded by a block in this figure shows such a node that appears plural times by adding $T3$. By cutting these nodes from the end it becomes the state of (b). Then the state (c) is the final state that is obtained by merging the $p2$ that is selected by the condition of the argument by disjunction. When two specification-tree that have the predicate $p2$ as a top node is merged, every domain for the variables in merged specification-tree is calculated again and modified.

Here is a problem of defining specification-tree having the same node as a top. The condition that it has the same predicate with the same number of arguments and the same input-output relation is mandatory. As well depending on whether the equality of the domains of the input-output variables that were obtained by tracing is added to the condition or not, the generated specification can be different. When two specification-trees with the different input

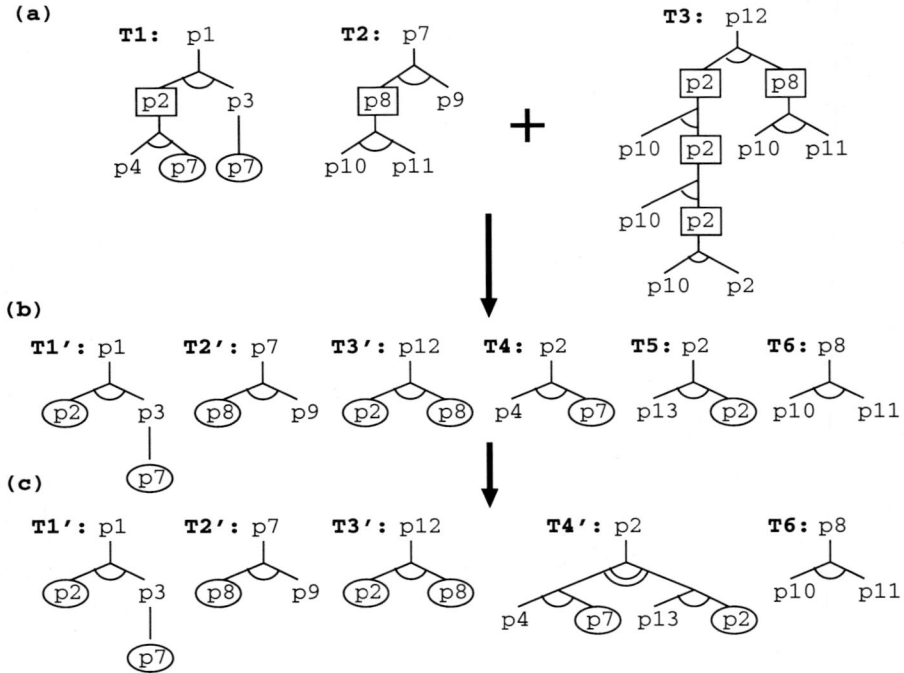

Fig. 4. program structure extraction

domains are merged, the generated program code can be shorter but its processing time can be longer because the number of determining the conditions increases. This becomes a problem for the program unit that is used frequently. This problem is resolved by handling the specification-trees including the different input domains as different until *generalizeTree* could be finished. That is, until *GeneralizedTree* could have been obtained, the structure is converted giving the processing speed the higher priority and, at the optimization stage, it is decided whether two specification-tree with the different input domains are to be merged or not. Since every structure has already been obtained at the optimization stage, there is no problem to merge two trees as long as the tree structure is the same even if the trees have the different input domains. When two tree structures are different to each other because of the difference of the input domains, there is no fixed basis to decide whether to reduce the program size by merging or to keep processing speed high by leaving the tree without being merged. The extra condition, for example by human decision, is necessary.

4.5 Branch Conditions

Finally a method to extract branch condition is discussed. Extracting branch condition is achieved after *GeneralizedTree* could have been generated and optimization would have been finished. Specification-tree at this stage includes information on an execution order of program components and function call conditions, branch condition at the disjunctive node is not described explicitly. It is necessary for identifying the branch condition. There are two cases in the inference operation to select rules that induce branch operation. One is the case in which different rules are selected by backtracking. The other is the case rule is selected by matching of variables. In case of using KAUS, rule selection by matching the domains can occur because it includes domain information explicitly.

Operation performed at extracting branch condition is identification of predicate that is defined to represent conditional operation and generating a decision tree for rule selection by matching between identified branch condition and variable. These methods are now being investigated. At the moment, such a decision tree is not made but the condition is given to every variable. Also the predicate that has been defined as representing conditional operation it is shown explicitly that it forms a branch condition. Since it is anticipated that the conditional operation is made intentionally when it is described as a procedural specification, it is possible to ask persons to describe it explicitly. But it is necessary in future to study on a method to detect conditional operation automatically.

5 Summary

A method of translating human activities into a program is discussed. To allow human-like representation and to replace the subject of programming from person to computer, a new modeling scheme is introduced. By using domain which is defined to variables of rules, the optimum program specification as specification-tree can be generated from rules. The system based on this method discussed here is now being developed. We already developed the basic part of this system and now implementing the method of domain tracing.

References

[OHS96] S. Ohsuga; Multi-Strata Modeling to Automate Problem Solving Including Human Activity, Proc.Sixth European-Japanese Seminar on Information Modelling and Knowledge Bases, 1996

[OHS98] S. Ohsuga; Toward truly intelligent information systems - from expert systems to automatic programming, Knowledge Based Systems, Vol.10, No.3 1998

[OHS99] S. Ohsuga, T. Aida; Externalization of Human Idae and Problem Description for Automatic Programming, Proceedings of ISMIS, pp163-171, 1999

[ROS97] F. H. Ross; Artificial Intelligence, What Works and What Doesn't? AI Magazine, Volume 18, No.2, 1997

PLAtestGA: A CNF-Satisfiability Problem for the Generation of Test Vector for Missing Faults in VLSI Circuits

Alfredo Cruz, PhD

Department of Electrical Engineering, Polytechnic University of
Puerto Rico, 377 Ponce de León Ave., San Juan, Puerto Rico, 00918
across@coqui.net

Abstract. An evolutionary algorithm (EA) approach is used in the development of a test vector generation application for single and multiple fault detection of growth faults in Programmable Logic Arrays (PLA). Evolutionary algorithms are search and optimization procedures that find their origin and inspiration in the biological world. In this paper, we apply the genetic operators to the CNF-satisfiability problem for the generation of test vectors for growth faults. CNF has several advantages, there are not dependencies between bits: any change would result in a legal (meaning) vector (either a minterm or a maxterm). Thus we can apply mutations and crossover without any need for decoders or repair algorithms. The crossover operation unlike previous operators used in PLA test generation, does not use lookups or backtracking.

1 Introduction

Recent Literature has addressed the problem of PLA test generation [1], [2], and [4]. Several algorithms have been also proposed for PLA testing using the sharp (T) operation or a modified version of this operation, but they tend to be computationally expensive [4], [5], and [6]. Smith [7] suggests simplifying the algorithm by generating a test for every fault. This results in considerably larger test vectors. Hence, a minimal test set is not guaranteed. Other approaches employ additional hardware, which means greater costs, and potential degradation of PLA performance. In [8] an overhead ranging from 20% to 50% has been reported for large PLAs using various such methods.

However, PLA testing based on genetic and evolutionary algorithms is in its earliest development. An algorithm for shrinkage faults using genetic algorithms is proposed recently in [3].

In this paper we present an algorithm for PLA test generation and its implementation using genetic operators to the CNF-satisfiability problem which shows that test pattern generation can be very efficient. This technique eliminates operations (such as backtracking and T operation) that can become computationally intractable with increasing PLA size.

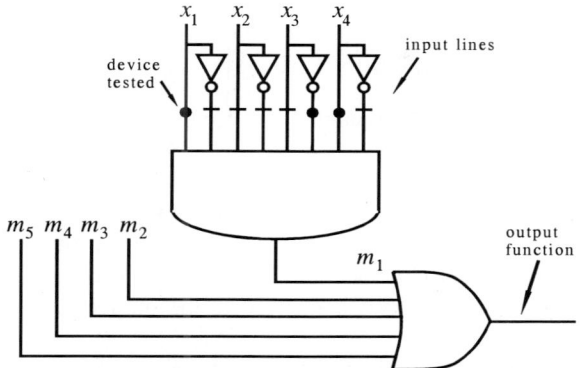

Fig. 1. Product Term Under Test

2 Fault Modeling

In testing digital circuits, the most commonly considered fault model is the stuck-at fault (i.e., s-a-0 or s-a-1). However, because of the PLA's array structure the stuck-at fault alone cannot adequately model all physical defects in a PLA [9]. The intersections between product lines and input bit lines or between output function lines and product term lines are called crosspoints. Each product line is used to realize an implicant (product term) of the given function by placing appropriate crosspoint devices into what is known as the AND plane. Therefore, a new fault class model, known as the crosspoint model is used. The unintentional presence or absence of a device in the PLA causes a crosspoint fault.

The focus of this paper is on the use of genetic algorithms for the generation of test vectors for growth faults.

3 The Growth Fault

Growth faults correspond to the removal of a literal, in the AND plane, from an implicant (product term) of the function which causes the growth of the implicant. A growth fault causes the ON-set (i.e., minterms) of a fault-free PLA to grow into the OFF-set (i.e., maxterms). To detect a fault in a PLA, it is important that the PLA output in the presence of the fault differs from the PLA output in the absence of the fault. The two requirements for fault detection are: Fault Sensitization and Fault Propagation.

$$\begin{array}{cccc} m_1 & 1 & X & 0 & 1 \\ m_2 & 0 & X & 1 & 0 \\ m_3 & 0 & 1 & 0 & X \\ m_4 & 0 & 0 & 0 & X \\ m_5 & 0 & X & 0 & 1 \end{array}$$

Fig. 2. Example PLA for Growth Faults

A missing device fault in the AND plane will be sensitized if and only if the implicant under testing carries a 0 when fault-free and if the implicant carries a 1 in the presence of the fault. Once a fault has been sensitized then a propagation path must be established, otherwise the fault is masked[1]. The propagation is done by deselecting all other product lines connected to the output except the product term under testing.

The procedure for deriving the growth test vectors is explained with aid of Fig. 1.

The product term is represented by an AND gate of 4 inputs. A dash '-' in the input lines indicates the absence of a device, whereas a circle 'O' on the input lines indicates the presence of a device.

For example, to detect a missing device at x1 of the implicant under consideration [1X01], a logic 0 must be applied to the input x1, while the care values (at input x3 x4) remain unchanged. Since the value of the literal was changed from 1 to 0, then a term from this set could detect a fault in the uncomplemented bit-line. To detect a fault in the complemented bit-line the literal must be changed from 0 to 1.

Now we should be able to sensitize this fault (if one exists) at the output of the AND gate under consideration. A value 1 at the output of the AND gate denotes a fault while the implicant under test carries a 0 in the absence of a fault. To generate a 0 on the product line (required for sensitization of growth faults) the input value connected to the target growth fault bit-line is toggled to the value opposite the value representing the used bit-line.

The small PLA of Fig. 2 is used as a running example for illustrating the test pattern generation for growth faults using genetic algorithms (GA). The function of Fig. 2 can be expressed both as a truth table or as sum-of-products as shown below:

$$f(x_1, x_2, x_3, x_4) = \sum (0,1,2,4,5,6,9,13)$$

---

[1] The necessary condition under which masking occurs in PLA is given by [10].

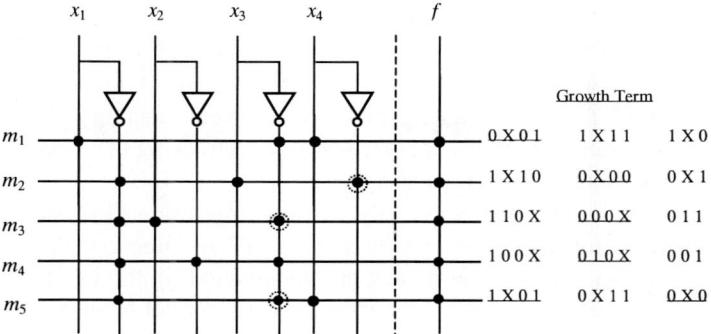

Fig. 3. The Growth Term of the PLA

3.1 Growth Term

The above discussion leads to the following rules that must be established for the generation of test vectors, for growth faults. A growth term stands for the set of extra terms contributed by a growth fault.

The growth term from a given set of product terms is derived as follows:

PROCEDURE 1:

For each product line mj
Do (n - log2Ω) times
 {
 Construct a growth term as follows:

 Scan the product term from left to right until an unmarked literal is found;
 Mark the literal and toggle its value from 0 to 1, or from 1 to 0;

 Leave the other components of the product term intact (both literal and don't care values). These extra terms correspond to the growth term.
 }

(n - log2Ω) is the number of literals on each product term [3].

The growth term for the sample PLA is derived using Procedure 1 (see Fig. 3). The fully redundant growth terms are underlined. A growth term is fully redundant when it is fully covered by one or more input product terms. A growth term may be partially redundant, i. e., partially covered by one or more input product terms.

A growth term may have terms in the ON-set function (i. e., minterns) and in the OFF-set function (i. e., maxterms). The terms in the ON-set function fail to select uniquely the product line on which the target is located, since it will also select the product terms that cover them. Therefore, the fault can not be propagated.

Furthermore, since a fault can be sensitized by a term from the ON-set function, it is necessary to delete those terms from the growth term. This procedure can be carried out by computing the intersection (denoted by ($\cap$)) between the growth term generated for each product with the complement function (OFF-set) [4], [6]. One of the disadvantages in this approach is the backtracking that could occur when the test is chosen and fails to propagate.

Another approach is to apply the sharp operation (T) between the growth term generated for each product term and the ON-set function. Bose in [1] and [5] uses this operation to find terms that are not covered by the ON-set function. However, terms partially covered by any input product can not be eliminated from the growth term without eliminating the growth term. That is, an invalid test vector may be generated after the Quine-McCluskey method is applied.

3.2 Conjunctive Normal Form

PLAtestGA uses the conjunctive normal form (CNF) logical expression, equivalent to the complement's function, to derive the test set for growth faults. The use of the CNF is supported by the De Morgan's theorem [11]. The terms complement and OFF-set function are equivalent.

The following logical expression of the example PLA of Fig. 2 is in CNF:

$$f'(x_1, x_2, x_3, x_4) = (x_1' \text{ OR } x_3 \text{ OR } x_4') \text{ AND}$$
$$(x_1 \text{ OR } x_3' \text{ OR } x_4) \text{ AND}$$
$$(x_1 \text{ OR } x_2' \text{ OR } x_3) \text{ AND}$$
$$(x_1 \text{ OR } x_2 \text{ OR } x_3) \text{ AND}$$
$$(x_1 \text{ OR } x_3 \text{ OR } x_4')$$

We apply a genetic algorithm to the CNF-satisfiability problem for the generation of test vectors for growth faults. The problem is to determine whether there exists a truth assignment for the variables in the expression, so that the CNF expression evaluates to TRUE. For example, the above CNF logical expression has several truth assignment (a valid candidate test vector) for which the whole expression evaluates to TRUE, e. g., any assignment with x3 = TRUE and x4 = TRUE. The CNF expression of the example PLA is made up of five clauses. That will allow us to rank potential bit pattern solutions in the range of 0 to 5, depending on the number of clauses that pattern satisfies. Table I shows the fitness of each element. When a pattern has a fitness of 5, a maxterm of the function is evaluated. A growth fault can be detected by this pattern if the intersection exists with a term(s) from the growth term set. The growth term set is the union ($\cup$) of the growth term generated by each product term (refer to Procedure 1). It is important to understand that an undetectable fault can not be detected by any pattern.

It is hard to imagine a problem with better suited representation: a binary vector of fixed length similar as the PLA physical layout should do the job. There are other several advantages, there are not dependencies between bits: any change would result in a legal (meaning) vector (either a minterm or a maxterm). Thus we can apply mutations and crossovers without any need for decoders or repair algorithms. Even other less frequently used genetic operators, such as the inversion (reversing the order of bits in the pattern) or exchange (interchanging two different bits in the pattern) leave the resulting bit pattern a legitimate possible solution [12], [13].

4 Test Generation Using Genetic Operators

The basic genetic algorithm, where P(t) is the population of strings at generation t is given below:

```
procedure genetic algorithm
{
  set time t := 0
  select an initial population P(t)
  while the termination condition is not met, do:
  {
      evaluate fitness of each member of P(t);
      select the fittest members from P(t);
      generate offspring of the fittest pairs (using genetic operators);
      replace the weakest members of P(t) by these offspring;
      set time t := t+1
  }
}
```

Selection is done on the basis of relative fitness and it probabilistically eliminates from the population those candidate test vectors which have relatively low fitness. Recombination, which consists of mutation and crossover, imitates sexual reproduction.

Crossover is performed with crossover probability Pcross between two selected strings, called parents, by exchanging parts of their genomes (i.e., encoding) to form two new individuals, called offspring. It is implemented by choosing a random point between 1 and the string length (δ) minus one [1, δ − 1] in the selected pair of parents and exchanging the substring defined by that point (i.e., swap the tail portion of the string) to produce new offspring. That is, all the information from one parent is copied from the start up to the crossover point, and then all the information from the other parent is copied from the crossover point to the end of the offspring (chromosome). The new chromosome thus gets the head of one parent's chromosome combined with the tail of the other. For example, consider strings 1 and 3 of Table I from our example initial population. See Fig. 4.

In choosing a random number between 1 and 4, we obtain a K = 2 (as indicated by the separator symbol |). The resulting crossover yields two new strings that are part of the generation. These offspring are: 0 1 1 0 and 0 0 0 1.

Crossover is both simple and efficient. This operation enables the evolutionary process to move towards optimal solutions in the search space. The usefulness of crossover is due to the combination of better than average substrings coming from different individuals [14].

Mutation probabilistically chooses a bit and flips it. Mutation is needed because if selection and crossover together search new solutions, they tend to cause rapid convergence and there is a danger of losing potentially useful genetic materials, such as 0s and 1s at particular location of the specified values of the candidate test vector under evolution. The usual interpretation of bit mutation rate is the following: for each string in the population and for each bit within the string generate a random number r between 0 and 1, if r ≤ Pmut flip the bit. This operator is applied to strings 1, 6, and 8 of Table I. For example, string 1 is changed from 0 1 1 0 to 0 1 1 1 after mutation.

The following definition applies to the discussions that follow.

Definition 1: Hamming distance

The number of bit positions in which two product terms hold non-don't care values that are different is called the Hamming distance, dH.

For example, in Fig. 2 the hamming distance between m3 and m4 is one, i.e. these terms differ in bit position x2.

The following GA parameters are used for testing growth faults of the example PLA of Fig. 2:

- Uniform Crossover Single Cut Point
- Number of generations : Until a minimal test set is found
- Size of Population : 8
- Crossover Probability : 1.0
- Mutation Probability : 0.1

PLAtestGA begins, at generation 0, with a population of 8 patterns. For each generation, each individual in the population is calculated as the number of clauses that pattern satisfies. A maximum value of 5 means that the pattern (candidate test vector) matches each clause of the CNF expression and consequently it is a valid candidate. For example, the fitness for the pattern [0001] of Table I is 3, while the fitness for the pattern [1110] is 5. The string [1110] in particular permits the detection of missing devices in product terms that are not activated, i. e., product terms that are not compatible with the pattern under consideration.

The patterns {[1110], [0111]} generated on Table I at the end of generation 0 are

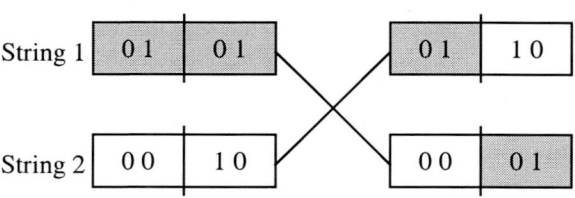

Fig. 4. The Crossover Operation

valid tests. Since genetic operators generate the pattern [1110], then it is a valid candidate test. This is necessary to assure propagation of the fault. The next step is to determine if there are product terms with a dH equal to 1 from the pattern (candidate test). The missing fault that can be detected by this pattern is the complement bit-line of the first input line of the product term m2.

Table 1. Generation 0

String	Population $X_1\ X_2\ X_3\ X_4$	Fitness $f(x_i)$	# of copies reproduced	Mate Pool (cross point site shown)	Mate #	Crossover $P_{cross}= 1.0$	Mutation $P_{mut}= 0.1$
(1)	0 0 0 1	3	0	0 1 | 0	13	0 1 1 ⓪	0 1 1 ①
(2)	0 1 0 1	4	1	1 ¦ 1 1	06	1 1 1 0	1 1 1 0
(3)	1 1 1 0	⑤	1	0 0 | 1	01	0 0 0 1	0 0 0 1
(4)	0 0 1 0	4	1	0 1 1 ¦ 0	05	0 1 1 0	0 1 1 0
(5)	0 1 1 0	4	1	1 1 1 ¦ 0	04	1 1 1 0	1 1 1 0
(6)	1 1 1 0	⑤	2	1 ¦ 1 1	02	1 ① 1 0	1 ⓪ 1 0
(7)	0 1 1 1	⑤	1	0 1 1 ¦	18	0 1 1 1	0 1 1 1
(8)	0 1 0 1	4	1	0 1 0 |	17	0 1 ⓪ 1	0 1 ① 0

Sum   35
Average 4.25
Max   5
Min   3

The following Lemma is necessary to the present discussion.

Lemma 1. A maxterm generated with the genetic operators with a dH equal to 1 from any product term is qualified to detect a missing device fault in that product.

The proof of this Lemma is supported by Procedure 1 and the CNF used for the pattern generation.

The string [0111] uses Lemma 1 to find product terms which are dissidents in one literal. These product terms are: m2, m3, and m5. Therefore, this pattern detects a growth fault in the product terms where they have the dissident bits (one Hamming distance away). The missing faults detected by the pattern [0111] (with fitness 5) with the aid of Lemma 1 are shown in Fig. 3. The faults detected are circled by a broken-line in their respective positions in the PLA.

The patterns {[0111], and [1110]} have a fitness of 5. The fittest members can be selected more than once. A bias roulette-wheel is used as the reproduction operator (see Fig. 5). To reproduce, we simply spin the weighted roulette wheel thus defined eight times, each with a size proportional to the pattern fitness. A probability is assigned to each pattern as follows:

$$p_i = \frac{f(x_i)}{Sum}$$

A cumulative probability is obtained for each pattern by adding up the fitness of the preceding population members:

$$c_i = \sum_{k}^{i} p_k, \quad i = 1, 2, \ldots, population\ Size$$

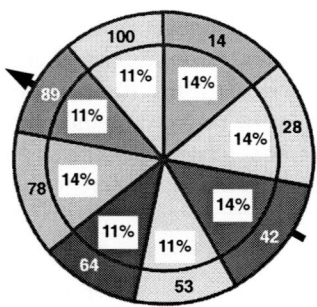

Fig. 5. Biased Roulette Wheel

In this way the selection of fittest members have proportionally more chances of being reproduced and patterns can be selected more than once. Table I shows how many times an individual is reproduced. Once the new population has been reproduced, strings are paired at random and recombined through crossover. The new individuals (patterns) will enter the new population in place of their parents. Crossover is applied with a frequency Pcross = 1.0.

After crossover, mutation is applied to the population members with a frequency Pmut = 0.1. It is interesting to note that after these genetic operators are applied in each generation the population average fitness continues to improve until the population becomes little differentiated and the fitness levels-off.

The final growth test set for the PLA under consideration were found after the second generation. The test vectors are: {[0011], [0111], [1100], [1110], [1111]}.

5 Conclusions

This article describes the use of genetic operator to the CNF-satisfiability problem for testing growth faults in PLAs. Existing methods tend to be computationally expensive.

The drawback of these methods is the backtracking that could occur when the test is chosen and fails to propagate. Our proposed algorithm overcomes this problem to generate good solutions efficiently. The CNF constraint satisfaction problem has several advantages over other approaches used for testing PLAs. It eliminates the possibility of intersecting a redundant growth term with a candidate test vector. Therefore, backtracking is not needed. Also, a minimal test set is guaranteed.

References

1. P. Bose, "A Novel Technique for Efficient Implementation of a Classical Logic/Fault Simulation Problem," IEEE Trans. on Computers, Vol. 37, pp.1569-1577, December 1984.

2. A.Cruz and R. Reilova, "A Hardware Performance Analysis for a CAD Tool for PLA Testing, " 39th Midwest Symposium on Circuits and Systems, 1997.
3. A. Cruz and S. Mukherjee, "PLAGA: A Highly Parallelizable Genetic Algorithm for Programmable Logic Arrays Test Pattern Generation," Congress on Evolutionary Computation, Vol. 2, July 6-9, 1999.
4. R. S. Wei and Sangiovanni-Vicentelli, "PLAtypus: A PLA Test Generation Tool." Trans. on Computer Aided Design, Vol. CAD-5, October 1986.
5. P. Bose, "Generation of Minimal and Near-Minimal Test Set for Programmable Logic Arrays," International Conference on Computers, December 1984.
6. M. Robinson and Rajski, "An Algorithm Branch and Bound Method For PLA Test Pattern Generation," International Test Conference, pp. 66-74, 1988.
7. J. E. Smith, "Detection of Faults in Programmable Logic Arrays," IEEE Transactions on Computers, Vol. C-28, No. 11, November 1979.
8. Hua and et al., "Built-in Tests for VLSI Finite State Machines," Dig. of Papers 14th Int'l Conf. on Fault Tolerant Computing, June 1984.
9. M. Abramoci and et al., "Digital Systems Testing and Testable Design," 41 Madison Avenue, New York, NY 10010: Computer Science Press, 1990.
10. V. K. Agarwal, "Multiple Fault Detection in Programmable Logic Arrays," IEEE Trans. on Computers, Vol. C-28, pp. 518-522, June 1980.
11. M. Mano and R. Kime, "Logic and Computer Design," 2nd Ed., Prentice Hall, 2000.
12. Z. Michalewicz, "Genetic Algorithms + Data Structures = Evolution Programs," Third, Revised Edition, Springer Verlag, 1996.
13. G. Lugger and W. A. Stubblefield, Artificial Intelligence: Structures and Strategies for Complex Problem Solving," 3rd Ed., Addison-Wesley, 1998.
14. G. Goldberg, "Genetic Algorithms in Search, Optimization and Machine Learning," Addison Wesley, Reading, MA, 198

Evaluating Migration Strategies for an Evolutionary Algorithm Based on the Constraint-Graph that Solves CSP

Arturo Núñez* and María-Cristina Riff**

Computer Science Department
Universidad Técnica Federico Santa María.
Valparaíso, Chile
{anunez,mcriff}@inf.utfsm.cl

Abstract. Constraint satisfaction problems (CSPs) occur widely in artificial intelligence. In the last twenty years, many algorithms and heuristics were developed to solve CSP. Recently, a constraint-graph based evolutionary algorithm was proposed to solve CSP, [17]. It shown that it is advantageous to take into account the knowledge of the constraint network to design genetic operators. On the other hand, recent publications indicate that parallel genetic algorithms (PGA's) with isolated evolving subpopulations (that exchange individuals from time to time) may offer advantages over sequential approaches, [1]. In this paper we examine the gain of the performance obtained using multiple populations - that evolve in parallel - of the constraint-graph based evolutionary algorithm with a migration policy. We show that a multiple populations approach outperforms a single population implementation when applying it to the 3-coloring problem.

1 Introduction

Constraint satisfaction problems (CSPs) occur widely in artificial intelligence. They involve finding values for problem variables subject to constraints on which combinations are acceptable. For simplicity we restrict our attention here to binary CSPs, where the constraints involve two variables. Binary constraints are binary relations. If a variable i has a domain of potential values D_i and a variable j has a domain of potential values D_j, the constraint on i and j, R_{ij}, is a subset of the Cartesian product of D_i and D_j. If the pair of values a for i and b for j is acceptable to the constraint R_{ij} between i and j, we will call the values consistent (with respect to R_{ij}). The entity involving the variables, the domains, and the constraints, is called constraint network. In the last twenty

---

* Partially supported by the Research Department Grant at University Santa María, Chile
** Supported by the Computer Science Department LIP6 at the Université Pierre et Marie Curie, France. e-mail:Maria-Cristina.Riff@lip6.fr

years, many algorithms and heuristics were developed to find a solution in constraint network [6], [10], [9]. Following these trend from the constraint research community in the evolutionary computation community some approaches was also proposed to tackle CSP with success [7], [8], [12], [18] in particular [16], [17] proposed an evolutionary algorithm based on the constraint network to solve CSP. Our motivations to present a parallel version for CSP are threefold. In our knowledge, all published evolutionary algorithms that address the constraint satisfaction problems are sequential approaches, i.e., one (or two in the case of co-evolution [14]) population evolves by means of genetic operators. However, recent publications indicate that parallel genetic algorithms (PGA's) with isolated evolving subpopulations (that exchange individuals from time to time) may offer advantages over sequential approaches, [1]. The contributions of this paper are:

- comparisons of the performance of the multiple populations evolving in parallel with previous sequential strategies.
- for this specific algorithm, an investigation on the influence of various parallelization parameters on its performance.

Throughout this work, we will use the term MpGA to describe a genetic algorithm with multiple populations (population structures) evolving in parallel. Accordingly, "sequential genetic algorithm" indicates a genetic algorithm with a single population. This usage is consistent with many previous papers. However, it is important to note that "parallel" and "sequential" refer to population structures, not the hardware on which the algorithms are implemented. In particular, the MpGA could be simulated on a single processor platform (as any discrete parallel process can) and the sequential genetic algorithm could be executed on a multiprocessor platform.

2 Problem Formulation

The problem at hand is that of constraint satisfaction problem (CSP) defined in the sense of Mackworth [11], which can be stated briefly as follows: We are given a set of variables, a domain of possible values for each variable, and a conjunction of constraints. Each constraint is a relation defined over a subset of the variables, limiting the combination of values that the variables in this subset can take. The goal is to find a consistent assignment of values to the variables so that all the constraints are satisfied simultaneously.

CSP's are, in general, NP-complete and some are NP-hard [4]. Thus, a general algorithm designed to solve any CSP will necessarily require exponential time in problem size in the worst case. Resulting from these considerations, three objectives are used in this work to asses the quality of the solution: A fitness function that takes into account the connection degree in the constraint network, a dynamic adaptation of the genetic operators, and a parallel migration between populations.

2.1 Notions on CSP

A *Constraint Satisfaction Problem* (CSP) is composed of a set of *variables* $V = \{X_1, \ldots, X_n\}$, their related *domains* $D_1, \ldots, D_n$ and a set θ containing η *constraints* on these variables. The domain of a variable is a set of values to which the variable may be instantiated. The domain sizes are $m_1, \ldots, m_n$, respectively, and we let **m** denote the maximum of the m_i. Each variable X_j is *relevant* (in the next we denote "being relevant for" by $\triangleright$), to a subset of constraints $C_{j_1}, \ldots, C_{j_k}$ where $\{j_1, \ldots, j_k\}$ is some subsequence of $\{1, 2, \ldots, \eta\}$. A constraint which has exactly one relevant variable is called a *unary constraint*. Similarly, a *binary constraint* has exactly two relevant variables. A binary CSP is associated with a constraint graph, where nodes represent variables and arcs represent constraints. If two values assigned to variables that share a constraint are not among the acceptable value-pairs of that constraint, this is an *inconsistency* or constraint violation.

Definition 2.1. *(Constraint Matrix)*
A Constraint Matrix **R** *is a $\eta \times n$ rectangular array, such that:*

$$\mathbf{R}_{\alpha j} = \mathbf{R}[\alpha, j] = \begin{cases} 1 & \text{if variable } X_j \triangleright C_\alpha \\ 0 & \text{otherwise} \end{cases}$$

Definition 2.2. *(Instantiation)*
An Instantiation **I** *is a mapping from a n-tuple of variables* $(X_1, \ldots, X_n) \to D_1 \times \ldots \times D_n$, *such that it assigns a value from its domain to each variable in* V.

Definition 2.3. *(Constraint Arity)*
We define the Constraint Arity *for a constraint* C_α, a_α, *as the number of relevant variables for* C_α.

Definition 2.4. *(Partial Instantiation)*
Given $V_p \subseteq V$, *a Partial Instantiation* $\mathbf{I_p}$ *is a mapping from a j-tuple of variables* $(X_{p_1}, \ldots, X_{p_j}) \to D_{p_1} \times \ldots \times D_{p_j}$, *such that it assigns a value from its domain to each variable in* V_p.
Note: For a given $\mathbf{I_p}$ *we will talk about satisfaction of* C_α *iff all of their relevant variables are instantiated.*

A solution to the CSP consists of an instantiation of all the variables which does not violate any constraint.

3 Network-Based Evolutionary Algorithm

The algorithm uses a non-binary genetic representation. The initial population is generated randomly. The variable values are selected from their domains with a uniform probability distribution. The selection algorithm is biased to the better evaluated individuals.

3.1 Fitness Function

In [15] we propose a fitness function specifically defined for CSP, which we describe briefly in the next two definitions.

Definition 3.1. *(Error-evaluation)*
For a binary CSP with a constraint matrix **R**, *an instantiation* **I**, *and a binary non-satisfied constraint* C_α *which has* X_k *and* X_l *as relevant variables (it has just two, exactly these two), we define the Error-evaluation* $\mathbf{e}(C_\alpha, I)$ *by:*

$\mathbf{e}(C_\alpha, I) = a_\alpha +$ *(Propagation Effect* X_k *and* X_l*)*
where Propagation Effect X_k *and* X_l *in a binary constraint network, is defined as the number of constraints* C_β, $\beta = 1, \ldots, \eta, \beta \neq \alpha$ *that have either* X_k *or* X_l *as relevant variables.*

Remark 3.1. If C_α is satisfied then $\mathbf{e}(C_\alpha, I)$ is equal to zero

The fitness function is the sum of the *Error-evaluations* (equation 3.1) of all constraints in the CSP, that is:

Definition 3.2. *(Fitness Function)*
For a binary CSP with constraint matrix **R** *and an instantiation* **I**, *and Error-evaluation* $\mathbf{e}(C_\alpha, I)$ *for each constraint* $C_\alpha, (\alpha = 1, \ldots, \eta)$, *the Fitness Function* $\mathbf{Z}(\mathbf{I})$ *is:*

$$\mathbf{Z}(\mathbf{I}) = \sum_{\alpha=1}^{\eta} \mathbf{e}(C_\alpha, I) \qquad (1)$$

The goal of the search is to minimize $\mathbf{Z}(\mathbf{I})$, which equals to zero when all constraints are satisfied.

3.2 Operator: Constraint Dynamic Adapting Crossover

Constraint Dynamic Adapting Crossover uses the idea that there are not fixed points to make crossover. It makes a crossover between two randomly selected individuals to create a new one. The child inherits its variable values using a greedy procedure, which analyzes each constraint (arc) according to a dynamic priority. The constraint dynamic priority not only takes into account the network structure, but also the current values of the parents. The priority is constructed using the next procedure: First, it identifies the *number of violations*, **nv**, that means, between both parents selected, how many are violating the current constraint. Second, it classifies the constraints in one of the following three categories : 0, 1 or 2 number of violations. Finally, within each category (0,1 or 2 number of violations), the constraints are ordered according to their contribution to the fitness function. To make crossover the operator uses two *partial fitness functions*. The first one is the *partial crossover fitness function*, **cff**, which allows us to guide the selection of a combination of variable values by constraint. The second one is the *partial mutation fitness function* **mff** for choosing a new variable value. The whole process is introduced, with details, in [17].

4 Model and Migration Policy

The Migration Model used in this work is shown in figure 1. The model is made up of a master node and i nodes. Each node has a population that evolves independently using the algorithm described in the previous section. The master goal is to send the initial parameters to each node (population size, seed, mutation and crossover probabilities). Once the nodes receive the parameters each node is ready to begin its evolution.

4.1 Migration Policy

We define a parameter called "migration rate", it specifies the number of iterations required before sending the best individuals to the neighboring node. This model also accepts another interesting parameter, this is the number of individuals migrating from each node. In the figure 1 dot lines shown the migration policy, that is:

- each node sends to its neighboring node its better individuals found until now
- each node receives the best individuals from its neighboring node, incorporating them to its population

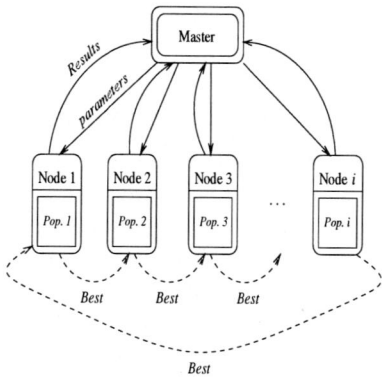

Fig. 1. Migration Model

Therefore our model needs the following parameters:

- i: number of nodes
- Mr : Migration Rate
- MtE: Members to Exchange
- $Popsize$: population size of each node

5 Tests

The aim of the experiments was to investigate the effect to incorporate multiple populations to the constraint-graph based evolutionary algorithm, and to compare it with the sequential approach. The algorithm has been tested by running experiments on randomly generated 3-coloring graphs, subject to the constraint that adjacent nodes must be colored differently. We used the Joe Culberson library, [5] to generate the random graphs. We have tested the algorithms for 3-coloring problems with solution, with a connectivity between [4.1..5.9]. For each connectivity we have generated 10000 random 3-coloring graph problems. In order to discard the "easy problems" we have applied DSATUR [3] to solve them. Thus, we have selected the problems not solved by DSATUR. DSATUR is one of the best algorithms to solve this kind of problems. The number of problems selected was 300 for each connectivity. It is important to remark that it is easy to find problems not solved by DSATUR in the hard zone [4], is not the case with others connectivities.

5.1 Hardware

The hardware platform for the experiments was a PC Pentium III-500 Mhz with 128 MB RAM under LINUX. For parallel support we have used PVM, [2]. PVM allows to use any computer as a virtual parallel machine with message-passing model. The algorithm was implemented in C using the PVM libraries.

5.2 Results

The single population algorithm and the multiple populations algorithm use the same parameters found for the algorithm introduced in [16], that is:

- Mutation probability = 0.2
- Crossover probability = 0.9

The parameters of the migration model are (i:3, Mr:50, MtE:1, $Popsize$:20)

Figure 2 and 3 shown the results obtained. The multiple populations algorithm was able to solve more than 88 % of the problems selected, even in the hard zone. Thus it works better than the sequential approach. The number of generations required was also reduced using the parallel approach, that is shown in the figure 3.

6 Discussion and Further Issues

We have obtained better results applying a model with multiple populations using the same sequential algorithm. Nevertheless, in order of being exact in the interpretation of the results we must consider that our new model works with

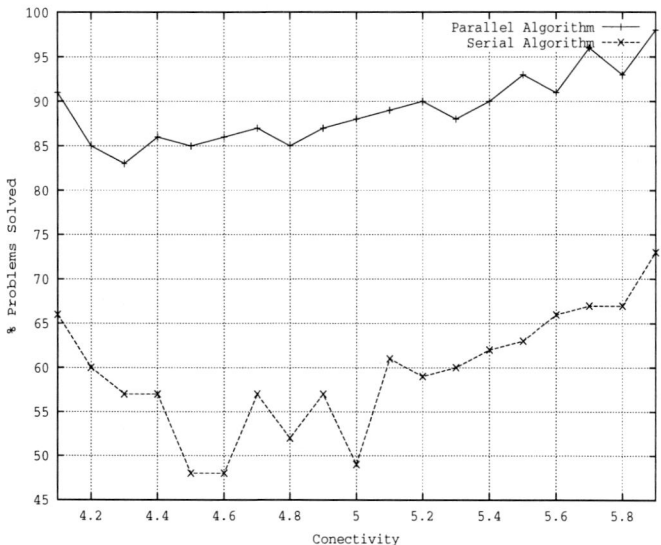

Fig. 2. Solved Problems by Multiple populations and single population

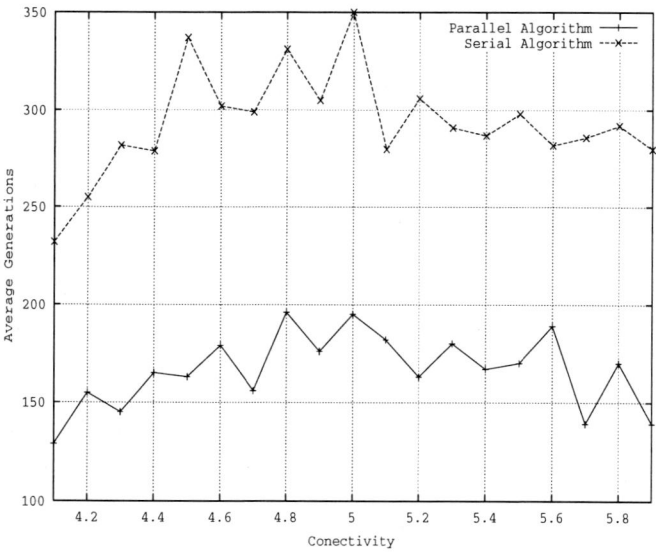

Fig. 3. Number of Generations by Multiple populatios and single population

three populations instead of one of the sequential model. For that, we define a measure of efficiency as:

$$\eta = \frac{\text{Generations serial model}}{\text{Generations parallel model} \times i} \times 100 \qquad (2)$$

Thus, we can conclude that the efficiency of the parallel model is approximately 60% better than the sequential one. It suggests that could be advisable to explore the behavior of the algorithm running in a parallel hardware platform.

7 Conclusion

A model based in multiple populations improves the performance of the graph-based evolutionary algorithm that solves CSP. Our research allows us to conclude that using an evolutionary algorithm with migration policy we are able to solve around 85% of the problems that are in the hard zone. The results suggest that our technique is a good option for solving CSPs.

There is a variety of ways in which the techniques presented here can be extended. The principal advantage of our method is that it is general, i.e., the approach is not related to a particular problem. Now our research is directed towards selecting parameters and testing in other hardware platforms.

Acknowledgments. We wish to gratefully acknowledge the discussions with Dr. Xavier Bonnaire (LIP6).

References

1. Adamis, Review of parallel genetic algorithms, Dept. Elect. Comp. Eng, Aristitele Univ. Thessaloniki, Greece, Tech. Rep. 1994.
2. A. Beguelin, J. J. Dongarra, G. A. Geist, R. Manchek, and V. S. Sunderam, A Users' Guide to PVM Parallel Virtual Machine, Oak Ridge National Laboratory, ORNL/TM-12187, September, 1994
3. Brelaz, New methods to color vertices of a graph. Communications of the ACM, 22,pp. 251-256, 1979.
4. Cheeseman P.,Kanefsky B., Taylor W., Where the Really Hard Problems Are. Proceedings of IJCAI-91, pp. 163-169, 1991
5. Culberson, J. http://web.cs.ualberta.ca/ joe/.
6. Dechter R., Enhancement schemes for constraint processing: backjumping, learning, and cutset decomposition. Artificial Intelligence 41, pp. 273-312, 1990.
7. G. Dozier, J. Bowen, and Homaifar, Solving Constraint Satisfaction Problems Using Hybrid Evolutionary Search, IEEE Transactions on Evolutionary Computation, Vol. 2, No. 1, 1998.
8. A.E. Eiben, J.I. van Hemert, E. Marchiori, A.G. Steenbeek. Solving Binary Constraint Satisfaction Problems using Evolutionary Algorithms with an Adaptive Fitness Function. Fifth International Conference on Parallel Problem Solving from Nature (PPSN-V), LNCS 1498, pp. 196-205, 1998.
9. Freuder E., The Many Paths to Satisfaction. Constraint Processing, Ed. Manfred Meyer, CAPringer-Verlag LNCS 923, pp. 103-119, 1995.
10. Kumar. Algorithms for constraint satisfaction problems:a survey. AI Magazine, 13(1):32-44, 1992.
11. Mackworth A.K., Consistency in network of relations. Artificial Intelligence, 8:99-118, 1977.

12. Marchiori E., Combining Constraint Processing and Genetic Algorithms for Constraint Satisfaction Problems. 7th International Conference on Genetic Algorithms (ICGA97), 1997.
13. Minton S., Integrating heuristics for constraint satisfaction problems: A case study. Proceedings of the Eleventh National Conference on Artificial Intelligence, 1993.
14. J. Paredis, Coevolutionary Algorithms, The Handbook of Evolutionary Computation, 1st supplement, BSck, T., Fogel, D., Michalewicz, Z. (eds.), Oxford University Press.
15. Riff M.-C., From Quasi-solutions to Solution: An Evolutionary Algorithm to Solve CSP. Constraint Processing (CP96), Ed. Eugene Freuder, pp. 367-381, 1996.
16. Riff M.-C., Evolutionary Search guided by the Constraint Network to solve CSP. Proc. of the Fourth IEEE Conf on Evolutionary Computation, Indianapolis, pp. 337-342, 1997.
17. Riff M.-C., A network-based adaptive evolutionary algorithm for CSP, In the book "Metaheuristics: Advances and Trends in Local Search Paradigms for Optimisation", Kluwer Academic Publisher, Chapter 22, pp. 325-339, 1998.
18. Tsang, E.P.K., Wang, C.J., Davenport, A., Voudouris, C., Lau,T.L., A family of stochastic methods for constraint satisfaction and optimization, The First International Conference on The Practical Application of Constraint Technologies and Logic Programming (PACLP), London, pp. 359-383, 1999

Relative Robustness: An Empirical Investigation of Behaviour Based and Plan Based Paradigms as Environmental Conditions Change

Jennifer Kashmirian and Lin Padgham

Dept. of Computer Science,
RMIT University, Melbourne, Australia.
linpa@cs.rmit.edu.au

Abstract. This paper compares a behaviour based architecture, and a plan based architecture for agents in multi-agent systems, with respect to the issue of robustness. The type of robustness investigated is stability of the systems as two aspects - unpredictability and rate of perception compared to speed of the environment - are modified. The comparison is done using a simulation scenario which was designed to be constrained, but to capture important qualities of the real world. The scenario was also chosen to have characteristics that could be favourable to both behaviour based and plan based paradigms. An analysis of the data collected for the two approaches provides strongly suggestive evidence that the plan-based system is more robust than the behaviour based one in some ways. There is no indication of better robustness in the behaviour based system with respect to the aspects of robustness investigated.

1 Introduction

Two fairly well established architectural paradigms in agent systems are the behaviour based paradigm and the plan-based paradigm. Both paradigms recognise the need for and incorporate reactivity as a fundamental quality of agent systems in dynamic environments, although they approach this in different ways. There are often claims from behaviour based proponents that behaviour based systems are more robust (e.g. [JF92,Bro86]), while plan-based proponents often argue that plans are necessary for addressing complex applications.

This work explores specifically a particular aspect of robustness, namely the stability of the agent or program behaviour as characteristics of the environment change. The particular environmental characteristics that we have manipulated are the predictability of the world and the speed of change in the world relative to the ability of the agent to perceive and act on those changes. Although the results are not entirely conclusive[1] the tendency is that plan-based systems appear to be more stable under environmental change than behaviour based systems.

---

[1] This was due to lack of sufficient data and hardware problems which made it impossible to collect further data.

The behaviour-based approach relies on low level parallel behaviours which react to sensed information about the current world situation without any modelling of or reasoning about either the world or the agent's actions in the world. Intelligent behaviour is seen to *emerge* from the combination of simple behaviours within a complex environment. There are several forms of behaviour based architectures of which Brooks' subsumption architecture [Bro86] is the most well known. Activation nets (e.g. [Mae95]) are another approach based on low level parallel behaviours.

BDI (*belief, desire, intention*) architectures are plan based systems which rely on a library of outline plans which indicate how to achieve particular goals in various situations. At execution time the details of the plan are filled out by the agent using sub-plans based on the actual state of the world when it is time to execute the relevant sub-plan. This approach makes practical reasoning tractable as the scope of deliberation is limited to choosing between competing matching plans [BIP88]. It also allows the agent to be reactive to environmental change by suspending or aborting execution of a plan in favour of a more relevant sub-plan or a plan reacting to a more important event. These systems are often referred to as *reactive planners* [Fir87,AC90]. Reaction is to both external events conveyed via sensors or to events based on agent modelling of the world and the expected results of the agent's own actions.

There has been a limited amount of work attempting to scientifically compare the applicability of these two paradigms to particular problem types. Drogoul [Dro95] has done some work in this direction, attempting to apply a behaviour based approach to chess and also looking at certain situations in Pengi[2]. In general Pengi is a game which is so dynamic and unpredictable that little or no planning is possible [AC87]. However, Drogoul identified some situations where it would appear that planning could be advantageous. Drogoul's findings were that although a modicum of success was achieved in behaviour based chess it was insufficient to allow successful competition against a deliberative chess program such as GNU chess. However, in the Pengi situation it was possible to simply provide agents with more complicated behaviours inorder to address the situations that it seemed would benefit from planning. From this work it appears that plan based systems are better for achieving intelligent behaviour in situations where strategy over time is critical. However, for applications which may be complex in terms of multiple competing goals and the need for reactivity, but do not allow for strategy because of the totally unpredictable nature of the environment, intelligent behaviour can be successfully obtained using a purely behaviour based approach.

Although robustness is one of the often claimed advantages of behaviour based systems, to our knowledge there has not been any work that attempts to define some aspect of robustness and actually test this claim. Robustness is essentially the ability of the system to continue functioning well over a wide

---

[2] Pengi is based on the arcade game Pengo where penguins must try to collect as many diamonds as possible while avoiding being being crushed by moving ice blocks or stung by roving bees.

range of environmental conditions. In this work we have considered two ways in which environmental conditions may change. The first is that the speed at which things happen in the agent's world may change relative to the speed of the agent itself (or its ability to perceive the change). One example of this is that a robot in Robocup[3] with a camera that perceives a fixed rate of frames per second may well perform differently depending on the speed of the other agents in the environment. If an agent's environment is made up of other systems then an upgrade of some of those systems may result in an effective speed up of the environment. Similarly an upgrade in the agent system hardware may result in the environment being perceived to be slower, relative to the speed of the agent system.

Another way in which we consider the environment changing is in its level of predictability. When a system is built it is often tailored in some way to the patterns of the environment. As times change these environmental patterns may change, resulting in the environment being less predictable than it was originally. Ideally one would like the system to learn new patterns, but an ability to degrade gracefully would also be useful. An argument is often made that because there is not the same level of explicit tailoring in behaviour based systems as there is in plan-based systems, they should be able to better handle this kind of environmental change.

In this paper we describe a simulation scenario where we modify two parameters of the scenario representing these aspects and measure the changes in the performance of the two types of agent systems. Because we are interested in robustness, i.e. stability under a wide range of conditions, we are not interested so much in whether one system is better than the other in any particular situation, but rather whether their rates of change (as the environment changes) differ in any significant way.

2 Description of project

We wished to establish a scenario that seemed suitable for either approach, and where parameters could be modified to simulate changing environmental conditions. We also wanted a scenario which could be viewed graphically to allow for the possibility of qualitative evaluation as well as quantitative.

The scenario developed was one where two sheep-dog agents needed to herd a flock of sheep through three different gates. The sheep mostly behaved in a predictable flocking manner, but would occasionally exhibit random breakaway behaviour at a frequency that was parameterised. Relative success of the agents was measured as time taken to achieve the goal of herding the sheep through all three gates. Time was measured in terms of world cycles.

We modelled levels of unpredictability by modifying how often a sheep would break away from the flock and move in a random direction, rather than with the flock. We modelled relative speed of the environment by having the dogs (or

---

[3] Robocup is an international forum where robots compete at playing soccer.

sheep) move less than every world cycle. For example if the sheep moved only every second world cycle, then the world would be changing less rapidly from the point of view of the dogs. If the dogs received input and had a move only every second world-cycle, then the environment would be changing more rapidly from the dog's point of view.

The base platform used for the simulation was the PAC system [PT97]. PAC attempts to provide an environment where scenarios can quickly and easily be built up, varying aspects of agent personality (or emotions) and agent cognition (plans and beliefs). The purpose of the PAC system is to allow experimentation with different combinations of agents in different worlds.

PAC consists of four modules - cognitive, emotional, behavioural and system management. The cognitive module manages plan-based agent behaviour using dMars, a descendant of PRS [GI89] and was used for implementing the plan-based dog agents. The behavioural module manages the graphical models for each agent plus their behavioural functions (e.g. walk forward, turn, etc.). The system management module integrates the other modules, manages scenario simulation and interfaces to the user interface for the interactive system. The behaviour based dog agents were implemented in C++ as a separate piece of code that was integrated with the system management module. The emotion module was not used in this work.

The sheep in the environment were implemented as simple agents in C++ which used a flocking algorithm [Rey87] to move. While this was visually not quite realistic for sheep flocking it provided appropriate predictable behaviour. Under normal circumstances sheep action was determined by a blending of four basic behaviours: obstacle avoidance, velocity matching, flock centering and avoidance of the dogs. At each world cycle a movement vector was calculated and then multiplied by priorities of 16, 4, 9 and 14 respectively.[4] The resulting vectors were then summed to provide a final movement vector. When breakaways were operational a sheep would make a totally random move some number of times every thousand world cycles at randomly determined intervals.

The plan based agents had plans to look for gates and for sheep, to move towards the flock and to return wayward sheep to the flock. They also had plans that enabled them to co-ordinate their behaviour to position themselves behind the sheep and then move towards the gate. The behaviour based agents had a number of behaviours such as moving towards the target, avoiding obstacles and moving away from the other dog if too close, which were then combined using behaviour blending in a similar way to the sheep. Some of the behaviours were disabled if other behaviours matched strongly. Both agent programs were developed to a point where they appeared to be herding the sheep successfully.

In the following sections we describe the specific experiments that were run and the results obtained.

---

[4] These priorities were determined by trial and error.

3 Experiments

One set of experiments varied the unpredictability in the environment by running scenarios with four different rates of breakaway of sheep. The rates used were 0, 4, 20 and 50 breakaways per 1000 world cycles.

The other set of experiments varied the rate at which the dogs could sense and act compared with the rate at which the sheep were moving. The ratios of dog action to sheep action used were 1:4; 1:3; 1:2; 1:1; 2:1; 3:1; 4:1. These experiments were run with a predictable environment (no breakaways) and with a moderately unpredictable environment (20 breakaways per 1000 cycles). Due to machine problems we were unable to obtain data for the behaviour based agents at ratios 1:4, 1:3 and 3:1.

For each set of parameters we ran thirty scenarios. Each scenario started with a random movement by the first sheep, which ensured that each scenario was unique. The dogs could herd the sheep through the gates sequentially in any order. When all sheep had been herded through a particular gate, the gate would shut and the number of moves required by the dogs to herd the sheep through that gate was recorded. When all three gates had shut the scenario was concluded. We recorded three basic pieces of data for each scenario:

- **Total World Cycles (TWC)**: the number of world cycles taken for the scenario from start to finish.
- **Goal World Cycles (GWC)**: the number of world cycles where the sheep were actively driving the sheep towards a gate, as opposed to rounding up breakaway sheep.
- **Breakaway World Cycles (BWC)**: the number of world cycles where one or more sheep were separated from the flock.

We then used this data to calculate what we refer to as an *average performance measure* and an *average efficiency measure*.

The *average performance measure* was calculated simply as the number of total world cycles averaged over the 30 runs. This gave an average "time" taken by the agent to achieve its task - the less time taken the better the performance.

The *average efficiency measure* was calculated using the formula
$\frac{100*GWC}{BWC+GWC}$
to obtain a score for each run. These scores were then averaged over the thirty runs. This was intended to capture how well the dogs were maintaining control over the sheep. The higher this number the better the dogs were effectively maintaining control over their world.

As well as collecting statistics we recorded observations for a number of randomly selected test runs. When these runs were short they were observed from start to finish. Longer runs (some over six hours duration) were observed at random intervals.

4 Results

In observing the test runs the plan-based dogs appeared to have a controlling behaviour, while the behaviour based dogs appeared opportunistic, reacting to opportunities that arose, rather than actively using a strategy to direct the sheep. As the breakaway frequency increased both sets of agents appeared to lose control of the sheep.

We show our analysis of the statistical data in terms of the two basic questions we are addressing: How adaptable are the two different agent types to an increasingly unpredictable world? How adaptable are the agents to changes in speed of the environment relative to their own speed?

4.1 Adaptability to unpredictability

As the unpredictability of the environment increased the performance of both plan-based and behaviour based agents decreased exponentially. Figure 1 shows the increasing number of world cycles taken to complete the task for both types of agents as numbers of breakaways increases. This figure also shows the performance of the control group of sheep without any dogs herding them. Clearly both experimental agent types are achieving better results than the sheep in the control group with no dog agents ($p < 0.005$). Both agent types appear to have almost identical deterioration patterns under decreasing predictability.

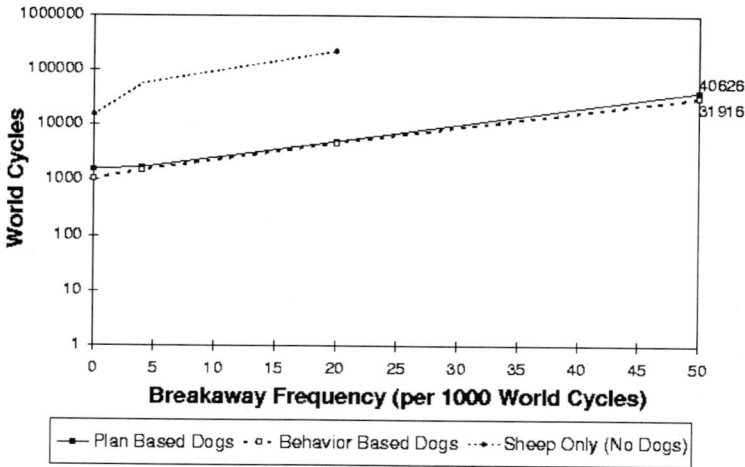

Fig. 1. Performance under varying unpredictability.

Looking at the initial part of the curve (0 - 4 breakaways per 1000 world cycles) it seems that both the agent types are also deteriorating less quickly than the control group, as seen in the slope of the line. However, this is less interesting than the fact that each of the agent types appear to be following a very similar pattern.

Using the measure of efficiency, rather than performance (i.e. what percentage of agent time was spent driving the sheep towards the gate, as opposed to doing the subsidiary task of managing the unpredictable breakaways), we see a linear, rather than exponential decline as the world becomes more unpredictable. This is shown in figure 2. In this case at both a breakaway rate of 4:1000 and at 20:1000 the behaviour based dogs are significantly less efficient than the plan-based dogs ($0.05 \leq p \geq 0.005$ for 4:1000 and $p \leq 0.005$ for 20:1000). At a breakaway rate of 50:1000 there is no longer any significant difference between the dog types, probably indicating that at this level of unpredictability chance is the overriding factor.

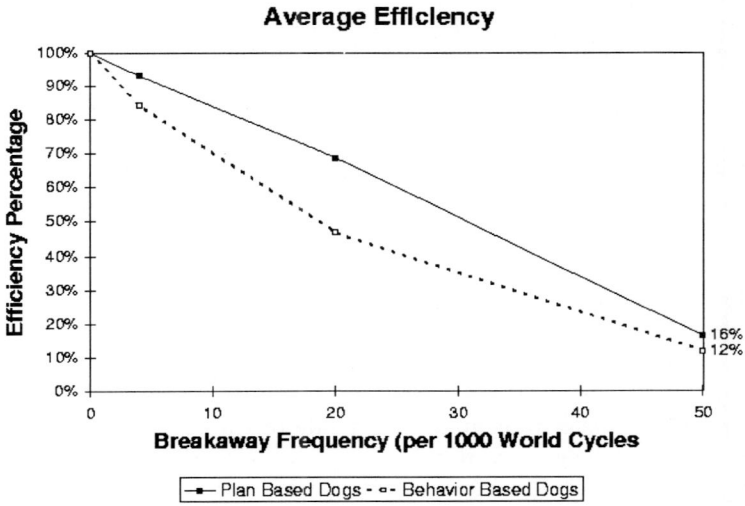

Fig. 2. Efficiency under varying unpredictability.

It seems odd that the difference in efficiency is not mirrored by a difference in performance. During observation we noticed that on a number of occasions the plan-based dogs would get caught in a circling behaviour as they tried to drive the sheep towards the gate. This was essentially a bug in the plan set which was not discovered until part way through the experimentation and we were unable to rerun the experiments with this bug fixed. This needless circling drives up the number of total world cycles which would cause the performance measure to decrease while causing the efficiency measure (which is a ratio of cycles where the dogs are driving the sheep towards a gate to the total number of world cycles) to increase.

Consequently we can conclude that in this scenario the behaviour based agents are not more robust than the plan-based agents under decreasing predictability of the environment. However, it is not possible to ascertain whether or not the plan-based agents are more robust.

4.2 Adaptability to relative speed fluctuations

Figure 3 shows the performance results as the speed of the environment changes with respect to the perception speed of the dogs. The most striking difference is between the plan-based and behaviour based dogs as the speed of the world increases with respect to the perception speed of the agents (see LHS of figure 3). The behaviour based agents appear to deteriorate exponentially for both predictable (0 breakaways) and somewhat unpredictable (20:1000 breakaways). Interestingly the plan-based agents in the predictable environment actually improve significantly when the environment speeds up by a factor of four ($0.05 \geq p \leq 0.005$), remaining relatively stable at speed up rates of two and three. With the somewhat unpredictable environment the plan-based agents maintain their performance at a doubling of world speed, but deteriorate as the world speeds up by factors of three and four.

As the world slows down relative to the speed of the dogs (see RHS of figure 3, we see a much more stable pattern. The plan-based agents at both predictability levels maintain a relatively constant performance as the world becomes up to four times slower. There is a significant improvement ($p \leq 0.005$) in the performance of the behaviour based dogs as the moderately unpredictable world (20:1000 breakaway rate) becomes slower.

Ideally one would like to analyse the data using regression analysis to determine whether the change function for the plan based dogs is significantly different to that of the behaviour based dogs as the environment speeds up and slows down at the different predictability levels. However, this would require a minimum of five measurements on each side of the center line and so is not possible to do with the current data.

Looking at the efficiency data in figure 4 we see that in the moderately unpredictable environment (20:1000 breakaways) both types of agents deteriorate as the world gets both slower and faster. However, it appears that the plan-based agents deteriorate less rapidly than the behaviour based agents as the world becomes faster, while the opposite effect is apparent as the world becomes slower. Once again the lack of sufficient data points makes it impossible to do regression analysis which would be the most appropriate form of statistical analysis. However, by shifting the data by a fixed amount it is possible to obtain a common start point at normal world speed and then use a paired two sample t-test. This indicates that the behaviour based dogs are significantly better ($p = 0.05$) which would suggest that the deterioration functions of the two dog types are significantly different. Comparing with the LHS of the graph we see that if the trend of the behaviour based agents continued with further data points beyond

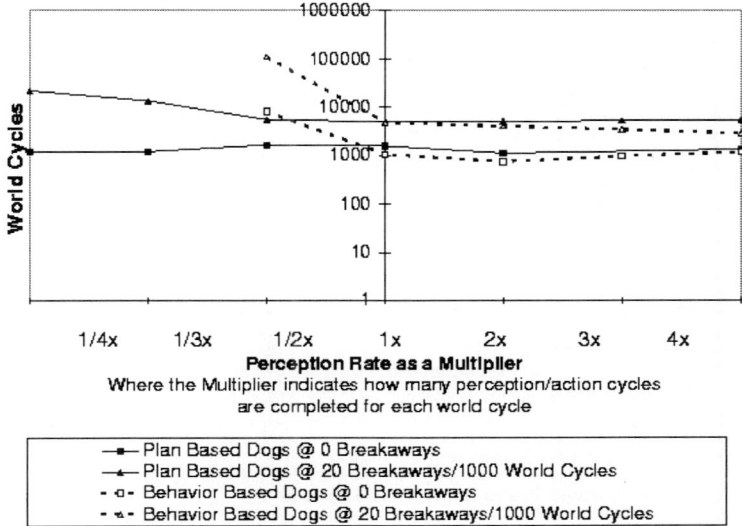

Fig. 3. Performance under varying rates of perception.

the two obtained, the difference would be greater than that on the right hand side as the lines diverge more quickly.

In a predictable world (0 breakaways) the plan-based agents appear to be stable until the world is speeded up by a factor of four, where they start to deteriorate. They are also stable as the world slows down. The behaviour based agents deteriorate as the world slows down by factors of two, three and four. The behaviour based agents in the speeded up world only have data for the speed up by two, and at this point they are stable.

In summary, there is a strong suggestion that as the world speeds up plan based agents are more stable and deteriorate less rapidly than behaviour based agents. As the world slows down the situation is less clear with respect to differences between the agent types and there appears to be far less effect than when the world speeds up. One interesting effect in a slowed down world is that if it is moderately unpredictable the behaviour based agents improve significantly when the world slows down by a factor of four.

5 Discussion and Conclusion

The results of this work do not lend support to the idea that behaviour based agents are more robust than plan-based agents if robust is defined in terms of

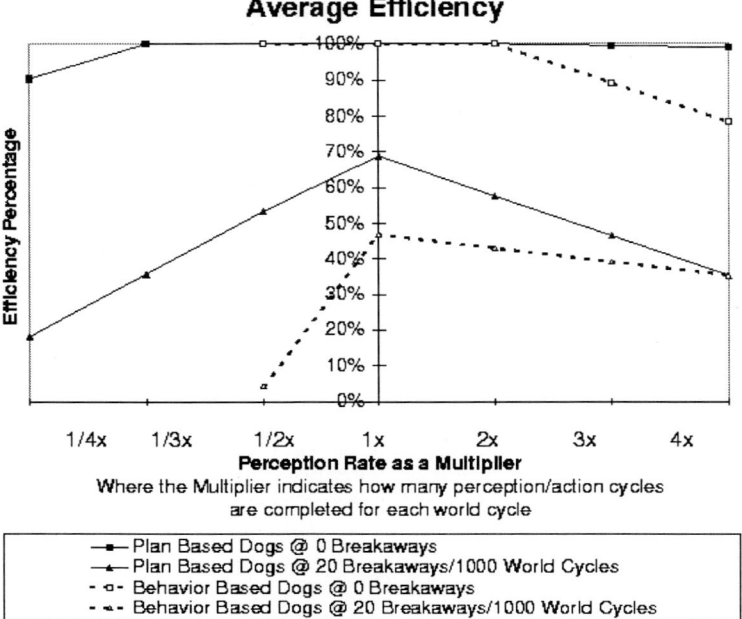

Fig. 4. Efficiency under varying rates of perception.

changes to the relative speed of the environment and to the unpredictability of the environment. In fact, the results suggest that plan-based agents may be more robust in this sense. The most surprising result is that as the world speeds up the behaviour-based agents appear to deteriorate rapidly while the plan-based agents remain stable or even improve slightly. Possibly this is related to the results obtained by Kinny [KG91] which showed that agents which were able to commit to a plan of action actually did better as the world became more dynamic, than agents that had no such ability.

It may also be the case that the co-ordination between the two agents that was required for successful herding of the sheep was more susceptible to environmental perturbations in the behaviour based program where it "emerged" due to combining of simple behaviours, than when it was more explicitly represented.

Questions about the relative robustness of different software paradigms as environments change are important to explore given that the digital world is changing extremely rapidly and that systems are increasingly likely to be interacting with other digital systems in a distributed manner.

Further work is needed to collect a larger number of data points in particular to allow regression analysis to determine whether the deterioration function differs significantly with different architectures. Work is also needed to explore these questions in an environment that is more closely related to application environments rather than a visual simulation.

Both systems exhibited similar degradation as unpredictability increased. However, it would be interesting to further explore this issue, perhaps identifying different sorts of unpredictability that may arise and whether the two approaches have different stability patterns with respect to differing types of unpredictability.

References

[AC87] Philip E. Agre and David Chapman. Pengi: An implementation of a theory of activity. In *Proceedings of AAAI-87*, pages 268–272, 1987.

[AC90] Philip E. Agre and David Chapman. What are plans for? *Robotics and Autonomous Systems*, 6:17–34, 1990.

[BIP88] Michael E. Bratman, David J. Israel, and Martha E. Pollack. Plans and resource bounded practical reasoning. *Computational Intelligence*, 4(4):349–355, 1988.

[Bro86] Rodney Brooks. A robust layered control system for a mobile robot. *IEEE Journal of Robotics and Automation*, 2(1):14–25, March 1986.

[Dro95] Alexis Drogoul. When ants play chess. In C. Castelfranchi and J.P. Müller, editors, *From Reaction to Cognition*, volume 957 of *LNAI*, pages 13–27. Springer-Verlag, Berlin-Heidelberg, 1995.

[Fir87] R. James Firby. An investigation into reactive planning in complex domains. In *Proceedings of AAAI-87*, pages 202–206, July 1987.

[GI89] M. Georgeff and F. Ingrand. Decision-making in an embedded reasoning system. In *Proceedings of the 10th International Conference on Artificial Intelligence*, pages 972–978, 1989.

[JF92] Alexis Drogoul Jaques Ferber. Using reactive multi-agent systems in simulation and problem solving. In *Distributed Artificial Intelligence: Theory and Praxis*, pages 53–80. ECSC-EEC-EAEC, Bruxelles et Luxembourg, 1992.

[KG91] David Kinny and Michael Georgeff. Commitment and effectiveness of situated agents. In *Proceedings of the 12th International Conference on Artificial Intelligence*, pages 82–88, Melbourne, Australia, 1991.

[Mae95] Pattie Maes. Modelling adaptive autonomous agents. In Christopher G. Langton, editor, *Artificial Life: An Overview*. MIT Press, Cambridge, 1995.

[PT97] Lin Padgham and Guy Taylor. Pac - personality and cognition: an interactive system for modelling agent scenarios. In *Proceedings of the 15th International Conference on Artificial Intelligence*, Tokyo, Japan, 1997.

[Rey87] Craig Reynolds. Flocks, herds and schools: A distributed behavioural model. In *Proceedings of SIGGRAPH*, pages 25–34, 1987.

A Heuristic for Domain Independent Planning and Its Use in an Enforced Hill-Climbing Algorithm

Jörg Hoffmann

Institute for Computer Science
Albert Ludwigs University
Am Flughafen 17
79110 Freiburg, Germany
hoffmann@informatik.uni-freiburg.de

Abstract. We present a new heuristic method to evaluate planning states, which is based on solving a relaxation of the planning problem. The solutions to the relaxed problem give a good estimate for the length of a real solution, and they can also be used to guide action selection during planning. Using these informations, we employ a search strategy that combines Hill-climbing with systematic search. The algorithm is complete on what we call *deadlock-free* domains. Though it does not guarantee the solution plans to be optimal, it does find close to optimal plans in most cases. Often, it solves the problems almost without any search at all. In particular, it outperforms all state-of-the-art planners on a large range of domains.

1 Introduction

The standard approach to obtain a heuristic is to relax the problem $\mathcal{P}$ at hand into some easier problem $\mathcal{P}'$. The optimal solution length to a situation in $\mathcal{P}'$ can then be used as an admissible estimate for the optimal solution length of the same situation in $\mathcal{P}$. An application of this idea to domain independent planning was first used in the HSP system [3]. The planning problem $\mathcal{P}$ is relaxed by simply ignoring the delete lists of all operators. However, computing the optimal solution length for a planning problem without delete lists is still NP-hard, as was first shown by Bylander [4]. Therefore, the HSP heuristic is only a rough estimate of the optimal relaxed solution length. In short, it is obtained by summing up the minimal distances of all atomic goals.

In this paper, we go one step further. We introduce a method that computes *some*, not necessarily optimal, solution to the relaxed problem. These solutions are helpful in two ways:

- their length provides an informative estimate for the difficulty of a situation;
- one can use them as a guidance for action selection.

The solution length estimates are used to control a local search strategy similar to Hill-climbing, which is combined with systematic breadth first search in order to escape local minima or plateaus. The guidance information is employed to cut down the branching factor during systematic search. The method shows good behavior over all domains that are commonly used in the planning community. In particular, we will see that it is complete on the class of problems we call *deadlock-free*. Performing local search, the method can not guarantee its solution plans to be optimal. In spite of this, it finds close to optimal plans in most cases. As a benefit from the severe restriction of its search space, it shows very competitive runtime behavior. For example, *logistics* problems are solved faster than by any other domain independent planning system known to the author at the time of writing.

2   Background

Throughout the paper, we consider simple STRIPS domains. We briefly review two standard notations. An *action o* has the form

$$o = \langle\, pre(o) \Rightarrow add(o), del(o)\,\rangle$$

where $pre(o)$, $add(o)$ and $del(o)$ are sets of ground facts. Plans P are sequences $P = \langle o_1, \ldots, o_n \rangle$ of actions, i.e., we consider only linear plans.

3   Heuristic

In this section, we introduce a method for heuristically evaluating planning states S. Basically, the method consists of two parts.

1. First, the *relaxed fixpoint* is built on S. This is a forward chaining process that determines in how many steps, at best, a fact can be reached from S, and with which actions.
2. Then, a *relaxed solution* is extracted from the fixpoint. This is a sequence of parallel action sets that achieves the goal from S, if their delete effects are ignored.

The first part corresponds directly to the heuristic method that is used in HSP [3]. The second part goes one step further: while in HSP, the heuristic is extracted as a side effect of the fixpoint, we invest some extra effort to find a relaxed plan, and use the plan to determine our heuristic value. The fixpoint process is depicted in Figure 1.

The algorithm can be seen as building a layered graph structure, where fact and action layers are interleaved in an alternating fashion. The process starts with the initial fact layer, which are the facts that are TRUE in S. Then, the first action layer comprises the actions whose preconditions are contained in S. The effects of these actions lead us to the second fact layer, which, in turn, determines

$F_0 := S$
$k := 0$
while $\mathcal{G} \not\subseteq F_k$ **do**
    $O_k := \{o \in \mathcal{O} \mid pre(o) \subseteq F_k\}$
    $F_{k+1} := F_k \cup \bigcup_{o \in O_k} add(o)$
    if $F_{k+1} = F_k$ **then**
        break
    endif
    $k := k+1$
endwhile
$max := k$

Fig. 1. Computing the relaxed fixpoint on a planning state S. $\mathcal{O}$ and $\mathcal{G}$ denote the action set and goal state of the problem at hand, respectively.

the next action layer and so on. The process terminates, and remembers the number max of the last layer, if all goals are reached or if the new fact layer is identical to the last one.

The crucial information that the fixpoint process gives us are the *levels* of all facts and actions. These are defined as the number of the first fact- or action layer they are members of.

$$\text{level}(f) := \begin{cases} min\{i \mid f \in F_i\} & \text{ex. } i : f \in F_i \\ \infty & \text{otherwise} \end{cases}$$

$$\text{level}(o) := \begin{cases} min\{i \mid o \in O_i\} & \text{ex. } i : o \in O_i \\ \infty & \text{otherwise} \end{cases}$$

We now show how to extract a relaxed plan from the fixpoint structure. This is done in a backward chaining manner, where we simply use any action with minimal level to make a goal TRUE. The exact algorithm is depicted in Figure 2. Note that we *do not need to search*, we can proceed right away to the initial state and are guaranteed to find a solution.

Before plan extraction starts, an array of goal sets G_i is initialized by inserting all goals with corresponding level. The mechanism then proceeds down from layer max to layer 1, and selects an action o for each goal g at the current layer i, incrementing the plan length counter h. No actions are selected for goals that are marked TRUE at the time being, as they are already added. The achiever o is required to have $\text{level}(o) = i - 1$. This is minimal as the goal g has level i, i.e., the first action that achieved g in the fixpoint came in at level $i - 1$. The preconditions of o are inserted as new goals into their corresponding goal sets. If the current layer is i, then the levels of o's preconditions are at most $i - 1$, so these new goals will be made TRUE later during the process.

for $i := 1, \ldots, max$ **do**
    $G_i := \{g \in \mathcal{G} \mid \text{level}(g) = i\}$
endfor
$h := 0$
for $i := max, \ldots, 1$ **do**
    for all $g \in G_i$, g not TRUE at i **do**
        select o with $g \in add(o)$ such that $\text{level}(o) = i - 1$
        $h := h + 1$
        for all $f \in pre(o)$, f not TRUE at $i - 1$ **do**
            $G_{\text{level}(f)} := G_{\text{level}(f)} \cup \{f\}$
        endfor
        for all $f \in add(o)$ **do**
            mark f as TRUE at $i - 1$ and i
        endfor
    endfor
endfor

Fig. 2. The algorithm that extracts a relaxed solution to a state S after the fixpoint has been built.

3.1 Goal Distance

To obtain the heuristic goal distance value $h(S)$ of a given planning state S, we now simply chain the two algorithms together. First, we perform the fixpoint computation from Figure 1. If the process terminates without reaching the goals, we set $h(S) := \infty$. Otherwise, we extract a relaxed plan, Figure 2, and use the plan length for evaluation, i.e., $h(S) := h$.

The overall structure of the relaxed planning process is quite similar to planning with planning graphs [1]. It amounts to a very special case, as no negative interactions at all occur between facts or actions in the relaxed problem.

3.2 Helpful Actions

We can also use the extracted plan to determine a set of actions that seem to be helpful in reaching the goal. To do this, we turn our look on the actions that are contained in the *first time step* of the relaxed solution, i.e., the actions that are selected at level 0. These are often the actions that are useful in the given situation. Let us see a simple example for that, taken from the *gripper* domain, as it was used in the 1998 AIPS planning systems competition. We do not repeat the exact definition of the domain here, as it is easily understood intuitively. There are two rooms, A and B, and a certain number of balls, which shall be moved from room A to room B. The planner changes rooms via the **move** operator, and controls two grippers which can **pick** or **drop** balls. Each gripper can only hold one ball at a time. We look at a small problem where 2 balls must be moved into room B. A relaxed solution to the initial state that our heuristic might extract is

< { **pick** ball1 A left,
   pick ball2 A left,
   move A B },
 { **drop** ball1 B left,
   drop ball2 B left } >

This is a parallel relaxed plan consisting of two time steps. Note that the **move** A B action is selected parallel to the **pick** actions, as the relaxed planner does not notice that it can not **pick** balls in room A anymore once it has moved into room B. In a similar fashion, both balls are picked with the left gripper. Nevertheless, two of the three actions in the first step are helpful in the given situation: both **pick** actions are starting actions of an optimal sequential solution. Thus, one might be tempted to define the set $H(S)$ of helpful actions as only those that are contained in the first time step of the relaxed plan. However, this is too restrictive in some cases. We therefore define our set $H(S)$ as follows.

$$H(S) := \{o \in O_0 \mid add(o) \cap G_1 \neq \emptyset\}$$

After plan extraction, O_0 contains the actions that are applicable in S, and G_1 contains the facts that were goals or subgoals at level 1. Thus, we consider as helpful those actions which add at least one fact that was a (sub)goal at the lowest time step of our relaxed solution.

4 Search

We now introduce a search algorithm that makes effective use of the heuristics we defined in the last section. The key observation that leads us to the method is the following. On some domains, like the *gripper* problems from the 1998 competition and Russel's *tyreworld*, it is sufficient to use our heuristic in a naive *Hill-climbing* strategy. In these problems, one can simply start in the initial state, pick, in each state, a best valued successor, and ends up with an optimal solution plan. This strategy is very efficient on the problems where it finds plans.

However, the naive method does *not* find plans on most problems. Usually, it runs into an infinite loop. To overcome this problem, one could employ standard Hill-climbing variations, like restarts, limited plateau moves, or a memory for repeated states. We use an *enforced* Hill-climbing method instead, see the definition in Figure 3.

The algorithm combines Hill-climbing with systematic breadth first search. Like standard Hill-climbing, it picks some successor of the current state at each stage of the search. Unlike in standard Hill-Climbing, this successor does not need to be a direct one, and, unlike in standard Hill-Climbing, we do not pick any best valued successor, but *enforce* the successor to be one that is *strictly better* than our current state.

More precisely, at each stage during search a successor state is found by performing breadth first search starting out from the current state S. For each search state S', all successors are generated and evaluated heuristically. Doubly

```
initialize the current plan to the empty plan <>
S := I
obtain h(S) by evaluating S
if h(S) = ∞ then
   output "No Solution", stop
endif
while h(S) ≠ 0 do
   breadth first search for a state S' with h(S') < h(S)
   if no such state can be found then
      output "No Solution", stop
   endif
   add the actions on the path to S' at the end of the current plan
   S := S'
endwhile
```

Fig. 3. The Enforced Hill-climbing algorithm. $\mathcal{I}$ denotes the initial state of the problem to be solved.

occuring states are pruned from the search by keeping a hashtable of past states in memory, and the search stops as soon as it has found a state S' that has a lower heuristic value than S. This way, the Hill-climbing search escapes plateaus and local minima by simply performing exhaustive search for an exit, i.e., a state with strictly better heuristic evaluation.

4.1 Helpful Actions

So far, we have only used the goal distance heuristic. We integrate the helpful actions heuristic into our search algorithm as follows. During breadth first search, we do not generate *all* successors of any search state S' anymore, but consider *only those* that are obtained by applying actions from $H(S')$. This way, the branching factor for the search is cut down. However, considering only the actions in $H(S')$ might make the search miss a goal state. If this happens, i.e., if the search can not reach any new states anymore when restricting the successors to $H(S')$, we simply switch back to complete breadth first search starting out from the current state S and generating *all* successors of search nodes.

5 Completeness

The Enforced Hill-climbing algorithm is complete on *deadlock-free* planning problems. We define a *deadlock* to be a state S that is reachable from the initial state $\mathcal{I}$, and from which the goal can not be reached anymore. A planning problem is called *deadlock-free*, if it does not contain any deadlock state. We remark that a deadlock-free problem is also solvable, cause otherwise the initial state itself would already be a deadlock.

Theorem 1. *Let $\mathcal{P}$ be a planning problem. If $\mathcal{P}$ is deadlock-free, then the* Enforced Hill-climbing *algorithm, as defined in Figure 3, will find a solution.*

Due to space restrictions, we do not show the (easy) proof of Theorem 1 here and refer the reader to [5]. In short, if the complete breadth first search starting from a state S can not reach a better evaluated state, then, in particular, it can not reach a goal state, which implies that the state S is a deadlock in contradiction to the assumption.

In [5], it is also shown that most of the currently used benchmark domains are in fact *deadlock-free*. Any solvable planning problem that is *invertible* in the sense that one can find, for each action sequence P, an action sequence $\overline{P}$ that undoes P's effects, does not contain deadlocks. One can always go back to the initial state first and execute an arbitrary solution thereafter. Moreover, planning problems that contain an *inverse action* $\overline{o}$ to each action o *are* invertible: simply undo all actions in the sequence P by executing the corresponding inverse actions. Finally, most of the current benchmark domains *do* contain inverse actions. For example in the *blocksworld*, we have **stack** and **unstack**. Similarly in domains that deal with logistics problems, for example *logistics*, *ferry*, *gripper* etc., one can often find inverse pairs of actions. If an action is not invertible, its role in the domain is often quite limited. A nice example is the **inflate** operator in the *tyreworld*, which can be used to inflate a spare wheel. Obviously, there is not much point in defining something like a **deflate** operator. More formally speaking, the operator does not destroy a goal or a precondition of any other operator in the domain. In particular, it does not lead into deadlocks.

6 Empirical Results

For empirical evaluation, we implemented the Enforced Hill-climbing algorithm, using relaxed plans to evaluate states and to determine helpful actions, in C. We call the resulting planning system FF, which is short for FAST-FORWARD planning system. All running times for FF are measured on a Sparc Ultra 10 running at 350 MHz, with a main memory of 256 M Bytes. Where possible, i.e., for those planners that are publicly available, the running times of other planners were measured on the same machine. We indicate run times taken from the Literature in the text. All planners were run with the default parameters, unless otherwise stated in the text, and all benchmark problems are the standard examples taken from the Literature. Some benchmark problems have been modified in order to show how planners scale to bigger instances. We explain the modifications made, if any, in the text. Dashes indicate that the corresponding planner failed to solve that problem within half an hour.

6.1 The *Logistics* Domain

This is a classical domain, involving the transportation of packets via trucks and airplanes. There are two well known test suites. One has been used in the

1998 AIPS planning systems competition, the other one is part of the BLACKBOX distribution. The problems in the competition suite are very hard. In fact, they are so hard that, up to date, no planner has been reported to solve them all. FAST-FORWARD is the first one that does. See Figure 4, showing also the results for GRT [12] and HSP-r [2], which are—as far as the author knows—the two best other domain independent *logistics* planners at the time being.[1]

problem	HSP-r time	HSP-r steps	GRT time	GRT steps	FF time	FF steps
prob-01	0.36	35	0.28	30	0.06	27
prob-02	3.13	36	1.32	34	0.19	32
prob-03	25.45	64	5.55	60	0.71	54
prob-04	50.13	63	19.28	69	0.98	58
prob-05	0.62	27	0.39	26	0.08	22
prob-06	293.60	83	14.39	80	1.95	73
prob-07	6.20	37	1.76	37	0.38	36
prob-08	-	-	16.37	48	2.04	41
prob-09	371.03	97	50.48	98	2.08	91
prob-10	287.64	121	23.13	117	3.20	103
prob-11	4.58	34	1.54	36	0.21	30
prob-12	-	-	43.06	48	2.01	41
prob-13	-	-	85.58	79	7.73	67
prob-14	-	-	60.20	104	6.97	98
prob-15	19.52	120	67.50	106	1.27	93
prob-16	92.75	69	31.58	62	1.23	55
prob-17	29.35	61	12.19	53	0.63	44
prob-18	-	-	335.05	193	50.76	167
prob-19	-	-	238.98	174	16.26	151
prob-20	-	-	324.12	169	24.40	139
prob-21	-	-	294.23	120	8.93	102
prob-22	-	-	-	-	246.05	282
prob-23	100.67	145	16.86	118	3.84	126
prob-24	-	-	98.54	49	4.17	40
prob-25	-	-	-	-	106.23	181
prob-26	-	-	-	-	71.15	183
prob-27	-	-	-	-	71.26	141
prob-28	-	-	-	-	679.43	265
prob-29	-	-	-	-	589.75	323
prob-30	-	-	-	-	62.4	131

Fig. 4. Results of the three domain independent planners best suited for *logistics* problems on the 1998 competition suite. Times are in seconds, *steps* counts the number of actions in a sequential plan. For HSP-r, the weighting factor W is set to 5, as was done in the experiments described by Bonet and Geffner in [2].

The times for GRT in Figure 4 are from the paper by Refanidis and Vlahavas [12], where they are measured on a Pentium 300 with 64 M Byte main memory. FF outperforms both HSP-r and GRT by an order of magnitude. Also, it finds shorter plans than the other planners.

We also ran FF on the benchmark problems from the BLACKBOX distribution suite, and it solved all of them in less than half a second. Compared to the results shown by Bonet and Geffner [2] for these problems, FF was between 2 and 10 times faster than HSP-r, finding shorter plans in all cases.

---

[1] It is important to distinct the results shown in Figure 4 from those reported earlier for HSP-r [2]. Those results were taken on the problems from the BLACKBOX distribution, while our results are taken on the 1998 competition test suite.

6.2 Mixed Classical Problems

FAST-FORWARD shows competitive behavior on all commonly used benchmark domains. To exemplify this, we show a table of running times on a variety of different domains in Figure 5, comparing FF against a collection of state-of-the-art planning systems: IPP [8], STAN [9], BLACKBOX [7], and HSP [3].

domain	problem	IPP time	IPP steps	STAN time	STAN steps	BLACKBOX time	BLACKBOX steps	HSP time	HSP steps	FF time	FF steps
tyreworld	fixit-1	0.04	19	0.10	19	0.43	19	0.35	23	0.04	19
tyreworld	fixit-2	11.29	30	1.25	30	114.32	30	-	-	0.09	30
tyreworld	fixit-3	-	-	-	-	933.14	41	-	-	0.20	41
tyreworld	fixit-4	-	-	-	-	-	-	-	-	0.42	52
hanoi	tower-3	0.03	7	0.03	7	0.23	7	0.31	7	0.01	7
hanoi	tower-5	0.11	31	0.27	31	680.6	31	2.04	31	0.09	31
hanoi	tower-7	1.93	127	6.10	127	-	-	23.18	163	0.52	127
hanoi	tower-9	39.31	511	230.20	511	-	-	-	-	6.45	511
sokoban	sokoban-1	1.15	25	1.51	25	1283.29	25	13.87	29	0.22	25
manhattan	mh-7	4.82	35	20.04	35	-	-	1.12	35	0.09	38
manhattan	mh-11	65.12	40	1013.96	40	-	-	13.31	40	0.26	43
manhattan	mh-15	-	-	-	-	-	-	-	-	0.64	59
manhattan	mh-19	-	-	-	-	-	-	-	-	1.53	87
blocksworld	bw-large-a	0.47	10	0.57	10	10.30	10	0.78	11	0.04	7
blocksworld	bw-large-b	2.20	14	4.04	14	160.14	14	1.54	13	0.10	10
blocksworld	bw-large-c	88.17	25	267.08	26	-	-	4.34	20	0.56	16
blocksworld	bw-large-d	362.19	33	-	-	-	-	11.36	27	1.42	20

Fig. 5. Running times and quality (in terms of number of actions) of plans for FF and state-of-the-art planners on various classical domains. All planners are run with the default parameters, except HSP, where loop checking needs to be turned on.

In Figure 5, the planning problems shown are the following. The *tyreworld* problem was originally formulated by Russell, and asks the planner to replace a flat tire. The problem is modified in a natural way so as to make the planner replace n flat tires. FF is the only planner that is capable of replacing more than three tires, scaling up to much bigger problems.

The *hanoi* problems make the planner solve the well known *Towers of Hanoi* problem, with n discs to be moved. FF also outperforms the other planners on these problems.

The *sokoban* problem encodes a small instance of a well known computer game, where a single stone must be pushed to its goal position. Although the problem contains deadlocks, FF has no difficulties in solving it.

The *manhattan* domain was first introduced by McDermott [10]. In these problems, the planner controls a robot which moves on a $n \times n$ grid world, and has to deal with different kinds of keys and locks. The original problem taken from [10] corresponds to the *mh-11* entry in Tabular 5, where the robot moves on a 11×11 grid. The other entries refer to problems that have been modified to encode 7×7, 15×15 and 19×19 grid worlds, respectively. FF easily handles all of them, finding slightly suboptimal plans.

Finally, the *blocksworld* problems in Figure 5 are benchmark examples taken from [6]. FF outperforms the other planners in terms of running time as well as in terms of solution length.

7 Related Work

The closest relative to the work described in this paper is, quite obviously, the HSP system [3]. In short, HSP does Hill-climbing search, with the heuristic function

$$h(S) := \sum_{g \in \mathcal{G}} weight_S(g)$$

The weight of a fact with respect to a state S is, roughly speaking, the minimum over the sums of the precondition weights of all actions that achieve it. The weights are obtained as a side effect of doing exactly the same fixpoint computation as we do. The main problem in HSP is that the heuristic needs to be recomputed for each single search state, which is very time consuming. Inspired by HSP, a few approaches have been developed that try to cope with this problem, like HSP-r [2] and the GRT-planner [12].

The authors of HSP themselves handle the problem by sticking to their heuristic, but changing the search direction, going backwards from the goal in HSP-r instead of forward from the initial state in HSP. This way, they need to compute a weight value for each fact only once, and simply sum the weights up for a state later during search.

The authors of [12] invert the direction of the HSP *heuristic* instead. While HSP computes distances by going towards the goal, GRT goes from the goal to each fact, and estimates its distance. The function that then extracts, for each state during forward search, the state's heuristic estimate, uses the pre-computed distances as well as some information on which facts will probably be achieved simultaneously.

For the FAST-FORWARD planning system, a somewhat paradoxical extension of HSP has been made. Instead of avoiding the major drawback of the HSP strategy, we even worsen it, at first sight: the heuristic keeps being fully recomputed for each search state, and we even put some extra effort on top of it, by extracting a relaxed solution. However, the overhead for extracting a relaxed solution is marginal, and the relaxed plans can be used to prune unpromising branches from the search tree.

To verify where the enormous run time advantages of FF compared to HSP come from, we ran HSP using Enforced Hill-climbing search with and without helpful actions pruning, as well as FF without helpful actions on the problems from our test suite. Due to space restrictions, we can not show our findings in detail here. It seems that the major steps forward are our variation of Hill-climbing search in contrast to the restart techniques employed in HSP, as well as the helpful actions heuristic, which prunes most of the search space on many problems. Our different heuristic distance estimates seem to result in shorter plans and slightly, about a factor two, better running times, when one compares FF to a version of HSP that uses Enforced Hill-climbing search and helpful actions pruning. We did not yet find the time to do these experiments the other way round, i.e., integrate our heuristic into the HSP search algorithm, as this would involve modifying the original HSP code, which means a lot of implementation work.

There has been at least one more approach in the Literature where goal distances are estimated by ignoring the delete lists of the operators. In [10], Greedy Regression-Match Graphs are introduced. In a nutshell, these estimate the goal distance of a state by backchaining from the goals until facts are reached that are TRUE in the current state, and then counting the estimated minimal number of steps that are needed to achieve the goal state.

To the best of our understanding, the action chains that lead to a state's heuristic estimate in [10] are similar to the relaxed plans that we extract. However, the backchaining process seems to be quite costly. For example, building the Greedy Regression-Match Graph for the initial state of the *manhattan* world 11×11 grid problem is reported to take 25 seconds on a Sparc 2 station. For comparison, we ran FF on a Sparc 4 station. Finding a relaxed plan for the initial state takes less than one hundredth of a second, i.e., the time measured is 0.00 CPU seconds.

The helpful actions heuristic shares some similarities with what is known as relevance from the literature [11]. The main difference is that relevance in the usual sense refers to what is useful for solving the whole problem. Being helpful, on the other hand, refers to something that is useful *in the next step*.

8 Conclusion and Outlook

In this paper, we presented two heuristics for domain independent STRIPS planning, one estimating the distance of a state to the goal, and one collecting a set of promising actions. Both are based on an extension of the heuristic that is used in the HSP system. We showed how these heuristics can be used in a variation of Hill-climbing search, and we have seen that the algorithm is complete on the class of deadlock-free domains. We collected empirical evidence that the resulting planning system is among the fastest planners in existence nowadays, outperforming the other state-of-the-art planners on quite a range of domains, like the *logistics*, *manhattan* and *tyreworld* problems.

To the author, the most exciting question is this: *Why* is the heuristic information obtained in this simple manner so good? It is not really difficult to construct abstract examples where the approach produces arbitrarily bad plans, or uses arbitrarily much time, so why does it almost never go wrong on the benchmark problems? Why is the relaxed solution always so close to a real solution, except for the *Tower of Hanoi* problems? Is it possible to define a notion of "simple" planning domains, where relaxed solutions have desirable properties?

First steps into that direction seem to indicate that, in fact, there might be some underlying theory in that sense. In particular, it can be proven that the Enforced Hill-climbing algorithm finds optimal solutions when the heuristic used is *goal-directed* in the following sense:

$$h(S) < h(S') \Rightarrow min(S) < min(S')$$

Here, $min(S)$ denotes the length of the shortest possible path from state S to a goal state, i.e., Enforced Hill-climbing is optimal when heuristically better evaluated states are really closer to the goal.

It can also be proven that the length of an *optimal* relaxed solution *is*, in fact, a goal-directed heuristic in the above sense on the problems from the *gripper* domain that was used in the 1998 planning systems competition. We have not yet, however, been able to identify some general structural property that implies goal-directedness of optimal relaxed solutions.

Apart from these theoretical investigations, we want to extend the algorithms to handle richer planning languages than STRIPS, in particular ADL and resource constrained problems.

Acknowledgments. The author thanks Bernhard Nebel for helpful discussions and suggestions on designing the paper.

References

1. A. Blum and M. Furst. Fast planning through planning graph analysis. *Artificial Intelligence*, 90(1-2):279-298, 1997.
2. B. Bonet and H. Geffner. Planning as heuristic search: New results. In *Proceedings of the 5th European Conference on Planning*, pages 359-371, 1999.
3. B. Bonet, G. Loerincs, and H. Geffner. A robust and fast action selection mechanism for planning. In *Proceedings of the 14th National Conference of the American Association for Artificial Intelligence*, pages 714-719, 1997.
4. T. Bylander. The computational complexity of propositional STRIPS planning. *Artificial Intelligence*, 69(1-2):165-204, 1994.
5. J. Hoffmann. A heuristic for domain independent planning and its use in a fast greedy planning algorithm. Technical Report 133, Albert-Ludwigs-University Freiburg, 2000.
6. H. Kautz and B. Selman. Pushing the envelope: Planning, propositional logic, and stochastic search. In *Proceedings of the 14th National Conference of the American Association for Artificial Intelligence*, pages 1194-1201, 1996.
7. H. Kautz and B. Selman. Unifying SAT-based and graph-based planning. In *Proceedings of the 16th International Joint Conference on Artificial Intelligence*, pages 318-325, 1999.
8. J. Koehler, B. Nebel, J. Hoffmann, and Y. Dimopoulos. Extending planning graphs to an ADL subset. In *Proceedings of the 4th European Conference on Planning*, pages 273-285, 1997.
9. D. Long and M. Fox. Efficient implementation of the plan graph in STAN. *Journal of Artificial Intelligence Research*, 10:87-115, 1999.
10. D. McDermott. A heuristic estimator for means-ends analysis in planning. In *Proceedings of the 3rd International Conference on Artificial Intelligence Planning Systems*, pages 142-149, 1996.
11. B. Nebel, Y. Dimopoulos, and J. Koehler. Ignoring irrelevant facts and operators in plan generation. In *Proceedings of the 4th European Conference on Planning*, pages 338-350, 1997.
12. I. Refanidis and I. Vlahavas. GRT: A domain independent heuristic for strips worlds based on greedy regression tables. In *Proceedings of the 5th European Conference on Planning*, pages 346-358, 1999.

Planning while Executing: A Constraint-Based Approach

R. Barruffi, M. Milano, and P. Torroni

DEIS, University of Bologna, Viale Risorgimento 2, 40136 Bologna, Italy,
Tel: 0039 051 2093086 Fax: 0039 051 2093073
{rbarruffi, mmilano, ptorroni}@deis.unibo.it

Abstract. We propose a planning architecture where the planner and the executor interact with each other in order to face dynamic changes of the application domain. According to the *deferred planning* strategy proposed in [14], a plan schema is produced off-line by a generative constraint based planner and refined at execution time by retrieving up-to-date information when that available is no longer valid. In this setting, both planning and execution can be seen as search processes in the space of partial plans. We exploit the Interactive Constraint Satisfaction framework [12] which represents an extension of the Constraint Satisfaction paradigm for dealing with incomplete knowledge. Given the uncertainty of the plan execution in dynamic environments, a backup and recovery mechanism is necessary in order to allow backtracking at execution time.

1 Introduction

In dynamic and changing environments, a plan produced *off-line* by a traditional generative planner can fail during execution due to the fact that the environment can change, often in unpredictable ways. In particular, our planner works in a networked computer system environment and assembles configuration plans. The information about system services and resources cannot be complete at planning time due to its vastity and dynamicity. In these cases it is impossible to produce a complete successful plan at plan generation time and there is need to sense correct and up-to-date information at execution time in order to refine the plan. In [14], the authors propose a classification followed by a deep analysis of the main planning strategies able to integrate execution time sensory data into the planning process. The strategy we follow is called *deferred planning* consisting in delaying until execution the decisions depending on sensing. As a consequence, there is need for a sensing mechanism for testing the environment and a procedure for plan refinement at execution time. We integrate a constraint-based planner aimed at producing a plan schema with an executor able to refine the plan before executing it. Both those components are able to sense the real world by means of a constraint based framework, called Interactive Constraint Satisfaction Problem (ICSP), proposed in [12]. The main point of this paper is to describe our architecture and how it implements the *deferred planning* strategy by exploiting the ICSP framework.

2 Different Strategies to Cope with Dynamicity

The enhanced complexity of traditional planning techniques when applied to dynamic environments is due to the facts that (*i*) typically the planner is not the only agent that causes changes on the system and (*ii*) often changes are not deterministic. This can lead to a failure of the plan execution, either because action preconditions are no longer verified at execution time, or because action effects are not those expected.

In [14], the authors present three different extensions to conventional planning techniques whose aim is to cope with uncertainty:

- *planning for all contingencies*, so that once sensing is performed, only the plan correspondent to the actual contingency will be executed [15,2];
- *making assumptions*, so that planning decisions will be based only on the assumed value of the sensing result [5,9];
- *deferring planning decisions* until information depending on sensors is available [14,5,8].

The appropriateness of the strategy depends on the application, and, in particular, on the criticality of mistakes, on the complexity of the domain, and on the acceptability of suspending execution to do more planning.

Our architecture follows the *deferred planning* strategy, as it will be described in the next section. The *deferred planning* approach aims at avoiding doing useless computation at planning time. Some portions of the plan requiring information which can be available only at execution time are left incomplete. In this way the planner could miss some important dependencies between the partial plans it is producing. This is why plan execution can fail and it is strongly required that the actions contained in partially specified plans are reversible.

3 An ICSP-Based Planning Architecture

Our planning architecture is in charge of computing configuration plans in a networked computer system [4,10]. The domain knowledge is composed by many different types of objects (e.g., machines, users, printers, services, files, processes), their attributes (e.g., sizes, availability, location) and relations among them (e.g., user u is logged on machine m). In this case, there is an enormous amount of knowledge to consider. In addition, this information can change during the system's life due to actions performed on the objects (e.g., removing or creating files, connecting or disconnecting machines, adding or deleting users, starting or killing processes). Thus, it is not convenient, if possible at all, to store all this information in advance and keep it up-to-date. We developed a planner able to deal with dynamic and incomplete knowledge. Our solution follows the *deferred planning* approach described in [14]. However, while in [14] the deferred decisions are represented by all the goals involving data that must be obtained through sensing, our planner does not defer until execution all the goals which require sensing since it is able to sense at planning time. In our approach deferred decisions are represented by:

- non deterministic variable bindings: since variable domain values represent alternative resources whose state can change during or after plan construction, we want to avoid as much as possible to commit to premature choices;
- acquisition of up-to-date information when that sensed at planning time is no longer valid.

Both the planner and the executor are able to sense the real system by means of a constraint based framework representing an extension of the Constraint Satisfaction paradigm and called Interactive Constraint Satisfaction Problem framework [12].

3.1 Preliminaries

Interactive Constraints (ICs) are declarative relations among variables whose domain (i.e., the set of values the variables can assume) is possibly partially or completely unknown. An interactive domain is defined as $D(X) = [List \cup Undef]$ where $List$ represents the set of known values for variable X, and $Undef$ is a domain variable itself representing (intensional) information which is not yet available for variable X. An Interactive Constraint Satisfaction Problem (ICSP) is defined on a set of variables ranging on interactive domains. Variables are linked by ICs that define (possibly partially known) combinations of values that can appear in a consistent solution. As for traditional Constraint Satisfaction Problems, a solution to an ICSP is found when all the variables are instantiated consistently with constraints. For a formal definition of the ICSP framework see [12]. ICs operational behaviour extends standard constraint propagation with a data acquisition mechanism devoted to retrieving consistent values for variable domains. In particular given a binary interactive constraint $IC(c(X,Y))$, its operational behaviour is the following:

1. **If** both variables are associated to a partially or completely unknown domain, the constraint is suspended;
2. **else, if** both variables range on a completely known domain, the constraint is propagated as in classical CSPs;
3. **else, if** one variable (say X) ranges on a fully known domain and the other (Y) is associated to a fully unknown domain a knowledge acquisition step is performed; this returns either a finite set of consistent values representing the domain of Y, or an empty set representing failure.
4. **else, if** X ranges on a fully known domain and Y is associated to a partially known one, Y domain is pruned from values non consistent with X. If Y domain becomes empty a new knowledge acquisition step is performed for Y driven by X.

This is a general framework which can be used in many applications. It is particularly suited for all the applications that process a large amount of constrained data provided by a lower level system, see for instance [11,12].

3.2 The Algorithm

According to the *deferred planning* strategy proposed in [14], a plan schema is produced off-line by a generative planning process and refined at execution time by retrieving up-to-date information when that available is no longer valid. In this setting, both planning and execution represent search processes in the space of partial plans. More precisely the plan execution can be seen as the second phase of the same search algorithm aimed both at producing and executing a plan. The generative phase of the algorithm represents a Partial Order Planner (POP)[16] interleaving open condition[1] achievement and conflict resolution steps. As far as the open condition achievement is concerned, three alternative cases are possible: (i) the open condition is already satisfied in the initial state, (ii) it can be satisfied by an action already in the plan, (iii) there is need of a new action in order to satisfy it.

The planning problem is mapped onto an ICSP so that the planner becomes able to both exploit constraint satisfaction techniques in order to reduce the search space and deal with incomplete knowledge. The method we propose embeds knowledge acquisition activity into the constraint solving mechanism, thus simplifying the planning process in two points. First of all, there is no need to add declarative sensing actions to the plan [1,6,10], we provide a sensing mechanism where no further declarative action is needed apart from the causal actions. Second, only significant information for the planner is retrieved. As a consequence, variable domains are significantly smaller than in the standard case.

Open conditions are treated as ICs. Variables appearing in ICs represent system resources, and domain values represent alternative instances. Variable domains contain all the known alternative resources; they can be either (i) completely known, containing objects which can be assigned to the corresponding variable; (ii) partially known, containing some values already at disposal and a variable representing intensional future acquisitions; (iii) totally unknown, when no information has already been retrieved for the variable. As soon as an open condition $p(X,Y)$ is selected, the constraint solver will propagate the corresponding $IC(p(X,Y))$ to test if there exists at least one value of X and Y that already satisfies p in the initial state. When variables (X, Y) range on known domains, traditional constraint propagation is performed in order to prune inconsistent values from domain, otherwise constraint propagation results in acquisition of domain values. In order to provide Interactive Constraints with the capability to sense the system we need to associate them with appropriate information gathering procedures, working as access modules to the real world. In our environment such procedures can be represented by simple UNIX sensing commands as well as by scripts when sensory requests involve setup activities. It is worth noting that when appropriate sensors are available, Interactive Constraint retrieve only information consistent with the context so as to simplify the task of pruning inconsistent alternatives. For instance, suppose that, during the planning process, we need to locate a file *mydoc* in a UNIX system, i.e., we need to propagate

---

[1] An open condition is indifferently represented by a precondition or a final goal conjunct still to be satisfied.

the interactive constraint $inDirectory(mydoc, Location)$. Suppose, also, that the file *mydoc* is initially contained in three different directories $dir1$, $dir2$ and $dir3$. If variable *Location* has an unknown domain, an acquisition step is performed and those three values are retrieved (through the $find$ Unix sensing command), otherwise the domain is pruned from not consistent values (e.g $dir4$).

If a constraint fails (i.e., a variable domain becomes empty), it means that the corresponding precondition is not satisfied in the initial state (i.e., there is need of an action in order to achieve it). On the other hand, when more than one value are left in a variable domain after all possible propagation, it means that all those values satisfy that constraint in the initial state. In a traditional CS_based approach, there is need for a non deterministic labelling step in order to find a final solution. In our architecture, the labelling step takes place at plan execution time so that at the end of the generative phase variables might be associated with a domain containing more than one value.

Given the plan schema produced by the generative phase, the executor selects the first action to be executed. An interactive constraint propagation activity checks the satisfiability of its preconditions in the real world. If precondition variables are already instantiated, the interaction with the underlying system results in a consistency check, while if those variables are associated to a domain, the domain can be pruned in order to remove values which are no longer consistent with the current state of the system. Value removal can trigger constraint propagation which, in turn, removes values from other variable domains, thus reducing the execution search space. If, after propagation, a domain is empty, meaning that values retrieved at planning time no longer verify the correspondent precondition p, a backtracking step is performed in order to select an alternative action or partial plan which satisfies p. When all the variables of the action range on non empty domains, necessary non deterministic labelling steps are performed and the action is executed. The same reasoning applies until all the actions are successfully executed.

3.3 An Example

Let us consider a network where a monitoring application ensures that certain processes are *up and running* (i.e. that their *status* is on). Once the status of the system is recognized as faulty, the planner is activated in order to provide a recovery plan.

Let us suppose that one of those processes, called Trigger, is off, and that for activating it the planner generates the plan P_1 of actions shown in Figure 1. Note that some domains are partially known, others are still completely unknown. TriggerStart is the daemon process in charge of activating the Trigger process, and its code is contained in the executable file TMAboot. When activating TriggerStart, TMAboot must be located in a directory (X) corresponding to the so-called *runlevel* (I) of the process. For instance, if TriggerStart is to be activated at runlevel '3', TMAboot must be in a directory called '/sbin/rl3'. The runlevel is a parameter of the machine which is set at boot-time. In particular, in order to achieve the goal of having processes TriggerStart and Trigger on, P_1

```
***** plan to be executed: *****

killProcess(TriggerStart)
copy(TMAboot, D1, X); X:: [Undef] D1:: [/sbin/rl1, /sbin/rl2, Undef]
onTriggerStart(I, X); X:: [Undef] I:: [3, Undef]
```

Fig. 1. Plan to be executed.

suggests that process `TriggerStart` is killed, that file `TMAboot` is copied from directory `D1` to directory `X` and that process `TriggerStart` is activated from the directory `X` corresponding to the runlevel `I`. At planning time only relevant

```
***** executing plan... *****

now checking preconditions...
          ---> condition status(TriggerStart, on) succeeded
...preconditions checked.
now doing labelling on preconditions...
labelling killProcess(TriggerStart)
...labelling on preconditions done.
now executing action 1: killProcess(TriggerStart)...
          ---> action killProcess(TriggerStart) succeeded
now checking preconditions...
          ---> condition inDirectory(TMAboot, D1) succeeded
...preconditions checked.
now doing labelling on preconditions...
labelling copy(TMAboot, D1, X)
...labelling on preconditions done.
now executing action 2: copy(TMAboot, /sbin/rl1, X)...
          ---> action copy(TMAboot, /sbin/rl1, /sbin/rl3) succeeded
now checking preconditions...
          ---> condition status(TriggerStart, off) succeeded
          ---> condition inDirectory(TMAboot, /sbin/rl3) succeeded
          ---> condition configDir(/sbin/rl3, I) succeeded
...preconditions checked.
now doing labelling on preconditions...
labelling onTriggerStart(I, /sbin/rl3)
...labelling on preconditions done.
now executing action 3: onTriggerStart(3, /sbin/rl3)...
          ---> action onTriggerStart(3, /sbin/rl3) succeeded

***** ...plan executed *****
```

Fig. 2. Output messages generated during the execution of the plan: case 1.

facts are retrieved from the world: in particular, the planner knows that the machine is on at runlevel 3, that four directories ('/sbin/rl0', '/sbin/rl1', '/sbin/rl2', '/sbin/rl3') exist and that they correspond to four different runlevels, that process `TriggerStart` is on and that process `Trigger` is off. Finally it knows that a copy of the file `TMAboot` is contained in two different directories ('/sbin/rl1', '/sbin/rl2'). If the world does not change the executor will instantiate variable `D1` either to '/sbin/rl1' or to '/sbin/rl2', variable

```
now checking preconditions...
        ---> condition inDirectory(TMAboot, D1) succeeded
...preconditions checked.
now doing labelling on preconditions...
labelling copy(TMAboot, D1, X)
...labelling on preconditions done.
now executing action 2: copy(TMAboot, /sbin/rl2, X)...
        ---> action copy(TMAboot, /sbin/rl2, /sbin/rl3) succeeded
```

Fig. 3. Output messages generated during the execution of action `copy`: case 2.

X to '/sbin/rl3' and I to 3. The output messages generated by the execution module are those of Figure 2. We can recognize different steps in the execution of each action: a first phase where the executor checks if the preconditions of the current action hold, a labelling phase where domains, if any, are labelled and eventually an execution phase, which modifies the state of the world.

If the actions are successfully performed, the world is led to a final state with all the relevant processes on. Now, let us suppose that before executing action copy(TMAboot, D1, X) some external agent in the world deletes file TMAboot from '/sbin/rl1'. The actual world contains only one instance of such file, in directory '/sbin/rl2'. Therefore the executor cannot label the plan in the same way as before (i.e., copy(TMAboot, /sbin/rl1, /sbin/rl3)). What it does, after checking in the world the domain of D1, is to choose one of the domain values which are actually left (i.e., '/sbin/rl2'). The execution proceeds as in the first case, with the only difference that the file is copied from a different source ('/sbin/rl2'). See Figure 3.

4 Non-monotonic Changes

Up to now, we have considered that values acquired during plan construction can be no longer available during plan execution. However, a more complex situation occurs when some new values are available during plan execution and have not been retrieved during plan construction. Standard CSPs do not deal with value insertion in variable domains since it implies reconsidering previously deleted values which can be supported by the newly inserted value. The ICSP framework can cope with non monotonic changes of variable domains thanks to the *open domains* data structure.

An *open domain* is represented by a set of known values and a variable representing the unknown domain parts, i.e., potential future acquisitions. Thus, if, during plan execution, the entire set of known values (those acquired during plan construction), is deleted because of precondition verification, a new acquisition can start aimed at retrieving new consistent values. If no values are available, backtracking is performed in order to explore the execution of other branches in the search space of partial plans. If the overall process fails (and only in this case), a re-planning is performed.

Dynamic Constraint Satisfaction (DCS) [13] has been proposed in order to deal with non monotonic changes. DCP solvers maintain proper data structures so as to tackle modifications of the constraint store. Thanks to the ICSP framework we do not need to store additional information for restoring the constraint store consistency as done by DCS approaches. On the other hand, our method makes the propagation we perform less powerful than that performed by dynamic approaches. In fact, if we consider a constraint between variables X and Y, the variable inserted in the domain of variable X represents a potential support for values in the domain of variable Y, which cannot be pruned until the domain of X becomes closed.

Example. Given the example above let us consider a third case. If the domain initially retrieved for D1 is completely wiped and an instance of file TMAboot is put in another directory, let us say '/sbin/rl0', it is necessary to perform more acquisition via the undefined part of the D1 domain. In particular, the constraints active on D1 can shape from the new world setting the correct domain for it at execution time and let the plan be once more successfully executed. Figure 4 shows the output generated by the execution of action copy.

```
now checking preconditions...
         ---> condition inDirectory(TMAboot, /sbin/rl0) succeeded
...preconditions checked.
now doing labelling on preconditions...
labelling 2
labelling copy(TMAboot, /sbin/rl0, X)
...labelling on preconditions done.
now executing action 2: copy(TMAboot, /sbin/rl0, X)...
         ---> action copy(TMAboot, /sbin/rl0, /sbin/rl3) succeeded
```

Fig. 4. Output messages generated during the execution of action copy: case 3.

5 Conclusion

This paper describes an approach to *deferred planning*, which represents one of the main planning strategy to plan in presence of dynamic environments. The idea is to delay some planning decisions regarding sensory data, as much as possible, in order to reduce the gap between the world as it is observed at planning time and the world the executor performs on. We exploit the Interactive Constraint Satisfaction framework [12], which represents an extension of the CS framework based on Interactive Constraints, in order to interact with the real world. Sensing is performed both at planning and at execution time.

The implementation of this architecture has been carried out by using the finite domain library of ECL^iPS^e [3] properly extended to cope with the interactive framework. ECL^iPS^e is a Constraint Logic Programming (CLP) [7] system merging all the features and advantages of Logic Programming and Constraint Satisfaction techniques. CLP on Finite Domains, CLP(FD), can be used to represent planning problems as CSPs.

A repair mechanism is currently under development in order to cope with failures and backtracking steps over already executed actions. The repair mechanism supports all cases in which the executor realises that the effects of the action are not those expected.

Acknowledgments. Authors' work has been partially supported by Hewlett Packard Laboratories of Bristol-UK (Internet Business Management Department) and CNR (Project 40%).

References

1. N. Ashish, C.A. Knoblock, and A. Levy. Information gathering plans with sensing actions. In *Proceedings of the 4th European Conference on Planning*, 1997.
2. D. Draper, S. Hanks, and D. Weld. Probabilistic planning with information gathering and contingent execution. In *Proceedings of AIPS-94*, 1994.
3. ECRC. ECL^iPS^e *User Manual Release 3.3*, 1992.
4. O. Etzioni, H. Levy, R. Segal, and C. Thekkath. Os agents: Using ai techniques in the operating system environment. Technical report, Univ. of Washington, 1993.
5. K. Golden. *Planning and Knowledge Representation for Softbots*. PhD thesis, University of Washington, 1997.
6. K. Golden and D. Weld. Representing sensing actions: The middle ground revisited. In *Proceedings of 5th Int. Conf. on Knowledge Representation and Reasoning*, 1996.
7. P. Van Hentenryck. *Constraint Satisfaction in Logic Programming*. MIT Press, 1989.
8. C.A. Knoblock. Planning, executiong, sensing, and replanning for information gathering. In *Proc. 14th IJCAI*, 1995.
9. N. Kushmerick, S. Hanks, and D. Weld. An algorithm for probabilistic least-commitment planning. In *Proceedings of AAAI-94*, 1994.
10. C.T. Kwock and D.S. Weld. Planning to gather information. Technical report, Department of Computer Science and Engineering University of Washington, 1996.
11. E. Lamma, M. Milano, R. Cucchiara, and P. Mello. An interactive constraint based system for selective attention in visual search. *Proceedings of the ISMIS'97*, 1997.
12. E. Lamma, M. Milano, P. Mello, R. Cucchiara, M. Gavanelli, and M. Piccardi. Constraint propagation and value acquisition: why we should do it interactively. *Proceedings of the IJCAI*, 1999.
13. S. Mittal and B. Falkenhainer. Dynamic constraint satisfaction problems. In *Proceedings of AAAI-90*, 1990.
14. D. Olawsky and M. Gini. Deferred planning and sensor use. In *Proceedings DARPA Workshop on Innovative Approaches to Planning, Scheduling, and Control*, 1990.
15. M. Peot and D. Smith. Conditional nonlinear planning. In J. Hendler, editor, *Proc. 1st AIPS*, pages 189–197, San Mateo, CA, 1992. Kaufmann.
16. D.S. Weld. An introduction to least commitment planning. *AI Magazine*, 15:27–61, 1994.

Problem Decomposition and Multi-agent System Creation for Distributed Problem Solving

Katsuaki Tanaka, Michiko Higashiyama, and Setsuo Ohsuga

Waseda University, Department of Information and Computer Science
3-4-1 Ohkubo Shinjuku-ku, Tokyo 169-8555, Japan

Abstract. As human society glows large and complex problems which human being must solve is also becoming large and complex. In many cases, a problem must be solved cooperatively by many people. There arise a problem of decomposing the problem into sub-problems, distributing these sub-problems to number of persons and organizing these people in such a way that the problem can be solved most efficiently. This organization is not universal but is made specific to the given problem. It is possible to create a multi-agent system to correspond to the cooperative work by persons. Here is a problem of creating an organization of the agents dynamically that is suited for coping with the specific problem. It is the major objective of this paper to discuss a way of generating a multi-agent system with examples.

1 Introduction

As social systems grow large and complex, problems which human being must solve are also becoming large and complex. In many cases, a problem must be solved cooperatively by many people. There arise a problem of decomposing the problem into sub-problems, distributing these sub-problems to number of persons and organizing these people in such a way that the problem can be solved most efficiently. Managing this process has been one of very important tasks by human being. But the growth of the scale of problems induces the increase of complexity in their management tasks and it is worrying that the current method of managing the process is inadequate for following up the growth of the problem scale. We are required today to develop a new method of management to resolve this problem.

One of the reasons that makes the current method inadequate is that large amount of decision were distributed to number of persons who join the process and made there without being recorded fully. Large part of these decisions remains in worker's brain, and the manager cannot follow the development process afterward for checking them. This problem remains unresolved as far as the main body of development is person because it is caused by an intrinsic nature of human being. An alternate way to improve this situation is to introduce computers in problem solving much more than ever before and let them record the history of its process, especially the history of decisions made there by persons in this process.

We introduce computers as software agents. It follows that agents replace persons in a organization for problem solving.

Many papers argue about problem solving system by software agents [1][2][8]. In this paper, Multi-strata model[5] is used to describe problem solving process, and multi-agent systems are created based on this model.

There are some problems. How is the multi-agent system organization generated and managed? How is the human decision recorded? How is the past record found and used for new problem solving? And so on.

Some of these issues, especially a way of generating a multi-agent system, are discussed in this paper. Every agent in this system is intelligent in a sense that not only it can solve problems autonomously based on a knowledge base but also it generates the other intelligent agents as needed.

2 Problem Solving

2.1 Problem Solving Scheme

Problems can be divided roughly into two types; design type and analysis type. Design type problem is define as those to obtain a structure of object with the required function while analysis type problem is to obtain the functionality of object based on a given structure. As will be described in the following, design type problem solving is defined as a repetitive operation including analysis type problem. In this paper therefore design type problem is mainly discussed.

A basic operation for design type problem solving is represented in this paper roughly as composed of three stages as follows and is shown in Fig. 1.

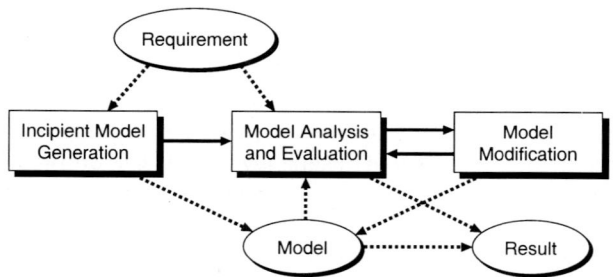

Fig. 1. Standard process of design problem solution

1. Make an incipient model as an embodiment of person's idea on problem solving. It includes this person's requirement to be satisfied.
2. Analyze the model to obtain its functionality and behavior, and evaluate it whether it satisfies the person's requests.
3. Based on the result of the analysis and evaluation, modify the model.

If the request is satisfied, stop the design process. The model represents a solution.

2.2 Solving Large Problem by Persons

When a problem is very large and is solved by person, this basic process cannot be applied directly but the problem must be decomposed to a set of smaller sub-problems and these sub-problems are distributed to the different persons. Since these sub-problems are not specified in advance but generated by decomposition after given the original problem, these persons cannot be assigned in advance but must be generated dynamically in parallel with decomposition process. This method is shown in the case of aircraft design as an example. It is shown in Fig. 2.

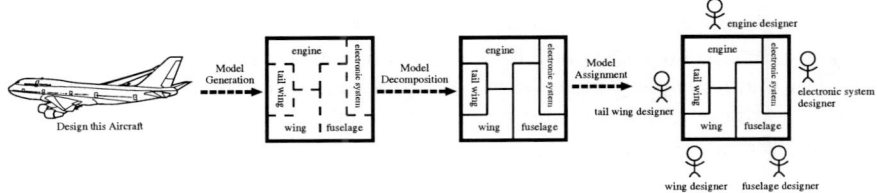

Fig. 2. Design process of an aircraft

The outline of the process is shown as follows.

1. **Model generation**
   Given requirement from client, a chief designer prepares to generate an incipient generous model of the whole airplane based on his/her experiences and referring to the case base. First, he/she creates a top node of a hierarchy to represent the object and gives it the design requirement.
2. **Model decomposition**
   Design starts top-down. In general, a complex object is decomposed to a number of assemblies and each assembly is further decomposed to a set of sub-assemblies and so on. In this way a hierarchical object is generated. It is required that the functionality of designed object meets the given functional requirement. The functionality of an object is decided by the functionality of its components and their structure. In this sense a model to represent an object, called an object model hereafter, is characterized by the bottom-up dependency. But it is difficult to design a large object bottom-up because of the combinatorial explosion of computations. Instead, a top-down design must be performed actually.
   The designer responsible to design an object decides tentatively an upper part structure of this object, i.e. main components and their structure. For example, an aircraft designer decide tentatively first main components (assemblies) such as engine, main wing, fuselage, ladder wing, tail wing, vertical wing, landing gear, wire-harness, electronic system, etc. Then their structural relation is defined. By assuming the functionality of these components, the functionality of the object can be estimated. If this tentative structure does

not satisfy the given requirement, the designer has to find another structure or changes the functionality of the components. The functionality becomes the requirement to the component design in the following process.

3. **Model assignment to component-design**
   If the designer satisfies the estimate, then he/she fixes tentatively this part of design and distributes a problem of designing each component to an expert of component design. For example, the engine design is assigned to an engine expert. After then, the similar process is performed for designing each component by the expert assigned in this way. Thus many people commit the design of the common objects. Since behavior of these components are related closely to each others, the components design cannot go independently from the others but needs very close interactions. Usually therefore those people are organized to assure easy communication and cooperation. That many people join the same design means that decisions are distributed to the different persons and remain there without being recorded. It causes the difficulty of tracing afterward the design for checking and maintenance. It will also be very much troubled in the document acquisition in modeling, if previous record is imperfect.

3 Outline of Autonomous Problem Solving

3.1 Problem Solving System Architecture

This human-centered process is replaced by computer-centered process. The computer-centered process means that a computer system manages a total process and persons join the problem solving in parts. In this computer system an object problem is represented by a knowledge representation language and a knowledge processing agents deal with the problem cooperatively. The agents is organized into a multi-agent system in which distributed agents are related each other in the best way to solve the given specific problem. The structure of the agents corresponds to the human organization in the human problem solving discussed in section 2. The major parts of the system are a global knowledge base, a distributed problem solving system composed of plural agents and user interface. The global knowledge base supports knowledge necessary for problem solving to every agent. The overall structure of agents for problem solving is shown in Fig. 3.

3.2 KAUS as Knowledge Representation Language

A language suited for representing this system is necessary. In order to cope with problem model of the form as will be discussed in section 2, it must be suited for representing predicate including data-structure as argument and also for describing meta-level operation such as knowledge for describing on other knowledge. KAUS (Knowledge Acquisition and Utilization Language) has been developed for the purpose[10]. In the following, some logical expressions appear as knowledge. In order to keep consistency and integrity of expressions throughout the

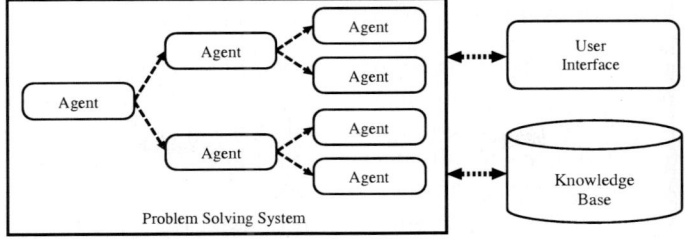

Fig. 3. Agent architecture of problem solving

whole system these must be written in KAUS language. However, these are not necessarily written in correct KAUS expressions but locally simplified. It is because KAUS syntax is not included in this volume and these locally simplified expressions are more comprehensive than correct expressions.

4 Multi-agent Problem Solving System

4.1 Design Principle of an Agent

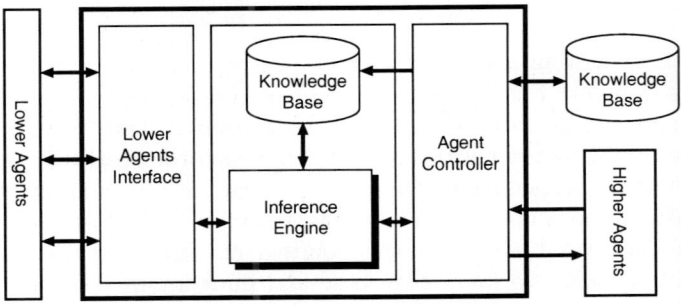

Fig. 4. Structure of an Agent

An autonomous problem solving system is designed as a multi-agent system. An agent in this multi-agent system is designed not as a special purpose agent to achieve a special role but a general-purpose problem solving system that can accept and cope with any problem as far as it is represented by a modeling scheme developed for the system. Every agent does not have any object knowledge (knowledge related to a specific object) beforehand. It retrieves necessary knowledge from global knowledge base when problem is assigned and just before to start problem solving. This agent has three layers: one for solving actually the given problem, second for generating the problem solving system by retrieving necessary knowledge from the global knowledge base and third for generating the other agent (Fig. 4). Throughout this paper, it is considered that an agent is a subject of an activity. Therefore, if some activity is defined in a computer system, there must be some agent to behave as the subject of the activity.

4.2 Behavior of Agents

The behavior of agents is similar to persons working together as has been discussed in section 2. Design-type problems are kept in mind. When a problem-solving task starts, a highest-level agent is prepared that corresponds to a human chief designer. Receiving user's requirement, it tries to decompose an object top-down as a first step to generate an object structure. It analyze and evaluate the structure and if it is decided that it succeeds in making the object structure at the highest level, then it generates and assigns an agent to every component of this structure. In this way the problem solving proceeds top-down. When it reaches the components at the bottom, that is, the components that are no more necessary to be decomposed, then the process stops in success.

```
(| (designA A) ~(designX X) ~(designY Y)
               ~(mergeModel A X Y)).
(designX 10).
(designY 5).
(| (mergeModel A X Y) ~($add A X Y)).
```

Fig. 5. Knowledge of designA

The behavior of an agent in this process is explained first using a simplest example as shown in Fig. 5. This is a problem of designing an imaginary object A. It is to merge the results from two designs for X and Y. In reality these designs are to obtain the entities X and Y of which the results are given as the number 10 and 5 respectively. The merge operation is to add the numbers. A design problem starts given the requirement (designA A)? First an agent at the highest level is created. It is given the requirement. The agent tries to satisfy the requirement. It refers to knowledge base and finds knowledge which says that the requirement can be satisfied by achieving two designs X and Y, and merging them. This means that three new activities are defined and accordingly that there should be three subjects for the activities. Traditionally these activities are combined into a program because each new activity is very simple. This is a case in which the same single computer represents the three subjects. But there can be different ways when sub-problems are large. For example, every subject can be different from each other and an agent is created to represent a subject. For the purpose of explanation, a design process is presented assuming this last case in the following. In the following explanation, the sentences headed by * shows the case of applying the general method to this specific example.

1. An agent receives the problem from user interface. The agent becomes the highest-level agent. The agent analyzes the problem, takes out some relating knowledge from the knowledge base, and saves it in the local knowledge base within the agent.
   * An agent receives '(designA A)?' Agent Controller in the agent retrieves related knowledge from Global Knowledge Base, and records it into Local

Knowledge Base. In this case it is '(| (designA A) ~(designX X) ~(designY Y) ~(mergeModel A X Y))'

2. The agent selects knowledge to use for problem solving from the local knowledge base, and analyzes its structure to decide whether the problem should be decomposed or not. If the object is to be decomposed, then an agent is created corresponding to every component. New requirement is given to every new agent.
   If there is no suitable knowledge in local knowledge base, the agent requires other knowledge via user interface.
   * Inference engine selects knowledge '(| (designA A) ~(designX X) ~(designY Y) ~(mergeModel A X Y))' in knowledge base. The object model A is decomposed to X and Y. Agents X and Y are created corresponding to the objects X and Y respectively.
3. Assign each problem to the lower agents. After then, the lower agents are activated.
   * The agent assigns the requirements '(designX X)?' and '(designY Y)?' to the agent X and Y respectively.
4. The higher-level agent receives solution from the lower level agent. Using the result problem solving continues there. If the solution is not obtained, the problem solving is carried out again using different knowledge.
   * The higher-level agent receives X=10 and Y=5 from the lower level agents. '(mergeModel A 10 5)?' is solved there to obtain A=15.
5. The solution is returned to the user interface. Inference engine replaces the variable part of the knowledge with solutions of sub-problems and problem, and Agent Controller registers it in Case Base.
   * 'A=15' is returned to user interface.

Interactions between agents are very important, but it is a very large problem. Our group still study about agents' interactions[7], this system has not yet realized interaction between agents that has same higher agent.

5 Knowledge Base

The knowledge base is required to retrieve an appropriate knowledge for the requests from agents in a short time for assuring the practicality of the system. Since large amount of knowledge from various types and domains is saved, the knowledge base must be well managed by a knowledge management system. The knowledge is divided into chunks by type information, domain information and the other information for aiding rapid retrieval of knowledge. These chunks are structured in the large knowledge base. The large knowledge base management system is itself a special agent. It accepts the request from the other agents, retrieve the required knowledge and send it back to the requesting agent.

6 Overall System Architecture

An overall system architecture is shown in Fig. 6.

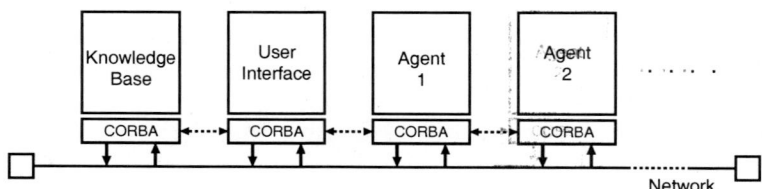

Fig. 6. System architecture

It consists of many computers on a network. The group of these computer systems is organized dynamically into a distributed, multi-agent system specified to a given problem. In principle, every computer is the same and symmetric to each other except a large knowledge-base management system that is specifically designed to achieve the specific role. In order for agents to cooperate to each other, CORBA is introduced. Since every agent works on KAUS system, a CORBA extension named KAUS-CORBA was developed.

7 Experiment

A slightly more complex problem than that used for explanation in section 5 has been solved as an example. This is to design a private house[3]. In this example, a house M is composed of such components as movementLine M0, equipment M1, livingSpace M2, privateRoom M3. A part of knowledge that corresponds to '(| (designA A) ~(designX X) ~(designY Y) ~(mergeModel A X Y))' used in the example before is given as follows.

(| (design-house M) ~(design-movementLine M0) ~(design-equipment M1)
      ~(design-livingSpace M2) ~(design-privateRoom M3)
      ~(merge M [M0 M1 M2 M3])).

―

(|(design-privateRoom Y) ~(wall Y0) ~(pdoor Y1) ~(pwindow Y2) ~(pfloor Y3)
      ~(pceiling Y4) ~(merge Y [Y0 Y1 Y2 Y3 Y4])).

―

(pfloor wooden). (pfloor plastic). (pfloor tatami). (pfloor carpet). (pfloor stone).

―

Fig. 7. A part of knowledge to design a house

This system was applied to design of the house. The arrangement of rooms is passed to top agent as requirement, agents select all parts of house.

The problem of design the house was divided into the sub-problems of design the parts; the different agents were assigned to the sub-problems and the designed the parts were merged to obtain the all of the house. Every sub-problem was solved by the different agents.

The knowledge base includes various alternative rules and it was confirmed that depending on the problem the way for decomposing the problem was also changed, the different organization of the agents is also generated and the results of the past trials were used effectively.

8 Conclusion

In this paper, it was discussed a way of solving large problems in a distributed multi-agent system. A problem is decomposed into sub-problems and, depending on this decomposition of the problem, agents are generated. These agents keep relation to each other for cooperation depending on the relations of sub-problems and therefore a multi-agent system is formed tailored to this specific problem. An agent is intelligent in the sense that it can solve various type of problem autonomously, and also it can create the other agent as needed. A basic idea, a way of problem solving, also a way of generating a multi-agent system, two experiments using very simple examples were included in this paper. This system is a part of a larger system the author's group is developing now. This is a very large system and there remain many problems to be solved. The part discussed in this paper is a central portion of the ideas on this system development.

Acknowledgement. This research was conducted sponsored by The Science and Technology Agency of Japanese Government. The authors would like to express sincere thanks to their support.

References

1. Caroline C. Hayes, Agents in a Nutshell - A very Brief Introduction, IEEE Transactions on Knowledge and Data Engineering, Vol. 11, No. 1, January/February 1999
2. M. Harandi and G. Rendon, "A Support Environment for Building Distributed Problem Solvers," Proc. IEEE Conf. Systems, Man, and Cybernetics, Oct. 1997.
3. M. Higashiyama, Construction Of Problem Model, Graduation Thesis of Waseda University, 2000
4. Y. Nishikawa, A Study on Construction of Large Scale Knowledge Base, Master Thesis of Waseda University, 2000
5. Setsuo Ohsuga, Toward truly intelligent information systems — from export systems to automatic programming, Knowledge-Based Systems, pp.363-396, Oct. 1998
6. Setsuo Ohsuga, Takumi Aida, Externalization of Human Idea and Problem Description for Automatic Programming, In Proceeding of the Eleventh International Symposium on Methodologies for Intelligent Systems, pp. 163-171, Jun. 1999
7. N. Shinkai, Negotiation and Strategy among Multi Agents, Graduation Thesis of Waseda University, 1999

8. G.W. Tan, C.C. Hayes, and M. Shaw, An Intelligent-Agent Framework for Concurrent Product Design and Planning, IEEE Trans. Eng. Management, vol.43, no.3, pp.297-306, Aug. 1996.
9. K. Tanaka, An Distributed Problem Solving System on the Use of Multi-agent, Master Thesis of Waseda University, 2000
10. Hiroyuki Yamamuchi, KAUS User's Manual Version 6.502, RCAST, University of Tokyo, 1999

A Comparative Study of Noncontextual and Contextual Dependencies

S.K.M. Wong[1] and C.J. Butz[2]

[1] Department of Computer Science, University of Regina
Regina, Saskatchewan, Canada S4S 0A2
E-mail: wong@cs.uregina.ca

[2] School of Information Technology & Engineering, University of Ottawa
Ottawa, Ontario, Canada K1N 6N5
E-mail: butz@site.uottawa.ca

Abstract. There is current interest in generalizing Bayesian networks by using dependencies which are more general than *probabilistic conditional independence* (CI). *Contextual* dependencies, such as *context-specific independence* (CSI), are used to decompose a subset of the joint distribution. We have introduced a more general contextual dependency than CSI, as well as a more general noncontextual dependency than CI. We developed these probabilistic dependencies based upon a new method of expressing database dependencies. By defining database dependencies using *equivalence relations*, the difference between the various contextual and noncontextual dependencies can be easily understood. Moreover, this new representation of dependencies provides a convenient tool to readily derive other results.

1 Introduction

Bayesian networks [5] have become an established framework for uncertainty management in artificial intelligence. Bayesian networks only use a single type of dependency, called *probabilistic conditional independence* (CI), to losslessly decompose a joint probability distribution. There is current interest, however, in generalizing Bayesian networks with more general dependencies. In [1], a *contextual* (horizontal) dependency, called *context-specific independence* (CSI), was introduced to capture CIs that only hold in some of the tuples in a joint distribution. In [8], we introduced a more general contextual dependency than CSI, as well as a more general noncontextual dependency than CI. The important point, however, is that our probabilistic dependencies were motivated by corresponding database dependencies.

Weak multivalued dependency (WMVD) [2,3] is a more general database dependency than *multivalued dependency* (MVD) [4]. Fischer and Van Gucht [2] gave several characterizations of WMVD. In this paper, we suggest a new characterization of both MVD and WMVD based on *equivalence relations*. In this framework, the difference between the various database dependencies can be easily understood. Moreover, this new representation of dependencies provides a convenient tool to readily derive other results.

This paper is organized as follows. In Section 2, we review some pertinent notions in the relational database model, and recall some notions about equivalence relations. We use this framework to express contextual and noncontextual dependencies in Section 3. In Section 4, we demonstrate the simplicity of our framework by showing the soundness of some known inference axioms. The conclusion is given in Section 5.

2 Basic Notions

2.1 Relational Databases

Here we review some notions used in the elegant relational database model [4].

A *relation scheme* $R = \{A_1, A_2, \ldots, A_m\}$ is a finite set of attributes. Corresponding to each attribute A_i is a nonempty finite set D_i, $1 \leq i \leq m$, called the *domain* of A_i. Let $D = D_1 \cup D_2 \ldots \cup D_m$. A *relation* r on the relation scheme R, written $r(R)$, is a finite set of mappings $\{t_1, t_2, \ldots, t_s\}$ from R to D with the restriction that for each mapping $t \in r$, $t(A_i)$ must be in D_i, $1 \leq i \leq k$, where $t(A_i)$ denotes the value obtained by restricting the mapping t to A_i. The mappings are called *tuples* and $t(A)$ is called the A-value of t. We use $t(X)$ in the obvious way and call it the X-value of t.

Mappings are used in our exposition to avoid any explicit ordering of the attributes in the relation scheme. To simplify the notation, however, we will henceforth denote relations by writing the attributes in a certain order and the tuples as lists of values in the same order. Furthermore, the following relational database conventions will be adopted for simplified notation. Uppercase letters A, B, C from the beginning of the alphabet may be used to denote attributes. A relation scheme $R = \{A_1, A_2, \ldots, A_m\}$ may be written as simply $A_1 A_2 \ldots A_m$. A relation r on scheme R is then written as either $r(R)$ or $r(A_1 A_2 \ldots A_m)$. The singleton set $\{A\}$ is sometimes written as A and concatenation XY may be used to denote set union $X \cup Y$.

The *select* σ, *project* π, and *natural join* $\bowtie$ operators are defined as follows.

When the *select* operator σ is applied to a relation r, it yields another relation that is a subset of tuples of r with a certain value on a specified attribute. Let r be a relation on scheme R, $A \in R$, and $a \in D_A$. Then

$$\sigma_{A=a}(r) = \{t \mid t \in r \text{ and } t(A) = a\}.$$

Whereas the select operator chooses a subset of tuples in a relation, the *project* operator π chooses a subset of attributes. Let r be a relation on R and X a subset of R. The *projection of* r *onto* X, written $\pi_X(r)$, is defined as

$$\pi_X(r) = \{ t(X) \mid t \in r \}. \tag{1}$$

The *natural join* of two relations $r_1(X)$ and $r_2(Y)$, written $r_1(X) \bowtie r_2(Y)$, is defined as

$$r_1(X) \bowtie r_2(Y)i = \{\ t(XY) \mid t(X) \in r_1(X) \text{ and } t(Y) \in r_2(Y)\ \}. \tag{2}$$

A fundamental database dependency, namely, *multivalued dependency* (MVD), can now be defined.

Definition 1. *Let X, Y, Z be pairwise disjoint subsets of scheme $R = XYZ$. A relation $r(XYZ)$ satisfies the multivalued dependency MVD(Y,X,Z), if for any two tuples t_1 and t_2 in r with $t_1(X) = t_2(X)$, there exists a tuple t_3 in r with $t_3(XY) = t_1(XY)$ and $t_3(Z) = t_2(Z)$.*

The multivalued dependency MVD(Y,X,Z) is a *necessary* and *sufficient* condition for $r(XYZ)$ to be losslessly decomposed as

$$r(XYZ) = \pi_{XY}(r) \bowtie \pi_{XZ}(r). \tag{3}$$

Example 1. The following relation $r_1(ABC)$ on the left satisfies the multivalued dependency MVD(A,B,C), since

$$r_1(ABC) = \pi_{AB}(r_1) \bowtie \pi_{BC}(r_1).$$

However, $r_2(ABC)$ on the right does *not* satisfy MVD(A,B,C) since

$$r_2(ABC) \neq \pi_{AB}(r_2) \bowtie \pi_{BC}(r_2).$$

A B C		A B		B C		A B C		A B		B C		A B C
$r_1(ABC) =$	0 0 0	$=$	0 0	$\bowtie$	0 0	, $r_2(ABC) =$	0 0 0	$\neq$	0 0	$\bowtie$	0 0	$=$	0 0 0
	0 0 1		1 0		0 1		1 0 1		1 0		0 1		0 0 1
	1 0 0		1 1		1 0		1 1 1		1 1		1 1		1 0 0
	1 0 1				1 1								1 0 1
	1 1 0												1 1 1
	1 1 1												

2.2 Properties of Equivalence Relations

Since we suggest that dependencies can be conveniently expressed using equivalence relations, we first recall some familiar notions about relations [6].

Given any subset $X \subseteq R$, we can define an *equivalence relation* $\theta(X)$ on r (a partition of r): for all $t_i, t_j \in r$,

$$t_i\ \theta(X)\ t_j, \quad \text{if } t_i(X) = t_j(X). \tag{4}$$

The *composition* operator $\circ$ is used to combine relations. Let $T = \{t_1, t_2, \ldots, t_s\}$ denote a finite set of objects. Consider two relations θ_1 and θ_2 on T. The binary operator $\circ$, called the *composition*, is defined by: for $t_i, t_k \in T$,

$$t_i(\theta_1 \circ \theta_2)t_k, \quad \text{if for some } t_j \in T \text{ both } t_i\theta_1 t_j \text{ and } t_j\theta_2 t_k. \tag{5}$$

It can be shown that the composition $\theta_1 \circ \theta_2$, of two individual equivalence relations θ_1 and θ_2, is itself an equivalence relation (a partition) if and only if $\theta_1 \circ \theta_2 = \theta_2 \circ \theta_1$.

We can then define MVD using equivalence relations as follows:

Definition 2. *Relation $r(XYZ)$ satisfies MVD(Y,X,Z), if*

$$\theta(X) \ = \ \theta(XY) \circ \theta(XZ) \ = \ \theta(XZ) \circ \theta(XY). \qquad (6)$$

3 Generalizing Multivalued Dependency

In this section, we generalize MVD with both *contextual* and *noncontextual* dependencies. Contextual dependencies only decompose a subset of the relation, while noncontextual dependencies decompose the entire relation.

3.1 Context Strong Multivalued Dependency (CSMVD)

Sometimes only a few tuples in a relation cause the violation of an MVD. In this section, we introduce *context strong multivalued dependency* (CSMVD) in order to losslessly decompose *part* (a subset) of a relation.

Consider the relation $r_1(ABC)$ in Figure 1. It can be verified that $r_1(ABC)$ does *not* satisfy MVD(A,B,C). The reason is because the definition of MVD requires that MVD(Y,X=x,Z) holds for *all* X-values x in relation $r(XYZ)$. In this example, this means that MVD(A,B=0,C) and MVD(A,B=1,C) must both hold. However, it can be seen that the MVD(A,B,C) holds when B=0, but *not* when B=1. The important point is that even though the entire relation $r_1(ABC)$ cannot be losslessly decomposed using MVD, namely,

$$r_1(ABC) \ \neq \ \pi_{AB}(r_1) \bowtie \pi_{BC}(r_1),$$

it is still possible to losslessly decompose the tuples $\sigma_{B=0}(r_1) = \{t_1, t_2, t_3, t_4\}$:

$$\sigma_{B=0}(r_1) \ = \ \pi_{AB}(\sigma_{B=0}(r_1)) \bowtie \pi_{BC}(\sigma_{B=0}(r_1)).$$

Definition 3. *Relation $r(XYZ)$ satisfies the context strong multivalued dependency CSMVD(Y,X=x,Z), if the equivalence class defined by $X = x$ in the equivalence relation $\theta(X)$ satisfies the following condition:*

$$\theta(X=x) \ = \ \theta(X=xY) \circ \theta(X=xZ) \ = \ \theta(X=xZ) \circ \theta(X=xY). \quad (7)$$

$$r_1(ABC) = \begin{array}{|ccc|} \hline A & B & C \\ \hline 0 & 0 & 0 \\ 0 & 0 & 1 \\ 1 & 0 & 0 \\ 1 & 0 & 1 \\ 0 & 1 & 0 \\ 0 & 1 & 1 \\ 1 & 1 & 1 \\ \hline \end{array}, \quad r_2(ABC) = \begin{array}{|ccc|} \hline A & B & C \\ \hline 0 & 0 & 0 \\ 0 & 0 & 1 \\ 1 & 0 & 0 \\ 1 & 0 & 1 \\ 2 & 0 & 2 \\ 3 & 0 & 2 \\ 0 & 1 & 0 \\ 1 & 1 & 0 \\ \hline \end{array}, \quad r_3(ABC) = \begin{array}{|ccc|} \hline A & B & C \\ \hline 0 & 0 & 0 \\ 0 & 0 & 1 \\ 1 & 0 & 0 \\ 1 & 0 & 1 \\ 2 & 0 & 2 \\ 2 & 0 & 3 \\ 3 & 0 & 2 \\ 1 & 1 & 1 \\ \hline \end{array}$$

Fig. 1. Relation $r_1(ABC)$ satisfies CSMVD(A,B=0,C). Relation $r_2(ABC)$ satisfies WMVD(A,B,C). Relation $r_3(ABC)$ satisfies CWMVD(A,B=0,C).

Example 2. Context Strong Multivalued Dependency. Let us verify that relation $r_1(ABC)$ in Figure 1 satisfies CSMVD(A,B=0,C). By Equation (4), we first obtain

$$\theta(B=0) = \{[t_1, t_2, t_3, t_4]\}.$$

By another application of Equation (4), we obtain the equivalence relations:

$$\theta(AB=0) = \{[t_1, t_2], [t_3, t_4]\},$$

and

$$\theta(B=0C) = \{[t_1, t_3], [t_2, t_4]\}.$$

Applying Equation (5) gives:

$$\theta(AB=0) \circ \theta(B=0C) = \{[t_1, t_2, t_3, t_4]\} = \theta(B=0C) \circ \theta(AB=0).$$

We have our desired result since:

$$\theta(B=0) = \theta(AB=0) \circ \theta(B=0C) = \theta(B=0C) \circ \theta(AB=0).$$

However, it can be similarly verified that CSMVD(A,B=1,C) is *not* satisfied.

CSMVD generalizes MVD by only decomposing *some* of the tuples in a relation. However, it is also possible to generalize MVD with a noncontextual dependency, called *weak multivalued dependency* (WMVD), which decomposes *all* of the tuples in a relation.

3.2 Weak Multivalued Dependency (WMVD)

Weak multivalued dependency (WMVD) [3] generalizes MVD(Y,X,Z) in Definition (2) by not requiring the equivalence relation $\theta(XY) \circ \theta(XZ) = \theta(XZ) \circ \theta(XY)$ to be equal to $\theta(X)$.

Definition 4. *Relation $r(XYZ)$ satisfies the weak multivalued dependency WMVD(Y,X,Z), if*

$$\theta(XY) \circ \theta(XZ) = \theta(XZ) \circ \theta(XY). \qquad (8)$$

Example 3. Weak Multivalued Dependency. Let us verify that relation $r_2(ABC)$ in Figure 1 satisfies WMVD(A,B,C). By Equation (4), we obtain:

$$\theta(AB) = \{[t_1, t_2], [t_3, t_4], [t_5], [t_6], [t_7], [t_8]\},$$

and

$$\theta(BC) = \{[t_1, t_3], [t_2, t_4], [t_5, t_6], [t_7, t_8]\}.$$

Applying Equation (5), we obtain our desired result since:

$$\theta(AB) \circ \theta(BC) = \{[t_1, t_2, t_3, t_4], [t_5, t_6], [t_7, t_8]\} = \theta(BC) \circ \theta(AB).$$

Thus, even though a relation does not satisfy MVD(Y,X,Z), it may still be possible to losslessly decompose the *entire* relation using WMVD(Y,X,Z).

3.3 Context Weak Multivalued Dependency (CWMVD)

We can introduce a contextual version of WMVD, called *context weak multivalued dependency* (CWMVD).

Definition 5. *Relation $r(XYZ)$ satisfies the context weak multivalued dependency CWMVD(Y,X=x,Z), if there exists a maximal disjoint compatibility class $\{t_i, \ldots, t_j\}$ in the relation $\theta(X = xY) \circ \theta(X = xZ)$.*

Definition 5 implies that $\{t_i, \ldots, t_j\}$ satisfies MVD(Y,X,Z).

Example 4. Context Weak Multivalued Dependency. To verify that relation $r_3(ABC)$ in Figure 1 satisfies WMVD(A,B=0,C), we first obtain:

$$\theta(AB = 0) = \{[t_1, t_2], [t_3, t_4], [t_5, t_6], [t_7]\},$$

and

$$\theta(B = 0C) = \{[t_1, t_3], [t_2, t_4], [t_5, t_7], [t_6]\}.$$

Applying Equation (5), we obtain $\mathcal{R} = \theta(AB = 0) \circ \theta(B = 0C)$:

$$\begin{aligned}\mathcal{R} = \{&t_1\mathcal{R}t_1, t_1\mathcal{R}t_2, t_1\mathcal{R}t_3, t_1\mathcal{R}t_4, t_2\mathcal{R}t_1, t_2\mathcal{R}t_2, t_2\mathcal{R}t_3, t_2\mathcal{R}t_4,\\ &t_3\mathcal{R}t_1, t_3\mathcal{R}t_2, t_3\mathcal{R}t_3, t_3\mathcal{R}t_4, t_4\mathcal{R}t_1, t_4\mathcal{R}t_2, t_4\mathcal{R}t_3, t_4\mathcal{R}t_4,\\ &t_5\mathcal{R}t_5, t_5\mathcal{R}t_6, t_5\mathcal{R}t_7, t_6\mathcal{R}t_5, t_6\mathcal{R}t_6, t_6\mathcal{R}t_7, t_7\mathcal{R}t_5, t_7\mathcal{R}t_7\}.\end{aligned}$$

Note that $t_7\mathcal{R}t_6$ is not a member in $\mathcal{R}$. Thus, $\{t_1, t_2, t_3, t_4\}$ is a maximal disjoint compatibility class, i.e., $\{t_1, t_2, t_3, t_4\}$ satisfies MVD(A,B,C). Therefore, relation $r(ABC)$ satisfies CWMVD(A,B=0,C).

4 Comparing Strong Versus Weak Dependencies

Our purpose in this section is show that weak dependencies are more general than strong dependencies.

Lemma 1. [2,3] MVD is a special case of WMVD.

Lemma 2. CSMVD is a special case of CWMVD.

Similarly, contextual dependencies are more general than their noncontextual counterparts.

Lemma 3. CSMVD is a more general dependency than MVD.

Lemma 4. CWMVD is a more general dependency than WMVD.

The relationships between all of these dependencies can be summarized as:

$$MVD \implies WMVD \implies CWMVD,$$

and

$$MVD \implies CSMVD \implies CWMVD.$$

It should be noted that WMVD does not logically imply CSMVD, and vice versa. For example, relation $r_2(ABC)$ in Figure 1 satisfies WMVD(A,B,C), but not CSMVD(A,B=0,C). On the other hand, relation $r_1(ABC)$ in Figure 1 satisfies CSMVD(A,B=0,C), but not WMVD(A,B,C).

5 Axiomatization of the Noncontextual Dependencies

By expressing dependencies using equivalence relations, it is straightforward to show the soundness of several inference axioms.

The following two axioms (MW1) and (MW2) are a sound and complete axiomatization for the mixture of MVD and WMVD [7]:

(MW1)   If MVD(Y,X,Z), then WMVD(Y,X,Z);
(MW2)   If WMVD(Y,XZ,W), WMVD(Y,XW,Z), and MVD(Z,XY,W), then WMVD(Y,X,ZW).

The soundness of axiom (MW1) follows directly from the definitions of MVD and WMVD. By definition, WMVD(Y,XZ,W), WMVD(Y,XW,Z), and MVD(Z, XY,W) imply:

$$\theta(XZY) \circ \theta(XZW) = \theta(XZW) \circ \theta(XZY), \tag{9}$$

$$\theta(XWY) \circ \theta(XZW) = \theta(XZW) \circ \theta(XWY), \tag{10}$$

and
$$\theta(XY) = \theta(XYZ) \circ \theta(XYW) = \theta(XYW) \circ \theta(XYZ), \quad (11)$$

respectively. Using Equations (9)-(11) it follows
$$\begin{aligned}\theta(XY) \circ \theta(XZW) &= \theta(XYZ) \circ \theta(XYW) \circ \theta(XZW) \\ &= \theta(XYZ) \circ \theta(XZW) \circ \theta(XYW) \\ &= \theta(XZW) \circ \theta(XYZ) \circ \theta(XYW) \\ &= \theta(XZW) \circ \theta(XY). \end{aligned} \quad (12)$$

Equation (12) indicates that WMVD(Y,X,ZW) as desired.

As a second example, the following four inference axioms (W1)-(W4) are a sound and complete axiomatization for WMVD [2]:

(W1)   If $U \subseteq X$, then WMVD(U,X,YZ);
(W2)   If WMVD(YX,XVZ,W), then WMVD(Y,XVZ,W) and WMVD(YXV,XVZ,W);
(W3)   If WMVD(Y,X,ZW), then WMVD(Y,XZ,W);
(W4)   If WMVD(Y,X,Z), then WMVD(Z,X,Y).

Two properties [6] of $U \subseteq X$ are that $\theta(XU) = \theta(X)$ and
$$\theta(U) \circ \theta(X) = \theta(X) \circ \theta(U) = \theta(X).$$

Thus, inference axiom (W1) is sound since $\theta(XU) \circ \theta(XYZ) = \theta(X) \circ \theta(XYZ) = \theta(XYZ) = \theta(XYZ) \circ \theta(X) = \theta(XYZ) \circ \theta(XU)$. Therefore, $WMVD(U, X, YZ)$.

To show the soundness of (W2), we are given:
$$\theta(XVZYX) \circ \theta(XVZW) = \theta(XVZW) \circ \theta(XVZYX),$$

or equivalently,
$$\theta(XVZY) \circ \theta(XVZW) = \theta(XVZW) \circ \theta(XVZY).$$

This is the definition of WMVD(Y,XVZ,W). Now consider
$$\begin{aligned}\theta(XVZYXV) \circ \theta(XVZW) &= \theta(XVZY) \circ \theta(XVZW) \\ &= \theta(XVZW) \circ \theta(XVZY). \end{aligned}$$

This is the definition of WMVD(YXV,XVZ,W).

In inference axiom (W3), we are initially given:
$$\theta(XY) \circ \theta(XZW) = \theta(XZW) \circ \theta(XY).$$

We want to show:
$$\theta(XYZ) \circ \theta(XZW) = \theta(XZW) \circ \theta(XYZ).$$

Consider
$$t_1\theta(XYZ)t_2 \text{ and } t_2\theta(XZW)t_3.$$
Since $t_1(XYZ) = t_2(XYZ)$, this implies that
$$t_1\theta(XY)t_2 \text{ and } t_2\theta(XZW)t_3.$$
By the given WMVD(Y,X,ZW), we obtain
$$t_1\theta(XZW)t_4 \text{ and } t_4\theta(XY)t_3.$$
What remains to be shown is that $t_4(XYZ) = t_3(XYZ)$, namely, $t_4(Z) = t_3(Z)$. Now $t_4(Z) = t_1(Z) = t_2(Z) = t_3(Z)$. Therefore, we have our desired result:
$$t_1\theta(XZW)t_4 \text{ and } t_4\theta(XYZ)t_3.$$

The soundness of (W4) follows directly from Definition 4.

6 Conclusion

In this paper, we have suggested a new characterization of MVD and WMVD based on equivalence relations. This characterization clearly exhibits the difference between not only these two database dependencies, but also their contextual counterparts. By expressing MVD and WMVD with equivalence relations, other results can be readily shown as we demonstrated by proving the soundness of the corresponding inference axioms. More importantly, the results here can be applied to the recent interest in contextual probabilistic conditional independence in Bayesian networks.

References

1. C. Boutilier, N. Friedman, M. Goldszmidt and D. Koller, Context-specific independence in Bayesian networks, *Proceedings of the Twelfth Conference on Uncertainty in Artificial Intelligence*, 115–123, 1996.
2. P. Fischer and D. Van Gucht, Weak multivalued dependencies, *Proceedings of the Third ACM SIGACT-SIGMOD Symposium on the Principles of Database Systems*, 266–274, 1984.
3. G. Jaeschke and H.J. Schek, Remarks on the algebra on non first normal form relations, *Proceedings of the First ACM SIGACT-SIGMOD Symposium on the Principles of Database Systems*, 124–138, 1982.
4. D. Maier, *The Theory of Relational Databases*, Computer Science Press, Rockville, Maryland, 1983.
5. J. Pearl, *Probabilistic Reasoning in Intelligent Systems: Networks of Plausible Inference*, Morgan Kaufmann, San Francisco, California, 1988.
6. F. Preparata and R. Yeh, *Introduction to Discrete Structures*, Addison-Wesley, Don Mills, Ontario, 1973.
7. D. Van Gucht and P. Fischer, MVDs, weak MVDs and nested relational structures, *Technical report*, CS-84-19, Vanderbilt University, October, 1984.
8. S.K.M. Wong and C.J. Butz, Contextual weak independence in Bayesian networks, *Proceedings of the Fifteenth Conference on Uncertainty in Artificial Intelligence*, 670–679, 1999.

Extended Query Answering Using Integrity Rules

Barry G.T. Lowden and Jerome Robinson

Department of Computer Science, The University of Essex,
Wivenhoe Park, Colchester CO4 3SQ, Essex,
United Kingdom
lowdb@essex.ac.uk

Abstract. The conventional use of databases is commonly restricted to the retrieval of factual data in the form of tuples or records. However most databases also contain metadata in the form of integrity rules which can provide a rich source of additional information not normally available to the user. Integrity rules define what data values and relationships may exist within the database and so their interrogation can provide answers as to whether a certain database state is possible. Our paper describes how this may be achieved and specifies a formal approach to implementing such an enquiry system.

1 Introduction

A database can be seen as comprising an extension, which is the set of tuples representing the current state of the database content, and the schema which describes the structure of and permitted relationships within the database. An important component of the latter is the set of integrity rules (constraints) which define the conditions which must apply to all entries within the extension [2,3,5]. These rules are enforced whenever changes are made to the database content. Most relational systems only permit query execution against the extension, the query result then consisting of a subset of the current database content. This means that query answers are limited to the current situation since they relate to the particular state of the database at the time of the query. Integrity rules, on the other hand, embody information which defines all legal states of the database, that is they specify what is possible.

Being able to access integrity information would permit the formulation of modal queries i.e. queries about what can be or must be the case. Examples of modal queries might be 'must all managers who earn more than £30,000' and work in London, be provided with a company car?', or 'under what conditions can Jones earn £25,000?'. Most current systems permit only limited access to the integrity rules and do not allow the user to formulate this type of query, even though the information is available to provide an answer. Note that this type of question is different from the 'what if' scenario of say financial planning [12] or AI reasoning [14] where the user is interested in the implications of some hypothetical update of the database which is known to be valid, see also earlier work [13]. We are here concerned with whether certain database states are permissible with respect to the integrity rules and, if not, what further conditions would make them permissible.

In the following sections we describe an approach to processing certain classes of modal query which can be used to enhance and augment an existing conventional query system. Section 2 outlines the basis for our approach and presents a restricted specification for the query constructs. In section 3 we show, through a series of examples, the algorithm for generating a modified database state which reflects the requirements of the modal query. The process of evaluating this changed state against the integrity rules is then described in section 4 while finally our conclusions are presented in section 5. This work is an extension to the programme of research, by the authors, into the generation, evaluation and application of rules in databases [6,7,8,11].

2 Representation of Modal Queries

Referring back to our second example above, we can see that were we to simulate a modification to the database so that Jones did indeed earn £25,000 and then determine whether the modified state violated any of the integrity rules then we would know the answer to the query. Any violations could be reported together with a brief explanation of their nature. If, however, there are no rule violations, associated with the hypothetical update, then it may be seen that there are no required conditions for the specified state to be permissible. The basis of our approach is therefore to generate a set of modifications to the existing database state which makes the query true, and then evaluate the integrity rules with respect to this modified state.

In relational systems, integrity rules are traditionally expressed in some form of the predicate calculus [4,9,10], however, in order not to restrict ourselves to any particular implementation, we will use a standard version of the Tuple Relational Calculus (TRC) to express both queries and integrity rules [4]. To simplify matters further we will concern ourselves only with the modal operator of possibility P, and make the assumption that modal expressions may contain only existential quantification, conjunction and the standard comparison operators. The syntax of our restricted TRC subset is therefore:

Modal queries: $P\phi$
Sentences $\phi, \eta ::= (\phi \wedge \eta) \mid (\exists v \in \tau \phi) \mid (x \theta y)$
Terms $x, y ::= \kappa \mid v.\alpha$
Rel $\theta ::= > \mid < \mid = \mid \neq \mid \geq \mid \leq$
    where $\tau \in$ *Type*, $\kappa \in$ *Constant*, $v \in$ *Variable*, $\alpha \in$ *Attribute*

A modal query of the form $P\phi$ would be checked initially to determine whether ϕ was true with respect to the current database state. If this were the case then the query processing mechanism terminates with a response to the user that the proposition is currently true. If, however, it is the case that ϕ does not hold with respect to the current database state then we need to find some modification to the existing state which makes $P\phi$ true and then check this modified state against the integrity rules.

3 Generating Hypothetical Modifications to the Database

The process of generating database modifications which make ϕ true involves building a representation of the basic requirements of the query and then instantiating this representation according to information currently in the database relations. The process will now be illustrated, by means of a simple example, with respect to the following database relation:

EMP(name, status, salary, department).

which contains the tuple:

(name = Jones, status = full-time, salary = 35K, department = Sales)

Let us assume that we wish to pose the query *'Can Jones work in Accounts?'*. This query translates to the TRC expression:

$\phi 1$: *P $\exists x \in$ EMP[x.name = Jones $\wedge$ x.department = Accounts]*

To begin with, we encode the requirements of the query using a procedure taking four arguments: the TRC expression ϕ, a variable binding environment E, an incoming list of tuples IN and an outgoing list of tuples OUT. The initial state is a full TRC expression, an empty environment and an empty list of incoming tuples, whilst the outgoing tuples are unknown.

Thus, for the above example:

TRC = $\exists x \in$ *EMP[x.name=Jones $\wedge$ x.department=Accounts]*
E = *NULL*
IN = *NULL*
OUT = ?

For each occurrence of an existential quantifier, in the query expression, we generate a unique tuple identifier together with an extended environment in which the quantified variable is associated with this identifier. For each associated relation, we also define a tuple template in which each attribute value is associated with a unique uninstantiated variable. The template(s), together with the tuple identifier(s), is then added to the incoming list, IN. We may now process the body of the quantified expression with respect to the new environment and the extended incoming list.
Thus:

TRC = *x.name=Jones $\wedge$ x.department=Accounts*
E = *[x:t1]*
IN = *[t1: (name=u1, status= u2, salary= u3, department= u4)$\in$EMP]*
OUT= ?

Conjunction is handled by processing each branch of the conjunct separately. Each conjunct will inherit the same environment, but the left conjunct is processed with respect to the IN list of the whole conjunction whilst the IN list of the right conjunct is set to the OUT list of the left conjunct.

We will consider the left conjunct first.

TRC = *x.name=Jones*
E = *[x:t1]*
IN = *[t1: (name= u1, status= u2, salary= u3, department= u4)∈EMP]*
OUT = *?*

This is the base case for the recursive process and we now unify [1] the objects denoted by each side of the equality. To do this we retrieve the value associated with x.name by noting that the variable x is associated with the tuple t1, and t1 identifies the tuple whose name attribute has the value u1. Unifying Jones with u1 therefore instantiates tuple t1's name attribute to the value Jones. In cases where there is a variable on each side of the equality, unification will ensure that the appropriate attributes take the same value, whilst if there is a constant on each side then the unification process will only succeed if their values are the same.

Having unified the appropriate objects we now simply copy the IN list to the OUT list and begin to unwind the recursion. We will therefore exit with the values:

TRC = *x.name=Jones*
E = *[x:t1]*
IN = *[t1: (name=Jones, status= u2, salary= u3, department= u4)∈EMP]*
OUT = *[t1: (name=Jones, status= u2, salary= u3, department= u4)∈EMP]*

We may now proceed to process the right branch of the conjunction, which inherits its IN list from the OUT list of the left conjunct.

TRC = *x.department=Accounts*
E = *[x:t1]*
IN = *[t1: (name=Jones, status= u2, salary= u3, department= u4)∈EMP]*
OUT = *?*

This is also a base case and, using the procedure described, we exit with the values:

TRC = *x.department=Accounts*
E = *[x:t1]*
IN = *[t1: (name=Jones, status= u2, salary= u3, department=Accounts)∈EMP]*
OUT = *[t1: (name=Jones, status= u2, salary= u3, department=Accounts)∈EMP]*

Each branch of the conjunction has now been processed and we exit by setting the OUT list of the whole conjunction to be the OUT list of the right conjunct, ie.

TRC = *x.name=Jones* ∧ *x.department=Accounts*
E = *[x:t1]*
IN = *[t1: (name=Jones, status= u2, salary= u3, department=Accounts)∈EMP]*
OUT = *[t1: (name=Jones, status= u2, salary= u3, department=Accounts)∈EMP]*

The final stage of our query processing algorithm is to set the OUT list of the existentially quantified expression to be that of the body of the expression (the conjunction):

$$TRC = \exists x \in EMP[x.name{=}Jones \land x.department{=}Accounts]$$
$$E = NULL$$
$$IN = NULL$$
$$OUT = [t1: (name{=}Jones,\ status{=}\ u2,\ salary{=}\ u3,\ department{=}Accounts) \in EMP]$$

and we see that, in order for the TRC expression associated with our query to be true, a tuple of the form:

$(name{=}Jones,\ status{=}\ u2,\ salary{=}\ u3,\ department{=}Accounts) \in EMP$

must be present in the database.

Before attempting to instantiate this with respect to the current database state we must merge any generated tuples which have the same key fields. For example, the TRC translation for 'Can Jones work in Accounts and earn 30K?' is:

$$P\{\exists x \in EMP[x.name{=}Jones \land x.department{=}Accounts]$$
$$\land\ \exists x \in EMP[x.name{=}Jones \land x.salary{=}\ 30K]$$

Applying the above algorithm, returns two tuples:

$(name{=}Jones,\ status{=}\ u2,\ salary{=}\ u3,\ department{=}Accounts\} \in EMP$
$(name{=}Jones,\ status{=}\ u2,\ salary{=}\ 30K,\ department{=}\ u4\} \in EMP$

Since name, however, is the key field of EMP, these tuples must be the same and we therefore merge them to produce a single tuple:

$(name{=}Jones,\ status{=}\ u2,\ salary{=}\ 30K,\ department{=}Accounts\} \in EMP$

The process, described above, can fail at two points, namely when we attempt to unify both sides of an equality which equates different values, and when we attempt to merge tuples whose keys are the same yet which differ in some other attribute. In both cases the underlying problem is that the query is about a contradictory database state, ie a state which would be impossible regardless of any constraints that might exist. In such cases we proceed no further, reporting this reason for failure to the user.

Having generated the query encoded tuples, the next stage involves hypothetically modifying the database so that it contains tuples which match their characteristics. Changes to a database may be effected in three main ways: inserting a new tuple, modifying an existing tuple or deleting an existing tuple. Our system makes use of inserts and updates, there being one of these operations for each query tuple generated by the query expression. The information contained in the tuple determines which of the above operations is applied. Where the key field of the query tuple is fully instantiated and matches the key field of an existing tuple, then the information in the query tuple will be used to update the existing tuple. If there is no key match then the query tuple will be used to insert a new tuple into the database.

Updates

We first consider modal queries which lead to updates only. Returning to our example, it is clear that a tuple of the form:

    t1(name=Jones, status= u2, salary= u3, department=Accounts) ∈*EMP*

must be present in the database for the query to be true. However, it is the case that the tuple:

    (name=Jones, status= 'full-time', salary= 35K, department=Sales)

already exists.

Thus we have the situation where the key of the query tuple is instantiated and matches the key field of an existing tuple. The other attributes of the query tuple may now be instantiated to provide a specification of the appropriate tuple. This is accomplished by setting each uninstantiated attribute to the value of the associated attribute in the existing tuple. Thus the query tuple, in our example, becomes:

    t1(name=Jones, status= full-time, salary= 35K, department=Accounts) ∈*EMP*

The next step is to hypothetically update the existing tuple to the new values specified and check the integrity rules, using a standard integrity enforcement mechanism, with respect to the modified state. If none are violated then the modified state is legal, and the system responds to the modal question in the affirmative. If there are any violations, then these can be reported together with an explanatory message indicating which rules were violated and thus the reasons why the state specified by the query is not possible. The database is then returned to its original state.

Ambiguous Queries

Unlike conventional queries, modal queries may involve ambiguity which is only made apparent when the query tuples are instantiated with respect to the current database. Consider, for example, the query 'Can Jones earn Smith's salary?' which translates to:

    $\phi 2$: $P \exists x \in EMP \exists y \in EMP[x.name= Jones \land y.name= Smith \land x.salary= y.salary]$

This generates the two query tuples:

    t1{name = Jones, status = u2, salary =u3, department = u4}) ∈ *EMP*
    t2{name = Smith, status = u6, salary = u3, department = u8}) ∈ *EMP*

where the salary attributes of both tuples, though uninstantiated, have the same value.

Let the EMP relation contain the following tuples:

    {name = Jones, status = full-time, salary = 35K, department = Sales}
    {name = Smith, status = part-time, salary = 40K, department = Engineering}

Note that both of the query tuples will result in updates, since their keys match existing tuples, but the resulting database state will be determined by which of the tuples we instantiate first. Choosing 'Jones' gives:

t1{name=Jones, status=full-time, salary= 35K, department=Sales} ∈EMP
t2{name=Smith, status= u6, salary= 35K, department= u8}) ∈EMP

and we see that instantiating Jones' tuple also sets the salary attribute of Smith's tuple. Instantiating this second tuple with respect to the database results in:

t1{name=Jones, status= full-time, salary=35K, department=Sales}) ∈EMP
t2{name=Smith', status= part-time, salary=35K, department=Engineering} ∈EMP

The overall effect then, is of leaving Jones' tuple unchanged while setting Smith's salary to that of Jones. Instantiating Smith's tuple first, however, gives:

t1{name = Jones, status = u2, salary = 40K, department = u4}) ∈ EMP
t2{name=Smith, status=part-time, salary=40K, department=Engineering} ∈EMP

then instantiating Jones' tuple produces the following query tuples:

t1{name=Jones, status= full-time, salary=40K, department=Sales}) ∈EMP
t2{name=Smith, status=part-time, salary=40K, department=Engineering} ∈EMP

In this latter case, the resulting database state would reflect that Smith's tuple was unchanged, and Jones' salary is set to Smith's. Whilst, intuitively, this modification seems more consistent with what was intended, both alternatives would appear to be acceptable. The system must therefore recognise that ambiguity can arise, in this type of situation, and be able to present the user with some means of choosing between the alternatives available.

Insertions

We now consider modal queries which generate tuple inserts to the database. Let us assume that we wish to check whether certain employment conditions are valid before making an offer of employment to a candidate named Walker. A possible query might be 'Can Walker have part-time working status and earn 35K and work in Engineering?' This translates to the following TRC query:

$\phi 3$: P $\exists x \in$ EMP[x.name=Walker ∧ x.status=part-time ∧ x.salary=35K
∧ x.department=Engineering]

resulting in the single query tuple:

t1(name=Walker, status=part-time, salary=35K, department=Engineering) ∈EMP

There is no existing tuple with a key field value of Walker, but all the attributes are instantiated. An attempt may therefore be made to insert this hypothetical tuple into the database. Should the transaction lead to an integrity violation then this would imply that the above conditions of employment were not valid with respect to the database.

Evaluation of 'must' type queries may be seen conceptually as the construction of two possibility sub-queries. Thus the query 'must all full-time salespersons earn at least 20K ?' would translate to:

(1) 'Can a person be full-time and work in Sales ?'

and if so (2) 'Can a person be full-time and work in Sales and earn less than 20K ?'

A positive response to (1) and a negative response to (2) would be required to establish the truth of the original query. It is necessary to establish the truth of the first sub-query since a negative response to the second alone may be attributable, for example, to violation of an integrity rule which requires that salespersons have to be either 'occasional' or 'part-time'.

In practice it is sufficient to pose the first sub-query and then let the system determine what assumptions need to be made regarding any unknown values in order for the modal record to be a legal transaction with respect to the integrity rules. Thus if one such assumption were that the employee earns 20K, or more, then this would imply a positive response to the original question. This process is discussed further in the next section.

4 Integrity Rule Evaluation

In cases where the database is modified using query tuples in which all attribute values are known, the program for checking the new database state against the integrity rules is relatively straightforward. Problems arise, however, when the query tuples contain unknown values ie. ones which have not been instantiated. In such cases it is necessary for the algorithm to make assumptions about these values based on the conditions imposed by the integrity rules. If a set of assumptions can be established which meets the criteria for satisfying all the integrity rules then the database state is deemed consistent and the query may be answered in the affirmative. If, however, no such set of assumptions can be constructed then the hypothetical update fails and the query conditions cannot be met.

For example, assume that our database relation is extended to include the attribute location and we wish to determine under what conditions a new employee can work in Engineering and also be located in London. This translates to the modal query:

$\phi 5$: P $\exists x \in$ EMP[x.department=Engineering $\wedge$ x.location=London]

resulting, since no further values may be instantiated, in the query tuple:

t1(name=u1, status=u2, salary=u3, department=Engineering, location=London)
$\in$ EMP

Assume further that we have the following integrity rules:

$\Psi 1$: $\forall x \in$ EMP[x.status = {part-time, full-time, occasional}]

$\Psi 2$: $\forall x \in$ EMP[x.status = occasional $\rightarrow$ x.salary < 10K]

$\Psi 3$: $\forall x \in$ EMP[x.department=Engineering $\rightarrow$ x.salary > 20K]

$\Psi 4$: $\forall x \in$ EMP[x.department=Engineering $\rightarrow$ x.location={Hull,London, Swindon}]

The evaluation algorithm processes through the rules as follows:

(i) $\Psi 1$ is true if $(u2 = \{part\text{-}time, full\text{-}time, occasional\})$
(ii) $\Psi 2$ is true if $(u2 = occasional) \wedge (u3 < 10K)$
(iii) $\Psi 3$ is true if $(u3 > 20K)$
(iv) $\Psi 4$ is true.

Clearly (ii) and (iii) are inconsistent and the algorithm now re-examines the necessary conditions which make these rules true. Simple analysis reveals that $\Psi 2$ is also satisfied by:

(a) $\neg(u2 = occasional) \wedge (u3 < 10K)$
or (b) $\neg(u2 = occasional) \wedge \neg(u3 < 10K)$

Whilst (a) still yields an inconsistency, it may be seen that (b) makes the query true with respect to the rules and so the necessary conditions for answering in the affirmative are that:

$(u2 = \{part\text{-}time, full\text{-}time, occasional\}) \wedge \neg(u2 = occasional)) \wedge (u3 > 20K)$

The system response is therefore that the database state implied by the original question is possible provided that the employee's status is 'part-time' or 'full-time' and that his/her salary is greater that 20K.

In practice the algorithm can be made to terminate when a set of necessary conditions is found or, alternatively, it may continue to find all possible sets of necessary conditions and report to the user that any one of these sets will result in an affirmative answer. This latter response enables the user to explore the complete range of possibilities for making the desired state true. Assumptions common to all possible sets imply a necessary condition which *must* be met.

Whilst, in the interests of clarity, we have limited ourselves to relatively simple examples in presenting the rule evaluation process, the working algorithm is equally suited to large and complex integrity rule sets.

5 Conclusions

In this paper, we have explained how it is possible to provide an extended database answering system which permits modal queries against the database integrity rules. An algorithm has been described which takes the query and from it constructs a set of database modification tuples which, when applied, bring about a changed state of the database which is consistent with the original question.

This changed state may then be checked against the integrity rules to determine whether the required conditions have been met and whether the question may be answered in the affirmative. We have also addressed the issue of ambiguity inherent in many questions concerning possibility and shown that our system can respond in a manner which permits the user to choose the interpretation closest to her intentions.

We have also examined the situation where the answers to modal questions are qualified by imposing one or more conditions and the way in which different condition scenarios may be presented to the user. This further enhances the user's understanding of the relationships inherent in the database which determine the system's response, thereby providing further information which is helpful in guiding the query process.

Work is currently being carried out on implementing a more comprehensive system to incorporate a greater range of query constructs and operators.

6 References

[1] Bundy A., 'The computer modeling of mathematical reasoning', Academic Press, London, 1986.

[2] Codd E.F., 'Domains, keys and referential integrity in relational databases', Info DB3, No.1, 1988

[3] Elmasri R. and Navathe S.B., 'Fundamentals of database systems', 3nd Edition, Addison-Wesley, 2000

[4] Freytag J. and Goodman N., 'On the translation of relational queries into iterative programs', ACM Trans. Database Syst. Vol.14, No.1, 1 – 27, 1989.

[5] Godfrey P., Grant J., Gryz J. and Minker J., 'Integrity Constraints: Semantics and applications', Logics for Databases and Information Systems, Kluwer, Ch.9, 1998.

[6] Lowden B.G.T. and Robinson J., 'A semantic query optimiser using automatic rule derivation', Proc. Fifth Annual Workshop on Information Technologies and Systems, Netherlands, 68-76, December 1995.

[7] Lowden B.G.T. and Robinson J., 'A statistical approach to rule selection in semantic query optimisation', Proc. 11th ISMIS International Symposium on Methodologies for Intelligent Systems, Warsaw, June 1999.

[8] Lowden B.G.T. and Robinson J., 'A fast method for ensuring the consistency of integrity constraints', Proc. 10th International DEXA conference on Database and Expert Systems Applications, Florence, August 1999

[9] Qian X., 'The expressive power of the bounded-iteration construct', Acta Inf. 28, 631 – 656, October 1991.

[10] Qian X., 'The deductive synthesis of database transactions', ACM Trans. Database Syst. 18, 4, 626 – 677, December 1993.

[11] Sayli A.and Lowden B.G.T., 'A fast transformation method for semantic query optimisation', Proc. IDEAS'97, IEEE, Montreal, 319-326, 1997.

[12] Sprague R.H. & Watson H.J., Decision support systems: putting theory into practice, Prentice Hall, 1989.

[13] Stonebraker M., Hypothetical databases as views, Proc. ACM Sigmod International Conference on Management of Data, 1981.

[14] Vielle L. et al., The EKS- V1 system, Proc. International Conference on Logic Programming & Automated Reasoning, Springer Verlag, 1992.

Finding Temporal Relations: Causal Bayesian Networks vs. C4.5

Kamran Karimi and Howard J. Hamilton

Department of Computer Science
University of Regina
Regina, SK
Canada S4S 0A2
{karimi,hamilton}@cs.uregina.ca

Abstract. Observing the world and finding trends and relations among the variables of interest is an important and common learning activity. In this paper we apply TETRAD, a program that uses Bayesian networks to discover causal rules, and C4.5, which creates decision trees, to the problem of discovering relations among a set of variables in the controlled environment of an Artificial Life simulator. All data in this environment are generated by a single entity over time. The rules in the domain are known, so we are able to assess the effectiveness of each method. The agent's sensings of its environment and its own actions are saved in data records over time. We first compare TETRAD and C4.5 in discovering the relations between variables in a single record. We next attempt to find temporal relations among the variables of consecutive records. Since both these programs disregard the passage of time among the records, we introduce the flattening operation as a way to span time and bring the variables of interest together in a new single record. We observe that flattening allows C4.5 to discover relations among variables over time, while it does not improve TETRAD's output.

1 Introduction

In this paper we consider the problem of discovering relations among a set of variables that represent the states of a single system as time progresses. The data are a sequence of temporally ordered records without a distinguished time variable. Our aim is identify as many cases as possible where two or more variables' values depend on each other. Knowing this would allow us to explain how the system may be working. We may also like to control some of the variables by changing other variables. We use data from a simple Artificial Life [6] domain because it allows us to verify the results, and thus compare the effectiveness of the algorithms.

Finding associations among the observed variables is considered a useful knowledge discovery activity. For example, if we observe that $(x = 5)$ is always true when $(y = 2)$, then we could predict the value of y as 2 when we see that x is 5. Alternatively, we could assume that we have the rule: **if** $\{(x = 5)\}$ **then** $(y = 2)$, and use it to set the value of y to 2 by setting the value of x to 5. Some researchers [1, 10] have tried to find the stronger notion of causality among the observed variables. In the previous example, they may call x a *cause* of y.

In this paper we consider two approaches to the problem of finding relations among variables. TETRAD [9] is a well-known causality miner that uses Bayesian networks [3] to find causal relations. One example of the type of rules discovered by TETRAD is $x \rightarrow y$, which means that x causes y. From the examples in [1, 10], it appears that Bayesian networks discover more causal relations than actually exist in the domain. Bayesian networks find causality even in domains where the existence of causal relations itself is a matter a debate. For words in political texts, Bayesian networks find rules such as `"Minister" is caused by "Prime"` [10]. This suggests that there is a considerable amount of disagreement about the concept of causality. There are ongoing debates about the suitability of using Bayesian networks for mining causality [2, 4, 5, 11]. Here we apply TETRAD to identify relationships between variables without claiming that all of them are causal relationships.

C4.5 [8] creates decision trees that can be used to predict the value of one variable from the values of a number of other variables. A decision tree can easily be converted to a number of rules of the form **if** $\{(x = \alpha) \text{ AND } (y = \beta)\}$ **then** $(z = \gamma)$. The variables x and y may be causing the value of y, or they may be associated together because of some other reason. C4.5 makes no claim about the nature of the relationship.

Both these programs ignore any temporal order among the records, while in the data that we use there does exist relations among the variables in consecutive records, in the sense that the values of some variables in a record affect the values of variables in later records. We will describe the method used to overcome this problem.

The rest of the paper is organized as follows. Section 2 describes the simple environment that we chose for testing the different methods. In Section 3 we first compare the results obtained from TETRAD and C4.5 when there is an association among the variables of a record, but no causality. After that we attempt to discover temporal relations among the records. Section 4 concludes the paper.

2 An Agent's View of Its Environment

We use an Artificial Life simulator called URAL [12] to generate data for the experiments. URAL is a discrete event simulator with well known rules that govern the artificial environment. There is little ambiguity about what causes what. This helps to judge the quality of the discovered rules.

The world in URAL is made of a two dimensional board with one or more agents (called *creatures* in Artificial Life literature) living in it. An agent moves around and if it finds food, eats it. Food is produced by the simulator and placed at positions that are randomly determined at the start of each run of the simulator. There is a maximum for the number of positions that may have food at any one time, so a position that was determined as capable of having food may or may not have food at a given time. The agent can sense its position and also the presence of food at its current position. At each time-step, it randomly chooses to move from its current position to Up, Down, Left, or Right. It cannot get out of the board, or go through the obstacles that are placed in the board by the simulator. In such cases, a move action will not change the agent's position. The agent can sense which action it takes in each situation. The aim is to learn the effects of its actions at each particular place.

URAL employs Situation Calculus [7] to build graphs with observed situations as the nodes and the actions as transition arcs between the situations. Agents use the graphs to store their observations of the world and to make plans for finding food. URAL differentiates between volatile and non-volatile properties of a situation. The x and y positions are non-volatile, and are used to distinguish among the situations. The presence of food is a volatile property, which means that the same situation can be in different states. The creature only keeps the last observed state of a situation. URAL was modified for this experiment to log each encountered situation in a file.

For our agent, time passes in discrete steps. At each time step, it takes a snapshot of its sensors and randomly decides which action it should perform. This results in records such as <x position, y position, is food here?, action>. C4.5 treats the last variable in a record as the decision attribute, so if necessary the variables are rearranged in the log file. Figure 1 shows two example sequences of records. In Figure 1(a) the last variable is the action, while in 1(b) it is the x position. Here time passes vertically, from top to bottom.

<x, y, f, a>
<1, 3, false, L>
<0, 3, false, L>
<0, 3, true, D>
<0, 4, false, U>
<0, 3, true, D>

(a)

<y, f, a, x>
<3, false, L, 1>
<3, false, L, 0>
<3, true, D, 0>
<4, false, U, 0>
<3, true, D, 0>

(b)

Fig. 1. Two example sequences of records

The agent, moving randomly around, can visit the same position more than once. Unlike the real-world data studied by many people [1, 2, 10], here we can reliably assume a temporal order among the saved records, as each observation follows the previous one in time. Considering the similarities between data gathered by an agent and the statistical observations done by people for real-world problems, it is interesting to see if we can use the same data mining techniques to extract knowledge about this environment.

The agent can move around in this very simple world, so it has a way of changing its position by performing a move action. Creating food, on the other hand, is completely beyond its power. Finding the effects of the agent's actions requires looking at more than one record at a time (two consecutive records in this environment), because in this environment the effects of an action always appears at a later time. Any algorithm that does not consider the passage of time is limiting itself in finding causal rules. So this domain contains causal relations detectable by the agents over time (the effects of moving), as well as relations that are not detectable by the agent (the place of food).

3 Experimental Results

The effectiveness of TETRAD and C4.5 at finding valid relations is assessed in two situations: within a single record and within consecutive pairs of records.

3.1 Experiment 1: Relationships within a Single Record

From the semantics of the domain, we know that no causal relationship exists within a single record. Causal relations appear across the records. However, there is an association between the x and y position of an agent and the presence of food at that position, as the simulator places food at only certain places. A position may or may not contain food at any given time. We created a log file of the first 1000 situations encountered by a single agent and used it for the experiments.

We first fed the log file to TETRAD version 3.1. In TETRAD's notation, $A \bullet \rightarrow B$ means that either A causes B, or they both have a hidden common cause, $A \bullet \!-\!\bullet B$ means that A causes B or B causes A, or they both have a hidden common cause, and $A \leftrightarrow B$ means that both A and B have a hidden common cause.

TETRAD would not accept more than 8 different values for each variable, so the world was limited to an 8 × 8 square with no obstacles. In the log file, x and y denote the agent's position, f denotes the presence of food, and a is the performed action. The presence of food and the actions were represented by numerical values to make them compatible with what TETRAD expects as input. The results were generated by the "Build" command. It did assume the existence of latent common causes, and used the exact algorithm. TETRAD's output is shown in Table 1.

Table 1. The rules discovered by TETRAD.

Case	Significance Level(s)	Discovered Rules	C	PC	W
1	0.0001	$y \bullet \rightarrow x$, f, $a \bullet \rightarrow x$	1	1	1
2	0.001, 0.005, 0.01	$a \bullet \rightarrow y$, $x \bullet \!-\!\bullet a$, $x \bullet \rightarrow y$, $f \bullet \rightarrow y$	0	2	2
3	0.05, 0.1	$x \bullet \!-\!\bullet y$, $f \bullet \rightarrow x$, $a \bullet \rightarrow x$, $f \bullet \rightarrow y$, $a \bullet \rightarrow y$	0	2	2
4	0.2	$x \bullet \!-\!\bullet y$, $x \bullet \!-\!\bullet f$, $x \bullet \!-\!\bullet a$, $y \bullet \!-\!\bullet f$, $y \bullet \!-\!\bullet a$, $f \bullet \!-\!\bullet a$	0	4	2

Based on the rules enforced by URAL in the artificial environment, the desired output in TEDRAD's notation would be the following relations: $x \bullet \!-\!\bullet f$ (x and f are associated), $y \bullet \!-\!\bullet f$ (y and f are associated), f (f has no cause and does not cause others). The relations that are not totally wrong are shown in bold. In the table the C column indicates the number of Correct rules, the PC column show the number of Partially Correct (non-conclusive) rules, and W shows the number of Wrong rules.

The appearance of food at a certain position depends on a random variable inside the URAL code, and there is no causal relation that the agent can discover, but there is an association between positions and the presence of food.

In case 1 the rule $a \bullet \rightarrow x$ correctly guesses that a may be a cause of x. However this is not conclusive in the sense that it considers it possible for a hidden common cause to exist. This is an important distinction. In the next rule f, TETRAD identifies f as something that does not cause anything, and is not caused by anything else. The other rule, $y \bullet \rightarrow x$, is wrong because the y coordinate of a position does not determine the x coordinate. Case 2 does not go wrong in the first two rules, even though none of them is conclusive. The other two rules are wrong. Case 3 does better in finding the relationships among a and x, and a and y, but the results are still not conclusive. Case 5 finds associations among x, y and f, but then wrongly does the same for a and f too.

As seen, TETRAD draws many wrong conclusions from the data, and with the exception of one rule (f), the rest are not conclusive. Notice that here we have been generously interpreting the rules involving a as if TETRAD is aware that an action will have an effect on the *next* value, and not the current value, of x or y.

We then tried the c4.5rules program of the C4.5 package in the default configuration, on the same data. We assigned the presence of food as the decision attribute. C4.5rules eliminates unneeded condition attributes when creating rules. For example, if the value of x_2 is sufficient to predict the outcome regardless of the value y_2, the generated rule will not include y_2. We are looking for rules of the form **if** $\{(x = \alpha)$ **AND** $(y = \beta)\}$ **then** $(f = \gamma)$, which means that there is a relation between the position (x and y) and the presence of food.

Table 2 shows the c4.5rules program's results for determining the value of f. Rules that are actually and useful for finding food are shown in bold. The rules predicting the *presence* of food correctly included both x and y. The rest of the rules deal with cases where no food was present.

Table 2. Attributes used in rules generated by C4.5 to determine the presence of food.

Decision Attribute	Condition Attribute(s)	Number of Rules	Example	Correctness
f	x	1	If$\{(x = 5)\}$ then $(f = 0)$	Correct
f	y	2	If$\{(y = 2)\}$ then $(f = 0)$	Correct
f	a	1	If$\{(a = L)\}$then $(f = 0)$	Wrong
f	**x, y**	**2**	**If$\{(x = 2)$ AND $(y = 3)\}$ then $(f = 1)$**	**Correct**

C4.5 was unable to find useful rules for determining the values of x or y, as they depend on the previous position and action, which are not available. The decision tree for x, for example, wrongly included the presence of food as a condition attribute.

3.2 Experiment 2: Relationships among Consecutive Records

As mentioned, the log file consists of temporally ordered records. Neither C4.5 nor TETRAD considers the temporal order and adding a simple discrete time stamp as an attribute will not allow them to find temporal relationships. With the semantics of our example domain in mind, the most one can hope for in the previous tests is finding a correct association between the agent's position and the presence of food.

The effects of an action will not be seen until later in time. Using a preprocessing step, a *flattened log file* is created with two or more consecutive records as a single record. Flattening the sequences in Figure 1(a) and 1(b) using a *time window* of size 2 gives the sequences sown in Figure 2(a) and 2(b) respectably, where time passes horizontally from left to right and also vertically from top to bottom. Here we have renamed the variables to remove name clashes. With the exception of the first and last records, every record appears twice, once as the second half (effect), and then as the first half (cause) of a combined record.

The appropriate size of the time window depends on the domain. It should be wide enough to include any cause and all its effects. If we suspect that the effects of an action will be seen in the next two records, then we may flatten Figure 1(a) to get records like: <1, 3, false, L, 0, 3, false, L, 0, 3, true, D>. The algorithms accepting the flattened records as input may not know about the passage of time, but flattening brings the causes and the effects together, and the resulting record then has the information about any changes in time.

$<x_1, y_1, f_1, a_1, x_2, y_2, f_2, a_2>$
$<1, 3,$ false, L, 0, 3, false, L $>$
$<0, 3,$ false, L, 0, 3, true, D $>$
$<0, 3,$ true, D, 0, 4, false, U $>$
$<0, 4,$ false, U, 0, 3, true, D $>$

(a)

$<y_1, f_1, a_1, x_1, y_2, f_2, a_2, x_2>$
$<3,$ false, L, 1, 3, false, L, 0 $>$
$<3,$ false, L, 0, 3, true, D, 0 $>$
$<3,$ true, D, 0, 4, false, U, 0 $>$
$<4,$ false, U, 0, 3, true, D, 0 $>$

(b)

Fig. 2. The flattened sequences of Figures 1(a) and 1(b).

For the next experiment, a window size of 2 was used because in the URAL domain the effects of an action are perceived by the agent in the next situation. In the combined record, x_1, y_1, f_1 and a_1 belong to the first record, and x_2, y_2, f_2 and a_2 belong to the next one. TETRAD's output for the resulting data is shown in Table 3.

Table 3. The rules discovered by TETRAD from the flattened records.

Case	Significance Level(s)	Discovered Rules	C	PC	W
1	0.0001	$y_1 \bullet\!\!\rightarrow x_1$, $a_1 \bullet\!\!\rightarrow x_1$, $x_1 \bullet\!\!-\!\!\bullet x_2$, $y_2 \bullet\!\!\rightarrow x_1$, $a_2 \bullet\!\!\rightarrow x_1$, $y_1 \bullet\!\!\rightarrow x_2$, $y_1 \bullet\!\!\rightarrow y_2$, $a_1 \bullet\!\!\rightarrow x_2$, $a_1 \leftrightarrow y_2$, $a_2 \bullet\!\!\rightarrow a_1$, $y_2 \bullet\!\!\rightarrow x_2$, $a_2 \bullet\!\!\rightarrow x_2$, f_1, f_2	2	4	8
2	0.0005, 0.001, 0.005, 0.01	$y_1 \bullet\!\!\rightarrow x_1$, $x_1 \bullet\!\!-\!\!\bullet a_1$, $x_1 \bullet\!\!-\!\!\bullet x_2$, $x_1 \bullet\!\!-\!\!\bullet y_2$, $a_2 \bullet\!\!\rightarrow x_1$, $y_1 \bullet\!\!\rightarrow a_1$, $y_1 \bullet\!\!\rightarrow x2$, $y_1 \bullet\!\!\rightarrow y_2$, $a_1 \bullet\!\!-\!\!\bullet x_2$, $a_1 \bullet\!\!-\!\!\bullet y_2$, $a_2 \bullet\!\!\rightarrow a_1$, $x_2 \bullet\!\!-\!\!\bullet y_2$, $a_2 \bullet\!\!\rightarrow x_2$, $a2 \bullet\!\!\rightarrow y2$, f_1, f_2	2	4	10
3	0.1	$x_1 \bullet\!\!\rightarrow y_1$, $x_1 \bullet\!\!-\!\!\bullet a_1$, $x_1 \bullet\!\!\rightarrow x_2$, $x_1 \bullet\!\!-\!\!\bullet y_2$, $x_1 \bullet\!\!-\!\!\bullet a_2$, $f_1 \bullet\!\!\rightarrow y_1$, $a_1 \bullet\!\!\rightarrow y_1$, $y_1 \leftrightarrow x_2$, $y_2 \bullet\!\!\rightarrow y_1$, $a_2 \bullet\!\!\rightarrow y_1$, $a_1 \bullet\!\!\rightarrow x_2$, $a_1 \bullet\!\!-\!\!\bullet y_2$, $a_1 \bullet\!\!-\!\!\bullet a_2$, $y_2 \bullet\!\!\rightarrow x_2$, $f_2 \bullet\!\!\rightarrow x_2$, $a_2 \bullet\!\!\rightarrow x_2$, $y_2 \bullet\!\!-\!\!\bullet a_2$	0	3	14
4	0.2	$x_1 \bullet\!\!-\!\!\bullet y_1$, $f_1 \bullet\!\!\rightarrow x_1$, $x_1 \leftrightarrow a_1$, $x_1 \leftrightarrow x_2$, $y_2 \bullet\!\!\rightarrow x_1$, $a_2 \bullet\!\!\rightarrow x_1$, $f_1 \bullet\!\!\rightarrow y1$, $y_1 \leftrightarrow a_1$, $y_1 \leftrightarrow x_2$, $y_2 \bullet\!\!\rightarrow y_1$, $a_2 \bullet\!\!\rightarrow y_1$, $a_1 \bullet\!\!-\!\!\bullet x_2$, $y_2 \bullet\!\!\rightarrow a_1$, $f_2 \bullet\!\!\rightarrow a_1$, $a_2 \bullet\!\!\rightarrow a_1$, $y_2 \bullet\!\!\rightarrow x_2$, $f_2 \bullet\!\!\rightarrow x_2$, $a_2 \bullet\!\!\rightarrow x_2$, $y_2 \bullet\!\!-\!\!\bullet a_2$	0	1	18

Here we are looking for the following relations: $x_1 \bullet\!\!-\!\!\bullet f_1$ (x_1 and f_1 are associated), $y_1 \bullet\!\!-\!\!\bullet f_1$ (y_1 and f_1 are associated), f_1, f_2 (f_1 and f_2 have no causes and do not cause anything), $x_2 \bullet\!\!-\!\!\bullet f_2$ (x_2 and f_2 are associated), $y_2 \bullet\!\!-\!\!\bullet f_2$ (y_2 and f_2 are associated), $a_1 \rightarrow x_2$ (a_1 causes x_2), $a_1 \rightarrow y_1$ (a_1 causes x_1), $x_1 \rightarrow x_2$ (x_1 causes x_2), and $y_1 \rightarrow y_2$ (y_1 causes y_2). The relations that are not totally wrong are shown in bold.

Most rules discovered by TETRAD on this data are wrong. In comparison with Experiment 1, the increased number of variables in Experiment 2 has resulted in an increase in the number of discovered rules, most of which are either wrong or not conclusive.

We applied C4.5 to the same data. The desired output rules are of the following forms: **if** $\{(x_2 = \alpha)$ AND $(y_2 = \beta)\}$ **then** $(f_2 = \gamma)$ (Association between x, y, and food), **if** $\{(x_1 = \alpha)$ AND $(a_1 = \beta)\}$ **then** $(x_2 = \gamma)$ (predicting the next value of x), and **if** $\{(y_1 = \alpha)$ AND $(a_1 = \beta)\}$ **then** $(y_2 = \gamma)$ (predicting the next value of y). The results produced by c4.5rules are shown in Table 4.

Table 4. C4.5's results after flattening the records.

Decision Attribute	Condition Attribute(s)	Number of Rules	Validity
f_2	x_2	1	Correct
f_2	y_2	2	Correct
f_2	a_2	1	Wrong
f_2	x_2, y_2	2	Correct
x_2	x_1, a_1	32	Correct
y_2	y_1, a_1	32	Correct

Rules that can actually be used for finding and reaching food are shown in bold. C4.5rules generated 32 correct rules for each of x_2 and y_2. In a two-dimensional space, there are 4 possible actions and 8 distinct values for x_1 and y_1. In this example the creature has explored all the world, and there are 32 (8 × 4) rules for predicting the next value of x_2 and y_2. There were no changes in the rules for f_2 (the actually useful rules that predict the presence of food still depend on both x_2 and y_2) even though C4.5 now has more variables to choose from. This is because the current value of f_2 is not determined by any temporal relationship. Overall, C4.5 did a much better job in pruning the irrelevant attributes than TETRAD.

4 Concluding Remarks

People interested in finding relations among observed variables usually gather data from different systems at the same time. Here we used data that represented the state of a single system over time. While there were relations among the variables in each state, some interesting temporal relations existed among the variables of different states.

We applied a causality miner that uses Bayesian networks to find relations in a very simple and well-defined domain. The results were similar to the real-world problems: both correct and wrong rules were found. Bayesian causality miners need a domain expert (a more powerful causal relation discoverer) to prune the output. Flattening the records to give the algorithm more relevant data resulted in many more irrelevant rules being discovered. We also tested C4.5 on the same data, and observed that it is very good in pruning non-relevant attributes of the records and finding temporal relations without actually claiming to be a causality discoverer. One consideration with C4.5 is that the user has to identify the decision attribute of interest.

We observed that flattening enabled C4.5 to discover new relations that it could not find otherwise. Flattening increases the number of variables in the resulting records. While creating a bigger search space for rule mining, flattening gives more information to the rule miner, and makes temporal relations explicit.

Acknowledgements

We thank Yang Xiang, Clark Glymour, and Peter Spirtes for their help. Funding was provided by a Strategic Grant from Natural Science Engineering Research Council of Canada.

References

1. Bowes, J., Neufeld, E., Greer, J. E. and Cooke, J., A Comparison of Association Rule Discovery and Bayesian Network Causal Inference Algorithms to Discover Relationships in Discrete Data, *Proceedings of the Thirteenth Canadian Artificial Intelligence Conference (AI'2000)*, Montreal, Canada, 2000.
2. Freedman, D. and Humphreys, P., *Are There Algorithms that Discover Causal Structure?*, Technical Report 514, Department of Statistics, University of California at Berkeley, 1998.
3. Heckerman, D., *A Bayesian Approach to Learning Causal Networks*, Microsoft Technical Report MSR-TR-95-04, Microsoft Corporation, May 1995.
4. Humphreys, P. and Freedman, D., The Grand Leap, *British Journal of the Philosophy of Science 47*, pp. 113-123, 1996.
5. Korb, K. B. and Wallace, C. S., In Search of Philosopher's Stone: Remarks on Humphreys and Freedman's Critique of Causal Discovery, *British Journal of the Philosophy of Science 48*, pp. 543- 553, 1997.
6. Levy, S., Artificial Life: A Quest for a New Creation, Pantheon Books, 1992.
7. McCarthy, J. and Hayes, P. C. Some Philosophical Problems from the Standpoint of Artificial Intelligence, *Machine Intelligence 4*, 1969.
8. Quinlan, J. R., *C4.5: Programs for Machine Learning*, Morgan Kaufmann, 1993.
9. Scheines R., Spirtes P., Glymour C. and Meek C., *Tetrad II: Tools for Causal Modeling*, Lawrence Erlbaum Associates, Hillsdale, NJ, 1994.
10. Silverstein, C., Brin, S., Motwani, R. and Ullman J., Scalable Techniques for Mining Causal Structures, *Proceedings of the 24th VLDB Conference*, pp. 594-605, New York, USA, 1998.
11. Spirtes, P. and Scheines, R., Reply to Freedman, V. McKim and S. Turner (editors), *Causality in Crisis*, University of Notre Dame Press, pp. 163-176, 1997.
12. ftp://orion.cs.uregina.ca/pub/ural/URAL.java

Learning Relational Clichés with Contextual LGG

Johanne Morin and Stan Matwin

School of Information Technology and Engineering, University of Ottawa
Ontario, K1N 6N5, Canada
jmorin@site.uottawa.ca

Abstract. Top-down learners suffer often from the plateau problem (or *myopia*) of their greedy search algorithms. One way to address this is to extend the top-down greedy search, which grows the clauses, with relational clichés. Using clichés the search is no longer constrained to adding one literal at a time: combinations of literals instantiating clichés are tried as well. The paper presents CLUSE: Clichés Learned and USEd, a system that learns clichés that are then used either within a domain, or across domains. CLUSE is a bottom-up learner, in which generalization proceeds according to Contextual LGG (CLGG). CLGG is an extension of LGG that takes into account the context in which a pair of literals is generalized. The paper defines CLGG, illustrates how clichés are learned, and shows that the complexity of this learning is polynomial.

1 Introduction

Inductive learners that use a first-order language to express examples, background knowledge and hypotheses (or concept descriptions) are called inductive relational learners. Because they induce hypotheses in the form of *logic programs* they are also called *inductive logic programming* (ILP) systems. Top-down inductive relational learners such as FOIL [13] and FOCL [10] learn Horn clauses adding one literal at a time using a greedy-search algorithm. At each step the coverage of the rule after adding a literal is tested on training examples. The literal that best discriminates the remaining positive and negative examples is added to the current clause. The clause is complete when it no longer covers negative examples. Top-down systems suffer from *myopia*, which arises when the best discrimination would be obtained by adding more than one literal at once. Solving the problem requires searching for combinations of literals rather than just single literals. Unfortunately, trying all possible combinations of literals can be intractable. A mechanism to search efficiently through the space of combinations of literals is needed. A learner can be provided with such a mechanism in form of a special-purpose bias.

We propose CLUSE (*Clichés Learned and Used*) [6] to learn combinations of literals automatically as a particular type of bias. These combinations of literals are called *relational clichés*. The underlying idea is to learn clichés from examples of a concept and to use them within and across domains (). Assuming that clichés express subconcepts common to a domain, and that in the same domain literals used to express different concepts overlap, then clichés learned from one concept should provide appropriate lookahead to learn other concepts in the same domain. On the other hand, these clichés probably have few literals in common with concepts in other domains, hence the need for more general clichés. To solve this, CLUSE learns two kinds of clichés: *Domain Dependent Clichés* (DDCs) expressed as a conjunction of literals specific to a domain, and *Domain Independent Clichés* (DICs) where literals have variable predicate symbols (hence they are notspecific to a domain). When DICs are transferred across domains they are instantiated with literals in the domain of the target concept.

CLUSE is a bottom-up inductive relational learner based on Relative Least General Generalization (RLGG) [12]remedy the inefficiency and the overgeneralization problems of RLGG, we have also developed a modified version of RLGG that exploits the context in which LGG is applied. The modified RLGG is called *Contextual Least General Generalization* (CLGG).

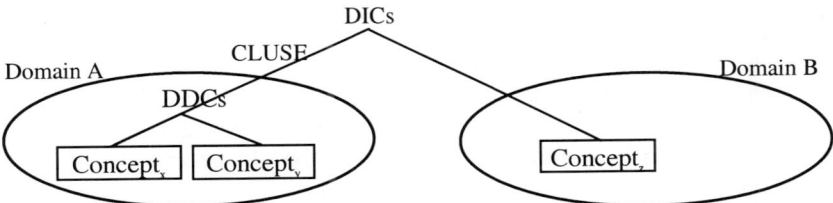

Figure 1. CLUSE learns relational clichés (DDCs and DICs) from examples of a concept in one domain. DDCs are useful to learn concept within the same domain, whereas DICs are useful across domains.

This paper describes the learning of relational clichés with CLUSE. It introduces CLGG, the similarity measures, and the notion of chains used in CLUSE. It describes CLUSE's algorithm and its complexity. How these clichés address the myopia problem of an inductive relational learner is described in [7].

2. CLGG: Contextual Least General Generalization

Much of the existing work in learning in the first-order logic setting is based on Plotkin's Least General Generalization LGG [11], and on its extension – called *Relative LGG* (RLGG) [12]. From the machine learning perspective, however, there are certain practical shortcomings of the LGG approach to generalization. First, in the worst case, the cost of applying LGG on two clauses is equal to the length of the first clause times the length of the second one. So the cost of applying LGG on a set of clauses is exponential in the number of literals to generalize. Second, additional knowledge (*e.g.* taxonomic hierarchies) is often available during generalization. Many learning methods take such knowledge into account in the generalization process [1, 10]. LGG does not use any background knowledge (BK) in the generalization process. RLGG is supposed to address knowledge-driven generalization, but since RLGG compiles all the knowledge during generalization in the form of additional literals, this compounds the efficiency problems of LGG.

The learning system GOLEM [9] is based on RLGG but reduces the cost by constraining the BK to a finite Herbrand model. Even using a finite model, the length of the LGGs is exponential in the number of given examples. By using restrictions like the *ij*-determinism and *syntactically generative*[1] background clauses, the length of the LGG of a set of examples no longer depends on the number of examples. ITOU [15], CLINT [3], and Kodratoff's system [4] encounter the same problem of efficiency as RLGG, generating the Herbrand models of the BK (by an exhaustive saturation process). Other learning systems like FOCL [16] and CIGOL [8] use BK, but these are not based on RLGG.

This section (see [6] for a full presentation) outlines an alternative to LGG that exploits the context in which LGG is applied. The *context* is meant here to include both the additional knowledge available during generalization, as well as the similarity of literals in the context of the clauses being generalized. Although CLGG is defined for literals with nested arguments involving functors, in this presentation we limit ourselves to simpler functor-free arguments.

To extend LGG so that context is taken into account during generalization, the similarity between every pair of constants (or bindings) occurring in the same relation in the two clauses being generalized is computed before the generalization. The similarity of constants takes into account their occurrences in clauses (see below). Constant bindings with a similarity higher than a threshold are bound to a variable. The constants and the variable constitute the *similarity bindings*, which are then passed to the CLGG to limit its search. When generalizing two clauses, literals that *match*[2] (even with multiple occurrences) must have at least one similarity binding to be generalized. Moreover, to take into account the context in which the generalization of two clauses takes place and to address the shortcomings of RLGG, the BK is used in a lazy manner. Only unmatched literals restricted with similarity bindings that find a generalization in the BK are generalized.

2.1 Similarity Measure Evaluates Bindings of Constants

We borrowed the similarity measure from Bisson [2] to evaluate the bindings of constants in clauses. For each constant a_i, a list of occurrences (denoted by $occ(a_i)$) is made of pairs (*predicate-a_i, position-of-a_i*) for each literal where the constant occurs. The *predicate-a_i* is the literal's predicate and the *position-of-a_i* is the term's position among the arguments of *predicate-a_i*.

Two constants match if they occur in two literals whose predicates are identical. The similarity between two constants from two clauses is the ratio of the length of the lists of common occurrences to the maximum length of constants' occurrences in the

---

[1] A clause is said to be *syntactically generative* if the variables in its head are a subset of the variables in its body.

[2] Two literals match if they have the same predicate and arity.

two clauses. It results in a value between [0..1], where the closer the value gets to 1, the more similar the constants are. The *similarity measure* formula is:

$$sim(a_i, a_j) = \frac{length(occ(a_i) \cap occ(a_j))}{MAX(length(occ(a_i)), length(occ(a_j)))}$$

The overall idea of this definition is that constants occurring in two clauses are similar if they occur in a similar enough context. It is required that occurrences of constants in relations (*i.e.* literals with more than one argument) match, otherwise their similarity binding is zero[3].

2.2 CLGG Relative to a BK

As in RLGG, the CLGG (in the functor-free case) of two clauses *C1* and *C2* (denoted *CLGG(C1, C2)*) is the least upper bound (or *least general generalization*) of *C1* and *C2* in the θ-subsumption lattice. Unlike RLGG, CLGG exploits an intensional BK in a lazy way. It first generalizes clauses and then uses the BK to generalize unmatched literals that have at least one similarity binding.

Figure 2 illustrates the extent to which similarity bindings limit CLGG search when there exist multiple examples of the same predicate in clauses (black/1). Unlike RLGG, CLGG limits the generalization of the predicate *black* to combinations with one similarity binding (*e.g.* V1 and V2). It results in a generalization (above(V1, V2), black(V1), black(V2)) and unmatched literals tri(a), rect(b), and small(b) for C1, and sq(c), sq(d) for C2. When a literal from the BK subsumes unmatched literals with a similarity binding, then the subsuming literal is added to the generalization and unmatched literals are discarded. For instance, Poly(V1) is added to the generalization because it subsumes tri(a) and sq(c).

C1: scene(a, b):- above(a, b), tri(a), rect(b), small(b), black(a), black(b).
C2: scene(c, d):- above(c, d), sq(c), sq(d), black(c), black(d).
CLGG(C1, C2) with BK:
⇒ scene(V1,V2):- above(V1, V2), black(V1), black(V2), Poly(V1), Rect(V2).
Generalized literals: Poly(V1) = (tri(a), sq(c))
Rect(V2) = (rect(b), sq(d))
Unmatched literals: small(b)
Bindings: V1 = (a, c), V2 = (b, d)

Figure 2. CLGG relative to the BK. Predicates with a capital letter are learned using a taxonomy of geometric forms.

---

[3] CLGG is used to learn relational clichés where relations are more important than attributes [6].

3 CLUSE Learns Relational Clichés

The generalization problem of learning clichés consists of finding common parts in examples. In relational domains, examples are represented with different relations and features, which makes it difficult to find a generalization process that will succeed in finding common parts of a set of examples. Moreover, in most of these domains, important concepts are represented by a small number of connections among constants defining examples. For these reasons, CLUSE splits examples into their shortest chains. This is similar to the idea of relational path finding [14]. Intuitively, a *chain* is a pattern showing how objects are related to one another and how their features are used in examples. So, each relation in an example (which provides the structural information between objects) and all features of the related objects form a chain. Every relation and feature of an example is preserved and some features may occur in more than one chain. A chain is thus defined as a connected conjunction of literals where one and only one literal is a relation.

```
E:  on(x, y), cir(x), rect(y), leftof(y, z), iso(z)
C1: on(x, y), cir(x), rect(y)
C2: leftof(y, z), rect(y), iso(z)
```

Figure 3. Example E expressed in terms of chains (C1 and C2).

Figure 3 shows the shortest chains `C1` and `C2` in the example `E`. They are connected combinations of literals with a single relation (`on(x, y)` and `leftof(y, z)`). Because relevant relations are not always connected to the head of examples, chains are used without the example's head. For instance, both relations `on(x, y)` and `leftof(y, z)` would be related to the head `scene(x, y)`. On the other hand, `leftof(y, z)` would not be connected if the head is `scene(x)`.

3.1 CLUSE's Algorithm

The general algorithm of learning DDCs and DICs with CLUSE is as follows. Positive and negative examples (and optionally) the BK are given to CLUSE. Examples are split into chains. At the beginning, positive chains are considered roots of the structure. CLUSE evaluates the similarity of each pair of roots. It chooses the two most similar roots and generalizes them using CLGG. The resulting generalization becomes the parent (and the new root) of the two generalized chains. The similarity of this root with each other root is evaluated. CLUSE repeats this process until no more generalizations are possible. When taxonomies are available to CLUSE, the most similar chains with the lowest cost are generalized first. The cost expresses the distance between predicates in the taxonomy. CLUSE uses the *similarity of clauses* to choose the two most similar chains to generalize first. The similarity measure of two clauses is computed from similarity bindings of constants. The formula is:

$$sim(C1,C2) = \prod_{i=1}^{n} \prod_{j=1}^{m} sim(a1_i, a2_j) \qquad \text{for } sim(a1_i, a2_j) \neq 0$$

where $a1_i \in C1$ and $a2_j \in C2$, n is the number of constants in $a1$, and m is the number of constants in $a2$.

A generalization is added to the structure as the parent of the two chains that it generalizes. To avoid a tree with duplicate generalizations (when many chains result in the same generalization), chains that are exactly subsumed by their generalization (*i.e.* they differ only by their argument names) are removed. Every unmatched literal (when combined with the generalization) that is exactly subsumed by the generalization it accompanies, is also removed.

Pruning the structure preserves generalizations (DDCs) with good coverage of examples and discards others. CLUSE traverses a structure depth-first and computes coverage frequencies for positive and negative examples. *Coverage frequencies* correspond to the number of positive (or negative) examples subsumed by the generalization divided by the total number of positive (or negative) examples[4]. A generalization is preserved when it covers fewer negative examples than the generalization that subsumes it, and satisfies the user-defined coverage thresholds. This way, the pruning eliminates generalizations with low recall and low precision.

DDCs are useful for learning concepts in the same domain where they are learned. On the other hand, clichés independent of the domain are useful for learning concepts in other domains. To learn such clichés, first-order predicates of DDCs are replaced with second-order predicates giving DICs. The information as to whether a predicate of a DDC is generalized using the BK or not is preserved within the predicate name in the DIC. A predicate that subsumes other predicates on the taxonomy in called an *intensional* predicate, whereas a predicate that belongs to examples is called an *extensional* predicate. Intensional predicates are generalized to a predicate variable of the form IntP# and extensional predicates to a predicate variable of the form ExtP#. So, When DICs are used to learn a concept in a new domain, this information can be recovered and used to instantiate second-order predicates with intensional or extensional predicates of the new domain.

The overall complexity of learning clichés is polynomial, due to the similarity evaluation of all chains $O(k^2 n^4)$ where k is the maximum number of arguments for any relation, and n the number of literals (see [6] for a full presentation).

3.2 Examples of Learning Relational Clichés

This section illustrates an example of CLUSE learning the concept *Scene* in the *blocks* domain[5]. It shows the chains used for learning, the structure of generalizations, the

---

[4] The frequency of coverage gives more flexibility than a measure like information gain [13]. Unlike information gain, the coverage frequencies of a generalization explicitly represent the proportion of subsumed positive and negative examples. This allows the user to fix two different coverage thresholds for choosing generalizations. For instance, the user may choose to preserve only generalizations that cover at least 50% of the positive and at most 25% of the negatives.

[5] CLUSE has also been used to learn clichés in the real-life domain of the Finite Element Mesh Design (see http://www.site.uottawa.ca/~jmorin/Programs/CLUSE/Output/Mesh/).

DDCs and the DICs learned. *Scene* is a disjunctive concept describing 1) an *ellipse above* a *rectangle*, which is *left of* an *isosceles triangle*; 2) an *ellipse above* an *isosceles triangle* and a *rectangle left of* the *triangle*. In both cases the ellipse may be *small*, the *rectangle* may be *large* and the *triangle* may be *red*. Moreover, the *ellipse* can also be a *circle*, the *rectangle* a *square*, and the *isosceles triangle* a *right-angled isosceles triangle*.

Figure 4 illustrates a part of the structure of generalizations built with CLUSE for the concept *Scene*. Each level includes: the generalization, the coverage frequencies of chains, unmatched literals (generalized or not) and the positive chains subsumed[6]. For instance, at the lowest level of the structure, generalization G46 subsumes chains 1, 7 and 17 with one unmatched literal large(x1) that belongs to the chain {1}. G49 subsumes G46 and chain 13. CLUSE knows from the BK (a taxonomy of geometric forms) that an *equilateral triangle* and an *isosceles right-angled triangle* are also *isosceles triangles*. So, equil(V12) from G46 and iso_rangl(x26) from chain 13 are generalized into Iso(V18) in G49.

The Structure of generalization	F+	F-
G58: leftof(V1,V2),Rect(V1), Iso(V2).	0.5	0.25
Rect(V1) -> (sq(x37), rect(V21)).		
Iso(V2) -> (equil(x38), iso(V22)).		
large(x29) {15}	0.05	0
Subs. chains: {1, 3, 5, 7, 9, 11, 13, 15, 17, 19}		
G51: leftof(V21,V22), Rect(V21), iso(V22), red(V22).	0.4	0
Rect(V21) -> (sq(V17), rect(V9)).		
Subs. chains: {1, 3, 5, 7, 9, 11, 13, 17}		
G49: leftof(V17,V18), sq(V17), Iso(V18), red(V18).	'	
Iso(V18) -> (equil(V12), iso_rangl(x26)).		
Subs. chains: {1, 7, 13, 17}		
G46: leftof(V11,V12), sq(V11), equil(V12), red(V12).		
large(x1) {1}		
Subs. chains: {1, 7, 17}		
G45: leftof(V9,V10), large(V9), rect(V9), Iso(V10), red(V10).		
Iso(V10) -> (iso(V1), iso_rangl(x6)).		
Subs. chains: {3, 5, 9}	'	

Figure 4. CLUSE generalizes chains into a structure and prunes this structure according to generalizations' coverage frequencies. Generalizations are identified by *G#* (and appear in bold), followed by predicate bindings, unmatched literals, and subsumed chains. Generalizations G45, G46, and G49 are pruned (shaded).

After the generalization, CLUSE prunes the structure in a top-down manner according to the coverage frequencies of generalizations. For instance, CLUSE preserves G58, which covers 50% of the positives and 25% of the negatives (F+ = 0.5 and F- = 0.25). The combination of the unmatched literal large(x29) with G58 covers 5% of the positives and none of the negatives. It covers fewer negatives than G58 itself, so

---

[6] Similarly for negative chains subsumed.

the unmatched literal is preserved. CLUSE continues with G51 and finds the coverage frequencies to be 40% for the positives and 0% for the negatives. G51 covers fewer negative examples than G58, so CLUSE preserves it. Since the coverage of negatives is already at the minimum and generalizations under G51 are more specific than G51 (and similarly if G51 had unmatched literals), CLUSE knows that no generalizations under G51 can cover fewer negatives than G51. Therefore, CLUSE prunes all descendants of G51 (*i.e.* G49, G46, and G45).

Generalizations left after pruning, are returned with their frequencies as learned DDCs (Table 1)[7]. A generalization with each of its unmatched literals makes a DDC. For instance, DDCs 1 and 2 are created from generalization G58: DDC 1 corresponds to G58 itself, DDC 2 corresponds to G58 with its unmatched literals large(Y). Table 1 also shows DICs generalized from each DDC.

Table 1. Learned DDCs with their coverage frequencies and their corresponding DIC.

G#	DDC	F+	F-	DIC
58	leftof(X,Y),Rect(X),Iso(Y)	0.5	0.25	ExtP1(X,Y),IntP1(X),IntP2(Y)
58	leftof(X,Y),Rect(X),large(X),Iso(Y)	0.05	0	ExtP1(X,Y),IntP1(X),Extp2(X),IntP2(Y)
51	leftof(X,Y),Rect(X),iso(Y),red(Y)	0.4	0	ExtP1(X,Y),IntP1(X),ExtP2(Y),ExtP3(Y)

4 Conclusion

The paper presented an extension of LGG/RLGG that exploits additional knowledge available during generalization and the similarity of literals in the context of the clauses being generalized. CLGG is less expensive to apply than LGG/RLGG. No literals are added to clauses prior to the generalization, and matching is restricted to literals with similarity bindings.

This paper showed the underlying algorithm to learn clichés with CLUSE and the algorithm's complexity. CLUSE uses CLGG and the notion of chains to learn relational clichés in a bottom-up manner into a hierarchy of generalizations. CLUSE prunes this hierarchy according to the generalizations' coverage frequencies of chains. Preserved generalizations and their coverages are returned as learned DDCs. DDCs are further generalized into DICs with variable predicates. DDCs are considered domain-dependent, since they are expressed with predicates specific to a domain, whereas DICs are domain-independent.

CLUSE could be used to create a library of concept hierarchies from different domains of application. Concept hierarchies (as described in Langley [5]) provide a better memory organization than flat lists of clichés and allow some pruning, giving a solution to the utility problem [5]. Classifying new instances with a concept hierarchy involves moving downward through the hierarchy. At each level, instantiate the cliché or use coverage frequencies on the alternative nodes to select one to expand, then recurse to the next level.

---

[7] For simplicity, variable names are changed.

Reference

1. BISSON, G. (1990). "KBG: A Knowledge Based Generalizer." *Proceedings of the Seventh International Conference on Machine Learning*, Austin, Texas, Morgan Kaufmann, 9-15.
2. BISSON, G. (1992). "Learning in FOL with a Similarity Measure." *Proceedings of the Tenth National Conference on Artificial Intelligence*, San Jose, CA, Morgan Kaufmann, 82-87.
3. DE RAEDT, L. AND M. BRUYNOOGHE (1992). "An Overview of the Interactive Concept-Learner and Theory Revisor CLINT." *Inductive Logic Programming*, Muggleton S. (ed.), Academic Press, 163-191.
4. KODRATOFF, Y. (1990). "Learning Expert Knowledge by Improving the Explanations Provided by the System." *Machine Learning: An Artificial Intelligence Approach-III*, Michalski R. S. and Y. Kodratoff (eds.), Morgan Kaufmann, 433-473.
5. LANGLEY, P. (1996). *Elements of Machine Learning*, Morgan Kaufmann.
6. MORIN, J. (1999). "Learning Relational Clichés with Contextual Generalization". PhD Thesis, School of Information Technology and Engineering, University of Ottawa.
7. MORIN, J. AND S. MATWIN (2000). "An Empirical Evaluation of Relational Clichés used Within and Across Domains." *Submitted*.
8. MUGGLETON, S. AND W. BUNTINE (1992). "Machine invention of first-order predicates by inverting resolution." *Inductive Logic Programming*, Muggleton S. (eds.), Academic-Press, 261-280.
9. MUGGLETON, S. AND C. FENG (1992). "Efficient Induction of Logic Programs." *Inductive Logic Programming*, Muggleton S. (eds.), Academic Press, 281-298.
10. PAZZANI, M. AND D. KIBLER (1992). "The Utility of Knowledge in Inductive Learning." *Machine Learning*, 9(1):57-94.
11. PLOTKIN, G. (1970). "A note on inductive generalization." *Machine Intelligence*, Meltzer, B. and D. Michie (eds.), Edinburg University Press, Edinburg, 5, 153-163.
12. PLOTKIN, G. (1971). "A Further Note on Inductive Generalization." *Machine Intelligence*, Meltzer B. and D. Michie (eds.), 6, 101-124, Edinburgh.
13. QUINLAN, R. (1990). "Learning Logical Definitions from Relations." *Machine Learning*, 5:239-266.
14. RICHARDS, B. L. AND R. J. MOONEY (1992). "Learning Relations by Pathfinding." *Proceedings of the Tenth National Conference on Artificial Intelligence*, San Jose, CA, AAAI Press, 50-55.
15. ROUVEIROL, C. (1991). "ITOU: Induction of First Order Theories." *Proceedings of the International Workshop on Inductive Logic Programming*, Vienna de Castelo, Portugal, 127-151.
16. SILVERSTEIN, G. AND M. PAZZANI (1991). "Relational Clichés: Constraining constructive induction during relational learning." *Proceedings of the Eighth International Conference on Machine Learning*, Evanston, Illinois, Morgan Kaufmann, 203-207.

Design of Rough Neurons: Rough Set Foundation and Petri Net Model

J.F. Peters[1], A. Skowron[2], Z. Suraj[3], L. Han[1], and S. Ramanna[1]

[1] Computer Engineering, Univ. of Manitoba, Winnipeg, MB R3T 5V6 Canada
[2] Institute of Mathematics, Warsaw Univ., Banacha 2, 02-097 Warsaw, Poland
[3] Institute of Mathematics, Pedagogical Univ., Rejtana 16A, 35-310 Rzeszów, Poland
{jfpeters,liting,ramanna}@ee.umanitoba.ca

Abstract. This paper introduces the design of rough neurons based on rough sets. Rough neurons instantiate approximate reasoning in assessing knowledge gleaned from input data. Each neuron constructs upper and lower approximations as an aid to classifying inputs. The particular form of rough neuron considered in this paper relies on what is known as a rough membership function in assessing the accuracy of a classification of input signals. The architecture of a rough neuron includes one or more input ports which filter inputs relative to selected bands of values and one or more output ports which produce measurements of the degree of overlap between an approximation set and a reference set of values in classifying neural stimuli. A class of Petri nets called rough Petri nets with guarded transitions is used to model a rough neuron. An application of rough neural computing is briefly considered in classifying the waveforms of power system faults. The contribution of this article is the presentation of a Petri net model which can be used to simulate and analyze rough neural computations.

1. Introduction

This paper considers the design of a rough neuron, which is based on rough set theory [1]-[3]. The study of rough neurons is part of a growing number of papers on neural networks based on rough sets. Rough-fuzzy multilayer perceptrons in knowledge encoding and classification were introduced in [4]. Rough-fuzzy neural networks have recently been also used in classifying the waveforms of power system faults [5]-[6]. Purely rough membership function neural networks were introduced in [7] in the context of rough sets and the recent introduction of rough membership functions [8]. There are two types of rough neurons: approximation neurons and rule-based decider neurons. An approximation neuron consists of a number of input ports governed by filters, a processing element which constructs a rough set, and one or more output ports which utilizes a rough membership function to compute the degree-of-accuracy of the approximate knowledge represented by the rough set derived by the neuron. The notion of an input port filter comes from signal processing. A filter is a device which transmits signals in a selected band of frequencies and rejects (or attenuates) signals in other bands [9]-[10]. Filters can be calibrated by adjusting the bandwidth of values, which can stimulate an approximation neuron. The contribution of this article is the presentation of a Petri net model of a rough neuron which can be used to simulate and analyze rough neural computations.

This paper is organized as follows. The basic concepts of rough sets, decision rules and rough membership functions underlying the design of rough neurons are presented in Section 2. The design of sample rough neurons is also presented in Section 2. A Petri net model of a rough neuron is given in Section 4.

2. Basic Concepts

A brief introduction to the basic concepts underlying the design of rough membership function neurons is given in this section.

2.1 Rough Sets

Rough set theory offers a systematic approach to set approximation [1]-[3], [8]. To begin, let $S = (U, A)$ be an information system where U is a non-empty finite set of objects and A is a non-empty finite set of attributes where $a:U \rightarrow V_a$ for every a ∈ A. For each $B \subseteq A$, there is associated an equivalence relation $\text{Ind}_A(B)$ such that

$$\text{Ind}_A(B) = \{(x,x') \in U^2 \mid \forall a \in B. \, a(x) = a(x')\} \tag{1}$$

If $(x, x') \in \text{Ind}_A(B)$, we say that objects x and x' are indiscernible from each other relative to attributes from B. The notation $[x]_B$ denotes equivalence classes of $\text{Ind}_A(B)$. For $X \subseteq U$, the set X can be approximated only from information contained in B by constructing a B-lower and B-upper approximation denoted by $\underline{B}X$ and $\overline{B}X$ respectively, where $\underline{B}X = \{ x \mid [x]_B \subseteq X \}$ and $\overline{B}X = \{ x \mid [x]_B \cap X \neq \emptyset \}$. The objects of $\underline{B}X$ can be classified as members of X with certainty, while the objects of $\overline{B}X$ can only be classified as possible members of X. Let $\text{BN}_B(X) = \overline{B}X - \underline{B}X$. A set X is rough if $\text{BN}_B(X)$ is not empty.

2.2 Rough Membership Functions

A rough membership function (rmf) makes it possible to measure the degree that any specified object with given attribute values belongs to a given set X [8], [16]. A rm function μ_x^B is defined relative to a set of attributes $B \subseteq A$ in information system $S = (U, A)$ and a given set of objects X. The equivalence class $[x]_B$ induces a partition of the universe. Let $B \subseteq A$, and let X be a set of observations of interest. The degree of overlap between X and $[x]_B$ containing x can be quantified with a rmf given in (2):

$$\mu_X^B : U \to [0,1] \text{ defined by } \mu_X^B(x) = \frac{|[x]_B \cap X|}{|[x]_B|} \qquad (2)$$

2.3 Example Rough Membership Function

A sample rough member function computation is given in this section (see Fig. 1(a)).

$$\mu_F^B(u) = \frac{\|\overline{BF} \cap [u]_B\|}{\|[u]_B\|} = \frac{1}{4}$$

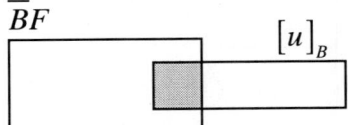

Fig. 1(a). Sample rmf value

Fig. 1(b). Overlapping regions

Let B be a set of attributes of waveforms of power system faults (e.g., b1 = phase current, b2 = maximum phase current, and so on). Let F be a set of fault signal files. Further, let $\overline{BF}$ = {f3,f4,f7,f8} be an upper approximation, and let $[u]_B$ = {f4, f9, f10, f15} be an equivalence class containing files representing a known fault. For overlapping regions shown in Fig. 1(b), the degree of overlap between $\overline{BF}$ and $[u]_B$ can be computed as in Fig. 1(a).

2.4 Design of Rough Neurons

Neural networks are collections of massively parallel computation units called neurons. A neuron is a processing element in a neural network. Two types of rough neurons have been identified: approximation and decider neurons [7]. Let $\overline{BX}$ be an upper approximaton relative a set of attributes B and reference set X. An approximation neuron η computes $y = \mu_U^B(\overline{BX}, [u]_B)$. A decider rough neuron implements a collection of decision rules by (i) constructing a condition vector c_{exp} from its inputs which are rm function values (ii) discovering the rule $c_i => d_i$ with a condition vector c_i which most closely matches an input condition vector c_{exp}, and (iii) outputs $\min(e_i d_i)$ where $d_i \in \{0,1\}$ and $e_i = \|c_{exp} - c_i\|/\|c_i\| \in [0,1]$. In cases where d = 0, then $y_{rule} = \min(e_i d_i) = 0$, and the classification is unsuccessful. If d = 1, then $y_{rule} = \min(e_i d_i) = e_i$ indicates the relative error in a successful classification.

2.5 Sample Rough Neural Network

A high voltage direct current (dc) transmission system connected between ac source and ac power distribution system has two converters. In the case where the flow of power is from the ac side to the dc side as in Fig. 2, then a converter acts as a rectifier in changing ac to dc. The inverter in Fig. 2 converts dc power to ac power at desired output voltage and frequency. The Dorsey Station in the Manitoba Hydro system, for example, acts as an inverter in converting dc to ac, which is distributed throughout North America.

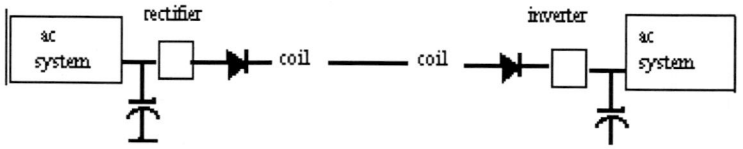

Fig. 2. dc Link Between ac Systems

A decision (d) to classify a waveform for a power transmission fault depends on an assessment of phase current (pc), current setting (cs), maximum phase current (max pc), ac voltage error (acve), pole line voltage (plvw) and phase current (pcw) waveforms. A sample commutation failure decision table is given next. In Table 1, d = 1 {0} indicates that the waveform for a fault represents {does not represent} a power system failure.

Table 1. Sample Power System Failure Decision Table

	acve	pc/cs	plvw	pcw	cs	max pc	d
file 1	0.059	0.069	0	0.0187	0	0	0
file 3	0.059	0.069	1	0.0187	0.1667	0.0856	1

Signal data needed to construct the condition granules in Table 1 come from files specified in column 1 of the table. Sample discretized rules derived from Table 1 using Rosetta [17] are given in (3) and (4).

$$\text{plvw}([*, 0.750)) \text{ AND } \text{cs}([0.111, *)) \text{ AND } \text{max-pc}([*, 0.043)) \Rightarrow d(no) \qquad (3)$$

$$\text{plvw}([0.750, *)) \text{ AND } \text{cs}([0.111, *)) \text{ AND } \text{max-pc}([0.043, *)) \Rightarrow d(yes) \qquad (4)$$

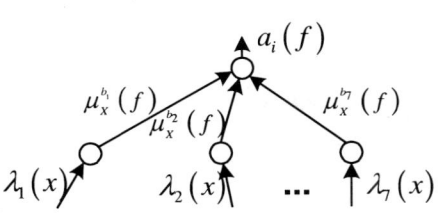

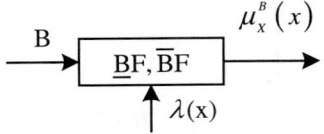

Fig. 3(b). Appoximation neuron

Fig. 3(a). Partial rough neural network

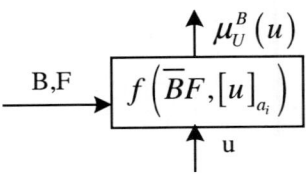

Fig. 3(c). Decider neuron

The basic structure of a rough neural network is given in Fig. 3(a)The decider neuron in Fig. 3(b) implements rules derived from Table 1. In the network in Fig. 3(a), the parameters to be tuned are represented by B, the set of relevant features. The goal of tuning is to improve the quality of concept (Fault) approximation.

2.6 Sample Verification

A comparison between the output from a rough neural network used to classify power system faults relative to 24 fault files and known classification of the sample fault data is given in Fig 4. In all of the cases considered in Fig. 4, there is a close match between the target faults and the faults identified the neural network.

3. Petri Net Model of a Rough Neuron

In what follows, it is assumed that the reader is familiar with classical Petri nets [19] and coloured Petri nets [20]. Rough Petri nets are derived from coloured and hierarchical Petri nets as well as from rough set theory [21]. A rough Petri net provides a basis for modeling, simulating and analyzing rough neurons, rough neural networks, and granular decision systems.

3.1 Rough Petri Nets

A rough Petri net (rPn) is a structure (Σ, P, T, A, N, C, G, E, I, W, $\Re$, ξ) where
- S is a finite set of non-empty data types called color sets.
- N is a 1-1 node function where N: A $\to$ (P x T) $\cup$ (T x P).

- C is a color function where C: P→Σ.
- G is a guard function where G: T → [0, 1].
- E is an arc expression function where E: A → Set_of_Expressions where E(a) is an expression of type C(p(a)) and p(a) is the place component of N(a).
- I is an initialization function where I: P→ Set_of_Closed_Expressions where I(p) is an expression of type C(p).
- W is a set of strengths-of-connections where ξ: A → W.
- $\Re = \{\rho_\xi \mid \rho \text{ constructs } \xi \in \{\text{rough set structure}\}\}$

Let U, S, A, d be a set of inputs, information system S, attributes of S, decision d, respectively. Examples of rough set structures constructed by ρ from information granules are the decision system S = (U, A ∪ {d}) and the set OPT(S) of all rules derived from reducts of a decisions system table for S. Borrowing from coloured Petri nets, a rough Petri net provides data typing (colour sets) and sets of values of a specified type for each place. The expression E(p, t) specifies the input associated with the arc from input place p to transition t, and the expression E(t, p') specifies a transformation (activity) performed by transition t on its inputs {E(p, t)} to produce an output for place p'.

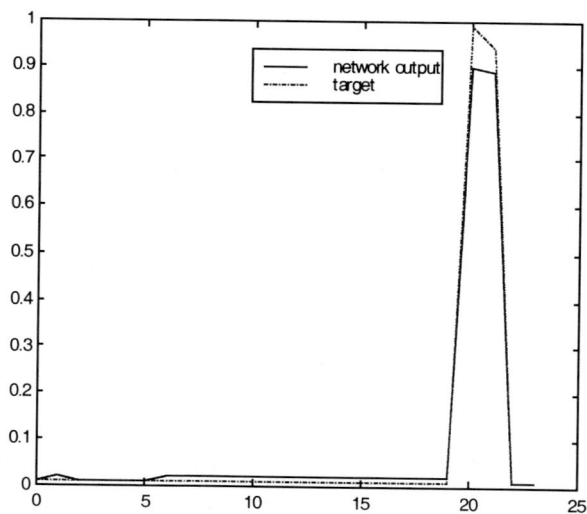

Fig. 4. Sample Verification

3.2 Guarded Transitions

In a rough Petri net, various families of guards can be defined which induce a level-of-enabling of transitions [21]. Consideration of level-of-enabling stems from guards named after Jan Lukasiewicz [22], who inaugurated the study of multivalued logic. Let U denote a universe of objects, and let $X \subseteq U$. Let $\lambda:U \rightarrow [0, 1]$.

Def. 1 *Lukasiewicz Guard.* A *Lukasiewicz guard* on transition t with input x is a higher order propositional function $P(\lambda(x))$ labeling the transition t with input x and output $\lambda(x)$. The guard $P(\lambda(x)) = \lambda(x) \in (0,1]$, where $0 < \lambda(x) \leq 1$ enables t.

With one exception, notice that $\lambda(x)$ can be used to model a filter on an input port of an approximation neuron, since there is interest in preventing input signals with zero strength from enabling an input transition. To complete the modeling of an input port filter, a restricted Lukasiewicz guard is needed.

Def. 2 *Restricted Lukasiewicz Guard.* A *restricted Lukasiewicz guard* on transition t with input x is a function $P(\lambda(x))$ labeling the transition t with input x and output $\lambda(x)$. The guard $P(\lambda(x)) = \lambda(x) \in (0,1]$, where $0 < \lambda(x) \leq 1$ enables t.

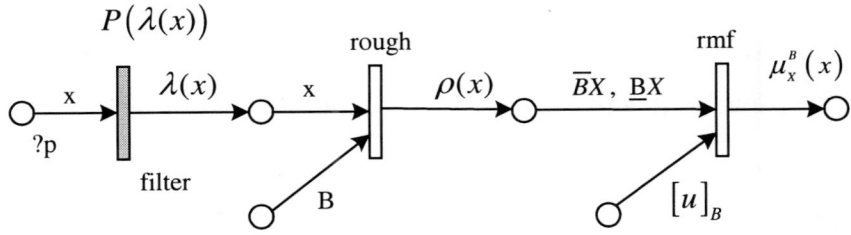

Fig. 5. Rough Neuron Petri Net Model

3.3 Petri Net Model of a Rough Neuron

Let η be an approximation neuron with a single input port p_i and single output port p_o.

Let X be a set of inputs for η, B a set of attributes, and λ a filter on p_i of η. Let ρ be a procedure which constructs $\overline{B}X$, $\underline{B}X$ and let $\mu_x^B(x)$ compute the output of η (see Fig. 5). The notation ?p indicates a receptor place which is always input ready. The filter $\lambda(x)$ returns x in cases where $\lambda(x) > 0$, $\lambda(x) \in [a, b] \subseteq (0, 1]$. The transition labeled "rough" in Fig. 5 is enabled by the input of signal x and set of attributes B. When this transition fires, $\rho(x)$ constructs $\overline{B}X$, $\underline{B}X$. The availability of $\overline{B}X$, $\underline{B}X$ and equivalence class $[u]_B$ enables the transition labeled "rmf" in Fig. 5. Whenever the rmf transition fires, $\mu_x^B(x)$ computes the degree of overlap between $[u]_B$ and $\overline{B}X$. The advantage in constructing a Petri net model of a rough neuron is facilitates a number of tests such as reachability of each of the transitions in the model and the action of the guard modeling a filter on a rough neuron input port.

4. Concluding Remarks

The basic features in the design of a particular kind of rough neuron called an approximation neuron are presented in this paper. The introduction of rough neurons has been motivated by the search for improved means of identifying and classifying features in a feature space. The output of an approximation neuron is a rough membership function value, which indicates the degree of overlap between an approximation region and some other set of interest in a classification effort. A Petri net model of an approximation neuron has also been given. The guarded transitions in a rough Petri net make it possible to model a filter on an input port of a rough neuron. A sample application of these neurons in a power system fault classification system has been given. Future work will entail a study of a more complete classification and design of rough neurons.

References

1. Z. Pawlak, Rough sets, Int. J. of Computer and Information Sciences, Vol. 11, 1982, 341-356.
2. Z. Pawlak, Rough Sets: Theoretical Aspects of Reasoning About Data, Boston, MA, Kluwer Academic Publishers.
3. Z. Pawlak. Reasoning about data--A rough set persepective. Lecture Notes in Artificial Intelligence 1424, L. Polkowski and A. Skowron (Eds.). Berlin, Springer-Verlag, 1998, 25-34.
4. M. Banerjee, S. Mitra, S.K. Pal, Rough fuzzy MLP: Knowledge encoding and classification, IEEE Trans. Neural Networks, vol. 9, 1998, 1203-1216.
5. 5 L. Han, J.F. Peters, S. Ramanna, R. Zhai, Classifying faults in high voltage power systems: A rough-fuzzy neural computational approach. In: N. Zhong, A. Skowron, S. Ohsuga (Eds.), New Directions in Rough Sets, Data Mining, and Granular-Soft Computing, Lecture Notes in Artificial Intelligence 1711. Berlin: Springer, 1999, 47-54.
6. L. Han, R. Menzies, J.F. Peters, L. Crowe, High voltage power fault-detection and analysis system: Design and implementation. Proc. CCECE99, 1253-1258.
7. J.F. Peters, A. Skowron, L. Han, S. Ramanna, Towards rough neural computing based on rough membership functions: Theory and Application, Rough Sets and Current Trends in Computing (RSCTC'2000) [submitted].
8. Z. Pawlak, A. Skowron, Rough membership functions. In: R. Yager, M. Fedrizzi, J. Kacprzyk (Eds.), Advances in the Dempster-Shafer Theory of Evidence, NY, John Wiley & Sons, 1994, 251-271.
9. A. J. Rosa, Filters (Passive). In: R.C. Dorf (Ed.), The Engineering Handbook. Boca Raton, FL: CRC Press, Inc., 1996, 1155-1166.
10. P. Bowron, F.W. Stephenson, Active Filters for Communication and Instrumentation. London: McGraw-Hill Book Co., 1979.
11. W. Pedrycz, J.F. Peters, Learning in fuzzy Petri nets, in *Fuzziness in Petri Nets* edited by J. Cardoso and H. Scarpelli. Physica Verlag, a division of Springer Verlag, 1998.

12 A. Skowron, J. Stepaniuk. Constructive information granules. In: Proc. of the 15th IMACS World Congress on Scientific Computation, Modelling and Applied Mathematics, Berlin, Germany, 24-29 August 1997. Artificial Intelligence and Computer Science 4, 1997, 625-630.

13 A. Skowron, C. Rauszer. The discernability matrices and functions in information systems. In: Intelligent Decision Support, Handbook of Applications and Advances of the Rough Sets Theory, Slowinski, R. (Ed.), Dordrecht, Kluwer Academic Publishers, 1992, 331-362.

14 A. Skowron, J. Stepaniuk, Information granules in distributed environment. In: N. Zhong, A. Skowron, S. Ohsuga (Eds.), New Directions in Rough Sets, Data Mining, and Granular-Soft Computing, Lecture Notes in Artificial Intelligence 1711. Berlin: Springer, 1999, 357-365.

15 A. Skowron, J. Stepaniuk. Constructive information granules. In: Proc. of the 15th IMACS World Congress on Scientific Computation, Modelling and Applied Mathematics, Berlin, Germany, 24-29 August 1997. Artificial Intelligence and Computer Science 4, 1997, 625-630.

16 J. Komorowski, Z. Pawlak, L. Polkowski, A. Skowron, Rough sets: A tutorial. In: S.K. Pal, A. Skowron (Eds.), Rough Fuzzy Hybridization: A New Trend in Decision-Making. Singapore: Springer-Verlag, 1999, 3-98.

17 Rosetta software system, http://www.idi.ntnu.no/~aleks/rosetta/

18 W. Pedrycz, Computational Intelligence: An Introduction, Boca Raton, CRC Press, 1998.

19 Petri, C.A., 1962, "Kommunikation mit Automaten", Schriften des IIM Nr. 3, Institut für Instrumentelle Mathematik, Bonn, West Germany.

20 Jensen, K., 1992, "Coloured Petri Nets--Basic Concepts, Analysis Methods and Practical Use 1", Berlin, Springer-Verlag.

21 J.F. Peters, A. Skowron, Z. Surai, S. Ramanna, Guarded transitions in rough Petri nets. In: Proc. of 7th European Congress on Intelligent Systems & Soft Computing (EUFIT'99), Sept. 1999, Aachen, Germany.

22 J. Lukasiewicz, O logice trojwartosciowej, Ruch Filozoficzny 5, 1920, 170-171. See "On three-valued logic", English translation in L. Borkowski (Ed.), Jan Lukasiewicz: Selected Works, Amsterdam, North-Holland, 1970, pp. 87-88.

Towards Musical Data Classification via Wavelet Analysis

Alicja Wieczorkowska

Polish – Japanese Institute of Information Technologies
ul. Koszykowa 86, 02-008 Warsaw, Poland
alicja@pjwstk.waw.pl

Abstract: In order to search through a sound database, information about the musical contents has to be attached to the file, otherwise the user has to look for the specific musical information by himself. Wavelet analysis is one of possible tools that can be used as a basis for automatic classification of musical data. In this paper, the author presents wavelet-based parameters extracted from sounds of musical instruments. These parameters have been used as a basis of automatic classification of musical instrument sounds. Tests evaluating the efficiency of such parameterization were performed by means of rough set based algorithms and decision trees. Results of these tests are presented in this paper.

Keywords: Knowledge Discovery and Data Mining, Soft Computing, Sound Analysis, Musical Sound Classification

1 Introduction

One of the main problems concerning sound databases is how to classify automatically the musical material, contained in a recording. For instance, if the information what musical instruments are playing in the piece is not attached to a file, it is not possible to extract such information automatically. Automatic classification makes possible automatic labeling of the content of the multimedia data, which is the aim of ISO/IEC standard MPEG-7 that is under development by MPEG (Moving Picture Experts Group).

Sounds (even singular sounds) that are to be processed by a classification algorithm cannot be represented as raw data, i.e. as a set of samples. Objects processed by a classifier should be described by a set of attributes (the less the better), so sound data need parameterization before classification. Additionally, since the same sound may be changed dramatically by musical interpretation and recording conditions, the appropriate parameterization is necessary as a preprocessing before classification.

Such a parameterization is based on sound analysis. Sounds can be analyzed both in temporal and spectral domain, using many methods, such as Fourier transform, wavelet transform, correlation, cepstral analysis, filtering, statistical methods and so on [1], [2], [3], [4]. Wavelet transform is especially useful for musical applications, because this time-frequency analysis divides the spectrum into frequency bands that are of equal width in a logarithmic scale, what is similar to the human hearing. Therefore, wavelet analysis can be used as a tool for the classification of musical instrument sounds and for labeling of the recordings.

The sound parameterization that is appropriate for instrument classification is a difficult task and usually requires very careful choice of attributes [6], [12]. The parameterization presented here is quite simple and the set of attributes contains 162 parameters. Many attributes are redundant and soft computing methods have been used to find the most useful attributes among them. In this paper, presented results were obtained using decision trees and rough set based methods, but other classifiers (such as neural networks, k-nearest neighbor etc.) can also be used [5], [9], [10], [12], [13].

2 Wavelet Analysis and Parameterization of Musical Data

Wavelet analysis applied in the presented work is based on the division of the spectrum into octave bands, using filter of the second order, proposed by Daubechies and Coifman (see Fig. 1). The wavelet transform of the function f is performed as a decomposition of f using a mother wavelet ψ and a scaling function φ in the following way [11]:

$$f = \sum_{j,k} \langle f, \psi_{j,k} \rangle \psi_{j,k},$$

where: <> - inner product,.
  j – resolution level,
  k – time instant,

$$\{\psi_{jk}(t) = 2^{j/2} \psi(2^j t - k)\},$$
$$\psi(t) = \sqrt{2} \sum_k g_k \varphi(2t - k),$$
$$\{\varphi_{jk}(t) = 2^{j/2} \varphi(2^j t - k)\},$$
$$\varphi(t) = \sqrt{2} \sum_k h_k \varphi(2t - k),$$

g_k, h_k – coefficients of highpass and lowpass filters,
$g_k = (-1)^k h_{1-k}.$

The above analysis gives good frequency resolution and poor time resolution for low frequency bands, and good time resolution and poor frequency resolution for high frequency bands.

Singular sound of any instrument can be quite long: it may last for some seconds, and its timbre may change with time. Since the most important parts of sound for the recognition by human is the beginning (starting transient, i.e. the attack) and the middle part of the sound (quasi-steady state), these parts have been taken into account during parameterization. Sounds from CDs [7], digitally recorded stereo with sampling frequency 44.1kHz and 16 bit resolution, have been analyzed (each channel separately) using wavelet transform with analyzing frame 4096 samples, taken from the attack and from the quasi-steady state of the sound. The calculated parameters are based on a part of each frame, containing the coefficient of the greatest energy. Exemplary result of wavelet analysis of a sound, with the area selected for parameterization marked by a black frame, is presented in Fig. 2.

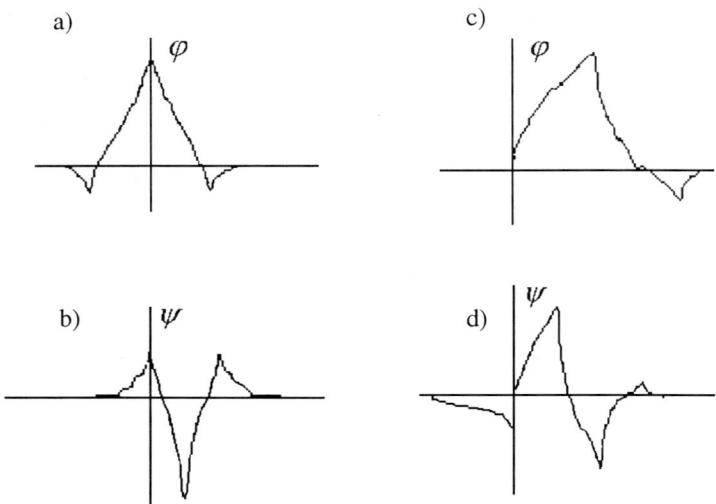

Fig. 1. Scaling functions φ and mother wavelets ψ for filters of order 2:
(a, b) proposed by Coifman
(c, d) proposed by Daubechies

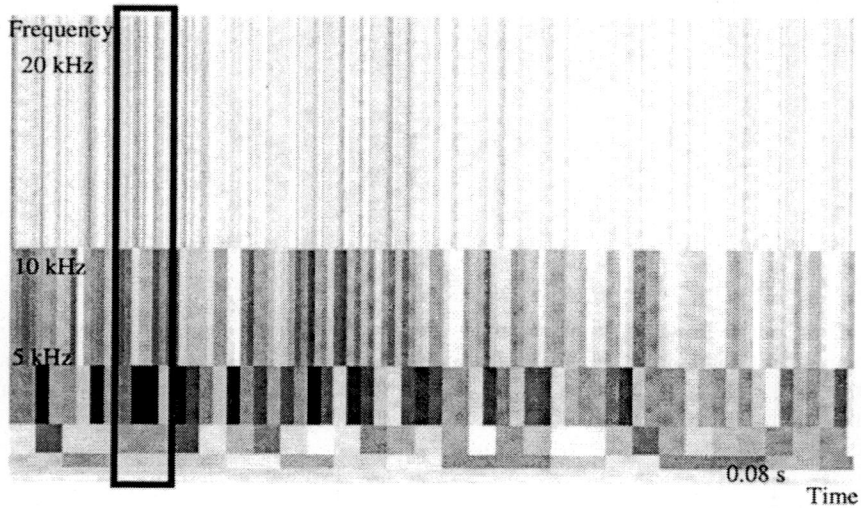

Fig. 2. Wavelet analysis (Daubechies φ and ψ) of the clarinet sound a^3 (1760 Hz), for sampling frequency 44.1 kHz and analyzing frame 4096 Sa; the darker the area, the grater the magnitude

The parameters calculated on the wavelet analysis are as follows [12]:
- $W_1, ..., W_{38}$ – energy of the parameterized spectrum bands for the Daubechies wavelet of order 2, in the middle of the attack;
  $W_i = E_i/E$, where:
  E – overall energy of the parameterized part of the frame; E_i – partial energy:
  i=23,...,38 – spectral components in the frequency band 11.025-22.05kHz,
  i=15,...,22 – spectral components in the frequency band 5.5125-11.025kHz,
  i=11,...,14 – spectral components in the frequency band 5.5125-11.025kHz,
  i=9,10 – spectral components in the frequency band 2.75626-5.5125kHz,
  i=1,...,8 – spectral components for lower frequency bands;
- $W_{39}, ..., W_{76}$ – energy for Daubechies wavelet of order 2, in the middle of the steady state;
- W_{77} - position of the middle of the attack, $W_{77} \in (0, 1)$;
- W_{78} - position of the middle of the steady state, $W_{78} \in (0, 1)$;
- $W_{79}, ..., W_{115}$ – energy of the parameterized spectrum bands for the Coifman wavelet of order 2, in the middle of the attack;
  $W_i = E_i/E$, where:
  i=100,...,115 – spectral components in the frequency band 11.025-22.05kHz,
  i=92,...,99 – spectral components in the frequency band 5.5125-11.025kHz,
  i=88,...,91 – spectral components in the frequency band 5.5125-11.025kHz,
  i=86,87 – spectral components in the frequency band 2.75626-5.5125kHz,
  i=79,...,85 – spectral component in lower frequency bands;
- $W_{116}, ..., W_{152}$ – energy for Coifman wavelet of order 2, in the middle of the steady state.

The set of parameters presented here is used to describe singular sound of a musical instrument. The data represent all available sounds of musical scale of the following instruments:
- bowed string instruments: violin, viola, cello and double bass;
- woodwinds: flute, oboe and clarinet;
- brass: trumpet, trombone, French horn and tuba.

Sounds of these instruments were recorded using various playing techniques, namely vibrato, pizzicato (when strings are plucked with fingers), with and without muting.

The investigated data have been grouped into classes in the following ways:
- 18 classes – each class contains objects representing sounds of one instrument, played using one technique: flute – vibrato, oboe – vibrato, B flat clarinet, C trumpet, C trumpet – muted, French horn, French horn – muted, tenor trombone, tenor trombone – muted, tuba, violin – vibrato, violin – pizzicato, viola – vibrato, viola – pizzicato, cello – vibrato, cello – pizzicato, double bass – vibrato, double bass – pizzicato;
- 11 classes, containing objects representing sounds of one instruments, played using various techniques: flute, oboe, B flat clarinet, C trumpet, French horn, tenor trombone, tuba, violin, viola, cello, double bass;
- 5 classes, containing objects representing sounds of family of instruments, played with the same technique: woodwinds, brass without muting, brass with muting, strings vibrato, strings pizzicato;

- 3 classes, containing objects representing sounds of family of instruments, played with various techniques: woodwinds, brass, strings;
- 4 classes, representing sounds of instruments played with the same technique: vibrato (strings, flute, oboe), pizzicato (strings), muting (trumpet, French horn, trombone), without vibrato or muting (clarinet, brass)
- 3 classes, representing sounds of instruments played with the same technique, but muting is not extracted into separate class: vibrato (strings, flute, oboe), without vibrato (clarinet, brass), pizzicato (strings).

The data set contains 1358 objects. The proposed set of parameters is probably superfluous, but classification algorithms can be applied as a tool of filtration of these attributes.

3 Classification Algorithms

As it was mentioned in the first section, automatic classification of data can be performed in many ways. In the described research, classification algorithms have been used not only to learn classification rules, but also to test the proposed set of parameters. The author decided to choose decision trees and rough set based algorithms, because they are quite fast and present the results in a way that is visually easy to interpret for a user.

Decision trees we use here are all binary, constructed from a root to leaves. Nodes are labeled with attributes (parameters), chosen by maximal gain ratio criterion [9]:

$$Gr = \frac{I(a \to d)}{H(a)}, \qquad \text{where}$$

$I(a \to d) = H(d) - H(d \mid a)$ - information gain for the attribute a and the class d

$H(d) = -\sum_{i=1}^{k} p(d_i) \cdot \log p(d_i)$ - entropy of the class d

$H(d \mid a) = -\sum_{j=1}^{l} p(a_j) \cdot \sum_{i=1}^{k} p(d_i \mid a_j) \cdot \log p(d_i \mid a_j)$ - conditional entropy for d and a

$p(v)$ – probability of the value v.

Edges are labeled with values of the attribute labeling the parent node. Real value data are quantized, and optimal cut point c is found on the basis of the entropy criterion. Attribute values x are divided into 2 sets: $x>c$ and $x \leq c$, and these sets are used to label the edges. The leaves of the tree represent classes with probability controlled by a user.

Since objects representing the investigated classes are mixed and some attributes in the data set are redundant, the created trees have been pruned, which results in the reduction of their number of attributes, their depth and the number of branches. Generally speaking, the pruning is driven by the admissible probability of incorrect classification of new coming objects.

Rough set [8] theory is based on a specific concept of membership function, describing elements x of a set X. In classical Cantor theory, the membership function $\mu_X(x)$ is defined as

$$\mu_X(x) = \begin{cases} 0 \text{ for } x \notin X \\ 1 \text{ for } x \in X \end{cases}$$

In rough set theory, we assume preliminary information $I(x)$ about elements $x \in U$ of the set $X \subseteq U$. On the basis of the information function $I: U \to 2^U$ such that ($\forall x \in U$) [$x \in I(x)$], the membership function $\mu_X(x) \in [0, 1]$ is defined as

$$\mu_X^I(x) = \frac{card(X \cap I(x))}{card\ I(x)}$$

Rough set based systems allow processing of imprecise or inconsistent data. The system used in the described research [10] includes quantization of real-value attributes. The domain of each attribute is divided into intervals of equal width; number of intervals (10 by default) can be selected by the user.

4 The Experiments

Both decision trees and rough set based systems have been used to extract rules describing musical instrument data. Exemplary decision tree describing the investigated data for 18 classes (a part only) is presented below:

```
W78 <= 0.5661 :
| W132 <= 0.00012145 :
| | W88 <= 0.0053521 :
| | | W51 > 0.0047161 : violin pizzicato
| | | W51 <= 0.0047161 :
| | | | W10 > 0.00305402 : viola pizzicato
| | | | W10 <= 0.00305402 :
| | | | | W43 <= 0.00090222 :
| | | | | | W47 <= 0.00113098 : double bass pizzicato
| | | | | | W47 > 0.00113098 : viola pizzicato
...
| | W88 > 0.0053521 :
| | | W3 <= 0.0003781 : viola pizzicato
| | | W3 > 0.0003781 : violin pizzicato
| W132 > 0.00012145 :
| | W43 > 0.117745 : cello pizzicato
| | W43 <= 0.117745 :
| | | W39 > 0.00088248 : trumpet muted
...
W78 > 0.5661 :
| W78 <= 0.9079 :
| | W122 <= 0.388391 :
| | | W45 <= 0.594423 :
...
```

The attributes in the tree describe both the attack and the steady state of the sound, using both Daubechies and Coifman wavelet. The root of the tree is labeled by one of temporal attributes, since this attribute allows easy discernment between sounds played pizzicato and vibrato. The same attribute labels the trees for each division of the data into classes.

Apart from decision trees, classification rules in a classical form have also been extracted for these data, using both C4.5 and DataLogic/R+. The obtained rules are of various length and accuracy. For example, some rules are based only on singular or very few attributes:

- $W_{152} > 0.0008401$ => trumpet muted (accuracy 96.7%),
- $W_{78} > 0.9087 \land W_{84} > 0.858949$ => flute (accuracy 79.4%),
- $W_{78} \leq 0.5661 \land W_{81} > 0.400169$ => double bass pizzicato (accuracy 85.7%),
- $W_{43} > 0.117745 \land W_{78} \leq 0.5661 \land W_{84} > 0.0127438$ => cello pizzicato (accuracy 85.5%).

Some of the obtained rules are longer, for example:

- $W_1 \leq 0.0001733 \land W_{78} \leq 0.5661 \land W_{88} > 0.00026218 \land W_{132} > 0.00012145$ => violin pizzicato (accuracy 96.6%),
- $W_6 \leq 0.782861 \land W_{13} > 0.00019162 \land W_{14} > 2.133e-05 \land W_{25} \leq 0.00513477 \land W_{45} > 0.00027319 \land 0.00041218 < W_{46} \leq 0.678519 \land W_{77} \leq 0.0258 \land 0.9079 < W_{78} \leq 0.949 \land W_{85} > 5.832e-05 \land W_{93} > 2.819e-05$ => clarinet (accuracy 93.4%),
- $W_{13} > 5.94e-06 \land W_{34} \leq 7.555e-05 \land W_{51} > 0.00057545 \land W_{59} \leq 0.00097323 \land W_{77} \leq 0.0583 \land 0.5661 < W_{78} \leq 0.8519 \land W_{121} > 0.56432 \land W_{127} \leq 0.0369227$ => French horn (accuracy 97.9%),
- $W_{73} > 5.054e-05 \land W_{74} \leq 3.476e-05 \land 0.5661 < W_{78} \leq 0.7812 \land W_{103} \leq 0.00116903 \land W_{122} > 0.380377$ => oboe (accuracy 82.0%),
- $W_{23} \leq 0.00010887 \land W_{59} \leq 0.00124617 \land 0.5661 < W_{78} \leq 0.9087 \land W_{121} > 0.56432 \land W_{127} > 0.0369227$ => trombone (accuracy 96.7%),
- $W_1 \leq 0.00370801 \land W_{28} \leq 0.00088872 \land W_{77} > 0.04505 \land W_{78} > 0.8276 \land W_{121} \leq 0.56432 \land W_{122} \leq 0.380377 \land W_{148} \leq 5.086e-05$ => tuba (accuracy 89.9%),
- $W_{14} \leq 0.00288078 \land W_{77} > 0.0457 \land W_{78} > 0.5661 \land W_{103} > 0.00010219 \land W_{121} > 0.56432 \land W_{127} <= 0.0369227$ => viola vibrato (accuracy 80.9%),

and so on.

Rules extracted both using DataLogic/R+ and C4.5 contain attributes calculated by means of both filters, from the attack and steady state of the sound. The constructed classifiers are based on about 60 attributes.

The obtained trees and rules have been tested, using 70% of the data as a training set and the remaining 30% as a test set. Rough sets based experiments have been performed for various settings of DataLogic/R+ [10] and the best results have been obtained for the following settings: roughness value 0.01, rule precision threshold 0.90, i.e. for quite precise rules. Decision trees have also been created for various settings of C4.5 [9], with quite good results for standard settings, i.e. with pruning confidence level 25% (and even better accuracy for the settings adjusted individually to each data set). The results are presented in Tab. 1.

Tab. 1. Percentage of correct classification for the musical instrument sound data

Data	Rough sets	Decision trees
18 classes	42.75%	57.9%
11 classes	46.68%	56.8%
5 classes	63.14%	75.8%
3 classes (woodwinds, brass, strings)	69.53%	80.1%
4 classes	70.27%	78.6%
3 classes (vibrato non-vibrato, pizzicato)	71.01%	79.5%

As we see, the results for decision trees are about 10% better on average, and obviously all results for 3-5 classes are better than for 11 or 18 classes. These results are worse than 70% of accuracy obtained for different methods of parameterization, i.e. not based on wavelet analysis, obtained by the author and another researchers [6], [12]. But this wavelet-based parameterization is very simple (and can be improved), whereas other parameterization methods require precise calculation of pitch and very careful analysis of spectrum of sounds. Additionally, in real recordings we usually have a sequence of sounds, i.e. a musical phrase, instead of singular sounds. Therefore, classification of such a phrase can be of higher accuracy. Of course, additional preprocessing is necessary in real recordings so as to extract singular sounds of instruments, namely separation of solo instrument from the musical background, and separation of consequent sounds in a phrase.

5 Conclusions

The contents-based search of audio and video data is one of main goals of multimedia research nowadays. Therefore, automatic classification of musical sounds is necessary as a tool of labeling of sound data. Parameterization of sounds that allows efficacious classification of musical instruments is quite difficult, because the timbre of sound depends on many circumstances. Additionally, the timbre changes within musical scale of an instrument, that makes classification (that should be correct independent on the pitch) even more difficult. That is why such a parameterization has to be done very carefully, involving quite sophisticated recipes of parameterization.

The next stage of this process is classification of the calculated data. The author decided to use rough set based algorithms and decision trees, since their outcomes are easy to interpret and they identify the most useful attributes in the proposed parameterization. These methods show the number of necessary attributes and allow evaluation of their importance.

Wavelet based parameterization described here is very simple, and the results are somewhat weaker than for more sophisticated methods, based on Fourier analysis and pitch calculation. Unfortunately, precise pitch extraction is usually elaborated for each instrument independently, and may introduce octave errors. Wavelet based parameterization does not require pitch calculation, and after some improvement, can be quite helpful in classifying musical instrument sounds.

Acknowledgements

This research has been partially performed at the Sound Engineering Department, Faculty of Electronics, Telecommunications and Informatics at the Technical University of Gdańsk, and also at the Computer Science Department, School of Information Technology at University of North Carolina, Charlotte.

References

[1] Ando S., Yamaguchi K., Statistical Study of Spectral Parameters in Musical Instrument Tones, *J. Acoust. Soc. of America*, 94, 1, July 1993, 37-45.
[2] Garnett G. E., Music, Signals, and Representations: A Survey, in: De Poli G., Piccialli A., Roads C. (ed.), Representations of Musical Signals, MIT Press, Cambridge, Massachusetts, 1991, 325-369.
[3] Ifeachor E. C., Jervis B. W., Digital signal processing: a practical approach, Addison–Wesley Publishing Co., Wokingham, England, 1995.
[4] Kronland-Martinet R., Grossmann A., Application of Time-Frequency and Time-Scale Methods (Wavelet Transforms) to the Analysis, Synthesis, and Transformation of Natural Sounds, in: De Poli G., Piccialli A., Roads C. (ed.), Representations of Musical Signals, MIT Press, Cambridge, Massachusetts, 1991, 45-85.
[5] Kubat M., Bratko I., Michalski R. S., A Review of Machine Learning Methods, in: Michalski R., Bratko I., Kubat M. (ed.), Machine Learning and Data Mining: Methods and Applications, John Wiley & Sons Ltd., Chichester 1998.
[6] Martin K. D., Kim Y. E., 2pMU9. Musical instrument identification: A pattern-recognition approach, Internet: ftp://sound.media.mit.edu/pub/Papers/kdm-asa98.pdf, presented at the 136th meeting of the Acoustical Society of America, October 13, 1998.
[7] Opolko F., Wapnick J., MUMS – McGill University Master Samples, compact discs, McGill University, Montreal, Canada, 1987.
[8] Pawlak Z., Hard and Soft Sets, *ICS Research Report* 10/94, Warsaw University of Technology, February 1994.
[9] Quinlan J. R., C4.5: Programs for Machine Learning, Morgan Kaufmann Publishers, San Mateo, California, 1993.
[10] Reduct Systems, Datalogic/R, 1990-95 Reduct Systems Inc., Regina, Saskatchewan, Canada.
[11] Wolfram Research, Mathematica, Wavelet Explorer, Champaign, Illinois, 1996.
[12] Wieczorkowska A., The recognition efficiency of musical instrument sounds depending on parameterization and type of a classifier (in Polish), Ph.D. Dissertation, Technical University of Gdańsk, Gdańsk, 1999.
[13] Zurada J., "Introduction to Artificial Neural Systems", West Publishing Company, St. Paul New York - Los Angeles - San Francisco, 1992.

Annotated Hyperresolution for Non-horn Regular Multiple-Valued Logics*

James J. Lu[1], Neil V. Murray[2], and Erik Rosenthal[3]

[1] Department of Computer Science, Bucknell University, Lewisburg, PA 17837.
U.S.A., jameslu@bucknell.edu
[2] Department of Computer Science, State University of New York, Albany, NY
12222, nvm@cs.albany.edu
[3] Department of Mathematics, University of New Haven, West Haven, CT 06516,
brodsky@charger.newhaven.edu

Abstract. This paper focuses on non-Horn formulas for the class of *regular* signed logics, also known as *annotated logics*. Resolution-based inference systems for these logics are not new, but most earlier work has concentrated on Horn formulas, to which the logic programming paradigm applies. Here a restriction of annotated resolution and reduction called *annotated hyperresolution* is introduced. The new rule is developed for arbitrary CNF formulas of regular signed logics and is shown to be complete.
Keywords: Logic for AI, ℧-resolution, hyperresolution, inference, multiple-valued logic, signed and annotated logic

1 Introduction

Hyperresolution is an example of a theorem proving technique that employs macro steps. Such inference rules usually impose significant restrictions on what choices are admissible in the search for a proof. Another important feature is that only the conclusions of the macro steps are retained, not the conclusions of the constituent steps. In this paper hyperresolution is extended to a class of multiple valued logics (MVL's).

Signed logics [13,4] provide a general[1] framework for reasoning about MVL's. They evolved from a variety of work on non-standard computational logics, including [2,4,9,11,15,20]. The key is the attachment of *signs*—subsets of the set of truth values—to formulas in the MVL. This approach is appealing because it facilitates the utilization of classical techniques for the analysis of non-standard logics, reflecting the classical nature of human reasoning. That is, regardless of the domain of truth values associated with a logic, at the meta-level, humans interpret statements about the logic to be either *true* or *false*.

This paper focuses on the class of *regular signed logics*. These logics are of interest in the knowledge representation and logic programming communities

---

* This research was supported in part by the National Science Foundation under grant CCR-9731893.
[1] Hähnle, R. and Escalada-Imaz, G. [7] have an excellent survey encompassing deductive techniques for a wide class of MVL's, including (properly) signed logics.

because they correspond to the class of paraconsistent logics known as *annotated logics*, introduced by Subrahmanian [19], Blair and Subrahmanian [2], and Kifer et al. [10,21]. In [13], they were also shown to capture fuzzy logics, but in this paper, regular signed logics will refer to annotated logics. In most of the work on annotated logics, the focus has been on Horn sets, widely applied within logic programming. The inference rule *annotated hyperresolution* is developed in this paper for arbitrary regular signed formulas in conjunctive normal form (CNF). Establishing completeness involves substantive reformulations of techniques developed for the classical counterparts (see [1,8]).

There has been related work by Sofronie-Stokkermans on adapting hyperresolution to logics with truth value sets based on finite distributive lattices [17, 18]. Under these assumptions, signs can be restricted to prime filters and their complements. This eliminates the need to resolve more than one positive literal against one negative literal. A similar situation occurs with regular signs when the truth value set is linearly ordered. Hähnle has exploited this to obtain resolution refinements in [5] and to introduce a version of hyperresolution under these conditions [6].

The next section is a summary of the basic ideas of signed formulas and annotated logics. Theorems 1–4 in Section 2.3 were proved in [12]. The main results are found in Section 3: The pure rule is adapted to signed and annotated logics in Section 3.1; annotated hyperresolution is developed in Section 3.2.

2 Signed Logics

Detailed descriptions of the basics of signed logics can be found in [15] and in [13]; the presentation in this section is brief.

Given a language Λ, let Δ be a complete lattice of truth values under some ordering $\preceq$ [2].

A *sign* is a subset of Δ, and a *signed formula* is an expression of the form $S:\mathcal{F}$, where S is a sign and $\mathcal{F}$ is a formula in Λ.

To answer arbitrary queries, we represent queries about formulas in Λ by formulas in a classical logic Λ_S, *the language of signed formulas*; it is defined as follows: The literals are signed formulas and the connectives are (classical) conjunction and disjunction. It should be emphasized that a signed formula $S:\mathcal{F}$ is a literal in Λ_S regardless of the size or complexity of $\mathcal{F}$ and thus has no component parts in the language Λ_S. The set of truth values is $\{true, false\}$. A formula in Λ_S is defined to be Λ-*atomic* if whenever $S:A$ is a literal in the formula, then A is an atom in Λ.

An arbitrary interpretation for Λ_S may make an assignment of $true$ or $false$ to any signed formula (i.e., to any literal) in the usual way. To focus attention only on those interpretations that relate to the sign in a signed formula, restrict attention to Λ-*consistent interpretations*. An interpretation I over Λ assigns to

---

[2] As usual, the greatest and least elements of Δ are denoted $\top$ and $\bot$, respectively, and Sup and Inf denote, respectively, the supremum (least upper bound) and infimum (greatest lower bound) of a subset of Δ.

each literal, and therefore to each formula $\mathcal{F}$, a truth value in Δ, and the corresponding Λ-consistent interpretation I_c is defined by $I_c(S:\mathcal{F}) = \textit{true}$ if $I(\mathcal{F}) \in S$; $I_c(S:\mathcal{F}) = \textit{false}$ if $I(\mathcal{F}) \notin S$.

The annotated hyperresolution rule developed in this paper lifts in the usual way; attention is mostly restricted to the ground case in this paper.

2.1 Signed Resolution

In this section, we review a method for adapting resolution to signed formulas. The inference rules $\mho$-resolution, introduced in [13], and annotated hyperresolution, defined later, are based on a generalized notion of complementary literals, which is characterized in the next lemma.

Lemma 1. (The Reduction Lemma) Let $S_1:A$ and $S_2:A$ be Λ-atomic atoms in Λ_S; then $S_1:A \wedge S_2:A \equiv_\Lambda (S_1 \cap S_2):A$ and $S_1:A \vee S_2:A \equiv_\Lambda (S_1 \cup S_2):A$. □

Consider a Λ-atomic formula $\mathcal{F}$ in Λ_S in conjunctive normal form (CNF). Let $C_j, 1 \leq j \leq r$, be clauses in $\mathcal{F}$ that contain, respectively, Λ-atomic literals $\{S_j:A\}$. Thus we may write $C_j = K_j \vee \{S_j:A\}$. Then the resolvent R of the C_j's is defined to be the clause

$$\left(\bigvee_{j=1}^{r} K_j\right) \vee \left(\left(\bigcap_{j=1}^{r} S_j\right):A\right).$$

The rightmost disjunct is called the *residue* of the resolution; observe that it is unsatisfiable if its sign is empty and satisfiable if it is not. In the former case, it may simply be deleted from R.

In clausal resolution systems, merging is crucial. (Consider unsatisfiable clause sets for which the minimal clause size is two.) In this paper, we treat clauses as sets, and merging is assumed.

Observe that if $S \subseteq S'$, and if two clauses are resolved on the literals $S:A$ and $S':A$, then the residue will be $S:A$ (after all, $S \cap S' = S$), so the clause containing $S:A$ must entail the resolvent. This proves

Lemma 2. The resolvent produced by resolving on two literals in which the sign of one contains the sign of the other is entailed by one of its parents. □

2.2 Regular Signed Formulas and Annotated Logics

Let $(P; \preceq)$ be any partially ordered set, and let $Q \subseteq P$. Then $\uparrow Q = \{y \in P | (\exists x \in Q)\ x \preceq y\}$. Note that $\uparrow Q$ is the smallest *upset* containing Q (see [3]). If Q is a singleton set $\{x\}$, then we simply write $\uparrow x$. We say that a subset Q of P is *regular* if for some $x \in P$, $Q = \uparrow x$ or $Q = (\uparrow x)'$ (the set complement of $\uparrow x$). We call x the *defining element* of the set. In the former case, we call Q *positive*, and in the latter *negative*. Observe that both Δ and $\emptyset$ are regular since $\Delta = \uparrow \bot$ and $\emptyset = \Delta'$. Observe also that if $z = \text{Sup}\{x, y\}$, then $\uparrow x \cap \uparrow y = \uparrow z$. A signed formula is regular if every sign that occurs in it is regular. Note that we may assume that no regular signed formulas have any signs of the form $(\uparrow\perp)'$.

An *annotated logic* is a signed logic in which only regular signs are allowed.

A regular sign is completely characterized by its defining element, say x, and its *polarity* (whether it is positive or negative). A regular signed atom may be written $\uparrow x : A$, while the complement is the set $(\uparrow x)' : A$. Observe that $(\uparrow x)' : A = \sim (\uparrow x : A)$; that is, the signed atoms are complementary with respect to Λ-consistent interpretations. With annotated logics, the most common notation is $\mathcal{F} : x$ and $\sim \mathcal{F} : x$. There is no particular advantage of one or the other, and it is perhaps unfortunate that both have arisen. We will follow the $x : \mathcal{F}$ convention when dealing with signed logics and use $\mathcal{F} : x$ for annotated logics.

2.3 Signed Resolution for Annotated Logics

A sound and complete resolution proof procedure was defined for clausal annotated logics in [9]. The procedure contains two inference rules that we will refer to as *annotated resolution* and *reduction*.[3] These two inference rules correspond to disjoint instances of signed resolution. Two annotated literals L_1 and L_2 are said to be *complementary* if they have the respective forms $A : \mu$ and $\sim (A : \rho)$, where $\mu \geq \rho$, and annotated resolution is defined as follows: Given the annotated clauses $(L_1 \vee D_1)$ and $(L_2 \vee D_2)$, where L_1 and L_2 are complementary, then the *annotated resolvent* of the two clauses on the annotated literals L_1 and L_2 is $D_1 \vee D_2$.

Two clauses can be so resolved only if the annotation of the positive annotated literal that is resolved upon is greater than or equal to the annotation of the negative literal resolved upon. In that case the two clauses are said to be *resolvable* on the annotated literals L_1 and L_2.

The reduction rule is defined when two occurrences of an atom have positive signs. Suppose $(A : \mu_1 \vee E_1)$ and $(A : \mu_2 \vee E_2)$ are two annotated clauses in which μ_1 and μ_2 are incomparable. Then the annotated clause $(A : Sup\{\mu_1, \mu_2\}) \vee E_1 \vee E_2$ is called a *reductant* of the two clauses, and we say that the two clauses are *reducible* on the annotated literals $A : \mu_1$ and $A : \mu_2$. A reduction step may be required to produce a positive sign that in turn enables an annotated resolution step.

It is straightforward to see that the two inference rules are both captured by signed resolution. In particular, annotated resolution corresponds to an application of signed resolution (to regular signed clauses) in which the signs of the selected literals are disjoint. Reduction on the other hand, corresponds to an application of signed resolution in which the signs of the selected literals are both positive and thus have a non-empty regular intersection.

Theorem 1. Suppose that $\mathcal{F}$ is a set of annotated clauses and that $\mathcal{D}$ is a deduction of $\mathcal{F}$ using annotated resolution and reduction. Then $\mathcal{D}$ is a signed

---

[3] Kifer and Lozinskii refer to their first inference rule simply as resolution. However, since we are working with several resolution rules in this paper, appropriate adjectives will be used to avoid ambiguity.

deduction of $\mathcal{F}$. In particular, if $\mathcal{F}$ is an unsatisfiable set of first order annotated clauses, then there is a signed refutation of $\mathcal{F}$. □

The viewpoint of signed logics provides insight into annotated logics. At the same time, the restriction to regular signs has practical advantages. The next theorem, which will be quite useful in Section 3, is an example.

Theorem 2. Suppose $S_1, \ldots, S_n$ are regular signs whose intersection is empty, and suppose that no proper subset of $\{S_1, \ldots, S_n\}$ has an empty intersection. Then exactly one sign is negative; i.e., for some $j, 1 \leq j \leq n, S_j = (\uparrow x_j)'$, and for $i \neq j, S_i = \uparrow x_i$, where $x_1, \ldots, x_n \in \Delta$. □

The intersection of a positive regular sign and a negative regular sign is regular if and only if it is empty, and two negative signs can have a regular intersection if and only if one is a subset of the other. In view of Lemma 2, the latter situation need never be considered, so a signed deduction is defined to be *regular* if every sign that appears in the deduction is regular and if no residue sign is produced by the intersection of two negative signs. The next two theorems are immediate. Theorem 4 states that the class of regular signed deductions is precisely the class of deductions using annotated resolution and reduction. As a result, restricting signed resolution to regular clauses captures annotated resolution and reduction without increasing the search space. Deductions obeying this restriction are called *regular*.

Theorem 3. A signed deduction of a regular formula is regular if and only if the sign of every satisfiable residue is produced by the intersection of two positive regular signs. □

Theorem 4. Let $\mathcal{D}$ be a sequence of annotated clauses. Then $\mathcal{D}$ is an annotated deduction if and only if $\mathcal{D}$ is a regular signed deduction. □

It follows from the theorem that regular signed resolution is complete.
Corollary. Suppose $\mathcal{F}$ is an unsatisfiable set of regular signed clauses. Then there is a regular signed deduction of the empty clause from $\mathcal{F}$. □

3 Regular Signed Deduction with Non-horn Sets

The ideas described in Section 2.3 were employed in [12] to develop $\mho$-resolution for annotated logic programs. One nice feature of $\mho$-resolution is that it allows simple SLD-style proof procedures for annotated logic programs over a large class of lattices. It does so eliminating the expensive reduction rule, yet it does not require irregular deductions. Moreover, for any deduction using annotated resolution and reduction, there is a corresponding deduction using $\mho$-resolution that is at least as short.

These advantages apply to the logic programming paradigm. In this paper, the more general non-Horn setting is addressed. *Every* deduced clause is cached, subject perhaps to certain deletion strategies. We begin by adapting the notion of purity from classical logic to signed and annotated logics.

3.1 Purity

Adapting the notion of purity to signed and annotated logics is not completely straightforward. In classical logic, a literal in a set of clauses is said to be *pure* if its complement does not occur in any other clause. In that case, the clause containing the pure literal is also said to be pure.[4]

The literal set (conjunction) $L = \{S_1 : A, S_2 : A, \ldots, S_m : A\}$ is unsatisfiable if $(\bigcap_{i=1}^{m} S_i) = \emptyset$; L is *minimally unsatisfiable* if the removal of any literal from L produces a satisfiable set. A literal in a set S of clauses is *pure* if it does not belong to any minimally unsatisfiable set of literals in which distinct literals occur in distinct clauses. In that case, the clause containing the pure literal is also said to be pure.

Observe that it is necessary to include minimally unsatisfiable in the definition: It is possible for $\{l\} \cup L$ to be unsatisfiable but for l not to be in any minimally unsatisfiable subset. A trivial example is for L to be any unsatisfiable literal set and for l to be $\Delta : A$. (In essence, l is the constant *true*.) Looked at another way, the definition assures that if l is *not* pure, its removal makes some unsatisfiable literal set in which it resides satisfiable.

Lemma 3 (Signed Pure Rule). Let S be a set of signed clauses in which the clause C contains pure literal $S : A$. Then S is unsatisfiable if and only if $S' = S - \{C\}$ is unsatisfiable. □

For annotated logics, the literal set $L = \{A : x_1, A : x_2, \ldots, A : x_m, \sim A : x_{m+1}, \ldots, \sim A : x_{m+r}\}$ is unsatisfiable if $m, r > 0$ and, for some j, $m+1 \leq j \leq m+r$, $\text{Sup}\{x_1, \ldots, x_m\} \geq x_j$. Then L is *minimally unsatisfiable* if the removal of any member of L results in a satisfiable literal set. In light of Theorem 2, $r = 1$ for any minimally unsatisfiable annotated literal set.

A literal in a set S of annotated clauses is said to be *pure* if it does not belong to any minimally unsatisfiable set of literals in which distinct literals occur in distinct clauses. In that case, the clause containing the pure literal is also said to be pure.

Lemma 4 (Annotated Pure Rule). Let S be a set of annotated clauses in which the clause C contains the pure literal l. Then S is unsatisfiable if and only if $S' = S - \{C\}$ is unsatisfiable. □

The next lemma is useful in proving the completeness of annotated hyperresolution (Section 3.2).

Lemma 5. Let $S = \{C_0, C_1, C_2, \ldots, C_k\}$ be a minimally unsatisfiable set of annotated clauses (i.e., no proper subset of S is unsatisfiable), and suppose $C_0 = \{l\} \cup \{l_1, l_2, \ldots, l_n\}$, $n \geq 0$. Obtain S' from S by deleting every occurrence of l in S. Then S' is unsatisfiable, and every minimally unsatisfiable subset of S' contains $C_0' = \{l_1, \ldots, l_n\}$. □

---

[4] The pure rule states that a set of clauses is unsatisfiable iff the set with all pure clauses removed is unsatisfiable. This rule does extend to signed logics, but the definition of purity must be properly formulated: A literal in a clause might not be pure even though no other clause contains a complementary literal.

3.2 Annotated Hyperresolution

Recall that annotated resolution consists of two inference rules, annotated resolution and reduction. Both of these are special cases of signed resolution restricted to regular signs. The annotated hyperresolution rule defined below can be thought of as a single rule that executes several regular signed resolution steps at once.

Let $\mathcal{S}$ be a set of annotated clauses. Clauses are defined to be *positive* or *negative* as they are for hyperresolution in classical logic [5]: A clause is *positive* if it does not contain any negative literals, and it is *negative* if it contains at least one negative literal. Note that if $\mathcal{S}$ is unsatisfiable, it must contain at least one clause with only negative literals and at least one clause with only positive literals.

Nucleus and satellite clauses are required to define annotated hyperresolution, as they are in the classical case. However, the definitions are a bit more complicated for the annotated case. Let

$$N = (\cup_{k=1}^{n} \sim b_k : \mu_k) \cup C$$

be a negative clause in $\mathcal{S}$, where C is positive, and select *sets* of positive clauses $\mathcal{B}_1, \mathcal{B}_2, ..., \mathcal{B}_n$ as follows.

For $1 \leq k \leq n$, $\mathcal{B}_k = \{B_{k1}, B_{k2}, ..., B_{kn_k}\}$, where $b_k : \beta_{kt} \in B_{kt}$, and $\mu_k \leq \text{Sup}\{\beta_{k1}, \ldots, \beta_{kn_k}\}$; let $\beta_k = \text{Sup}\{\beta_{k1}, \ldots, \beta_{kn_k}\}$. Intuitively, for each $\sim b_k : \mu_k \in N$, $\mathcal{B}_k$ consists of n_k positive clauses, and each clause contains an annotated atom of the form $b_k : \beta_{kt}$. Furthermore, the members of $\mathcal{B}_k$ can be used to form a reductant $\mathcal{B}_k^{\mathcal{R}}$ containing the annotated atom $b_k : \beta_k$, and this reductant resolves against the literal $\sim b_k : \mu_k$ in the nucleus clause.

It turns out that the following additional condition, which further restricts the search, is useful: For any $b' : \beta' \in B_{kt} - \{b_k : \beta_{kt}\}$, $b' : \beta' \neq b_j : \beta_{jl}, 1 \leq j \leq n, 1 \leq l \leq n_j$; $b' : \beta'$ is not one of the atoms contributing to the residue of some $B_k^{\mathcal{R}}$. (Intuitively, all such atoms are ideally resolved away by $\sim b_k : \mu_k$, and we prohibit them from being reintroduced by other satellites.) This condition is referred to as the *satellite redundancy condition* in the proof of Theorem 6.

Then the clause

$$R = (\cup_{k=1}^{n}(\mathcal{B}_k^{\mathcal{R}} - \{b_k : \beta_k\})) \cup C$$

is the *annotated hyperresolvent* of the *nucleus clause* N and the *satellite clauses* $B_{11}, B_{1,2}, \ldots, B_{nn_n}$.

Obviously, R is a positive clause. That R can be soundly inferred from N and the $\mathcal{B}_k$'s can easily be seen by noting that a sequence of binary annotated resolutions between N and the $\mathcal{B}_k^{\mathcal{R}}$'s produces R (and the $\mathcal{B}_k^{\mathcal{R}}$'s are the result of a sequence of reductions). Semantically, any interpretation I that satisfies the parent clauses either satisfies one literal in C (and thus R) or satisfies some $\sim b_k : \mu_k$. But then I falsifies $b_k : \mu_k$ and thus falsifies $b_k : \beta_k$ as well. Hence, some $b_k : \beta_{kj}$ is falsified in B_{kj}. Since B_{kj} is satisfied by I, so is some literal in $B_{kj} - \{b_k : \beta_{kj}\}$, i.e., some literal in R. This proves

---
[5] This is Robinson's original terminology [16]; others have used mixed and negative to describe non-positive clauses

Theorem 5. Annotated Hyperresolution is a sound rule of inference for annotated logic. □

Consider an example over the lattice SIX, which is a six element lattice containing two chains: $\bot < \text{lt} < \text{t} < \top$ and $\bot < \text{lf} < \text{f} < \top$.

Suppose we have the following unsatisfiable set of annotated clauses: (1) $\{p:\text{t},\ \sim q:\top,\ \sim r:\text{t},\ \sim s:\text{f}\}$, (2) $\{q:\text{lt},\ p:\text{f}\}$, (3) $\{q:\text{lf}\}$, (4) $\{r:\text{t}\}$, (5) $\{s:\text{t},\ p:\text{t}\}$, (6) $\{s:\text{lf}\}$, (7) $\{\sim p:\text{lt}\}$, and (8) $\{\sim p:\text{lf}\}$.

An annotated hyperresolution proof is obtained by using clause (1) as nucleus for satellites (2) through (6) to produce $p:\text{t} \lor p:\text{f}$. It then serves as the only required satellite in the remainder of the deduction, in which clauses (7) and (8) serve as nuclei.

Annotated hyperresolution is complete; the proof is not trivial.

Theorem 6. Annotated hyperresolution is refutation complete for propositional annotated logic.

Proof. Let $\mathcal{S} = \{C_1, C_2, \ldots, C_m\}$ be an unsatisfiable set of annotated clauses. Assume that $\mathcal{S}$ is minimally unsatisfiable; otherwise, restrict attention to a minimally unsatisfiable subset. We must show that there is a refutation of $\mathcal{S}$ using annotated hyperresolution.

Proceed by induction on the number n of literal occurrences in $\mathcal{S}$. If there are none, then $\mathcal{S}$ is $\{\emptyset\}$, and we are done. Note that $\mathcal{S}$ cannot contain exactly one literal occurrence.

So suppose that all minimally unsatisfiable annotated clause sets with at most n literal occurrences can be refuted with annotated hyperresolution, and assume that $\mathcal{S}$ has size $n + 1$. Let q be a predicate occurring in $\mathcal{S}$. Note that since $\mathcal{S}$ is minimal, the Pure Rule implies that q cannot be pure, so there are minimally unsatisfiable literal sets, whose members have q as the predicate and are taken from distinct clauses.

Consider the set $\{q:x_1, q:x_2, \ldots, q:x_{r^+}\}$ of all positive q-literals in $\mathcal{S}$. We shall first show that for each i, $1 \leq i \leq r^+$, the unit clause $\{q:x_i\}$ can be derived by annotated hyperresolution. To do so, remove all occurrences of $q:x_i$ from $\mathcal{S}$ to produce $\mathcal{S}'_i$. This formula is unsatisfiable by Lemma 5. Consider a minimally unsatisfiable subset; by the same lemma, every clause that came from a clause in $\mathcal{S}$ containing $q:x_i$ is in this set. Since the number of literals in $\mathcal{S}'_i$ is at most n, the induction hypothesis applies, and there is a refutation $\mathcal{R}_{q_i}$ by annotated hyperresolution.

Now apply that refutation to $\mathcal{S}$; that is, construct a deduction in $\mathcal{S}$ by doing the identical annotated hyperresolution steps with the exception that each deleted $q:x_i$ is included. Observe that the effect is that whenever a clause, whether nucleus or a satellite, contains $q:x_i$, that literal is added to the resolvent. Note also that the satellite redundancy condition is obeyed: Satellites do not reintroduce any of the positive literals that, collectively, resolve against the negative nucleus literal. The deduction in $\mathcal{S}'$ has this property by the induction hypothesis, and $q:x_i$ does not occur in $\mathcal{S}'$.

Call the resulting deduction (it may no longer be a refutation) $\mathcal{R}'_{q_i}$ which, with merging, may produce the unit clause $\{q:x_i\}$ rather than the empty clause.

Nevertheless, each step is an annotated hyperresolution step. The reason is that by reintroducing positive occurrences of $q\!:\!x_i$, the status of all clauses as nucleus or as satellite in an inference step is unchanged. Thus each of the r^+ unit clauses $\{q\!:\!x_i\}$, $1 \leq i \leq r^+$ may be derived with annotated hyperresolution.

Consider the negative occurrences $\sim q\!:\!y_j$, $1 \leq j \leq r^-$ of q in $\mathcal{S}$. Note that, since none is pure, $\operatorname{Sup}\{x_1,\ldots,x_{r^+}\} \geq y_j, 1 \leq j \leq r^-$. (Otherwise, $\sim q:y_j$ would not be in an unsatisfiable literal set.) Thus some subset of the positive units $\{q : x_1, ..., q : x_{r^+}\}$ will suffice to resolve away any particular negative literal containing q.

Now delete from $\mathcal{S}$ all occurrences of $\sim q : y_j$ for some j. The resulting formula is unsatisfiable, and a refutation $\mathcal{R}_{\sim q_j}$ by annotated hyperresolution can be found. Let the proof that results from applying $\mathcal{R}_{\sim q_j}$ to $\mathcal{S}$ be denoted by $\mathcal{R}'_{\sim q_j}$. This proof yields either the empty clause or the unit $\{\sim q\!:\!y_j\}$. However, it may fail to be an annotated hyperresolution proof in two ways. Reintroducing $\sim q_j : y_j$ into a clause may add to the negative literals in a nucleus clause or may convert a positive (satellite) clause into a negative cause. In the first case, the nucleus would have a negative literal not resolved away, and in the second case, a clause that cannot act as a satellite would be produced. In either case, an annotated hyperresolution proof can be constructed from $\mathcal{R}'_{\sim q_j}$ using the units $\{q\!:\!x_i\}$ produced by the deductions $\mathcal{R}'_{q_i}$, as we show below.

Suppose first that a step in $\mathcal{R}_{\sim q_j}$ with nucleus $N = \{\sim b_1 : y_1, \ldots, \sim b_n : y_n, c_1, ..., c_m\}$ adds $\sim q : y_j$ to N. As noted earlier, some subset of the derived unit clauses $\{q\!:\!x_i\}$ resolve away $\sim q : y_j$. Employing these units as additional satellites results in an annotated hyperresolution step in which the deduced (positive) clause is exactly the same as in $\mathcal{R}_{\sim q_j}$.

Suppose now that $\sim q_j : y_j$ is added to a satellite clause B, producing a negative clause B' with one negative literal. If B' is used as a nucleus clause, and if the derived unit clauses that resolve away $\sim q_j : y_j$ are used as satellite clauses, then B is the annotated hyperresolvent. Note that this construction assures that the last step that produced $\sim q : y_j$ in $\mathcal{R}'_{\sim q_j}$ now produces the empty clause.

Again the satellite redundancy condition is obeyed; in $\mathcal{R}_{\sim q_j}$ by the induction hypothesis, and in $\mathcal{R}'_{\sim q_j}$ since the only changes introduced involve additional collections of unit satellites.

Finally, combining the deductions $\mathcal{R}'_{q_i}$ and the modified deduction $\mathcal{R}'_{\sim q_j}$ produces the required annotated hyperresolution refutation. □

References

1. Anderson, R. and Bledsoe, W. A linear format for resolution with merging and a new technique for establishing completeness, *J. ACM* 17(3) (1970), 525–534.
2. Blair, H.A. and Subrahmanian, V.S., Paraconsistent logic programming, *Theoretical Computer Science*, 68:135–154, 1989.
3. Davey, B.A., and Priestley, H.A., *Introduction to Lattices and Order*, Cambridge Mathematical Textbooks, (1990)
4. Hähnle, R., *Automated Deduction in Multiple-Valued Logics*, International Series of Monographs on Computer Science, vol. 10. Oxford University Press, 1993.

5. Hähnle, R., Exploiting data dependencies in many-valued logics, *Journal of Applied Non-Classical Logics* **6**(1): 49-69, 1996.
6. Hähnle, R., Transformation between signed and classical clause logic, *Proceedings* of *ISMVL-99*, 248-255, 1999.
7. Hähnle, R. and Escalada-Imaz, G., Deduction in many-valued logics: a survey, *Mathware & Soft Computing*, IV(2), 69–97, 1997.
8. Hähnle, R., Murray, N.V. and Rosenthal E., Some Remarks on Completeness, Connection Graph Resolution and Link Deletion, In *Proceedings* of the *International Conference TABLEAUX'98 - Analytic Tableaux and Related Methods*, Oisterwijk, The Netherlands, May 1998. In *Lecture Notes in Artificial Intelligence* (H. de Swart, Ed.), Springer-Verlag, Vol. 1397, 172-186.
9. Kifer, M., and Lozinskii, E., A logic for reasoning with inconsistency, *J. of Automated Reasoning* 9, 179–215, 1992.
10. Kifer, M., and Lozinskii, E.L., RI: A Logic for Reasoning in Inconsistency, *Proceedings of the Fourth Symposium of Logic in Computer Science*, Asilomar, 253-262, 1989.
11. Kifer, M., and Subrahmanian, V.S., Theory of generalized annotated logic programming and its applications, the *J. of Logic Programming* 12, 335–367, 1992.
12. Leach, S.M., Lu, J.J., Murray, N.V., and Rosenthal, E., Ʊ-resolution: an inference for regular multiple-valued logics. *Proceedings* of *JELIA'98*. IBFI Schloss Dagstuhl (International Conference and Research Center for Computer Science), October 1998. In *Lecture Notes in Artificial Intelligence*, Springer-Verlag, Vol. 1489, 154-168.
13. Lu, J.J., Murray, N.V., and Rosenthal, E., A Framework for Automated Reasoning in Multiple-Valued Logics, *J. of Automated Reasoning* **21**,*1* 39–67, 1998.
14. Murray, N.V., and Rosenthal, E., Improving tableaux deductions in multiple-valued logic, *Proceedings* of the 21st *International Symposium on Multiple-Valued Logic*, Victoria, B.C., Canada, May 26-29, 1991, 230-237.
15. Murray, N.V., and Rosenthal, E., Adapting classical inference techniques to multiple-valued logics using signed formulas, *Fundamenta Informaticae* 21:237–253, 1994.
16. Robinson, J.A., Automatic deduction with hyper-resolution, *International Journal of Computer Mathematics*, **1** (1965), 227-234.
17. Sofronie-Stokkermans, V., On translation of finitely-valued logics to classical first-order logic, *Proceedings* of the *13th ECAI*, 1998.
18. Sofronie-Stokkermans, V., Automated theorem proving by resolution for finitely-valued logics based on distributive lattices with operators, *Multiple Valued Logic Journal*, to appear.
19. Subrahmanian, V.S., On the Semantics of Quantitative Logic Programs, in: *Proceedings of the 4th IEEE Symposium on Logic Programming*, Computer Society Press, 1987.
20. Subrahmanian, V.S., Paraconsistent Disjunctive Databases, *Theoretical Computer Science*, 93, 115–141, 1992.
21. Thirunarayan, K., and Kifer, M., A Theory of Nonmonotonic Inheritance Based on Annotated Logic, *Artificial Intelligence*, 60(1):23–50, 1993.

Fundamental Properties on Axioms of Kleene Algebra

Tomoko Ninomiya[1] and Masao Mukaidono[2]

[1] Tamagawa University, Dept. of English and American
6-1-1 Tamagawagakuen, Machida-Shi, Japan
ninomiya@lit.tamagawa.ac.jp
[2] Meiji University, Dept. of Computer Science
1-1-1 Higashimita, Tama-Ku, Kawasaki-Shi, Japan
masao@cs.meiji.ac.jp

Abstract. The research works on axioms of Kleene algebra are surveyed and the fundamental properties of axioms of Kleene algebra are clarified through the method of indeterminate coefficients.. Especially the algorithm for checking a given axiom of Kleene algebra is independent or not from the other axioms is shown. Finally, all finite models of Kleene algebra of 8 elements are derived as an example.

1 Introduction

Kleene algebra was firstly proposed and investigated by J. A. Kalmann[1] under the name of "a normal i-lattice", and in the book of R. Balbes and P. Dwinger[2] it was described that the name "Kleene algebra" was found in the paper of D. Brignole and A. Monteiro[3].

On the other hand, after fuzzy logic was proposed by L. A. Zadeh[4] to treat ambiguous states or phenomena exist in the real world, many researchers investigated the algebraic structures of fuzzy logic. Almost all of them characterized the fuzzy logic as De Morgan algebra. Among them, only F. P. Preparata and R. T. Yeh[5] and M. Mukaidono[6] pointed out that Kleene's law $(\sim a \vee a) \geq (\sim b \bullet b)$ holds in fuzzy logic. In such situations, M. Mukaidono[7] declared firstly that fuzzy logic is a model of "Kleene algebra" and found out the independent and complete axioms for Kleene algebra, and after that he clarified the canonical forms of free Kleene algebra under the name of fuzzy switching functions[8]~[10] (some times in the literatures this algebra was called also as "fuzzy algebra" or "soft algebra"). Kleene algebra is a weaker algebra than Boolean algebra and a stronger algebra than De Morgan algebra because Kleene algebra is De Morgan algebra satisfying Kleene's laws and Kleene's laws are weaker version of the law of excluded middle that is the essential law in Boolean algebra. It was already shown that Kleene algebra is essentially 3-valued[2],[10].

Recently, Kleene algebra appears in many fields and plays essential roles to represent ambiguous or uncertainty states especially in the field of intelligent systems. In this paper the research works on axioms of Kleene algebra are surveyed and the fundamental properties of axioms of Kleene algebra are clarified through the method of indeterminate coefficients[11], where the method of indeterminate coefficients is a strong tool to derive all finite models satisfying the given a set of all axioms. Especially we show a given axiom of Kleene algebra is independent or not from the other axioms, and through these investigations we can clarify the properties of each axiom of Kleene algebra and a set of independent and complete axioms of Kleene algebra is shown. Finally, all finite models of Kleene algebra, for the number of elements being 8 are derived as an example.

2 Axioms of Kleene Algebra

In the fuzzy theory[4], at first, three kinds of set operations $A \cup B, A \cap B$ and A^c are defined as follows:
$$\mu_{A \cup B}(a) = \mu_A(a) \vee \mu_B(a)$$
$$\mu_{A \cap B}(a) = \mu_A(a) \bullet \mu_B(a)$$
$$\mu_{A^c}(a) = \sim \mu_A(a),$$
where $\mu_A(a)$ and $\mu_B(a)$ are membership values for an element a to belong to the fuzzy set A and B, respectively, and take any values of the unit interval [0,1], and $\vee$, $\bullet$ and $\sim$ are logic operations and defined as

[Definition 1]
$a \vee b = \max(a, b)$
$a \bullet b = \min(a, b)$
$\sim a = 1-a$
where a, b are elements of [0,1].

The logic operations $\vee$ and $\bullet$ in Definition 1 are afterward generalized into t-conorm and t-norm, respectively, but the above definitions of logic operations OR($\vee$), AND($\bullet$) and NOT($\sim$) are always fundamental and essential in fuzzy logic.

It is easily shown that the above definitions of a set of logic operations $\{\vee, \bullet, \sim\}$ satisfy the following equalities listed in Table 1 in the closed interval [0,1].

If an algebraic system satisfies the equations [1]~[4] then it is a lattice, if the equations [1]~[7] then it is a bounded distributed lattice, and if the equations [1] ~[9] then it is De Morgan algebra.

[Definition 2] The bounded distributive lattice satisfying [8], [9] and [10] is called Kleene algebra.

That is, Kleene algebra is De Morgan algebra satisfying [10]Kleene's laws, where Kleene's laws are weaker conditions of
[10]' The Complementary laws: $\sim a \bullet a = 0$, $\sim a \vee a = 1$.
The above complementary laws correspond to the law of excluded middle and the law of contradiction, which are essential parts of Boolean algebra or two-valued logic.

Table 1. Axioms of Kleene Algebra

[1] Commutative laws: $a \vee b = b \vee a$ ----[1-1], $\quad a \bullet b = b \bullet a$ ---------------- [1-2]
[2] Idempotent laws: $a \vee a = a$ ----------- [2-1], $\quad a \bullet a = a$ ------------------- [2-2]
[3] Absorption laws: $a \vee (a \bullet b) = a$ -----[3-1], $\quad a \bullet (a \vee b) = a$ ------------- [3-2]
[4] Associative laws: $a \vee (b \vee c) = (a \vee b) \vee c$ -[4-1], $\quad a \bullet (b \bullet c) = (a \bullet b) \bullet c$ ---- [4-2]
[5] Distributive laws: $a \bullet (b \vee c) = (a \bullet b) \vee (a \bullet c)$ -[5-1], $a \vee (b \bullet c) = (a \vee b) \bullet (a \vee c)$ - [5-2]
[6] The least element: $0 \vee a = a$ -------[6-1], $\quad 0 \bullet a = 0$ ------------------- [6-2]
[7] The greatest element: $1 \vee a = 1$ -------[7-1], $\quad 1 \bullet a = a$ ------------------- [7-2]
[8] Double negations law: $\sim(\sim a) = a$ -----[8]
[9] De Morgan's laws: $\sim(a \vee b) = \sim a \bullet \sim b$- [9-1], $\quad \sim(a \bullet b) = \sim a \vee \sim b$ ------- [9-2]
[10] Kleene's laws □ $(\sim a \vee a) \vee (\sim b \bullet b) = \sim a \vee a$ -[10-1],
     $(\sim a \vee a) \bullet (\sim b \bullet b) = \sim b \bullet b$ -- [10-2]
where a, b ∈ [0,1].

[Theorem 1] An algebra of fuzzy logic <[0,1], $\vee$, $\bullet$, $\sim$> is Kleene algebra.

3 A Finite Model Showing an Axiom Being Independent from Others

One of the interesting problems concerning the above axioms of Kleene algebra is whether each axiom is independent or not from the others in the set of axioms. This problem was considered firstly by M. Mukaidono[7] and he showed a set of axioms of Kleene algebra in which each axiom is independent from the others. Recently, this problem was investigated again in more thoroughly by the authors[12] by using the method of indeterminate coefficients[11]. Here in this paper we will explain this topic in more detail. Before that, we have to explain what is the method of indeterminate coefficients. The method of indeterminate coefficients was developed firstly by M. Goto[11] to find out many-valued truth tables for undefined operators in axioms by a computer. In axioms of an algebra described by equations, in general, there are some operators in the equations such as $\vee$, $\bullet$, $\sim$. In the method of indeterminate coefficients, these operators are regarded as undefined and the axioms are regarded as constrained conditions that these operators should be satisfied. As an output, the algorithm based on the method of indeterminate coefficients gives truth tables of operators, that is, finite models of the algebra that satisfy all given axioms. In finding finite models, at first, we have to designate the number N of elements (corresponds to the number of truth values in N valued logic) in the model.

[The algorithm based on the method of indeterminate coefficients]
Input:
(1) N: number of elements
(2) A set of axioms
Output:
Truth tables of undefined operators satisfying all given axioms

The method of indeterminate coefficients is described briefly in appendix. If you are interested in the algorithm in more detail, please see the reference (11) or (12).

[Example]
Input:
(1) N=3
(2)[2-1]Idempotent law: $a \vee a = a$
Output:
Table 2:

Table 2. $a \vee b$

```
        a       1   2   0
   b  |----------
   1  |     1   *   *
   2  |     *   2   *
   0  |     *   *   0
```

In the above example, we designate the number of truth values as 3, that is, {0,1,2}, and a set of axioms as only one equation of the idempotent law: $a \vee a = a$. We obtain one solution as shown in Table 2. Here please notice that in the table there are symbols ∗ which means don't care, that is, any element of {0,1,2} is admissible in ∗. In this sense, the number of solution is only one but the number of truth tables (finite models) satisfying the idempotent law is 3x3x3x3x3x3=729 in three-valued because the number of ∗ is 6 in Table 2.

By using the method of indeterminate coefficients we can clarify the properties and the power of each axiom to determine the solutions. Especially, we can examine whether an axiom is independent or not from the other axioms. That is, at first we obtain the truth tables satisfying the other axioms and finally we add the given axiom and obtain the truth tables again satisfying all axioms. If the number of truth tables is reduced, it is proofed that the given axiom is independent from the other axioms and one of the disappeared truth tables is an example (a finite model) that shows the given axiom is independent from the others. Every one of the disappeared truth tables is a counter example showing the axiom is independent.

3-1 Double Negations Law Is Independent

At first we show that
[8]Double negations law: $\sim(\sim a) = a$ -----[8]
is independent from the others. In Table 3, the number of solutions satisfying all axioms before the axiom including it self is listed in turn when N=3. The upper line shows the axiom number and the lower line shows the number of solutions that satisfy the axiom with all axioms before that. Although the order of axioms to be examined is appropriate in principle, it is selected such that the number of solutions does not become so huge on the way in practice. Please notice again the numbers in the lower line are of solutions and not of truth tables. The final solution is obtained as only one solution which is a truth table described in Table 4, which is only one three valued

Fundamental Properties on Axioms of Kleene Algebra 315

model of Kleene algebra. In Table 3 until the axiom [10-2] there exist 10 solutions, but at the axiom[8], 9 solutions have disappeared, all of which are models showing [8] Double negations law is independent from the others. Table 5 is one example of them, that is, this truth table satisfies all axioms of a set axioms of Kleene algebra except [8] Double negations law.

Table 3. Solutions of [8]Double negations law in three valued

Axiom #	[6-1]	[7-1]	[9-1]	[3-1]	[2-1]	[4-1]	[5-1]	[10-1]	[1-1]	[1-2]
N.of solut.	1	1	316	4350	3453	2033	287	173	107	25

	[4-2]	[3-2]	[5-2]	[2-2]	[9-2]	[6-2]	[7-2]	[10-2]	[8]
	24	10	10	10	10	10	10	10	1

Table 4. A model of Kleene algebra in three valued

```
   a 1 2 0           a 1 2 0              a|
b |---------      b |---------   1|0
1 |1 1 1          1 |1 2 0       2|2
2 |1 2 2          2 |2 2 0       0|1
0 |1 2 0          0 |0 0 0
   a ∨ b             a • b          ~a
```

Table 5. A model showing [8] double negations law is independent

```
   a 1 2 0           a 1 2 0              a|
b |---------      b |---------   1|1
1 |1 1 1          1 |1 2 0       2|1
2 |1 2 2          2 |2 2 0       0|1
0 |1 2 0          0 |0 0 0
   a ∨ b             a • b          ~a
```

3-2 De Morgan's Laws Are Independent

To show one of
[9]De Morgan's laws: ~(a ∨ b)= ~a • ~b---[9-1]
                      ~(a • b)= ~a ∨ ~b---[9-2]
is independent from others, we will choose [9-1] as the last axiom and apply the algorithm of the method of indeterminate coefficients. If let N=3, that is three valued, then we can get the result shown in Table 6. In this case the truth table is decided uniquely at the axiom [1-2], which means that we cannot decide [9-1]De Morgan's laws is independent or not from the other axioms in the scope of three valued. In general, even if the number of solution is not decreased at the final stage in N valued, it is not proof that the last axiom is not independent from the set of former axioms because we have a possibility to find out a counter example in the N+1 valued. Indeed in this case letting N=4, we can show that [9-1]De Morgan's law ~(a ∨ b)= ~a • ~b is independent as shown in Table 7. An example of truth tables disappeared lastly in Table 7 is shown in Table 8, which is a model showing that [9-1]De Morgan's laws is independent. The situation is same if [9-2]De Morgan's law: ~(a • b)= ~a ∨ ~b is located lastly instead of [9-1]De Morgan's law ~(a ∨ b)= ~a • ~b.

Table 6. Solution of [9-1]De Morgan's law in three valued

Axiom #	[8]	[6-1]	[7-1]	[1-1]	[10-1]	[3-2]	[5-1]	[1-2]	[4-1]	[3-1]
N.of solut.	4	4	4	12	52	5	10	1	1	1

	[2-1]	[4-2]	[5-2]	[2-2]	[10-2]	[6-2]	[7-2]	[9-1]
	1	1	1	1	1	1	1	1

Table 7. Solution of [9-1]De Morgan's law in four valued

Axiom #	[8]	[6-1]	[7-1]	[1-1]	[10-1]	[3-2]	[5-1]	[1-2]	[4-1]	[3-1]
N.of solut.	0	10	10	640	9490	985	984	5	5	5
	[2-1]	[4-2]	[5-2]	[2-2]	[10-2]	[6-2]	[7-2]	[9-1]		
	5	5	5	5	5	5	5	3		

Table 8. A model showing [9-1]De Morgan's law is independent

```
  a  1  2  3  0          a  1  2  3  0
b |---------------     b |---------------         a |
1 | 1  1  1  1         1 | 1  2  3  0         1 | 2
2 | 1  2  3  2         2 | 2  2  2  0         2 | 1
3 | 1  3  3  3         3 | 3  2  3  0         3 | 0
0 | 1  2  3  0         0 | 0  0  0  0         0 | 3
     a ∨ b                   a • b                 ~a
```

3-3 Commutative, Distributive and Kleenes, Laws Are Independent

In the similar manner we can show that each one of
Commutative laws: a ∨ b=b ∨ a--------[1-1]
                      a • b=b • a---------[1-2]
Distributive laws: a • (b ∨ c)=(a • b) ∨ (a • c)---[5-1]
                      a ∨ (b • c)=(a ∨ b) • (a ∨ c)---[5-2], and
Kleene's laws :   (~a ∨ a) ∨ (~b • b)=~a ∨ a---[10-1]
                      (~a ∨ a) • (~B • b)= ~b • b---[10-2]

is independent from others, respectively. In fact, Table 9 is a model showing that [1-1]Commutative law is, Table 10 is a model showing that [5-1]Distributive law is and Table 11 is a model showing that [10-1]Kleene's law is independent, respectively. The situations are same for [1-2]Commutative law:a • b=b • a, [5-2]Distributive law: a ∨ (b • c)=(a ∨ b) • (a ∨ c)   and   [10-2]Kleene's law (~a ∨ a) • (~b • b)=~b • b, respectively.

Table 9. A model showing [1-1]Commutative law is independent

```
  a  1  2  3  0          a  1  2  3  0
b |---------------     b |---------------         a |
1 | 1  1  1  1         1 | 1  2  3  0         1 | 0
2 | 1  2  2  2         2 | 1  2  3  0         2 | 3
3 | 1  2  3  0         3 | 3  3  3  3         3 | 2
0 | 1  2  3  0         0 | 0  0  0  0         0 | 1
     a ∨ b                   a • b                 ~a
```

Table 10 A model showing [5-1]Distributive law is independent

a∨b	1	2	3	4	0		a•b	1	2	3	4	0		~a	
b							b							a	
1	1	1	1	1	1		1	1	2	3	4	0		1	0
2	1	2	1	1	2		2	2	2	0	0	0		2	4
3	1	1	3	1	3		3	3	0	3	0	0		3	3
4	1	1	1	4	4		4	4	0	0	4	0		4	2
0	1	2	3	4	0		0	0	0	0	0	0		0	1

Table 11 A model showing [10-1] Kleene's law is independent

a∨b	1	2	3	0		a•b	1	2	3	0		~a	
b						b						a	
1	1	1	1	1		1	1	2	3	0		1	0
2	1	2	1	2		2	2	2	0	0		2	3
3	1	1	3	3		3	3	0	3	0		3	2
0	1	2	3	0		0	0	0	0	0		0	1

4 Independent and Complete Axioms of Kleene Algebra

From the above considerations, we can see that in the set of axioms of Kleene algebra, at least, the following axioms have to be included:
(1)[8]Double negations law: ~(~a)=a-----[8]
(2)One of [9]De Morgan's laws: ~(a ∨ b)= ~a • ~b---[9-1]
                                 ~(a • b)= ~a ∨ ~b---[9-2]
(3)One of [1]Commutative laws: a ∨ b=b ∨ a--------[1-1]
                                 a • b=b • a---------[1-2]
(4)One of [5]Distributive laws: a • (b ∨ c)=(a • b) ∨ (a • c)---[5-1]
                                 a ∨ (b • c)=(a ∨ b) • (a ∨ c)---[5-2]
(5)One of [10] Kleene's laws☐   (~a ∨ a) ∨ (~b • b)=~a ∨ a---[10-1]
                                 (~a ∨ a) • (~B • b)= ~b • b---[10-2]
Similarly it is easily shown that
(6)One of [6]The least element: 0 ∨ a=a---[6-1]   0 • a=0----[6-2]
       [7]The greatest element: 1 ∨ a=1---[7-1]   1 • a=a---[7-2]
is independent from the others by defining 1=~0

   It was shown that the each of the above axioms is independent each other in the set of axioms of Kleene algebra. Next question is what is the complete set of Kleene algebra. For obtaining such complete set of axioms, we have to show that every axioms of Kleene algebra ([1-1] ~[10-2]) is derived from the complete set of axioms. To show that an axiom is derived from the other axioms (that is, the axiom is not independent from the other axioms), we cannot use the method of indeterminate coefficients, because even if the number of solutions is not decreased when it was located as last axiom, it is not proof the axiom is not independent from the others as described in Section 3-2. The method of indeterminate coefficients only finds the candidates for the complete set of axioms. So, we have to show formally that every axioms of Kleene algebra can be derived from the complete set of axioms, where the above six axioms are candidates for the element of the complete set of axioms. In deed, by M. Mukaidono[7] it has been shown the following six axioms listed in Table 12, which are all selected among the above six candidates, are the complete axioms, which are, of course, also independent each other, of Kleeen algebra.

Table 12 An independent and complete axioms of Kleene algebra[7]
[1]Commutative laws: a ∨ b=b ∨ a--------[1-1]
[5]Distributive laws: a • (b ∨ c)=(a • b) ∨ (a • c)---[5-1]
[6]The least element: 0 ∨ a=a---[6-1]
[8]Double negations law: ~(~a)=a-----[8]
[9]De Morgan's laws: ~(a ∨ b)= ~a • ~b---[9-1]
[10] Kleene's laws:(~a ∨ a) ∨ (~b • b)=~a ∨ a---[10-1]

By using the above facts, another sets of independent and complete axioms of Kleene algebra were reported recently[12], and this technique is applied to axioms of Boolean algebra and 64 deferent kind sets of the independent and complete axioms of Booean algebra are discovered[13].

5 Finite Models of Kleene Algebra

As obtained a three-valued mode of Kleene algebra in Table 4, we can derive all finite models satisfying the given set of axioms by using the method of indeterminate coefficients in any N values in the scope of permitted time. In figure 1, are illustrated Hasse diagrams of all finite models of Kleene algebra in N=8 for an example. For obtaining these models based on the method of indeterminate coefficients we used the set of axioms of Kleene algebra listed in Table 12, which are independent and complete axioms, because it is easer to derive all models if the number of axioms is smaller.

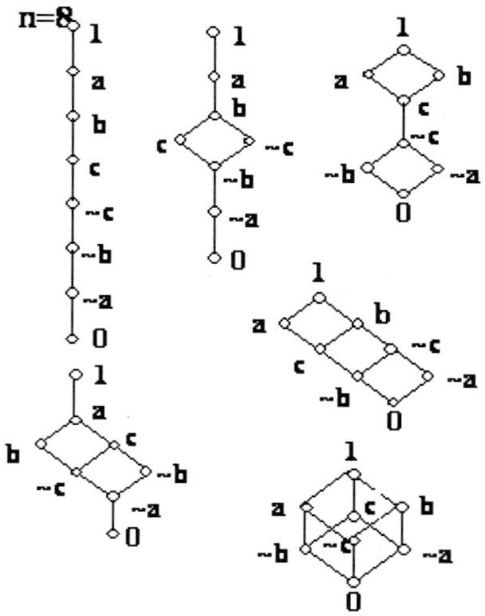

Fig. 1. All models of 8-valued Kleene Algebra

6 Conclusions

The properties and roles of each axioms of Kleene algebra are clarified and the independence of each axiom is examined through the method of indeterminate coefficients. As sub-products we can show that a set of independent and complete axioms of Kleene algebra and all finite models of Kleene algebra in some finite cases.

Reference

1. J. A. Kalman, Lattice with involution, Trans. Amer. Math. Soc. 87, pp.485-491, 1958
2. R. Balbes and P. Dwinger, Distributive Lattices, University of Missouri Press, pp.215, 1974
3. D. Brignole and A. Monteiro, Caracterisation des algebras de Nelson par des egalites, Notas de Logica Matematica, Instituto de Matematica Universidad del sur Bahia Blanca, 20, 1964
4. L. A. Zadeh, Fuzzy sets, Information Control, 8, pp.338, 1965
5. F. P. Preparata and R. T. Yeh, Continuously valued logic, J. of Computer and Systems Science, 6. pp.397, 1972
6. M. Mukaidono, On some properties of fuzzy logic, Systems •Computers •Controls, Vol.6, No.2, 1975
7. M. Mukaidono, A set of independent and complete axioms for a fuzzy algebra (Kleene algebra), Proceedings of The 8th International Symposium on Multiple-Valued Logic, IEEE, 1981
8. M. Mukaidono, New canonical forms and their applications to enumerating fuzzy switching functions, Proceedings of the 12th International Symposium on Multiple-Valued Logic,IEEE,pp.275~279,1982
9. J.Bermann and M.Mukaidono, Enumerating Fuzzy Switching Functions and Free Kleene Algebras, International Journal of Computer and Mathematics with Applications, Pergamon Press, Vol.10, No.1, pp.25-35, 1984
10. Masao Mukaidono, The representation and minimization of fuzzy switching functions, The Analysis of Fuzzy Information,Vol.1, edited by J. C. Bezdek, CRC Press, pp.213~229, 1987
11. M. Goto, S. Kao and T. Ninomiya, Determination of many-valued truth tables for undefined operators in axioms by a computer and their applications, Proceedings of the 7th International Symposium on Multiple-Valued Logic, IEEE, 1977
12. T. Ninomiya and M. Mukaidonbo, Clarifying the axioms of Kleene algebra based on the method of indeterminate coefficients, Proceedings of the 29th International Symposium on Multiple-Valued Logic, IEEE, 1999
13. T. Ninomiya and M. Mukaidonbo, Independence of the axioms of Boolean algebra in multiple-valued logic, Proceedings of the 30th International Symposium on Multiple-Valued Logic, IEEE, 2000

Appendix

A Brief Description of the Method of Indeterminate Coefficients

Let f(x,y) be an undefined operator appeared in a set of axioms $\{S_1,--,S_k,---,S_n\}$.
Then we can write $f(x, y) = \bigvee_{i,j=0}^{N-1} X_{ij} \bullet x^i \bullet y^j$,

where $x^i=1$ if $x=i$ and $x^i=0$ if $x \neq i$ and $X_{i,j}$ is an indeterminate coefficient and will takes a value of $\{0,1,\text{---}.N-1\}$ if the undefined operator is defined uniquely, but in the sequel $X_{i,j}(k,m)$ $(m=1,\text{--},m_k)$ $(i, j=0,1,\text{---},N-1)$ will takes a subset of $\{0,1,\text{---}.N-1\}$ where $X_{i,j}(k,m)$ means a m-th partial solution in k-th step.

(1) $X_{i,j}(0,1)=* =\{0,\text{---},N-1\}$ $(i, j=0,1,\text{---},N-1)$ for a starting partial solution. Let $m_0=1$ and $k=1$.

(2) Drive all partial solutions $X_{i,j}(k,m)(m=1,\text{--},m_k)$ satisfying the axiom S_k from the initial conditions $Xi,j(k-1,m)(m=1,\text{--},m_{k-1})$

(3) Repeat the above step $(k=1,\text{---},n)$ until S_n.

(4) The final partial solutions are the general solutions satisfying all given axioms $\{S_1,\text{--},S_k,\text{---},S_n\}$.and if $m_n=1$ and $X_{i,j}(n,1)$ takes an element of $\{0,\text{---},N-1\}$ for all i and j, then the truth table of $f(x,y)$ is determined uniquely.

Extending Entity-Relationship Models with Higher-Order Operators

Antonio Badia

University of Arkansas
abadia@godel.uark.edu

Abstract. The concept of Generalized Quantifier (GQ) was introduced to query languages in [9] and, independently, in [10]. The present paper shows how GQs can be used in Conceptual Modeling, specifically how they can be incorporated into Entity-Relationship diagrams ([4]) to increase their expressive power. A language to express E-R models is defined and given formal semantics. It is them shown how GQs can be easily added to this framework. Several GQs that have natural, intuitive interpretations in the context of conceptual modeling are defined; their use is shown through examples.

1 Introduction

Conceptual modeling is one of the most important steps in the creation of an Information System. Modeling is a notoriously difficult activity; it cannot be treated algorithmically, and it requires ingenuity and experience. One of the main tools for the task is the use of *conceptual models*, semiformal specifications of how to structure and express information. The Entity-Relationship (E-R) model is one of the most successful models ([4]); it is simple, intuitive, yet relatively powerful. However, its limitations are well-known. Many developments in conceptual modeling assume that logic methods are too inflexible, too limited and too unintuitive to be useful for modeling. This paper is a starting point for work that counters this assumptions and gives logic a place in conceptual models. In particular, it shows how to overcome many of the limitations of the E-R model by using higher-order operators (intuitively, relations on relations). Reasoning with higher-order concepts may be complex from both a computational complexity point of view and a conceptual point of view. We propose to use the framework of *Generalized Quantifiers (GQs)* to attack the problem. GQs are declarative, powerful, high-order operators that have a natural graphical representation.

In the next section, we review the basics of Entity-Relationship models, give a formal description for them, and introduce the concept of GQ. In section 3 we show some of the problems that this approach is trying to solve by giving some examples of situations in which information is hard or impossible to capture in a traditional E-R model. In section 4 we formalize E-R models and extend the formalization with a selected set of GQs; we give examples of how the extension solves the problems of section 3. Finally, we mention some related work and close with some conclusions and comments on further work.

2 Background

In this section we give some preliminary definitions for the rest of the paper. For completeness, the next subsection briefly introduces the basic ideas of Entity-Relationship models, while subsection 2.2 introduces the concept of Generalized Quantifier, together with some examples.

2.1 Entity-Relationship Models

An *E-R* model is a data model with three basic concepts: *entities*, *attributes* and *relationships*. Entities represent things either real or conceptual. They denote sets of objects, not particular objects; in this respect they are close to *classes* in object-oriented models. The set of objects modeled by an entity are called its *extension*.

Relationships are connections among entities. The *arity* of a relationship is the number of entities involved: unary, binary, ternary, and so on. Binary relationships are the most common. Two kinds of constraints are associated with relationships. The *participation constraint* tells us whether all objects in the extension of an entity are involved in the relationship, or whether some may not be. For example, entities Office and Employee may have a relationship works-on between them. If all offices have employees, then participation of Office in works-on is total (otherwise is partial). If all employees are based in an office, then participation of Employee is also total. The *cardinality constraint* tells us how many times an object in the entity's extension may be involved in a relationship, and allows us to classify the relationship as **one-to-one**, **one-to-many** and **many-to-many**. We note that *recursive relationships* are allowed: they relate one entity to itself. For example, the relationship Child-of relates the entity Person to itself. To distinguish the ways in which Person participates in this relationship, *roles* are added to the entity (*father* and *son*, in this example).

Entities and relationships have attributes, which are properties with a *value*. Attributes convey characteristics or descriptive information about the entity to which they belong. Attributes may be simple or composite, single or multivalued, primitive or derived.

2.2 Generalized Quantifiers

Generalized Quantifiers were first introduced in logical studies ([13], [12]). The concept has attracted attention lately for its uses, among others, in query languages ([9],[10]) and other languages like description logics ([3]).

Given a set M, a *Generalized Quantifier (GQ)* on M is a relation among subsets of relations on M.

Definition 1. *Let a* type *be a finite sequence of positive numbers, which will be written* $[k_1, \ldots, k_n]$. *Then a* generalized quantifier *of type* $[k_1, \ldots, k_n]$ *on* M *is an n-ary relation between subsets of* $M^{k_1}, \ldots, M^{k_n}$ *(i.e. between elements of* $\mathcal{P}(M^{k_1}) \times \ldots \times \mathcal{P}(M^{k_n})$).

Not every relation between subsets of the domain is considered a GQ. Intuitively, we would like a GQ to behave as a *logical* operator, in the sense that it should not distinguish between elements in the domain. Thus, many authors pose the following constraint on the definition:

Definition 2. (PERM) A quantifier Q follows PERM if, whenever f is a permutation on M, then $Q_M(A_1, \ldots, A_n)$ iff $Q_M(f[A_1], \ldots, f[A_n])$.

In the context of database query languages, this constraint ensures that quantifiers are *generic* operations ([1]).

The following are examples of GQs. A universe M is fixed. We use Q as a variable over GQs, and write $Q(A_1, \ldots, A_n)$ to indicate that sets $A_1, \ldots, A_n$ belong to the extension of Q, i.e. that they are in the relation denoted by Q.

all $= \{X, Y \subseteq M | X \subseteq Y\}$
some $= \{X, Y \subseteq M | X \cap Y \neq \emptyset\}$
no $= \{X, Y \subseteq M | X \cap Y = \emptyset\}$
at least n $= \{X, Y \subseteq M | |X \cap Y| \geq n\}$
at most n $= \{X, Y \subseteq M | |X \cap Y| \leq n\}$
I $= \{X, Y \subseteq M | \ |X| = |Y|\ \}$ (Hartig's quantifier)
$\mathbf{Q_R} = \{X \subseteq M | \ |X| > |M - X|\ \}$ (Rescher's quantifier)
$\mathbf{H} = \{R \subseteq M^4 \mid \exists f: M \to M \ \exists g: M \to M \ \forall a, b \in M \ < a, f(a), b, g(b) > \in R\}$
$\mathbf{W^R} = \{X \subseteq M, \ R \subseteq M^2 \mid R \text{ well-orders } X\}$

All the above quantifiers are of type [1,1], except $\mathbf{Q_R}$ (type [1]), $\mathbf{H}$ (type [4]) and $\mathbf{W^R}$ (type [1,2]). Note that the first five quantifiers are first-order definable; the last four are not.

3 Limitations of the E-R Model

E-R models have a *layered approach* to organizing information, in the sense that only entities and relationships can have attributes and only entities can be involved in relationships. Thus, it is not possible for attributes to have attributes, or to be involved in relationships; and it is not possible for relationships to be involved in other relationships. This results in limitations on what can be expressed in the model. We call limitations due to the first rule *constraints*: they usually express some condition on the values that some attribute(s) can take. We will not deal with them in this paper. Here we concentrate on overcoming the second class of limitations by proposing higher-order operators (intuitively, relations on relations). To give an idea of the problems that E-R models face, we give several examples of situations in which the model is not able to capture necessary information.

One such situation is the *connection traps*: let E_1, E_2, E_3 be entities and $R_1 \subseteq E_1 \times E_2$, $R_2 \subseteq R_2 \times R_3$ be relationships. The *connection trap* problem is that of inferring properties of a possible connection between E_1 and E_3 based on R_1 and R_2. Sometimes the relationship may not exist at all, sometimes the relationship may exist but it is not determined by composing R_1 and R_2. The following examples are taken from [5].

Example 1. Let entities `Division`, `Staff` and `Branch` be related through relationships `IsAllocated`, between `Division` and `Staff`, and `Operates`, between `Division` and `Branch`. Participation in `IsAllocated` is total, 1 on the `Division` side and many on the `Staff` side (i.e. all staff members are allocated to one and only one division, and each division is allocated one or more staff members). Participation in `Operates` is total, 1 on the Division side and many on the branch side (i.e. all divisions are assigned one or more branches, and all branches are assigned to one and only one division). One could assume that staff members work at particular branches, and that such a relationship between elements of `Staff` and `Branch` can be inferred from the two explicit relationships. However, we note that both relationships are 1-M, with the 1 on their common domain (`Division`). Therefore some element in `Staff` may be related to more than one branch, even if one one would expect that each staff member works in only one branch. This is called the *fan trap* in [5]. Another, different problem is called the *chasm trap*. The chasm trap would come up if any of the two relationships were partial instead of total. Assume the situation this time involved entities `Branch`, `Staff` and `PropertyForRent`, related as follows: `Branch` and `Staff` are related by `IsAllocated`, as before, and `Staff` and `PropertyForRent` are related through relationship `Oversees`. Participation in `Oversees` is partial (i.e. not all people in the staff oversees a property for rent, and not all property for rent has someone from `Staff` assigned to oversee it). It would seem that one could connect each branch with the property or properties that someone at that branch oversees. However, there is no guarantee that all branches can be related to at least some property for rent (since participation in `Oversees` is partial).

Example 2. Assume an E-R model for a university. The model contains entities `Teacher`, `Class` and `Department`. There are relationships `Teaches` between `Teacher` and `Class`, `OfferedBy`, between classes and departments, and `Faculty`, between teachers and departments, with the obvious interpretation. A university rule is that teachers can only teach classes offered by the department in which they are faculty. This rule cannot be enforced in the E-R model.

Example 3. Assume an E-R model for a company, which contains entities `Client` and `Representative`. The representatives are employees whose mission is to interact with and attend to clients; a business rule is that every client must have a representative, and that every representative must attend to several clients. Note that this relationship, which is one-to-many and total, induces a partition on the set of clients. Many properties of such partitions cannot be expressed in the E-R model; for instance, a rule stipulating that all representatives must have the same number of clients (i.e. all sets in the partition have the same size).

Example 4. An example of a recursive relationship is the relationship `ManagerOf` on the entity `Employee`. Two roles are associated with `Employee` through this relationship: *manager* and *managee*. Such relationship is partial on the *manager* role (not all employees are managers) and total on the *managee* role (all employees have a manager). This relationship has several properties which cannot be expressed in the model: it is an irreflexible relationship (no one can be his or her own manager). In most situations, it will also be asymmetric (if employee **a** is the manager of employee **b**, it cannot be that employee **b** is, in turn, a manager of **a**) and transitive (managers have higher-level managers, and so on). Such information can be used by a system to check insertions in the relationship for correctness, but cannot be represented in a E-R model.

4 Extending the E-R Model

We formalize E-R models in a general framework that will allow the integration of GQs. First we define signatures and give a formal semantics to E-R models. We then show how to extend the model with GQs and define several GQs which are useful in extending the semantic expressivity of E-R models in the context of conceptual modeling. Finally, we give several examples of how our extension deals with the problems introduced in the previous section.

4.1 Formalization of E-R Models

Usually an E-R model is displayed graphically by an *E-R diagram*. An E-R diagram is a graph where entities are represented by nodes, relationships (including IS-A relationships) are represented by edges; attributes are depicted next to the entity or relationship they belong to. There are variations between authors in the way the information is represented graphically. In order to be able to define our extensions of E-R models without depending on a particular graphical representation, and to develop a rigorous framework, we define a formal language in which to express the model. We use a lisp-like syntax, with expressions always in balanced parenthesis. In the following, $^+$ means one or more, and $[A]$ means A is optional. Thus, (<attr-name>)$^+$ means a list of one or more of the objects of type <attr-name>.

Definition 3. *A signature S is a triple $< \mathcal{E}, \mathcal{R}, \mathcal{A} >$, where $\mathcal{E}$ (denoted by $ent(S)$) is a set of entity names; $\mathcal{R}$ (denoted by $rel(S)$) is a set of relationship names; and $\mathcal{A}$ (denoted by $att(S)$) is a set of attribute names. Each set is disjoint from the other two.*

Definition 4. *An E-R model* D *for a signature S is a set of sentences, where each sentence is either*

- *of the form* (E <entity-name> (<attr-name>$^+$))*, with each*<entity-name> $\in ent(S)$ *and each* <attr-name> $\in attr(S)$*; or*
- *of the form* (R <relationship-name> (<entity-name>:<role><part-constraint> <card-constraint>)$^+$ (<attr-name>)$^+$), *where* <part-constraint> *is one of* total *or* partial, <card-constraint> *is one of* 1 *or* M, *each*<relationship-name> $\in rel(S)$, *each* <entity-name> $\in ent(S)$, *and each* <attr-name> $\in attr(S)$*; or*

- of the form (A <attr-name> <attr-type> [(<attr-name>)$^+$]), where each <attr-name> $\in$ attr($\mathcal{S}$), <attr-type> is one of simple or complex plus one of single or multivalued, and the optional list of attribute names is used if the attribute is complex[1].

Definition 5. *Given signature $\mathcal{S}$, diagram D, for each $r \in rel(\mathcal{S})$, $comp(r) \subseteq ent(\mathcal{S})$ is the set of entities involved in the relationship denoted by r, that is $\{e_1, \ldots, e_n \mid$ (R r ($e_1 : r_1$ p_1 c_1) $\ldots$ ($e_n : r_n$ p_n c_n) (attr$_i$)) $\in D\}$, where each r_i is a role, each p_i a participation constraint, each c_i a cardinality constraint and each attr$_i$ a set of attributes, for $1 \leq i \leq n$.*

Definition 6. *Given signature $\mathcal{S}$, diagram D, for each $a \in attr(\mathcal{S})$, $edom(a) = \{e \in ent(\mathcal{S}) \mid$ (E e attr) $\in D \wedge a \in$ attr$\}$, where attr is a list of attribute names, and $rdom(a) = \{r \in rel(\mathcal{S}) \mid$ (R r el attr) $\in D \wedge a \in$ attr$\}$, where el is a list of entity components and attr is a list of attribute names. Thus $edom(a)$ is the list of entities e such that a is an attribute of e, and $rdom(a)$ is the list of relationships r such that a is an attribute of r[2].*

We give formal semantics to the language by defining a *conceptual structure* as follows.

Definition 7. *Given an E-R model $\mathcal{D}$ over signature $\mathcal{S}$, a* conceptual structure *is a tuple $< M, V, I >$, where M and V are disjoint, nonempty sets; and I is an* interpretation function *from the elements of $\mathcal{S}$ to $M \cup V$ with the following characteristics:*

- For each $e \in ent(\mathcal{S})$, $I(e) \subseteq M$.
- For each $r \in rel(\mathcal{S})$, such that $comp(r) = \{e_1, \ldots, e_n\}$, $I(r) \subseteq I(e_1) \times \ldots \times I(e_n)$, and
  - if participation constraint is total for entity $e_i \in \{e_1, \ldots, e_n\} = comp(r)$, then $r[e_i] = I(e_i)$[3], and
  - if cardinality constraint is 1 for entity $e_i \in \{e_1, \ldots, e_n\} = comp(r)$, then $r < e_i >$ is a function[4].
- For each $a \in att(\mathcal{S})$, $I(attr)$ is either
  - a function $f : M \rightarrow V$, such that for each $e \in edom(a)$ $I(e) \subseteq dom(f)$, or
  - a function $f : M^n \rightarrow V$, such that for each $r \in adom(a)$, then $I(r) \subseteq dom(f)$.

---

[1] For simplicity we will assume from now on that all attributes are simple and single. This simplifies notation while not subtracting anything substantial from the model.
[2] In the following, we will assume for simplicity that every attribute applies only to entities or to relationships. Again, this simplifies notation without affecting expressive power.
[3] $r[e_i] = \{x_i \in I(e_i) \mid \bigwedge_{1 \leq j \neq i \leq n} \exists x_j \in I(e_j)\ r(x_1, \ldots, x, \ldots, x_n)\}$, that is, the elements in the extension of e_i that are related to other elements by relationship r.
[4] $r < e_i >= \{< e_i, < e_1, \ldots, e_{i-1}, e_{i+1}, \ldots, e_n >>|< e_1, \ldots, e_{i-1}, e_i, e_{i+1}, \ldots, e_n >\in I(r)\}$, that is, the binary function obtained from $I(r)$ by considering entity e_i and the combination of values that e_i is related to by relation r.

Intuitively, M is the domain of objects which we are trying to model, while V is a set of values[5]; each entity e is assigned by the structure a set of objects $I(e)$ as its extension; likewise, relationships are assigned relations over the entities involved in the relationship, and attributes are considered functions relating values to entities and relationships.

4.2 Extensions of the Model

It has become common to extend E-R models with ideas from object oriented models, like IS-A (class/subclass) relationships (see [14] for a developed (and complex!) example of an object oriented modeling framework). Our formal language for E-R models did not have notation for IS-A relationships, or other extensions. Our strategy is to start with a very simple E-R model, to incorporate GQs into the model and use their power to model the extensions that are needed to increase the modeling power of the initial model.

Definition 8. Given a *conceptual structure* $<M, V, I>$, an *extended conceptual structure* is a tuple $<\mathcal{Q}, M, V, I>$, where M, V, I are as before and $\mathcal{Q}$ is a set of GQs defined over M.

Thus, the GQs defined may have as arguments entities (which are represented by sets in the structure) and relationships (which are represented by relations in the structure). The obvious issue is to find a set of GQs that will be helpful in the conceptual modeling task. We introduce several such quantifiers next. In the following definitions, it will be assumed implicitly that all sets and relations come from M.

First we note a form of useful quantification is in the form of *structural (typeless) quantifiers*. Since GQs can capture relationships among sets, they can capture the semantics of IS-A relationships, which usually correspond to simple set inclusion. We point out, however, that there is more information about class/subclass relationships than mere inclusion. Advanced models classify IS-A relationships at least along two dimensions; first, *disjointness/overlap* of the subclasses (i.e. whether the subclasses are allowed to have any common elements -this corresponds to inclusive and exclusive choices), and second, *coverage* of the superclass by the subclasses (i.e. whether the superclass is the union of the subclasses or not). These variants can all be captured by generalized quantification:

$Q_I(A, A_1, \ldots, A_n) = \{\bigwedge_{i \in 1, \ldots, n} A_i \subseteq A\}$
$Q_{IC}(A, A_1, \ldots, A_n) = \{\bigwedge_{i \in 1, \ldots, n} A_i \subseteq A \wedge \bigcup_{i \in 1, \ldots, n} A_i = A\}$
$Q_{ID}(A, A_1, \ldots, A_n) = \{\bigwedge_{i \in 1, \ldots, n} A_i \subseteq A \wedge \bigwedge_{i,j \in 1, \ldots, n, i \neq j} A_i \cap A_j = \emptyset \}$
$Q_{IDC}(A, A_1, \ldots, A_n) = \{\bigwedge_{i \in 1, \ldots, n} A_i \subseteq A \wedge \bigwedge_{i,j \in 1, \ldots, n, i \neq j} A_i \cap A_j = \emptyset \wedge \bigcup_{i \in 1, \ldots, n} A_i = A\}$

Clearly, $Q_I(A, A_1, \ldots, A_n)$ indicates that $A_1, \ldots, A_n$ are subclasses of A, with not further restrictions; Q_{IC} further constraints the relationship so that $A_1, \ldots, A_n$ *cover* A (that is, all the elements in A are in one of the subclasses);

---
[5] Separating values for attributes from the entities makes the model simpler and agrees with standard practice in building (semantic) data models ([11]).

Q_{ID}, on the other hand, constraints the relationship so that all subclasses are disjoint. Finally, Q_{IDC} adds both constraints[6].

This does not give us more expressive power than we already had in some extensions of E-R models. However, it is easy to define GQs that give well-known properties of relations which are not expressible in E-R models, even extended ones. A few, simple examples are:

$Q_R(R) = \{R \subseteq M^2 \mid R \text{ is reflexive}\}$
$Q_T(R) = \{R \subseteq M^2 \mid R \text{ is transitive}\}$
$Q_S(R) = \{R \subseteq M^2 \mid R \text{ is symmetric}\}$

and their respective negations, Q_{NR}, Q_{NT} and Q_{NS}[7].

The above GQs involve only one relation. More complex examples may involve more than one relationship, and determine whether they can be composed or not; the following is called *pseudo-transitivity*, for the obvious reasons:

$Q_{PT}(R_1, R_2, R_3) = \{R_1 \subseteq M^2, R_2 \subseteq M^2, R_3 \subseteq M^2 \mid \forall x, y, z R_1(x, y) \wedge R_2(x, z) \to R_3(y, z)\}$

The following is called *compositionality*, as it determines when it is possible *in principle* to compose two relations with a common entity:

$Q_c(A, B, C, R_1, R_2) = \{A, B, C \subseteq M, R_1 \subseteq A \times B, R_2 \subseteq B \times C \mid R_1[B] = R_2[B]\}$

The following is called *functionality* as it states that two relations not only can be composed, but furthermore their composition is a function:

$Q_f(R_1, R_2) = \{R_1, R_2 \subseteq M^2 \mid \forall x \, \forall y \, \exists! z R_1(x, y) \wedge R_2(y, z)\}$[8]

Finally, some GQs may involve relations and sets (relationships and entities) and express some properties that the relation determines on the set. As stated above, any time there is a one-to-many relationship R on entities E_1, E_2, the relationship on the many side (say E_2) has a partition created on it by R (strictly speaking, this only happens if participation of E_2 in R is total; otherwise, a set of elements not in $R[E_2]$ must be considered). Different properties of such partition can be expressed by GQs[9]:

$Q_{PC}(A, B, R) = \{A, B \subseteq M, R \subseteq A \times B \mid \forall y, z \in B \mid (\{x \mid R(x, y)\}\mid = \mid\{x \mid R(x, z)\}\mid)\}$

We show the applicability of these GQs by using them to solve the problems introduced before.

Example 5. Recall example 1 about connection traps. $Q_f(\text{IsAllocated}, \text{Operates})$ states that the relationships can be composed in a functional manner. Thus, fan traps are avoided. $Q_c(\text{Branch}, \text{Staff}, \text{PropertyForRent}, \text{IsAllocated}, \text{Oversees})$ states that the extensions of the relationships coincide on their common entity. Thus, chasm traps are avoided. Note that Q_C does not constraint the relationships to be total or partial, leaving the analyst free to combine this and other properties.

---

[6] In the quantifier name, I stands for *inheritance*, C for *cover* and D for *disjointness*.
[7] In the quantifier name, N stands for *not*, as in NR for *not reflexible*, and so on.
[8] The notation $\exists! z$ is a shortcut for *there exists a unique z*. Note that this condition is less restrictive than asking that both R_1 and R_2 are functional.
[9] $|A|$ denotes the *cardinality* of set A.

Example 6. Recall example 2 about teachers, classes and departments. Then $Q_{PT}(\text{Teaches}, \text{OfferedBy}, \text{Faculty})$ enforces the restriction that every professor can only teach classes offered by the department where (s)he is faculty, as desired.

Example 7. Recall example 3 about clients and representatives. Then $Q_P(\text{Client}, \text{Representative}, \text{Attends})$ states that all representatives attend the same number of clients.

Example 8. Recall example 4 about the recursive relationships `ManagerOf` on the entity `employee`. We can express the desired properties of `ManagerOf` now as follows: $Q_{NR}(\text{ManagerOf})$, $Q_{NS}(\text{ManagerOf})$, $Q_T(\text{ManagerOf})$.

5 Related Work

The seminal paper ([4]) introduced Entity-Relationship modeling. Although widely used because of its balance of simplicity and expressive power, the limitations of the model have been noted for quite some time and given rise to several extensions: [2] proposed to add facilities to model the concept of *transaction*; [6] proposed the addition of data types; [7] propose the addition of more functionality by describing entity behavior. The paper [8] adds several powerful concepts, including abstract data types, arbitrarily complex structures, and defines a powerful query language for E-R diagrams. It must be pointed out that not all the cited work provides a formal foundation in the form of well-defined, formal semantics ([8] is an exception). Our work is different in that we extend E-R diagrams as little as possible, by introducing only one new category (that of GQ), while at the same time capturing a rich class of semantic information which is of interest for conceptual modeling (i.e. our goal is not to define a query language or data types, but to assist the analyst in capturing mode domain information, therefore helping to restrict possible interpretations of the model). By defining a formal language and giving a formal semantics to a very basic E-R model, the work presented here is independent of notational variations and extensions of the model, and has a formal semantics.

6 Conclusion and Further Research

This paper is a starting point for work that uses logical methods in conceptual modeling. It was argued that GQs are a good fit for conceptual modeling as they are *declarative* and *high-level*. They are also an extremely rich and powerful category; the challenge is to define relevant sets of GQs for the goal at hand. We note that, even though we have worked with a formal language for several reasons, to incorporate GQs into E-R diagrams is easy because GQs have an intuitive graphic depiction, as it was shown in [15]. Therefore, modelers could actually work with a diagrammatic representation of the ideas introduced here.

Some issues that deserve further attention include *reasoning* with GQs. Some sort of limited deduction may allow analysts to check properties of the model; introductory work has already been carried out ([16],[17]) but it does not seem to be well known outside logical circles. We hope the present work will help disseminate potentially helpful work from pure logic to more applied enterprises.

References

1. Abiteboul, S., Hull, R. and Vianu, V., *Foundations of Databases*, Addison-Wesley, 1995
2. Atzeni, P., Batini, C., Lenzerini, M., Villanelli, F. *INCOD: A system for conceptual design of data and transactions in the entity-relationship model*, in Proceedings of the 2nd International Conference on Entity-Relationship Approach to Information Modeling and Analysis, 1981.
3. Badia, A. *Extending Description Logics with Generalized Quantification*, in Proceedings of the 11th International Symposium on Methodologies for Intelligent Systems, LNAI, number 1609, Ras and Skowron, editors, Springer-Verlag, 1999.
4. Chen, P. *The Entity-Relationship Model -Towards a Unified View of Data*, ACM Transactions on Database Systems, v. 1, n. 1, 1976.
5. Connolly, T., Begg, C. and Strachan, A. *Database Systems*, Addison-Wesley, 1999.
6. Dos Santos, C. S., Neuhold, E. J. and Furtado, A. L. *A Data Type Approach to the Entity-Relationship Model*, in Proceedings of the 1st International Conference on Entity-Relationship Approach to Software Engineering, 1980.
7. Eder, J., Kappel, G., Tjoa, A. and Wagner, R. *BIER: The Behavior Integrated Entity Relationship Approach*, in Proceedings of the 5th International Conference on Entity-Relationship Approach, 1986.
8. Gogolla, M. and Hohestein, U. *Towards a Semantic View of en Extended Entity-Relationship Diagram*, ACM Transactions on Database Systems, v. 16, n. 3, 1991.
9. Gyssens, M., Van Gucht, D. and Badia, A., *Query Languages with Generalized Quantifiers*, in *Application of Logic Databases*, Ramakrishnan, Ragu ed., Kluwer Academic Publishers, 1995
10. Hsu, P. Y. and Parker, D. S., *Improving SQL with Generalized Quantifiers*, in Proceedings of the Tenth International Conference on Data Engineering, 1995
11. Hull, R. and King, R. *Semantic Database Modeling: Survey, Applications and Research Issues*, ACM Computing Surveys, vol. 19, n. 19, 1987.
12. Lindstrom, P., *First Order Predicate Logic with Generalized Quantifiers*, Theoria, volume 32, 1966
13. Mostowski, A., *On a Generalization of Quantifiers*, Fundamenta Mathematica, volume 44, 1957
14. Rumbaugh, J., Jacobson, I. and Booch, G. *The Unified Modeling Language Reference Manual*, Addison-Wesley, 1999.
15. Sarathy, V. and Van Gucht, D. and Badia, A., *Extended query graphs for declarative specification of set-oriented queries*, in Workshop on Combining Declarative and Object-Oriented Databases (in conjunction with SIGMOD 93), Washington, D.C.
16. Van Benthem, J. *Questions about Quantifiers*, Journal of Symbolic Logic, v. 49, 1984.
17. Westerstahl, D., *Quantifiers in Formal and Natural Languages*, in *Handbook of Philosophical Logic*, Reidel Publishing Company, Gabbay, D. and Guenther, F., editors, vol. IV, 1989.

Combining Description Logics with Stratified Logic Programs in Knowledge Representation

Jianhua Chen
Computer Science Department
Louisiana State University
Baton Rouge, LA 70803-4020
jianhua@bit.csc.lsu.edu

Abstract. Hybrid knowledge representations that combine description logics with logic programs are considered. Previous works combine description logics with Horn logic programs. In this paper, the expressive power of such hybrid systems is extended by allowing the combination of function-free, non-recursive *stratified* logic programs with description logics. Two model-theoretic definitions for the semantics of the hybrid knowledge representation are presented. It is shown that the inference problem based on the second semantics is decidable. When the logic program is Horn, the two semantics defined in this paper coincide with the semantics in [4] with regard to the inference problem.

1. Introduction

The use of hybrid representations is an important research problem in knowledge representation and reasoning. Several recent papers [2,4,5,6] addressed various aspects of hybrid knowledge representations. In [6], a framework based on unification of constraint logic programming, annotated logic programming, and stable model semantics, was developed to handle the multiple modes of reasoning in hybrid knowledge bases. The work in [2] investigated the connections between description logics and predicate logics, and compared their relative expressiveness. Combinations of description logics with Horn logic programs were investigated [3,4]. In particular, it was shown in [4] that the inference problem is *decidable* in a hybrid system combining the description logic *ALCNR* with a non-recursive Horn logic program.

A natural extension in similar direction is to consider the integration of description logics with *stratified* logic programs for knowledge representation. Stratified logic programs are extensions of Horn logic programs which allow certain restricted form of negations in the antecedent of program rules, and thus are more expressive. The ability to combine stratified programs and description logics within one hybrid system significantly enhances the system's expressive power.

In this paper, we present such an extension. The proposed hybrid system combines the description logic *ALCNR* with a stratified logic program. We present two model-theoretic semantics for the hybrid knowledge representation system: The preferred model semantics, and the preferred-canonical model semantics. We show that the inference problem under the preferred-canonical model semantics is decidable based on a straight forward application of the decidability result in [4].

2. Preliminaries

In the new hybrid representation, a knowledge base $\Delta = \langle T, \pi \rangle$ consists of two components: a terminology T in the description logic *ALCNR*, and a logic program π which is function-free, non-recursive, and *stratified*. The concepts and roles from the terminology T can occur (positively) in the antecedents of rules in π.

2.1. The Terminological Component

A description logic language contains a set of unary relations called concepts that represent sets of objects in the domain of discourse, and binary relations called roles that represent relationships between these objects. Composite formulas in a description logic are built from primitive concepts and roles by using a set of *constructors* in the logic. Here as in [4], the description logic component in a hybrid knowledge base can be any subset of the *ALCNR* language. Descriptions in *ALCNR* are built in the following way: Each primitive concept A is a concept description; the special concepts $\top$ (truth) and $\bot$ (falsity) are concept descriptions; let C and E be concept descriptions and let R be a role description, then $C \cap E$, $C \cup E$, $\neg C$, $\exists R.C$, $\forall R.C$, $(\geq n\, R)$ and $(\leq n\, R)$ are concept descriptions; each role description R is of the form $R_1 \cap ... \cap R_m$ where each R_j is a primitive role.

The general terminological part of a terminology T is a set of sentences in *ALCNR*, where a sentence is either a concept definition, a concept inclusion, or a role definition. A concept definition is of the form $C := E$ where C is a concept name and E is a concept description. A concept inclusion is of the form $C \subseteq E$ where both C and E are concept descriptions. A role definition is of the form $P := R$ where P is a role name and R is a role description. We do allow recursive concept definitions. The assertional part of T is a set of ground atoms of the form $C(a), R(a, b)$.

The meaning of a terminology T is determined by a model-theoretic semantics. We define an interpretation I to be a non-empty domain O, a mapping from the set of constants in T to O such that $a^I \neq b^I$ if $a \neq b$, a mapping from each concept name C to a unary relation C^I in O, and a mapping from each role name R to a binary relation $R^I \subseteq O \times O$. The mappings of I can be naturally extended to the composite descriptions in a straight forward way.

An interpretation I satisfies a concept instance $C(a)$ if $a^I \in C^I$. I satisfies a role instance $R(a, b)$ if $(a^I, b^I) \in R^I$. I satisfies a concept definition $C := E$ if $C^I = E^I$, it satisfies a concept inclusion $C \subseteq E$ if $C^I \subseteq E^I$, and it satisfies a role definition $P := R$ if $P^I = R^I$. I is a model of a terminology T if I satisfies each sentence in T.

2.2. Stratified Logic Programs

A function-free (normal) logic program is a set of program rules of the form
$$\text{r: } B_1(X_1) \wedge ... \wedge B_m(X_m) \wedge \neg C_1(Y_1) \wedge ... \wedge \neg C_n(Y_n) \rightarrow A(X). \tag{1}$$

Here each B_i, C_j and A is an atom, m, n $\geq$ 0. The conjunction on the left hand side of the arrow "→" is called the *antecedent* (*body*) of the rule, and $A(X)$ is the *consequent* (*head*) of the rule. X, X_i and Y_j are tuples of variables and constants. In this study we consider only *safe* programs, i.e., each variable appearing in the head of r or in a negative literal in the body of r must occur in a positive literal in the body of r. Moreover we consider only *non-recursive* programs.

An interpretation I for a program π consists of a non-empty set D (the domain of I), a mapping from the constants in π to D such that $a^I \neq b^I$ if $a \neq b$, and a mapping from each n-ary predicate P to a subset of D^n. Here an n-tuple $\alpha \in P^I$ for the n-ary predicate P means I assigns $P(\alpha)$ to be "true". For a fixed domain D, one can equivalently replace a program π by the (possibly infinite) set of ground rules which are obtained from π by substituting the variables in each rule r by the objects in D. Let us call this instantiated program π_D. An interpretation I is said to be a *model* of a program π if I satisfies the instantiated program π_D.

In logic programming, the class of *stratified* programs [1, 8] received a lot of attention. A normal logic program π is said to be *stratified* if there is a stratification Σ which partitions the predicates in π
$$\Sigma = S_1 \cup S_2 \cup ... \cup S_k$$
such that for each rule r of the form (1) in π, we have stratum$(B_i) \leq$ stratum(A) for $1 \leq i \leq m$, and stratum$(C_j) <$ stratum(A) for $1 \leq j \leq n$. Here stratum$(q) = j$ if $q \in S_j$. From a stratification Σ of π, we can equivalently define a partition of the rules in π
$$\pi = \pi_1 \cup \pi_2 \cup ... \cup \pi_k$$
such that the rules with head atom in S_j form the set π_j.

Without loss of generality, we assume in this paper that the stratification considered for a program π is the tightest stratification in the sense that for each atom A in S_j ($j \geq 2$), there exists an atom q in S_{j-1} such that $\neg q$ occurs in the body of a rule with A as the rule head. We remark that in case some concepts and roles from the terminological part T appear in the body of rules in the stratified program π, these concepts and roles will always belong to S_1 in the tightest stratification of π. Note that π_1 may be empty for the tightest stratification (see Example 1).

Let π be a stratified program with stratification $\Sigma = S_1 \cup S_2 \cup ... \cup S_k$. Let M, N be two models of π based on the same domain D. We say that M is more preferable to N, written as M $\leq$ N, if for each ground atom $P(\alpha) \in$ M $-$ N ($P(\alpha) =$ true in M, and $P(\alpha) =$ false in N), there is an atom $Q(\beta) \in$ N $-$ M such that stratum$(Q) <$ stratum(P). We write M $<$ N if M $\leq$ N and M $\neq$ N. Clearly, for a stratified program π, the preference relation "$\leq$" is a partial order. A model M is said to be a *most preferred model* of π, if there is no model N of π such that N $<$ M. Such a most preferred model is called a *perfect* model by Przymusinski [7].

In this paper, we always consider perfect models of a stratified program to be its designated models. Moreover, whenever we discuss the models of a program π with respect to a fixed domain D, we will always replace π by its instantiated program π_D, which is equivalent to a propositional logic program. It has been shown that a stratified propositional program has a unique perfect model which can be obtained iteratively using iterative Horn program consequences and program reduction.

Let π be a propositional program with negations allowed in the body of rules. Let I be a partial interpretation of the form $I = \langle Pos, Neg \rangle$ where Pos is the set of atoms assigned to be true in I, Neg is the set of atoms assigned to be false in I, $Pos \cap Neg = \emptyset$. The atoms occurring in π but not in $Pos \cup Neg$ are *undefined* in I. The reduction of π w.r.t. I is a program π/I which is obtained from π by the following two operations:

(1) For each rule of the form
$$B_1 \wedge ... \wedge B_m \wedge \neg C_1 \wedge ... \wedge \neg C_n \rightarrow A,$$
delete the rule if some $B_i \in Neg$ or $C_j \in Pos$.

(2) For each remaining rule, remove from the body any occurrence of literals whose atom occur in $Pos \cup Neg$.

Example 1. Consider the hybrid knowledge base $\Delta = \langle T, \pi \rangle$. The terminological part T has the following sentences:

author := ∃ write.paper
author ⊆ student ∪ professor
student ∩ professor ⊆ ⊥
author(John)

The concepts *student, professor, and paper* are primitive ones, whereas the concept *author* is a derived concept. The first inclusion states authors are either professors or students, and the second inclusion states that professors and students are disjoint. The assertion "author(John)" implies that John has a filler of the role *write*.

The logic program π consists of the following rules:

(1) author(x) ∧ ¬previous-author(x) → new-author(x)
(2) student(x) ∧ write(x, y) ∧ paper(y) → eligible(x)
(3) professor(x) ∧ write(x, y) ∧ paper(y) ∧ new-author(x) → eligible(x)

Here in the tightest stratification of π, $S_1 = \{$student, professor, author, paper, write, previous-author$\}$ and $S_2 = \{$new-author, eligible$\}$. Also note that in the partition of the program π, we have $\pi_1 = \{\}$ and $\pi_2 = \pi$. Imagine that the program π formalizes a professional conference organizer's rules to decide the eligibility of authors for travel support. Intuitively, the rules say that student authors are eligible for the travel support, and that professor authors are also eligible if they have not previously submitted papers to this conference series. ♣

2.3 Preferred Models of A Hybrid Knowledge Base

Recall that a hybrid knowledge base Δ consists of two components T and π, where T is a terminology and π is a stratified logic program. The concepts and roles in T are allowed to occur positively in the *body* of rules in π. On the other hand, no concepts or roles are allowed in the *head* of any rule in π, because the terminology T is supposed to specify a *complete* definition of the concepts and roles. The predicates in π but not in T are called *ordinary* predicates, which can be of any arity.

The meaning of a hybrid knowledge base $\Delta = \langle T, \pi \rangle$ is given by a model-theoretic semantics. First, an interpretation I for the hybrid knowledge base consists of a nonempty domain D, a mapping from the constants in T and π to D, and a mapping from each predicate P of arity n (including the concepts, roles and ordinary predicates) to P^I which is a subset of D^n. Now, how do we define the models of a hybrid knowledge base Δ? A naive way would be to define a model of Δ as a model of both of its components, namely, an interpretation I is a model of Δ if the restriction of I on predicates in T is a model of T and the restriction of I on the predicates in π is a perfect model of π. However, this has some undesirable consequences. Since concepts and roles occur only positively in the rule antecedents, any perfect model of π (with respect to all predicates in π) will assign "false" to each ground instance of such concepts and roles. However, this assignment may not be consistent with any of the models of T. Moreover, the notion of minimization and perfect models should really be applied to *ordinary predicates only*. Thus the desirable definition should keep the extension of concepts and roles *fixed* (as in parallel circumscription) according to the models of T, and then based on these fixed truth assignments to concepts and roles, define perfect models of π with respect to the ordinary predicates. Thus we develop the following definition:

Definition 1 (Preferred Models).

Let $\Delta = \langle T, \pi \rangle$ be a hybrid knowledge base. Let $I = I_T \cup I_\pi$ be an interpretation of Δ, where I_T defines the mapping for concepts and roles, I_π defines the mapping for ordinary predicates. Let D be the domain of I. We say that I is a *preferred model* of Δ if it satisfied the following two conditions:

(1) I_T is a model of T.

(2) I_π is a perfect model of π_D/I_T.

The collection of preferred models of Δ is denoted as $Pref(\Delta) = \{I: I$ is an interpretation and a preferred model of $\Delta\}$. A ground atom $P(a)$ is entailed by Δ (under this preferred model semantics), written $\Delta \models_{Pref} P(a)$. if $P(a)$ is true in each model in $Pref(\Delta)$. ♣

The reasoning problem under the preferred model semantics is precisely to determine whether we have $\Delta \models_{Pref} P(a)$ where P is an ordinary predicate and a is a tuple of constants. We do not have a decision procedure for this general problem when the program π is not Horn. This is because the terminology T may have *infinitely many* models, each corresponding to a preferred model of Δ, and finding decision procedures for Δ in this case may be difficult. However, if we modify the preferred model semantics and focus on an interesting, *finite* subset of models of T, we will consider only finitely many models of Δ as designated ones. Hence the inference problem under the modified semantics becomes decidable.

What are the models of T that belong to the finite subset of interest in the modified semantics? In [4], it was shown that one can always find a finite subset Ω from the models of T such that for each ground atom $P(\alpha)$ (P is an ordinary predicate and α is a tuple of constants), $T \cup \pi \models p(\alpha)$ if and only if $I \cup \pi \models P(\alpha)$ for each $I \in \Omega$. We will focus on precisely these models of T in defining the modified semantics. According to [4], the set Ω can be obtained as follows: First, we build an initial constraint system S_T from T, which is *equivalent* to T in the sense that it has the same models as T. A constraint system is a non-empty set of constraints of the form $s: C$,

$s\ R\ t$, $\forall x. x:C$ and $s \neq t$, where s and t are either constants or variables, C is a concept description, R is a primitive role name. Second, we expand S_T by repeatedly applying a set of *propagation rules*. This will produce a finite set of completions which are constraint systems such that no propagation rule is applicable to them. Some of the completions may contain a *clash*, i.e., a contradiction. For each of the clash-free completions, a unique canonical model can be constructed. All such canonical models of clash-free completions together form the set Ω.

A naive application of the propagation rules in expansion of S_T may not terminate because of the *generating rules* which can introduce new variables to the constraint system. Thus the n-tree equivalence condition was developed in [4] to assure expansion termination in finitely many steps. In [4], each completion S is obtained from S_T by application of propagation rules with the $U(\Delta)$-tree equivalence termination condition. Here for a knowledge base $\Delta = \langle T, \pi \rangle$ where T is a terminology and π is a Horn program, $U(\Delta)$ is the *maximal size* (number of literals in the antecedent) of a rule derivable by chaining the rules in π. In our case of non-Horn stratified programs, $U(\Delta)$ is defined to be the *maximal size* of a rule derivable by chaining the rules in π_1, which is the first stratum of the program rules in π. When π_1 is empty, $U(\Delta)$ is defined to be zero.

Now we are ready to present a modified preferred model semantics as follows: Instead of considering all models I_T for a terminology T when defining models of $\langle T, \pi \rangle$, an interpretation I is considered only when I_T is the *canonical model* of a clash-free completion S, which is obtained from S_T based on the $U(\Delta)$-tree equivalence condition. The formal definition follows:

Definition 2 (Preferred-Canonical Models).

Let $\Delta = \langle T, \pi \rangle$ be a hybrid knowledge base. Let $I = I_T \cup I_\pi$ be an interpretation of Δ, where I_T defines the mapping for concepts and roles in T, and I_π defines the mapping for ordinary predicates. Let D be the domain of I. We say that I is a *preferred-canonical model* of Δ if it satisfied the following two conditions:

(1) I_T is the canonical model of S, which is a clash-free completion obtained from S_T under the $U(\Delta)$-tree equivalence condition.

(2) I_π is a perfect model of π_D/I_T.

The collection of preferred-canonical models of Δ is denoted as *Pref-Cano*$(\Delta) = \{I : I$ is an interpretation and a preferred-canonical model of $\Delta\}$. A ground atom $P(a)$ is entailed by Δ (under this preferred-canonical model semantics), written $\Delta \models_{Pref-Cano} P(a)$, if $P(a)$ is true in each model in *Pref-Cano*(Δ). ♣

Note that when the program π is a Horn program which is a special case of stratified programs, the preferred model semantics and the preferred-canonical model semantics coincide with respect to the entailment of ground atoms. Moreover, as far as entailment of ground atom is concerned, these semantics also coincide with the semantics in [4] defined for combination of an *ALCNR* terminology and a Horn program. Here we use the notation $\Delta \models P(\alpha)$ to denote the entailment of $P(\alpha)$ by Δ under the semantics defined in [4].

Theorem 1.

Let $\Delta = \langle T, \pi \rangle$ be a hybrid knowledge base, where T is a terminology in the ALCNR logic and π is a non-recursive Horn program. Let $P(\alpha)$ be a ground atom where P is an ordinary predicate in π. Then
$$\Delta \models_{Pref} P(\alpha) \Leftrightarrow \Delta \models_{Pref-Cano} P(\alpha) \Leftrightarrow \Delta \models = P(\alpha).$$

3. Decidable Reasoning with Description Logics and Stratified Logic Programs

From the discussions in Section 2, we see that for a hybrid system $\Delta = \langle T, \pi \rangle$, the terminology T has only finitely many canonical models I_S which are constructed from clash-free completions S of S_T. Each such canonical model I_S is finite. By Definition 2, the preferred-canonical models of the hybrid system $\Delta = \langle T, \pi \rangle$ consider precisely these canonical models I_S as the designated models of T. Moreover, for each such canonical model I_S, there is a unique perfect model I_π for the reduced program π_D/I_S. It follows immediately that there are finitely many preferred-canonical models of Δ, each is finite. Thus we get the following theorem by a straight forward application of the decidability result from [4]:

Theorem 2.

Let $\Delta = \langle T, \pi \rangle$ be a hybrid knowledge base, where T is a terminology in the ALCNR logic and π is a stratified logic program. Let $P(\alpha)$ be a ground atom where P is an ordinary predicate in π. Then the problem of determining whether $\Delta \models_{pref-cano} P(\alpha)$ is decidable.

The following outlines the algorithm for the reasoning problem $\Delta \models_{Pref-Cano} P(\alpha)$.

(1) Build the initial constraint system S_T starting with all ground atoms in T. Compute $U(\Delta)$ using the tightest stratification of π.

(2) Apply the propagation rules to S_T and obtain clash-free completions S with the $U(\Delta)$-tree equivalence condition for termination. Let W be the set of clash-free completions obtained.

(3) For each $S \in W$, construct the canonical model I_S for S. Let Ω be the set of such canonical models.

(4) For each $I_S \in \Omega$, construct the unique perfect model I_π of the reduced program π_D / I_S using the "Construct" algorithm below. Here D is the domain of I_S extended by constants appearing in π but not in T.

(5). If $P(\alpha)$ is true in every I_π obtained in step (4), then output answer "yes", otherwise output answer "no".

Before we present the algorithm for constructing the unique perfect model I_π of π with respect to a model I_S of the terminology T, we need to introduce several notations. Recall that for a given stratified program π, we can partition $\pi = \pi_1 \cup \pi_2 \cup \ldots \cup \pi_k$ according to its tightest stratification. For a non-empty set D which is the domain of an interpretation of Δ, we define U_j to be the set of ground atoms of the form $A(\alpha)$ where A is an ordinary predicate in stratum j, and α is a tuple of objects in D. For a given interpretation I_S of the terminology T, we use π' to denote the reduced program π_D/I_S. Clearly, π' can be represented as $\pi' = \pi'_1 \cup \pi'_2 \cup ... \cup \pi'_k$, where each π'_j for $1 \leq j \leq k$ is the reduction of $(\pi_j)_D$ by I_S.

The "Construct" algorithm mentioned above is the following:

Algorithm Construct

Input: I_S, the canonical model of a clash-free completion S and π, a stratified logic program.

Output:
  I_π, the unique perfect model of the reduced program π_D/I_S.

(1) Let D be the domain of I_S extended by adding all constants in π but not in T. Let the program π' be the reduced program π_D/I_S. Let $Mod_0 = \langle \emptyset, \emptyset \rangle = \langle P_0, N_0 \rangle$.

(2) For $i = 1$ to k do

   (2.1) $Pos_i = Neg_i = \emptyset$.

   (2.2) While there is a ground rule in π'_i of the form $B_1 \wedge B_2 \wedge ... \wedge B_m \to A$, such that A is not in Pos_i and each B_j is in Pos_i, add A to Pos_i.

   (2.3) Let $Neg_i = U_i - Pos_i$ and $T_i = \langle Pos_i, Neg_i \rangle$.

   (2.3) Let $P_i = P_{i-1} \cup Pos_i$, $N_i = N_{i-1} \cup Neg_i$. Let $Mod_i = \langle P_i, N_i \rangle = Mod_{i-1} \cup T_i$.

   (2.4) If $i < k$, then $\pi'_{i+1} = \pi'_{i+1}/Mod_i$.

(3) Output $I_\pi = Mod_k$ as the result.

Example 2. Consider the hybrid knowledge base Δ in Example 1. Here $U(\Delta) = 0$. This knowledge base has two preferred-canonical models I_1 and I_2, with the same domain $D = \{John, v_1\}$. Both I_1 and I_2 contain the ground atoms $\{author(John), write(John, v_1), paper(v_1), eligible(John), \neg previous\text{-}author(John), new\text{-}author(John)\}$. I_1 contains $student(John)$ and $\neg professor(John)$, while I_2 contains $professor(John)$ and $\neg student(John)$. Since both models contain $eligible(John)$, it follows that $\Delta \mid =_{Pref-cano} eligible(John)$.

To derive $eligible(John)$, program rule (2) is used in building I_1, and program rules (1) and (3) are used in constructing I_2. Note the use of *negation as failure* in constructing I_2. The reduced program π' consists of two ground rules "$\neg previous\text{-}author(John) \to new\text{-}author(John)$" and "$new\text{-}author(John) \to eligible(John)$". In applying the "Construct" algorithm, we get $\neg previous\text{-}author(John)$ in Mod_1 by negation as failure, and thus subsequently we get $new\text{-}author(John)$ and $eligible(John)$ in Mod_2. We can not infer $eligible(John)$ if we do not use the perfect model semantics which sanctions the negation as failure inference illustrated above.
♣

4. Conclusions

In this paper, we present a new hybrid knowledge representation system which allows the combination of description logics with stratified logic programs. Two semantics are defined for the new hybrid knowledge representation. It is shown that under the preferred-canonical model semantics, the inference problem is decidable. Algorithms for performing such inferences are also presented.

The work reported here is still quite preliminary, and further studies are needed to investigate the decidability of reasoning under the preferred model semantics when the terminological cycles are allowed.

Acknowledgment

The author thanks Kevin P. Grant for useful discussions related to this work.

References

[1] K.R. Apt, H. Blair, A. Walker, Towards a Theory of Declarative Knowledge, In: *Foundations of Deductive Databases and Logic Programming* (J. Minker, Ed.), Morgan Kaufmann Publishers, Los Altos, CA, 1988, pp. 89-148.

[2] A. Borgida, On the Relative Expressiveness of Description Logics and Predicate Logics, *Artificial Intelligence*, **82**(1996), pp. 353-367.

[3] M. Buchheit, F.M. Donini, A. Schaerf, Decidable Reasoning in Terminological Knowledge Representation Systems, *Journal of Artificial Intelligence Research*, **1** (1993), pp. 109-138.

[4] A. Y. Levy, M.-C. Rousset, Combining Horn Rules and Description Logics in CARIN, *Artificial Intelligence*, **104**(1998), pp. 165-209.

[5] A. Y. Levy, M.-C. Rousset, CARIN: A Representation Language Integrating Rules and Description Logics, *Proc. of European Conf. on Artificial Intelligence*, Budapest, Hungary, 1996.

[6] J. Lu, A. Nerode, V.S. Subramanian, Hybrid Knowledge Bases, *IEEE Transactions on Knowledge and Data Engineering*, **8**5), 1996, pp. 773-785.

[7] T. Przymusinski, On the Declarative Semantics of Deductive Databases and Logic Programs, In: *Foundations of Deductive Databases and Logic Programming* (J. Minker, Ed.), Morgan Kaufmann Publishers, Los Altos, CA, 1988, pp. 193-216.

[8] A. Van Gelder, Negation as Failure Using Tight Derivations for General Logic Programs, In: *Foundations of Deductive Databases and Logic Programming* (J. Minker, Ed.), Morgan Kaufmann Publishers, Los Altos, CA, 1988, pp. 149-176.

Emergence Measurement and Analyzes of Conceptual Abstractions during Evolution Simulation in OOD

Mourad OUSSALAH, Dalila TAMZALIT

IRIN, 2, rue de la Houssinière - BP 92208 44322 Nantes cedex 03 France
{Mourad.Oussalah, Dalila Tamzalit}@irin.univ-nantes.fr
Tel : 02-51-12-58-47   Fax : 02-51-12-58-12

Abstract. When a designer has the delicate task to integrate new or badly specified needs, it is not easy specially within engineering applications having significant class hierarchies with bulky object bases. He is only sure about the changes to bring punctually, more explicitly on instances. We propose to this kind of designer a simulation tool of class evolution according to structural evolution of instances. It has to provoke emerge of new and adapted conceptual abstractions and to detect their position in the class hierarchy. The objective of this article is to analyze emergent abstractions by using metrics.

1. Introduction

A designer has a great number of strategies to manage the evolution of engineering applications [6]. A designer can use, according to his needs, one or several strategies. However, experience gained in OO systems and applications [10] has brought to light that new needs appear more often during their manipulation, so during manipulating instances.

In order to face unforeseen changes for complex and bulky engineering applications, we propose to a designer a simulation tool of class evolution. This tool searches and releases several possible directions of evolution of specifications thanks to dynamic evolution of instance structure. The general principle is to provoke the emergence of conceptual abstractions more adapted. This emergence is based on these newly expressed requirements and one those already present in the database. Hereafter, the simulation tool has to detect the location of these new abstractions in the hierarchy and to determine possible impac4ts.

2. The object evolution: a state of the art

In a general way, to prepare a system or an application to evolve, it is necessary to be able:

1. to formulate changes in order to achieve the pursued goal, namely the model after evolution;
2. to manage the impacts generated by these changes;

3. to define the link between the starting model and the arrival one (it's the same but in two different stages).

OO defined systems and programming languages propose strategies and mechanisms to manage evolution. Experience gained in OO design and development outlines, as well at design level as implementation one, lacks of actual evolutionary approaches. Existing strategies are varied and meet partially requirements. We propose to examine these strategies according to different viewpoints or *facets* of evolution:

2.1 The three Facets of Evolution

Rather than classifying some evolutionary strategies according to some common but not exhaustive criteria, we propose a classification resting on own evolution's criteria. We consider that evolution of any OO system presents three facets:

- **Type of evolution:** when needs are taken into account during the analysis and design phases, the evolutionary strategy is *preventive* or *anticipated*. When an evolutionary strategy can face new or badly specified needs, the evolution is said *curative* or *unanticipated*.
- **Object of Evolution:** the evolution can concern be the *product* (code, class, schema...) or the *process* (a part of reasoning, a development process of an application...).
- **Process of Evolution:** we distinguish two kinds of evolutionary processes: *development* and *emergence*. The development concerns classes and their impacts on corresponding sub-classes and instances. The emergence concerns instance evolution and their impacts on corresponding classes.

Each facet is represented by an axis. The combination of the three axis gives a three-dimension representation of the object evolution.

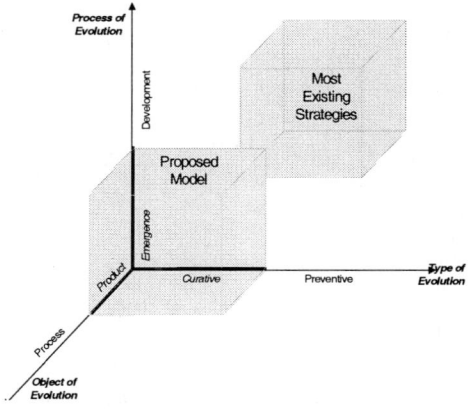

Fig. 1. Three facets of the evolution object

In order to classify a strategy, we have just to answer these three questions:

1. What kind of evolution this strategy propose (curative or preventive)?
2. What does it process on (product or process)?
3. What kind of evolutionary process does it allow (development or emergence)?

By responding to these questions for each studied strategy [9], we notice that most of them propose a *curative* evolution, that they principally work on the *product* by considering essentially *development* process. Our research work aims to complete these preventive strategies. For that, our model responds to the three aforesaid questions as follows: 1: *curative* – 2: *product* – 3: *principally* *emergence*.

2.2 Object Evolution problematic under the product viewpoint

We restrict Fig.1 to a two-dimension figure because we only consider the product. For space restriction reason, we only situate evolutionary strategies on the Fig. 2 in order to position them according to the object evolution problematic. Fig.2 shows that most of OO evolutionary strategies are preventives and allow development processes. Only categorization [7] allows emergence but in a preventive way with a break in the life cycle. We note that the principal lack of the existing evolutionary strategies are their inability to cope with unexpected or poorly specified needs and incomplete data. Moreover, instance evolution is always limited by class' one. This situation constitute a restrictive and unnatural aspect of their evolution. Our model leads with this aspect, principally with the emergence.

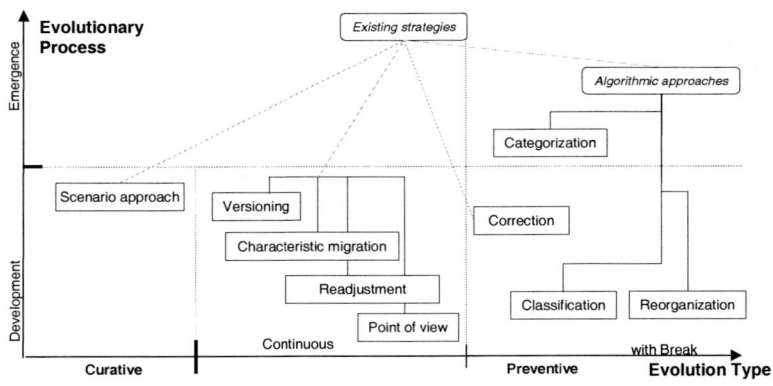

Fig. 2. Principal strategies according to the process and the type of evolution

3. The proposed model

Because instances are the object representatives of real entities, we consider them as full *individuals*. This leads de facto to make an analogy with leaving individuals, principally in their evolving and adapting feature according to their environment. Since we use some principles and concepts of Artificial Evolution, we give a brief presentation of Artificial Life and Genetic Algorithms. For more details, see [9].

3.1 Artificial Evolution: our inspiration source

Even if they have been defined and are used in different scientific areas, Artificial Life [4] and Genetic Algorithms [3] toke their inspiration from biology in order to simulate evolutionary biological mechanisms. From evolutions and mutations, newly well adapted parts of information arise. We have been attracted by this principle. Artificial Life uses concepts of GTYPE and PTYPE (by analogy to *genotype* and *phenotype* of biology). They evolve by unceasingly interacting through *development* and *emergence* processes. Genetic Algorithms [3][5] inspire us in their mechanical operating and used operators.

3.2 Concepts

1. Basic concepts: population, Instance-PTYPE and Class-GTYPE:

- *Population and Genetic patrimony:* a group of classes representing various abstractions of one and the same entity forms a *population* (like the population of members of a university). All the attributes constitute its *Genetic Patrimony*.
- *Instance-PTYPE:* instances are the phenotype and represent entities called upon to evolve.
- *Class-GTYPE[1]:* classes define instances features, their genetic code.

We present an example which will be taken again and unrolled all along the article :

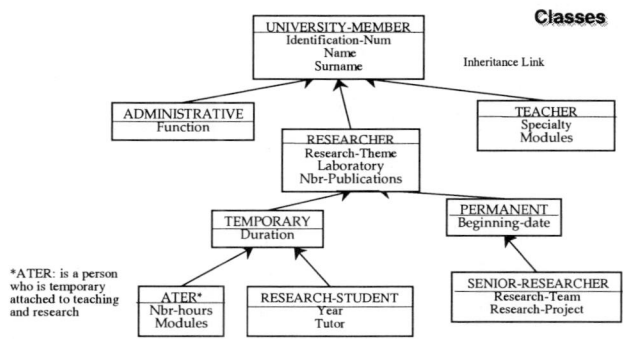

Fig. 3. Members of a university described at the class level

2. Advanced concepts: Fundamental, Inherited and Specific Genotypes: In a class, not every gene plays the same role or has the same prevalence. We consider that any class is entirely specified through three types of genotypes:
- *Fundamental Genotype or FG:* any object presents fundamental features, represented by particular genes representing the minimal semantics inherent to all classes of a same population.

---

[1] In order to simplify, we will use the classical term of class (respectively, instance) in place of Class-GTYPE (respectively, Instance-PTYPE).

☐ *Inherited Genotype or IG:* properties inherited by a class from its super-class constitute the *Inherited Genotype*.
☐ *Specific Genotype or SG:* it consists of properties locally defined within a class, specific to it.

3. **The scheme:** the scheme expresses in a simple and concise way, genes in the attributes and methods form. It has the same genetic structure as the represented entity. Each gene is represented by 0, 1 or #: 0 for absence of the gene, 1 for its presence and # for its indifference. The scheme is a simple and powerful means to model groups of individuals. We consider two kinds of schemes: the *permanent scheme*, associated with each specified class and having the same structure, and the *temporary scheme*, which is a selection unit of one or a group of entities (instances or classes).

The example's model presented in Fig.3 is described in Fig. 4 taking into account the aforesaid concepts.

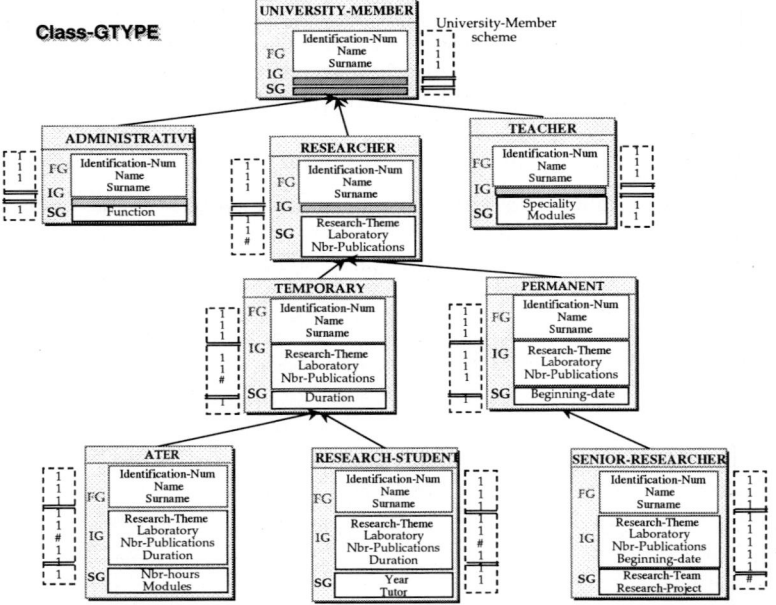

Fig. 4. the same example in our model

3.3 Evolutionary Processes

An evolutionary process is triggered when a change, envisaged or not, appears in the model. The process must be able to detect this change, find entities implicated in the evolution and reflect this change adequately:

1. **Phases:** we consider that an instance's evolutionary process is carried out in three phases: an *extraction* phase, an *exploration* phase, and finally an *exploitation* phase:

- *Extraction Phase:* extracts the object's genetic code within a temporary scheme.
- *Exploration phase:* explores all classes to locate adapted, even partially, ones. First it selects set of concerned populations, then it carries out the search in that set. Selection is the operator used thanks to the calculation of the adaptation values Avs (section 2).
- *Exploitation phase:* manages the impacts by development or emergence way. The *development process* represents the impact of class evolution on instances, while the *emergence processes* concern any emergence of new conceptual information, by way of impacts on classes. There are two possible outcomes: *local emergence* is related to the emergence of new information within existing class(es). The genetic code of the object has mutated and this can force mutation in its class; and the *global emergence* related to the emergence of a new conceptual entity.

2. Object operators: it is necessary to define basic operators to handle instances and classes. The two most important are those of selection and crossing-over:

- *Selection:* is defined to determine, after structural evolution of an instance, which class holds part or all of its specification.
- *Crossing-over:* works on two entities via their scheme to interchange their genes to define a new group of genes. It constitutes the core of the emergence process (took from [8]). It amounts to granting a weight relating to parents for genes transmission to children. We add to that a significant constraint: a permanent scheme presents at most two significant blocks (after FG): IG's genes and SG' genes. When processing the crossing-over, these blocks must be respected. It is the constraint of *bocks of genes*. The crossing-over is guided by the block constraints in order to ensure a minimal coherence for emergent schemes.
- *Adaptation value Av:* calculate the semantic distance between the evolved object and classes. Denoting the evolved object's scheme by Sch_{obj} and the close class's scheme by Sch_{param}, the adaptive function is defined, using the operator $\wedge$ (and_logic): $Av(Sch_{param}) = \sum_{(i=1 \to n)} \{Sch_{obj}[i] \wedge Sch_{param}[i]\}/n$, where n is number of genes specified in the evolved object; i is the variable index from 1 to n, defining, at each stage, the position of two respective genes of the analyzed schemes.
- *Semantic Distance sd :* is the value which expresses the semantic proximity between an emergent scheme and one of its ancestor. It helps to choose the super-class of the new abstraction. We use the same adaptive function defined for the calculation of Avs.

3. Examples: we consider following instances which not only evolve in their first structure, but they also introduce new attributes to become:

Genetic Patrimony		Instances become				
		O_1	O_2	O_3	O_4	O_5
FG	Identification-Num (1)	#3	#8	#4	#7	#1
	Name (2)	N1	N2	N3	N4	N5
	Surname (3)	P1	P2	P3	P4	P5
	Research-theme (4)	Automatic	-Mathematics	Object	Constraints	Object
	Laboratory (5)	L1	- L2	L3	L1	L5
	Nbr-Publications (6)		- 1	4	6	
	Specialty (7)					
	Modules (8)	+ Segmentation				DSI
	duration (9)				1	- 1
	Beginning-date (10)	01-10-2000		01-10-2000		01-09-2000
	Nbr-hours (11)					- 96
	Year (12)	- 1st	- 2nd	- 3rd		
	Tutor (13)	- Dupont	- Durant	- Dupond		
	Research-Team (14)	+ Vision		+ Object		
	Project-Team (15)					
	Rank	++ Professor				
	Position		++ Engineer			
	Responsibility			++ Supervisor		
	Responsible					++ Mr. X

attribute : existing attribute   -attribute Lost attribute   +attribute : gained attribute   ++attribute : new attribute

☐ Exploration phase: we calculate the Adaptation Value Av for each temporary scheme with each existing class:

Entity	Scheme (IG and SG)												AvO$_1$	AvO$_2$	AvO$_3$	AvO$_4$	AvO$_5$
Object O$_1$	1	1	#	#	1	0	1	0	0	0	1	#					
Object O$_2$	0	0	0	0	0	0	0	0	0	0	0	0					
Object O$_3$	1	1	1	0	0	0	1	#	0	0	1	#					
Object O$_4$	1	1	1	#	0	1	0	0	#	0	#	#					
Object O$_5$	1	1	#	#	1	0	1	0	0	0	#	#					
University-Member	0	0	0	0	0	0	0	0	0	0	0	0	0/5	0	0/5	0/4	0/4
Researcher	1	1	#	0	0	0	0	0	0	0	0	0	2/5	0	3/5	3/4	2/4
Teacher	0	0	0	1	1	0	0	0	0	0	0	0	**1/5**	0	0/5	0/4	**1/4**
Temporary	1	1	#	0	0	1	0	0	0	0	0	0	2/5	0	4/5	**4/4**	2/4
Permanent	1	1	1	0	0	0	1	0	0	0	0	0	3/5	0	4/5	3/4	3/4
Ater	1	1	#	0	1	1	0	1	0	0	0	0	**3/5**	0	3/5	**4/4**	**3/4**
Research-Student	1	1	#	0	0	1	0	0	1	1	0	0	**2/5**	0	3/5	**4/4**	2/4
Senior-Researcher	1	1	1	0	0	0	1	0	0	0	1	#	**4/5**	0	5/5	3/4	3/4
with :	IG				SG												

Conclusions for each object: O_1 : partially adapted classes: Teacher, Ater, Research-Student and Senior-Researcher. They are candidates for crossing-over. O_2 : no adapted class. O_3 : one completely adapted class: Senior-Researcher. O_4 : three completely adapted classes: Temporary, Ater and Research-Student. O_5 : partially adapted classes: Teacher, Ater, Research-Student and Senior-Researcher. They are candidates for crossing-over

☐ Exploitation phase

☐ O_1: *crossing-over* on Teacher, Ater and Senior-Researcher' schemes. Avp is an adaptation value which is pondered with the other Avs.

N° Population	Crossing-over												
	O_1	1	1	#	#	1	0	1	0	0	0	1	#	
1 : Teacher Ater Sr-Researcher	Random bonds	0.34	0.59	0.42	0.70	0.15	0.60	0.50	0.43			0.10	0.90	Avp
	Teacher	0	0	0	1	1	0	0	0			0	0	1/4
	Ater	1	1	#	0	1	1	0	1			0	0	3/4
	E1	1	1	#	1	1	0	0	0			0	0	3/4
	E2	0	0	0	1	1	0	0	0			0	0	1/4
	O_1	1	1	#	#	1	0	1	0	0	0	1	#	
2 : E1 E2 Sr-Researcher	Random bonds	0.05	0.12	0.20	0.10	0.49	0.79	0.37	0.25			0.85	0.22	Avp
	E2	0	0	0	1	1	0	0	0			0	0	5/21
	S'Researcher	1	1	1	0	0	0	1	0			1	#	16/21
	E3	0	0	0	1	0	0	1	0			1	0	2/5
	E4	1	1	1	0	1	0	1	0			1	#	5/5
	O_1	1	1	#	#	1	0	1	0	0	0	1	#	
3 : E1 E3 E4	Random bonds	0.34	0.19	0.02	0.70	0.15	0.60	0.50	0.33			0.80	0.90	Avp
	E1	1	1	#	1	1	0	0	0			0	0	8/23
	E3	0	0	0	1	0	0	1	0			1	0	15/23
	E5	1	1	#	1	1	0	1	0			1	0	5/5
	E6	1	1	1	1	0	0	0	0			0	0	2/5

Scheme	Selected scheme
0	Absent gene from all schemes are ignored in the crossing-over since it has no significance

Crossing-over ends because we have two emergent schemes: E4 and E5.

- O_2: *global emergence by direct creation of the abstraction - place of insertion?* As a sub-class of University-Member because O_2 has only the FG which is common with all other classes.

- O_3: *local emergence* of the new attribute 'Responsibility' in Senior-researcher class because it is the unique completely adapted class.

- O_4: *global emergence by direct creation of the abstraction - place of insertion?* Three Avs are equal to 1. Among the three well-adapted classes, Temporary is the most eligible because it's the representative abstraction of the temporary subpopulation. But as it's an abstract class, O_4 represents another abstraction of temporary researchers. So, it provokes emergence of a new sub-class of temporary.

- O_5: *crossing-over* on the same population as O_1. Crossing-over steps are the same. The final step stops with the emergent scheme E5.

3.4 Discussion on emergence

After the structural evolution of O_1, ..., O_2, the emergent processes permit to detect the kind of emergence, the abstractions concerned and also the location of changes and insertion of new classes (there are more details in [9]). But sometimes, we can just conclude that the emergent scheme is an abstraction of the permanent sub-population, but not it's precise location in this sub-population. We propose to enrich the emergence process in order to control it and to choose in a better way by using metrics.

4. Model of metrics

We apply the GQM technique [1] in order to determine where metrics are useful in the emergence process. We take our inspiration from class context metrics [2]. As we want to analyze any emergent abstraction in an internal and external way, we identify two contexts for metrics:

Question	Sub-questions	Metric	Comments
Intra-Abstraction Context	the most *significant*	S=difference between scheme and instance	nearest scheme to instance
	the less *contradictory* ?	C_t= number of contradictions between attributes and attribute blocs	weaker =less contradictions
	the most *coherent*?	C_h =Σsimple attributes +Σ attribute contradictions + Σ blocs + Σ blocs contradiction	weaker =less incoherence
Inter-Abstraction Context	*parents* ?	P_s = detection of super-class(es)	Sd
	Contradiction with parents?	C_d=Σcontradictory attributes+Σcontradictory blocs	weaker=less contradiction
	Coupling	C_c = Σreferences towards and from other classes	weaker=minimal coupling
	Depth in hierarchy?	P_d = position inside the hierarchy	weaker=lessreorganization

Since we apply these metrics, we could better choose and apply emergence results.

5. Conclusion

The first objective of our research work is to allow a designer to attempt to apprehend and to anticipate the future changes and requirements of complex and bulky OO applications. This is possible by simulating several evolution ways by expressing new requirements on instances. We have seen in fact that several ways of class evolution can emerge from structural instance evolution. In this paper, we propose metrics in order to analyze and control what emerge, how it can change class specifications and the possible impacts. We propose two kinds of metrics: *intra-abstraction context metrics* and *inter-abstraction context metrics*. Metrics are not systematically applied if the designer precise invariants to respect during evolution and emergence. All this is done in order to offer a simulation tool of application evolution to help the designer for evolution and maintenance of complex applications.

Bibliography

[1] Basili, V.R., Weiss, D.,"A methodology for collecting valid software engineering data", IEEE TSE, Nov.1984 p728-738.
[2] Chidamber S.R Kemerer C.F"A Metrics Suite for OOD"IEEE TSE vol.20(6) 94, p.476-493.
[3] Goldberg D.E. "Algorithmes Génétiques", Edition Addison-Wesley, 1994
[4] Heudin J.C. "La Vie Artificielle", Edition Hermès, 1994.
[5] Holland J "Adaptation in Natural and Artificial Systems", University of Michigan Press, 75.
[6] Kim W. "Introduction to OODB", MIT Press, Cambridge Massachussetts, 1990.
[7] Napoli A. "Représentation à objets et raisonnement par classification en I.A", Thèse 1992.
[8] Syswerda G. "Uniform Crossover in Genetic Algorithms", ICGA, 1989, p.2-9.
[9] Tamzalit Oussalah"From Object Evolution to Object Emergence ACM CIKM'99 p.514-521
[10] Wei Li, J. Talburt "Empirically Analyzing OOsoftware Evolution" JOOP Sept 98, p.15-19

Using Intelligent Systems in Predictions of the Bacterial Causative Agent of an Infection

Diana R. Cundell[1], Randy S. Silibovsky[2], Robyn Sanders[2], and Les M. Sztandera[1]

[1] School of Science and Health, Philadelphia University, Philadelphia, PA 19144, USA
[2] Department of Infectious Diseases, Albert Einstein Medical Center, Philadelphia, PA 19104, USA

Abstract. In this study, we designed a fuzzy logic system to examine the influence of the demographic variables of age, blood type, gender and race on bacterial infection rates using a medical database assembled over 17 months from patients presenting to Albert Einstein Medical Center. The intelligent system was created using 155 patients, randomly selected from the database, and consisted of four input categories of demographic variables and four output categories of bacterial infection ("streptococci", "staphylococci", *"Escherichia coli"* and "non-*E. coli* gram negative rods"). The remaining 32 patients were used to assess the program's ability to correctly determine bacterial infection when provided only with demographic data. Our intelligent system correctly assigned the bacterial output group in 27 of these 32 patients, giving an overall correlation of 84.4%. These studies suggest that demographic variables are major factors influencing bacterial infection. Such a system may, therefore, hold promise as a diagnostic tool.

1 Introduction

The ability of physicians to diagnose bacterial infections is currently dependent on the use of a series of developed algorithms in which the most likely etiological agent is determined based on the patient's symptoms, previous history and predisposing physiological factors. Improvement of this diagnostic tool would involve identifying additional variables, which might serve as risk factors for infection by a particular bacterial species or genus.

Previous studies have indicated that the demographic variables of age and blood type might act as predisposing agents in bacterial infection [1-3]. Indeed, advanced age has been shown to be a risk factor for pneumococcal infection [1] and expression of blood types A or AB appears to predispose individuals to tuberculosis [2] or cholera [3] infection. In the case of tuberculosis infection blood type expression may even be a major risk factor; a study of the Innuit showed that the infection was three times more common in individuals of blood types A and AB than any other group [2]. Using traditional statistical methods, these studies have been able to show a putative role for individual demographic variables, as risk factors in bacterial infections but shed no further light as to how these variables might be involved in the course of bacterial disease. The development of the fields of artificial intelligence (fuzzy logic) and genetic algorithms now allow the creation of computer programs in which complex associations between several variables can be "learned" and used predict the outcome of given situations [4,5]. Fuzzy logic programming has proven to be of particular use for the development of models of biological and medical systems since these frequently include "shades of gray or maybe" interactions between several variables which can not be efficiently analyzed using traditional computerized statistical methods [4,5].

The association between demographic factors and bacterial infections represents a highly suitable system to model using fuzzy logic, as all variables are definable using well-established parameters. A prospective investigation was therefore undertaken to examine the association between blood type, age, gender and race and bacterial infection rates, using a medical database obtained over a 17-month period from 187 patients presenting to the wards of Albert Einstein Medical Center. To investigate how closely the variables of blood type, age, gender and race were associated with bacterial infection, the fuzzy logic program generated from these patients' data was tested for the ability to correctly ascribe bacterial infection when given only the demographic data of 32 randomly selected patients.

2 Methods

2.1 Data Collection

Collection of data was governed under the rules of a collaborative, expedited IRB agreement (HN-2092) made between Philadelphia University and Albert Einstein Medical Center. Each patient provided written consent, permitting confidential use of their data, and was admitted into the study if their infection resulted from a single, identifiable pathogen. All bacteriology data was obtained courtesy of the Clinical Laboratories of Albert Einstein Medical Center. Patients not included in the study included those with underlying disease predisposing to infection, pregnancy, mental disability or minor patients.

2.2 Medical Database and Fuzzy System

The data set investigated consisted of a real medical database comprising 187 patients. Patient data was randomly assigned into two categories; training data (155 patients) and test data (32 patients; 8 patients within each of the four bacterial output groups). The intelligent system was modeled using the 155 patient training data using four input classes (demographic variables of age, blood type, gender and race) and four output classes (bacterial infections with the species "staphylococci" (*S. aureus* and *S. epidermidis*), "streptococci" (*S. pneumoniae,* and groups B and D streptococci), "*Escherichia coli*" and "non-*E. coli* gram negative rods" (species of *Klebsiella, Serratia, Bacteroides, Morganella, Prevotella, Pseudomonas* and *Proteus*). The four output and input spaces were divided into several fuzzy subsets and assigned linguistic terms. Decision surfaces for two inputs; age and blood type and the output bacterial classes are shown in Figure 1. Each region was then assigned a fuzzy membership function. A triangular shape was selected with height 1 at the center of the region and 50% overlap between neighboring sets (for the input parameters). Fuzzy sets for the output parameters are shown in Figure 2. Fuzzy rules were generated, using "IF... AND....THEN", where "IF...AND" were generated from the input parameters and "THEN" from the output parameters. The system generated 173 rules, 159 of which received the highest count and were retained. A fuzzy interference engine was executed and mapping made based on the 159 remaining rules using correlation product interference. Defuzzification of the data was based on the center of gravity method, which was sensitive to all remaining rules. The 32 patients constituting the "test" set, in that they had been previously unseen by the system, consisted of 8 patients clinically defined as belonging to each of the four created output groups.

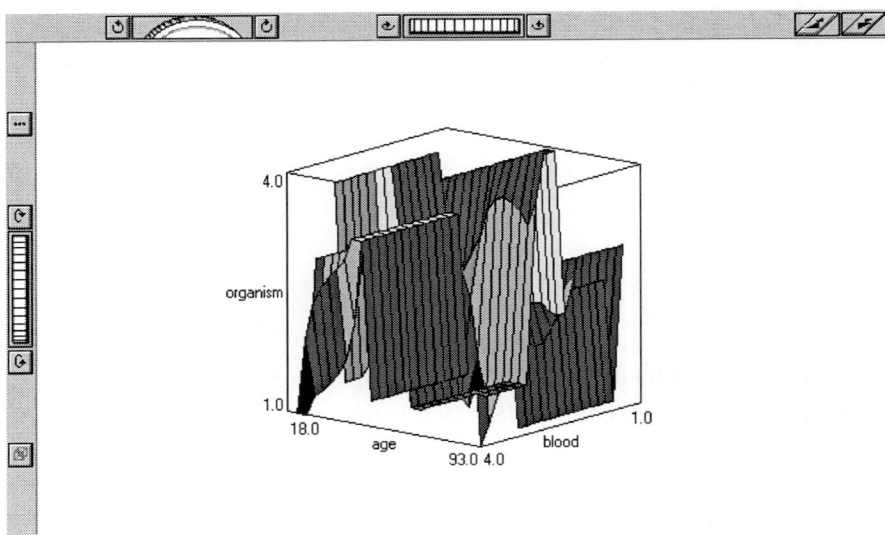

Fig. 1 – Decision surfaces for two inputs (age and blood type) and the output (bacteria) demonstrating correlation between blood type/ patient's age and the type of bacteria

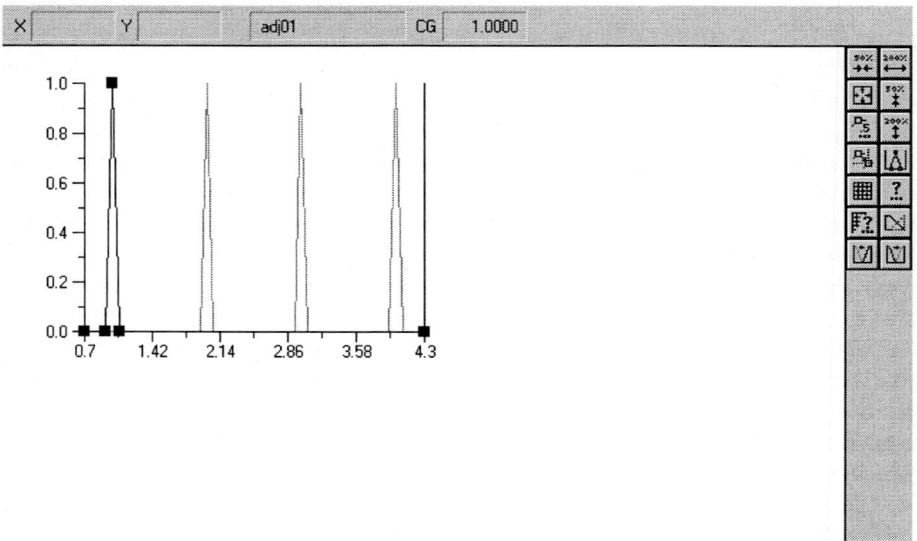

Fig. 2 Fuzzy sets representing output bacterial classes. The four output bacterial classes of staphylococci, streptococci, *E. coli* and non-*E. coli* gram negative rods were each described, in the order shown, using a triangular shape with height 1 at the center of the region and no overlap between classes

3 Results

3.1 Age, Blood Type, Gender and Race Distribution for the Four Output Classes of Bacterial Infection within the Patient Database

The demographic and bacterial infection variables of the patients in the medical database generated for this study are shown in Table 1. Of the 187 patients, 64 were infected by staphylococcal species (34% of total infections). Patients infected with staphylococcal species *per se* demonstrated more than a two-fold increase in the frequency of blood type B (22%) in comparison with that normally observed in the general population (10%) with a decrease in the frequency of type A (31% compared

Table 1. Age, blood type, gender and race distribution for the four output classes of bacterial infection Frequency in each output class (%)

Input	Staphylococci (n = 64)	Streptococci (n = 40)	E.coli (n= 28)	Non-E. coli GNR (n= 55)	Total (n = 187)
Age					
20-40	6 (9%)	5 (12%)	3 (11%)	8 (14%)	22 (12%)
41-60	15 (23%)	18 (45%)	4 (14%)	13 (24%)	50 (27%)
61-80	33 (52%)	10 (25%)	16 (57%)	23 (42%)	82 (44%)
81-100	10 (16%)	7 (18%)	5 (18%)	11 (20%)	33 (17%)
Blood type					
A	20 (31%)	15 (38%)	9 (32%)	17 (31%)	61 (33%)
AB	2 (3%)	4 (10%)	1 (4%)	3 (5%)	10 (5%)
B	14 (22%)	9 (22%)	2 (8%)	5 (9%)	30 (16%)
O	28 (44%)	12 (30%)	16 (56%)	30 (55%)	86 (46%)
Gender					
Male	38 (59%)	27 (68%)	12 (43%)	31 (56%)	108 (58%)
Female	26 (41%)	13 (37%)	16 (57%)	24 (44%)	79 (42%)
Race					
African-American	34 (53%)	23 (58%)	14 (50%)	35 (64%)	106 (57%)
Asian	2 (3%)	1 (2%)	2 (7%)	0	5 (3%)
Caucasian	26 (41%)	13 (33%)	11 (39%)	17 (31%)	67 (36%)
Hispanic	2 (3%)	3 (7%)	1 (4%)	3 (5%)	9 (4%)

Groups of bacterial species defining output classes used in the Table were defined according to standardized microbiological classifications [6-8]
to a normal frequency of 40%). Frequencies of types AB and O were similar to the general population.

The 40 patients who were infected by streptococcal species demonstrated more than a two-fold increase in the frequency of both blood types AB (10%) and B (22%) in comparison with the general healthy population (5% and 10%, respectively) with a decrease in the frequency of type O (30% compared with 45%). The frequency of type A was similar to that expected in the general population at large.

E. coli infections accounted for 28 of the 187 patients in the study (15% of total infections). There was an increase in the frequency of type O in these patients of 10% compared with the expected frequency from the general population and a commensurate decrease in frequency of type A.

Infections with non-*E. coli* GNR constituted 55 of the 187 patient admissions (29%) and involved several species of bacteria: *Klebsiella pneumoniae* (n=14), *Proteus mirabilis* (n=11), *Bacteroides* species (n=10), *Serratia* species (n=7), *Enterobacter cloacae* (n=4), *Pseudomonas* species (n=4), *Morganella morgani* (n=3) and *Prevotella* species (n=2). Interestingly, the blood group distribution for this broad-spectrum category was similar as that for *E. coli*.

In all output groups except one, the gender distribution was similar to that of the general patient population with slightly more males than females (58% compared with 42% in each group). Interestingly, in the *E. coli* output group, the reverse was true with only one-third of those infected being male and two-thirds female. Three groups also showed a similar age distribution (staphylococci, *E. coli* and non-*E. coli* GNR), with the majority of patients being over 60 (68, 75 and 62%, respectively). In contrast, the streptococcal group was somewhat younger with 57% of patients being 60 or less. Race distribution was found to be fairly homogeneous across the four output groups.

3.2 Intelligent System as a Predictor of Bacterial Output Class

The novel fuzzy logic system generated from 155 of the 187 patient database was also tested for its ability to correctly determine the bacterial infectious agent of the remaining 32 patients when provided solely with demographic data. The program was able to correctly assign all patients with streptococcal infections (8/8), 7 out of 8 patients with *E. coli* infections and 6 out of 8 patients with either staphylococcal or non-*E. coli* gram negative rod infections (non-*E. coli* GNR) to their output groups. This gave an overall prediction rate for the patient sample of 27/32 or 84.38%.

The 5 patients who were incorrectly classified by the program were placed into the following output groups: 2 patients with non-*E. coli* GNR were assigned to the *E. coli* category, 2 patients with staphylococcal infections and 1 patient with *E. coli* infection were assigned to the streptococcal category. This system is currently under development and, as such, no clinical evaluation of its efficacy at diagnosing bacterial infection in the absence of standard clinical algorithms has yet been performed.

4 Discussion

These data suggest that, with training, bacterial infection may be predicted with relative efficiency by inputting a patient's data for blood type, gender, age and race into a fuzzy logic program. There have been previous studies in the literature, which have suggested a putative correlation between demographic variables and bacterial infection [1-3], but it is not until now, with the advent of this powerful analytical tool that the dynamics between these factors can be interpreted.

By designing a fuzzy logic system, using a real medical database, we were able to correctly diagnose 27 patients from the 32 patient test group using only the demographic variables of age, blood type, gender and race, for four separate groups of infectious agents, namely staphylococci, streptococci, *Escherichia coli* and non-*E. coli* gram negative rods. Of these four output groups, the program was able to correctly assign all patients with streptococcal infections (8/8), 7 out of 8 patients with *E. coli* infections and 6 out of 8 patients with either staphylococcal or non-*E. coli* gram negative rod infections (non-*E. coli* GNR) to their output groups. This gave an overall prediction rate for the patient sample of 27/32 or 84.38%. The 5 patients who were incorrectly classified by the program were placed into the following output groups: 2 patients with non-*E. coli* GNR were assigned to the *E. coli* category , 2 patients with staphylococcal infections and 1 patient with *E. coli* infection was assigned to the streptococcal category.

This finding is particularly impressive since staphylococci and streptococci (output groups 1 and 2) have a number of features in common [9,10]. Both organisms inhabit the same microbial niches, being chiefly residents of the skin and mucous membranes, have a similar structure to their outer cell wall (gram-positive), produce closely similar virulence factors for invasion and produce similar spectra of disease [9,10]. As a result, they are often difficult for clinicians to distinguish based on clinical algorithms alone and in the absence of microbiological laboratory data. Interestingly the remaining two output groups, namely the *E. coli* and the non-*E. coli* GNR categories, also share a number of features. Both *E. coli* and non-*E. coli* GNR bacteria inhabit similar microbial niches, being chiefly residents of the gastrointestinal tracts of humans and animals, are gram negative organisms and produce similar spectra of disease[11]. The finding that *E. coli* was incorrectly assigned in two individuals is therefore not a surprising one since as well as sharing the habitats of these opportunists, *E. coli* frequently behaves as one in individuals whose immune systems are compromised due to surgery, tracheostomy, catheterization or renal dialysis [11].

Examination of the distribution of the demographic variables across the four output groups (Table 1) demonstrated that there were subtle differences in age, blood type and gender between the four output groups, which had clearly allowed the intelligent system to differentiate between them. Of these variables, blood type distribution was the most significantly different within the patient population and between the four output groups (Table 1). In the general, healthy human population, regardless of racial origin, approximately 40% of individuals will be of blood type A, 5% of type AB, 10% of type B and 45% of type O [12]. In contrast, the hospitalized population demonstrated a decrease in expected frequency of blood type A (40% to 33%) and an increase in frequency of blood type B (10 to 16%), with types AB and O having values as expected (Table 1).

Differential distributions of blood type frequency were also observed when the output classes of staphylococcal and streptococcal infection were compared (Table 1). These infections constituted the majority of cases in the medical database (104/187; 56%) and thus would be expected to provide the most permutations in terms of demographic variable combinations. In addition, the groups were both microbiologically similar, since both are gram-positive organisms, with similar demographic variable distributions of gender and race. Only one variable, apart from blood type, was different between the two classes; the streptococcal-infected patients represented a somewhat younger group being mostly below the age of 60 (Table 1). Both groups demonstrated an increased frequency of blood type B (22%) above that expected for both the patient population (16%) and the general healthy population (10%) and a decrease in frequency of blood type A commensurate with the general population (33%; Table 1). Staphylococci-infected patients had similar frequencies of blood types AB (3%) and O (44%) as both the general patient and normal healthy populations. In contrast, patients with streptococcal infections demonstrated an increased frequency of individuals with type AB (10%) and a decrease in the frequency of type O (30%; Table 1)

Patients with *E. coli* and non-*E. coli* GNR infections were also found to have a differential blood type distribution from the general patient population (Table 1). The organisms from these two output groups shared microbiological features in common but differed demographically since the *E. coli* group were significantly older than the general patient population distribution (71% over 60 years) and were predominantly female (67%) (Table 1). In spite of differences in other demographic variables, both groups showed a similar blood type distribution with an increased frequency of blood type O, when compared with both the general healthy population and the patient population as a whole (53-55% compared with 45%; Table 1).

Bacterial infection is a highly selective and dynamic process in which the host is targeted based on a complex interplay of many factors, which are only now gradually beginning to be understood. The results of this study suggest that the demographic variables of blood type, gender, age and race may be involved in bacterial host selection, and that such differential targeting may allow us to use these variables as a predictor of disease. In this study, the size of our population did not allow us to establish which variables were the most strongly associated with the bacterial infectious agent. An increased patient bank of data, which is currently being generated courtesy of an Einstein Society Award from Albert Einstein Medical Center, would allow further tuning and development of the program to eliminate any variables, which prove to be redundant. More patient data will also allow the subdivision of the current bacterial infection categories to single species, making the program more specific. Indeed, it is anticipated that the combination of currently used clinical algorithms with a user-friendly, simplified version of the current program might allow its eventual use by all physicians to make more accurate initial predictions of the bacterial causative agent of an infection.

References

1. Nester, E.W., Roberts C.E., Nester M.T.: Interactions between humans and microorganisms. In: Nester, E.W., Roberts C.E., Nester M.T (eds.): Microbiology, A human perspective. First edn. Wm. C. Brown Publishers, Dubuque (1995) 353-354.
2. Overfield T., Klauber M.: Prevention of tuberculosis in Eskimos. Human Biology, 52 (1980) 87-91.
3. Minkoff C., Baker P.J.: Variation among human populations. In: Schank D, Connell T. (eds.): Biology Today an Issues Approach. First edn. McGraw-Hill Publishers, New York (1996) 146-158.

4. Sztandera L.M., Goodenday L.S., Cios K.J.: A Neuro-Fuzzy Algorithm for Diagnosis of Coronary Artery Stenosis. Computers in Biology and Medicine Journal, 26 (1996) 97-107.
5. Kosko, B.: Fuzzy logic and engineering. In: Kosko, B. (ed.) Fuzzy Engineering. First edn. Prentice Hall, Upper Saddle River (1997) 3-37.
6. Holt, J.G. Krieg, N.R., Sneath, P.H.A., Staley, J.T.,Williams, S.T.: Group 17: Gram Positive Cocci. In: Holt, J.G., Krieg N.R., Sneath P.H.A., Staley, J.T., Williams, S.T. (eds.): Ninth edn. Bergey's Manual of Determinative Bacteriology, Williams and Wilkins, Baltimore (1994) 528.
7. Gilchrist, M.J.: Enterobacteriacae: Opportunistic pathogens and other genera. In: Murray, P.R., Baron, E.J., Pfaller, M.A., Tenover, F.C., Yolken R.H. (eds.): Sixth edn. Manual of Clinical Microbiology, ASM Press, Washington, (1995) 457-464.
8. Jousimies-Somer, H.R., Summanen, P.H., Finegold S.M.: *Bacteroides, Porphyromonas, Prevotella, Fusobacterium* and other anaerobic gram-negative bacteria. In: P.R. Murray, E.J. Baron, M.A. Pfaller, F.C. Tenover, R.H. Yolken (eds.): Sixth edn. Manual of Clinical Microbiology, ASM Press, Washington (1995) 603-620.
9. Kloos, W.E., Bannerman, T.L.: Staphylococcus and Micrococcus. In: Murray, P.R., Baron, E.J., Pfaller, M.A., Tenover, F.C., Yolken R.H. (eds.): Sixth edn. Manual of Clinical Microbiology, ASM Press, Washington, (1995) 284-291.
10. Ruoff, K.L.: Streptococcus. In: Murray, P.R., Baron, E.J., Pfaller, M.A., Tenover, F.C., Yolken R.H. (eds.): Sixth edn. Manual of Clinical Microbiology, ASM Press, Washington, (1995) 299-302.
11. Talaro, K., Talaro, A.: The Gram-Negative Bacilli of Medical Importance. In: Talaro, K, Talaro, A. (eds.): Fourth Edn. Foundations of Microbiology, W.C. Brown Publishers, Boston (1999) 630-640.
12. Audesirk T., Audesirk G.: Patterns of Inheritance. In: Minkoff E.C., Baker P.J. (eds): First edn. Biology, Life on Earth, Prentice Hall Publishers, Upper Saddle River (1999) 213.

An Intelligent Lessons Learned Process

Rosina Weber[1], David W. Aha[2], Hector Muñoz-Ávila[3], and Leonard A. Breslow[2]

[1]Department of Computer Science, University of Wyoming, Laramie, WY 82071-3682
[2]Navy Center for Applied Research in Artificial Intelligence,
Naval Research Laboratory (Code 5515), Washington, DC 20375
surname@aic.nrl.navy.mil
[3]Department of Computer Science, University of Maryland, College Park, MD 20742-3255
surname@cs.umd.edu

Abstract. A learned lesson, in the context of a pre-defined organizational process, summarizes an experience that should be used to modify that process, under the conditions for which that lesson applies. To promote lesson reuse, many organizations employ *lessons learned processes*, which define how to collect, validate, store, and disseminate lessons among their personnel, typically by using a standalone retrieval tool. However, these processes are problematic: they do not address lesson reuse effectively. We demonstrate how reuse can be facilitated through a representation that highlights reuse conditions (and other features) in the context of lessons learned systems embedded in targeted decision-making processes. We describe a case-based reasoning implementation of this concept for a decision support tool and detail an example.

1 Lessons Learned Process

Lessons learned (LL) processes (Weber et al., 2000b) are knowledge management (KM) solutions for sharing and reusing knowledge gained through experience (i.e., *lessons*) among an organization's members. LL systems are motivated by the need to preserve an organization's knowledge and convert individual knowledge into organizational knowledge so that, when experts become unavailable; other employees who encounter conditions that closely match some lesson's context may benefit from applying it. Therefore, a lesson learned is a validated working experience that, when applied, can positively impact an organization's processes. While some organizations can quickly update the processes targeted by lessons, thus eliminating the need for a repository of lessons, other organizations (e.g., the US military, the Department of Energy) do not have this luxury (i.e., they cannot easily update their processes), which necessitates using LL systems to explicitly store and retrieve lessons.

LL systems are ubiquitous; we easily located[1] over 40 of them on the WWW, are aware that many others are used in private industry, and discovered that they rarely succeed in promoting knowledge reuse/sharing for two reasons (Weber et al., 2000b). First, the selected representations of lessons typically are not designed to facilitate reuse, either because they do not clearly identify the process to which the lesson applies, its contribution to that process, or its pre-conditions for application. Second,

---

[1] Our compiled findings are posted at www.aic.nrl.navy.mil/~aha/lessons.

these systems are usually not integrated into an organization's decision-making process, which is the primary requirement for any solution to successfully contribute to KM activities (Reimer, 1998; Leake et al., 1999; Aha, 1999).

KM solutions usually involve both organizational dynamics and technological components. We propose a *technological* solution to designing LL systems that includes a lesson representation chosen to potentiate knowledge sharing in an embedded system in which lessons are proactively brought to the attention of users. In the remainder of this paper we summarize research on LL systems, introduce a representation for lessons that promotes knowledge sharing, discuss the lessons learned process, describe the design of an *active lessons delivery* system, and detail an example of its use as a module in HICAP[2] (Muñoz-Avila et al., 1999), a decision support tool for interactive plan authoring.

2 Related Work

Although dozens of lessons learned centers and their respective systems exist, few researchers have addressed LL systems, and almost none in artificial intelligence.[3] This is somewhat surprising, given that their developers and users overwhelmingly agree that current LL systems are insufficient. That is, there are several unanswered research issues regarding intelligent LL systems that need to be addressed.

Several KM publications have reported on issues related to lessons learned systems (van Heijst et al., 1996; O'Leary, 1998; Secchi, 1999; Habbel et al. 1999, SELLS, 1999). However, few of these discussed topics related to *intelligent systems* (e.g., van Heijst et al. (1997) stress the relationship between case-based reasoning (CBR) and LL systems). The only *deployed* application that uses CBR technology is NASA's RECALL system (Sary & Mackey, 1995), although three research groups have recently *proposed* CBR approaches that promote knowledge sharing.

First, the Air Campaign Planning Advisor (ACPA) (Johnson et al., 2000) disseminates videotaped stories (e.g., best practices) in a planning environment. However, ACPA does not reason on the stories, nor highlight reuse components or conditions. Thus, the user must decide whether or not to apply the memory captured in the story according to their interpretation of it.

Second, CALVIN (Leake et al., 2000) captures lessons concerning which online information resources should be searched for a given research topic. The subject and research results are used to index lessons so that when a user starts a search, previously stored results are proactively brought to the user's attention. Unlike most LL systems, CALVIN is task-specific rather than organization-specific.

Finally, we propose the Active Lessons Delivery System (ALDS), whose implementation in HICAP is discussed and exemplified in this paper. Users can interact with HICAP to author plans by iteratively decomposing complex tasks into primitive actions. ALDS monitors changes in the plan and plan state (i.e., described by a set of <question, answer> pairs), and triggers a lesson when its applicable task

---

[2] For more information and demonstrations of HICAP and ALDS, both developed in Java 1.2, please see http://www.aic.nrl.navy.mil/hicap.

[3] This motivated us to organize the AAAI'00 Intelligent Lessons Learned Workshop, whose homepage is www.aic.nrl.navy.mil/AAAI00-ILLS-Workshop.

matches a task in the (evolving) plan and its conditions closely match the plan state. ALDS differs from the previous two embedded architectures in that (1) it focuses specifically on organizational lessons in the context of planning tasks, (2) it automatically determines a triggered lesson's interpretation for the evolving plan, and (3) it allows users to automatically implement a lesson by pressing a button.

3 Lessons Learned Knowledge Representation

In our survey of LL systems (Weber et al., 2000b), we found that lessons are often represented inadequately, preventing them from being easily reused or understood. For example, recorded lessons often do not highlight the task for which they apply, or precisely specify their triggering conditions. Also, free text representations, which are used in all the deployed LL systems we have found, complicate reuse because this text has to be correctly interpreted to ensure proper lesson reuse.

A lesson is derived from an experience in which the *result* derived from applying an *originating action* yields significant new knowledge (i.e., a *contribution*), due to a success or failure, that can, and should, be taught to others. A lesson's *conditions for reuse* are the relevant state variables that existed when the originating action occurred. An ideal, validated lesson facilitates its dissemination by clearly stating its contribution and the *decision*, *task*, or *process*[4] for which, by applying its *recommended response action* (i.e., a *suggestion*), a user can reduce or eliminate the potential for failures or mishaps, or reinforce a positive result. In more detail, the features of a lesson that target improvements to planning tasks are:

Originating action: The action taken in the lesson's initiating experience.

Result: This indicates whether the experience was positive or negative, and helps to determine whether to recommend repeating or avoiding the same experience.

Lesson contribution: This is the crucial feature (e.g., a set of constraints) that characterizes the originating action and is responsible for the result of the original experience. The lesson's contribution is the element that should be repeated, in conjunction with the originating action, when the experience has a positive result, and it should be avoided when the result is negative.

Applicable task: This is a pre-defined task in an organization's targeted planning process. The lesson author must identify the task to which the lesson is applicable.

Conditions for reuse: These are the values of the state that, when matched closely, will cause a lesson to be reused. Knowledge for identifying and assessing similarity between conditions and state variables must be elicited from domain experts.

Suggestion: This is the recommended response action. It is entailed by a lesson's other features (i.e., a negative experience should be avoided) and provided by the lesson author.

---

[4] In decision-making systems, lessons are applicable to decisions. In planning, lessons are applicable to tasks.

We illustrate this representation with a lesson from the Joint Unified Lessons Learned System[5] concerning non-combatant evacuation operations (Section 5.2). This lesson refers to the step in which non-combatants had to be registered prior to evacuation in a disaster relief operation after the April 1991 eruption of Mt. Pinatubo in the Philippines. The lesson's summary is: *The evacuee registration process was very time consuming and contributed significantly to delays in throughput and to evacuee discomfort under tropical conditions.* Our representation for this lesson is as follows:

Originating action ➔ *Evacuee registration*
Action result ➔ *Delays, time consuming, and evacuee discomfort* ➔ *negative*
Contribution ➔ *Triple registration process is problematic*
Applicable task ➔ *Evacuee registration*
Conditions ➔ *Under tropical conditions*
Suggestion ➔ *Locate an INS (Immigration and Naturalization Service) screening station at the initial evacuation processing site. Evacuees are required to clear INS procedures prior to reporting to the evacuation processing center.*

This lesson refers to a negative outcome (e.g., evacuee discomfort). The expression "under tropical conditions" is a condition for reuse. In this lesson, the applicable task is the same as the originating action, although this is not true for all lessons. The lesson recommends an alternative method of registration that is not time consuming, which defines its suggestion.

Because a lesson may still be applicable even when its *conditions* are not perfectly matched by the state, reusing lessons using a CBR approach is appropriate. The similarity assessment between conditions and state variables is modeled using elicited expert knowledge. Adaptation (e.g., replacing *tropical condition*s with *winter conditions*) is not supported because the user must decide whether to apply the recommended suggestion. The only feature that can be inferred is the *suggestion*, from information embedded in the *originating action, lesson contribution,* and *result*.

In the implementation of ALDS in HICAP, the applicable planning task and conditions are used for indexing a lesson.. To improve retrieval and consequently improve reuse, an effective indexing should anticipate the end users' needs and indexing style (Kolodner, 1993). Therefore, a different indexing strategy is required to facilitate retrieval of lessons that target technical decision making. This indexing strategy should use an expert's model so that technicians can identify the model component targeted by the lesson (instead of identifying an applicable task) and other features (e.g., the problem, its causes, and the symptoms associated with that component).

4 Lessons Learned Process

In Section 1 we identified two problems with traditional lessons dissemination approaches: lesson representations that do not promote reuse and standalone retrieval tools. In Section 3 we proposed a representation that facilitates lesson reuse. This section focuses on embedding LL systems in their targeted processes.

An organization's lesson learned process typically involves the following tasks: *collecting, validating, storing, disseminating,* and *reuse*. For example, military

---

[5] https://www-secure.jwfc.acom.mil/protected/jcll.

organizations request their members, after completing a mission, to submit lessons to a *LL center*, where they are analyzed, indexed according to a task list specific to that branch of the armed services, validated, and stored in a repository. Lesson repositories are provided to military personnel, and are accessible on the secure military network SIPRNET and also on CD-ROMs. An accompanying search engine is used to submit queries in the hope of retrieving relevant lessons. Thus, LL centers are responsible for *collecting, validating, storing*, and *disseminating* lessons so potential users can *reuse* them. These five steps summarize the standard LL process, which varies slightly among LL centers.

Most systems for lesson retrieval are standalone and passive, and thus ill suited for promoting lesson dissemination and reuse because they require users to master a new process (i.e., search for relevant lessons in a separate standalone LL retrieval tool) that is independent of their problem-solving task. In fact, this process makes several unrealistic assumptions: it assumes that a user is reminded of the potential utility of a LL system whenever it may be useful, knows that the system exists, knows where to find it, has the time and the skills to use it, and can correctly interpret and reuse retrieved lessons.

We identified two desired characteristics of a LL process for facilitating knowledge sharing. First, it must deliver lesson knowledge during process execution (e.g., business, planning) to support decision-making. Second, it must be embedded in the process targeted by a lesson. An embedded LL system should *monitor* this process, *identify* changes in the plan state, *recognize* when a lesson is applicable to the current decision or task (i.e., when the conditions of the lesson and plan state match), and *proactively highlight* relevant lessons to the user (Figure 1). This process will allow a user to incorporate a relevant lesson's suggestion, which can potentially modify the user's decision-making. Thus, this *active delivery* process promotes embedding knowledge reuse into the decision-making process.

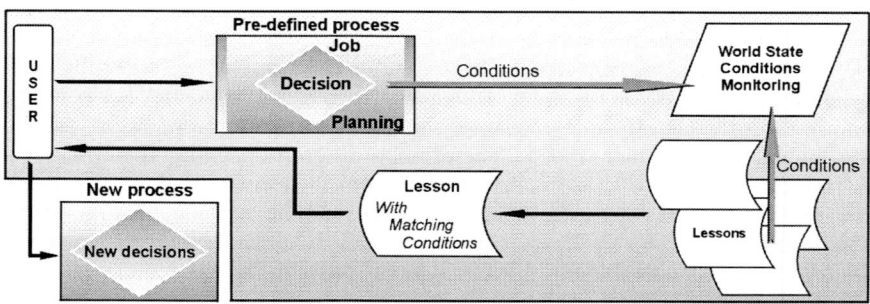

Fig. 1 Proposed lessons learned process.

These observations motivated us to design an active lessons delivery approach, to be embedded in a user's decision support tool. By automatically bringing relevant lessons to the user's attention, it promotes lesson reuse by reducing the burden on the user. In particular, this process can clarify how a lesson is relevant to the user's current decision-making task by reducing or eliminating problems of lesson interpretation and selection, does not require the user to consult a separate LL system, should increase the precision and recall of lesson retrieval, and should allow users to automatically incorporate a triggered lesson's suggestion into the evolving plan.

5 An Active Lessons Delivery System

An embedded active lessons delivery module monitors a decision-making process, bringing lessons to the user's attention when they become relevant. The primary constraint on the embedding decision support tool is that it represents and maintains information on this process that can be used to index appropriate lessons (e.g., lesson's task and triggering conditions). We illustrate this active lessons delivery approach for a military planning process in the context of HICAP. The following subsections introduce HICAP and detail an example that illustrates the use of ALDS.

5.1 The Decision Support Tool: HICAP

HICAP (<u>H</u>ierarchical <u>I</u>nteractive <u>C</u>ase-based <u>A</u>rchitecture for <u>P</u>lanning) (Breslow et al., 2000) helps users to formulate a hierarchical plan, which is represented as a tuple $P = \{T,R,A\}$. $T=\{T,<,\wedge\}$ is a *hierarchical task network* (HTN), where each task $t \in T$ is defined by its name t_n and duration t_d, the relation $<$ defines a (partial) temporal ordering on tasks, and $t \wedge t'$ means that t is a parent of t' in T. The leaves of T comprise the primitive actions to be included in the plan. R, which is also represented using an HTN, is the plan's set of resources. Finally, A is a set of assignments between the plan's tasks and resources. Also of interest is $S=\{<q,a>^+\}$, which denotes state information in the form of a set of <question,answer> pairs.

HICAP's modules include, among others, a *Hierarchical* Task Editor (HTE) that allows users to edit a plan, a conversational case retriever (NaCoDAE/HTN) that allows users to *interactively* select a stored decomposition to apply to a task in T, and a generative planner (JSHOP) that can be selected to *automatically* decompose tasks in T into subtasks. The plan state S is updated by direct user input, through user interactions with NaCoDAE/HTN, or by JSHOP.

5.2 The Task Domain: Noncombatant Evacuation Operations (NEOs)

We initially designed HICAP for deliberative NEO planning; no AI system has been deployed to assist military experts to plan NEOs. NEOs (DoD, 1994) are performed by the US military to assist in the evacuation of non-combatants, non-essential military personnel, and others (e.g., host nation citizens) whose lives are in danger (e.g., due to political insurgencies, volcanic eruptions) from an endangered location (e.g., a beleaguered US embassy) to an appropriate safe haven.

Each lesson in HICAP is indexed by its *applicable task* and *conditions*. For example, one such lesson for the NEO planning domain is:

Originating action ➔ *Assign conventional use of air wing*
Action result ➔ *Increases the risk to detection of clandestine SOF* ➔ *negative*
Contribution ➔ *Conventional (low visibility) air wing increases SOF risk*
Applicable task ➔ *Assign air wing*
Conditions ➔ *Q: Is it necessary to use covert SOF helicopters? A: Yes*
Suggestion ➔ *Assign high visibility to conventional air wing*

A lesson's *conditions* are represented as <question,answer> pairs so their similarity with state variables can be easily assessed. The user decides whether and how the lesson's suggestion will be implemented, as illustrated below.

5.3 Active Lessons Delivery Module: An Example

For this example, we use the fictitious *Terror in the Jungle* NEO scenario, obtained from the DISA *Adaptive Courses of Action* ACTD.[6] Some tasks in its task hierarchy can be further decomposed using interactive case retrieval. After the user selects a task to expand, NaCoDAE/HTN displays alternative expansions that could apply, along with questions that, if answered by the user, could help determine which case's conditions best matches *S*. The task being expanded here is *Rescue mission*, which concerns how to safely evacuate the evacuees.

After answering some questions and thus updating the state, the case retriever then displays the question *Is it necessary to use covert SOF helicopters?* The user answers *Yes*, yielding a perfect match with a task decomposition case that expands to the subtasks *Use ground support* and *Assign conventional use of air* wing.

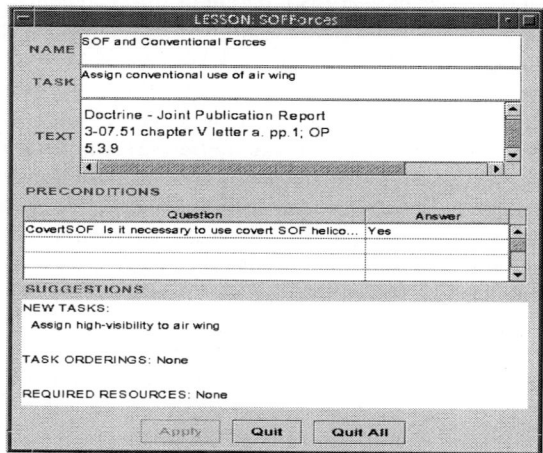

Fig. 2. A lesson pertaining to camouflaging special operations forces.

When expanding these tasks, ALDS recognizes that a lesson applies (i.e., the user had indicated the need to use Special Operations Forces (SOF) helicopters for the evacuation) and displays it (Figure 2). This lesson, which is applicable to the task *Assign conventional use of air wing*, suggests replacing this task with *Assign high visibility to air wing*. Figure 3 displays the resulting task hierarchy. The meaning of this lesson is that military protocol dictates that SOF forces should be made less conspicuous whenever they are deployed. In this example, a high-visibility air wing, composed of conventional forces, will more easily hide the SOF forces.

---

[6] http://www.les.disa.mil/insert/acoa/index.htm

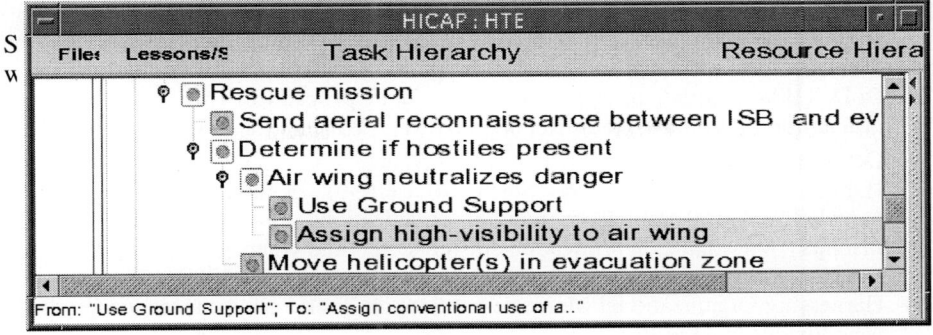

Fig. 3. A subset of the task hierarchy after applying the lesson shown in Figure 2.

6 Concluding Remarks and Future Work

In this paper we focused on the reuse of lessons learned. We identified two problems that interfere with lesson reuse: inadequate lesson representations (e.g., how different features should be highlighted to enable interpretation) and system architecture (i.e., how lessons learned systems should be embedded into the decision-making process). We then proposed an active lessons delivery approach (ALDS) to address these problems and exemplified its use in HICAP, a plan authoring tool.

We have not yet evaluated the utility of ALDS in NEO exercises, and instead developed a simple travel planning domain for evaluating the impact of ALDS (Weber et al., 2000a). In future work, we will examine how to use HICAP to guide interactive lesson elicitation, demonstrate the utility of active lessons delivery for other decision support tasks, and transition HICAP to the ACOA ACTD project.

Acknowledgements

This research was supported by grants from the Office of Naval Research, the Naval Research Laboratory, and the University of Wyoming.

References

1. Aha, D.W. (1999). The AAAI-99 KM/CBR workshop: Summary of contributions. In C. Gresse von Wagenheim & C. Tautz (Eds.) *Proceedings of the ICCBR-99 Workshop on Practical Case-Based Reasoning Strategies for Building and Maintaining Corporate Memories.* Munich: Unpublished.

2. Breslow, L.A., Muñoz-Avila, H., Aha, D.W., Weber, R., & Nau, D. (2000). *HICAP: Hierarchical interactive case-based architecture for planning* (Technical Report AIC-00-006). Washington, DC: NRL, NCARAI.
3. DoD (1994). *Joint tactics, techniques and procedures for noncombatant evacuation operations* (Joint Report 3-07.51). Washington, DC: Department of Defense.
4. Habbel, R., Harter, G., & Stech, M. (1999). Knowledge management: knowledge-critical capital of modern organizations. *Booz Allen & Hamilton Insights*. [www.bah.com/viewpoints/insights/cmt_knowmanage_2.html]
5. Johnson, C., Birnbaum, L., Bareiss, R., & Hinrichs, T. (2000). War Stories: Harnessing Organizational Memories to Support Task Performance. *Intelligence: New Visions of AI in Practice*, *11*(1), 17-31.
6. Kolodner, J. (1993). *Case-based reasoning*. San Mateo, CA: Morgan Kaufmann.
7. Leake, D.B., Bauer, T., Maguitman, A., & Wilson, D.C. (2000). Capture, storage, and reuse of lessons about information resources: Supporting task-based information search. In D.W. Aha & R. Weber (Eds.) *Intelligent Lessons Learned Systems: Proceedings of the AAAI Workshop* (Technical Report WS-00-08). Menlo Park, CA: AAAI Press.
8. Leake, D., Birnbaum, L, Hammond, K., Marlow, C., & Yang, H. (1999). Task-based knowledge management. In D.W. Aha, I. Becerra-Fernandez, F. Maurer, & H. Muñoz-Avila (Eds.) *Exploring Synergies of Knowledge Management and Case-Based Reasoning: Proceedings of the AAAI-99 Workshop* (Technical Report WS-99-10). Orlando, FL: AAAI Press.
9. Muñoz-Avila, H., McFarlane, D., Aha, D.W., Ballas, J., Breslow, L.A., & Nau, D. (1999). Using guidelines to constrain interactive case-based HTN planning. *Proceedings of the Third International Conference on Case-Based Reasoning* (pp. 288-302). Munich: Springer.
10. O'Leary, D.E. (1998). Enterprise Knowledge Management. *Computer*, *31*(3), 54-61.
11. Reimer, U. (1998). Knowledge integration for building organisational memories. *Proceedings of the Eleventh Banff Workshop on Knowledge Acquisition.*. [http://ksi.cpsc.ucalgary.ca/KAW/KAW98/KAW98Proc.html]
12. Sampson, M. (1999). NASA parts advisories – nine years of experience, and counting. In (Secchi, 1999).
13. Sary, C., & Mackey, W. (1995). A case-based reasoning approach for the access and reuse of lessons learned. *Proceedings of the Fifth Annual International Symposium of the National Council on Systems Engineering* (pp. 249-256). St. Louis, Missouri: NCOSE.
14. Secchi, P. (Ed.) (1999). *Proceedings of Alerts and Lessons Learned: An Effective way to prevent failures and problems* (Technical Report WPP-167). Noordwijk, The Netherlands: ESTEC.
15. SELLS (1999). *Proceedings of the Society for Effective Lessons Learned Sharing Spring Meeting*. Las Vegas, NV: Unpublished. [www.tis.eh.doe.gov/ll/sells]
16. van Heijst, G., Hofman, M., Kruizinga, E., & van der Spek, R. (1997). AI-techniques and the knowledge pump. In B. Gaines & R. Uthursamy (Eds.) *Artificial Intelligence in Knowledge Management: Proceedings of the 1997 Spring Symposium* (Technical Report SS-97-01). Menlo Park, CA: AAAI Press.

17. van Heijst, G., van der Spek, R., & Kruizinga, E. (1996). Organizing corporate memories. *Proceedings of the Tenth Banff Workshop on Knowledge Acquisition.* Banff, Canada. [ksi.cpsc.ucalgary.ca/KAW/KAW96/KAW96Proc.html]
18. Weber, R., Aha, D.W., Muñoz-Avila, H., & Breslow, L.A. (2000a). Active delivery for lessons learned systems. To appear in *Proceedings of the Fifth European Workshop on Case-Based Reasoning.* Trento, Italy: Springer.
19. Weber, R., Aha, D.W., & Becerra-Fernandez, I. (2000b). *Intelligent lessons learned systems.* To appear in *International Journal of Expert Systems Research & Applications.*

What the Logs Can Tell You: Mediation to Implement Feedback in Training

David A. Maluf*, Gio Wiederhold**

* Research Institute for Advanced Computer Science, Computational Science Division, NASA Ames. maluf@ptolemy.arc.nasa.gov
** Department of Computer Science Stanford University, Stanford.

Abstract. The problem addressed by Mediation to Implement Feedback in Training (MIFT) is to customize the feedback from training exercises by exploiting knowledge about the training scenario, training objectives, and specific student/teacher needs. We achieve this by inserting an intelligent mediation layer into the information flow from observations collected during training exercises to the display and user interface. Knowledge about training objectives, scenarios, and tasks is maintained in the mediating layer. A designer constraint is that domain experts must be able to extend mediators by adding domain-specific knowledge that supports additional aggregations, abstractions, and views of the results of training exercises.

The MIFT mediation concept is intended to be integrated with existing military training exercise management tools and reduce the cost of developing and maintaining separate feedback and evaluation tools for every training simulator and every set of customer needs. The MIFT Architecture is designed as a set of independently reusable components which interact with each other through standardized formalisms such as the Knowledge Interchange Format (KIF) and Knowledge Query and Manipulation Language (KQML).

1 Mediation applied to military exercise management

The initial application of MIFT is the Exercise Analysis and Feedback phase of military exercise management as schematically shown in Figure 1. More precisely, the focus is on simulation-based army training exercises [1]. MIFT handles some of the information flows involved in training exercise management. The intent of MIFT is to supplement the flow of information from simulations to evaluation and review and complete a feedback loop by supplying information to plan and tailor future training exercises.

MIFT processes the data that is logged during training exercises and uses scenario information and domain knowledge to organize the data from the exercises in ways that are meaningful and useful for the Observer/Controllers (O/Cs) managing the exercises, trainees, commanders, exercise evaluators, and others interested in the results of training exercises. MIFT is designed to feed information to other software systems that generate training scenarios and help commanders plan future training exercises tailored to the needs of their trainees. The MIFT design is intended to integrate with other exercise management applications (see Figure 3) and achieve two key application goals for exercise feedback:

1. The software is easy to use by using domain-specific exercise concepts and terminology.
2. Domain experts are able to extend feedback software and tailor it to domain-specific and local needs.

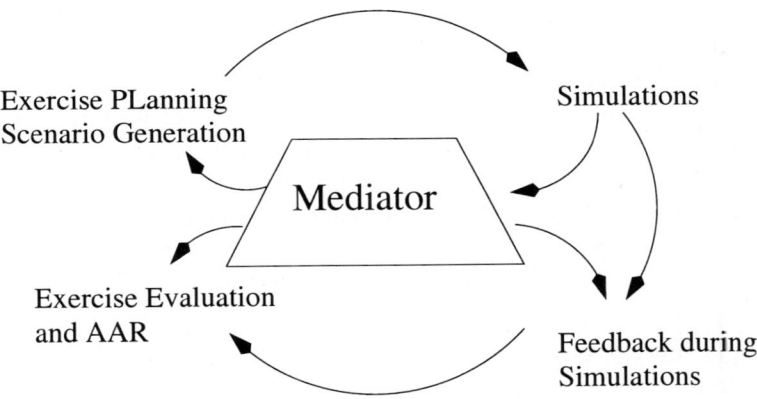

Fig. 1. MIFT's mediators supplement the flow of information from simulations to evaluation and review and complete the feedback loop by supplying information to plan and tailor future training exercises.

MIFT will achieve the first goal by incorporating knowledge about the scenario objectives and the task and subtasks to be trained. MIFT uses this scenario knowledge to relate simulation results to the objectives and tasks to be trained so that O/Cs, trainees, and commanders can query the simulation results using scenario-based terminology. For example, rather than forcing the O/C to formulate a query to "select all enemy detections of Alpha company before an assault," the O/C can simply ask whether Alpha company achieved its scenario subtask of remaining hidden until the beginning of the attack. The mediator will know that enemy detections before the attack are evidence that the unit was not successful in remaining hidden. In general, MIFT produces results tailored to the needs of exercise planners, weapons designers, and tactics developers. The second goal of a mediator-based architecture is to enable military training and support personnel to tailor and extend analysis and feedback software to meet their own local needs [6]. Figure 1 illustrates MIFT's mediators to supplement the flow of information from simulations to evaluation and review and complete the feedback loop by supplying information to plan and tailor future training exercises.

2 Mediation Technology

A mediator is a software module that exploits encoded knowledge about certain sets or subsets of data to create information for a higher layer of applications [8] [9]. It should be small and simple, so that it can be maintained by one expert or, at most, a small and coherent group of experts. The first step in developing a mediation architecture for training feedback is to isolate the mediators from lower-level data sources and from higher-level user interface and application code [5]. This will enable mediators to achieve the role of a reusable middleware. Mediators interact with each other through the standardized knowledge exchange and communications protocols. We have used standard knowledge exchange and communications protocols based on Knowledge Interchange Format (KIF) [4] and Knowledge Query and Manipulation Language (KQML) [3] so that mediators can work with data from multiple knowledge sources and supply information that is reusable in multiple roles. The MIFT mediation architecture combines plug-in components at three levels:

1. User interfaces that accept information from mediators and provide a standard set of display options.
2. Mediators that use scenario-based knowledge to analyze, transform, query, and present simulation results. A mediator supports numerous modules which are relatively small components. A module is a collection of rules reflecting the domain knowledge functionalities. Domain experts extend the analysis functionality by adding domain knowledge to mediators or by plugging in additional modules.
3. Wrappers connect MIFT with the output formats of operational simulators. Currently wrappers are tailored for JANUS and SimNet/LEAF data.

3 Implementation and Functionalities

The current MIFT user interface is built on Web browsers, hence enabling a multiple platform execution. In other words, the MIFT user interface can run at any location that supports Web browsing; the user does not have to download the simulation data. An innovation of the user interface is that it is designed to display information received from a mediator. Users connect to MIFT and the underlying exercise results by using a Java-capable browser. Building the user interface in a browser has several advantages:

1. Users can access exercise results in the same way they access other information from local and remote sources. The user interface will be increasingly familiar to O/Cs and trainees.
2. The exercise data may be local or remote. Startup and initialization is simple. Users do not have to download and manage the exercise data.

A key benefit of mediators for military training applications is that they avoid the need for each simulation program having to build from scratch and maintain a separate set of analysis and feedback software packages.

The operations referenced by the mediator can be layered in the direction of the data-to-knowledge aggregation as shown in Figure 2. For example, the first two levels in the mediator perform standard aggregations, selections, and analyses on the data sources. We have implemented these two levels to provide a basic level of functionality for higher levels. The third level in the mediator uses knowledge of the training scenario so that O/Cs and trainees can obtain feedback about how well specific scenario tasks have been performed. The mediator allows users to obtain specific feedback without having to understand the structure of the underlying data. A planned fourth level in the mediator will use domain-specific models about the exercise, the scenario, and causal relationships in the exercise to analyze the data for its probable significance and automatically call the users' attention to what it perceives as the more relevant exercise results. It is useful to think of the mediator as composed of three parts:

1. Data from disparate sources are converted into object instances over which inferences can be performed.
2. Knowledge about the application domain is maintained in declarative representations.
3. An inference engine processes the knowledge and data sources to produce higher level information that is passed to other mediators or to the user interface in a standardized form.

One of the MIFT functionalities is that an Observer/Controller (O/C) will depend upon it during an After Action Review (AAR) or that a

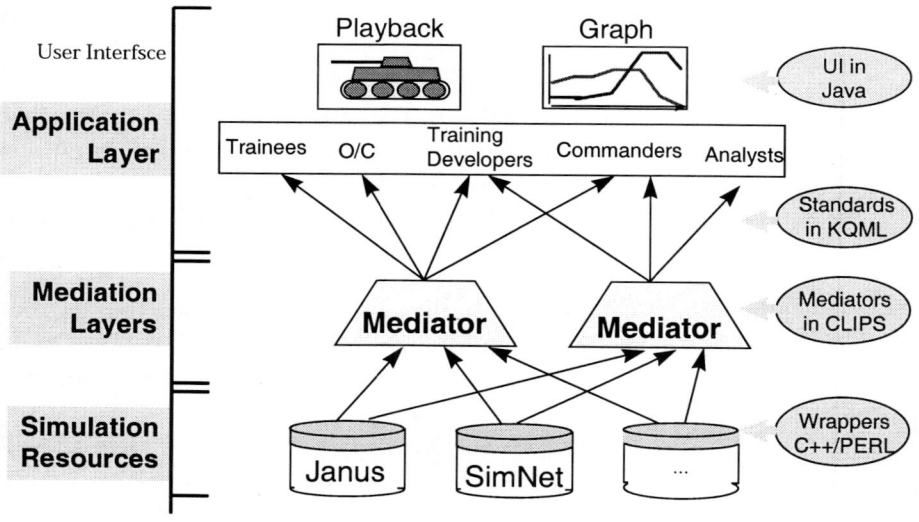

Fig. 2. The operations envisaged from the mediator can be layered in the direction of the data-to-knowledge aggregation.

trainee will use it after the AAR. As similar MIFT functionality will be useful to commanders, exercise evaluators, weapons designers, and others, but each of these other users is likely to want a different user interface and additional mediator functionality.

MIFT uses wrappers to isolate the mediators from the specific data formats and other differences between simulator outputs. When a mediator needs additional information, it calls the appropriate wrapper. The wrapper accesses the data and creates instances of the appropriate objects. The current implementation includes wrappers that process the outputs of Janus simulation runs, and LEAF[1] formated data from SimNet results. We believe that MIFT functionality can be made available

---

[1] Janus simulation databases.

for additional simulators by writing the appropriate wrapper to process simulator outputs. Writing additional wrappers requires programming expertise, but it is not a major undertaking. Using MIFT on a different simulation may also require additional modules and/or user interfaces to provide new functionality appropriate for that simulation. For example, the mediator that creates force ratios is more useful for simulations at the battalion or higher level and might not have been developed for analysis of simulations at the company level.

3.1 Implementing a Programmable Mediator

The architecture used for the MIFT mediator was based on a system that can sustain minimal first order logic inference capability. To further minimize development cost, the Mediator is finally written in parts in Clips [7], a widely-available and easily portable expert system shell. With little and careful programming, Clips was capable of supporting networking [2], a forward or backward chainer, a unifier, an in-memory object oriented database and a knowledge base that accepts and translate knowledge in the form of objects, rules and facts. The major function of the architecture is to allow a temporal hyper-graph construction that triggers modules to it which will perform their assigned tasks. The major five modules were as follows: A *Conflict Resolver* to maintain the truth values in the system, *Domain Modules* or the processes revolving around the domain knowledge of the main requirements, a *Report Agent* in which reports are generated and wrapped in KQML after the main requirements are accomplished, a *Maintenance Module* once some processes have terminated after the data and finally *Data Wrappers* which perform the necessary wrapping to maintain a correct syntax for the language in use. Hence, template structures are not violated. This reduces tremendously the amount of data to be loaded in comparision to the amount that will be used. Typically most databases are collections of instance events which have a time stamp associated with them and hence the wrappers are capable of playing back the databases as a function of time. Wrappers are mostly written in C++ to suit the variety and embedded complexity of the original databases.

Programming the MIFT mediator as a reusable system from task to task is performed by changing the domain module. Although attempt was made to make the conflict resolver generic in its functionalities among the tasks, domain specific rules are used in the module. The major goal of the conflict resolver is to identify the knowledge which might be disruptive to the overall mediator operation. The domain expert rules were divided under

1. Cyclic behavior: where asserted events result in cyclic effects in the process of inference.
2. Repetition and redundancy: where asserted events are redundant in the databases.
3. Constrained Space: where asserted events who's truth value conflicts with prior asserted events. For example a stated destroyed tank appearing later on in the simulation as a functional unit. Conflicts were generically sorted out using deduction rules which eliminates the erroneous event.

4 Conclusion

This paper describes Mediation to Implement Feedback in Training to customize the feedback from training exercises by exploiting knowledge about the training scenario, training objectives, and specific student/teacher needs. We plan to achieve this by inserting intelligent mediators into the information flow from observations collected during training exercises to the display and user interface functionality. Knowledge about training objectives, scenerios, and tasks is maintained in the mediators. A technical constraint is that domain experts must be able to extend mediators by adding domain-specific knowledge that supports additional aggregations, abstractions, and views of the results of training exercises.

MIFT is intended to allow analysis and evaluation software to be reused by all of the different consumers of simulation results. In addition to trainees, O/C, and commanders, others who need to analyze and evaluate simulation results include exercise planners, training managers, weapons designers, tactics developers, and doctrine writers. MIFT can also provide results to other software applications; for example, software used to assist in exercise planning and preparation can use MIFT analyses of previous exercises to identify the tasks and subtasks that need to be emphasized in additional training. Thus MIFT contributes to completing the feedback loop from the results of one simulation run into the planning and preparation for future training.

The Mediator is currently written in Clips 6.0 [7], a widely-available and easily portable expert system shell. Since user interface functions and data access functions are separated out into other components, the module implementations are quite small. For example, the force ratio computation for any set and/or combination of units is only four rules for a total of 12 lines. Most other mediators at the current stage are smaller. We believe that some domain experts will be able to write modules in Clips.

5 Acknowledgments

This research was supported by a grant from DARPA under the CAETI-EXMAN program and under contract N66001-95-C08618; Kirstie Bellman is the Program Manager. Input from Julia Loughran from IDA has been particularly helpful. This research has tremendously benefited from Ted Linden (Project Manager, Myriad Software) and Priya Panchapagesan (Stanford Graduate Research Assistant).

References

1. M. Crissey, G. Stone, D. Briggs and M. Mollaghasemi, "Training Exercise Planing: Leveraging Data and Technologies"; Proceedings of the 16th Interservice/Industry Training System and Education Conference. Washington, DC: national Security Industrial Association.
2. Clips Knowledge Networking Protocol www-db.stanford.edu/~maluf/cknp/
3. T. Finin, J. Weber, G. Wiederhold, M. Genesereth, R. Fritzson, J. McGuire, S. Shapiro, and C. Beck, "Specification of the KQML Agent-Communication Language"; Stanford February 1994.
4. M. Genesereth."Knowledge Interchange Format"; Technical Report Logic-92-1, Stanford University, 1992.
5. D. Maluf, G. Wiederhold, T. Linden and P. Panchapagesan, "Mediation to Implement Feedback in Training"; CrossTalk: Journal of Defense Software Engineering Software Technology Support Center, Department of Defense, 1997.
6. MIFT Home page www-db.stanford.edu/mift
7. Riley, G. "CLIPS: An Expert System Building Tool"; Proceedings of the Technology 2001 Conference, San Jose, CA, December 1991.
8. G. Wiederhold, "Mediators in the Architecture of Future Information Systems"; IEEE Computer, March 1992, pages 38-49.
9. Y. Papakonstantinou, H. Garcia-Molina and J. Widom, "Object Exchange Across Heterogeneous Information Sources"; International Conference on Data Engineering, 1995.

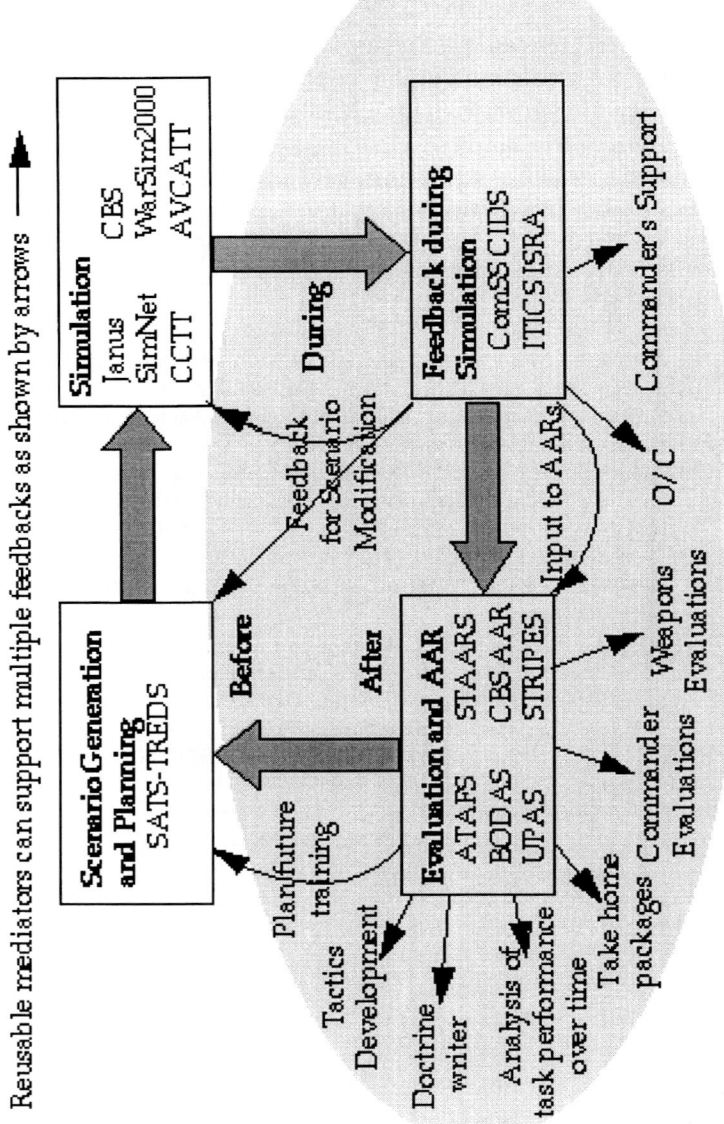

Fig. 3. The application Mediation to Implement Feedback in Training (MIFT) is the Exercise Analysis and Feedback phase of military exercise management. This figure illustrates the many different simulation results and the roles that MIFT can play by implementing reusable mediators that aggregate, summarize, and analyze simulation results and deliver them to various consumers in terms tailored to their individual needs.

Top-Down Query Processing in First Order Deductive Databases under the DWFS

C A Johnson

Computer Science Department, University of Keele, Staffs, ST5 5BG, England

Abstract. A top-down query processing method for first order deductive databases under the disjunctive well-founded semantics (DWFS) is presented. The method is based upon a characterisation of the DWFS in terms of the Gelfond-Lifschitz transformation, and employs a hyper-resolution like operator and quasi cyclic trees to handle minimal model processing. The method is correct and complete, and can be guaranteed to terminate given certain mild constraints on the format of database rules. The efficiency of the method may be enhanced by the application of partial compilation, subgoal re-ordering, and further constraints on the format of database rules. For finite propositional databases the method runs in polynomial space.

1 Introduction

Over the last few years there has been a great deal of interest in the study of semantics for deductive databases and logic programs [5], one of the most prominent to emerge being the disjunctive well-founded semantics (DWFS). This was introduced by Brass and Dix [1-4] as the weakest semantics which satisfies certain desirable properties, including the generalised principle of partial evaluation. In [2,3] an extension of the Gelfond-Lifschitz transformation was employed to give a bottom-up characterisation of DWFS. This was then used in the DISLOP project [1] to develop a bottom-up method of computing DWFS, using the (bottom-up) methods of [12] to handle minimal model reasoning.

In [10] we presented a characterisation of the DWFS directly in terms of the Gelfond-Lifschitz transformation, and using this derived a top-down method of testing DWFS membership in propositional countable logic programs. In this paper we extend these techniques to provide a top-down query processing method for first order deductive databases under the DWFS. Our method is correct and complete, and can be guaranteed to terminate given certain mild constraints on the format of database rules. We also consider how our method can be made more efficient by the application of partial compilation, subgoal re-ordering and further restrictions on database rule format.

In Section 3 we restate the bottom-up characterisation of DWFS given in [2,3] and its re-characterisation in terms of the Gelfond-Lifschitz transformation [10]. In Section 4 we (re)introduce the concept of a deduction tree [7,8] which is based on a hyperresolution-like operator, and facilitates top-down query processing in positive databases. In Section 5 we (re)introduce quasi cyclic trees [10], which are variants of cyclic trees [6-9], and enable us to perform top-down minimal model reasoning in databases resulting from applications of the Gelfond-Lifschitz transformation. In Section 6 we combine deduction and quasi cyclic trees to form our top-down method, which is presented by means of an example. Sections 7 and 8 then examine the construction (traversal) of deduction and quasi cyclic trees, and the related termination and efficiency issues. Finally Section 9 contains our conclusions and suggestions for further research.

2 Terminology

Throughout $\mathcal{L} = \{P_0, P_1, \ldots, P_n, c_1, c_2, \ldots, c_m\}$ denotes a finite function free first order language. We assume that $\{P_1, P_2, \ldots, P_n\}$ is the disjoint union of EXT($\mathcal{L}$) (the *extensional* predicates) and INT($\mathcal{L}$) (the *intensional* predicates). A positive (negative) atom is a formula of the form $P(\mathbf{x})$ ($\neg P(\mathbf{x})$), and informally we write $P(\mathbf{x}) \in$ EXT($\mathcal{L}$) when $P \in$ EXT($\mathcal{L}$), etc. $\mathcal{H}$ denotes the set of positive ground atoms, ie., the Herbrand base. If I is a set of atoms, then $\overline{I} = \{\neg K : K \in I\}$. If θ is a first order formula, then $VAR(\theta)$ denotes the variables in θ.

A *deductive database* T is a finite set of rules C of the form $A_1 \wedge A_2 \wedge \ldots \wedge A_n \wedge \neg A_{n+1} \wedge \neg A_{n+2} \wedge \ldots \wedge \neg A_{n+h} \to B_1 \vee B_2 \vee \ldots \vee B_r$, where each A_i, B_j is a positive atom and $r > 0$. antec(C) $= \{A_1, A_2, \ldots, A_n\}$, conseq($C$) $= \{B_1, B_2, \ldots, B_r\}$ and $\mathcal{N}(C) = \{A_{n+1}, A_{n+2}, \ldots, A_{n+h}\}$. We make the standard assumptions that $VAR(\text{conseq}(C) \cup \mathcal{N}(C)) \subseteq VAR(\text{antec}(C))$, and that if $\{B_1, B_2, \ldots, B_r\} \cap$ EXT($\mathcal{L}$) $\neq \emptyset$, then $\{B_1, B_2, \ldots, B_r\} \subseteq$ EXT($\mathcal{L}$) and antec(C) $\cup \mathcal{N}(C) = \emptyset$.

T is regarded as representing the set of its ground instances, which we denote by $gr(T)$. T is *positive* iff $\mathcal{N}(C) = \emptyset$ for each $C \in T$.

As in [6-8] we assume the existence of a set of *semi-definite* predicates SD($\mathcal{L}$) $\subseteq$ INT($\mathcal{L}$) such that for each rule C, if $P \in$ SD($\mathcal{L}$) appears in conseq(C), then C is definite (ie., $|\text{conseq}(C)| = 1$), and each predicate appearing in the body of C is in EXT($\mathcal{L}$) $\cup$ SD($\mathcal{L}$). SD(T) $= \{C \in T : \text{conseq}(C) = \{B\}, B \in SD(\mathcal{L})\}$, EXT($T$) $= \{C \in T : \text{conseq}(C) \subseteq$ EXT($\mathcal{L}$)$\}$, and INT(T) $= T -$ EXT(T). Notice that EXT(T) consists of disjuncts of ground positive extensional atoms.

3 The disjunctive well-founded semantics

Definition 3.1. If C is a ground rule, then $pos(C) = \bigwedge \text{antec}(C) \to \bigvee \text{conseq}(C)$. If T is ground, and $N \subseteq \mathcal{H}$, then let $T|_g N$ denote the *Gelfond-Lifschitz transformation*, $T|_g N = \{pos(C) : C \in T, \mathcal{N}(C) \cap N = \emptyset\}$.

Theorem 3.2 [2,3]. Let $D_0 = \emptyset$ and $D_{\alpha+1} = D^+_{\alpha+1} \cup D^-_{\alpha+1}$, where
$D^+_{\alpha+1} = \{\bigvee \mathcal{P} : \mathcal{P} \subseteq \mathcal{H}, (gr(T)/D_\alpha)|_g \mathcal{H} \models \bigvee \mathcal{P}\}$,
$D^-_{\alpha+1} = \{\neg Q : Q \in \mathcal{H}, (\forall N \subseteq \mathcal{H})(N \models D^+_{\alpha+1} \Longrightarrow (gr(T)/D_\alpha)|_g N \models_{min} \neg Q)\}$,
and $gr(T)/D_\alpha$ is formed from $gr(T)$ by (i) removing any rule C for which $\overline{D^+_\alpha} \models \bigvee \mathcal{N}(C)$, and (ii) for each remaining rule C, replacing $\mathcal{N}(C)$ by $\mathcal{N}(C) - \overline{D^-_\alpha}$.

Then $D_0 \subseteq D_1 \subseteq D_2 \subseteq \ldots$ grows monotonically and $\text{DWFS} = \bigcup_\alpha D_\alpha$.

Recall that $\overline{D_\alpha^-} = \{K : \neg K \in D_\alpha^-\}$. $T \models_{min} \neg Q$ iff Q is false in all minimal models of T. Note that no clause for limit ordinals is needed since $\mathcal{H}$ is finite.

The construction of DWFS given above is based upon computing a set of disjuncts and negative atoms, using these to reduce the database with the /-operator, and then repeating the process. This is ideal for a bottom-up computation, but for a top-down approach it is essential to express each of the sets $D_{\alpha+1}$ directly in terms of $gr(T)$ rather than $gr(T)/D_\alpha$.

Theorem 3.3 [10]. $D_{\alpha+1}^+ = \{\bigvee \mathcal{P} : \mathcal{P} \subseteq \mathcal{H}, gr(T)|_g(\mathcal{H} - \overline{D_\alpha^-}) \models \bigvee \mathcal{P}\}$, and $D_{\alpha+1}^- = \{\neg Q : Q \in \mathcal{H}, (\forall N \subseteq \mathcal{H} - \overline{D_\alpha^-})(N \models D_{\alpha+1}^+ \implies gr(T)|_g N \models_{min} \neg Q)\}$.

4 Computation in positive databases

In Theorem 3.3 we saw that the computation of $D_{\alpha+1}^+$ employs the positive database $gr(T)|_g(\mathcal{H} - \overline{D_\alpha^-})$. Derivability in positive databases can be characterised using deduction trees [7,8,10] which in turn are based upon a hyperresolution-like operator akin to that employed in SLO-resolution [13].

Definition 4.1. Suppose that T is positive and that $\mathcal{P}$ is a set of positive atoms. A *deduction tree* for $\mathcal{P}$ in T is a finite tree containing predicate nodes and rule nodes satisfying the following conditions.
(i) The root node (at the top of the tree) is a predicate node labelled with $\mathcal{P}$. All other predicate nodes are labelled with a positive atom.
(ii) If N is a predicate node, then let $\text{ACT}(N)$ denote the set of atoms labelling predicate nodes at or above N (on the current branch). N has a single child node which is a rule node labelled with an instance $C\theta$ of a rule $C \in T$ (written $RN_{C\theta}$) such that $\text{conseq}(C\theta) \subseteq \text{ACT}(N)$. For each $K \in \text{antec}(C\theta)$, $RN_{C\theta}$ has a (predicate) child node labelled with K.
(iii) Each leaf node is a rule node.

An *instance* of $\mathcal{T}$ is formed by applying some substitution to all labels (including rules) within the tree. Clearly if $\mathcal{T}$ is a deduction tree in T, then so is any instance of $\mathcal{T}$. A predicate node N is *redundant* iff there is a predicate node $N' > N$ such that $lab(N)$ equals, or is contained in, $lab(N')$ [7,8].

Theorem 4.2 [7,8]. If T is positive and $\mathcal{P} \subseteq \mathcal{H}$, then $T \models \bigvee \mathcal{P}$ iff $\mathcal{P}$ has a deduction tree in $gr(T)$ in which
(i) no predicate node is redundant, and
(ii) if N is a predicate node with $lab(N) \in SD(\mathcal{L})$ and $RN_{C\theta}$ is the child node of N, then $\text{conseq}(C\theta) = \{lab(N)\}$.

Thus we see that such trees may be constructed using ancestor pruning (condition (i)) and that semi-definite atoms are expanded in a linear fashion. The construction of deduction trees is discussed in Section 7.

5 Quasi cyclic trees

In [6-8] we introduced the notion of a cyclic tree as a means of testing minimal model membership. The following variant of this notion (adapted from [10]) enables us to characterise minimal models of databases of the form $T|_g N$. Motivation for the following definition appears in [6-10].

Definition 5.1. A *quasi cyclic tree* $\mathcal{T}$ for $P(\mathbf{t})$ in T is a finite tree $\mathcal{T}$ containing predicate nodes and rule nodes satisfying the following conditions.
(a) The root node (at the top of $\mathcal{T}$) is a predicate node labelled with $P(\mathbf{t})$.
(b) Each predicate node N is labelled with a positive atom, denoted by $lab(N)$, and $\text{CYC}(N) = \{lab(N') : N'$ is a predicate node, $N' \geq N$, and $\exists N'' \geq N', N''$ is a predicate node, $lab(N'') = lab(N)\}$. Let $Pred(\mathcal{T}) = \{lab(N) : N$ is a predicate node in $\mathcal{T}\}$.
(c) A predicate node N has at most a single child node which (if it exists) is a rule node labelled with an instance $C\theta$ of a rule $C \in T$ (written $RN_{C\theta}$) such that $\text{conseq}(C\theta) \cap \text{CYC}(N) \neq \emptyset$, $\text{antec}(C\theta) \cap \text{CYC}(N) = \emptyset$, and $\mathcal{O}(RN_{C\theta}) = (\text{conseq}(C\theta) - \text{CYC}(N))$ is disjoint from $Pred(\mathcal{T})$. For each $K \in \text{antec}(C\theta)$, $RN_{C\theta}$ has a (predicate) child node labelled with K.
(d) If N is a predicate node with $lab(N) \in SD(\mathcal{L})$, then N is not redundant.

We define $\mathcal{N}(\mathcal{T}) = \bigcup\{\mathcal{N}(C\theta) : RN_{C\theta}$ is a rule node in $\mathcal{T}\}$, and $\mathcal{O}(\mathcal{T}) = \bigcup\{\mathcal{O}(RN_{C\theta}) : RN_{C\theta}$ is a rule node in $\mathcal{T}\}$. As in [6-10], an *unfactored* quasi cyclic (UQC) tree is a quasi cyclic tree in which each leaf node is a rule node.

Notice that condition (c) is inherently top-down. Also note that if $lab(N) \in \text{EXT}(\mathcal{L}) \cup SD(\mathcal{L})$, then $\text{CYC}(N) = \{lab(N)\}$. Let $RN_{C\theta}$ be the child node of N. If $lab(N) \in SD(\mathcal{L})$, then $\text{conseq}(C\theta) = \{lab(N)\}$ and $C \in SD(T)$. If $lab(N) \in \text{EXT}(\mathcal{L})$, then $\theta = \emptyset$, $lab(N) \in \text{conseq}(C)$ and $C \in \text{EXT}(T)$.

The following theorem details the basic properties of UQC trees.

Theorem 5.2 [6,7,10]. *Let* $N \subseteq \mathcal{H}$.
(a) *If* $\mathcal{T}$ *is a UQC tree, then all labels in* $\mathcal{T}$ *are ground.*
(b) *Suppose that* M *is a minimal model of* $gr(T)|_g N$ *with* $P(\mathbf{a}) \in M$. *Then we may find a UQC tree* $\mathcal{T}$ *for* $P(\mathbf{a})$ *in* T *such that* $Pred(\mathcal{T}) \subseteq M \subseteq \mathcal{H} - \mathcal{O}(\mathcal{T})$ *and* $\mathcal{N}(\mathcal{T}) \cap N = \emptyset$.
(c) *Suppose that* $\mathcal{T}$ *is a UQC tree in* T, $\mathcal{N}(\mathcal{T}) \cap N = \emptyset$, *and* $M \models gr(T)|_g N$ *with* $M \cap \mathcal{O}(\mathcal{T}) = \emptyset$. *Then* $Pred(\mathcal{T}) \subseteq M$.

In order to perform top-down testing of membership in $D^-_{\alpha+1}$, we will need the following characterisation.

Theorem 5.3 [10]. $\neg Q \in D_{\alpha+1}^-$ *iff for each UQC tree* $\mathcal{T}$ *for* Q *in* T, *either* $\bigvee \mathcal{N}(\mathcal{T}) \in D_{\alpha+1}^+$ *or* $gr(T)|_g(\mathcal{H} - \overline{\mathcal{N}(\mathcal{T}) \cup D_\alpha^-}) \models \bigvee \mathcal{O}(\mathcal{T})$.

Note that the two clauses on the right hand side of Theorem 5.3 are actually quite similar, since $\bigvee \mathcal{N}(\mathcal{T}) \in D_{\alpha+1}^+$ iff $gr(T)|_g(\mathcal{H} - \overline{D_\alpha^-}) \models \bigvee \mathcal{N}(\mathcal{T})$.

Note also that if $\mathcal{T}_1, \mathcal{T}_2$ are UQC trees for Q with $\mathcal{O}(\mathcal{T}_1) \subseteq \mathcal{O}(\mathcal{T}_2)$ and $\mathcal{N}(\mathcal{T}_1) \subseteq \mathcal{N}(\mathcal{T}_2)$, then $\mathcal{T}_2$ is redundant as far as Theorem 5.3 is concerned, since $\bigvee \mathcal{N}(\mathcal{T}_1) \in D_{\alpha+1}^+$ implies $\bigvee \mathcal{N}(\mathcal{T}_2) \in D_{\alpha+1}^+$, and $gr(T)|_g(\mathcal{H}-\overline{\mathcal{N}(\mathcal{T}_1) \cup D_\alpha^-}) \models \bigvee \mathcal{O}(\mathcal{T}_1)$ implies $gr(T)|_g(\mathcal{H} - \overline{\mathcal{N}(\mathcal{T}_2) \cup D_\alpha^-}) \models \bigvee \mathcal{O}(\mathcal{T}_2)$.

6 Top-down query processing

We can now combine the results of previous sections to develop a top-down method of query processing under DWFS. We illustrate the method by means of an example.

Example 6.1. Let T consist of the following rules.
1. $\neg P(y) \land S(y,z) \land Q(z,y) \rightarrow Q(a,y) \lor Q(y,b)$
2. $\neg F(w,w) \land F(w,x) \land F(x,y) \rightarrow S(x,y)$ 3. $E(a,b)$ 4. $E(a,a)$
5. $\neg P(x) \land Q(x,x) \rightarrow R(x)$ 6. $F(a,b)$ 7. $F(b,a)$
8. $D(a,b)$ 9. $D(b,a)$ 10. $\neg E(y,x) \land D(y,x) \rightarrow R(x)$
11. $\neg E(x,y) \land D(y,x) \rightarrow P(y) \lor R(y)$ 12. $Q(a,b) \lor Q(b,a)$
where x,y,z,w are variables and a,b are constants.

Suppose that we wish to test whether $Q(a,a) \lor Q(a,b)$ is in DWFS. In order to show that $Q(a,a) \lor Q(a,b) \in D_{\alpha+1}^+$, we need to develop a deduction tree for $\{Q(a,a), Q(a,b)\}$ in $gr(T)|_g(\mathcal{H} - \overline{D_\alpha^-})$. The only rule whose head unifies with a subset of $\{Q(a,a), Q(a,b)\}$ is rule 1, with the unifier $\{y \rightarrow a\}$, thus resulting in the partial deduction tree $\mathcal{T}_1$ in Figure 6.1(i).

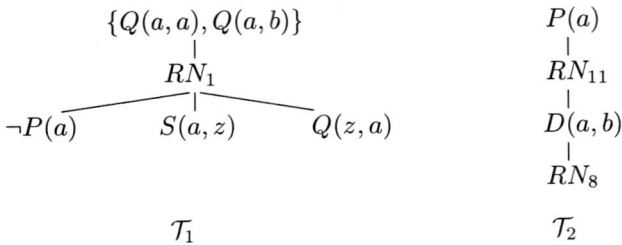

Figure 6.1(i)

Thus if $\neg P(a) \in D_\alpha^-$, then $Q(a,a) \lor Q(a,b) \in D_{\alpha+1}^+$ iff $S(a,z) \lor Q(a,a) \lor Q(a,b) \in D_{\alpha+1}^+$ and $Q(z,a) \lor Q(a,a) \lor Q(a,b) \in D_{\alpha+1}^+$ (for some z). Note however that if $\neg P(a) \notin D_\alpha^-$, then the fact that $S(a,z) \lor Q(a,a) \lor Q(a,b) \in D_{\alpha+1}^+$ and

$Q(z,a) \vee Q(a,a) \vee Q(a,b) \in D_{\alpha+1}^+$ (for some z) is of no help whatsoever. We thus attack the negative subgoal $\neg P(a)$ first.

The only UQC tree for $P(a)$ is depicted as $\mathcal{T}_2$ in Figure 6.1(i), with $\mathcal{O}(\mathcal{T}_2) = \{R(a)\}$ and $\mathcal{N}(\mathcal{T}_2) = \{E(b,a)\}$. By Theorem 5.3 we are thus required to show that either $E(b,a) \in D_\alpha^+$ or that $gr(T)|_g(\mathcal{H} - \{E(b,a)\} \cup \overline{D_{\alpha-1}^-}) \models R(a)$. In general we therefore need to pursue one of these subgoals, returning to the other if the first fails. In this particular case it is evident that $E(b,a) \notin D_\alpha^+$, since no rule has a consequent that unifies with $E(b,a)$.

We thus set about trying to show that $gr(T)|_g(\mathcal{H} - \{E(b,a)\} \cup \overline{D_{\alpha-1}^-}) \models R(a)$. This resembles our original problem, and is depicted in Figure 6.1(ii) via the node $(\{R(a)\}:\{E(b,a)\})$. The "rule node" $RN_{\mathcal{T}_2}$ simply indicates the computation of the UQC tree.

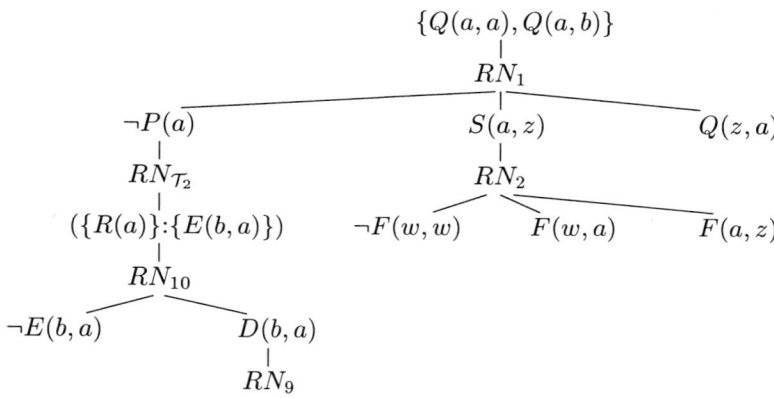

Figure 6.1(ii)

We thus look for a rule whose head unifies with $R(a)$. Notice that rule 5 should not be applied since it would introduce a duplicate subgoal ($\neg P(a)$) which leads to the circular argument: $\neg P(a) \in D_\alpha^-$ if $\neg P(a) \in D_{\alpha-1}^-$. Hence the only applicable rule is rule 10 yielding the two subgoals $\neg E(y,a)$ and $D(y,a)$. For the former, we need to look for an instance of $E(y,a)$ which is in $\{E(b,a)\} \cup \overline{D_{\alpha-1}^-}$. In this case we set y equal to b (Figure 6.1(ii)), whence $D(b,a)$ is solved by the application of rule 9. Thus $\neg P(a) \in D_1^-$.

Returning to the subgoal $S(a,z)$, we apply rule 2 to yield three child nodes, $\neg F(w,w)$, $F(w,a)$ and $F(a,z)$. Again for the first of these we are required to generate UQC trees for instances of $F(w,w)$ in order to find an instance of $\neg F(w,w)$ that is contained in D_α^-. Since $F(a,a)$ and $F(b,b)$ have no UQC trees, it is trivially the case that both $\neg F(a,a)$ and $\neg F(b,b)$ belong to D_1^-. Setting w equal to a would not allow the next subgoal to be solved, thus we set w equal to b. $F(a,z)$ is then solved via $\{z \to b\}$ and rule 6. Thus $S(a,b) \in D_2^+$.

Finally rule 12 shows that $Q(a,a) \vee Q(a,b) \vee Q(b,a) \in D_1^+$, whence $Q(a,a) \vee Q(a,b) \in D_2^+$.

Notes. Our method thus employs a combination of deduction tree and UQC tree constructions, and can be viewed as lifting the methods of [10] to the first order level. At each stage, if the current leaf node is a positive atom (or a set there-of), then we extend the current branch by applying some database rule (unifying the consequent with some of the atoms on the branch). If the current leaf node is a negative atom $\neg Q(\mathbf{x})$, then we may compute UQC trees in order to (try to) find an instance of $\neg Q(\mathbf{x})$ in the relevant D_β^-. As with $E(y, a)$, if the most recent UQC tree gave rise to a goal of the form $(\mathcal{P}{:}\mathcal{Q})$, then we have a second option, which is to try to find an instance of $Q(\mathbf{x})$ in $\mathcal{Q}$.

Each UQC tree yields a choice of two subgoals, the processing of which is independent of the other parts of the tree, with the exception that we may need to examine earlier nodes on the current branch to prevent duplicates.

The correctness and completeness of our method follows from Theorems 3.3, 4.2, 5.3 and the correctness and completeness of our constructions (below) for deduction and UQC trees. In addition, if these tree constructions terminate, then the method as a whole terminates, since the tree developed passes *down* the D_α hierarchy (in particular disallowing duplicate negative subgoals).

7 Constructing deduction trees

Constructing deduction trees at the ground level is trivial: at each stage we extend the current predicate leaf node N by a rule $C \in T$ such that $\mathrm{conseq}(C) \subseteq \mathrm{ACT}(N)$ and $\mathrm{antec}(C) \cap \mathrm{ACT}(N) = \emptyset$. This process must terminate since duplicate atoms do not appear along any branch. Testing derivability in positive databases can be achieved "branch at a time" [7,8], and therefore operates in space which is linear in $|\mathcal{H}|$.

The first order method of constructing deduction trees employed informally in Example 6.1 is taken from [7,8]. The method is top-down, left-to-right and depth-first, and yields a correct and complete method of testing derivability in first order positive deductive databases [7,8]. The obvious difference at the first order level is the use of unifying substitutions, which have the effect of enlarging the search space. In addition, termination is far more difficult to guarantee, and negative subgoals may need special attention, again in order to limit the size of the search space. These issues are discussed in Sections 7.1-7.3 below.

7.1 Termination

During the construction of deduction trees we can (by Theorem 4.2) employ ancestor pruning at will. Consequently, it is the introduction of new variables into the construction which threatens termination. In [7,8] we showed that termination can be guaranteed if we apply ancestor pruning, adopt a linear expansion of semi-definite atoms (as per Theorem 4.2), and assume the existence of a level function $\ell : \{P_1, P_2, \ldots, P_n\} \to \{0, 1, 2, \ldots, n+1\}$ such that
(a) $\mathrm{EXT}(\mathcal{L}) = \{P \in \mathcal{L} : \ell(P) = 0\}$ and $SD(\mathcal{L}) = \{P \in \mathcal{L} : 1 \leq \ell(P) \leq n\}$. (We may therefore define $\ell(C) = \ell(P)$ for any P occurring in $\mathrm{conseq}(C)$.)

(b) If $C \in T$ and $x \in VAR(\text{antec}(C)) - VAR(\text{conseq}(C))$, then we may find an $R(\mathbf{t}) \in \text{antec}(C)$ such that $\ell(R) < \ell(C)$ and x appears in $\mathbf{t}$.

Note that the use of ancestor pruning within our termination argument is dependent upon the use of a function free language.

7.2 Groundedness and yes/no answers

We saw in Example 6.1 that the subgoal $S(a, z)$ becomes grounded as a result of the expansion of the tree below this subgoal. In [7,8] it is shown that in effect this always occurs, thus allowing us to perform answer extraction in the case when the top-level goal is not grounded.

In addition, if we assume the conditions of Section 7.1, and attack semi-definite subgoals first, then we can ensure that if $P(\mathbf{t}) \in \text{INT}(\mathcal{L}) - SD(\mathcal{L})$ occurs on the current branch, then each variable in $\mathbf{t}$ occurs in the root node (whence again, ancestor pruning limits the number of such atoms). This has the important consequence of limiting the search space when applying the disjunctive rules in $\text{INT}(T) - SD(T)$. (As an aside, note that the search space is already limited when applying rules in $SD(T)$ (by linearity and definiteness) and when applying rules in $\text{EXT}(T)$ (by groundedness).)

Note in particular that if the root node is ground (as it is in a yes/no query and in the subtree below a UQC tree), then all positive atoms in $\text{INT}(\mathcal{L}) - SD(\mathcal{L})$ lying on the current branch are ground. In this case, the unifying substitution when applying a rule $C \in \text{INT}(T) - SD(T)$ is simply a ground instantiation of $\text{conseq}(C)$.

7.3 Negative subgoals and subgoal re-ordering

Generating UQC trees for instances of $Q(\mathbf{x})$ will (by Theorem 5.3) tell us which instances of $\neg Q(\mathbf{x})$ are contained in the relevant D_β^-. It will not however tell us which of these instances to pick, hence the choice is somewhat arbitrary. This is not desirable, since in resolution based reasoning the unifications are expected to perform the necessary instantiations. Two solutions which guarantee that $\mathbf{x}$ becomes grounded before we attack $\neg Q(\mathbf{x})$ are as follows.

Since $VAR(\mathcal{N}(C)) \subseteq VAR(\text{antec}(C))$ we could re-order the subgoal $\neg Q(\mathbf{x})$ so that it appears to the right of (ie., after) other positive siblings containing the variables in $\mathbf{x}$. Since such siblings become grounded as a result of their expansion, this would have the desired effect. On the other hand the re-ordering itself is somewhat undesirable since (as explained in Example 6.1) we would prefer to attack negative subgoals first.

An alternative solution is to make a further assumption:

(*) If $C \in \text{INT}(T) - SD(T)$ and $Q(\mathbf{x}) \in \mathcal{N}(C)$ with $Q \in \text{INT}(\mathcal{L}) - SD(\mathcal{L})$, then $VAR(\mathbf{x}) \subseteq VAR(\text{conseq}(C))$.

As in Section 7.2, in the case when the root is grounded, condition (*) has the effect of ensuring that a negative subgoal $\neg Q(\mathbf{x})$ (with $Q \in \text{INT}(\mathcal{L}) - SD(\mathcal{L})$) will be grounded as soon as it enters the tree. This still leaves the problem of ungrounded negative subgoals $\neg Q(\mathbf{x})$ when $Q \in \text{EXT}(\mathcal{L}) \cup SD(\mathcal{L})$, but this is no loss since condition (b) of Section 7.1 still requires us to handle such subgoals below ungrounded positive semi-definite subgoals (as with $\neg F(w,w)$ in Example 6.1). Moreover the relevant UQC trees are far easier to compute (ie., in a purely linear fashion). In addition, they also allow a simpler testing of $\neg Q(\mathbf{x}) \in D^-_{\beta+1}$, since if $Q(\mathbf{a}) \in \text{EXT}(\mathcal{L}) \cup SD(\mathcal{L})$ and $\mathcal{T}$ is a UQC tree for $Q(\mathbf{a})$, then $\mathcal{N}(\mathcal{T}) \subseteq \text{EXT}(\mathcal{L}) \cup SD(\mathcal{L})$, and $\mathcal{O}(\mathcal{T}) \subseteq \text{EXT}(\mathcal{L})$ (whence $gr(T)|_g(\mathcal{H} - \mathcal{N}(\mathcal{T}) \cup \overline{D^-_\beta}) \models \bigvee \mathcal{O}(\mathcal{T})$ iff $\mathcal{O}(\mathcal{T})$ contains some disjunct from $\text{EXT}(T)$).

8 Computing UQC trees

Let $P(\mathbf{t}) \in \text{EXT}(\mathcal{L}) \cup SD(\mathcal{L})$. In [6] we showed that cyclic (and hence UQC) trees for instance of $P(\mathbf{t})$ can be computed via a top-down, left-to-right, depth-first and linear construction. In order to guarantee termination, ancestor pruning is employed (as dictated by condition (d) of Definition 5.1) and we assumed the existence of a level function ℓ as in Section 7.1 such that

(†) If $C \in SD(T)$ and $x \in VAR(\text{antec}(C)) - VAR(\text{conseq}(C))$, then we may find an $R(\mathbf{t}) \in \text{antec}(C)$ such that $\ell(R) < \ell(C)$ and $x \in \mathbf{t}$.

Note that (†) is subsumed by condition (b) of Section 7.1, hence it is no further imposition. In [6,7] we presented a construction of UQC trees for atoms $P(\mathbf{t}) \in \text{INT}(\mathcal{L}) - SD(\mathcal{L})$ under conditions (†) and (♯):

(♯) If $C \in \text{INT}(T) - SD(T)$ and $x \in VAR(C)$, then we may find an $R(\mathbf{t}) \in \text{antec}(C)$ such that $\ell(R) < \ell(C)$ and $x \in \mathbf{t}$.

However (♯) is a far more stringent condition than (†). We can eliminate the need for (♯) by partially compiling the construction of UQC trees for atoms in $\text{INT}(\mathcal{L}) - SD(\mathcal{L})$ using semi-factored quasi cyclic trees (see below). Such compilation is motivated and justified by the fact that we generally expect $\text{INT}(T)$ to be relatively static (in comparison to $\text{EXT}(T)$). Pre-processing of cyclic trees has also been employed in [9] to facilitate query compilation in propositional stratified databases.

8.1 Semi-factored quasi cyclic trees

Definition 8.1.1. Let $P(\mathbf{t}) \in \text{INT}(\mathcal{L}) - SD(\mathcal{L})$. A *semi-factored* quasi cyclic (SQC) tree for $P(\mathbf{t})$ is a quasi cyclic tree $\mathcal{T}$ in which for each predicate node N, N is a leaf node iff $lab(N) \in \text{EXT}(\mathcal{L}) \cup SD(\mathcal{L})$.
Let $leaf(\mathcal{T}) = \{N : N \text{ is a predicate leaf node in } \mathcal{T}\}$.

UQC trees are constructed by extending SQC trees in the obvious way.

Theorem 8.1.2. Let $P(\mathbf{a}) \in \mathcal{H} \cap (\text{INT}(\mathcal{L}) - SD(\mathcal{L}))$. Then $\mathcal{T}$ is a UQC for $P(\mathbf{a})$ in T iff we may find an SQC tree $\mathcal{T}'$ for $P(\mathbf{a})$ in $gr(T)$, and for each $N \in leaf(\mathcal{T}')$ we can find a UQC tree $\mathcal{S}_N$ for $lab(N)$ in T such that
(a) $\mathcal{T}'$ is an initial segment of $\mathcal{T}$,

(b) *for each* $N \in leaf(\mathcal{T}')$, $\mathcal{S}_N$ *equals the subtree of* $\mathcal{T}$ *below* N, *and*
(c) $\bigcup\{Pred(\mathcal{S}_N) : N \in leaf(\mathcal{T}')\} \cap \bigcup\{\mathcal{O}(\mathcal{S}_N) : N \in leaf(\mathcal{T}')\} = \emptyset$.

Moreover if conditions (a), (b) and (c) are satisfied, then $Pred(\mathcal{T})$, $\mathcal{O}(\mathcal{T})$ *and* $\mathcal{N}(\mathcal{T})$ *are the union of the corresponding sets from* $\mathcal{T}'$ *and* $\{\mathcal{S}_N : N \in leaf(\mathcal{T}')\}$.

8.2 Partial compilation of UQC trees

We will partially compile the computation of UQC trees for atoms in $INT(\mathcal{L}) - SD(\mathcal{L})$. The compilation phase will consist of computing a set of SQC trees, and the run-time computation of UQC trees will involve extending these SQC trees to UQC trees.

8.2.1 Computing SQC trees. As the construction of a first order quasi cyclic tree proceeds, existing branches are extended by the application of rules and unifying substitutions. The main problem is to ensure that these unifying substitutions do not violate the conditions of Definition 5.1. This is complicated still further by the fact that the unifying substitutions can cause the CYC sets to be altered. In [6,7] we showed that when computing UQC trees, these problems can be overcome by the use of conditions (†) and (♯), since these cause sufficiently large parts of the tree to become grounded as the construction proceeds.

In this section we wish to compute SQC trees without resorting to (♯). ((†) is of no help, since SQC trees employ only rules in $INT(T) - SD(T)$.) Thus in order to overcome the above-mentioned problems, we will, as the construction proceeds, develop a set of constraints $\mathcal{I}$ which will guarantee the conditions of Definition 5.1, and also ensure that the CYC sets do not change. Termination of our construction is guaranteed by the fact that the length of any branch through an SQC tree in $gr(T)$ is bounded by $1 + |\mathcal{H}'| * (|\mathcal{H}'| + 1)/2$, where $\mathcal{H}' = \mathcal{H} \cap (INT(\mathcal{L}) - SD(\mathcal{L}))$ [6]. Of course $\mathcal{H}'$ is likely to be large, but this compilation only needs to be performed once (unless $INT(T) - SD(T)$ changes).

Suppose that we have constructed a partial SQC tree $\mathcal{T}$ with existing constraints $\mathcal{I}$. A *valid extension* of $(\mathcal{T}, \mathcal{I})$ is constructed as follows. Pick a branch through $\mathcal{T}$, $(root(\mathcal{T}) = P_1(\mathbf{x}_1), P_2(\mathbf{x}_2), \ldots, N = P_r(\mathbf{x}_r))$, where $P_r \in INT(\mathcal{L}) - SD(\mathcal{L})$ and $r \leq |\mathcal{H}'| * (|\mathcal{H}'| + 1)/2$. $\mathcal{I}$ will contain $\{\mathbf{x}_i \neq \mathbf{x}_j : i < j < r, P_i(\mathbf{x}_i)$ and $P_j(\mathbf{x}_j)$ are unifiable, but $\mathbf{x}_i \neq \mathbf{x}_j\}$. First we need to fix CYC(N).

If $P_r(\mathbf{x}_r) \in \{P_i(\mathbf{x}_i) : i < r\}$, then $\mathcal{I}$ already ensures that CYC(N) is fixed, thus suppose that $P_r(\mathbf{x}_r) \notin \{P_i(\mathbf{x}_i) : i < r\}$. If $P_r(\mathbf{x}_r)$ is not unifiable with any $P_i(\mathbf{x}_i)$ ($i < r$), then CYC(N) = $\{lab(N)\}$ is fixed. Finally if $P_r(\mathbf{x}_r)$ is unifiable with some $P_i(\mathbf{x}_i)$ ($i < r$) then we either (i) pick some such i such that the most general unifier $\mu = mgu\{\mathbf{x}_r, \mathbf{x}_i\}$ does not violate $\mathcal{I}$, and apply μ to $\mathcal{T}$ and $\mathcal{I}$, or (ii) add $\{\mathbf{x}_r \neq \mathbf{x}_i : i < r, P_r(\mathbf{x}_r)$ and $P_i(\mathbf{x}_i)$ are unifiable$\}$ to $\mathcal{I}$.

We can then extend our branch with some rule $C \in T$. We need to pick a most general unifier η for a subset of conseq(C) and a subset of CYC(N) such that η does not violate $\mathcal{I}$, antec($C\eta$)$\cap$CYC(N)$\eta = \emptyset$ and CYC(N)η−conseq($C\eta$) is disjoint from $Pred(\mathcal{T}\eta) \cup$ antec($C\eta$) (cf., condition (c) of Definition 5.1). η is applied to $\mathcal{T}$ and $\mathcal{I}$, and $\mathcal{T}\eta$ is then extended with $RN_{C\eta}$ (and the corresponding child nodes of $RN_{C\eta}$). Let $\mathcal{T}'$ be the extended tree.

For each $Q(\mathbf{x}) \in \mathcal{O}(RN_{C\eta})$, if $Q(\mathbf{x})$ is unifiable with some $Q(\mathbf{x}') \in Pred(\mathcal{T}')$, then we add $\mathbf{x} \neq \mathbf{x}'$ to $\mathcal{I}$. Similarly if $Q(\mathbf{x}) \in \text{antec}(C\eta)$ is unifiable with some $Q(\mathbf{x}') \in \text{CYC}(N)$, then we add $\mathbf{x} \neq \mathbf{x}'$ to $\mathcal{I}$.

A pair $(\mathcal{T}_m, \mathcal{I}_m)$ is *complete* iff $\mathcal{T}_m$ is an SQC tree and there is a sequence $(\mathcal{T}_0, \mathcal{I}_0), (\mathcal{T}_1, \mathcal{I}_1), \ldots, (\mathcal{T}_m, \mathcal{I}_m)$ such that $\mathcal{T}_0$ consists of the single root $P(x_1, x_2, \ldots, x_k)$, $\mathcal{I}_0 = \emptyset$, and each $(\mathcal{T}_{i+1}, \mathcal{I}_{i+1})$ is a valid extension of $(\mathcal{T}_i, \mathcal{I}_i)$.

Theorem 8.2.2 *Let $P(\mathbf{a}) \in \mathcal{H} \cap (\text{INT}(\mathcal{L}) - SD(\mathcal{L}))$. Then $\mathcal{T}$ is an SQC tree for $P(\mathbf{a})$ in $gr(T)$ iff there is complete pair $(\mathcal{T}_m, \mathcal{I}_m)$ and a substitution θ such that θ that does not violate $\mathcal{I}_m$, $\mathcal{T} = \mathcal{T}_m \theta$ and $lab(root(\mathcal{T})) = P(\mathbf{a})$.*

After constructing a complete pair $(\mathcal{T}, \mathcal{I})$, we can then discard the tree $\mathcal{T}$ itself: we simply need to keep $(root(\mathcal{T}), leaf(\mathcal{T}), \mathcal{N}(\mathcal{T}), \mathcal{O}(\mathcal{T}), \mathcal{I})$. As indicated at the end of Section 5, we can also eliminate redundant SQC trees. Specifically if we can find a substitution θ such that $root(\mathcal{T})\theta = root(\mathcal{T}')$, $leaf(\mathcal{T})\theta \subseteq leaf(\mathcal{T}')$, $\mathcal{N}(\mathcal{T})\theta \subseteq \mathcal{N}(\mathcal{T}')$, $\mathcal{O}(\mathcal{T})\theta \subseteq \mathcal{O}(\mathcal{T}')$ and $\mathcal{I}\theta \subseteq \mathcal{I}'$, then we can discard $(root(\mathcal{T}'), leaf(\mathcal{T}'), \mathcal{N}(\mathcal{T}'), \mathcal{O}(\mathcal{T}'), \mathcal{I}')$.

For example suppose that $\text{INT}(T) - SD(T) = \{S(x) \to Q(x) \vee R(x), S'(x,y) \wedge Q(y) \to Q(x) \vee P(x)\}$, with $\text{INT}(\mathcal{L}) - SD(\mathcal{L}) = \{P, Q, R\}$. SQC trees for $Q(x)$ can be developed by m applications of the second rule followed by a single application of the first. For $m > 0$ the tree developed is subsumed, in the above sense, by the tree developed for $m = 0$.

8.2.3 Computing UQC trees from SQC trees. By Theorems 8.1.2 and 8.2.2, if $P(\mathbf{a}) \in \text{INT}(\mathcal{L}) - SD(\mathcal{L})$ then $\mathcal{T}$ is a UQC trees for $P(\mathbf{a})$ iff we can find a complete pair $(\mathcal{T}_m, \mathcal{I}_m)$ and for each $N \in leaf(\mathcal{T}_m)$ a UQC tree $\mathcal{S}_N$ for some instance $lab(N)\theta$ of $lab(N)$ such that θ does not violate $\mathcal{I}_m$, $lab(root(\mathcal{T}_m))\theta = P(\mathbf{a})$ and $\mathcal{T}_m \theta$ and $\{\mathcal{S}_N : N \in leaf(\mathcal{T}_m)\}$ satisfy the conditions of Theorem 8.1.2.

9 Conclusions and further research

We have presented a top-down correct and complete query processing method for first order deductive databases under the DWFS. We have also investigated termination and efficiency aspects of our method by examining partial compilation, subgoal re-ordering and restrictions on database rule format. Our method is based upon a branch by branch tree traversal, and therefore (in common with the methods of [1]) for propositional databases operates in space that is polynomial in the size of the underlying language. The techniques presented in this paper have also been applied, suitably modified to the perfect and the disjunctive stable model semantics [6-9,11].

The following open questions are worthy of further investigation.

(i) Can our techniques handle query compilation in DWFS, in which we preprocess a query using INT(T) so that its run-time processing then involves EXT(T) only?
(ii) To what extent can we employ search space pruning techniques in order to make the tree constructions more efficient. In particular, can we use information from a UQC tree construction (say) to prune the resulting deduction tree constructions?
(iii) Can our methods be extended to languages containing function symbols?
(iv) Could we guarantee termination with weaker constraints and/or promote efficiency either by combining our methods with the bottom-up methods of [1], or by making further "natural" assumptions about deductive databases (eg., with respect to the amount of disjunctive or recursive information)?

References

1. C. Aravindan, J. Dix and I. Niemelä, DISLOP: A research project on disjunctive logic programming, AI communications, vol. 10 (1997), 151-165.
2. S. Brass, J. Dix, I. Niemelä and T. C. Przymusinski, A comparison of the static and disjunctive well-founded semantics, in: A.G. Cohn, L.K. Schubert and S.C. Shapiro (eds.), Proceedings of the 6th International Conference on Principles of Knowledge Representation and Reasoning (Morgan Kaufmann, 1998).
3. S. Brass and J. Dix, Characterisations of the disjunctive well-founded semantics: confluent calculi and iterated GCWA, J. Automated Reasoning, vol. 20 (1998), 143-165.
4. S. Brass and J. Dix, Semantics of (disjunctive) logic programs based upon partial evaluation, J. Logic Programming, vol. 40 (1999), 1-46.
5. J. Dix, Semantics of logic programs: Their intuitions and formal properties, in A. Fuhrman and H. Rott (eds.), Logic, Action and Information, Essays on Logic in Philosophy and Artificial Intelligence (DeGruyter, 1995), 241-327.
6. C. A. Johnson, On computing minimal and perfect model membership, Data and Knowledge Engineering, vol. 18 (1996), 225-276.
7. C. A. Johnson, Extended deduction trees and query processing, Computer Science technical report TR98-07, Keele University (1997).
8. C. A. Johnson, Top-down query processing in indefinite stratified databases, Data and Knowledge Engineering, vol. 26 (1998), 1-36. (Extracted from [7].)
9. C. A. Johnson, On cyclic covers and perfect models, Data and Knowledge Engineering, vol. 31 (1999), 25-65.
10. C. A. Johnson, On the computation of the disjunctive well-founded semantics, submitted, J. Automated Reasoning.
11. C. A. Johnson, Processing deductive databases under the disjunctive stable model semantics, Fundamenta Informaticae, vol. 41 (1999), 31-51.
12. I. Niemelä, A tableau calculus for minimal model reasoning, in: Proceedings of the 5th Workshop on Theorem Proving with Analytic Tableaux and Related Methods, Terrasini, Italy (Springer, 1996), 278-294.
13. A. Rajasekar, Semantics for Disjunctive Logic Programs, PhD thesis, University of Maryland (1989).

Discovering and Resolving User Intent in Heterogeneous Databases

Chris Fernandes[1] and Lawrence Henschen[2]

[1] Department of Computer Science,
Colby College, Waterville, ME 04901
[2] Department of Electrical and Computer Engineering,
Northwestern University, Evanston, IL 60201
henschen@ece.northwestern.edu

Abstract. We propose a system whereby subtle semantic ambiguity found in queries of distributed heterogeneous database systems can be resolved by considering the user's intentions. Through the use of domain-specific knowledge embedded within a mediator-based architecture, subtleties in meaning can be explicitly modeled. Through the use of dynamic profiles and active dialogue, the system can discover user intent, providing more satisfying query answers.

1 Introduction and Problem Statement

Modern heterogeneous database systems generally require users to issue queries via a global query language. This is because, typically, no one member database of the distributed system has all the concepts of the whole system. Moreover, two identical terms in two different local schemas may have slight semantic differences in the context of the entire network. To avoid the ambiguity caused by this heterogeneity, many systems have a global language containing a vocabulary of terms with exact pre-specified meanings. Wrappers are then used to translate global terms into their meanings within a local schema's context.

This global approach places excess burden on the users to first understand the semantic differences that terms may have in different databases and then to very carefully express queries to precisely reflect their desired semantics. First, note that global concepts are often expressed as generalizations that are broad enough to cover the varying meanings of that concept at the local levels. The user has to be aware of the different nuances that terms might have in the other local repositories and add appropriate constraints to be sure his/her intent is specified. Second, as new databases join the global network, existing terms may take on even more nuances, new terms and concepts may enter the global language, and perhaps even the global schema itself may change. Users will then need to become familiar with the changes, and so user training in a dynamic network is not a one-time affair like many researchers claim. Similarly, applications that had been written for the local database and been modified to adapt to the network at a particular point in time may also need additional updating as the network changes. For the same reason, commonly run queries that were written and saved under an older incarnation of the global schema may also need to be updated.

The opposite approach allows users to specify queries in their own local database language and provides translators to map those queries to a global form that reflects the semantics of that local database. This allows users to issue distributed queries and to correctly interpret answers without having to learn a new language and be continually retrained. This would be a distinct advantage for non-sophisticated users, and it is reasonable to expect a large number of database users to fit into this category. For example, a distributed information system might include car rental companies and meteorological databases, both of which use the term "map" with related but distinct meanings. A counter agent at a car rental location may never care about seeing meteorological maps of the region, only street maps to give to customers renting cars. Such counter agents are probably not database experts and should not be required to know about all the other kinds of maps that may be available and how to specify just street maps when issuing a query. On the other hand, the local-query approach forces any user who wants to take advantage of the variety of related information to again learn the global language and continuously keep up to date as the network evolves. A travel planner in the "maps" example may very well want one or both kinds of maps when planning a group tour to account for both the actual transportation as well as possible weather-related contingencies. Even more, the planner may issue the same query, for example "get a map of the Boston area", at different times and want different kinds of maps based on what aspect of the tour is being planned at that moment. But, again, travel planners are not likely to be database experts, and it would be advantageous to develop some kind of system that would help such a user cope with a semantically diverse and evolving distributed knowledge base.

We will describe in this paper a system that attempts to merge the best features of each approach and to help users of both kinds. Queries are expressed in local database languages, and the query-processing algorithm uses the query and a global knowledge base in combination with information from an individual user profile and even user dialogue to translate the local query into a global one. Through the use of the profile and dialogue, our system tries to discover the real intent of the user query. In addition, our system attempts to help the user specify the intent when it can't be guessed or when the user has requested help. The development and use of individual user profiles and the nature of the dialogue are the primary foci of this paper. They are described in Section 3. In Section 2, we differentiate our work from other research in this area, but space limitations preclude any extensive discussion of prior and related work. Section 5 presents concluding remarks and the direction we would like to take in the future.

2 Related Work

We have chosen a mediator-based system for our implementation since it allows for the two most important criteria we desire: the ability to ask queries through local member databases and the presence of a global knowledge base to perform translation on a query-by-query basis. Since intent can change over time, having this latter property is critical. Many other prototypes are also mediated systems, but most do not consider the idea of intent as we have stated it.

Most mainstream prototypes use a single global language to represent queries. These include rule-based languages such as the one used by the HERMES system [7], description logic-based languages such as Loom used by SIMS [1], clause-based languages such as that used by Information Manifold [4], and others such as OQL used by DISCO [8]. Using these global languages provides certain advantages. HERMES, for example, is able to incorporate a degree of probability into its rules, while Information Manifold's use of logical clauses provides a robust translation from its global language into local subqueries. However, they all prevent the user from using a local language with which s/he may already be familiar, and they all prevent the use of applications which already utilize that local language.

A means of adjusting to different user intents is another aspect found in only a few prototype systems. One such system, TSIMMIS [2], is web-based and allows for a limited type of intent clarification by including hypertext in a query result. By this hypertext the user can see more general or more specific information concerning the objects involved in the query answer. This could be extended so that the user could reissue a query after specifying a more appropriate level of object detail or after delineating ambiguous concepts. We propose a dialogue system which would clearly define user intent on a level not seen in these prototypes.

Another important aspect which we include in our system, but which others do not, is conditional attribute equivalence. Essentially, this term means that any two attributes which are considered to be equivalent or at least synonymous for one query may not be considered equivalent or synonymous for another. Consider a distributed system dealing with universities, where one local database keeps track of an undergraduate-only institution while the other holds data of a university with graduates and undergraduates. Two "student" attributes in these two local databases may be considered equivalent for a query dealing with counting up all students but quite dissimilar for a query dealing solely with graduate students. Existing systems such as OBSERVER [5] may use a single ontological term to relate a priori to multiple local attributes. As a result, there is no easy way for these equivalences to change from query to query. Our approach will use constructs called annotations [3,6] to accomplish conditional attribute equivalence. These annotations will provide for query-dependent processing, which we believe is essential for determining user intent.

3 Discovering a User's Intentions

Our system discovers different possible query interpretations at run-time. Due to space limitations we will not explain our query processing system in detail; the interested reader is referred to [3,6]. When necessary, we will explain those

Person		
Name	Phone	State
Fred	111-1111	VA
Sally	222-2222	IL
Tom	333-3333	HI

```
SELECT name, phone
FROM Person
WHERE state="IL"
```

Fig. 1. Sample query in an Employee Database

features of the query processor that intersect with our techniques for determining the user's true intentions.

3.1 A Motivating Example

We first illustrate that subtle heterogeneity can creep into even the most innocuous of queries. Figure 1 shows a table from an employee database showing each worker's name, phone number, and place of residence. The accompanying query is asking for the names and phone numbers of those employees residing in IL. While the semantics for such a query would be clear for any one relational database, it becomes more complex if this is a distributed query over many employee tables. Figure 2 shows two local databases containing identically structured employee tables, but differing in their interpretation of the **phone** attribute. In DB1, the phones are all cellular phones. They do not have a permanent location, and they are probably kept with the employee most of the time. However, in DB2, the phones are regular permanent phones, located most likely at the employee's place of residence. This difference is not reflected in either of the two tables, and it produces two different query interpretations. Did the user specify "IL" because s/he intended to call there, perhaps looking for the employee's family? Or did the user intend to contact *people* based in IL, regardless of where their phones are? One's initial solution in a small example like this might be to return both sets of phone numbers so that the user could decide. But this solution is not feasible in general, especially where scalability and time are factors. In an emergency situation such as a plane crash, passenger manifests need to be used to notify family members quickly. In this scenario, permanent phone numbers would be desired, not cell phone numbers. Having the system return hundreds of extraneous tuples that the user must sift through is not practical.

3.2 Finding Ambiguities in Intent

In order to choose the desired interpretations of a given query, a system must be able to explicitly model subtle semantic differences like the one given here.

DB1: cellular phones

Person		
Name	Phone	State
Fred	111-1111	VA
Sally	222-2222	IL
Tom	333-3333	HI

DB2: regular phones

Empl		
Name	Regphone	Place
Sally	444-4444	IL
Carl	555-5555	IL
Beth	666-6666	VA

Fig. 2. Subtle Heterogeneity in Logically Identical Tables

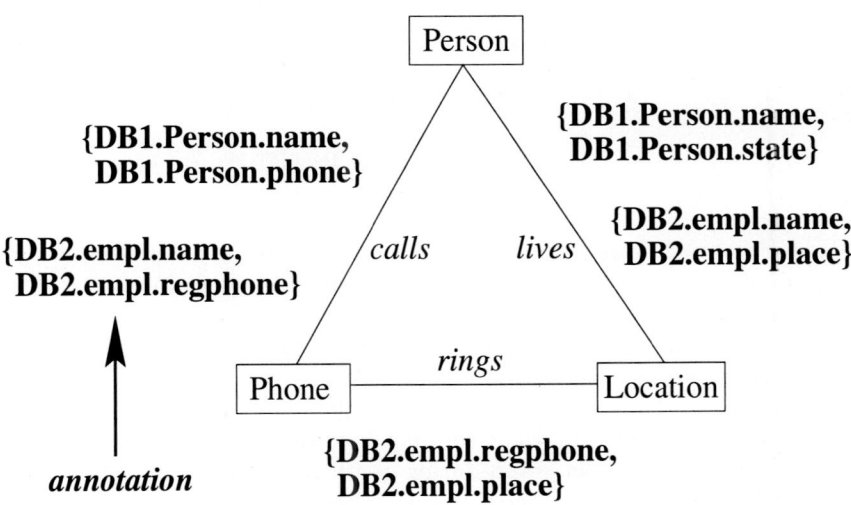

Fig. 3. Section of Mediator used for Phone Example

The ambiguity in the phone example stems from the fact that there are distinct relationships between the **phone** and **state/place** attributes in DB1 and DB2. We have developed a graph-based mediator [3,6] that precisely and explicitly maps out concepts and relationships at the global level and relates these to the concepts and relationships in the individual local databases. Although the global representation and the related algorithms are not the focus of this paper (see [3,6] for full discussions of these), we illustrate briefly how our system models semantic variations.

Figure 3 shows a portion of the mediator for our phone example. Vertices represent concepts, while edges represent relationships. Knowledge is embedded in the mediator in the form of **annotations**. An annotation is an ordered pair

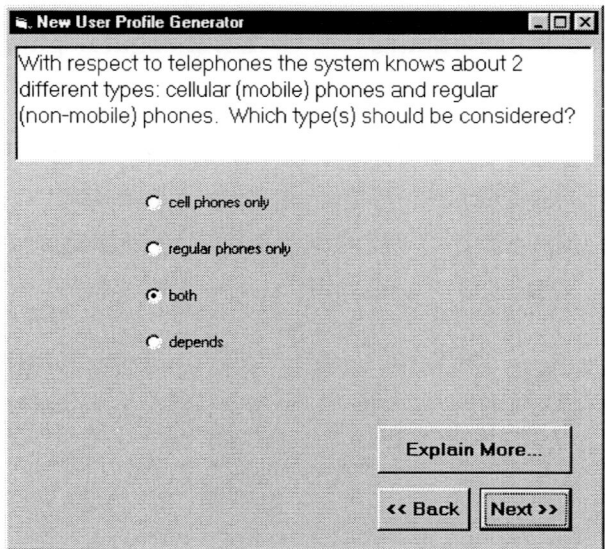

Fig. 4. Creating a User Profile

$\{x, y\}$ where x and y are local database attributes. If an annotation $\{x, y\}$ is associated with a relationship R, then R **must** hold between local attributes x and y. Thus, the annotation $\{DB2.empl.regphone, DB2.empl.place\}$ shown in Figure 3 means that the relationship *rings* must hold between the **Phone** and **Location** concepts in local database DB2. That is, for DB2, the phone must ring in the specified place. The lack of a similar annotation for DB1 for the *rings* relationship means that *rings* does not hold or is unknown in DB1. When analyzing this section of the graph during query processing, our algorithm can use the presence or absence of certain annotations to detect possible differences in intent.

3.3 Resolving Ambiguities in Intent

Once the source of competing interpretations has been found, the system can then move to the task of resolving the ambiguity. We have developed two methods for determining what the user truly intended: individual user profiles and on-the-fly dialogue.

A **user profile** is a list of preconceived notions that the user has about what local query concepts mean. A profile may contain domain-related information, e.g. the concept *student* means just undergraduates. It may also contain general query processing preferences, e.g. to exclude a particular local database from contributing answers or to allow separate databases to participate in providing partial answers that can be combined into a complete whole (i.e. inter-database joins.)

A profile can be created and modified before query run-time. In the system we are developing, a profile is initially created when a new user logs into the system for the first time. At that point, the user is asked a series of questions about how s/he interprets certain local concepts. A screenshot in our system for this process is shown in Figure 4. The result is a personal knowledge base that the query processor can consult to help determine a user's intent. In our phone example, the query processor could examine this profile when it considered the *rings* relationship shown in Figure 3. If the user had indicated that only non-mobile phones are of interest, then DB1, containing only cell phones, would be eliminated from the search space.

Profiles need not exist only for individual users, however. Profiles for specific classes of individuals could also be created. For example, all incoming business school students at a university or all new secretaries in a corporation could have a class profile that would ensure a uniform query vocabulary. The fact that a local query language is being used would provide a foundational interpretation of terms for creating such a class profile. Similarly, a profile could be set up for an application, extending its access from the original database to a distributed federation of databases. Moreover, as the global schema changes in response to the addition or deletion of member databases, administrators could modify these class profiles to keep end users and applications up-to-date. These changes would be transparent to non-sophisticated users while still allowing more advanced users to tweak their own profiles if they so desire.

In addition to a profile generator, we have also implemented a DBA toolkit by which profile generators can be created quickly. This toolkit is illustrated in Figure 5. We envision that both local and global database administrators are in the best position for understanding the subtle semantic differences between the syntactically identical concepts across individual member databases. The administrators can then utilize this tool to form the questions which will make up profile generators. These generators can be tailored by each local DBA so that questions are best expressed to match the level of user expertise and "lingo" in use at that local site.

A query-independent mechanism like a user profile is not sufficient by itself, however. A user may be aware that his/her interpretation of a term may change depending on the query asked. The profile generator accounts for this by allowing the user to pick a "depends" option as shown in Figure 4. This tells the system that when a particular concept is used in a query, its meaning will need to be determined at run-time. In this way, a user would not need to continually update his/her profile just because s/he wishes to execute a different query. Instead the "depends" option signals the query processor to ask the user (say, via a set of choices in a dialogue box) for the user's interpretation of the concept for this query. This type of dialogue is called **on-the-fly dialogue**.

Consider the mediator subset in Figure 6 dealing with an ecommerce network. Here a single member database, DB3, allows two possible interpretations for queries dealing with buyers. The Buyer concept can either represent those who have actually made purchases of a given product (kept in the *Product* ta-

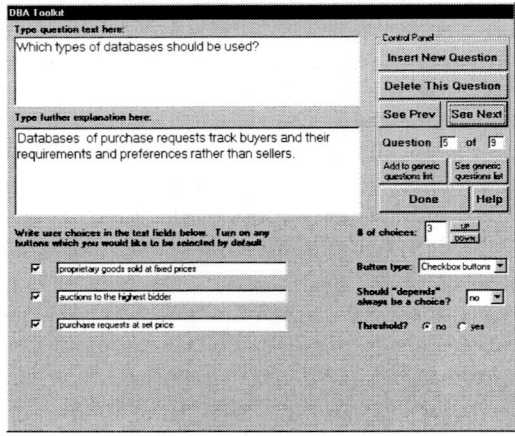

Fig. 5. DBA Toolkit for Creating Profile Generators

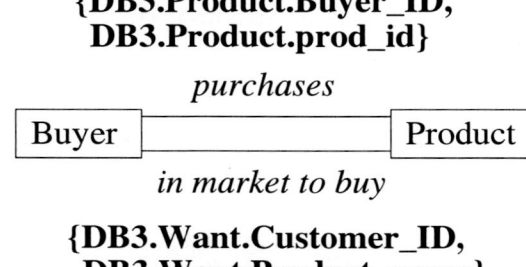

Fig. 6. Subset of Mediator used in an Ecommerce Network

ble of DB3), or it can represent those who are just in the market to buy said product (kept in the *Want* table of DB3.) Such a distinction would be made, for example, in a name-your-own-price website such as *priceline.com*. During processing of a query dealing with buyers, the mediator would detect the two different relationships possible within DB3 and consult the user's profile to find a preference. If one is not found (if the user chose "depends" as an answer, for example) then the user would need to be asked to specify which relationship s/he wanted. Tuples from the correct table in DB3 can then be retrieved.

The gravest potential pitfall of on-the-fly dialogue is its overuse. Since it is an interruption during query processing, we wish to eliminate unnecessary interaction. We have several ways by which our system can deduce a correct interpretation. First, the profiles of other users of the same local database can be used as a guide. If most users interpret a term in the same way, the system

DB1: Dan's Good & Services
```
Transaction(PURCHASE #, date, buyer
            product_code, quantity)

Product(PRODUCT_CODE, product_name,
        description, amt_in_stock, unit_price)
```

DB2: Al's Auctions
```
Product(PRODUCT_ID, description
        opening_bid)

Auction(AUCTION_ID, seller_id, product_id,
        high_bid, buyer_id, closing_time, num_bids)

Seller(SELLER_ID, seller_name, seller_email)

Buyer(BUYER_ID, buyer_name, buyer_email,
      current_bid, product_id)
```

DB3: Sam's Shopbot Site
```
Buyer(SAM_ID#, name, email)

Want(SAM_ID#, PRODUCT_NAME, PRICE)

Product(PROD_ID, SAM_ID#, product_name, seller,
        actual_price, qty, time_to_find)
```

Fig. 7. Schemas Used in Experimental Ecommerce Network (keys are in CAPS)

could prevent an interruption by assuming the same interpretation. Second, the result of an on-the-fly dialogue, no matter what the user answers, may end up being moot. In our phone example, a local database may be eliminated from the search space depending on which phone type the user specified in the dialogue. If that local repository has already been eliminated due to other query constraints, then asking the user for clarification is fruitless. Third, by using the log data of past queries, it is possible to detect when the result of an on-the-fly dialogue would only result in an insignificant increase in the number of returned tuples. In this situation, the interpretation which produced the greater number of answers could be assumed with insignificant effect on processing time or user time when sorting through the results. In general, on-the-fly dialogue could be bypassed if the effect of "guessing wrong" was negligible.

4 Preliminary Results and Future Work

We have already implemented our profile generators, DBA toolkit for making generators, and our annotated mediator. We have also completed preliminary experiments using concepts and queries from an ecommerce domain. In these experiments, both novice and expert computer users issued queries to a set of three databases that each performed on-line buying and selling in a different way. The first sold goods and services at fixed prices, the second auctioned off goods

Table 1. Results from Experiment

Query	User Level	Number of Matches	Number of Mismatches	Probability of Independence
Query 1	novice	3	23	.0008
	expert	0	7	
Query 2	novice	12	14	.7001
	expert	3	4	
Query 3	novice	15	11	.3296
	expert	5	2	
Query 4	novice	9	17	.2538
	expert	4	3	
Query 5	novice	16	10	.1154
	expert	6	1	

to the highest bidder, and the third was a shopbot-based database that worked in a similar fashion to *priceline.com*. Their schemas are shown in Figure 7.

Users were first asked to create a profile which recorded their own opinions concerning common terms used in this domain. For example, a user was asked to define the term "Buyer" given the two interpretations presented previously in Section 3.3 and shown in Figure 6. Then, each user was shown the same set of test queries and asked to profess what they considered to be "correct" answers based on their profiles and responses to on-the-fly dialogue. The users then compared the answers returned by the system to the answers they thought would be "correct" to see if the system could properly discern their interpretation of the query. Chi-Square analysis was performed on the raw data to test for a relationship between the two user groups. The results, displayed in Table 1, show the number of users whose query interpretations matched those of the system alongside those whose interpretations did not match.

Several conclusions can be drawn from the data. The low number of matches for the first query is not surprising since the idea of multiple query interpretations is a new one to most users. This query represents a training period for the user. As time went on, the user's interpretation of the query was correctly reflected by the system's interpretation with increasing accuracy. The continually-reducing probability of independence between the two user groups indicates that expert users tended to become adept at using the system in a shorter amount of time than novices. Because non-experts are the target audience for a system of this type, we hope in the future to adjust the wording of questions in the profile generator and to provide a more robust training period using more sample queries. It is hoped that these will provide a ramp to proficiency that is neither too steep nor too long for either type of user.

5 Conclusion

We believe that the discovery of intent in queries is an important aspect in determining if returned answers are indeed "correct" or not. If a system returns tuples based on its own static interpretation of query terms, regardless of how broadly encompassing those terms may be, the lack of consideration for the user's expectations will always leave room for misinterpretation.

We have developed and implemented a system that attempts to take the user's world assumptions into account by using profiles and on-the-fly dialogue to guide the query translation process. Using preferences specified both before and during run-time, a single query can return different sets of answers that correspond to what different users had in mind. Preliminary empirical evidence is favorable, though we hope to do more experiments to increase proficiency and decrease training time.

References

1. Arens, Hsu, and Knoblock. Query processing in the SIMS information mediiator. In Austin Tate, editor, *Advanced Planning Technology*, Menlo Park, CA, 1996. AAAI Press.
2. Garcia-Molina, Papakonstantinou, Quass, Rajaraman, Sagiv, Ullman, and Widom. The TSIMMIS approach to mediation: Data models and languages (extended abstract). *JIIS*, 1997.
3. Henschen, Neild, and Fernandes. An object-oriented graph traversal algorithm for data mediation. In *Proceedings of the 2nd Americas Conference on Information Systems (AIS96)*, Phoenix, AZ, August 1996.
4. Levy, Rajaraman, and Ordille. Query-answering algorithms for information agents. In *Proceedings of the Thirteenth National Conference on Artificial Intelligence (AAAI-96)*, Portland, OR, December 1996.
5. Mena, Kashyap, Sheth, and Illarramendi. OBSERVER: An approach for query processing in global information systems based on interoperation across pre-existing ontologies. In *COOPIS*, Brussels, Belgium, June 1996.
6. Tania Neild. *The Virtual Data Integrator: An Object-Oriented Mediator for Heterogeeous Database Integration*. PhD thesis, Northwestern University, June 1999.
7. Subrahmanian, Adah, Brink, Emery, Lu, Rajput, Rogers, Ross, and Ward. HERMES: A heterogeneous reasoning and mediator system. submitted for publication.
8. Tomasic, Raschid, and Valduriez. Scaling heterogeneous databases and the design of DISCO. In *Proceedings of the International Conference on Distributed Computer Systems*, 1996.

Discovering and Matching Elastic Rules from Sequence Databases

Sanghyun Park and Wesley W. Chu

Department of Computer Science
University of California, Los Angeles
Los Angeles, CA 90095, USA
{shpark, wwc}@cs.ucla.edu

Abstract. This paper presents techniques for discovering and matching rules with *elastic patterns*. Elastic patterns are ordered lists of elements that can be stretched along the time axis. Elastic patterns are useful for discovering rules from data sequences with different sampling rates. For fast discovery of rules whose heads (left-hand sides) and bodies (right-hand sides) are elastic patterns, we construct a trimmed suffix tree from succinct forms of data sequences and keep the tree as a compact representation of rules. The trimmed suffix tree is also used as an index structure for finding rules matched to a target head sequence. When matched rules cannot be found, the concept of *rule relaxation* is introduced. Using a cluster hierarchy and relaxation error as a new distance function, we find the least relaxed rules that provide the most specific information on a target head sequence. Experiments on synthetic data sequences reveal the effectiveness of our proposed approach.

1 Introduction

Rule discovery from sequential data is a data mining technique for trend prediction [3][7]. There have been several approaches [1][6][9][14] to discover useful rules from patterns occurring frequently in data sequences. A pattern is defined as a partially ordered collection of elements. According to the constraints on the arrangement of elements, patterns can be classified as serial patterns and parallel patterns [9].

As a subset of serial patterns, we can think of *elastic patterns* where elements can be stretched along the time axis by replicating themselves. Elastic patterns AB and ABC are interpreted as A^+B^+ and $A^+B^+C^+$, respectively, using the notation of a regular expression. $\langle A, B \rangle$ and $\langle A, A, B, B, B \rangle$ are instances of an elastic pattern AB while $\langle A, C, B \rangle$ is not. Elastic patterns are useful for discovering rules from data sequences whose sampling rates may vary. For example, consider medical data sequences that record the body temperatures of patients. Some data sequences may have temperature values taken every day while others may have values taken every week. Furthermore, even within a single data sequence, time intervals between neighboring temperature values can vary non-linearly. These sequences cannot be compared directly without considering stretches or compressions of elements along the time axis.

The rules whose heads (left-hand sides) and bodies (right-hand sides) are elastic patterns are called *elastic rules*. Given elastic patterns α and β, elastic rules have the format '$\alpha \to \beta$' that is interpreted as "if there occurs a sequence which is an instance of α, then it will be followed by a sequence which is an instance of β". Time intervals are not associated with elastic patterns because these patterns are flexible on the time axis.

There are many techniques [1][6][9][14] to discover rules with serial patterns. Many of them use the relationship between patterns and their sub-patterns. Given the serial patterns AB occurring 200 times and $ABAC$ occurring 150 times, they extract the rule '$AB \to AC$ with confidence $\frac{150}{200}(= 0.75)$'. Infrequent patterns whose numbers of occurrences are below a threshold are ignored because infrequent patterns are considered insignificant. To find frequent patterns, they first find short frequent patterns and then combine them to generate longer candidate patterns. Candidate patterns are checked whether they are frequent or not. These combining and checking steps are repeated until all frequent patterns are found. Therefore, repeated readings of data sequences are unavoidable.

Once rules are discovered from data sequences, they may be used to predict the future trend of a target head sequence $\vec{q}$ via the process of rule matching. We say that a rule is matched to $\vec{q}$ when each element of the rule head is equal to the corresponding element of $\vec{q}$. However, if there are large number of rules, it is not a trivial task to find rules efficiently that are matched to $\vec{q}$.

There are some occasions when we fail to find rules matched to a target head sequence $\vec{q}$. This failure often occurs when $\vec{q}$ is not a frequent sequence. For those infrequent target head sequences, we can introduce the concept of *rule relaxation*. Based on a cluster hierarchy, a rule R is relaxed to R' by replacing some elements of R with elements denoting more general concepts or broader range values. Given a target head sequence $\vec{q}$ and a rule R that is not matched to $\vec{q}$, we can relax R to R' so that R' *covers* $\vec{q}$. We say that a rule covers $\vec{q}$ when each element of the rule head represents the same range as or broader range than the one represented by the corresponding element of $\vec{q}$. Among many relaxed rules that can cover $\vec{q}$, we are interested in finding the least relaxed rules since they describe $\vec{q}$ more accurately than the other relaxed rules.

In this paper, we investigate the problems of discovering and matching elastic rules for data sequences with different sample rates. An efficient rule discovering algorithm is developed and algorithms for exact and relaxed rule matching algorithms are presented.

2 Background

2.1 Suffix Tree

A suffix tree [13] is an index structure that has been used as a fast access method to locate substrings (or subsequences) that are exactly matched to a target string (or a target sequence). The suffix tree structure is based on *tries* and *suffix tries*. A trie is an indexing structure used for indexing sets of keywords of varying sizes. A suffix trie is a trie whose set of keywords comprises the suffixes of sequences. Nodes of a suffix trie with a single outgoing edge can be collapsed, yielding a suffix tree. We use the notation PN for the parent node of N, and the notation $label(N_i, N_j)$ for the labels on the path connecting nodes N_i and N_j.

2.2 Type Abstraction Hierarchy

Type Abstraction Hierarchy (TAH) [5] is a data-driven multi-level cluster hierarchy that uses relaxation error as a goodness measure for generating clusters. For a cluster $C = \{x_1, x_2, ..., x_n\}$ of n elements, the relaxation error for C is defined as $RE(C) = \sum_{i=1}^{n} \sum_{j=1}^{n} P(x_i)P(x_j) \mid x_i - x_j \mid$ where $P(x_i)$ and $P(x_j)$ are occurring probabilities of x_i and x_j, respectively. The algorithms for generating binary and n-ary TAHs are given in [5]. TAH is easier to implement than the maximum-entropy clustering method and generates more accurate clusters than the equal-length interval clustering method. Figure 1 shows an example TAH built from a distribution of data sequences whose elements take values within the range of $[0, 7.0)$. The relaxation error and the value range are stored in each node, and the nodes are labeled with unique symbols.

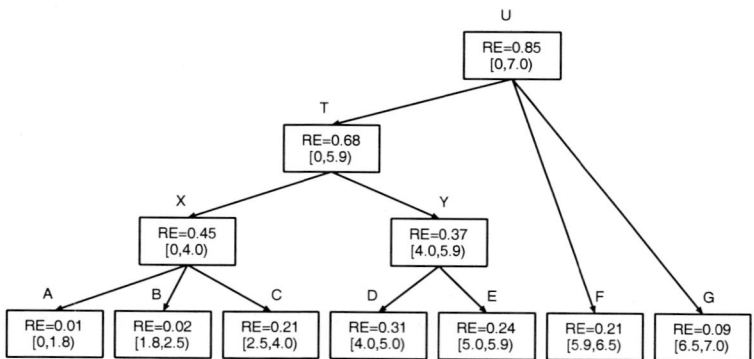

Fig. 1. An example of TAH. Each node is labeled with a unique symbol. The value range and the corresponding relaxation error are stored at each node.

3 Rule Discovery

In this section, we propose an efficient method to discover elastic rules from data sequences via a suffix tree. We assume that the TAH has been generated from data sequences and distinct symbols have been assigned to the TAH nodes.

The support value of the pattern α is defined as the number of suffixes having α as their prefixes. SUP_{min} is the minimum support value that is used to filter out infrequent patterns. We also define the relative support value of the pattern α as $RSUP(\alpha)$ = (the number of suffixes having α as their prefixes) / (the total number of suffixes). For the applications where the total number of data sequences and their lengths may vary, the relative support is better than the (absolute) support.

The problem of elastic rule discovery is defined as follows: Given a database with M sequences $\vec{x}_1, \vec{x}_2, ..., \vec{x}_M$ and the minimum support value SUP_{min}, discover rules composed of elastic patterns whose supports are at least SUP_{min}. This elastic rule discovery consists of the following five steps.

Step 1. Converting numeric elements to symbol elements: We convert each numeric element of data sequences into the symbol of the corresponding leaf node of the TAH. The symbolized representation of $\vec{x}$ is denoted as $S(\vec{x})$. For example, based on the TAH in Figure 1, a data sequence $\vec{x} = \langle 3.4, 3.0, 3.7, 2.3, 2.1, 4.3 \rangle$ is converted to $S(\vec{x}) = \langle C, C, C, B, B, D \rangle$.

Step 2. Compaction: We convert the symbolized data sequence $S(\vec{x})$ into the compact representation $C(S(\vec{x}))$ by replacing consecutive elements that have the same value with a single element of that value. This step is for considering the property of elastic patterns. For example, $S(\vec{x}) = \langle C, C, C, B, B, D \rangle$ is converted to $C(S(\vec{x})) = \langle C, B, D \rangle$. We use the notation $\vec{X}$ for $C(S(\vec{x}))$.

Step 3. Suffix tree construction: From the set of M converted data sequences $\vec{X}_1, ..., \vec{X}_M$, we build a suffix tree using either McCreight's algorithm [8] or incremental disk-based algorithm [4].

Step 4. Trimming: We compute the support values of the nodes and trim out the nodes whose support values are less than SUP_{min}. The support values of internal nodes are obtained by summing up the support values of their children nodes. The support values of the leaf nodes are the same as the number of suffixes represented by the leaf nodes. The trimmed suffix tree is called the *rule tree*.

Step 5. Rule extraction: We compute the confidence values of nodes and then extract rules. The expression for computing the confidence value of the node N is $confidence(N) = Support(N)/Support(PN)$ where PN is the parent node of N. If the number of labels on the path from PN to N is L, we extract L rules as shown in the following:

$R_1 : label(rootNode, PN) \to label(PN, N)$
$R_2 : label(rootNode, PN) \bullet (label(PN, N)[1:1]) \to label(PN, N)[2:L]$
$R_3 : label(rootNode, PN) \bullet (label(PN, N)[1:2]) \to label(PN, N)[3:L]$
...
$R_L : label(rootNode, PN) \bullet (label(PN, N)[1:L-1]) \to label(PN, N)[L:L]$

where $label(PN, N)[i : j]$ is the subsequence of $label(PN, N)$ including elements in positions i through j $(i \leq j \leq L)$, and '•' is the binary operator for concatenating two sequences. If N is the root node, then $label(rootNode, PN)$ becomes the empty sequence $\langle \rangle$. The confidence of R_1 is the same as confidence(N) while the confidences of R_2, R_3, ..., and R_L are 1. Figure 2 shows a part of a rule tree and the rules extracted from the tree. The values in the nodes represent their support values.

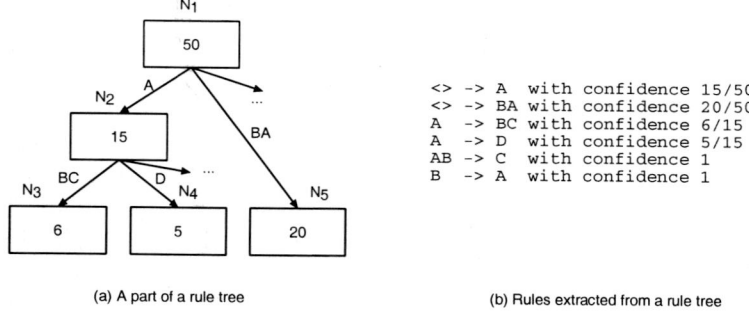

(a) A part of a rule tree   (b) Rules extracted from a rule tree

Fig. 2. An example of a rule tree and the corresponding rules.

4 Exact Rule Matching

Exact rule matching is defined as follows: Given a rule tree, a type abstraction hierarchy, and a target head sequence $\vec{q}$, find the rules matched to $\vec{q}$. Our approach for exact rule matching consists of two steps.

Step 1. Search for exactly matched rule head: Using the rule tree as an index structure, we find the rule head $\vec{h}$ that is exactly matched to a target head sequence. Algorithm 1 shows the exact matching algorithm RTI-E (Rule Tree Index for Exact matching). We use the notation $\vec{Q}$ for $C(S(\vec{q}))$. The algorithm traverses the rule tree and returns a pair (N, p) that represents the matched rule head $\vec{h} = label(rootNode, PN) \bullet (label(PN, N)[1 : p])$. The first call to the algorithm has two arguments: rootNode and $\vec{Q}$.

Step 2. Rule extraction from exactly matched rule head: Using the relationship between the exactly matched rule head and its following subsequences, we extract the rules. Let us assume that RTI-E has returned the pair (N, p) and the length of $label(PN, N)$ is L. If $p < L$, then the matched rule is '$label(rootNode, PN) \bullet (label(PN, N)[1 : p]) \rightarrow label(PN, N)[p + 1 : L]$ with confidence 1'. Otherwise, the number of matched rules is the same as the number of children of N. For each child node CN of N, the matched rule is '$label(rootNode, N) \rightarrow label(N, CN)$ with confidence(CN)'.

> **Input** : node N, target head sequence $\vec{Q}$
> **Output**: child node CN, length of matched prefix
>
> Visit the node N;
> Select the child node, CN, where $label(N, CN)$ is matched to the prefix of $\vec{Q}$;
> Remove the matched prefix from $\vec{Q}$;
> **if** $\vec{Q}$ *becomes empty* **then** return a pair (CN, the length of a matched prefix);
> **else** call RTI-E($CN, \vec{Q}$);

Algorithm 1: Exact matching algorithm RTI-E

5 Relaxed Rule Matching

Relaxed rule matching is defined as follows: Given a rule tree, a type abstraction hierarchy, and a target head sequence $\vec{q}$, find the least relaxed rules that cover $\vec{q}$.

Since a rule head $\vec{h}$ whose length is not equal to $|\vec{q}|$ may be stretched and relaxed to cover $\vec{q}$, we propose a relaxation-error based time warping distance function $D_{RE}(\vec{h},\vec{q})$ as a similarity measure of $\vec{h}$ and $\vec{q}$. $D_{RE}(\vec{h},\vec{q})$ stretches $\vec{h}$ and $\vec{q}$ non-linearly to find the best element mappings that minimize the difference of $\vec{h}$ and $\vec{q}$. Let $\vec{h}[i]$ be mapped to $\vec{q}[j]$. Then, the distance of this mapping is defined as $RE(\vec{h}[i],\vec{q}[j]) - RE(\vec{h}[i])$ where $RE(\vec{h}[i],\vec{q}[j]))$ is the relaxation error of the lowest node containing both $\vec{h}[i]$ and $\vec{q}[j]$, and $RE(\vec{h}[i])$ is the relaxation error of the lowest node containing $\vec{h}[i]$. The detailed description of $D_{RE}(\vec{h},\vec{q})$ is given in [10]

$D_{RE}(\vec{h},\vec{q})$ can be calculated efficiently by the dynamic programming technique [2] based on the recurrence relation $r(x,y)$. ($x = 1, ..., |\vec{h}|, y = 1, ..., |\vec{q}|$). The final cumulative distance, $r(|\vec{h}|, |\vec{q}|)$, is the amount of relaxation needed for $\vec{h}$ to cover $\vec{q}$. Using the cluster hierarchy (Figure 1), Figure 3 shows the cumulative distance table T for $D_{RE}(\vec{h} = \langle C, A, E, D, A \rangle, \vec{q} = \langle C, E, A \rangle)$ and the element mappings after time warping and relaxation. In the following, we present the two-step approach for relaxed rule matching.

Step 1. Search for the nearest rule head: To generate the least relaxed rules, we first traverse the rule tree to find the rule head $\vec{h}$ that requires the least relaxation to cover a target head sequence $\vec{q}$. The similarity matching algorithm RTI-S (Rule Tree Index for Similarity matching) is given in Algorithm 2. Note that a target head sequence $\vec{q}$ is converted to the compact representation $\vec{Q}$ before beginning the search process. The algorithm maintains three global variables during its execution: $\vec{Q}$, the nearest rule head $\vec{h}$ found so far, and its distance $MinDist$ from $\vec{Q}$. The first call to the algorithm has two arguments: rootNode and emptyTable. RTI-S reduces the search time by applying the branch-pruning approach [11] and by allowing the cumulative distance table to be shared by rule heads that have the same prefix.

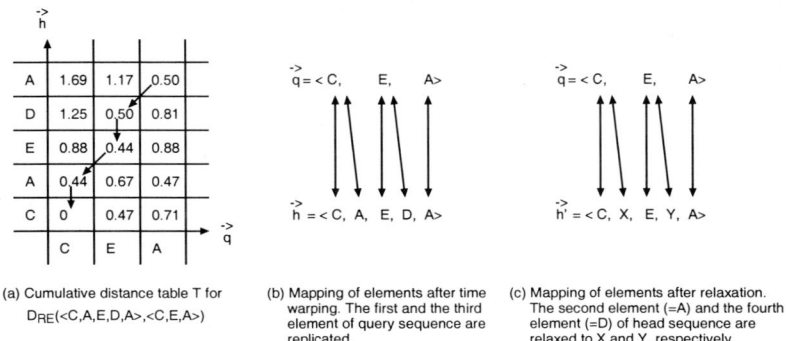

(a) Cumulative distance table T for $D_{RE}(<C,A,E,D,A>,<C,E,A>)$

(b) Mapping of elements after time warping. The first and the third element of query sequence are replicated.

(c) Mapping of elements after relaxation. The second element (=A) and the fourth element (=D) of head sequence are relaxed to X and Y, respectively.

Fig. 3. Cumulative distance table T for $D_{RE}(\vec{h} = \langle C, A, E, D, A \rangle, \vec{q} = \langle C, E, A \rangle)$ and the mapping of elements after time warping and relaxation.

Input : node N, cumulative distance table T

Visit the node N;
for each child node CN **do**
　Build a new cumulative distance table $newT$, by adding rows corresponding to $label(N, CN)$ on T;
　Find a nearer rule head $\vec{h}$ from $newT$ and update $MinDist$;
　If further traverse-down the tree is necessary, call RTI-S($CN, newT$);

Algorithm 2: Similarity matching algorithm RTI-S

Step 2. Rule extraction from the nearest rule head: After finding the rule head $\vec{h}$ most similar to $\vec{Q}$, we generate the least relaxed rules from $\vec{h}$ and its following subsequences. This step begins with extracting the rules from $\vec{h}$ using the method explained in Section 4. Then, we convert symbols of rule heads and bodies into their relaxed symbols according to the mapping of $\vec{h}$ and $\vec{Q}$, and get the compact representations of rules. Finally, the rules having the same head and body are merged and their confidence values are recomputed.

6 Experiments

To study the effectiveness of our proposed methods, we performed experiments on the *random walk* synthetic data sequences [10]. We used the relative minimum support value $RSUP_{min}$ to control the number of discovered rules.

6.1 Rule Discovery

We used the total elapsed time as a performance measure of our rule discovery algorithm. First, we increased the number of sequences from 100 to 10,000 while keeping their average length constant at 200. Then, we changed the average length of sequences from 100 to 1,000 while maintaining the number of sequences at 500. As shown in Figures 4 and 5, the total elapsed times increase linearly as the number of and the average length of data sequences grow. The figures also show that the linearity is maintained with different $RSUP_{min}$ values.

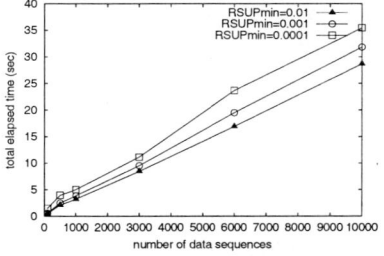 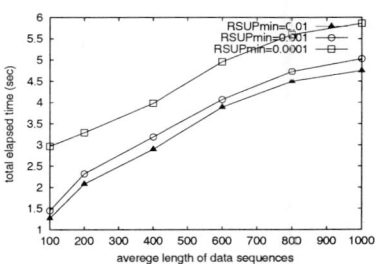

Fig. 4. Total elapsed time for discovering elastic rules with selected numbers of data sequences.

Fig. 5. Total elapsed time for discovering elastic rules with selected average length of data sequences.

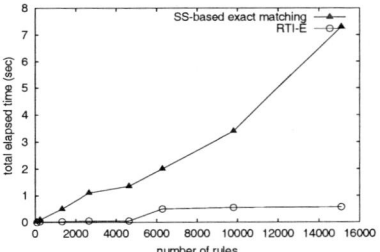 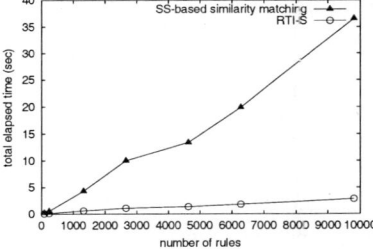

Fig. 6. Performance comparison between sequential scanning and RTI-E for exact rule matching.

Fig. 7. Performance comparison between sequential scanning and RTI-S for relaxed rule matching.

6.2 Rule Matching

For 500 data sequences with the average length 400, Figure 6 shows the average search times of RTI-E and SS(Sequential- Scanning)-based exact matching algorithm for increasing numbers of rules. The search times of SS-based exact matching algorithm increase linearly with the number of rules while the search times of RTI-E remain relatively constant. Figure 7 shows the average search times of RTI-S and SS-based similarity matching algorithm. The performance gain of RTI-S increases as the number of rules increases.

7 Conclusion

In this paper, we proposed a method to discover elastic rules from sequence databases. We also presented efficient techniques to find matched rules and to derive the least relaxed rules. We focused on data sequences consisted of univariate numeric values. If elements are non-numeric, we employ an encoding scheme that converts non-numeric elements to numeric elements.

Experiments on synthetic data sequences revealed that: 1) our rule discovering algorithm is linear to both the total number of and the average length of data sequences, and 2) our exact and relaxed rule matching algorithms are several orders of magnitude faster than sequential scanning.

References

1. R. Agrawal, and R. Srikant, "Mining Sequential Patterns", *Proc. IEEE ICDE*, 1995.
2. D. J. Berndt, and J. Clifford, "Finding Patterns in Time Series: A Dynamic Programming Approach", *Advances in Knowledge Discovery and Data Mining*, AAAI/MIT, 1996.
3. P. S. Bradley, U. M. Fayyad, and O. L. Mangasarian, "Data Mining: Overview and Optimization Opportunities", *Microsoft Research Report MSR-TR-98-04*, 1998.
4. P. Bieganski, J. Riedl, and J. V. Carlis, "Generalized Suffix Trees for Biological Sequence Data: Applications and Implementation", *Proc. Hawaii Int'l Conf. on System Sciences*, 1994.
5. W. W. Chu, and K. Chiang, "Abstraction of High Level Concepts from Numerical Values in Databases", *Proc. of AAAI Workshop on Knowledge Discovery in Databases*, 1994.
6. G. Das, K. Lin, H. Mannila, G. Renganathan, and P. Smyth, "Rule Discovery from Time Series", *Proc. International Conference on Knowledge Discovery and Data Mining*, 1998.
7. U. M. Fayyad, "Mining Databases: Toward Algorithms for Knowledge Discovery", *Data Engineering Bulletin 21(1)*, 1998.
8. E. M. McCreight, "A Space-Economical Suffix Tree Construction Algorithm", *Journal of the ACM*, Vol. 23, No. 2, 1976
9. H. Mannila, H. Toivonen, and A. I. Verkamo, "Discovering Frequent Episodes in Sequences", *Proc. International Conference on Knowledge Discovery and Data Mining*, 1995.
10. S. Park and W. W. Chu, "Discovering and Matching Elastic Rules from Sequence Databases", *UCLA Technical Report UCLA-CS-TR-200012*, 2000.
11. S. Park, W. W. Chu, J. Yoon, and C. Hsu, "Efficient Searches for Similar Subsequences of Different Lengths in Sequence Databases", *Proc. IEEE ICDE*, 2000.
12. L. Rabinar, and B. Juang. *Fundamentals of Speech Recognition*, Prentice Hall, 1993.
13. G. A. Stephen, *String Searching Algorithms*, World Scientific Publishing, 1994.
14. J. T.-L. Wang, G.-W. Chirn, T. G. Marr, B. Shapiro, D. Shasha, and K. Zhang, "Combinatorial Pattern Discovery for Scientific Data: Some Preliminary Results", *Proc. ACM SIGMOD*, 1994.

Perception-Based Granularity Levels in Concept Representation

Lorenza Saitta[1] and Jean-Daniel Zucker[2]

[1] Univ. del Piemonte Orientale Dip. di Scienze e Tecnologie Avanzate
Corso Borsalino 54, 15100 Alessandria (Italy)
saitta@di.unito.it

[2] Université Paris VI – CNRS, Laboratoire d'Informatique de Paris 6
4, Place Jussieu, F-75252 Paris (France)
Jean-Daniel.Zucker@lip6.fr

Abstract. In this paper we propose a perception-based view of abstraction, which originates from the observation that conceptualization of a domain involves entities belonging to several epistemological levels. The fundamental level corresponds to the *perception* of a world. For memorization purposes, some kind of *structure* is needed, in order to organize objects and relations perceived in the world into coherent ensembles. To communicate with others, a *language* must be invented, and, finally, a *theory* makes it possible to reason about the world. After discussing suitable properties abstraction should have to be useful for concept representation, examples of abstraction operators, designed to perform the abstraction process in practice, will be introduced.

1 Introduction

Abstraction, intended as the ability to forget irrelevant details and to find simpler descriptions, has been mainly investigated in problem solving [16, 20, 5, 3, 8], and in problem reformulation [12, 1, 18, 24]. In this paper we are interested, instead, in the role played by abstraction in a phase preceding problem solving, namely the phase of conceptualizing a domain, when a set of appropriate, possibly interrelated concepts is defined.

In problem solving and problem reformulation, abstraction consists of a transformation of the representation language that allows a theorem to be proved (or a problem to be solved) more easily, i.e., with a reduced computational effort. This pragmatic view of abstraction, which proved very useful to its intended goal, may not be sufficient for concept definition, where computational issues, even though important, are subsequent to the establishing of meaningful relations between the "concepts" and their referents in the world. In concept representation, in fact, the role of abstraction seems more related to "making sense" of the perception of the world, by transforming it into a set of meaningful "concepts", prior than to an efficient use of them. Abstraction is thus a fundamental mechanism for saving cognitive efforts, by offering us a "higher" level view of our physical and intellectual environment. Goldstone and Barsalou [6] have recently advocated a stricter link between perception and conceptualization in Cognitive Science. We think that their approach offers a cognitive foundation to our model of abstraction.

2 Related Work

Plaisted [16] has provided foundations of theorem proving with abstraction, seen as a function from a set of clauses to another one that satisfies some properties related to the resolution principle. Tenenberg [20] starts from Plaisted's work to define an abstraction as a *mapping* between predicates that preserves logical consistency. He defines an abstraction either as a *predicate mapping,* or a mapping between clauses based on predicate mappings, where only consistent clauses are considered.

Giunchiglia and Walsh [5] reviewed most of the work done in reasoning with abstraction. Extending Tenenberg's work, Nayak and Levy [14] proposed a semantic theory of abstraction. This theory defines abstraction as a *model level mapping* rather than *predicate mapping*. This semantic theory yields *model increasing abstractions* that are weaker than the base theory, i.e., they are strictly a subset of the "theorem decreasing" abstractions introduced in [5].

Abstraction has also been studied in relation with change of problem representation [2,1,18,19,11,12]. The notion of granularity is related to the analysis of possible links between levels of abstractions [7, 9, 15].

3 Definition of Perception-Based Abstraction

The novel perspective on abstraction that we propose originates from the observation that the conceptualization of a domain involves at least four different levels. Underlying any source of experience there is the *world* W. We consider a <u>fixed</u> part of the world, and we assume further that it does not change over time. The world is not really known, because we only have a mediated access to it through our perception. An actual world perception P is obtained through a process $\mathcal{P}$ of signal/information acquisition from/about the world:

$$P = \mathcal{P}(W)$$

Even though the primary source of information is the flow of sensory perceptions from the world, we cannot go back to them every time we approach a new task. Then, when we consider a world, we can think of using a perception system $\mathcal{P}$ more complex than the one sufficient to capture basic signals; $\mathcal{P}$ detects objects, properties and relations specified by P through mechanisms that we leave implicit. More precisely, in the world, both *atomic* and *compound* objects can be perceived. Atomic objects do not have parts, whereas compound objects do have parts that are themselves objects: a *part-of* hierarchy relates compound objects to their constituents. Single objects (both atomic and compound ones) have properties, which we call *attributes*. Other types of properties involve groups of objects, resulting in *functions* and *relations*. The percepts in P, provided by $\mathcal{P}$(W), can be grouped into four classes:

$$P = < OBJ, ATT, FUNC, REL >.$$

OBJ is a set of objects, *ATT* is a set of object attributes, *FUNC* is a set of functional links, and *REL* is a set of relations.

At the perception level, the percepts "exist" only for the observer, and only during the act of perceiving. In order to let the stimuli become available over time, for retrieval and further reasoning, they must be memorized and organized into a *structure* S [22]. This structure is an *extensional* representation of the perceived world, in which stimuli perceptually related one to another are stored together. In an artificial system, storage occurs in a relational database, manipulated via relational algebra operators [21]. We will denote by $\mathcal{M}$ the memorization process:

$$S = \mathcal{M}(P)$$

Finally, in order to describe in a symbolic way the perceived world, and to communicate with other agents, a *language* L is needed. L allows the perceived world to be described *intensionally*. Assigning "names" to the tables in the structure S and to their entries is a process of description:

$$L = \mathcal{D}(S)$$

Finally, a theory allows reasoning about the world. The theory may also contain general knowledge, which does not belong to any specific domain. At the theory level inference rules are used. We call *formalization* the process of expressing the theory in the language L (possibly enriched to accommodate domain-independent background knowledge):

$$T = \mathcal{F}(L)$$

The four levels are ordered as in Figure 1. The meaning of an arrow X → Y and, consequently, of the functional denotation Y = $\mathcal{g}$(X), indicates that the syntactic and semantic definition of level Y must take into account the content of level X.

As no world is totally isolated, a body of background knowledge provides, in principle, input to each level, especially to the theory, where general laws and domain-independent facts may be needed.

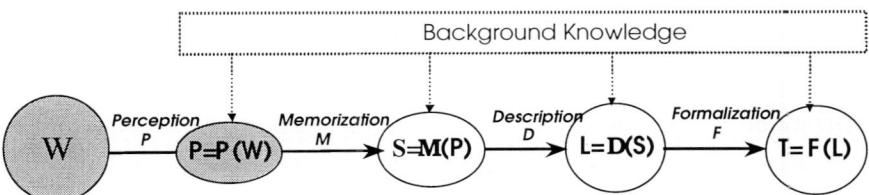

Fig. 1. Levels involved in representing and reasoning about a world W. P denotes a perception of objects and their physical links in W. S is a set of tables, each one grouping objects sharing some property. L is a formal language, whose semantics is evaluated on the tables of S. Finally, T is a theory formulated using L, in which properties of the world and general knowledge are embedded. General background knowledge may provide inputs to any level.

In this paper we consider the levels as given, and we do not discuss the nature of the $\mathcal{P}$, $\mathcal{M}$, $\mathcal{D}$, and $\mathcal{F}$ processes. Instead, we concentrate on representational issues and define a *Description Framework* D(W), over a world W, as the 4-tuple D(W) = (P, S, L, T). Given a world W, let P be a perception of W resulting from a process $\mathcal{P}$ that uses a set of sensors, each one tailored to capture a particular signal. Each sensor has a resolution threshold that establishes the minimum difference between two signals in order to consider them as distinct. A set of values provided by the sensors in $\mathcal{P}$ is called a *signal pattern* or a *configuration*. Let Γ be the set of possible configurations detectable by $\mathcal{P}$. In order to capture the intuitive notion that a more abstract configuration should be described in a simpler way, and that its simplicity should not depend upon the tool used for the description, but should be intrinsic to the configuration itself, we can exploit the notion of *algorithmic complexity* introduced by Kolmogorov [13].

Given $\mathcal{P}$, and, hence, Γ, any configuration $\gamma \in \Gamma$ can be described by a program π on a universal computer $\mathcal{U}$. Then, a complexity measure can be associated to Γ according to the following definition:

Definition 1 – Given a perception system $\mathcal{P}$ and the associated set Γ, let $\gamma \in \Gamma$ be any configuration. Given a universal machine $\mathcal{U}$ let

$$K(\gamma) = \min_{U(\pi) = \gamma} \ell(\pi) + C$$

be the Kolmogorov complexity of γ, defined as the length $\ell(\pi)$ of the shortest program π that output γ on the universal machine. The complexity of Γ can be defined as:

$$K(\Gamma) = \max_{\gamma \in \Gamma} K(\gamma) + C \qquad \square$$

The above definition has the advantage to be an "absolute" measure of the complexity of Γ, because it is machine-independent up to an additive constant. The additive constant C can be interpreted as the length of the program that describes the common structure of the configurations, i.e., the length of the program necessary for the machine $\mathcal{U}$ to interpret the descriptions π. Kolmogorov complexity is not computable. Nevertheless, we can use computable approximations of , K sufficient for practical purposes, provided that the same approximation is used uniformly over all the Γ's.

Definition 2 – Given a world W, let $\mathcal{P}_1$ and $\mathcal{P}_2$ be two perception processes generating P_1 and P_2, respectively. Let Γ_1 and Γ_2 be the corresponding configuration sets. The perceived world P_2 will be said *simpler*, according to Definition 1, than P_1 iff:

$$K(\Gamma_2) \leq K(\Gamma_1) \qquad \square$$

The above definition has the advantage of linking simplicity to its semantic meaning of cognitive effort of the information processing, rather to its syntactic expression. Obviously, syntactic complexity may have an effect on the simplicity of a perceived world, when a higher syntactic complexity implies more work to handle the conveyed information.

Definition 3 – Given a world W, let $D_g(W) = (P_g, S_g, L_g, T_g)$ and $D_a(W) = (P_a, S_a, L_a, T_a)$ be two description frameworks over the same world W, which we conventionally label as *ground* and *abstract,* respectively. An *Abstraction* is a mapping

$$\mathcal{A}: D_g(W) \to D_a(W)$$

such that P_a is "simpler" than P_g. □

Definition 3 is very general, and does not impose any "semantic" link between Γ and Γ'; it only states that Γ' is simpler to describe. Abstraction is, of course, a transitive relation [10, 23], and chains of abstraction mapping can be considered. Notice that abstraction is not concerned with the probability distribution of the configurations belonging to Γ: just one configuration, the observed one γ, is relevant. Also, the target problem is not *recognizing* the configuration, but *describing* it.

Assuming that a ground representation framework $D_g(W) = (P_g, S_g, L_g, T_g)$ has been defined over a world W, we will now investigate the relations between D_g and a more abstract representation framework $D_a(W) = (P_a, S_a, L_a, T_a)$. These relations are schematically illustrated in Figure 2.

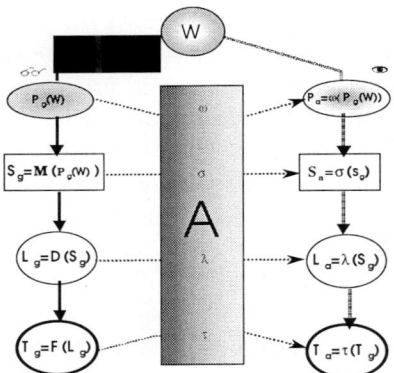

Fig. 2. Abstraction mapping between a ground representation framework $D_g(W) = (P_g, S_g, L_g, T_g)$ and a more abstract one $D_a(W) = (P_a, S_a, L_a, T_a)$.

The symbols ω, σ, λ and τ denote *abstraction operators* belonging to four sets Ω_W, Ω_S, Ω_L, and Ω_T, respectively acting at the four considered levels.

In Figure 2 two dimensions appear: the vertical one, along which the *nature* of the representation changes, and the horizontal one, along which *abstraction,* for each type of description, occurs. As abstraction is, in the proposed model, grounded on perception, the P-operators are the ones that have to be defined first, whereas the other operators must reflect, if possible, the perceptual transformations. If a perceptual transformation linking $\mathcal{P}$ and $\mathcal{P}'$ exists, no problems of consistency should appear [6]; instead, it is possible that the language (and/or the theory) becomes unable to describe the new perceived world. In the next section, more details about the abstraction operators will be given.

4 Abstraction Operators

The previous section provides a formal definition of abstraction as a functional mapping $\mathcal{A}$ between a given perception of a world and a simpler one. Given $\mathcal{A}$, we would like to introduce operators that can actually operate the transformation.

Definition 4 – An abstraction P-*operator* ω denotes a procedure that takes as input a perception $P_g(W)$ of a world W and outputs a *simpler* perception $P_a(W)$ of the same world. The *domain of application* of a P-operator represents the subset of the world perceptions to which it applies. □

Many P-operators can be defined, depending on the domain and task.

Given a P-operator ω, in order to obtain the structure S_a that *memorizes* the abstract world perception P_a, it would be enough to apply the $\mathcal{M}$ process to it, namely to define $S_a = \mathcal{M}(P_a)$. However, building the structure S_a in this way requires first to explicitly build the abstract world perception $P_a = \omega(P_g(W))$ and then to *memorize* it. Because the memorization step is difficult to automate, and the abstract perception could be difficult to acquire, it would be more useful to have a transformation that works directly on the ground structure S_g. Therefore, we shall define the notion of a structure level operator σ (S-operator), compatible with a P-operator ω. This compatibility provides the semantic foundation to the abstract operators at the structure level.

Definition 5 – An S-operator σ, applicable at the structure level, is *compatible* with a P-operator ω, at the world perception level, iff:

$$\mathcal{M}(\omega(P_g(W))) = \sigma(\mathcal{M}(P_g(W)))$$ □

For any given P-operator ω, there is no guarantee that a corresponding, effective S-operator, compatible with ω, exists, nor that, if it exists, it is unique.

Operators that directly modify languages have been predominant in modeling abstraction [16, 20, 5, 14]. Given a P-operator ω and a compatible S-operator σ, in order to obtain the abstract language L_a that *describes* the abstract structure $S_a = \sigma(S_g)$, it would be enough to apply $\mathcal{D}$ to the abstract structure, i.e., to define $L_a = \mathcal{D}(\sigma(S_g))$. However, the same considerations made for the S-operators hold also for the language level: we would like to define L-operators that operate directly on L_g.

Definition 6 – Given an S-operator σ, an L-operator λ at the language level is *compatible* with σ iff $\lambda(\mathcal{D}(S_g)) = \mathcal{D}(\sigma(S_g))$. □

Given a ground language L_g, a theory T_g is expressed as a set of formulas formalizing a body of knowledge. These formulas use predicates from L_g, the ones that have an operational semantics (their interpretations are in the structure), as well as other predicates symbols. Given a compatible L-operator λ, the abstract theory T_a

may be built up by *formalizing*, using the abstracted language $\lambda(L_g)$, the same knowledge that was formalized using language L_g, i.e., $T_a = \mathcal{F}(\lambda(T_g))$.

Definition 7 – Given a language L_g and an L-operator λ, a T-operator τ at the theory level is *compatible with* λ iff $\tau(\mathcal{F}(L_g)) = \mathcal{F}(\lambda(L_g))$. ◻

The definition of operators at the theory level is the most difficult step in the whole process of abstraction, the one that requires background knowledge to be performed. A preliminary solution to theory abstraction has been proposed by Giordana, Saitta and Roverso [4].

5 Conclusions

In this paper we have presented a new model of abstraction, which advocates the primary role played by perception in the conceptualization of a domain. The same claim, put forward recently by Goldstone and Barsalou [1998], provides the cognitive grounds to this model.

The presented work outlines only general properties of abstraction. We have also introduced the notion of *operator,* which, at each considered representation level, maps a ground to an "abstract" level. Specific operators may also be defined in connection with particular application and/or tasks.

Given the complexity of the problem, we are well aware that this work leaves open many fundamental questions. One of them is related to the "compatibility" notion and the concrete definition of compatible operators.

References

1. Amarel, S. (1983). "Representation in problem solving". In *Methods of Heuristics*, Lawrence Erlbaum, Palo Alto, CA, pp. 131-171.
2. Benjamin D.P., Dorst L., Mandhyan I., and Rosar M. (1990). "An Algebraic Approach to Abstraction and Representation Change". In *Proc. AAAI Workshop on Automatic Generation of Approximations and Abstractions* (Boston, MA), pp. 47-52.
3. Ellman T. (1993). "Synthesis of Abstraction Hierarchies for Constraint Satisfaction by Clustering Approximately Equivalent Objects". In *Proc. of 10th Int. Conf. on Machine Learning* (Amherst, MA), pp. 104-111.
4. Giordana A., Saitta L., Roverso D. (1991). "Abstracting Concepts with Inverse Resolution", In *Proc. 8th Int. Machine Learning Workshop*, pp. 142-146.
5. Giunchiglia F., and Walsh T. (1992). "A Theory of Abstraction". *Artificial Intelligence*, 56, 323-390.
6. Goldstone R., and Barsalou L. 1998). "Reuniting Perception and Conception". *Cognition*, 65, 231–262

7. Hobbs J. (1985). "Granularity". In *Proc. IJCAI-85* (Vancouver, USA), pp. 432-435.
8. Holte R.C., Mkadmi T., Zimmer R.M., and MacDonald A.J. (1996). "Speeding Up Problem-Solving by Abstraction: A Graph-Oriented Approach". *Artificial Intelligence, 85,* 321-361.
9. Imielinski T. (1987). "Domain Abstraction and Limited Reasoning". In *Proc.IJCAI-87* (Milano, Italy, pp. 997-1003.
10. Iwasaski Y. (1990). "Reasoning with Multiple Abstraction Models". In *Proc. AAAI Workshop on Automatic Generation of Approximations and Abstractions* (Boston, MA), pp. 122-133.
11. Korf, R. E. (1980). "Towards a Model for Representation Change". *Artificial Intelligence, 14,* 41-78.
12. Lowry, M. (1987). "The Abstraction/Implementation Model of Problem Reformulation". In *Proc. IJCAI-87* (Milano, Italy), pp. 1004-1010.
13. Li M., and Vitányi P. (1997). *An Introduction to Kolmogorov Complexity and Its Applications,* Springer-Verlag, New York, NY.
14. Nayak P.P., and Levy A.Y. (1995). "A Semantic Theory of Abstraction". In *Proc. IJCAI-95* (Montréal, Canada), pp. 196-202.
15. Pawlak Z. (1991). *Rough Sets : Theoretical Aspects of Reasoning about Data,* Kluwer Academic Publ., Norwell, MA.
16. Plaisted, D. (1981). "Theorem Proving with Abstraction". *Artificial Intelligence, 16,* 47-108.
17. Sacerdoti, E. (1973). Planning in a hierarchy of abstraction spaces. In *Proc. of the 3rd Int. Joint Conf. on Artificial Intelligence,* pp. 412-422.
18. Subramanian, D. (1990). "A Theory of Justified Reformulations". In D. P. Benjamin (Ed.),*Change of Representation and Inductive Bias,* Kluwer Academic Pub., Boston, MA, pp. 147-167.
19. Subramanian D., Greiner R., and Pearl J. (Eds.) (1997). *Artificial Intelligence, 97 (1-2).* Special Issue on Relevance.
20. Tenenberg J. (1987). "Preserving Consistency across Abstraction Mappings". In *Proc. IJCAI-87* (Milan, Italy), pp. 1011-1014.
21. Ullman, J. D. (1983). *Principles of Database Systems.* Computer Science Press.
22. Van Dalen, D. (1983). *Logic and Structure,* Springer-Verlag, Berlin, Germany.
23. Yoshida K. and Motoda H. (1990). "Towards Automatic Generation of Hierarchical Knowledge Bases". In *Proc. AAAI Workshop on Automatic Generation of Approximations and Abstractions* (Boston, MA), pp. 98-109.
24. Zucker J-D., and Ganascia J-G. (1996). "Changes of Representation for Efficient Learning in Structural Domains". In *Proc. 13th Int. Conf. on Machine Learning* (Bari, Italy), pp. 543-551.

Local Feature Selection with Dynamic Integration of Classifiers

Alexey Tsymbal and Seppo Puuronen

University of Jyväskylä, P.O.Box 35, FIN-40351 Jyväskylä, Finland
{Alexey, Sepi}@jytko.jyu.fi

Abstract. Multidimensional data is often feature-space heterogeneous so that different features have different importance in different subareas of the whole space. In this paper we suggest a technique that searches for a strategic splitting of the feature space identifying the best subsets of features for each instance. Our technique is based on the wrapper approach where a classification algorithm is used as the evaluation function to differentiate between several feature subsets. In order to make the feature selection local, we apply the recently developed technique for dynamic integration of classifiers. It allows us to determine what classifier and with what feature subset should be applied for each new instance. In order to restrict the number of feature combinations being analyzed we propose to use decision trees. For each test instance we consider only those feature combinations that include features present in the path taken by the test instance in the decision tree built on the whole feature set. We evaluate our technique on datasets from the UCI machine learning repository. In our experiments, we use the C4.5 algorithm as the learning algorithm for base classifiers and for decision trees that guide the local feature selection. The experiments show advantages of the local feature selection in comparison with the selection of one feature subset for the whole space.

1 Introduction

Current electronic data repositories contain enormous amount of data including also unknown and potentially interesting patterns and relations, which are tried to be revealed using knowledge discovery and data mining methods [8]. One approach commonly used is supervised machine learning, in which a set of training instances is used to train one or more classifiers that map the space formed by different features of the instances into the set of class values [1]. Each training instance is usually represented by a vector of the values of the features and the class label of the instance. An induction algorithm is used to learn a classifier, which maps the space of feature values into the set of class values.

The multidimensional data is sometimes feature-space heterogeneous so that different features have different importance in different subareas of the whole space. Many methods have been proposed for the purpose of feature selection, but almost all of them ignore the fact that some features may be relevant only in context (i.e. in some regions of the instance space) [7].

In this paper we describe a technique that searches for a division of the feature space identifying the best subsets of features for each instance. To make the feature selection local, we apply the recently developed technique for dynamic integration of classifiers to determine what classifier and with what feature subset is applied for each new instance [12]. We make experiments with well-known datasets of the UCI machine learning repository using ensembles of quite simple base classifiers [4].

In Chapter 2 we consider the dynamic integration of classifiers. Chapter 3 discusses local feature selection with the dynamic classifier integration. In the next chapter we consider our experiments with the local feature selection technique on different datasets. We conclude briefly in Chapter 5 with a summary and further research topics.

2 Dynamic Integration of Classifiers

In this chapter, the dynamic integration of classifiers is discussed, and a variation of stacked generalization, which uses a metric to locally estimate the errors of the base classifiers, is considered.

There are two main approaches to the integration. First, *combination approach*, where the base classifiers produce their classifications. The simplest method of combining classifiers is voting [1]. Examples of more complex algorithms are weighted voting (WV) and stacked generalization [12].

Second, *selection approach*, where one of the classifiers is selected and the final result is the result produced by it. One very popular but simple static selection approach is CVM (Cross-Validation Majority) [9]. And an example of more sophisticated dynamic selection approach predicts the correctness of the base classifiers for a new instance [11]. We have elaborated a dynamic approach that estimates the local accuracy of the base classifiers by analyzing the accuracy in near-by instances [12].

The dynamic integration approach contains two phases [12]. In the learning phase (procedure *learning_phase* in Fig. 1), the training set **T** is partitioned into v folds. The cross-validation technique is used to estimate the errors of the base classifiers $E_j(\mathbf{x}^*)$ on the training set and the meta-level training set $\mathbf{T}^*$ is formed. It contains the attributes of the training instances $\mathbf{x}_i$ and the estimates of the errors of the base classifiers on those instances $E_j(\mathbf{x}^*)$. Several cross-validation runs can be used in order to obtain more accurate estimates of the base classifiers' errors. Then each estimated error will be equal to the number of times that an instance was incorrectly predicted by the classifier when it appeared as a test example in a cross-validation run. The learning phase finishes with training the base classifiers C_j on the whole training set.

In the application phase, the combining classifier (either the function *DS_application_phase* or the function *DV_application_phase* in Fig. 1) is used to predict the performance of each base classifier for a new instance. Two different functions implementing the application phase were considered [12]. In the DS (Dynamic Selection) application phase the classification error E_j^* is predicted for each base classifier C_j using a nearest neighbor procedure and a classifier with the smallest error (with the least global error in the case of ties) is selected to make the final classification. In the DV (Dynamic Voting) application phase each base classifier C_j receives a weight W_j that depends on the local classifier's performance and the final classification is conducted by voting classifier predictions $C_j(\mathbf{x})$ with their weights W_j.

T_i i-th fold of the training set T
T^* meta-level training set for the combining algorithm
$c(x)$ classification of the instance with attributes x
C set of base classifiers
C_j j-th base classifier
$C_j(x)$ prediction produced by C_j on instance x
$E_j(x)$ estimation of error of C_j on instance x
$E^*_j(x)$ prediction of error of C_j on instance x
m number of base classifiers
W vector of weights for base classifiers
nn number of near-by instances for error prediction
W_{NNi} weight of i-th near-by instance
procedure *learning_phase*(T,C)
    begin {fills in the meta-level training set T^*}
        partition T into v folds
        loop for $T_i \subset T$, $i = 1,\ldots, v$
          loop for j **from** 1 **to** m train(C_j, $T-T_i$)
          loop for $x \in T_i$
            loop for j **from** 1 **to** m
              compare $C_j(x)$ with $c(x)$ and derive $E_j(x)$
            collect ($x, E_1(x),\ldots, E_m(x)$) into T^*
        loop for j **from** 1 **to** m train(C_j, T)
    end
function *DS_application_phase*(T^*,C,x) **returns** class of x
    begin
        loop for j **from** 1 **to** m
$$E^*_j \leftarrow \frac{1}{nn}\sum_{i=1}^{nn} W_{NN_i} \cdot E_j(x_{NN_i}) \quad \{\text{NN estimation}\}$$
$$l \leftarrow \arg\min_j E^*_j \quad \{\text{number of cl-er with min. } E^*_j\}$$
        {with the least global error in the case of ties}
        return $C_j(x)$
    end
function *DV_application_phase*(T^*,C,x) **returns** class of x
    begin
        loop for j **from** 1 **to** m
$$W_j \leftarrow 1 - \frac{1}{nn}\sum_{i=1}^{nn} W_{NN_i} \cdot E_j(x_{NN_i}) \quad \{\text{NN estimation}\}$$
        return Weighted_Voting($W, C_1(x),\ldots, C_m(x)$)
    end

Fig. 1. The algorithm for dynamic integration of classifiers [12]

3 Local Feature Selection

In data mining the object of processing is usually multidimensional data presented by a number of features. Commonly there are present also a number of irrelevant features. In this chapter the feature selection problem is discussed based on local considerations of the relevance of each feature.

The feature space is often heterogeneous, where the features that are important for data mining are different in different regions of the feature space [2]. Of the two main approaches to manage this [3] we apply the approach where the data mining problem is divided into subproblems and the solution of the whole classification task is guided by the heterogeneity of the feature space.

First, a feature selection algorithm can be based on a heuristic measure acting as a *filter* extracting features from a feature set before its use by the main algorithm. Second, a feature selection algorithm based on the actual accuracy acts as a *wrapper* around the main algorithm [6]. We propose to apply the wrapper approach for dynamic integration of classifiers [12] using local classification accuracy [15].

In this paper we consider an advanced version of the method presented in [15]. We previously analyzed all possible combinations of features to build the ensemble of base classifiers. However, this can be very expensive computationally and can lead to overfitting. The big number of feature subsets and, consequently, integrated base classifiers dramatically increases the number of degrees of freedom in the training process, leading to increased variance of predictions and an increased risk of overfitting the data. To reduce the risk, we propose to limit the number of considered feature subsets in our local wrapper technique. The base classifiers in the ensemble can be built using combinations of only potentially locally relevant features, discarding features that are definitely irrelevant at that region.

Some recursive partitioning techniques or some heuristic measures can be used to discard features that are locally irrelevant with a high probability. Thus, we propose to combine our wrapper-based method with a filter approach, using it in advance for restricting the possible feature combinations.

In [5] a decision tree was proposed to be used for local feature selection, where for a test case only those features are considered to be locally relevant, which lie on the path taken by the test case in the decision tree. We propose to use a decision tree approach [5] as a filter for feature selection with our method.

4 Experiments

In this chapter, we present our experiments with the use of the C4.5 decision tree algorithm to guide the local feature selection process. First we describe the experimental setting and then present results of our experiments. We conduct the experiments on eight datasets taken from the UCI machine learning repository [4] and on the Dystonia dataset considered in [16]. Previously we have experimentally evaluated the dynamic classifier integration [12] and the unguided local feature selection [15]. Here we use a similar experimental environment, but this time the algorithm builds C4.5 decision trees with and without pruning [14] at the end of the training phase for local feature filtering at the beginning of the application phase. For each test instance we consider only those feature combinations that include features present in the path taken by the test instance in the decision tree.

For each dataset 30 test runs are made. In each run 30 percent of the instances of the dataset are by random sampling picked up to the test set. The rest 70 percent of the instances form the training set which is passed to the learning algorithm. This training set is divided into 10 folds using stratified random sampling because we apply 10-fold cross-validation to build the cross-validation history for the dynamic integration of the base classifiers [12]. The base classifiers themselves are learned using the C4.5 decision tree algorithm with pruning [14] on feature subsets including only exactly one or two features. In the estimation of the classification errors of the base classifiers for a new instance we use the collected classification information about classification errors for seven nearest neighbors of the new instance [12]. Based on the comparisons between different distance functions for dynamic integration presented in [13] we decided to use the Heterogeneous Euclidean-Overlap Metric, which produced good test results. The test environment was implemented within the MLC++ framework (the machine learning library in C++) [10].

Table 1 presents accuracy values and numbers of analyzed features for our algorithm of local feature selection when considered feature subsets consisted of exactly one feature, and Table 2 presents the same results when the feature subsets consisted of two features each. The left-hand side (until the bold line) of the tables shows the averages of the classification accuracies over the 30 runs. The first five columns include the average of the minimum accuracies of the base classifiers (min), the average of average accuracies of the base classifiers (aver), the average of the maximum accuracies of the base classifiers (max), the average percentage of test instances where all the base classifiers managed to produce the right classification (agree), and the average amount of test instances where at least one base classifier during each run managed to produce the right classification (cover). The next four columns of the left-hand side of Table 1 and Table 2 include the accuracies for the four types of integration of the base classifiers.

These are: (1) CVM – Cross-Validated Majority, (2) WV – Weighted Voting, (3) DS – Dynamic Selection, and (4) DV – Dynamic Voting. The three columns of the right-hand side include the minimum (min), average (aver), and maximum (max) number of features used to classify test instances. All the above columns are averaged over the 30 test runs.

In vertical direction the main body of the tables is divided into groups of three rows each corresponding to one dataset. The first row contains accuracies received with unguided local feature selection, the second row contains accuracies received when the feature selection is guided by the C4.5 with pruning, and the third one contains accuracies received when the feature selection is guided by the C4.5 without pruning. The last group in the tables shows accuracies averaged over all the datasets.

When the numbers of features are compared inside the three lines of the groups corresponding to all the datasets, one can see that the use of the C4.5 algorithm with or without pruning for local feature selection significantly reduces the number of locally analyzed features. When the corresponding accuracies achieved are compared one can see that this has often happened without a loss in the classification accuracy. For many datasets, the number of locally analyzed features is several times less in average than the total number of features. For example, in the Dystonia dataset there are totally 7 features, whereas the average number of locally analyzed features in the case when one-feature subsets were analyzed (Table 1) is 1.133 for pruned C4.5 and the same number for C4.5 without pruning (the trees generated were already too simple to prune).

Table 1. Accuracy values and numbers of analyzed features for our algorithm of local feature selection with feature subsets including one feature

DB	Base classifiers: C4.5 with pruning					Integration of classifiers				Features		
	min	aver	max	agree	cover	CVM	WV	DS	DV	min	aver	max
Breast	0.673	0.706	0.737	0.473	0.887	0.695	0.709	0.709	0.709	9	9	9
	0.332	0.669	0.906	0.551	0.831	0.697	0.723	0.700	0.730	1.833	3.294	6.000
	0.561	0.720	0.875	0.524	0.840	0.695	0.725	0.697	0.725	1.933	4.432	7.433
Dystonia	0.892	0.977	1.000	0.847	1.000	1.000	1.000	1.000	1.000	7	7	7
	1.000	1.000	1.000	1.000	1.000	1.000	1.000	1.000	1.000	1.000	1.133	1.200
	1.000	1.000	1.000	1.000	1.000	1.000	1.000	1.000	1.000	1.000	1.133	1.200
Glass	0.274	0.430	0.570	0.004	0.858	0.517	0.590	0.530	0.570	9	9	9
	0.256	0.440	0.631	0.046	0.810	0.506	0.535	0.502	0.548	2.500	6.085	7.867
	0.278	0.465	0.713	0.042	0.812	0.506	0.542	0.512	0.556	2.500	6.435	8.533
Heart	0.493	0.641	0.793	0.013	0.999	0.720	0.770	0.641	0.741	13	13	13
	0.364	0.641	0.843	0.145	0.983	0.715	0.732	0.621	0.704	2.533	5.933	8.233
	0.354	0.640	0.840	0.138	0.984	0.716	0.730	0.621	0.703	2.533	6.032	8.233
Iris	0.502	0.775	0.965	0.351	0.986	0.942	0.935	0.930	0.934	4	4	4
	0.735	0.879	0.965	0.890	0.974	0.941	0.945	0.948	0.945	1.000	1.578	2.300
	0.412	0.764	0.966	0.848	0.978	0.941	0.942	0.939	0.934	1.000	1.752	3.000
Liver	0.496	0.557	0.623	0.066	0.967	0.563	0.596	0.597	0.616	6	6	6
	0.500	0.565	0.628	0.099	0.965	0.574	0.618	0.595	0.629	1.133	5.659	6.000
	0.500	0.565	0.628	0.099	0.965	0.574	0.618	0.595	0.629	1.133	5.659	6.000
MONK-1	0.403	0.491	0.757	0.091	0.957	0.757	0.509	0.453	0.462	6	6	6
	0.308	0.490	0.762	0.338	0.927	0.757	0.612	0.623	0.575	1.000	2.963	4.967
	0.314	0.490	0.757	0.332	0.937	0.757	0.613	0.625	0.577	1.000	3.053	5.167
MONK-2	0.669	0.669	0.669	0.669	0.669	0.669	0.669	0.669	0.669	6	6	6
	0.566	0.646	0.701	0.669	0.669	0.669	0.669	0.669	0.669	2.767	4.509	6.000
	0.572	0.643	0.692	0.669	0.669	0.669	0.669	0.669	0.669	2.967	4.778	6.000
MONK-3	0.473	0.604	0.813	0.290	1.000	0.781	0.570	0.787	0.566	6	6	6
	0.469	0.706	0.872	0.591	1.000	0.781	0.763	0.786	0.943	1.967	2.104	3.000
	0.469	0.706	0.872	0.591	1.000	0.781	0.763	0.786	0.943	1.967	2.104	3.000
Average	0.542	0.650	0.770	0.312	0.925	0.738	0.705	0.702	0.696			
	0.503	0.671	0.812	0.481	0.907	0.738	0.733	0.716	0.749			
	0.496	0.666	0.816	0.471	0.909	0.738	0.734	0.716	0.748			

For many datasets and on average, the classification accuracy is even higher with the guided feature selection. Thus, sometimes the guided feature selection helps to improve the classification accuracy, as on the Breast, MONK-1, and MONK-3 datasets. However, some datasets are complex, containing many locally relevant features, even more than there are levels in the decision tree. In those cases, the guided local feature selection usually produces slightly lower accuracy than the unguided feature selection, as on the Glass dataset.

Table 2. Accuracy values and numbers of analyzed features for our algorithm of local feature selection with feature subsets each including two features

DB	Base classifiers: C4.5 with pruning					Integration of classifiers				Features		
	min	aver	max	agree	cover	CVM	WV	DS	DV	min	aver	max
Breast	0.641	0.725	0.799	0.378	0.935	0.737	0.740	0.732	0.719	9	9	9
	0.196	0.725	1.000	0.617	0.819	0.744	0.744	0.739	0.745	2.100	3.549	6.333
	0.381	0.773	0.988	0.549	0.854	0.744	0.746	0.738	0.747	2.267	4.637	7.500
Dystonia	0.951	0.997	1.000	0.950	1.000	1.000	1.000	1.000	1.000	7	7	7
	1.000	1.000	1.000	1.000	1.000	1.000	1.000	1.000	1.000	1	1.18	1.267
	1.000	1.000	1.000	1.000	1.000	1.000	1.000	1.000	1.000	1	1.18	1.267
Glass	0.318	0.523	0.686	0.003	0.965	0.619	0.675	0.635	0.677	9	9	9
	0.166	0.500	0.765	0.110	0.908	0.599	0.604	0.545	0.584	2.367	5.938	7.600
	0.188	0.528	0.801	0.106	0.909	0.607	0.619	0.558	0.600	2.367	6.344	8.633
Heart	0.457	0.667	0.826	0.003	1.000	0.749	0.823	0.763	0.809	13	13	13
	0.155	0.651	0.979	0.128	0.972	0.754	0.762	0.711	0.758	2.167	5.817	8.100
	0.136	0.657	0.987	0.114	0.973	0.754	0.762	0.709	0.756	2.167	5.952	8.200
Iris	0.703	0.905	0.971	0.669	0.986	0.933	0.954	0.945	0.950	4	4	4
	0.838	0.939	0.984	0.932	0.954	0.940	0.949	0.943	0.948	1.000	1.607	2.267
	0.716	0.902	0.986	0.926	0.961	0.940	0.949	0.942	0.947	1.000	1.692	2.667
Liver	0.469	0.574	0.673	0.010	1.000	0.575	0.645	0.609	0.658	6	6	6
	0.468	0.574	0.677	0.058	0.990	0.582	0.646	0.615	0.653	1.100	5.648	6.000
	0.468	0.574	0.677	0.058	0.990	0.582	0.646	0.615	0.653	1.100	5.648	6.000
MONK-1	0.368	0.549	0.826	0.028	1.000	0.824	0.711	0.824	0.575	6	6	6
	0.204	0.618	1.000	0.490	0.962	0.950	0.748	0.944	0.800	1.000	3.062	5.467
	0.215	0.605	1.000	0.482	0.980	0.964	0.748	0.959	0.804	1.000	3.162	5.600
MONK-2	0.686	0.686	0.686	0.686	0.686	0.686	0.686	0.686	0.686	6	6	6
	0.547	0.646	0.724	0.686	0.686	0.686	0.686	0.686	0.686	2.767	4.541	6.000
	0.549	0.641	0.716	0.686	0.686	0.686	0.686	0.686	0.686	3.000	4.796	6.000
MONK-3	0.431	0.675	0.969	0.151	1.000	0.969	0.746	0.972	0.687	6	6	6
	0.426	0.730	0.968	0.902	1.000	0.969	0.969	0.981	0.963	1.900	2.088	3.000
	0.426	0.730	0.968	0.902	1.000	0.969	0.969	0.981	0.963	1.900	2.088	3.000
Average	0.509	0.663	0.805	0.241	0.947	0.762	0.748	0.771	0.720			
	0.375	0.673	0.887	0.490	0.911	0.778	0.764	0.771	0.767			
	0.385	0.676	0.890	0.478	0.919	0.781	0.766	0.774	0.770			

For some datasets, the dynamic integration (local feature selection) is clearly better than the static approaches, as on the Liver and MONK-3 datasets. The MONK-3 dataset is a good example of the benefit from the local feature selection guided by a decision tree. On this dataset, the guided feature selection is better than the unguided one, and the dynamic integration is better than the static integration. The dynamic integration of classifiers built on different feature subsets guided by the C4.5 decision tree is the best choice for this domain. The C4.5 decision tree clearly helps to reject the three irrelevant features present in the dataset. One can see that at maximum, just three relevant features are selected as locally relevant with the C4.5 on this dataset.

According to the last groups of the tables, the guided feature selection is better on average than the unguided one. The C4.5 without pruning naturally generates bigger trees, and it leads to greater average number of locally relevant features selected. However, the accuracies of the algorithm with C4.5 with and without pruning usually are almost undistinguishable. Only on the Glass dataset pruning gives clearly better results. When the two tables are compared, one could see, that Table 2 contains naturally greater accuracies (because subsets with 2 features were analyzed whereas only one-feature subsets were analyzed in Table 1). However, the difference is less than one could expect. Integration of classifiers based on only one feature gives surprisingly good results. And these results can be only slightly improved when more features are considered.

5 Conclusion

In this paper we described a technique that searches for a division of the feature space identifying the best subsets of features for each instance. The technique is based on the local wrapper approach, and uses the method for dynamic integration of classifiers to determine what classifier and with what feature subset is applied for each new instance. At the application phase, in order to restrict the number of feature combinations being analyzed, we used the C4.5 decision tree built on the whole feature set as a feature filter. For each test instance we considered only those feature combinations that included features present in the path taken by the test instance in the decision tree. Our technique can be applied in the case of implicit heterogeneity when the regions of heterogeneity cannot be easily defined by a simple dependency.

We conducted experiments on datasets of the UCI machine learning repository using ensembles of simple base classifiers each generated on either one or two features. The results achieved are promising and show that the local feature selection in comparison with selecting only one feature set for the whole space can be advantageous.

Further experiments can be conducted to make deeper analysis of applying recursive partitioning and the dynamic integration of classifiers for local feature selection (and particularly to define the dependency between the parameters of local feature selection, characteristics of a domain, and the data mining accuracy). Decision trees built on the whole instance set were used in our experiments. Use of other feature filters can be tested in future experiments. Another potentially interesting topic for further research is the analysis of feature subsets without the restriction on their size. Also it would be interesting to consider an application of this technique to a hard real-world problem.

Acknowledgments: This research is partly supported by the COMAS Graduate School of the University of Jyväskylä. We would like to thank the UCI machine learning repository of databases, domain theories and data generators for the datasets, and the machine learning library in C++ for the source code used in this study. We are grateful to the anonymous referees for their valuable comments and constructive criticism.

References

1. Aivazyan, S.A.: Applied Statistics: Classification and Dimension Reduction. Finance and Statistics, Moscow (1989).
2. Apte, C., Hong, S.J., Hosking, J.R.M., Lepre, J., Pednault, E.P.D., Rosen, B.K.: Decomposition of Heterogeneous Classification Problems. In: X. Hiu, P.Cohen, M. Berthold (eds.), Advances in Intelligent Data Analysis (IDA-97), Springer-Verlag, London (1997) 17-28.
3. Atkeson, C.G., Moore, A.W., Schaal, S.: Locally Weighted Learning. Artificial Intelligence Review, Vol. 11, Ns. 1-5 (1997) 11-73.
4. Blake, C.L., Merz, C.J.: UCI Repository of Machine Learning Databases [http://www.ics.uci.edu/ ~mlearn/ MLRepository.html]. Dep-t of Information and CS, Un-ty of California, Irvine CA (1998).
5. Cardie, C., Howe, N.: Improving Minority Class Prediction Using Case-Specific Feature Weights. In: Proc. 14th Int. Conf. on Machine Learning, Morgan Kaufmann (1997) 57-65.
6. Dash, M., Liu, H.: Feature Selection for Classification. Intelligent Data Analysis, Vol. 1, No. 3, Elsevier Science (1997).
7. Domingos, P.: Context-Sensitive Feature Selection for Lazy Learners. J. of AI Review, Vol. 11, Ns. 1-5 (1997) 227-253.
8. Fayyad, U., Piatetsky-Shapiro, G., Smyth, P., Uthurusamy, R.: Advances in Knowledge Discovery and Data Mining. AAAI/ MIT Press (1997).
9. Kohavi, R.: A Study of Cross-Validation and Bootstrap for Accuracy Estimation and Model Selection. In: C. Mellish (ed.), Proceedings of IJCAI'95, Morgan Kaufmann (1995).
10. Kohavi, R., Sommerfield, D., Dougherty, J.: Data Mining Using MLC++: A Machine Learning Library in C++. Tools with Artificial Intelligence, IEEE CS Press (1996) 234-245.
11. Merz, C.: Dynamical Selection of Learning Algorithms. In: D.Fisher, H.-J.Lenz (eds.), Learning from Data, Artificial Intelligence and Statistics, Springer-Verlag, NY (1996).
12. Puuronen, S., Terziyan, V., Tsymbal, A.: A Dynamic Integration Algorithm for an Ensemble of Classifiers. In: Z.W. Ras, A. Skowron (eds.), Foundations of Intelligent Systems: ISMIS'99, Lecture Notes in AI, Vol. 1609, Springer-Verlag, Warsaw (1999) 592-600.
13. Puuronen, S., Tsymbal, A., Terziyan, V.: Distance Functions in Dynamic Integration of Data Mining Techniques. In: B.V. Dasarathy (ed.), Data Mining and Knowledge Discovery: Theory, Tools and *Technology II, Proceedings of SPIE, Vol.4057*, USA, 2000, pp.22-32.
14. Quinlan, J.R.: C4.5 Programs for Machine Learning. Morgan Kaufmann, San Mateo, CA (1993).
15. Skrypnik, I., Terziyan, V., Puuronen, S., Tsymbal, A.: Learning Feature Selection for Medical Databases. In: Proc. 12th IEEE Symp. on Computer-Based Medical Systems CBMS'99, IEEE CS Press, Stamford, CT (1999) 53-58.
16. Terziyan, V., Tsymbal, A., Puuronen, S.: The Decision Support System for Telemedicine Based on Multiple Expertise. Int. J. of Medical Informatics, Vol. 49, No. 2 (1998) 217-229.

Prediction of Ordinal Classes Using Regression Trees

Stefan Kramer[1], Gerhard Widmer[2,3], Bernhard Pfahringer[4]
and Michael de Groeve[5]

[1] Department of Computer Science, Albert-Ludwigs-University Freiburg,
Am Flughafen 17, D-79110 Freiburg, Germany
[2] Department of Medical Cybernetics and Artificial Intelligence,
University of Vienna, Freyung 6/2, A-1010 Vienna, Austria
[3] Austrian Research Institute for Artificial Intelligence,
Schottengasse 3, A-1010 Vienna, Austria
[4] Department of Computer Science, University of Waikato, Hamilton, New Zealand
[5] Department of Computer Science, Katholieke Universiteit Leuven,
Celestijnenlaan 200A, B-3001 Heverlee, Belgium

Abstract. This paper is devoted to the problem of learning to predict ordinal (i.e., ordered discrete) classes using classification and regression trees. We start with S-CART, a tree induction algorithm, and study various ways of transforming it into a learner for ordinal classification tasks. These algorithm variants are compared on a number of benchmark data sets to verify the relative strengths and weaknesses of the strategies and to study the trade-off between optimal categorical classification accuracy (hit rate) and minimum distance-based error. Preliminary results indicate that this is a promising avenue towards algorithms that combine aspects of classification and regression.

1 Introduction

Learning to predict discrete classes or numerical values from preclassified examples has long been, and continues to be, a central research topic in Machine Learning (e.g., [Breiman et al., 1984, Quinlan, 1992, Quinlan 1993]). A class of problems between classification and regression, learning to predict *ordinal classes*, i.e., discrete classes with a linear ordering, has not received much attention so far, which seems somewhat surprising, as there are many classification problems in the real world that fall into that category.

Recently, [Potharst & Bioch, 1999] presented a tree-based algorithm for the prediction of ordinal classes. [Potharst & Bioch, 1999] assume that the independent variables are ordered as well, which implies that the predictions made should be consistent with the order of the attribute values in the decision nodes. So, the authors present "repair strategies" correcting inconsistent trees in case these consistency constraints are violated, as well as an algorithm for constructing consistent trees in the first place.

Other machine learning research that seems relevant to the problem of predicting ordinal classes is work on *cost-sensitive learning*. In the domain of propositional learning, some induction algorithms have been proposed that can take into account matrices of misclassification costs (e.g., [Schiffers, 1997,Turney, 1995]. Such cost matrices can be used to express relative distances between classes. In the area of statistics, there are methods directly relevant to our problem (e.g., *Ordinal Logistic Regression* [McCullagh, 1980]); some of these have also been studied in the field of neural networks (e.g., [Mathieson, 1996]). However, our goal is to develop induction algorithms that produce interpretable, symbolic models. Moreover, our algorithm S-CART, to be presented below, can learn in both propositional and relational domains.

The purpose of the research described in this paper is to study ways of learning to predict ordinal classes using regression trees. We will start with an algorithm for the induction of regression trees and turn it into an ordinal learner by some simple modifications. This seems a natural strategy because regression algorithms by definition have a notion of relative distance of target values, while classification algorithms usually do not. More precisely, we start with the algorithm S-CART (*Structural Classification and Regression Trees*) [Kramer 1996, Kramer 1999] and study several modifications of the basic algorithm that turn it into a distance-sensitive classification learner. Several variants of this algorithm are compared on a number of data sets to verify the relative strengths and weaknesses of the strategies and to study the trade-off between optimal categorical classification accuracy (hit rate) and minimum distance-based error.

2 The Basic Learning Algorithm: S-CART (Structural Classification and Regression Trees)

Structural Classification and Regression Trees (S-CART) [Kramer 1996, Kramer 1999] is an algorithm that learns a first-order theory for the prediction of either discrete classes or numerical values from examples and relational background knowledge. The algorithm constructs a tree containing a positive literal or a conjunction of literals in each node, and assigns a discrete class or a numeric value to each leaf. S-CART is a full-fledged relational version of CART [Breiman et al., 1984]. After the tree growing phase, the tree is pruned using so-called error-complexity pruning for regression or cost-complexity pruning for classification [Breiman et al., 1984]. These types of pruning are based on a separate "prune set" of examples or on cross-validation.

For the construction of a tree, S-CART follows the general procedure of top-down decision tree induction algorithms [Quinlan, 1993]. It recursively builds a binary tree, selecting a positive literal or a conjunction of literals (as defined by user-defined schemata [Silverstein & Pazzani, 1991]) in each node of the tree until a stopping criterion is fulfilled. The algorithm keeps track of the examples in each node and the positive literals or conjunctions of literals in each path leading to the respective nodes. This information can be turned into a clausal theory (i.e., a set of first-order classification or regression rules).

As a regression algorithm, S-CART is designed to predict a numeric (real) value in each node and, in particular, in each leaf. In the original version of the algorithm the target value predicted in a node (let us call this the *center value* from now on) is simply the mean of the numeric class values of the instances covered by the node. A natural choice for the *evaluation measure* for rating candidate splits during tree construction is then the *Mean Squared Error (MSE)* of the example values relative to the means in the two new nodes created by the split:

$$MSE = \frac{1}{n_1 + n_2} \sum_{i=1}^{2} \sum_{j=1}^{n_i} (y_{ij} - \bar{y}_i)^2 \qquad (1)$$

where n_i is the number of instances covered by branch i, y_{ij} is the value of the dependent variable of training instance e_j in branch i, and $\bar{y}_i$ is the mean of the target values of all training instances in branch i.

In constructing a single tree, the simplest possible stopping criterion is used to decide whether the tree should be further refined: S-CART stops extending the tree given some node when no literal(s) can be found that produce(s) two partitions of the training instances in the node with a required minimum cardinality. The post-pruning strategy then takes care of reducing the tree to an appropriate size.

S-CART has been shown to be competitive with other regression algorithms. Its main advantages are that it offers the full power and flexibility of first-order (Horn clause) logic, provides a rich vocabulary for the user to explicitly represent a suitable language bias (e.g. through the provision of schemata), and produces trees that are interpretable as well as good predictors.

As our goal is to predict discrete ordered classes, S-CART cannot be used directly for this task. We will, however, include results with standard S-CART in the experimental section to find out how paying attention to ordinal classes influences the mean squared error achievable by a learner.

3 Inducing Trees for the Prediction of Ordinal Classes

In the following, we describe a few simple modifications that turn S-CART into a learning algorithm for ordinal classification problems. In section 3.2, we consider some pre-processing methods that also might improve the results.

3.1 Adapting S-CART to Ordinal Class Prediction

The most straightforward way of adapting a regression algorithm like S-CART to classification tasks is to simply run the algorithm on the given data as if the ordinal classes (represented by integers) were real values, and then to apply some sort of *post-processing* to the resulting rules or regression tree that translates real-valued predictions into discrete class labels.

An obvious post-processing method is *rounding*. S-CART is run on the training data, producing a regular regression tree. The real values predicted in the leaves of the tree are then simply rounded to the nearest of the ordinal classes (not to the nearest integer, as the classes may be discontiguous; after preprocessing, they might indeed be non-integers — see section 3.2 below).

More complex methods for mapping predicted real values to symbolic (ordinal) class labels are conceivable. In fact, we did experiments with an algorithm that greedily searches for a mapping, within a defined class of functions, that minimizes the mean squared error of the resulting (mapped) predictions on the training set. Initial experiments were rather inconclusive; in fact, there were indications of the algorithm overfitting the training data. However, more sophisticated methods might turn out to be useful. This is one of the goals of our future research.

An alternative to post-processing is to modify the way S-CART computes the target values in the nodes of the tree *during tree construction*. We can force S-CART to always predict integer values (or more generally: a valid class from the given set of ordinal classes) in any node of the tree. The leaf values will thus automatically be valid classes, and no post-processing is necessary.

It is a simple matter to modify S-CART so that instead of the *mean* of the class values of instances covered by a node (which will in general not be a valid class value), it chooses one of the class values represented in the examples covered by the node as the center value that is predicted by the node, and relative to which the node evaluation measure (e.g., the mean squared error, see Section 2 above) is computed. Note that in this way, we modify S-CART's *evaluation heuristic* and thus its *bias*.

There are many possible ways of choosing a center value; we have implemented three: the *median*, the *rounded mean*, and the *mode*, i.e., the most frequent class. Let E_i be the set of training examples covered by node N_i during tree construction and C_i the multiset of the class labels of the examples in E_i, with $|E_i| = |C_i| = n$. In the MEDIAN strategy, S-CART selects the class $\hat{c}_i$ as center value that is the median of the class labels in C_i; in other words, if we assume that the example set E_i is sorted with respect to the class values of the examples, MEDIAN chooses the class of the $(n/2)^{th}$ example.[1] In contrast, the ROUNDEDMEANTOCLASS strategy chooses the class closest to the (real-valued) mean $\bar{c}$ of the class values in C_i. Finally, in the MODE strategy the center value $\hat{c}_i$ for node N_i is chosen to be the class with the highest frequency in C_i.

Table 1 summarizes the variants of S-CART that will be put to the test in Section 4 below.

3.2 Preprocessing

The results of regression algorithms can often be improved by applying various transformations to the raw input data before learning. The basic idea underly-

---

[1] In this case the *Mean Absolute Deviation (MAD)* is used as distance metric instead of the Mean Squared Error, because the former measure is the one that is known to be minimized by the median.

Table 1. Variants of S-CART for learning ordinal classes.

Name	Formula
POSTPROC. ROUND	$\hat{c}_i$ = mean of the $c_{ij} \in C_i$; real values in leaves of learned tree are rounded to nearest class in C_i
MEDIAN	$\hat{c}_i$ = median of class labels in multiset C_i
ROUNDEDMEANTOCLASS	$\bar{c}$ = mean of the $c_{ij} \in C_i$, $\hat{c}_i = \bar{c}$ rounded to nearest class $c_{ij} \in C_i$
MODE	$\hat{c}_i$ = most frequent class in C_i

ing different data transformations is that numbers may represent fundamentally different types of measurements. [Mosteller & Tukey, 1977] distinguish, among others, the broad classes of *amounts and counts* (which cannot be negative), *ranks* (e.g., 1 = smallest, 2 = next-to-smallest, ...), and *grades* (ordered labels, as in A, B, C, D, E). They suggest the following types of pre-processing transformations: for amounts and counts, translate value v to $tv = \log(v + c)$; for ranks, $tv = \log((v - 1/3)/(N - v + 2/3))$, where N is the maximum rank; and for grades, $tv = (\phi(P) - \phi(p))/(P - p)$, where P is the fraction of observed values that are at least as big as v, p is the fraction of values $> v$, and $\phi(x) = x \log x + (1 - x) \log(1 - x)$. We have tentatively implemented these three pre-processing methods in our experimental system and applied the appropriate transformation to the respective learning problem in our experiments. Table 2 summarizes them in succinct form, in the notational frame of our learning problem.

Note that these transformations do not by themselves contribute to the goal of learning rules for ordinal classes. They are simply tested here as additional enhancements to the methods described above. In fact, pre-processing usually transforms the original ordinal classes into real numbers. That is no problem, however, as the number of distinct values remains unchanged. Thus, the trans-

Table 2. Pre-processing types (c = original class value; tc = transformed class value)

Name	Formula
RAW	No pre-processing ($tc = c$)
COUNTS	$tc = \log(c + 1 - \min(Classes))$
RANKS	$tc = \log((c - 1/3)/(N - c + 2/3))$, where $N = \max(Classes)$
GRADES	$tc = (\phi(P) - \phi(p))/(P - p)$, where $\phi(x) = x \log x + (1 - x) \log(1 - x)$, P = fraction of observed class values $\geq c$, p = fraction of observed class values $> c$

formed values can still be treated as discrete class values without changing the learning algorithms.

In the experiments, we applied only the one type of pre-processing that we considered suitable for the dependent variable of the given learning problem. Subsequently, we rounded to the next class in the "transformed space" and mapped this prediction back. As it turned out, the dependent variable was a "grade" in all four application domains. So, due to the nature of the data, we actually use only one transformation in the experiments. Although we conjecture that the other transformations might give good results as well, this has to be confirmed (or refuted) in further experiments in other domains.

4 Experiments

4.1 Algorithms compared

In the following, we experimentally compare the S-CART variants and preprocessing methods on several benchmark datasets. Three quantities will be measured:

1. *Classification Accuracy* as the percentage of exact class hits,
2. the *Root Mean Squared Error (RMSE)* $\sqrt{1/n \sum_{i=1}^{n}(c_i - \hat{c})^2}$ of the predictions on the test set, as a measure of the average distance of the algorithms' predictions from the true class, and
3. the *Spearman rank correlation coefficient* with a correction for ties, which is a measure for the concordance of actual and predicted ranks.

As ordinal class prediction is somewhere 'between' classification and regression, we additionally include two 'extreme' algorithms in the experimental comparison. One, S-CART_CLASS, is a variant of S-CART designed for categorical classification. S-CART_CLASS chooses the most frequent class in a node as center value and uses the *Gini index of diversity* [Breiman et al., 1984] as evaluation measure; it does not pay attention to the distance between classes. The other extreme, called S-CART_REGRESS, is simply the original S-CART as a regression algorithm that acts as if the task were to predict real values; we are interested in finding out how much paying attention to the discreteness of the classes costs in terms of achievable RMSE. (Of course, the percentage of exact class hits achieved by S-CART_REGRESS cannot be expected to be high.) Finally, we will also list the *Default* or *Baseline Accuracy* for each algorithm on each data set and the corresponding *Baseline RMSE*.

4.2 Data sets

The algorithms were compared on four datasets that are characterized by a clear linear ordering among the classes. Three of the data sets were taken from the UCI repository: Balance, Cars and Nursery. The fourth dataset, the Biodegradability data set [Džeroski et al., 1999], describes 328 chemical substances in the familiar "atoms and bonds" representation [Srinivasan et al., 1995]. The task is to predict the half-rate of surface water aerobic aqueous biodegradation in hours. For previous experiments, we had already discretized this quantity and mapped it to the four classes *fast*, *moderate*, *slow*, and *resistant*, represented as 1, 2, 3, and 4.

4.3 Results

In Table 4.3, we summarize the results (RMSE and classification accuracy) of 10-fold stratified cross-validation runs on these data sets.

The first and most fundamental obervation we make is that the learners improve upon the baseline values in almost all cases, both in terms of RMSE and in terms of classification accuracy. In other words, they really learn something.

As expected, there seems to be a fundamental tradeoff between the two goals of error minimization and accuracy maximization. This tradeoff shows most clearly in the results of the 'extreme' algorithms S-CART_REGRESS and S-CART_CLASS: S-CART_CLASS, which solely seeks to optimize the hit rate during tree construction but has no notion of class distance, is among the best class predictors in all four domains, but among the worst in terms of RMSE. S-CART_REGRESS, on the other hand, is rather successful as a minimizer of the RMSE, but unusable as a classifier.

Interestingly, neither of the two solves its particular problem optimally: some ordinal learners beat S-CART_CLASS in terms of accuracy, and some beat the regression "specialist" S-CART_REGRESS in terms of the RMSE.

For Balance and for Cars, both the pre-processing and the simple post-processing method are able to achieve good predictive accuracy while at the same time keeping an eye on the class-distance-weighted error. These methods also perform favorably in terms of the Spearman rank correlation coefficient.

For Nursery, the results are less pronounced. Still, both methods improve upon the classification result of S-CART_REGRESS. Post-processing leads to a slight improvement in terms of the RMSE and the Spearman rank correlation coefficient (note that these are results for 12961 examples).

For Balance, Cars and Nursery, methods modifying the center value during tree construction (MEDIAN, ROUNDEDMEANTOCLASS and MODE) do not seem to perform as well. In particular for Nursery, these methods perform drastically worse than PREPROC. GRADES and POSTPROC. ROUND.

Results for Biodegradability appear to be different from the other results. The biodegradability domain is different from the other domains in several respects: It has fewer examples, it is known to have class noise and it is essentially relational. Here, methods modifying the center value during tree construction perform better, but not good enough to be competitive with either the classification method or the regression method. Still, it should be noted that the RMSE of these methods is between the RMSE of the classification "specialist" and the one of the regression "specialist".

Drawing more general conclusions from these limited experimental data seems unwarranted. Our results so far show that tree learning algorithms for predicting ordinal classes can be naturally derived from regression tree algorithms, but more extensive experiments with larger data sets from diverse areas will be needed to establish the precise capabilities and relative advantages of these algorithms.

Table 3. Results from 10-fold cross-validation for Balance (625 examples, 4 attributes), Cars (1278 examples, 6 attributes), Nursery (12961 examples, 8 attributes) and Biodegradability (328 examples)

	Balance			Cars		
Approach	Accuracy	RMSE	Spearm.	Accuracy	RMSE	Spearm.
BASELINE	46.1%	1.39	-	70.0%	0.84	-
S-CART_CLASS	77.8%	0.80	0.676	94.6%	0.29	0.933
S-CART_REGRESS	6.4%	0.69	0.677	78.7%	0.23	0.948
PREPROC. GRADES	77.0%	0.70	0.732	94.7%	0.24	0.942
POSTPROC. ROUND	77.6%	0.70	0.733	94.7%	0.25	0.946
MEDIAN	76.5%	0.76	0.692	91.5%	0.34	0.866
ROUNDEDMEANTOCLASS	75.0%	0.67	0.747	93.0%	0.29	0.913
MODE	78.9%	0.75	0.710	88.9%	0.40	0.818
	Nursery			Biodegradability		
Approach	Accuracy	RMSE	Spearm.	Accuracy	RMSE	Spearm.
BASELINE	33.3%	2.84	-	36.6%	1.01	-
S-CART_CLASS	98.4%	0.14	0.988	57.9%	0.84	0.561
S-CART_REGRESS	88.0%	0.11	0.984	1.2%	0.76	0.537
PREPROC. GRADES	97.7%	0.14	0.986	43.3%	0.84	0.436
POSTPROC. ROUND	98.2%	0.13	0.990	48.8%	0.82	0.489
MEDIAN	92.8%	0.27	0.963	50.3%	0.79	0.538
ROUNDEDMEANTOCLASS	93.1%	0.26	0.964	47.3%	0.80	0.510
MODE	92.3%	0.27	0.962	50.3%	0.82	0.506

5 Further Work and Conclusion

Further work will be to perform other experiments including the other transformations suggested by Mosteller and Tukey. It also would be interesting to build tree induction algorithms that do not enforce the prediction of "legal" classes during tree construction, but deal with this problem in the pruning phase. Moreover, it might be rewarding to experiment with tree learners that optimize some other measure (such as the Spearman rank correlation coefficient) for the prediction of ordinal classes.

In this paper, we have taken first steps towards effective methods for learning to predict ordinal classes using regression trees. We have shown how algorithms for learning ordered discrete classes can be derived by simple modifications to a basic regression tree algorithm. Preliminary experiments in four benchmark domains have shown that, in some cases, the resulting algorithms are able to achieve good predictive accuracy while at the same time keeping the class-distance-weighted error low.

Acknowledgments

This research is part of the project "Carcinogenicity Detection by Machine Learning", supported by the Austrian Federal Ministry of Science and Transport. It was also partly supported by the FWF project P12645-INF. Michael de Groeve was supported by a SOCRATES (ERASMUS) grant.

References

[Breiman et al., 1984] Breiman, L., Friedman, J.H., Olshen, R.A., & Stone, C.J. (1984). *Classification and Regression Trees*. Belmont, CA: Wadsworth International Group.

[Džeroski et al., 1999] Dzeroski, S., Blockeel, H., Kompare, B., Kramer, S., Pfahringer, B., Van Laer, W. (1999). Experiments in Predicting Biodegradability, in: *ILP-99: Proceedings Ninth International Workshop on Inductive Logic Programming*, Springer, 1999.

[Kramer, 1996] Kramer, S. (1996). Structural Regression Trees. In *Proceedings of the Thirteenth National Conference on Artificial Intelligence (AAAI-96)*. Cambridge, MA: AAAI Press/MIT Press.

[Kramer, 1999] Kramer, S. (1999). *Relational Learning vs. Propositionalization: Investigations in Inductive Logic Programming and Propositional Machine Learning*. Ph.D. Thesis, Vienna University of Technology.

[Mathieson, 1996] Mathieson, M. (1996). Ordered Classes and Incomplete Examples in Classification. In M. Mozer et al. (eds.), *Advances in Neural Information Processing Systems 9*. Cambridge, MA: MIT Press.

[McCullagh, 1980] McCullagh, P. (1980). Regression Models for Ordinal Data. *Journal of the Royal Statistical Society Series B* 42, 109–142.

[Mosteller & Tukey, 1977] Mosteller, F. & Tukey, J.W. (1977). *Data Analysis and Regression - A Second Course in Statistics*. Reading, MA: Addison-Wesley.

[Potharst & Bioch, 1999] Potharst, R. and Bioch, J.C. (1999). A Decision Tree Algorithm for Ordinal Classification. In *Proceedings of the Third Symposium on Intelligent Data Analysis (IDA-99)*. Berlin: Springer Verlag.

[Quinlan, 1992] Quinlan, J.R. (1992). Learning with Continuous Classes. In *Proceedings AI'92*. Singapore: World Scientific.

[Quinlan, 1993] Quinlan, J.R. (1993). *C4.5: Programs for Machine Learning*. San Mateo, CA: Morgan Kaufmann.

[Schiffers, 1997] Schiffers, J. (1997). A Classification Approach Incorporating Misclassification Costs. *Intelligent Data Analysis* 1(1).

[Silverstein & Pazzani, 1991] Silverstein, G. & Pazzani, M.J. (1991). Relational Clichés: Constraining Constructive Induction During Relational Learning. In *Proceedings of the 8th International Workshop on Machine Learning (ML-91)*. San Mateo, CA: Morgan Kaufmann.

[Srinivasan et al., 1995] Srinivasan, A., Muggleton, S., and King, R.D. (1995). Comparing the use of background knowledge by Inductive Logic Programming systems. In *Proceedings ILP-95*, Katholieke Universiteit Leuven, Belgium.

[Turney, 1995] Turney, P.D. (1995). Cost-Sensitive Classification: Empirical Evaluation of a Hybrid Genetic Decision Tree Induction Algorithm. *Journal of Artificial Intelligence Research* 2, 369–409.

Optimal Queries in Information Filtering*

Ali H. Alsaffar[1], Jitender Deogun[1], and Hayri Sever[2]

[1] The Department of Computer Science & Engineering
University of Nebraska-Lincoln
Lincoln, NE 68588, USA
aalsaffa,deogun@cse.unl.edu

[2] Department of Computer Science & Engineering
Hacettepe University
06532 Beytepe, Ankara, Turkey
sever@hacettepe.edu.tr

Abstract. Information filtering has become an important component of modern information systems due to significant increase in its applications. The objective of an information filtering is to classify/categorize documents as they arrive into the system. In this paper, we investigate an information filtering method based on steepest descent induction algorithm combined with a two-level preference relation on user ranking. The performance of the proposed algorithm is experimentally evaluated. The experiments are conducted using Reuters-21578 data collection. A micro-average breakeven effectiveness measure is used for performance evaluation. The best size of negative data employed in the training set is empirically determined and the effect of R_{norm} factor on the learning process is evaluated. Finally, we demonstrate effectiveness of proposed method by comparing experimental results to other inductive methods.

1 Preliminaries

We propose a framework for the information filtering problem based on *Steepest Descent Algorithm* SDA [8], which is an inductive learning method, combined with a two-level preference relation on user ranking. This SDA method is trained on pre-judged documents and an optimal query is formulated with respect to the proposed framework. The reason for choosing SDA is two folds. First, it has been shown by Wong [8] that SDA is a higher order approximation, whereas Rocchios' method (Relevance Feedback in Vector Space Model) is only a first-order approximation and does not necessarily provide a solution vector. Second, to best of our knowledge no investigation has been reported for SDA in the information filtering, especially when preference relation is involved. In the remaining of this section we briefly provide preliminaries necessary to capture essence of the framework.

---

* This work is supported by the Army Research Office of USA, Grant No. DAAH04-96-1-0325, under DEPSCoR program of Advanced Research Projects Agency, Department of Defense, and by University of Bahrain.

A typical information retrieval (IR) systems S can be defined as a 5-tuple, $S = (T, D, Q, V, f)$, where: T is a set of index terms, D is a set of documents, Q is a set of queries, V is a subset of real numbers, and $f : D \times Q \Rightarrow V$ is a retrieval function between a query and a document. IR systems based on the vector processing model represent documents by vectors of term values of the form $d = (t_1, w_{d_1}; t_2, w_{d_2}; \ldots; t_n, w_{d_n})$, where t_i is an index term in d (i.e. $t_i \in T \cap d$) and w_{d_i} is the weight of t_i that reflects relative importance of t_i in d. Similarly, for a query $q \in Q$, it is represented as $q = (q_1, w_{q_1}; q_2, w_{q_2}, \ldots; q_m, w_{q_m})$, where $q_i \in T$ is an index term in q (i.e. $q_i \in T \cap q$) and w_{q_i} is the weight of query term q_i that reflects relative importance of q_i in q.

Our objective is to formulate an optimal query, Q_{opt}, that discriminates more preferred documents from less preferred documents. With this objective in mind, we define a preference relation $\succeq$ on a set of documents, Δ, in a retrieval (ranking) output as follows. For $d, d' \in \Delta$, $d \succeq d'$ is interpreted as d is preferred to or is equally good as d'. It is assumed that the user's preference relation on Δ yields a weak order where the following conditions are hold [8]:

$d \succeq d'$ or $d' \succeq d$.
$d \succeq d'$ and $d' \succeq d'' \Rightarrow d \succeq d''$.

The essential motivation is that Q_{opt} provides an acceptable retrieval output; that is, for all $d, d' \in \Delta$, there exists $Q_{opt} \in Q$ such that $d \succeq d' \Rightarrow f(Q_{opt}, d) > f(Q_{opt}, d')$.

Given the *user ranking* as a preference relation defined on a set of documents, a system that produces a *system ranking* closer to the user ranking is better than a system that produces a ranking that is further away. To quantify this idea, a performance measure may be derived by using the distance between a user ranking and a system ranking. A possible evaluation measure is R_{norm} as suggested by Bollmann and Wong [2], other measures have also been proposed [9]. Let $(D, \succeq)$ be a document space, where D is a finite set of documents and $\succeq$ be a preference relation as defined above. Let Δ be some ranking of D given by a retrieval system. Then R_{norm} is defined as

$$R_{norm}(\Delta) = \frac{1}{2}\left(1 + \frac{S^+ - S^-}{S^+_{max}}\right) \qquad (1)$$

where S^+ is the number of document pairs where a preferred document is ranked higher than non-preferred document, S^- is the number of document pairs where a non-preferred document is ranked higher than preferred one, and S^+_{max} is the maximal number of S^+.

Example 1. Consider the ranking $\Delta = (rnrn \mid rnnnnn)$ for a two level preference relation, where r stands for a relevant document and n for a non-relevant document. So, $S^+_{max} = 21, S^+ = 13, S^- = 2$, and $R_{norm} = 0.762$. □

In this paper, Q_{opt} is formulated inductively by SDA as described in [8]. Let $B = \{b = d - d' : d \succeq d'\}$ be the set of difference vectors in an output ranking. To obtain Q_{opt} from any query Q, we solve $f(Q, b) > 0$ for all $b \in B$. It is assumed here that $f(Q, d) = Q^T d$, which is the cross product, and for

$f(Q,d) > f(Q,d') \Rightarrow Q^T d > Q^T d' \Rightarrow Q^T(d-d') > 0 \Rightarrow f(Q,b) > 0$. The steps of the algorithm are as follows.

1. Choose a starting query vector Q_0; let $k = 0$.
2. Let Q_k be a query vector at the start of the (k+1)th iteration; identify the following set of difference vectors:

$$\Gamma(Q_k) = \{b = d - d' : d \succeq d' \text{ and } f(Q_k, b) \leq 0\};$$

 if $\Gamma(Q_k) = \emptyset$, $Q_{opt} = Q_k$ is a solution and exit, otherwise,
3. Let

$$Q_{k+1} = Q_k + \sum_{b \in \Gamma(Q_k)} b$$

4. $k = k + 1$; go back to Step (2).

Theoretically it is known that SDA terminates only if the set of retrieved documents is linearly separable. Therefore, a practical implementation of the algorithm should guarantee that the algorithm terminates whether or not the retrieved set is linearly separable. In this paper, we use a pre-defined iteration number and R_{norm} measure for this purpose. The algorithm is terminated if the iteration number reaches the pre-defined limit or the R_{norm} value of the current query is higher than or equal to some pre-defined value. Within the algorithm loop we continually update the query, Q_k, that yields the highest R_{norm} value in order to return Q_k as optimal query.

2 Effectiveness Measures

In order to measure the performance of a classifier,[1] we use text categorization effectiveness measures. There are a number of effectiveness measures employed in evaluating text categorization algorithms. Most of these measures are based on the *contingency table* model. Consider a system that is required to categorize n documents by a query, the result is an outcome of n binary (or two-level) decisions. From these decisions a dichotomous table is constructed as in Figure 1(a). Each entry in the table specifies the number of documents with the specified outcome label. For example, a is the number of documents whose predicted and actual labels agree upon being relevant.

In our experiment, the performance measures are based on *precision* and *recall* whose values are computed as $a/(a + b)$ and $a/(a + c)$ in Fig 1(a), respectively. Usually a single composite recall-precision graph is reported reflecting the average performance of all individual queries in the system. Two average effectiveness measures, widely used in the literature, are: *Macro-average* and *Micro-average* [4]. In information retrieval, Macro-average is preferred in evaluating

---

[1] In this paper, a classifier is defined as a query. Therefore, we will use query and classifier interchangeably.

Predicted Label	Actual Label	
	relevant	non-relevant
Relevant	a	b
Non-relevant	c	d

(a)

	Training	Test
With at least one topic	7,775	3,019
With no topic	1,828	280
Total	9,603	3,299

(b)

Fig. 1. (a) Measures of system effectiveness. (b) Number of documents in the collection.

query-driven retrieval, while in text categorization Micro-average is preferred. Consider a system with n documents and q queries. Then there are q dichotomous tables each of which is similar to the one in Figure 1(a) representing the outcomes of two-level decisions (relevant or nonrelevant) by the filtering system (predicted label) and the user/expert (actual label) when a query is evaluated against all n documents. Macro-average computes precision and recall separately from the dichotomous tables for each query, and then computes the mean of these values. Micro-average, on the other hand, adds up the q dichotomous tables all together, and then precision and recall are computed.

For the purpose of plotting a single summary figure for recall versus precision values, an adjustable parameter is used to control assignment of documents to profiles (or categories in text categorization). Furthermore recall and precision values at different parameter settings are computed to show trade-off between recall and precision. This single summary figure is then used to compute what is called *breakeven* point, which is the point at which recall is approximately equal to precision [4]. It is possible to use linear interpolation to compute the breakeven point between recall and precision points.

3 The Experiment

In this section, we describe the experimental set up in detail. First, we describe how the Reuters-21578 dataset is parsed and the vocabulary for indexing is constructed. Upon discussion of our approach to training, the experimental results are presented.

3.1 Reuters-21578 Data Set and Text Representation

To experimentally evaluate the proposed information filtering method, we have used the corpus of Distribution 1.0 of Reuters-21578 text categorization test collection [2]. This collection consists of 21,578 documents selected from Reuters newswire stories. The documents of this collection are divided into training and test sets. Each document has five category tags, namely, *EXCHANGES, ORGS, PEOPLE, PLACES,* and *TOPICS*. Each category consists of a number of topics that are used for document assignment. We restrict our study to only *TOPICS*

---

[2] Reuters-21578 collection is available at: http://www.research.att.com/~lewis.

category. To be more specific, we have used the Modified Apte split of Reuters-21578 corpus that has 9,603 training documents, 3,299 test documents, and 8,676 unused documents.

The training set was reduced to 7,775 documents as a result of screening out training documents with empty value of *TOPICS* category. There are 135 topics in the *TOPICS* category, with 118 of these topics occurring at least once in the training and test documents[3]. Each of the three topics out of 118 ones has been assigned to only one document in the test set. We have chosen to experiment with all of these 118 topics despite the fact that three topic categories with no occurrence of training set automatically degrades system performance. Figure 1(b) shows some statistics about the number of documents in the collection.

We have produced a dictionary of single words excluding numbers as a result of pre-processing the corpus including performing parsing and tokenizing the text portion of the title as well as the body of both training and unused documents. We have used a universal list of 343 stop words to eliminate functional words from the dictionary [4]. The Porter stemmer algorithm was employed to reduce each remaining words to word-stems form [5]. Since any word occurring only a few times is statistically unreliable [6], the words occurring less than five times were eliminated. The remaining words were then sorted in descending order of frequency.

Our filtering framework is based on the Vector Space Model (VSM) in which documents and queries are represented as vectors of weighted terms. Let t_{jk} be a jth term of document with identity of k in a collection. One common function employed for computing document term weight, say w_{jk}, is to multiply term frequency (indicating the frequency of the term in a document) by the inverse document frequency of that term which can be formulated as $w_{jk} = t_{jk} \times \log(N/n_j)$ [6], where t_{jk} is the term frequency, N is the total number of documents in the collection, and n_j is the number of documents containing t_{jk}. We use a normalized version of this function (i.e., making magnitudes of document vectors one). A document is assigned to a topic by a particular classifier if the *cosine similarity* measure between this document and the query is greater than or equal to an externally supplied *threshold* value, called adjustable parameter previously.

3.2 Training

In contrast to information retrieval systems, in text categorization systems, we have neither a retrieval output nor a user query. Instead, we have a number of topics and for each topic the document collection is partitioned into training

---

[3] In the description of the Reuters-21578 read-me file it was stated that the number of topics with one or more occurrences in *TOPICS* category is 120, but we have found only 118. The missing two topics were assigned to unused documents.
[4] The stop list is available at: http://www.iiasa.ac.at/docs/R_Library/libsrchs.html.
[5] The source code for the Porter Algorithm is found at: http://ils.unc.edu/keyes/java/porter/index.html.

Table 1. Top 16 topics with more than 100 positive examples.

Name	Train	Test	Name	Train	Test
earn	2877	1087	ship	197	89
acq	1650	719	corn	182	56
money-fx	538	179	money-supply	140	34
grain	433	149	dlr	131	44
crude	389	189	sugar	126	36
trade	369	118	oilseed	124	47
interest	347	131	coffee	111	28
wheat	212	71	gnp	101	35

and test cases. The training set contains only positive examples of a topic. In this sense, the training set is not a counterpart of the retrieval output due to the fact that we do not have any negative examples. We can, however, construct a training set for a topic that consists of positive and negative examples, under the plausible assumption that any document considered as positive example for the other topics and not in the set of positive examples of the topic at hand is a candidate for being a negative example of this topic.

The maximum number of positive examples per topic in the corpus is 2877 and the average is 84. The size and especially the quality of the training set is an important issue in generating an induction rule set. In an information routing study [7], the learning method was not applied to the full training set but rather to the set of documents in the local region for each query. The local region for a query was defined as the 2000 documents nearest to the query, where similarity was measured using the inner product score to the query expansion of the initial query. Also, in [1] the rules for text categorization were obtained by creating local dictionaries for each classification topic. Only single words found in documents on the given topic were entered in the local dictionary.

In our experiment, the training set for each topic consists of all positive examples while the negative data is sampled from other topics. The reason for including the entire set of positive examples is that SDA is an enhanced version of a relevance feedback algorithm and thus a larger number of positive examples makes the algorithm produce more efficient induction rules. Additionally, the result published by Dumais et al. [3] for the Reuters-21578 data shows that with respect to micro-averaged score of the SVM (Support Vector Machine) over multiple random samples of training sets for the top 10 categories with varying sample size, but keeping size of negative data the same, performance of the SVM was degraded from 92% to 72.6% while the size was reduced from whole training set down to 1%. Another important finding reported in that study shows that performance of the SVM becomes somewhat unstable when a category has fewer than 5 positive training examples. Here we have investigated the size of training set from a different perspective and tried to estimate the best size for the negative data in proportion to positive ones, which is described in the remaining part of this section.

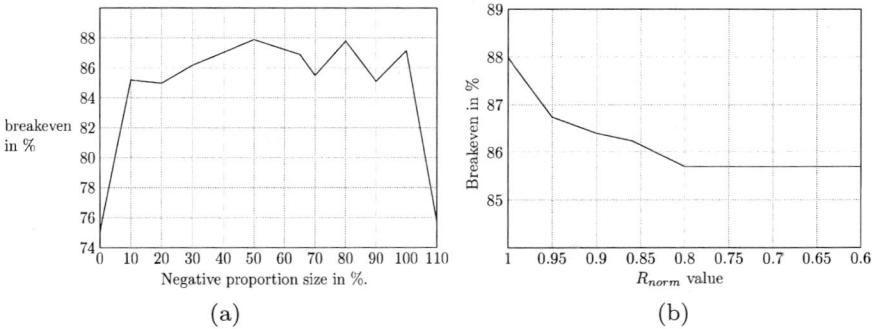

Fig. 2. (a) Performance of the top 16 topics at various negative-to-positive percentages. (b) R_{norm} value versus breakeven on the top 10 topics.

To estimate the best size for negative data in the training set, we have experimented with the top 16 topics of the Reuters-21578 data as shown in Table 1 in which training and testing sizes of positive data are given. We have trained each of these topics as follows: 1) Consider all positive data, 2) take a $\phi\%$ of the positive data as the sample size of the negative data, 3) apply SDA on this training set and compute breakeven point, 4) repeat the above steps but selecting a different negative sample in Step 2. This process is continued until all negative data (i.e., varying sample sizes of negative data for each of the 16 top topics) is exhausted. Figure 2(a) shows the performance for $\phi = 10, 20, \cdots, 110$. The initial query of SDA is first set to the mean of the positive examples and for subsequent iterations the initial query is simply set to the query obtained in the previous iteration. We set R_{norm} value to its maximum value (i.e. 1.0) and assert on maximum of 150 iterations in case the value of R_{norm} is not reached.

As indicated in Figure 2(a), the quality of the induction is effectively enhanced best when the proportion of negative data over positive data becomes 10% (local maximal point) with respect to ratio of increase in breakeven point over the one in size of negative data. Besides the maximum point is reached (i.e., the best performance in the absolute sense) when proportion of negative data becomes 50% or 80%. It is worth stating that the performance is abruptly degraded when the size of negative data exceeds that of positive data. For the concern of obtaining the best quality of induction, we fixed the size of negative sample to 50% of the positive set. For example, for 'grain' topic in Table 1(b), we considered all the 433 training examples as positive data and 216 as negative data sampled from other topics.

The choice of R_{norm} value used in terminating criteria of the SDA algorithm is important in the learning process. This is because there is an application-dependent trade off between quality of a query and processing overhead Figure 2(b) shows the performance trade offs on various values of R_{norm} on the top 10 topics. As the value of R_{norm} decreases, the system performance also decreases until a point where the query produced for subsequent values of R_{norm}

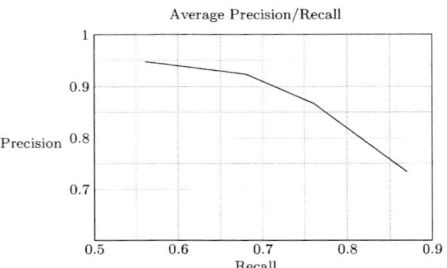

Fig. 3. (c) Average Precision/Recall on all 118 topics with breakeven point of 81.28%.

remains unchanged. This is because on the first run, the algorithm yields a query with a higher R_{norm} value than whatever the supplied R_{norm} value is.

3.3 Results

The average precision/recall graph of the 118 topics is shown in Figure 3. The graph is magnified and plotted around the area where recall and precision are approximately equal (i.e. breakeven point.) From the graph, the breakeven is approximately 81.28%. As a comparative study, Table 2 presents results of SDA and other five inductive algorithms that were recently experimented on Reuters-21578 dataset [3]. *Findsim* method is a variant of Rocchio's method for relevance feedback. The weight of each term is the average (or centroid) of its weight in positive instances of the topic. We may compare SDA with this method since it is a first-order approximation. The list of names such as *NBayes, BayesNets, Trees,* and *SVM* in Table 2 stand for Naive Bayes, Bayes Nets, Decision Trees, and Linear Support Vector Machines methods, respectively. For further details of these methods the reader is referred to [3].

SDA outperforms Findsim, NBayes, and BayesNet, and is almost as good as the Decision Trees method. It is however outperformed by the Linear SVM method due to the fact that relevance feedback methods (including SDA) require as large size of positive data as possible for drifting the query towards the solution region. Therefore, for topics with small number of positive examples, which represent the majority of the topics in the Reuters-21578 data, the optimal query which is close to the solution region is hard to find and the performance of the algorithm SDA is the same as Findsim method on these topics. Nevertheless, on average it outperforms the Findsim method by a significant margin which upholds the plausible fact that higher-order approximation methods, such as SDA, outperforms their counterpart first-order approximation methods, such as Findsim.

Table 2. Comparing results with other five inductive algorithms. Breakeven is computed on top 10 topics and on overall 118 topics.

Topic	Findsim	NBayes	BayesNets	Trees	SDA	SVM
earn	92.9%	95.9%	95.8%	97.8%	98.5%	98.0%
acq	64.7%	87.8%	88.3%	89.7%	95.9%	93.6%
money-fx	46.7%	56.6%	58.8%	66.2%	78.4%	74.5%
grain	67.5%	78.8%	81.4%	85.0%	90.6%	94.6%
crude	70.1%	79.5%	79.6%	85.0%	86.5%	88.9%
trade	65.1%	63.9%	69.0%	72.5%	76.06%	75.9%
interest	63.4%	64.9%	71.3%	67.1%	77.29%	77.7%
wheat	68.9%	69.7%	82.7%	92.5%	82.2%	91.9%
ship	49.2%	85.4%	84.4%	74.2%	88.1%	85.6%
corn	48.2%	65.3%	76.4%	91.8%	82.36%	90.3%
Avg. Top 10	64.6%	81.5%	85.0%	88.4%	87.99%	92.0%
Avg All Cat.	61.7%	75.2%	80.0%	N/A	81.28%	87.0%

References

1. Apt C., Damerau F., and Weiss S. M. Automated learning of decision rules for text categorization. *ACM Transactions on Information Systems*, 12(3):233-251, July 1994.
2. Bollman, P. and Wong, S. K. M. Adaptive linear information retrieval models. In *Proc. of the 10th Int. ACM SIGIR Conference on Research and Development in Information Retrieval*, pages 157-163, New Orleans, LA., June 1987.
3. Dumais, S., Platt, J., Heckerman, D., and Sahami, M. Inductive learning algorithms and representations for text categorization. In *Proceedings of ACM-CIKM98*, Nov. 1998.
4. Lewis, David D. Evaluating text categorization. In *Proceedings of Speech and Natural Language Workshop*, pages 312-318, February, 1991.
5. Raghavan, V. V., and Sever H. On the reuse of past optimal queries. In *Proc. of the 18_{th} Int. ACM-SIGIR Conference on Research and Development of Information Retrieval*, pages 344-350, 1995.
6. Salton, G, *Automatic Text Processing, The Transformation, Analysis, and Retrieval of Information by Computer*. Addison-Wesely, 1988.
7. Schütze, H., Hull, D. A., and Pederson, J. O. A comparison of classifiers and document representations for the routing problem. *In Proc. of the 18th Annual Int. ACM-SIGIR Conference on Research and Development of Information Retrieval* (Seattle, Washington, 1995) 229-237.
8. Wong, S. K. M., and Yao, Y. Y. Query formulation in linear retrieval models. *Journal of the American Society for Information Science*, 41, 5 (1990), 334-341.
9. Wong, S. K. M. Measuring retrieval effectiveness based on user preference of documents. *Journal of the American Society for Information Science*, 46(2):133-145, 1995.

Automatic Semantic Header Generator

Bipin C. Desai, Sami S. Haddad, and Abdelbaset Ali

Department of Computer Science,
Concordia University,
1455 de Maisonneuve Blvd West,
Montréal,
CANADA H3G 1M8
Email contact: bcdesai@cs.concordia.ca
http://www.cs.concordia.ca/faculty/bcdesai

Abstract. As the mounds of information and the number of Internet users grow, the problem of indexing and retrieving of electronic information resources becomes more critical. The existing search systems tend to generate misses and false hits due to the fact that they attempt to match the specified search terms without proper context in the target information resource. In environments that contain many different types of data, content indexing requires type-specific processing to extract indexing information effectively. The COncordia INdexing and DIscovery (Cindi) system is a system devised to support the registration of indexing meta-data for information resources and provide a convenient system for search and discovery. The Semantic Header, containing the semantic contents of information resources stored in the Cindi system, provides a useful tool to facilitate the searching for documents based on a number of commonly used criteria. This paper presents an automatic tool for the extraction and storage of some of the meta-information in a Semantic Header and the classification scheme used for generating the subject headings.

1 Introduction

Rapid growth in data volume, user base and data diversity render Internet-accessible information increasingly difficult to use effectively. The number of information sources, both public and private, available on the Internet are increasing almost exponentially. They include text, computer programs, books, electronic journals, newspapers, organisational, local and national directories of various types, sound and voice recordings, images, video clips, scientific data, and private information services such as price lists and quotations, databases of products and services, and speciality newsletters [3]. There is a need for an automated search system that allows easy search for and access to relevant resources available on the Internet which in turn requires proper indexing of the available information. The semantics of the resource are exploited in the current system to extract and summarise the relevant meta-information(Semantic

Headers [2]) to support its discovery. Specialised databases maintain archives of these Semantic Headers(SH) which could be searched by another component of Cindi which features cooperating distributed expert systems and helps users in locating pertinent documents.

The Cindi system provides mechanisms to register, search and manage the SHs, with the help of easy to use graphical user interfaces. Cindi avoids problems caused by differences in semantics and representation as well as incomplete and incorrect data cataloguing by using a standardized subject heading hierarchy. This meta-information could be entered by the primary resource provider with the help of an Automatic Semantic Header Generator (ASHG) described in this paper. ASHG is a software that assists the authors of documents to semi-automatically generate many of the fields of the SH and hence assist them in the registration of their documents in the Cindi system. One of the main tasks of ASHG is to classify a document under a list of subject headings as described herein. As the author is required to verify and complete the ASHG generated Semantic Header entry, the potential for its accuracy is high.

The paper is organized as follows: in section 2, we introduce the Cindi system. Section 3 covers our approach to the building of the thesaurus used in ASHG system and section 4 describes its components. Following this, we give the results of our tests to generate the SH on a set of documents prepared in the HTML, LaTeX, RTF and plain text format and our conclusions.

2 The Cindi System

Attempts to provide easy search of relevant documents has led to a number of systems [1,5,7,8,13,15,18,19,20]. However, the problem with many of these is that their selectivity of documents is often poor [3]. The chances of getting inappropriate documents and missing relevant information because of poor choice of search terms are great. Hence, there is a need for the development of a system which allows easy search for and access to resources available on the Internet. Using a standard index structure and building an expert system based bibliographic system using standardised control definitions and terms can alleviate the problem and provide fast, efficient and easy access to the Web documents. For cataloguing and searching, Cindi uses a meta-data description called SH[4] to describe an information resource. The SH includes those elements that are most often used in the search for an information resource. Since the majority of searches begin with a title, name of the authors (70%), subject and sub-subject (50%) [6], Cindi requires the entry for these elements in the SH. Similarly, the abstract and annotations are relevant in deciding whether or not a resource is useful, so they are included too[3]. The components of the SH are: Title, Alt-title, Language, Character Set, Keyword, Identifier, Date, Version, Classification, Coverage, System Requirements, Genre, Source and Reference, Cost, Abstract, Annotations and User ID, Password.

Preparing the primary source's SH requires identifying it as to its subject, title, author, keywords, abstract, etc. These problems are addressed by Cindi, which provides a mechanism to register, manage and search the bibliographic information.

The overall Cindi system uses knowledge bases and expert sub-systems to help the user in the registering and search processes. The index generation and maintenance sub-system uses Cindi's thesaurus to help the provider of the resource select the most-appropriate standard terms for items such as subject, sub-subject and keywords. Similarly, another expert sub-system is used to help the user in the search for appropriate information resources [2].

The SH information entered by the provider of the resource using a graphical interface is relayed from the user's workstation by a client process to the database server process at one of the nodes of a distributed database system (SHDDB). The node is chosen based on its proximity to the workstation or on the subject of the index record. From the point of view of the users of the system, the underlying database may be considered to be a monolithic system. In reality, it would be distributed and replicated allowing for reliable and failure-tolerant operations. The interface hides the distributed and replicated nature of the database. On receipt of the information, the server verifies the correctness and authenticity of the information and on finding everything in order, sends an acknowledgment to the client. The server node is responsible for locating the partitions of the SHDDB where the entry should be stored and forwards the replicated information to appropriate nodes. The various sites of the database work in a cooperating mode to maintain consistency of the replicated portion. The replicated nature of the database also ensures distribution of load and ensures continued access to the bibliography when one or more sites are temporarily nonfunctional.

Cindi search sub-system guides the user in entering the various search items in a graphical interface similar to the one used by the index entry system. Once the user has entered a search request, the client process communicates with the nearest SHDDB catalogue to determine the appropriate site of the SHDDB database. Subsequently, the client process communicates with this database and retrieves one or more SHs. The result of the query could then be collected and sent to the user's workstation. The contents of these headers are displayed, on demand, to the user who may decide to access one or more of the actual resources.

3 ASHG's Thesaurus

ACM[17], INSPEC[14] and Library of Congress Subject Headings (LCSH)[16] were the main building blocks of Cindi's three level Subject Hierarchy which currently is limited to the domains of Computer Science and Electrical Engineering. ASHG's computer science subject hierarchy uses ACM's subject hierarchy as the starting point, and electrical engineering subject hierarchy is based on that of INSPEC's. We have exploited LCSH's subject headings relations to refine both hierarchies. LCSH contained relations between subject headings such as BT (Broader Term), NT (Narrow Term), UF (Used For), and RT (Related To). In order to augment ACM and INSPEC subject hierarchies, a search for an

ACM or INSPEC subject heading was made in LCSH. If a match was found, the narrow terms found in LCSH under the matched subject were added to the list of subjects or terms under the ACM or INSPEC's matched subject heading. This augmentation produced a hierarchy composed of five or six levels. Since Cindi's subject hierarchy was limited to only three levels, the following rules illustrated in Figure 1, were applied to merge these subject headings. The ($Level\_0$) subject is *Computer Science* or Electrical Engineering. Some of the subject headings found in the $Level\_1$ and $Level\_2$ augmented subject hierarchies were concatenated to form the Cindi's $Level\_1$ subject heading. The same rule was applied on subject headings at $Level\_3$ and $Level\_4$ to yield Cindi's $Level\_2$ subject heading. The $Level\_5$ and $Level\_6$ subjects were used as controlled terms associated with Cindi's $Level\_2$ subject headings.

The resulting subject hierarchy has three levels and a set of control terms associated with the lowest level subject headings.

The reason behind the Control Term Subject association is to extract or classify the primary source under a number of subject headings by comparing the significant list of words contained in the document with the list of control terms. An association between the control terms and their corresponding subject headings is created.

Each control term has three lists of subject headings attached to it. The control terms are based on the terms found in ASHG's subject hierarchy and the additional terms that are associated with $Level\_2$ subject headings. For each subject heading and the additional controlled terms, we use their constituent English none noise words as their corresponding control terms. For example, the control term *compute* will be associated with *Computer Science* general subject heading. Similarly, the control term *hardware* will be associated with *Hardware integrated circuits* and *Hardware performance and reliability* level_1 subject headings and *Hardware Simulation Design Aids* level_2 subject heading. Each controlled term is associated with one or more subject headings.

Mapping ASHG's subject heading terms into control terms involves: removing noise (stop) words; stemming the remaining words to find the the root and associating the root with the corresponding subject heading.

4 ASHG Implementation

In this section, we present the implementation details of the Automatic Semantic Header Generator (ASHG) of the Cindi system. This is an important step in providing the author of a document a draft SH with an initial set of subject classifications and a number of components of the SH for the document. The ASHG scheme takes into account both the occurrence frequency and positional weight of keywords found in the document. Based on the document's keywords, ASHG assigns a list of subject headings by matching those keywords with the controlled terms found in the controlled term subject association. The ASHG also extracts some of the meta-information from a document such as title, abstract, keywords, dates, author, author's information, size and type.

ASHG uses the syntax of documents in HTML, LaTex, RTF or text to extract the document's meta-information. ASHG extracts summary information, such as the title, keywords, dates of creation, author, author's information, abstract and size. In tagged documents, the author might explicitly tag some of the fields to be extracted. In case these fields are not explicitly tagged, ASHG attempts to extract them using heuristics. However, if the explicit keywords were not found in the document, then words found in the title, abstract and other tagged words would be used to extract an implicit list of keywords.

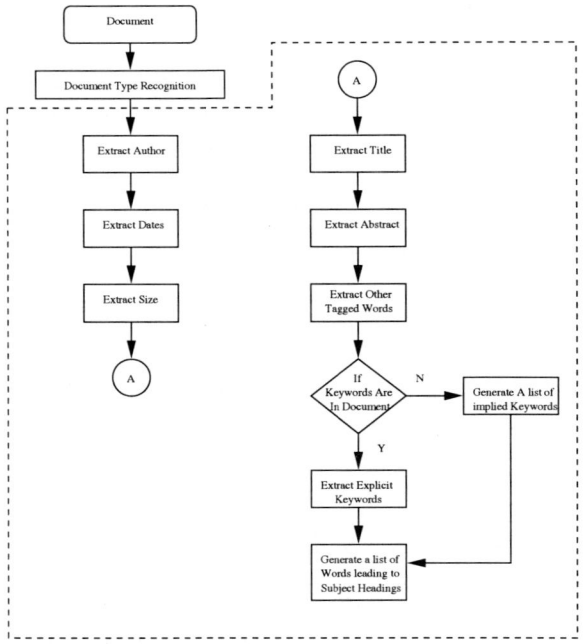

Fig. 1. ASHG's extraction steps

Generating an Implicit List of Keywords and Words Used in Document Classification

ASHG generates an implicit list of keywords in case explicit keywords were not found in the document, the system derives a list of words from the words found in the title, abstract, and other tagged fields. This list of derived words will also be used in classifying the document. However, if the keywords were explicitly stated in the document, then ASHG will augment them with a list of words from the words found in the title, abstract, and other tagged fields.

Generating both lists of words relies on the stemming process that will map the words into their root words, the stemmed word frequency of occurrence and the word location in the document. Because the terms are not equally useful for content representation, it is important to introduce a term weighting system that assigns high weights for important terms and low weight for the less important terms [11]. The weight assignment uses the following scheme:

If the keywords are explicitly included in the document, they convey some important concepts and hence are assigned the highest weight of five. Usually, words found in the abstract are the second most important words, and are assigned a weight of four. The words in the title, are assigned a weight of three. The word appearing in the other tagged fields, are assigned a weight of two.

Each numeric weight is a class by itself defining the words' location. The range of class weight generated will be between two and 14, depending on the positions where a word appears.

For each class, we set the maximum class frequency to be the frequency of occurrence of a term found most often in that class. For instance, if, in class four, we had three terms having two, four and six as frequencies, the system would select six as the maximum class four frequency. The words' frequencies are compared with their corresponding maximum class frequency. For low weighted classes such as two and three, significant terms have the maximum class frequencies. Thus, limiting the number of significant terms. However, all terms found in class eight and more are significant regardless of their frequency of occurrence.

Table 1. Weight and Frequency numbers used in extracting terms

Term Weight	Term Frequency
2	Maximum Class 2 frequency
3	Maximum Class 3 frequency
4	Greater or equal to Maximum Class 4 frequency minus 1
5	Greater or equal to Maximum Class 5 frequency minus 1
6	Greater or equal to Maximum Class 6 frequency minus 2
7	Greater or equal to Maximum Class 7 frequency minus 3
≥ 8	All

Two lists of words are generated. The first one containing only the root words or control terms found in Cindi's thesaurus. This list of control terms is used in the document's subject classification scheme. The second list contains the most significant root words not found in Cindi's thesaurus. If no keywords were found in the document, ASHG extracts words having a term weight more than four and their corresponding frequencies of occurrence is the same as the ones tabulated. These words are the document's keywords. In generating a list of control terms used to classify the document, terms having weight of two or more are extracted. The extracted words have the frequencies of occurrence as tabulated in Table 1.

ASHG's Document Subject Headings Classification Scheme

An important step in constructing the draft SH is to automatically assign subject headings to the documents. The title, explicitly stated keywords, and abstract are not enough by themselves to convey the ideas or subjects of the document.

Since the author tries to convey or to summarise his ideas in the previously mentioned fields, there is a need to use all none noise words found in those fields. To assign the subject headings, ASHG uses the resulting list of significant words generated from the previous section and the control term to subject association. The subject heading classification scheme relies on passing weights from the significant terms to their associated subjects, and selecting the highest weighted subject headings. The following algorithm is used to construct the three levels of subject headings:

1. For each term found in both Cindi's control terms and the generated list of words, the system traces the control term's attached list of subjects (list of *level0, level1 and level2*) headings, and adds the subject headings to their corresponding list of possible subject headings.
2. Weights are also assigned to the subject hierarchies. The weight for a subject is given according to where the term matching its controlled term was found. A subject heading having a term or set of terms occurring in both title and abstract, for instance, gets a weight of seven. The matched terms' weights are passed to their subject headings.
3. The system extracts *Level_2, Level_1* and *Level_0* subject headings having the highest weights from the three lists of possible subject headings.
4. After building the three lists for the three level subject headings, the system selects the subjects using the bottom-up scheme:
   a) Having selected the highest weighted *level_2* subject headings, the system derives their *level_1* parent subject headings.
   b) An intersection is made between the derived *level_1* subject headings and the list of the highest weighted *level_1* subject headings. The common *level_1* subjects are the document's *level_1* subject headings.
   c) The system uses the same procedure in selecting *level_0* subject headings.

Once the process of extracting the meta-information is terminated, the SH is displayed for the source provider to modify, add or remove some of the attributes. Once the provider finishes, the semantic header can be registered in the Cindi database.

5 Analysis of ASHG's Results and Conclusions

The experiments described here are designed to test the accuracy of the generated index and the subject headings classification results. After applying the ASHG on a set of documents, the generated index fields such as title, keywords, abstract and author are compared with those that are found in the document.

The experiments were conducted on a number of documents[21]. These documents dealt with computer science and electrical engineering subjects. Each of these documents was rendered manually in the four formats. ASHG was able to extract all the explicitly stated fields such as title, abstract, keywords, and author's information with a hundred percent accuracy. If the abstract was not explicitly stated, ASHG was able to automatically generate an abstract that would describe the paper. However, ASHG's implicit keyword extraction generated a list of words which included some words that were insignificant. These insignificant words in turn lead to the diversion in subject classification.

The ASHG's automatic subject headings classification results are compared with the INSPEC's classification and with what the papers' authors would regard as good subject classifications and poor ones. For the former we consulted the authors about the subject heading generated by ASHG system for their documents. The results are tabulated in Table 2. which shows a greater than 50% of acceptable subject headings. Some of the ASHG's subject classifications had different words than INSPEC's even though they described the same subject. That was due to the fact that our computer science subject classification was built from ACM and not from INSPEC.

Table 2. Summary of ASHG's tests

Document Type	Avg. Number of Subject Headings Generated	Avg. Number of Acceptable Subject Headings	Percent of Inspec Heading Discovered
HTML	4.9	66.1%	74%
LaTex	4.4	63%	80%
RTF	4.8	60.6%	65%
Text	5.9	57.0%	80%

ASHG's was able to generate between 65% and 80% of the subject heading that were generated by professional catalogers. However, since ASHG produced, on the average more classifications, the accuracy was lower at about 22%. Since our system was only based on the frequency and location of words in a document to determine the document's keywords and subject classification, it missed the importance of the word senses and the relationship between words in a sentence. The simplistic system did not capture the concepts behind the documents, or the ideas that the author was trying to convey. Our results support the idea that word frequency and location are not enough in information retrieval. However, since the ASHG's result will be used as a starting point by the author, he/she has the opportunity to correct the errors and include fields of the SH not given before registering it. Further work is required in refining the subject classification to reduce the number of poor classifications.

In conclusion, we believe that resolving word senses and determining the relationships that those words have to one another will have the greatest impact on refining the ASHG's subject classification scheme. Therefore, we plan to pursue semantic level language processing in the future.

References

1. De Bra, P., Houben, G-J., & Kornatzky, Y., *Search in the World-Wide Web*, http://www.win.tue.nl/help/doc/demo.ps

2. Desai B. C., *Cover page aka Semantic Header*, http://www.cs.concordia.ca/~faculty/bcdesai/semantic-header.html, July 1994, revised version, August 1994.
3. Desai B. C., *The Semantic Header Indexing and Searching on the internet*, Department of Computer Science, Concordia University. Montreal, Canada, February 1995.
http://www.cs.concordia.ca/~faculty/bcdesai/cindi-system-1.1.html
4. Desai, Bipin C., *Supporting Discovery in Virtual Libraries*, Journal of the American Society of Information Science(JASIS), 48-3, pp. 190-204, 1997.
5. Fletcher, J. 1993., Jumpstation,
http://www.stir.ac.uk/jsbin/js
6. Katz, W. A., *Introduction to Reference Work*, Vol. 1-2 McGraw-Hill, New York, NY.
7. Koster, M., *ALIWEB(Archie Like Indexing the WEB)*,
http://web.nexor.co.uk/aliweb/doc/aliweb.html
8. McBryan, Oliver A., *World Wide Web Worm*,
http://www.cs.colorado.edu/home/mcbryan/WWWW.html
9. Paice C. D., *Automatic Generation of Literature Abstracts - An Approach Based on the identification of self indicating phrases, in information retrieval research*, R.N. Oddy, S.E. Robertson, C.J. van Rijsbergen and P.W. Williams, editors, Butterworths, London, pp. 172-191, 1981.
10. Salton G., Allen J. , Buckley O. , *Automatic Structuring and Retrieval of Large Text Files*, Department of Computer Science, Cornell University. 1992.
11. Salton G., Allan J. , Buckley C., and Singhal A. , *Automatic Analysis, Theme Generation, and Summarization of Machine-Readable Texts*, Science, Vol264, pp. 1421-1426, June 1994.
12. Shayan N., *CINDI: Concordia INdexing and DIscovery system*, Department of Computer Science, Concordia University, Montreal, Canada, 1997.
13. Thau, R., *SiteIndex Transducer*,
http://www.ai.mit.edu/tools/site-index.html
14. Computer and Control Abstracts, Produced by INSPEC, No. 10, October 1997.
15. *Experimental Search Engine Meta-Index*,
http://www.ncsa.uiuc.edu/SDG/Software/Mosaic/Demo/metaindex.html
16. Library of Congress Subject Headings, September 1996.
17. http://www.acm.org/class/1998/ccs98.txt.
18. Search WWW document full text,
http://rbse.jsc.nasa.gov/eichmann/urlsearch.html
19. WebCrawler,
http://www.biotech.washington.edu/WebCrawler/WebQuery.html
20. World Wide Web Catalog,
http://cuiwww.unige.ch/cgi-bin/w3catalog
21. http://www.cs.concordia.ca/~faculty/bcdesai/cindi/listofpapers.html

On Modeling of Concept Based Retrieval in Generalized Vector Spaces*

Minkoo Kim[1+], Ali H. Alsaffar[2], Jitender S. Deogun[2], and Vijay V. Raghavan[1]

[1]The Center for Advanced Computer Studies
University of Louisiana
Lafayette, LA 70504, USA
{mkim, raghavan}@cacs.louisiana.edu
[2]Department of Computer Science & Engineering
University of Nebraska
Lincoln, NE 68588, USA

Abstract. One of the main issues in the field of information retrieval is to bridge the terminological gap existing between the way in which users specify their information needs and the way in which queries are expressed. One of the approaches for this purpose, called Rule Based Information Retrieval by Computer (RUBRIC), involves the use of production rules to capture user query concepts (or topics). In RUBRIC, a set of related production rules is represented as an AND/OR tree. The retrieval output is determined by Boolean evaluation of the AND/OR tree. However, since the Boolean evaluation ignores the term-term association unless it is explicitly represented in the tree, the terminological gap between users' queries and their information needs can still remain. To solve this problem, we adopt the generalized vector space model (GVSM) in which the term-term association is well established, and extend the RUBRIC model based on GVSM. Experiments have been performed on some variations of the extended RUBRIC model, and the results have also been compared to the original RUBRIC model based on recall-precision.

1 Introduction

Many intelligent retrieval approaches have been studied to meet the users' individual preferences properly [2, 6, 7]. However, there always exists a terminological gap between the way in defining queries and the way in representing documents. One of the approaches proposed in the literature for this purpose involves the use of production rules to capture user query concepts (or topics). The central ideas of such an approach were introduced in the context of a system called Rule Based Information Retrieval by Computer (RUBRIC) [6]. In RUBRIC, a set of related production rules is represented as an AND/OR tree, called a rule-base tree. RUBRIC allows the definition of detailed queries starting at a conceptual level. The retrieval output is

---

* This work is supported in part by the US Army Research Office, by Grant No.\ DAAH04-96-1-0325, under DEPSCoR program of Advanced Research
  Projects Agency, Department of Defense, and in part by the U.S. Department of Energy, Grant No. DE-FG02-97ER1220, and by the University of Bahrain.
[+] On leave from the Department of Computer Engineering in Ajou University, Korea.

determined by Boolean evaluation of the AND/OR tree. The RUBRIC concepts were incorporated into a commercial system called TOPIC. Though the system was not popular due to difficulties in generating rule-bases, recent research has developed ways to automate the creation and update (using relevance feedback) of rule-bases [4, 3]. Other challenges in successful implementation of RUBRIC deal with efficiency issues and the method of evaluating rule-bases. The efficiency issue has been addressed in [1, 5]. However, the method of evaluation, using Max and Min for OR and AND, respectively, still has limitations.

To solve the second problem, in this paper, we adopt the generalized vector space model (GVSM) [10, 11]. In GVSM, term-term associations are computed as an integral part of the automatic indexing process. We propose a way to integrate the ideas of concept based retrieval in RUBRIC with the generalized vector space model. Experiments have been conducted on some variations of the integrated model. The results show that the integrated model is more effective than the original one in terms of recall-precision.

2 Review of RUBRIC

In RUBRIC, concepts of interest are formulated using a top-down refinement strategy. In a top-down strategy, the first step is to express a given request as a single concept. The next step is to refine the initial concept by decomposing it into a set of component parts that are related through either the AND or OR logical operator. The individual components may take the form of a new concept defined at a different abstraction level, a text expression, or a single index term. In each case, a weight value is assigned to the individual concept-component pairs that are formed during the decomposition process. The assigned weight value represents the user's belief in the degree to which a given component characterizes the related concept.

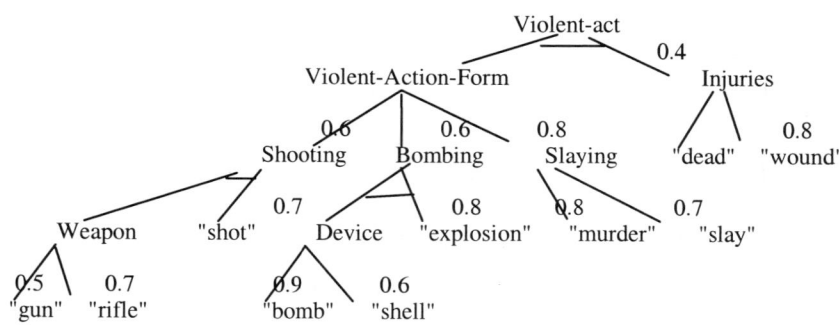

Fig. 1. Rule-base tree for concept **Violent-act**

Fig. 1 shows the rule-base tree for concept **Violent-act** where the leaf nodes are index terms and are enclosed by double quotations, the internal nodes are concepts and the weights are displayed along the edges connecting concepts and components. The concept **Violent-act** is first being decomposed into two component concepts,

Violent-Action-Form and **Injuries**, which are related to **Violent-act** with AND operator. The AND operator is denoted by drawing a line between its branches. If there is no line connecting the branches, the relationship is OR.

The evaluation of the relevancy (RSV: Retrieval Status Value) of a document to concept **Violent-act** could be processed by a bottom-up strategy. For example, if a document contains the words "gun," "shot," "bomb," "slay," and "dead" and no other words in the example rule base are referred to, then the index term nodes "gun," "shot," "bomb," "slay," and "dead" will receive a weight of 1.0 and all other index term nodes will receive a weight of 0. For example, the concept **Weapon** is composed of two component "gun" and "rifle," and the weight for "gun" is 1.0 and "rifle" is 0. Since the operator for **Weapon** is OR, the relevancy to concept **Weapon** is assigned to be the max value of the product of the weights of its components and the corresponding weights connecting its components. In this particular case, the result is Max (1.0*0.5, 0*0.7) = 0.5. In a similar way, finally we can get the relevancy (RSV) of the document to the concept **Violent-act**, which comes out to be 0.3.

In order to make the computation more efficient, Minimal Term Sets (MTSs) should be generated through static analysis of rule-base [1, 5]. A minimal term set consists of index terms that are necessary to make sure that the retrieval status value of a concept is larger than 0. That means if any of the index terms in the MTS is taken out, then the RSV would be 0. Here, we list all the MTSs and their RSVs for the concept **Violent-act** in Table 1.

Table 1. MTSs and RSVs for concept **Violent-act**

	MTS	RSV
MTS 1	{"gun," "shot," "dead"}	0.3
MTS 2	{"gun," "shot," "wound"}	0.3
MTS 3	{"rifle," "shot," "dead"}	0.4
MTS 4	{"rifle," "shot," "wound"}	0.32
MTS 5	{"bomb," "explosion," "dead"}	0.4
MTS 6	{"bomb," "explosion," "wound"}	0.32
MTS 7	{"shell," "explosion," "dead"}	0.36
MTS 8	{"shell," "explosion," "wound"}	0.32
MTS 9	{"murder," "dead"}	0.4
MTS 10	{"murder," "wound"}	0.32
MTS 11	{"slay," "dead"}	0.4
MTS 12	{"slay," "wound"}	0.32

3 Review of Generalized Vector Space Model (GVSM)

In information retrieval, it is common to model index terms and documents as vectors in a suitably defined vector spaces [8, 9]. This approach is usually called Vector Space Model (VSM). In VSM, all the items of interest to the information retrieval system are modeled as elements of a vector space. Let $t_1, t_2, ..., t_n$ be the terms used to index the documents in a collection. For each term, there is supposed to be a corresponding vector $\mathbf{t}_i$ in a vector space. Those vectors $\{\mathbf{t}_i \mid i=1,..., n\}$ are considered as the generating set of the subspace and therefore all the items of interests can be represented as a linear combination of $\mathbf{t}_i$ s.

Let $d_1, d_2, ..., d_m$ be the documents in a collection. Then, we can consider each of those documents as a vector in the form of

$$\mathbf{d}_\alpha = \sum_{i=1,n} a_{\alpha_i} \mathbf{t}_i \tag{3.1}$$

where a_{α_i} is the component of $\mathbf{d}_\alpha$ along the direction of the term $\mathbf{t}_i$. Similarly, we can consider a query $\mathbf{q}$ as a linear combination of the $\mathbf{t}_i$'s as follows.

$$\mathbf{q} = \sum_{j=1,n} q_j \mathbf{t}_j \tag{3.2}$$

where q_j is the component of $\mathbf{d}_\alpha$ along the direction of the term $\mathbf{t}_i$.

Then, the similarity of a document and a query could be acquired by computing the similarity of the document vector and the query vector in the vector space. Generally, the cosine similarity function is often to be used as the measure of similarity, that is,

$$\mathbf{d}_\alpha \bullet \mathbf{q} = \sum_{j=1,n} \sum_{i=1,n} a_{\alpha_i} q_j \, \mathbf{t}_i \bullet \mathbf{t}_j \tag{3.3}$$

The documents in a collection can be ranked by the Retrieval Status Values (RSVs) given by similarity function values between the documents and the query. For this purpose, we need to know a_{α_i}'s and $\mathbf{t}_i \bullet \mathbf{t}_j$. Note that sometimes we may or may not know the vector representation for $\mathbf{t}_i$ explicitly. Therefore earlier researchers made the assumption that $\mathbf{t}_i$ and $\mathbf{t}_j$ are orthogonal if $i \neq j$. That is $\mathbf{t}_i \bullet \mathbf{t}_j = 0$ if $i \neq j$. Unfortunately such an assumption does not hold in the real world. To solve this problem, the Generalized Vector Space Model (GVSM) is proposed.

The main idea of GVSM is to incorporate the representation of elements in a Boolean algebra to into a vector space. In this mapping, terms are represented as a linear combination of vectors associated with the atomic expressions (or concepts) that are pairwise orthogonal. Let $t_1, t_2, ..., t_n$ denote the terms that are used to index the documents in a collection. An atomic expression m_k, called a min-term, in the n literals $t_1, t_2, ..., t_n$ is a conjunction of the literals where each t_i appears exactly once and is either in complemented or uncomplemented form. That is, $m_k = x_1$ AND x_2 AND ... AND x_n where x_i has the form either of t_i or $\neg t_i$. Since the number of all possible min-terms is 2^n and the conjunction of any two different min-terms is always *zero* (*false*), we can map the 2^n min-terms into the orthogonal bases of the vector space R^{2^n} as follows:

$$\mathbf{m}_1 = (1,0,0,...,0), \ \mathbf{m}_2 = (0,1,0,...,0), \ ...., \ \mathbf{m}_{2^n} = (0,0,0,...,1) \tag{3.4}$$

Since each t_i is itself an element of the Boolean algebra generated, t_i can be expressed in its disjunctive normal form:

$$t_i = m_{i_1} \text{ OR } m_{i_2} \text{ OR} \ldots \text{OR } m_{i_r} \tag{3.5}$$

where the m_i's are those min-terms in which t_i is uncomplemented. Let the set of min-terms in Equation (3.5) be denoted by $\{m\}^i$. We can now define basis vectors analogous to Equation (3.4) and term t_i can be written in the vector notation as

$$\mathbf{t}_i = \sum_{k=1}^{2^n} c_{ik} \mathbf{m}_k \tag{3.6}$$

where unnomalized form of c_{ik} is given by

$$c_{ik} = \sum_{d_\alpha \in D_{mk}} w_{\alpha_i}$$

where D_{mk} is a set $\{d_\alpha \mid d_\alpha$ contains all the non-negated index terms in m_k and also excludes all the negated index terms in $m_k \}$. In the above, w_{α_i} is the importance of each term t_i in document d_α. In Wong and et al.'s work [11], they compute these weights from the term co-occurrence frequency.

4 Modeling RUBRIC Concepts in GVSM

In RUBRIC, the rule-base tree describes certain relationships among concepts. Specification of a concept as a rule-base tree not only helps the user to describe his/her retrieval request more accurately and more flexibly, but also makes the resulting rule-base tree more understandable to other users. However, in order to retrieve a document, RUBRIC requires that the document contain all the index terms in one MTS. It is a strong provision for information retrieval systems, even if each conjunction presents an alternative specification of a query topic. To overcome this problem, we should consider term correlations.

In GVSM, however, term correlations are well established based on the co-occurrence frequency. However GVSM itself does not provide a way for describing a query at a conceptual level, while the RUBRIC model does. Consequently, it is a natural way to incorporate these two models together. The main issue here is to construct a query vector in GVSM for a concept defined in a rule base tree.

4.1 Mapping MTSs to Query Vectors in GVSM

Now, we are trying to map MTSs into query vectors in GVSM. For this purpose, we consider *potential weights* of index terms to a given concept. The *potential weight* means the importance of the index term to the given concept based on the assumption that the index term is used as a term in the corresponding query to the concept. To compute the *potential weight* of an index term, we just follow the paths in the rule-base tree from the index term up to the root that denotes the given concept and multiply the weights in the paths. For example, in rule-base tree shown as Fig. 1, the *potential weight* of index term "gun" to concept **Violent-act** will be 0.5*1.0*0.6*1.0 = 0.3. Based on *the potential weights*, we can consider *weighted MTS* for a concept *c* as a MTS in which each term t_i is assigned the *potential weight* of t_i to *c*. For example, the *weighted MTS*1 corresponding to MTS1 in Table 1 is {("gun",0.3), ("shot",0.42), ("dead",0.4)}.

To complete the mapping of MTSs into query vectors in GVSM, it is necessary to define vector operations that correspond to OR and AND operators, respectively. In this paper, we adopt more general definitions for these operators as follows. Given $\mathbf{t}_1 = c_{11}\mathbf{m}_1 + c_{12}\mathbf{m}_2 + \ldots + c_{1p}\mathbf{m}_p$ and $\mathbf{t}_2 = c_{21}\mathbf{m}_1 + c_{22}\mathbf{m}_2 + \ldots + c_{2p}\mathbf{m}_p$, where $p = 2^n$, $\mathbf{t}_1 \oplus \mathbf{t}_2 = \Sigma \max(c_{1k}, c_{2k})\mathbf{m}_k$ and $\mathbf{t}_1 \otimes \mathbf{t}_2 = \Sigma$ and-max $(c_{1k}, c_{2k})\mathbf{m}_k$, where and-max(x, y) is max(x, y) if x and y are not zero, and otherwise zero. These are analogous to $\oplus$ and $\otimes$ introduced in Wong and *et al.* [12].

4.2 A Complete Example

In this subsection, we give an example to show how the extended RUBRIC model works. Suppose the rule-base tree is shown as Fig. 2, where **Shooting** and **Injuries** are defined as the same concepts as shown in Fig. 1. The *weighted MTS*s for concept **Killing** are listed as:

{"gun"(0.5*1.0*0.6=0.3), "shot"(0.7*0.6=0.42), "dead"(1.0*0.4=0.4)}
{"gun" "(0.5*1.0*0.6=0.3), "shot" (0.7*0.6=0.42), "wound" (0.8*0.4=0.32)}
{"rifle" "(0.7*1.0*0.6=0.42), "shot" (0.7*0.6=0.42), "dead" (1.0*0.4=0.4)}
{"rifle" "(0.7*1.0*0.6=0.42), "shot" (0.7*0.6=0.42), "wound" (0.8*0.4=0.32)}

Killing

0.6   0.4

Shooting   Injuries

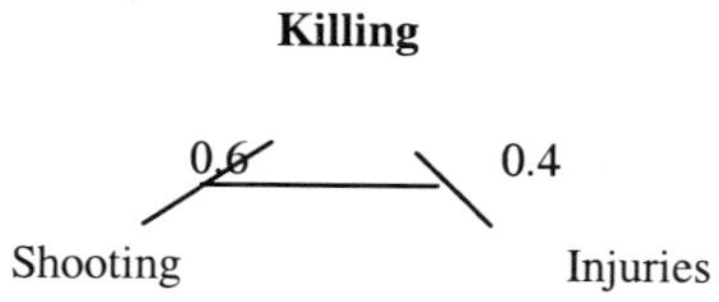

Fig. 2. An example rule-base tree

Suppose the Document-term matrix and the basis vectors are shown as Table 2 and Table 3, respectively.

Table 2. Document-term matrix

	wound	dead	shot	rifle	gun
d1	2	0	0	1	3
d2	1	2	2	0	0
d3	4	0	0	1	1
d4	0	2	3	4	0
d5	1	2	2	2	4
d6	2	1	4	0	1

Table 3. Basis vectors

	wound	dead	shot	rifle	gun	basis vector
m_0	0	0	0	0	0	m_0
m_1	0	0	0	0	1	m_1
.	.	.	.	.	.	.
m_{31}	1	1	1	1	1	m_{31}

Term vectors can be calculated as:

wound = $0.92\, m_{19} + 0.15\, m_{28} + 0.30\, m_{29} + 0.15\, m_{31}$
dead = $0.55\, m_{14} + 0.55\, m_{28} + 0.27\, m_{29} + 0.55\, m_{31}$
shot = $0.52\, m_{14} + 0.34\, m_{28} + 0.69\, m_{29} + 0.34\, m_{31}$
rifle = $0.81\, m_{14} + 0.40\, m_{19} + 0.40\, m_{31}$ **gun** = $0.69\, m_{19} + 0.17\, m_{29} + 0.69\, m_{31}$

Using the Equation (3.6), we can compute the term vectors. For example, we can get the coefficient of m_{19} in term vector **wound** as follows. Since the Boolean pattern corresponding to m_{19} is 10011, only documents d1 and d3 are relevant to this pattern. Therefore, m_{19}'s unnormalized coefficient is 2 + 4 = 6. Similarly, we can get the unnormalized coefficients of m_{28}, m_{29}, and m_{31} are 1, 2, and 1, respectively. Now, we normalize the coefficient of m_{19} as $6 / (6^2 + 1^2 + 2^2 + 1^2) = 0.92$. Using these term vectors, we can compute document vectors. For example, since d_1 = 2 **wound** + 1 **rifle** + 3 **gun** from the document-term matrix in Table 2, $d_1 = 0.81\, m_{14} + 4.31\, m_{19} + 0.30\, m_{28} + 1.11\, m_{29} + 2.77\, m_{31}$.

For expressing queries, we use the operators $\oplus$ and $\otimes$ introduced in section 4.1. For concept **Killing**, if we choose MTS1 = {(gun, 0.3), (shot, 0.42), (dead, 0.4)} as a query, then we can construct the query vector as follows.

$q = [0.3*(0.69\, m_{19} + 0.17\, m_{29} + 0.69\, m_{31})] \otimes$
$[0.42*(0.52\, m_{14} + 0.34\, m_{28} + 0.69\, m_{29} + 0.34\, m_{31})] \otimes$
$[0.4*(0.55\, m_{14} + 0.55\, m_{28} + 0.27\, m_{29} + 0.55\, m_{31})] = 0.29\, m_{29} + 0.22\, m_{31}$

For concept **Killing**, if we choose the following two MTSs as a query:

MTS1 = {(gun, 0.3), (shot, 0.42), (dead, 0.4)},
MTS2 = {(rifle, 0.42), (shot, 0.42), (dead, 0.4)}

Then the corresponding vector **q** is computed as follows. $q = q_1 \oplus q_2$, where q_1 and q_2 are query vectors for MTS1 and MTS2, repectively.

$q_1 = 0.29\, m_{29} + 0.22\, m_{31}$
$q_2 = \{[0.42*(0.81\, m_{14} + 0.40\, m_{19} + 0.40\, m_{31})] \otimes$
$[0.42*(0.52\, m_{14} + 0.34\, m_{28} + 0.69\, m_{29} + 0.34\, m_{31})] \otimes$
$[0.4*(0.55\, m_{14} + 0.55\, m_{28} + 0.27\, m_{29} + 0.55\, m_{31})]\} = 0.34\, m_{14} + 0.22\, m_{31}$
$q_1 \oplus q_2 = 0.34\, m_{14} + 0.29\, m_{29} + 0.22\, m_{31}$

5 Experimental Design and Results

For the experimental test, we choose *Google* search engine in Internet [13]. It uses a conjunction of keywords as queries and returns a set of links to web pages. We use the rule base tree shown in Fig. 1 for our experimental test. From Fig. 1, we can compute 12 *weighted MTSs* and construct a query for each MTS. For each query, we choose the top 20 links retrieved by *Google*. Since some links to the web pages might have been changed or moved after being indexed by *Google*, we must eliminate those links. Finally, we got a collection of 196 documents. The relevance judgments, for evaluation purpose, are determined by looking through each document and choosing those document related to the concept **Violent-act**. From the collected documents, we can construct the document-term matrix. Using the matrix, we can also compute the term vectors and document vectors. Since, with respect to each MTS, RSV (similarity between a query and a document) is generated for each document, further processing is done to combine them into a single RSV for each document satisfying more than one MTS. For this purpose, the user can be given the option of performing the disjuction operation over some or all the MTSs. Thus, the processing done by our system for ranking requires post-processing of results from Google.

In our experiment, we select all the MTSs and compute the performance in terms of recall-precision. We conduct experiments using two different disjunction operators, namly + and ⊕. We call the former *Extended Rubric version 1* (ER1) and the latter *Extended Rubric version 2* (ER2).

Table 4. Recall-Precision using the original RUBRIC approaches (R1 and R2) and the extended RUBRIC approaches (ER1 and ER2).

Recall	Precision			
	R1	R2	ER1	ER2
0.1	0.7143	0.5882	0.6250	0.7143
0.2	0.6250	0.5882	0.5714	0.6667
0.3	0.5179	0.5686	0.6170	0.7073
0.4	0.5672	0.5672	0.6032	0.6909
0.5	0.6026	0.5875	0.6184	0.6528
0.6	0.6364	0.6154	0.5895	0.6087
0.7	0.6019	0.5603	0.5328	0.5462
0.8	0.5362	0.5175	0.5175	0.5034
0.9	0.5030	0.5390	0.5155	0.5092
1.0	0.4742	0.4742	0.4769	0.4792

In order to compare the extended versions with the original RUBRIC approaches, we conduct the experiment using the versions of the original RUBRIC, denoted by R1 and R2, respectively. In R1, we adopt the original idea of RUBRIC. That is, we use max operator and min operator for disjunction and conjunction, respectively. In R2, we use the same operator (that is, min operator) for conjunction and use an operator different from max operator for disjunction. In R2, we adopt the weighted operator for disjunction [5]. Suppose that document d is satisfied by three *weighted MTSs* M_1, M_2, and M_3. And suppose that weights w_1, w_2, and w_3 are assigned to M_1, M_2, and M_3, respectively, and w_2 is the maximum value among w_1, w_2, and w_3. Then, we compute the RSV of d in the following weighted manner: RSV = $w_1 + 2w_2 + w_3$. The performance of ER1, ER2, R1, and R2 are given in Table 4.

6 Conclusions

From the experiments, we can reach to the following conclusions about the method of evaluating documents relative to a rule-base.

(1) The variation of RUBRIC (R2) does not improve the accuracy of RSVs.
(2) The extended approach ER1 shows a similar performance as the original RUBRIC approach R1
(3) ER2 shows a better performance than ER1.

Our approach can be extended in several ways. First of all, the term weight is not necessarily frequency information. For example, some existing search engines provide weights or scores for retrieved documents. In that case, if a single term is sent as the query to the search engine, the weight or ranking assigned to a document can be treated as the weight of the term in the document. There are other interesting issues currently being investigated, related to the RUBRIC approach. For example, based on relevance feedback, we can adjust the weights both in the rule base tree and the document-term matrix [3]. Furthermore, with the help of some knowledge bases, the user can construct the rule base tree more automatically [4].

References

1. Alsaffar, A. H., Deogun, J. S., Raghavan, V. V., and Sever, H. Concept-based retrieval with minimal term sets. In Z. W. Ras and A. Skowon, editors, Foundations of Intelligent Systems: Eleventh Int'l Symposium, ISMIS'99 proceedings, pp. 114-122, Springer, Warsaw, Poland, Jun, 1999.
2. Croft, W. B. Approaches to intelligent information retrieval. *Information Processing and Management*, 1987, Vol. 23, No. 4, pp. 249-254.
3. Kim, M. and Raghavan, V. V. Adaptive Concept-based Retrieval Using a Neural Network. In Proceedings of ACM SIGIR Workshop on Mathematical/Formal Methods in Information Retrieval, July 28, 2000, Athens, Greece.
4. Kim, M., Lu, F., and Raghavan, V. V. Automatic Construction of Rule-based Trees for Conceptual Retrieval. In Proceedings of SPIRE2000, September 27-29, 2000, A Coruna, Spain, IEEE Computer Society Press.

5. Lu, F., Johnsten, T., Raghavan, V. V., and Dennis Traylor, Enhancing Internet Search Engines to Achieve Concept-based Retrieval, In Proceeding of Inforum'99, Oakridge, USA
6. McCune, B. P., Tong, R. M., Dean, J. S., and Shapiro, D. G. RUBRIC: A System for Rule-Based Information Retrieval, *IEEE Transaction on Software Engineering*, Vol. SE-11, No. 9, September 1985.
7. Resnik, P. Using information content to evaluate semantic similarity in a taxonomy. In Proceedings of the 14th International Joint Conference on Artificial Intelligence, pp. 448-453, 1995.
8. Salton, G. and Lesk, M. E. Computer Evaluation of Indexing and Text Processing. *ACM* 15, 1 (Jan, 1968), pp. 8-36.
9. Salton, G. and McGill, M. J. *Introduction to Modern Information Retrieval*. McGraw Hill, New York, 1983.
10. Wong, S.K.M., Ziarko, W., and Wong, P. C. N. Generalized Vector Space Model in Information Retrieval. In Proceedings of the 8th Annual International *ACM-SIGIR* Conference, 1985, New York, pp. 18-25.
11. Wong, S.K.M., Ziarko, W., Raghavan,V., and Wong, P. C. N. On Modeling of Information Retrieval Concepts in Vector Spaces, ACM Transaction on Database System, Vol. 12, No. 2, June 1987. pp. 299-321.
12. Wong, S.K.M., Ziarko, W., Raghavan,V., and Wong, P. C. N. Extended Boolean Query Processing in the Generalized Vector Space Model, *Information Systems* Vol. 14, No. 1, pp. 47-63, 1989.
13. "Google Search Engine," http://www.google.com

Template Generation for Identifying Text Patterns

Cécile Boisson and Nahid Shahmehri

Department of Computer and Information Science, Linköpings universitet, Sweden
{cecbo,nahsh}@ida.liu.se

Abstract. It is common that a text document contains information that can be interpreted as instructions to pursue a given task. This information called, *pattern*, can be seen as the triggering mechanism for a set of predefined operations. We are interested in automating the recognition of these patterns for repetitive tasks. We introduce the notion of template generation which allows for the recognition of new patterns that trigger operations. We implemented an algorithm for template generation and we tested it in an electronic publishing application. The tests show that some characteristics of the processed text can be used to adapt the generation process and obtain templates that provide better precision and recall.

1 Introduction

Today, most operations on text such as classification or editing, are still done manually. Some of these operations are repeatedly performed on a daily basis, e.g. in e-publishing or document management. In these cases, it is common that a text document contains information that can be interpreted as instructions to pursue a given task. This information can be seen as the triggering mechanism for a set of predefined operations. We use the common name of pattern for this information. A pattern can appear in different forms such as word, phrase, sentence, etc. The following are examples of repetitive tasks on text which contain implicit or explicit instructions:

- Establishing an interaction in natural language with a computer. For example, a command like "`turn on the light in the kitchen`" is a written instruction that could be given on a regular basis to an intelligent device in charge of the environmental control of a house.
- Classifying/routing documents. For example, an electronic message that contains the pattern "`Mr Hill wants to change the carburetor of his car`" is forwarded to the team in charge of engine repair, while the message which contains the pattern "`Mrs Jones wants to change the door of her van`" is forwarded to the team in charge of chassis repair.
- Transforming textual information into data. For example, the string "`let's meet at 3 pm on Wednesday in my office`" written in a message can be an implicit instruction to create and insert a meeting appointment in a calendar.

– Performing an editing operation to fit a policy. For example, in an electronic journal, the references to articles like "in the IJCAI paper of Harris" will be transformed into a link to the actual article.

Typically, the properties shared by these repetitive tasks are:

– No initial corpus is provided for training. For example, to produce a training corpus, the editor of a journal would need to accumulate hundreds of articles and thus would produce tens of issues before getting any automatic support.
– The set of patterns to be identified can change. This can be due to the modification of the users habits or needs, to the modification of the way to express instructions in the texts, etc.
– Some background knowledge is needed to perform the tasks. The background knowledge can be domain specific concepts or information on the structure of the texts[1]. For example, the editor of ETAI [10], an electronic journal, uses a knowledge base of authors and references to articles from which an ontology can be built. In this paper we assume the existence of a basic set of concepts for each task.

Automating the operations on text by establishing a human-computer collaboration would save much human effort and material resources. Our work aims at automating the generation of templates which represent and are used for the identification of patterns that trigger the same operation. The means to recognize patterns range from the recognition of syntactic characteristics of the patterns (e.g. keywords, grammar rules) to the recognition of their semantical characteristics which implies reasoning about the meaning of words and their context (e.g. conceptualization, sentence understanding). In this paper we introduce a method for template generation to support the semantical identification of textual patterns in repetitive tasks. Template generation is performed after each execution of a given task to incrementally build the set of templates that identify all patterns triggering an operation.

Given a pattern p, template generation, TG, creates a template t by substituting specific information in p with less specific information (i.e. concepts). The substitutions are made through a number of rewritings according to the syntactic and semantical rules of a grammar G. A concept describes a characteristic of a piece of pattern and can be of different categories like semantic, grammatical role, language, etc. Thus, given the (semantic) concepts "<action>", "<device>" and "<room>", and the pattern p = "turn on the light in the kitchen", TG can rewrite p into t = "<action> the <device> in the <room>". Notice that it is possible to generate several templates from a given pattern. For example, TG can also create the template t' = "<action> the light in the <room>" from p. Given a pattern p which triggers the operation o, we denote by $T_p = \{t_1, ..., t_n\}$ the set of templates that can be created by TG, where each template identifies a different set of patterns $P_{t_i} = P_{t_i}^+ \cup P_{t_i}^-$. The elements of $P_{t_i}^+$ trigger o while

---

[1] Knowledge acquisition techniques such as data mining or interview of experts can support the construction of domain specific knowledge.

the elements of $P_{t_i}^-$ do not trigger o. In this paper we consider the case where p is rewritten into a unique template t such that (1) t identifies only patterns which trigger o (i.e. $P_t = P_t^+$)[2], (2) t is more general or equal to the template that identifies only p (i.e. $t \in T_p$), and (3) t is one of the most general templates that satisfy (1) and (2).

In section 2 we describe the characteristics of the ontology of concepts usable by TG. In section 3, we introduce our solution for template generation: pattern generalization (PG). In section 4 we give the task scenario that we use to evaluate our method. In section 5 we describe our tests results. In section 6 we relate template generation to other work. The paper concludes in section 7.

2 An Ontology for Template Generation

TG is based on the possibility to capture the semantics of a pattern. This is done by substituting pieces of the pattern with relevant concepts. For example, given the pattern $p =$ "the article published in IJCAI-99", and the concept <conference>, "IJCAI-99" can be substituted by <conference> to produce the template $t =$ "the article published in <conference>". It is then possible to recognize all the patterns syntactically similar to p which express a reference to a paper published in any conference. If some of these patterns do not trigger the creation of a link, the concept maybe too general and has to be replaced by a more specific one (e.g. <AI-conference>). Such substitutions are possible if there is an ontology which describes the concepts expressed in the processed texts. In [5] Guarino and Giaretta point out that the word "ontology" is ambiguous. We adopt the definition commonly used in knowledge engineering [1]: an ontology is a set of specifications of conceptualizations, where a conceptualization is a set of definitions of elements (concepts) in a domain. So, to describe an ontology, we should provide: (1) the descriptions of the concepts, and (2) the relations between the concepts (examples of common relations are is-a, part-of, consist-of, etc). The concepts and the relations aim at describing the piece of the real world that a given application needs to know. The is-a relation is important because it allows comparing the concepts in terms of generality. When it comes to providing definitions of concepts, we notice (1) the existence of distinct categories of concepts to characterize pieces of text (i.e. semantic, grammatical role, presentation, language, etc.) and (2) the possibility to describe concepts in several ways (i.e. natural language description, value enumeration, grammar, process, etc.). The existence of several categories of concepts augments the number of potential substitutions for a given piece of pattern and complicates the process of the creation of a template. Moreover, to be able to perform the substitution of a piece of pattern by a concept, TG should be able to understand the different means of descriptions used by the ontology. This problem is outside the scope of this paper. Figure 1 gives an example of ontology usable by TG.

---

[2] There is always a solution since the template that only identifies the pattern p fulfills the requirements.

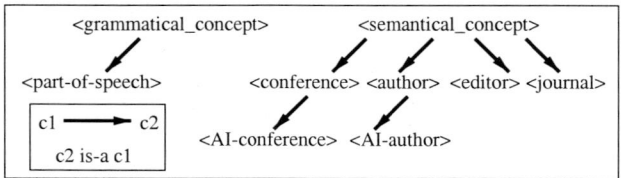

Fig. 1. An example of ontology for a task of journal editing

3 The Process of Pattern Generalization

In this section, we describe the process of pattern generalization (i.e. PG), our solution for template generation. PG consists of rewriting pieces of a pattern p into concepts according to the rewrite rules of a grammar.

- T_p is the set of templates that can be created from a pattern p.
- We define the order $>_g$ on T_p such that:
  $\forall t_i \forall t_j : P_{t_i} \supset P_{t_j} \wedge P_{t_i} \neq P_{t_j} \Rightarrow t_i >_g t_j$.
  If $t_i >_g t_j$ we say that t_i is more general than t_j and that t_j is more specific than t_i.
- The process of PG is the function PG(C, Txt, p, G) return t where
  - C is the ontology (cf. section 2).
  - Txt is the set of texts already processed where the pieces of text that are not patterns are annotated.
  - p is a pattern which triggers an operation o.
  - t is a more general template of T_p that can be used to recognize only patterns that trigger o: $P_t = P_t^+$.
  - G is a grammar (cf. table 3). A pattern is a string which forms either a sentence (S_struct) or a phrase (P)[3], and a template is a sequence of strings and concepts. G is composed of three kinds of rules:
    1. "SubPart -> Concept" defines the non terminal SubPart as a piece of pattern that can be substituted by a concept.
    2. "X -> X_struct | SubPart" expresses the possibility of transforming a piece of pattern X into either a "legal English structure" of X or a SubPart.
    3. "X_struct -> ... | ... " describes the possible "legal English structures" of X.[4]
    The grammar is ambiguous at two levels: (1) it is possible to create several parse trees (i.e. chains of rewrite rules) for a given pattern p, and (2) it is possible to associate several concepts to the same SubPart.

---

[3] In the grammar we restrict the patterns to the forms of "legal English structures" of sentences or phrases to allow the usage of NLP analyzers which require grammatically correct input. The principle of PG can handle any piece of text.
[4] These rules can be completed to parse more legal English structures of sentences, phrases, noun phrases, and verb phrases.

```
S_struct -> P | P PREP P | NP VP | ...
P -> P_struct | SubPart
P_struct -> NP | VP | ...
NP -> NP_struct | SubPart
NP_struct -> N | ADJ N | ART ADJ N | ...
VP -> VP_struct | SubPart
VP_struct -> V | V NP | ...
N -> N_struct | SubPart
V -> V_Struct | SubPart
ADJ -> ADJ_Struct | SubPart
ART -> ART_Struct | SubPart
SubPart -> Concept
```

S_struct: sentence, P: phrase, PREP: preposition, ADJ: adjective, ART: article, N: noun, V: Verb.

Concept is a terminal. Its instances belong to C.

N_Struct, V_Struct, ADJ_Struct, ART_Struct are terminals. Their values are given by a lexicon.

Table 1. Ambiguous grammar for rewriting a pattern into a template

The process of computing PG() is composed of the two following steps:
1. BuildGeneralTemplate(p, C) return t_g,
where t_g is such that: $(t_g \in T_p) \land (\forall t_i \in T_p : \neg(t_i >_g t_g))$
Building t_g implies the choice of the derivation which rewrites p into a most general template. Table 2 gives the rules to make this choice[5].
2. RefineTemplate(t_g, C, Txt) return t such that $(P_t = P_t^+)$.
RefineTemplate() = {For each $p^- \in P_{t_g}^-$ Do Refine(t_g, p, p^-, C)}.
Refine() applies on t_g the most specific derivation which provides the most general template that does not match p^-. It is done in three steps.
2.1 DifferentSubparts(t_g, p, p^-) return S, where S is the set of subparts of t_g such that $\forall s \in S$ the form of s is a concept in t_g, and the form of s in p and in p^- are different[6]. E.g. s = "<author>" in t_g, s = "Harris" in p and s = "Smith" in p^-.
2.2 OrderedDerivations(S, C) return D.
$\forall s \in S$, s is a concept and there are two kinds of derivations possible to make s more specific: (d1) $s \Rightarrow$ concept where concept is-a s and, (d2) $s \Rightarrow X \Rightarrow$ X_struct $\Rightarrow s' \Rightarrow g$ where g is the most general form of s'.
Table 3 gives the rules to choose the most specific derivations for a given subpart (i.e. a concept). D is the set of combinations of derivations on s where the derivations are stored from more general to more specific.
2.3. Enumerate(t_g, S, D, C, p^-) return t = {For each d in D Do $t \leftarrow$ ApplyDerivation(t_g, d); If $p^- \notin P_t$ then return t}

---

[5] Currently, for the cases 1 and 3 of rule 2, we use the first heuristic. More investigations need to be done on the second heuristic.
[6] This selection limits the space of the derivations to those which focus on the elimination of p^- from P_{t_g}.

1. X -> X_struct | SubPart
   Heuristics: the conceptualization of a subpart is more general than the sequence of conceptualization of its subparts. Thus,
   $(p \xRightarrow{n} A\ X\ B \Longrightarrow A\ \text{SubPart}\ B \xRightarrow{*} t_1) \wedge (p \xRightarrow{n} A\ X\ B \Longrightarrow A\ X\_\text{struct}\ B \xRightarrow{*} t_2) \Rightarrow t_1 >_g t_2$.
   E.g. given X = "the Oxford University Press",
   $(p \xRightarrow{n} A\ X\ B \Longrightarrow A\ \text{SubPart}\ B \Longrightarrow A\ \text{<editor>}\ B \xRightarrow{*} t_1) \wedge$
   $(p \xRightarrow{n} A\ X\ B \Longrightarrow A\ X\_\text{struct}\ B \xRightarrow{*} A\ \text{"the <university> Press"}\ B \xRightarrow{*} t_2)$
   $\Rightarrow t_1 >_g t_2$.

2. SubPart -> Concept
   Given the set $C_s = \{c_1, c_2, ..., c_n\}$ of concepts associated to the subpart s, the rewrite rule becomes s -> c_1 | c_2 | ... | c_n
   Case 1: some concepts belong to different categories of s.
   E.g. s = "Harris" $\wedge c_1$ = <author> $\wedge c_2$ = <personal-noun> $\wedge$ c_1 is-a <semantic_concept> $\wedge c_2$ is-a <grammatical_concept>.
   Two heuristics: (1) Defining an order among the categories (E.g. <semantic_concept> $>_g$ <grammatical_concept>), and (2) Consider all possible orders and generate a template for each.
   Case 2: there is an is-a relation between some concepts.
   $\exists (c_i, c_j) \in C^2 : (c_i\ \text{is-a}\ c_j) \wedge (p \xRightarrow{n} A\ \text{Subpart}\ B \Longrightarrow A\ c_i\ B \xRightarrow{*} t_1) \wedge$
   $(p \xRightarrow{n} A\ \text{Subpart}\ B \Longrightarrow A\ c_j\ B \xRightarrow{*} t_2) \Rightarrow (t_2 >_g t_1)$.
   E.g. s = "Harris" $\wedge c_1$ = <AI-author> $\wedge c_2$ = <author> $\wedge$
   (<AI-author> is-a <author>) $\wedge$
   $(p \xRightarrow{n} A\ s\ B \Longrightarrow A\ \text{<AI-author>}\ B \xRightarrow{*} t_1) \wedge$
   $(p \xRightarrow{n} A\ s\ B \Longrightarrow A\ \text{<author>}\ B \xRightarrow{*} t_2) \Rightarrow (t_2 >_g t_1)$.
   Case 3: some concepts belong to the same category and there is no relation is-a between them. E.g. s = "Brown", c_1 = <author> and c_2 = <university> $\wedge$ c_1 is-a <semantic_concept> $\wedge c_2$ is-a <semantic_concept>.
   There is only one concept which is true.
   Two heuristics: (1) Picking one concept randomly to generate t, and (2) Generating one template for each concept.

Table 2. Rules to choose a most general derivation

1. $\forall c \in C$: ApplyDerivation(c, d2) $>_g$ ApplyDerivation(c, d1).
2. $\forall (c_i, c_j) \in C^2$:
   c_i is-a c_j $\Rightarrow$ ApplyDerivation(c, d1(c_j)) $>_g$ ApplyDerivation(c, d1(c_i))
   where ApplyDerivation(c_1, d1(c_2)) replace c_1 with c_2.
3. $\forall c \in C : \exists s_1 : s_1$ = ApplyDerivation(c, d2) $\wedge$
   $\exists s_2 : s_2$ = ApplyDerivation(c, d2) $\wedge (s_1 \xRightarrow{n} s_2) \Rightarrow s_1 >_g s_2$
   E.g. (c = <editor> $\wedge s_1$ = <university> Press $\wedge s_2$ = <town> university Press $\wedge s_3$ = Oxford university Press) $\Rightarrow (s_1 >_g s_2 >_g s_3)$

Table 3. Rules to choose the most specific derivation

4 Test Platform

We identified the need for support for pattern identification in the task of the editor of ETAI [10], an electronic journal for artificial intelligence. While our method is adaptable to other domains, it is on this task that we have performed the evaluation of our methods of template creation and management (cf. section 5). Most of the examples which have illustrated our explanations are also examples from this task. The policy of the journal includes the public and collaborative reviewing of the articles submitted for publication. The task of the editor is to compile the email messages written by researchers to review a given article, into a web page. Each message is processed in its order of arrival. The messages can contain references to articles. One of the operations is to transform the pieces of text that refer to articles into links to the actual articles. Thus in this task, the patterns to recognize are the references to articles, and the texts to inspect are bodies of messages. Example of patterns are "in this paper" and "in the IJCAI paper of Harris". The ontology for this task, as described in figure 1, is composed of several <semantical_concept> which appear in references to articles (i.e. <conference>, <author>, <editor>, <journal>, <university>), and of one <grammatical_concept>: the <part-of-speech>. Each word of the patterns is associated to a <part-of-speech>. Some words, alone or in sequence, are also associated to a <semantical_concept>.

5 Practical Analysis of Pattern Generalization

Characteristics of the ontology (e.g. the order on the categories of concepts) and of the patterns (e.g. the similarity between the patterns to identify and the patterns to not identify) influence the granularity of the generated templates, and thus their performances in terms of precision and recall. Our first test illustrates these characteristics. In a second test, we evaluate the algorithm in terms of precision and recall and show that the generalization of templates provides a better recall than a method which incrementally collects the patterns and does not perform any generalization.

First test:
We have implemented the algorithm for pattern generalization which assumes the existence of a predefined order "$>_g$" on the categories of concepts (cf. heuristic 1 for case 1 of the rule 2 given in table 2). In our test ontology, there are two categories of concepts, <semantic_concept> and <grammatical_concept>, thus their are two possible orders. The purpose of this test is to observe how the order of the categories of concepts influences the form of the generated templates. From the observation we propose heuristics to choose one order that will contribute to a quicker generation of templates that provide good recall and precision.
Given two sets of patterns $P^- = \{p_1^-, ..., p_{10}^-\}$ and $P^+ = \{p_1, ..., p_{10}\}$, we performed PG and got the set of templates $T = \{t_{p_1}, ..., t_{p_{10}}\}$. PG was performed twice with two different biases on the ontology. Since the 10 templates produced

had the same characteristics, we illustrate the generalization for a single pattern:
$p = $ "the results by Hill and Smith".
Bias 1: <grammar_concept> $>_g$ <semantic_concept>.
Result: The elements of T are mainly sequences of part of speech tags.
$t_{bias1} = $ "<DT> <NNS> <IN> <NNP> <CC> <NNP>".
Bias 2: <semantic_concept> $>_g$ <grammar_concept>. Result: The elements of T are sequences of semantical concepts and part of speech. The part of speech are mostly conceptualizations of articles and verbs.
$t_{bias2} = $ "<DT> <NNS> <IN> <author> <CC> <author>". The templates created with Bias 2 are more specific to the task and likely to be more accurate. This test highlights some new parameters which influence the construction of the templates:

- The order between the categories of concepts.
- The specificity of the concepts to the task. For example, <grammar_concept> is less specific to the task than <semantic_concept>.
- The number of refinements performed on the templates which identify patterns that should not be identified.
- The syntactic and semantic similarity between elements of P^+ and elements of P^-: the more the elements of P^+ look like the elements of P^-, the higher the granularity of the ontology should be.

Given automatic techniques to observe these parameters, one can design an algorithm which adapts PG to the domain of the task.

Second test: evaluation of PG in terms of recall and precision.
Problem description:

- The texts processed are two e-mail discussions. The first discussion contains 14 mails and the second contains 8 mails.
- the order $>_g$ on the concepts of the ontology is completed with the bias 1: <grammar_concept> $>_g$ <semantic_concept>.
- The set of templates created during the processing of the first discussion continues its evolution during the processing of the second discussion.

Evaluation data:

- N_i^+: the number of elements of P^+ recognized with the templates while processing the ith message. In brackets: the number of patterns which are copies of pattern encountered in previous messages.
- N_i^-: the number of elements of P^- recognized with the templates.
- N_i^f: the number of element of P^+ that have not been recognized with the templates.
- N_i^{new}: the number of new patterns encountered in the ith message.

Results:

- Table 4 gives the evaluation data and the recall and precision rates for each discussion.

- For the first discussion, $N_i^- = 0$, thus the templates are never too general. Recall does not exceed 21.4% because there is a large syntactic difference between the patterns.
- For the second discussion, the high Recall is not due to the generalization process but to the high usage of identical patterns in the discussion. For example, the pattern "the paper" occurs 6 times. The low precision is due to the high level of generality (sequences of part of speech) of some templates. For example the template "<DT> <NN>" generated from the pattern "the paper" after the processing of the 4th message is responsible for most of the 8 wrong recognitions performed in the 5th message. The templates in fault are immediately adapted: 100% of the wrong recognition on this discussion is done on the 5th mail of the discussion, but none is done on later processing, while correct recognitions continue to take place.
- By comparison with a system which incrementally builds a set of patterns to recognize and perform the automatic recognition of exact copies in new texts processed, PG does augment the coverage of recognition of 66% in the first discussion, and of 18% in the second. This difference of performance is explained by the fact that the semantical concepts are well adapted to the semantic carried by the pattern of the first discussion, but not to the second. For the second discussion, the addition of a concept <publication> corresponding to words like "paper", "journal" or "article" would have increased the performance of the generalization.

	Discussion 1														Discussion 2								
Messages	0	1	2	3	4	5	6	7	8	9	10	11	12	13	Total	1	2	3	4	5	6	7	8	Total
N_i^+	0	0	0	0	0	1	1	0	0	1 (1)	0	0	0	0	3 (1)	0	0	0	2	3 (3)	3 (3)	3 (3)	0	11 (9)
N_i^-	0	0	0	0	0	0	0	0	0	0	0	0	0	0	0	0	0	0	0	8	0	0	0	8
N_i^f	1	3	0	0	0	2	2	0	1	1	0	0	0	1	11	0	1	1	0	0	1	8	0	11
N_i^{new}	1	3	0	0	0	3	3	0	1	1	0	0	0	1	13	0	1	1	0	0	1	8	0	11

	Recall	Recall due to PG()	Precision	Precision due to PG()
Discussion 1	21.4%	14.3%	100%	66.6%
Discussion 2	50%	4.5%	57.9%	5.26%

Table 4. Tests: Evaluation data, recall and precision for the discussions

To conclude, the experiment shows that the generalization is immediately useful if some parameters are properly set. The most important settings being (1) having a fairly complete ontology and (2) having a management of the categories of concepts adapted to the domain of the task. We believe that these needs can be met to some reasonable extent.

6 Related Work

Information Extraction [2] systems ([6], [7], [3], etc) are using techniques to collect sets of templates which identify interesting pieces of texts in large corpus. The creation of the template is done via a training on a part of the corpus. Some attempts have been made to automate the creation of the templates (Autoslog-TS [11]). These attempts simplify the training, but do not eliminate it. Moreover it is not possible to learn new templates on the fly. In our case, a requirement as having a large number of texts to train on before hand, is not acceptable. While IE techniques are very inspiring (templates, natural language analysis), they are not directly applicable to the support for repetitive tasks. In [4], R. Grishman and J. Sterling describe the generalization of acquired semantic patterns. Their approach is based on the computation of the frequency of each semantic pattern in a corpus. Once again it is an approach that we cannot adopt since we do not have any corpus before hand. When it comes to providing support to repetitive tasks, the APE project [12] describes a technique to learn incrementally the set of recurrent sequences of commands typed or clicked by users in a programming environment for Smalltalk. This project shows the usefulness of incremental learning in a system which records the actions of the user and analyses them to identify patterns. However, their domain of command sequences is less complex to reason about than the domain of natural language. Other projects explore the domain of personal assistance on repetitive tasks. CAL [8] refines the principles initiated for [9]: "learning through experience" a set of rules to manage a calendar. The concepts are predefined, fixed, and limited in number (we counted 22). We have no limitation on the number of concepts and manage several granularities and categories.

7 Conclusion and Future Work

We propose a method for template generation to support the identification of important pieces of texts with respect to a given task. This method learns incrementally a set of patterns. The set is completed and refined after each new execution of the task, thus providing a better support for the next execution. We described a solution for template generation: pattern generalization. Given a pattern which triggers the operation o, pattern generalization generates a most general template that recognizes only patterns that trigger o. We have tested the pattern generalization on a set of patterns extracted from a real life task. The results show that the performance depends on the completeness and granularity of the ontology, and on the relations between the categories of concepts. As further work, we plan to investigate the management of the templates. The current system already handles the refinement of templates after each execution of the task. It should also be able to recognize relations between the templates to identify inconsistencies or templates that cover the same sets of patterns. We also plan to investigate methods to incrementally adapt the ontology and to produce a cooperation between the semantic and grammatical categories of concepts. We intend to establish a system to evaluate the performances of templates. This system would allow the addition of new rules to choose the most suitable template among the most general templates.

Acknowledgements: This research has been supported by the Swedish Research Council for Engineering Sciences (TFR). The authors would like to thank Patrick Lambrix and Lena Strömbäck for their valuable comments on earlier versions of this paper.

References

1. A. Bernaras, I. Laresgoiti, N. Bartolome, and J. Corera. An ontology for fault diagnosis in electrical networks. In *International Conference on Intelligent Systems Application to Power Systems*, pages 199–203, 1996.
2. J. Cowie and W. Lehnert. Information extraction. *CACM*, 39(1):80–91, Jan. 1996.
3. R. Gaizauskas, T. Wakao, K. Humphreys, H. Cunningham, and Y. Wilks. Description of the lasie system as used for muc-6. In *6th Message Understanding Conference (MUC-6)*, pages 207–220, 1995.
4. R. Grishman and J. Sterling. Generalizing automatically generated selectional patterns. In *15th International Conference on Computational Linguistics (COLING 94)*, pages 742–747, 1994.
5. N. Guarino and P. Giaretta. Ontologies and knowledge bases: Toward a terminological clarification. In NJI Mars, editor, *Toward Very Large Knowledge Bases: Knowledge Building and Knowledge Sharing*, pages 25–32. 1995.
6. J. Hobbs, D. Appelt, J. Bear, D. Israel, M. Kameyama, M. Stickel, and M. Tyson. Fastus: a cascaded finite-state transducer for extracting information from natural-language text. In E. Roche and Y. Schabes, editors, *Finite State Devices for Natural Language Processing*. Cambridge MA: MIT Press, 1996.
7. W. Lehnert, J. McCarthy, S. Soderland, E. Riloff, C. Cardie, J. Peterson, F. Feng, C.Dolan, and S. Goldman. Description of the circus system as used for muc-5. In *5th Message Understanding Conference (MUC-5)*, pages 277–291, 1993.
8. T. M. Mitchel, R. Caruana, D. Freitag, J. McDermott, and D. Zabowski. Experience with a learning personal assistant. *CACM*, 37(7):80–91, July 1994.
9. T. M. Mitchell, S. Mahadevan, and L. I. Steinberg. Leap: A learning apprentice for vlsi design. In *IJCAI-85*, pages 573–580, 1985.
10. Electronic Transactions on AI (ETAI). http://www.ida.liu.se/ext/etai/, 1997.
11. E. Riloff. Automatically generating extraction patterns from untagged text. In *AAAI-96*, pages 1044–1049, 1996.
12. J-D. Ruvini and C. Dony. Learning users habits: The ape project. In *Learning About Users, IJCAI'99 Workshop*, July 1999.

Qualitative Discovery in Medical Databases

David A. Maluf *, Jiming Liu **

* Research Institute for Advanced Computer Science, Computational Science Division, NASA Ames. maluf@ptolemy.arc.nasa.gov
** Department of Computing Studies, Hong Kong Baptist University

Abstract. Implication rules have been used in uncertainty reasoning systems to confirm and draw hypotheses or conclusions. However a major bottleneck in developing such systems lies in the elicitation of these rules. This paper empirically examines the performance of evidential inferencing with implication networks generated using a rule induction tool called KAT. KAT utilizes an algorithm for the statistical analysis of empirical case data, and hence reduces the knowledge engineering efforts and biases in subjective implication certainty assignment. The paper describes several experiments in which real-world diagnostic problems were investigated; namely, medical diagnostics. In particular, it attempts to show that (1) with a limited number of case samples, KAT is capable of inducing implication networks useful for making evidential inferences based on partial observations, and (2) observation driven by a network entropy optimization mechanism is effective in reducing the uncertainty of predicted events.

1 Introduction

One of the important aspects of using expert systems technology to solve real-world problems lies in the management of domain-knowledge uncertainty. Several methods of reasoning under uncertainty have been proposed in the past [1] [11] [13] [15]. All these approaches require a representation of domain knowledge. Generally speaking, constructing a valid knowledge representation is a time-consuming task and often subject to opinion biases or semantics invalidity if it is built purely based on human heuristics. To overcome the difficulties in knowledge acquisition, several investigations have been carried out in recent years to explore the effectiveness and validity of automated means such as algorithms to perform this task.

Pitas *et al.* [14] have proposed a method of learning general rules from specific instances based on a minimal entropy criterion. Geiger [7] has formulated a learning algorithm for uncovering a Bayesian conditional dependence tree. This algorithm combines entropy optimization with Heckerman's similarity networks modeling scheme [8]. Cooper and Herskovits [2] have developed an algorithmic method of empirically inducing probabilistic networks, which utilizes a Bayesian framework to assess the probability of a network topology given a distribution of cases. A heuristic technique is provided to optimize the search for probable topologies. Simulation results have shown that a small 37-node, 46-link network can be derived with 3,000 cases.

In this paper, we present a new rule-learning algorithm for inducing implication relations based on a small number of empirical data samples. The major difference between Cooper and Herskovits' approach and ours is that their approach focuses on topological induction accuracy while ours is concerned with the accuracy of inferences based on an induced network, without regards to the topological uniqueness. Our approach to implication network induction has been implemented in a tool box called KAT, which contains several components; namely empirical data acquisition, implication rule elicitation module, network validation module, optimal observation determination module, and embedded diagnostic inferencing engine which implements uncertainty reasoning schemes.

Our approach to implication induction draws on the previous work on empirical construction of inference networks [4]. The present study further extends the earlier work by augmenting the implications with certainty measures. Another related work is the development of a *prediction logic* based on a contingency-table of probabilities, as proposed by Hildebrand et al. [9]. In their work, the emphasis was on the definition and computation of precision and accuracy of propositions represented. An analogy was made between contingency table based prediction logic and formal proposition logic. To validate the implication networks generated from KAT, we have conducted a series of empirical experiments to examine the performance of evidential inferencing with the induced networks. The chosen problem domain is *medical diagnosis*; this task shares many commonalities with other real-world problems as described in [1] [6] and has been in part inspired by earlier studies on *knowledge space theory* (KST) by Doignon and Falmagne [5]. The KST presents an interesting set-theory interpretation of knowledge states as well as its mathematical foundations. In our present framework, unlike the one by Doignon and Falmagne, the interdependencies among knowledge units are the closures under union and intersection, which can be correctly represented with a directed inference network. Hence, our implication networks representation (i.e., an instance of implication networks) can be viewed as a proper subset of the knowledge space representation.

In this paper, we examine the effectiveness and exactness of inferences with statistically induced networks. Our claim is that the proposed network induction method is capable of generating logically and empirically sound implication-based domain representations useful in predicting unobserved events upon receiving certain partial information. While validating the networks in several real-world task domains, we attempt to demonstrate the generality of the algorithmic rule induction and reasoning approach in solving problems where a complete set of events is too difficult to observe or the diagnostic judgments are subject to human errors.

2 Implication Network Induction

In the present work, we refer the term *implication network* to a directed acyclic graph in which the nodes represent individual event variables or hypotheses, and the arcs signify the existence of direct implication (e.g., influence) among the nodes. The value taken on by one event variable is dependent on the values taken on by all variables that influence it. Each value indicates the likelihood of an unobserved event. The value is updated every time new information is obtained (e.g., some symptom is observed). The strengths of the event interdependencies are quantified by functions (e.g., belief functions), as weights associated with the arcs.

Formally, an implication network can be represented as an ordered quadruple:

$$Net = \langle \mathcal{N}, \mathcal{I}, \alpha_c, p_{min} \rangle, \qquad (1)$$

where $\mathcal{N}$ is a finite set of nodes and $\mathcal{I}$ is a finite set of arcs. α_c is the network induction error and p_{min} is the minimal conditional probability to be estimated in the arcs. Furthermore, each induced implication rule can be specified by the following quadruple:

$$Imp = \langle N_{ant}, N_{concl}, W_I, \tilde{W}_I \rangle, \qquad (2)$$

where W_I and $\tilde{W}_I$ are weight functions that map the pairs of antecedent-consequent nodes, i.e., N_{ant} and N_{concl}, and their negations to a real number between 0 and 1, respectively. That is,

$$W_I : \quad N_{ant} \times N_{concl} \to [0,1]. \qquad (3)$$
$$\tilde{W}_I : \neg N_{concl} \times \neg N_{ant} \to [0,1]. \qquad (4)$$

	B	$\neg B$
A	$N_{A \wedge B}$	$N_{A \wedge \neg B}$
$\neg A$	$N_{\neg A \wedge B}$	$N_{\neg A \wedge \neg B}$

Fig. 1. contingency table where cells indicate the number of co-occurrences.

2.1 The Rule-Elicitation Algorithm

The basic idea behind the *empirical construction* is that in an ideal case, if there is an implication relation A ⇒ B, then we would never expect to find the co-occurrences as in Figure 1 that event A is true but not event B, from the empirical data samples. This translates into the following two conditions:

$$P(B|A) = 1 \qquad (5)$$

$$P(\neg A|\neg B) = 1 \qquad (6)$$

In reality, however, due to noise such as sampling errors, we have to relax Conditions 5 and 6. KE takes into account the imprecise/inexact nature of implications and verifies the above conditions by computing the lower bound of a $(1 - \alpha_{\text{error}})$

confidence interval around the measured conditional probabilities. If the verification succeeds, an implication relation between the two events is asserted. Two weights are associated with the relation[1], which correspond to the relations' conditional probabilities $P(B|A)$ and $P(\neg A|\neg B)$. In fact, these weights together express the degree of certainty in the implication. Once an implication relation can be determined, another logical operator "$-$" is readily defined as follows:

$$(A \Rightarrow B) \Rightarrow ((B \Rightarrow A) \Rightarrow (B - A)) \tag{7}$$

The elicitation of dependences among the nodes requires considering the existence (or nonexistence) of direct relationships between pairs of random variables in a domain model. In theory, there exist six possible types of implications between any two nodes or events.

The implication rule elicitation algorithm can be stated as follows:

The Rule-Elicitation Algorithm

Begin
set an arbitrary level α_c and a minimal conditional probability p_{min} (this test can be repeated for different α_c and p_{min}. An example is $\alpha_c = 0.05$ and $p_{min} = 0.5$).
for node$_i$, $i \in [0, n_{max} - 1]$ and node$_j$, $j \in [i+1, n_{max}]$

    for all empirical case samples

    compute a contingency table $T_{ij} = \begin{vmatrix} N_{11} & N_{12} \\ N_{21} & N_{22} \end{vmatrix}$

        for each rule type k out of the six possible cases.
        test the following inequality:

$$P(x \leq N_{\text{error\_cell}}) < \alpha_c \tag{8}$$

        based on the two lower tails of binomial distributions $Bin(N, p_{min})$ and $Bin(\tilde{N}, p_{min})$, where N and $\tilde{N}$ denote the occurrences of antecedent satisfactions in the two inferences using a type k implication rule, i.e., in *modus ponens* and *modus tollens*, respectively. α_c is the alpha error (or significance level) of the conditional probability test.
        if the test succeeds
            return a type k implication rule.
        endif, endfor
    endfor
endfor
End

Here it is assumed that the conditional probability is p in each sample, and all n samples are independent. If X is the frequency of the occurrence, then X

---

[1] With respect to the two directions of the inference, i.e. *modus ponens* vs. *modus tollens*.

satisfies a binomial distribution, i.e., $X \sim Bin(n,p)$, whose probability function $p_X(k)$ and distribution function $F_X(k)$ are given below:

$$p_X(k) = \binom{n}{k} p^k q^{n-k} \tag{9}$$

$$F_X(k) = p(X \leq k) = \sum_{j=0}^{k} \binom{n}{k} p^k q^{n-k}; \quad p = 1-q \tag{10}$$

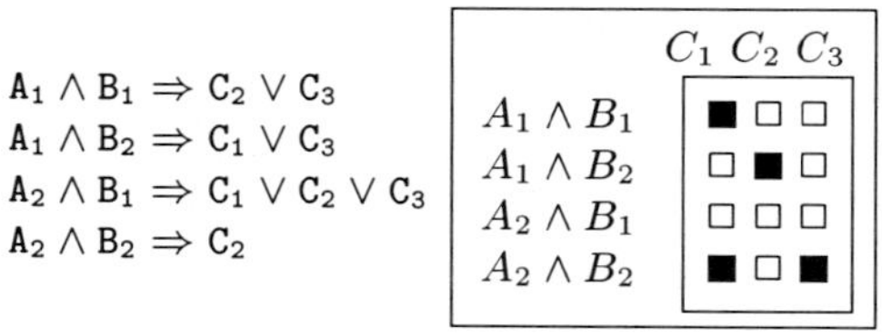

Fig. 2. A contingency table where cells indicate the number of co-occurrences in the case of multivariate implications.

Thus, the test of hypothesis for $A \Rightarrow B$ can be obtained by computing by a lower tail confidence interval over a binomial function:

$$p(X \leq N_{A \wedge \neg B}) = \sum_{i=0}^{N_{A \wedge \neg B}} \binom{n}{i} p^{n-i}(1-p)^i \tag{11}$$

where n has the same definition as above, and where p is set to the desired minimal conditional probability. This formula represents the probability that as small a number as X of unpredicted results would be observed if the true probability of a predicted result were exactly p. The smaller the probability given by the formula is, the less likely it is that the true probability of a predicted result is *less than p*.

From a theoretical point of view, we could increase the dimensionality of the distribution to incorporate all variables relevant to the problem in question and allow the variables to be multivariate as illustrated in Figure 2. In such a case, the probability function to be considered becomes: $p_{X_1,...,X_r}(k_1,...,k_r) = \frac{n!}{k_1!...k_r!} p_1^{k_1} \ldots p_r^{k_r}$. From a practical point of view, this would also introduce exponential computational complexity. In the present study, we concentrate on *bivariate* variables *pairwise*, which reduces the scope of problem for which probabilities have to be elicited. Often this is known as *naive Bayes*.

2.2 An Example of Positive Implication Induction

The following section illustrates how the above algorithm is used to verify the existence of a *positive* implication rule: $A \Rightarrow B$.

In the first step of positive implication rule induction, a two-dimensional contingency table for variables A and B is compiled. As computed from an empirical data set, the cells in the contingency table contain the observed joint occurrences for the respective four possible combinations of values. Table 1 shows an example of the contingency table with respective co-occurrences of variables A and B in a hypothetical data set.

	B	$\neg B$
A	20 ($N_{A \wedge B}$)	1 ($N_{A \wedge \neg B}$)
$\neg A$	8 ($N_{\neg A \wedge B}$)	1 ($N_{\neg A \wedge \neg B}$)

Table 1. Distribution of observed occurrences

where $N_{\bullet\bullet}$ denotes the occurrences of the respective situations. The total numbers of A and $\neg B$ can be derived accordingly as follows:

$$N_A = N_{A \wedge B} + N_{A \wedge \neg B} = 21$$
$$N_{\neg B} = N_{A \wedge \neg B} + N_{\neg A \wedge \neg B} = 2$$

Statistical Tests for Implication Existence: The second step of our induction method consists of an assessment of the numerical constraints imposed by $A \Rightarrow B$. More specifically, the assessment is based on the lower tails of binomial distributions $Bin(N_A, p_{min})$ and $Bin(N_{\neg B}, p_{min})$ to test measured conditional probabilities $P(B \mid A)$ and $P(\neg A \mid \neg B)$, where $N_A = N_{A \wedge B} + N_{A \wedge \neg B}$, $N_{\neg B} = N_{A \wedge \neg B} + N_{\neg A \wedge \neg B}$, and p_{min} is an arbitrary number chosen as the *minimal conditional probability* for an implication relation. For each of the two binomial distributions, we check to see whether Inequality 8 can be satisfied.

Suppose that in this example, $p_{min} = 0.85$; $\alpha_c = 0.20$. Accordingly the binomial distribution for testing $P(B \mid A)$ can be written as: $Bin(21, 0.85)$. The computation of the lower bound proceeds as follows:

$$\begin{aligned} P(x \leq N_{A \wedge \neg B}) &= P(x \leq 1) \\ &= P(x = 0) + P(x = 1) \\ &= \binom{21}{0} 0.85^{21}\, 0.15^0 + \binom{21}{1} 0.85^{20}\, 0.15^1 \\ &= 0.155 \end{aligned}$$

hence $P(x \leq N_{A \wedge \neg B}) < \alpha_c$ where symbol $\binom{j}{k}$ represents the number of combinations of k in j. The inference with $A \Rightarrow B$ in the *modus ponens* direction is significant with confidence level $(1 - \alpha_c)$. In a similar way, given $Bin(2, 0.85)$, the test for $P(\neg A \mid \neg B)$ yields:

$$P(x \leq N_{A \wedge \neg B}) = \binom{2}{0} 0.85^2\, 0.15^0 + \binom{2}{1} 0.85^1\, 0.15^1 = 0.98$$

hence,
$$P(x \leq N_{A \wedge \neg B}) \not< \alpha_c$$

Since Inequality 8 for the test of $P(\neg A|\neg B)$ is not satisfied, A ⇒ B cannot be used for *modus tollens* inference. Hence, the positive implication rule A ⇒ B is rejected. The overall, worst-case time complexity of inducing an implication network with the above algorithm is $O(n_{max}^2)$ where n_{max} is the number of nodes for modeling the domain.

3 Empirical Cases

This section describes the empirical data used in a series of experiments aimed to investigate the effectiveness and exactness of induced implication networks in diagnostic reasoning. The selected task domain is medical diagnosis.

In the current study, we model the different possible knowledge states by a *partial order*. Although this formalism could not fully represent all possible knowledge states, it captures a large part of the constraints on the ordering among KU and can be used for the purpose of automatic knowledge assessment [3]. The data used to induce implication networks for medical diagnosis consists of a set of attributes which are continuous variables. In order to build a network, these attributes were first transformed into bivariate (i.e., binary) values using thresholds.

The medical diagnostic method developed in this work was first validated using the empirical cancer data samples collected from 69 healthy people and 31 cancer patients. Each sample contains the information on 22 chemical residues (i.e., attributes) found in a bioposy. In order to build the network, we first transformed the ordered continuous variables, i.e., trace element concentrations, into two-valued Boolean variables, by means of thresholding.

Zn ⇒ Mg	0.7826 0.7959	Cd ⇒ Zn	0.7096 0.8333	Mg ⇒ Ca	0.8823 0.8775
Zn ⇒ Ca	0.8695 0.8775	Cd ⇒ Ni	0.8064 0.8846	Mg ⇒ Cu	0.7058 0.7272
Zn ⇒ Cu	0.6956 0.7454	Cd ⇒ Co	0.7096 0.8571	V ⇒ Ni	0.7058 0.8076
Co ⇒ Ni	0.7297 0.8076	Cd ⇒ Cu	0.8064 0.8909	Cu ⇒ Ca	0.7555 0.7755

Table 2. The original trace concentration data samples.

The derived data set was used to induce the network. Tables 3 and 4 show a few examples of the original and the derived data set samples, respectively. Table 3 presents a subset of the induced implication network in the form of pairwise gradation relations.

4 Evidential Inferences

To validate the accuracy of the evidential inferences generated from implication networks, we have conducted a series of experiments in simulated diagnostic task

Zn	Pb	Ni	Co	Cd	Mn	Cr	Mg	V	Al	Ca	Cu	Ti	Se	Categ.
237.84	8.50	1.532	1.045	0.590	1.953	1.717	223.62	1.696	0.010	1806.75	8.71	0.732	0.001	1
203.15	12.70	2.362	1.707	0.898	1.347	1.204	46.33	0.811	4.189	405.20	13.92	0.689	0.001	1
266.34	4.44	0.085	1.013	0.382	2.151	0.340	47.73	0.010	13.137	367.92	17.10	2.896	0.003	2

Table 3. The transformed trace concentration data samples (subset).

Zn	Pb	Ni	Co	Cd	Mn	Cr	Mg	V	Al	Ca	Cu	Ti	Se	Category
01.00	01.00	01.000	01.000	01.000	1.000	1.000	01.00	1.000	0.000	01.00	00.00	1.000	0.000	1
01.00	01.00	01.000	01.000	01.000	1.000	1.000	00.00	1.000	0.000	00.00	01.00	0.000	0.000	1
01.00	01.00	00.000	01.000	01.000	1.000	1.000	00.00	0.000	1.000	00.00	01.00	1.000	0.000	2

Table 4. Examples of the induced positive implication rules (subset).

settings. In particular, we used constructed implication networks as the basis for evidential inferences. Each simulation run consisted of selecting a portion of a subject's sample data and propagating evidential supports throughout the network.

4.1 Experimental Method

There exist various interpretations of the imprecision measure associated with an implication rule [11]. Each interpretation dictates the way in which inferences are to be performed. Bayesian inference is based on the mapping of an implication relation into conditional probabilities [13]. Taking an implication A ⇒ B for example, updating the probability would be based upon $P(B \mid A)$, which should approach 1.0 if the implication is strong. The difficulty with this scheme stems from the fact that if further observation of C is obtained and if there is a relation C ⇒ B, then there is a need to update the value of B based upon $P(B \mid A, C)$, and so on. As more observations occurs, the conditional probabilities become practically impossible to estimate, whether subjectively or from sample data. To address this difficulty in a Bayesian belief network, the assumption of independence is made between individual implication relations. In the present work, we have applied the Dempster-Shafer (D-S) method of evidential reasoning to propagate supports (whether confirming or disconfirming) throughout the implication network. The D-S inferencing scheme may be regarded as a complex theoretical deviation from the Bayesian theory. According to the D-S scheme, the set of possible outcomes of a node is called the *frame of discernment*, denoted by Θ. If the antecedents of a rule confirm a conclusion with degree m, the rule's effect on belief in the subsets of Θ can be represented by so-called probability masses. In our bivariate case of knowledge assessment, there are only two possible outcomes for each node, q_i, that is, $\Theta = \{known, \neg known\}$.

The D-S scheme provides a means for combining beliefs from distinct sources, known as *Dempster's rule of combination*. This rule states that two assignments, corresponding to two independent sources of evidence, may be combined to yield a new one, that is,

$$m(X) = k \sum_{X_i \cap X_j = X} m_1(X_i) m_2(X_j) \qquad (12)$$

where k is a normalization factor. Another evidential inference methodology, called *Certainty Factors* (CF) as previously implemented in MYCIN [1], was also applied in this study. This approach may be viewed as a special case of the D-S evidential reasoning. The two approaches differ from each other only in combining two opposite beliefs (i.e., one confirming and the other disconfirming).

4.2 Results in Medical Diagnosis

This section presents the empirical results of evidential inferences using the databases of cancer diagnosis instances as mentioned in Section 3. In each of the two experiments, the numeric-valued attributes were first discretized into binary values which were then used for both network induction and inferencing validation.

In the case of cancer diagnosis, 40 patient samples were compiled to induce the implication network with $p_{min} > 0.5$ and $\alpha_c < 0.30$. The generated network contains 87 implication relations. Another set of 60 patient samples was used to validate the evidential inferencing.

During the validation, a certain percentage of attributes in each test cases were randomly sampled, and the rest of the attributes were inferred from the implications. Upon the completion of inferencing, a pair of thresholds $\langle u, v \rangle$ (i.e., bi-directional thresholds) was defined to filter the numeric-valued weights. That is, if a specific node has a weight $w > v$, then the node is believed to be TRUE. On the other hand, if $w < u$, the node is believed to be FALSE (i.e., the corresponding attribute does not exist). The resulting filtered predictions were compared with the actual values in the test samples.

4.3 Experiment E-5 Cancer Diagnosis

Globally speaking, given the distributions of evidentially predicted *weights* and initial *weights* with respect to various bi-directional thresholds, it can be observed that in the guessing case, both the correctly predicted nodes and the errors were almost the linear functions of the observation rate. However, in the evidential inferencing case, the shapes of these two rate profiles were changed, indicating that as the observation increased, additional nodes were added to both the correct predictions and the errors. It should also be noted that the error rates in the inferencing case were quickly stablized after the amount of observation exceeded a certain percentage.

To further compare the results of inference-based prediction and initial weight-based guessing, a pair of bi-directional thresholds was picked up from each of the two figures such that the selected two cases would have similar error rates. At 0% sampling, the inferencing case predicted about 45% due to its conservative thresholding. However, as the observation increased, correct predictions were quickly added along with some wrong predictions. The evidential inferencing resulted in a consistently better performance in evaluating the unobserved nodes when the observation sampling exceeded 18%, as compared to the pure initial weight based guessing. For instance, at 45% observation, the inferencing method correctly predicted 4% more attributes than the guessing method.

5 Entropy-Driven (ED) Diagnosis Based on Induced Networks

In diagnostic reasoning, various rules may be applied to determine which node is to be observed next. One approach is to randomly choose symptom nodes from a complete symptom set that spans all the symptoms in the diagnostic structure, as studied in the previous section. Another approach is to apply entropy optimization and choose the most informative node. This section investigates the performance of entropy-driven (ED) evidential inferences based on the induced implication networks.

In the following experiments, the expected information yield of each individual node over all the possible outcomes is computed and weighted by the likelihood of each outcome. The node that has the maximum expected information yield is chosen as the potentially most informative one, which is to be observed next. Formally, the expected information yield of an observation is defined as follows:

$$\Delta I_i = E_{cur}(net) - E_{exp}(net)$$
$$= E_{cur}(net) - [p_i E(net \mid \texttt{node}_i = \texttt{TRUE}) + (1-p_i) E(net \mid \texttt{node}_i = \texttt{FALSE})]$$
$$= p_i \sum_{k=1}^{n_{max}} (p'_k \log p'_k + p''_k \log p''_k) - \sum_{k=1}^{n_{max}} (p_k \log p_k + p''_k \log p''_k)$$

where $E_{current}(net)$ denotes the current network entropy. $E(net \mid \bullet)$ denotes the updated network entropy having observed $\texttt{node}_i$. p_i is the current probability of $\texttt{node}_i = \texttt{TRUE}$. p'_k and p''_k are the updated probabilities of a network $\texttt{node}_k$, having observed that $\texttt{node}_i = \texttt{TRUE}$ and $\texttt{node}_i = \texttt{FALSE}$, respectively.

In what follows, we examine the diagnostic performance at the level of individual nodes. The performance is analyzed with respect to three observation modes, which are:

(I) inferences based on the E.D. observation: nodes are given initial probabilities (i.e., averaged weights). If a node is observed to be TRUE, it is assigned 0.9 and 0.1 otherwise, taking into account the random error in the original data. Inference propagation is performed based on that observed node;

(II) inferences based on random observation (as in the previous section): same as (I) but nodes are chosen at random; and,

(III) no inference condition (or guessing): same as (II) but no inference propagation is performed.

Since the comparison between the D-S and Certainty-Factors approaches, as presented in the preceding section, does not reveal any significant performance difference, here we shall focus on the methods of observation *with the D-S evidential inferencing only*.

5.1 Experiment E-11 Cancer Diagnosis

This section examines the performance of evidential inferences under E.D. observation mode. The networks to be tested in the following two experiments are the same as the ones used in the random mode observation as described in Section 4.2. During the validation, the inferred attributes were accepted based upon a pair of thresholds $\langle u, v \rangle$ for filtering the numeric-valued weights, and then compared with the actual discretized attribute values in the samples.

Unlike the distribution in experiment E-5, the distributions of the *weight-based-guessing-only mode* have become non-linear to the observation rate. This indicates that the E.D. observation tends to pick up the nodes with relatively higher uncertainty. At the same time, the inferences with E.D. observation added more information than the purely weight-based guessing with the same observation mode, revealing that the selection of the nodes was not based purely on the present weights of the nodes but also their connectivities in the network.

A main result may be stated that if the E.D. observation sampling is more than 13%, the performance of inferencing is consistently better than that of guessing. For instance, at 45% observation, the inferencing scheme produces 11% additional correct predictions as compared to the pure guesses.

For evidential inferencing with two different observation modes, i.e., E.D. vs. random, the results are significantly different. In the former observation mode, the correctly predicted nodes at 45% observation can reach up to 87%, whereas the latter produces around 81% given the same amount of observation. In the random mode, it requires at least 18% observation in order for the inferencing scheme to show better performance. In the present E.D. mode, this percentage is further lowered to 13%.

6 Conclusion

In this paper, we have described a series of empirical validation experiments which examined the performance of evidential inferences based on implication networks that were induced by a rule learning tool (KAT). In the experiments, building implication networks for evidential inferencing in various real-world diagnostic task domains (as shown in the experiments, some may have less strong implications than the others) is translated into the task of statistically induction, *from a small number of individual instances or empirical data samples* (e.g., the sizes of the samples for the experiments are respectively 47, 20, 40, and 153). Generally speaking, evidential inferencing with such induced networks is effective in generating valid predictions about unobserved events such as knowledge units and diagnostic attribute values.

This study also explored an E.D. diagnostic method and compared its performance with a random sampling method. The result of comparisons has shown that while both the random and the minimum-entropy-based methods are desirable, the latter is in general far better for reducing uncertainties, especially when the observation rate is more than 13% (e.g., as shown in Experiments 7, 11, and 14).

As validated in the cancer experiments, the binary representation of diagnostic attributes enables the induction of valid implication networks, which are useful not only in the predictions of unobserved attributes but also in patient diagnostic classification. The conducted experiments also reveal that the implication network is less sensitive to the particular inferencing scheme performed. In addition to the D-S and Certainty Factors schemes of evidential inferencing, we have also implemented and applied other schemes such as the Bayesian approach, with very little variation in the performance.

References

1. B. G. Buchanan and E. H. Shortliffe, Rule-Based Expert Systems: The MYCIN Experiments of the Stanford Heuristic Programming Project, Addison-Wesley, Reading, MA, 1984.
2. G. Cooper and E. Herskovits, A Bayesian method for the induction of probabilistic networks from data, Machine Learning, 9 1992, pp. 309-347.
3. E. Degreef, J.-P. Doignon, A. Ducamp, and J.-C. Falmagne, Languages for the assessment of knowledge, Journal of Mathematical Psychology, 30 1986, pp. 243-256.
4. M. C. Desmarais and J. Liu, Experimental results on user knowledge assesment with an evidential reasoning methodology, in Proceedings of International Workshop on Intelligent User Interfaces, Orlando, FL 1993.
5. J.-P. Doignon and J.-C. Falmagne, Spaces for the assessment of knowledge, International Journal of Man-Machine Studies, 23 1985, pp. 175-196.
6. T. D. Garvey, J. D. Lowrance, and M. A. Fischler, An inference technique for integrating knowledge from disparate sources, in Proceedings of IJCAI, 1981, pp. 319-325.
7. D. Geiger, An entropy-based learning algorithm of bayesian conditional trees, in Proceedings of the Eighth Conference on Uncertainty in Artificial Intelligence, Stanford University, July 17-19 1992, pp. 92-97.
8. D. Heckerman, Probabilistic Similarity Networks, The MIT Press, Cambrige, MA, 1991.
9. D. K. Hildebrand, J. D. Laing, and H. Rosenthal, Prediction Analysis of Cross Classications, John Wiley and Sons, New York, NY, 1977.
10. K. G. Joreskog and H. Wold, Systems Under Indirect Observation, North Holland, Amsterdam, 1982.
11. H. E. Kyburg, Bayesian and non-bayesian evidential updating, Artificial Intelligence, 1987, pp. 271-293.
12. S. K. Murthy, S. Kasif, and S. Salzberg, A system for induction of oblique decision trees, Journal of Artificial Intelligence Research, 2 1994, pp. 1-32.
13. J. Pearl, Probabilistic Reasoning in Intelligent Systems: Networks of Plausible Inference, Morgan Kaufmann, San Mateo, CA, 1988.
14. I. Pitas, E. Milios, and A. N. Venetsanopoulos, A minimum entropy approach to rule learning from examples, IEEE Transactions on Systems, Man, and Cybernetics, pp. 621-635.
15. G. Shafer, A Mathematical Theory of Evidence, Princeton University Press, Princeton, N. J., 1976.

Finding Association Rules Using Fast Bit Computation: Machine-Oriented Modeling

Eric Louie[1] and Tsau Young Lin[1,2]

[1] Department of Mathematics and Computer Science
San Jose State University, San Jose, California 95192
[2] Berkeley Initiative in Soft Computing
Department of Electrical Engineering and Computer Science
University of California, Berkeley, California 94720
tylin@cs.sjsu.edu; tylin@cs.berkely.edu

Abstract. This paper continue the study of machine oriented models initiated by the second author. An attribute value is regarded as a name of the collection (called granule) of the entities that have the same property (specified by the attribute value). The relational model uses these granules (e.g., bit representation of subsets) as attribute values is called machine oriented data model. The model transforms data mining, particularly finding association rules, into Boolean operations. This paper show that this approach speed up data mining process tremendously; in the experiments, it is approximately 50 times faster, the pre-processing time was included).
Keywords: data mining, association rules, Boolean operation, machine oriented model, granular computing

1 Introduction

Data mining is the search for hidden information from the stored data. Typically, it investigates the relationship of attribute values among tuples by *machine*. In other words, the meaning (to human) of attribute values plays no role in the processing. So we replace meaningful attribute value by granules (the set of entities that has the property specified by the attribute value). Let U be the set of entities and R the relation (representing U) under consideration. Each granule (as a subset of the universe U) is represented by bit patterns. This paper is a continuations of [4], [5]. We will explain how this data model is represented and how this model can be applied to speed up data mining. The fundamental techniques here is the computing of granules-granular computing. In this paper, the granulation is partition; So it is an extended rough set theory. We will discuss the technique to implement a database using equivalence classes (granules) and how is applicable in mining association rules.

2 Representations and Equivalence Classes (Granules)

Let U be the universe of discourse, a classical set. Its elements will be referred to as entities or objects. In relational database theory, U is the set of entities that are represented faithfully by a relation (not a relation scheme) [2]. Attribute values correspond to the properties of entities. We will refer to them as elementary concepts. The collection of elementary concepts will be denoted by C. Since the correspondence between entities and tuples is one to one in both directions. So we can also regard U as the relation (= the set of tuples), or more precisely, the set of tuple-ids or tuple-names.

A binary relation is an equivalence relation if it satisfies the three properties, reflexivity, symmetry, and transitivity. Values of the attribute A will be abbreviated as A-values. The attribute A induces an equivalence relation on U as follows: Two tuples (entities) are equivalent iff the corresponding A-values (property) are the same. The equivalence relation, by abuse of notation, denoted by A again, partitions U into mutually disjoint equivalence classes; we may refer to them as *granules* for short. Mathematically, the attribute A induces a projection from the relation to A-component. The projection induces an isomorphism $U/A \equiv Dom(A)$. One can regard each attribute value as the "meaningful name or id" of the granule via this isomorphism. A granule is a subset of U, at the same time, it is an element of U/A. We will regard the latter as the *canonical name* of the former. The canonical name of a granule may be represented by an explicit list of tuple-ids or binary representation of the granule as a subset of U. In binary form, the value, 1 or 0 at certain position indicates whether the certain tuple holds the particular attribute value or not.

Table 1. A Two Column Relation

Car Ids	Car Type
ID_1	Sedan
ID_2	Sport Utility
ID_3	Sedan
ID_4	Mini Van
ID_5	Coupe
ID_6	Mini Van
ID_7	Sport Utility
ID_8	Sedan
ID_9	Sedan
ID_{10}	Sport Utility
ID_{11}	Station Wagon
ID_{12}	Sedan

An example relation on vehicles consists of two columns, vehicle ID and vehicle type, is used to illustrate the idea; see Table 1. The values in the ID-column, identify uniquely the tuples (entities). This characteristic of the column, where attribute values are all unique, is not useful in data mining since the column values has the one-to-one correspondence to the tuple itself. But it will be used here to reference tuples (as names) in the relation. The vehicle type, however, is not unique. This column is refer to certain property of vehicles, a potentially useful attribute for data mining algorithms. Table 1 contains the following vehicles: six sedans, two mini vans, three sport utility, one coupe, and one station wagon. The quotient set is {Sedan, Sport Utility, Mini Van, Coupe, Station Wagon }. Each element is the meaningful name of a granule (equivalence class). Table 2 displays the partition in two forms, list and binary representations. Both forms represent the elements in the quotient set (or subsets of U). Its first column contains the name of each element in quotient set, on the second column, the binary representation, and on the third column, the list representation.

Note that each element in the quotient set is a subset of U. The number of elements in the quotient set is much more smaller than the number of rows; the quotient set is the domain, to be precise, the active domain [2]. Roughly, we have reduced the problem from the universe to the quotient sets (the domain); this is an essential idea in granular computing. A table of multiple columns then is a family of partitions. In the next section, we will discuss how data mining algorithm can run faster using these binary representations.

Table 2. Partition by Attribute Values; list representation

Meaningful Names of Granules	List Representation of Granules	Binary Representation of Granules
Sedan	$= \{ID_1, ID_3, ID_8, ID_9, ID_{12}\}$	$= 101000011001$
Sport Utility	$= \{ID_2, ID_7, ID_{10}\}$	$= 010000100100$
Mini Van	$= \{ID_4, ID_6\}$	$= 000101000000$
Coupe	$= \{ID_5\}$	$= 000010000000$
Station Wagon	$= \{ID_{11}\}$	$= 000000000010$

3 Using Equivalence Relations in Data Mining Algorithms

One common data mining question is to find all association rules in a given relation. Let X and Y be two elementary concepts (attribute values) in the relation. An association rule, (X, Y), exists if it appears in the $s\%$ of a given relation. In other words, $s\%$ of the tuples contains (X, Y) as sub-tuples.

There are many papers reporting various algorithms to discover association rules [6], [7]. In general, the procedure to find association rule is, first, to generate possible combinations (patterns) of attribute values, second, to count the number of times each possible patterns appears in the relation, and finally to declare the ones that meet or exceed the $s\%$ of tuples in the relation an association rule. Generally, the goal of these algorithms lies in reducing the number of checking the possible combinations in the relation. Below we will demonstrate the use of the granules (equivalence classes) to find the association rules. In terms of granules, the pattern, (X, Y), is an association rule if the bit count of the intersection $X \cap Y$ is equal or greater than the percentage, $s\%$, of the relation [4] The bit count is the number of 1's appearing in the bit stream that represents the granule as a subset of U. For simplicity, we will use $X \cap Y$ to mean the association rule (X, Y). The pattern $X \cap Y$ can be generalized to the form of intersecting multiple granules: $X_1 \cap X_2 \cap X_3 \cap \ldots \ldots X_n \cap Y_1 \cap Y_2 \cap Y_3 \cap \ldots Y_n$.

Table 3. A Car Relation

Car Ids	Car Type	Color Type	Cost Type
ID_1	Sedan	Red	Moderate
ID_2	Sport Utility	Blue	Expensive
ID_3	Sedan	Green	Expensive
ID_4	Mini Van	White	Moderate
ID_5	Coupe	Red	Cheap
ID_6	Mini Van	Red	Expensive
ID_7	Sport Utility	Black	Moderate
ID_8	Sedan	Blue	Expensive
ID_9	Sedan	Green	Expensive
ID_{10}	Sport Utility	Black	Moderate
ID_{11}	Station Wagon	Green	Moderate
ID_{12}	Sedan	Blue	Expensive

Table 3 is an extension from the previous vehicle relation. Two more columns, Color Type and Cost Type, are added. Each column has its own domain of attribute values as well as its own granules. The quotient sets for the last two columns are displayed on Figure 1.

* Color Type Quotient Set
    1. Red = $\{ID1, ID5, ID6\}$
    2. Blue = $\{ID2, ID8, ID12\}$
    3. Green = $\{ID3, ID9, ID11\}$
    4. White = $\{ID4\}$
    5. Black = $\{ID7, ID10\}$
* Cost Type Quotient Set
    1. Cheap = $\{ID5\}$
    2. Moderate = $\{ID1, ID4, ID7, ID10, ID11\}$
    3. Expensive = $\{ID2, ID3, ID6, ID8, ID9, ID12\}$
        Figure 1 Quotient Sets

From this small Table 3, one could conjecture that almost all mini vans are white, or 66% of the expensive vehicles are sedans. Or, one could ask the question what type of expensive vehicle exists in more than 10% of the data. To answer this last question, each granule of the quotient set, car type, must be checked with the granule, expensive, from the quotient set, cost type, to determine which these combinations (sub-tuples of the Cartesian product of quotient sets) are an association rule. Each combinations requires the Boolean AND operation between the two class types. There are five combinations. Table 4 shows the operations on those five combinations, the results on the intersection, and the bit counts on the results.

Table 4. AND operation of granules

Combination	Binary Form	Result	Count
Sport utility AND expensive	{010000100100} ∩ {011001011001}	{010000000000}	1
Mini van AND expensive	{000101000000} ∩ {011001011001}	{000001000000}	1
Coupe AND expensive	{000010000000} ∩ {011001011001}	{000000000000}	0
Station wagon AND expensive	{000000000010} ∩ {011001011001}	{000000000000}	0
Sedan AND expensive	{101000011001} ∩ {011001011001}	{001000011001}	4

The combination, sedan and expensive, is the only one that meet or exceed 10% of the rows in the table, vehicle. So, there is only one association rule out of those five combinations. The number of combinations to check is small compared to the more general question of finding all combinations between the car type and cost type that represents 10% of data. For this case, there are 25 combinations to check, (5-car type times 5-color type).

Determining association rules can be a very computational process when there are many quotient sets to consider. For two columns, the combinations are the pairs (2-tuples) of Cartesian product of the two quotient sets, $Q_1 \times Q_2$. The number of combinations is the product, Card (Q_1)Card (Q_2), of the two cardinal numbers. In general, the number of combinations is huge if the question touches on many columns (quotient sets) and each column has many elements in it. Below, we explain the methods in conjunction with the granules to lessen the number of possible combinations and to reduce the computation in determining whether a particular combination is an association rule.

4 The Algorithms and Comparisons

Let $A_i, i = 1, 2, \ldots$ be the attributes of a relation R. Each column A_i gives rise to a quotient set $Q_i = U/A_i$. Each tuple in R induces a tuple in the Cartesian product of $Q_1 \times Q_2$ For simplicity and clarity, the term tuple will be reserved for the relation R, and the *tuple* and its sub-tuples in the Cartesian product of quotient sets will be referred to as combinations of granules from $Q_i, i = 1, 2, \ldots$. A A combination of length q is called q-combination. A granules is said to be *large*, if it meets $s\%$ percentage, that is, $\text{Card}(Q_i)/\text{Card}(U) \geq s\%$. A q-combination (of granules) is said to be large, if the intersection of the granules in the combination is large.

- A large q-combination forms an association rule (of length q).

The task in this section is to develop an efficient algorithms to determine which combinations are association rules. The task involves two issues: First is to select the combination, which we treat at Section 4.1. The second issue is how to evaluate if the combination is large; this is treated at Section 4.2 We consider both together at Section 4.3.

4.1 The Outer Loop: Generating Potential Combinations to Test for Association Rules

One obvious way to reduce the complexity is to remove those granules in the quotient set, Q_n, that do not meet the percentage, $s\%$. These granules will be called lean granules. It should be clear any removal of those lean granules, L, also means the removal of any combinations that include those lean granules, L. With that in mind, generated large combinations must have all its sub-combinations large too.

- q-combinations should be generated by large $q-1$-combinations.
- All sub $(q-1)$combinations of q-combination are large.

This step is essential the same as Apriori Algorithms.

4.2 The Inner Loop: The Computation Cost of Determining a Generated Combination Is an Association Rule

In this section, we will explain how to identify a q-combination is large; and we will compare our approach with Apriori.

Counting bits. We will explain the logarithmic approach, counts the bits by partitioning the word. For each word, it does the count by every 2 bits, then by every 4 bits, next by every 8 bits, and finally by every 16 bits. Here is the sample (we use pseudo C syntax).

1. 2 bits count: odd bits are added to the preceding even bits: Let B be the bit pattern and $Count32$ is a variable holding the "new" bit pattern.

   $Count32 = (B\&0x55555555) + ((B >> 1)\&0x55555555);$

2. 4 bits count: odd 2-bits are added to even 2-bits (each 2-bit represents the sum of even and odd bits)

   $Count32 = (Count32\&0x33333333) + ((Count32 >> 2)\&0x33333333);$

3. 8 bits count: odd 4-bits are added to even 4-bits (each 4-bit represents the sum of even and odd 2-bits)

   $Count32 = (Count32 \& 0x0F0F0F0F) + ((Count32 >> 4) \& 0x0F0F0F0F);$

4. 16 bits count; odd 8-bits are add to even 8-bits

   $Count32 = (Count32 \& 0x00FF00FF) + ((Count32 >> 8) \& 0x00FF00FF);$

5. 32 bits count; odd 16-bits are add to the even 16-bits

   $TotalCount = (Count32 \& 0x0000FFFF) + (Count32 >> 16);$

The first four statements take 28 instructions, 4 * (3 move + 2 and + 1 add + 1 shr), and the last statement takes 7 instructions, (3 move + 1 add + 2 add, + 1 shr). The total for this approach is 35 instructions to count one 32-bit word. This can be faster than the usual shift method in the order of magnitude.

The Instructions Counts. The instruction count in computing a q-combination being large is simply the AND instructions between elements of quotient sets in binary form and the instructions to count the number of 1's in the result. Since computers process the size of a word at a time, the number of instructions is approximately the following:

1. Set $length$ = The length of the bit string representing a granule. This is the cardinality of the universe, Card(U).
2. Set w = Number of bits in a word; in this paper, w is 32.
3. Set $count$ = Number of instructions per word =
   $length/w * ((q-1) AND$ operations + cost of counting bits.)
4. The total instruction count to determine, in the worst case, whether a q-combination is large is the following:
   $count$ = Number of instructions per word = $length/w * ((q-1) + 35)$ instructions

For example, for a combination of length 10 with 2^{20} tuples (a "million" tuples)in the relation, 1,441,792 instructions will be executed.

$$(2^{20}/32) * ((10-1) + 35) = 215 * (44) = 1,441,792 \text{ instructions}.$$

Note: We may stop once the count meets $s\%$ of the data.

4.3 Inner and Outer Loop: Grouping Combinations

As explained in Section 4.1, there are two strategies in carrying out the computations: one is the full computations of the intersection; the other stop at the $s\%$ condition.

1. For $q = 1$, the inner loop identifies large granules, say $l_1, l_2, \ldots, l_n$, by counting the bits using the method of Section 4.2; note that the counting will stop at $s\%$. For uniformity in terminology, they will be referred to as large 1-combination.
2. For $q = 2$, the inner loop computes the 2-combination by $ANDing$ two 1-combinations and counting bit patterns of the result, say, $l_1 \cap l_2$. There are three approach in this computations:
   a) full computation: We compute the intersection and save the results. One needs memory and storage management, since the intersection can not be kept in main memory.
   b) partial computation: The computation of the intersection stop at the point $s\%$ reaches. No intersections of bit patterns are saved. The disadvantage is we may do some computations *repeatedly*. But in the case that s is well smaller than the length of bit strings, this is right approach.
   c) partial computation and partial save: Compute at $s\%$ and save the partial intersections. This could delay the storage management to a much later stage, so one might be able to avoid it in *not extremely large* databases.
3. For general q, the loop computes the q-combination by $ANDing$ two $q-1$-combinations and counting the bit pattern of the result. Again, we can use one of the three approaches.

5 Computational Data

The relation consists of 128K rows = 131072, 16 Columns, the support requires 8192, and memory is 10 megabytes:

Table 5. Experimental Results

Length of combination	# of Candidates	Association rules	Granule(Full Computation)	Granule Partial	Apriori Hybrid	Apriori 199	Apriori Tid
1	101	90	1.000s	0.968s	1.110s	1.110s	3.813s
2	3694	381	2.610s	2.547s	26.078s	25.765s	36.671s
3	1419	246	1.187s	1.172s	46.609s	37.938s	362.657s
4	137	39	0.141s	0.141s	25.938s	30.187s	26.312s
5	1	1	0.000s	0.000s	0.343s	74.094s	0.360s
6	0	0	0.000s	0.000s	0.000s	0.000s	0.000s
			4.938	4.828s	100.078s	169.094s	429.813

The program for Apriori, AporiTid and AprioriHybrid are our honest implementations of the algorithms in [6], [7]. In the implementation, we use some buffer scheme to speedup read/write for all algorithms.

Comparison with Apriori. In the outer loop, Apriori and granular approach is the same. For outer loop, Apriori goes through each transaction and use subset function to verify, if a q-combination belongs to that transaction (for a fixed q and fixed combination). It takes more than one instruction to hash through the hash tree. For granular approach, it involves only AND-operations, computed q times, which takes only $q/32$ (32=wordsize) instruction(s). So our computation is much faster. This is from computational point of view. From I/O point of view, the machine model often transforms database to a more compact form, so the reading of total databases takes less time too. We should like to point out that if attribute values are continuous or too many values one has to perform discretization, partitioning, granulation on the (active) attribute domain first. The so called concept hierarchy method can be applied [1].

6 Conclusions

It is clear why granular approach is faster than Apriori in our experiments. It is less clear why it is still faster than AprioriTid and AprioriHybrid; it may be due to the fact that the "Tid and Hybrid" algorithms only improve the Apriori at the latter phases and those phases do not play the decisive role. Further analysis and experiments are necessary.

References

1. Y.D. Cai, N. Cercone, and J. Han. "Attribute-oriented induction in relational databases," in Knowledge Discovery in Databases, pages 213-228. AAAI/MIT Press, Cambridge, MA, 1991.
2. Maier,D.: The Theory of Relational Databases. Computer Science Press, 1983 (6th printing 1988).
3. Pawlak, Z.: Rough Sets: Theoretical Aspects of Reasoning about Data, Kluwer Academic, Dordrecht (1991)
4. T. Y. Lin.,"Data Mining and Machine Oriented Modeling: A Granular Computing Approach," Journal of Applied Intelligence, Kluwer, in print
5. Eric Louie and T.Y. Lin, "A Data Mining Approach using Machine Oriented Modeling: Finding Association Rules using Canonical Names.". In: Proceeding of 14th Annual International Symposium Aerospace/Defense Sensing, Simulation, and Controls , SPIE Vol 4057, Orlando, April 24-28, 2000
6. Agrawal, R., T. Imielinski, and A. Swami, "Mining Association Rules Between Sets of Items in Large Databases," in Proceeding of ACM-SIGMOD International Conference on Management of Data, pp. 207-216, Washington, DC, June, 1993.
7. Agrawal, R., R. Srikant, "Fast Algorithms for Mining Association Rules," in Proceeding of 20th VLDB Conference San Tiago, Chile, 1994.

Using Closed Itemsets for Discovering Representative Association Rules *

Jamil Saquer[1] and Jitender S. Deogun[2]

[1] Computer Science Department,
Southwest Missouri State University,
Springfield, MO 65804, USA
jms481f@mail.smsu.edu,
[2] Computer Science & Engineering Department,
University of Nebraska, Lincoln, NE 68588, USA
deogun@cse.unl.edu

Abstract. A set of association rules is called representative if it is a minimal set of rules from which all association rules can be generated. The existing algorithms for generating representative association rules use all the frequent itemsets as input. In this paper, we present a new approach for generating representative association rules that uses only a subset of the set of frequent itemsets called frequent closed itemsets. This results in a big reduction in the input size and, therefore, faster algorithms for generating representative association rules. Our approach uses ideas from formal concept analysis to find frequent closed itemsets.

1 Introduction

Mining association rules is an important data mining problem that was introduced in [1]. The problem was first defined in the context of the market basket data to identify customers' buying habits. For example, it is of interest to a supermarket manager to find out that 80% of the customers who buy bagels also buy cream-cheese and 5% of all customers buy both bagels and cream-cheese. Here the association rule is the rule baggies $\Rightarrow$ cream-cheese, 80% is the confidence of the rule and 5% is its support. There has been a great deal of research in developing efficient algorithms for discovering association rules that satisfy user-specified constraints such as minimum support and minimum confidence [2, 3,12,14].

The number of discovered association rules is usually huge which makes it very difficult for an expert to analyse the rules and identify the interesting ones. Lately, there has been an interest in identifying the association rules that are of special importance to a user and in decreasing the number of discovered association rules [4,9,10,13]. Most of these approaches introduce additional measures

---

* This research was supported in part by the Army Research Office, Grant No. DAAH04-96-1-0325, under DEPSCoR program of Advanced Research Projects Agency, Department of Defense and by the U.S. Department of Energy, Grant No. DE-FG02-9 7ER1220.

for interestingness of a rule and prune the rules that do not satisfy the additional measures. A set of representative association rules, on the other hand, is a minimal set of rules from which all association rules can be generated. The number of representative association rules is much smaller than the number of all association rules. Furthermore, we do not need any additional measure for determining the representative association rules.

Algorithms for discovering representative association rules are given in [7, 8]. These algorithms use all the frequent itemsets to find the representative association rules. In this paper, we present a different approach for generating representative association rules. Our approach uses only a subset of the set of frequent itemsets which we call frequent closed itemsets. This results in reducing the input size and, therefore, results in faster algorithms for generating representative association rules. We use ideas from formal concept analysis to find the frequent closed itemsets.

2 Association Rules and Representative Association Rule

The problem of discovering association rules was first introduced in [1]. It can be described formally as follows [1,2]. Let $\mathcal{I} = \{i_1, i_2, \cdots, i_m\}$ be a set of m literals, called items. Let $\mathcal{D} = \{t_1, t_2, \cdots, t_n\}$ be a database of n transactions where each transaction is a subset of $\mathcal{I}$. Any subset of items X is called a k-itemset if the number of items in X equals k. The support of an itemset X, denoted by $sup(X)$, is the percentage of transactions in the database $\mathcal{D}$ that contain X. An itemset is called frequent if its support is greater than or equal to a user specified threshold value.

An association rule r is a rule of the form $X \Rightarrow Y$ where both X and Y are nonempty subsets of $\mathcal{I}$ and $X \cap Y = \emptyset$. X is called the antecedent of r and Y is called its consequent. The support and confidence of the association rule $r : X \Rightarrow Y$ are denoted by $sup(r)$ and $conf(r)$, respectively, and defined as

$$sup(r) = sup(X \cup Y) \quad \text{and} \quad conf(r) = sup(X \cup Y)/sup(X).$$

Support of $r : X \Rightarrow Y$ is simply a measure of its statistical significance and confidence of r is a measure of the conditional probability that a transaction contains Y given that it contains X.

The task of the association data mining problem is to find all association rules with support and confidence greater than user specified minimum support and minimum confidence threshold values. Throughout this paper, we will use the notation $AR(s, c)$ to denote the set of all association rules with minimum support s and minimum confidence c. We also write AR instead of $AR(s, c)$, when s and c are understood.

The number of association rules is usually huge. Representative association rules (RAR) were introduced in [7] to overcome this problem and to reduce the number of rules presented to a user. The user can mine around the RAR. For example, the user may ask to be presented with all the rules that are covered (or represented) by a certain rule of interest to him/her. Informally, the cover of a rule $r : X \Rightarrow Y$, denoted by $C(r)$, is the set of association rules that can be generated from r. Formally,

$$C(r : X \Rightarrow Y) = \{X \cup U \Rightarrow V \mid U, V \subseteq Y, \ U \cap V = \emptyset, \ \text{and} \ V \neq \emptyset\}.$$

An important property of the cover operator is that if an association rule r has support s and confidence c, then every rule $r' \in C(r)$ has support at least s and confidence at least c [7]. This property means that C is a well defined inference operator for association rules.

Using the cover operator, a set of representative association rules with minimum support s and minimum confidence c, $RAR(s,c)$, is defined as follows:

$$RAR(s,c) = \{r \in AR(s,c) \mid \nexists r' \in AR(s,c), \ r \neq r' \ \text{and} \ r \in C(r')\}.$$

That is, a set of representative association rules is a least set of association rules that cover all the association rules and from which all association rules can be generated. Clearly, $AR(s,c) = \bigcup \{C(r) \mid r \in RAR(s,c)\}$.

Let length of $X \Rightarrow Y$ be the number of items in $X \cup Y$. The following are important properties of RAR [7,8]:

Property 1. *Let $r : X \Rightarrow Y$ and $r' : X' \Rightarrow Y'$ be two different association rules, then*

1. *If r is longer than r', then $r \notin C(r')$.*
2. *If r is shorter than r', then $r \in C(r')$ iff $X \cup Y \subset X' \cup Y'$ and $X \supseteq X'$.*
3. *If r and r' are of the same length, then $r \in C(r')$ iff $X \cup Y = X' \cup Y'$ and $X \supset X'$.*

Property 2. *Let $r : X \Rightarrow Z \setminus X \in AR(s,c)$ and let $maxSup = max(\{sup(Z') \mid Z \subset Z' \subseteq I\} \cup \{0\})$. Then, $r \in RAR(s,c)$ if the following two conditions are satisfied:*

i. *$maxSup < s$ or $maxSup/sup(X) < c$.*
ii. *$\nexists X', \ \emptyset \subset X' \subset X$ such that $X' \Rightarrow Z \setminus X' \in AR(s,c)$.*

The first condition guarantees that r does not belong in the cover of any association rule with length greater than the length of r. The second condition guarantee that r does not belong in the cover of any association rule that has the same length as r.

Property 3. *Let $\emptyset \neq X \subset Z \subset Z' \subseteq I$ and $sup(Z) = sup(Z')$. Then, there is no rule $r : X \Rightarrow Z \setminus X \in AR(s,c)$ such that $r \in RAR(s,c)$.*

Property 3 holds because $r \in C(X \Rightarrow Z' \setminus X)$. The above properties led to the development of the algorithms *GenAllRepresentatives* and *FastGenAllRepresentatives* for discovering representative association rules [7,8]. Both algorithms use all frequent itemsets generated from applying the Apriori algorithm to the database $\mathcal{D}$ [2]. Our approach is different; we use only a subset of the frequent itemsets which we call *frequent closed itemsets*. Frequent closed itemsets are found by using methods from formal concept analysis. In the next section we develop the necessary theory for that.

3 Closed Itemsets

In this section, we develop theoretical results that lead to the development of our algorithm for RAR generation. These results are directly related to formal concept analysis [15]. Our notion of a closed itemset is similar to that of a concept. Informally, a concept is a pair of two sets: set of objects (transactions or itemsets) and set of features (items) common to all the objects. Using the framework of formal concept analysis, concepts are structured in the form of a lattice called the *concept lattice*. The concept lattice has proved to be a useful tool for knowledge representation and knowledge discovery [6].

Definition 1. *A data mining context* is defined as a triple $(T, \mathcal{I}, R)$ where T is a set of transactions, $\mathcal{I}$ is a set of items and $R \subseteq T \times \mathcal{I}$.

A data mining context is a formal definition of a database. The set T is the set of all transactions in the database and the set $\mathcal{I}$ is the set of all items in the database. For $t \in T$ and $i \in \mathcal{I}$ we write $(t, i) \in R$ to mean that the transaction t contains the item i. An example of a data mining context is shown in Table 1 where an X is placed in the tth row and ith column to indicate that $(t, i) \in R$. This example is generated from the database given in [7] which we use in this paper for comparison.

Table 1. Example of a Data Mining Context

	A	B	C	D	E	F	G	H
t_1	X	X	X	X	X			
t_2	X	X	X	X	X	X		
t_3	X	X	X	X	X		X	X
t_4	X	X			X			
t_5		X	X	X	X		X	X

Two dual mappings α and β are defined between the power sets of T and $\mathcal{I}$ as follows.

Definition 2. Let $(T, \mathcal{I}, R)$ be a data mining context, $X \subseteq T$, and $Y \subseteq \mathcal{I}$. Define the mappings α, β as follows:

$$\beta : 2^T \to 2^\mathcal{I}, \quad \beta(X) = \{i \in \mathcal{I} \mid (t, i) \in R \ \forall \, t \in X\},$$

$$\alpha : 2^\mathcal{I} \to 2^T, \quad \alpha(Y) = \{t \in T \mid (t, i) \in R \ \forall \, i \in Y\}.$$

The mapping $\beta(X)$ associates with X the set of items that are common to all the transactions in X. Similarly, the mapping $\alpha(Y)$ associates with Y the set of all transactions having all the items in Y. Intuitively, $\beta(X)$ is the maximum set of items shared by all transactions in X and $\alpha(Y)$ is the maximum set of transactions possessing all the attributes in Y.

Example 1. Consider the database presented in Table 1. Let $X = \{t_1, t_2\}$ and $Y = \{A, B, C\}$. Then, $\beta(X) = \{A, B, C, D, E\}$, $\alpha(Y) = \{t_1, t_2, t_3\}$, $\alpha(\beta(X)) = \alpha(\{A, B, C, D, E\}) = \{t_1, t_2, t_3\}$, and $\beta(\alpha(Y)) = \beta(\{t_1, t_2, t_3\}) = \{A, B, C, D, E\}$.

It is clear from this example that, in general, $\beta(\alpha(Y)) \neq Y$ where Y is an itemset. This leads to the following definition.

Definition 3. *An itemset Y that satisfies the condition $\beta(\alpha(Y)) = Y$ is called a closed itemset.*

Closed itemsets are important because all members of the concept lattice of a data mining context satisfy the condition $\beta(\alpha(Y)) = Y$ [15]. This step of considering only closed itemsets can be considered as a first step in pruning the itemset lattice.

Example 2. Let $Y = \{A, B, C, D, E\}$. $\beta(\alpha(Y)) = Y$. Therefore, Y is a closed itemset. On the other hand, the itemset $\{A, B, C\}$ given in Example 1 is not closed because $\beta(\alpha(\{A, B, C\})) = \{A, B, C, D, E\} \neq \{A, B, C\}$.

The concept lattice can be pruned more by considering only closed itemsets with support greater than minimum support which we call frequent closed itemsets. This leads to the following definition.

Definition 4. *A frequent closed itemset is a closed itemset which is also frequent. That is, it has support greater than or equals to user-specified value for minimum support.*

4 Algorithms

Our approach is to first generate the set of all frequent closed itemsets, FCI, and then to use FCI to generate the set of all representative association rules.

4.1 Generating Frequent Closed Itemsets

Let $\mathcal{D} = (T, \mathcal{I}, R)$ be a database mining context. The algorithm we use to generate FCI is a slight modification of the *Close* algorithm mentioned in [11] which we call *Close-FCI*. Both algorithms are similar to the *Apriori* algorithm [2].

Assume that the items in $\mathcal{I}$ are sorted in lexicographic order. The data structure used consists of two sets. Set of candidate frequent closed itemsets, FCC, and set of frequent closed itemsets, FC. The notations FCC_i and FC_i are used to indicate candidate frequent closed itemsets and frequent closed itemsets of size i, respectively. Each element in FCC_i and FC_i has three components. An itemset component, a closure component and a support component.

Closure of an itemset $X \subseteq \mathcal{I}$, denoted by $closure(X)$, is the smallest closed itemset containing X and is equal to the intersection of all itemsets containing X. It is also shown in [11] that Support(X) = Support(Closure(X)). The *Close-FCI* algorithm is given below.

Algorithm 1 *Close-FCI(D)*
1) $FCC_1.itemsets = \{\text{1-itemsets}\}$;
2) for (k = 1; $FCC_k \neq \emptyset$; k++) do begin
3)   forall $X \in FCC_k$ do begin
4)   X.closure = $\emptyset$;
5)   X.support = 0;
6)   end forall
7) FCC_k = Generate-Closures(FCC_k);
8)   forall candidate closed itemset $X \in FCC_k$ do begin
9)   if ((X.support $\geq$ minSupport) and (X.closure $\notin FC_k$)) then
10) $FC_k \leftarrow FC_k \cup \{X\}$;
11)   end if
12) FCC_{k+1} = Generate-Candidates(FC_k);
13) endfor
14) return $\bigcup_{i=1}^{k-1}\{FC_i.closure \text{ and } FC_i.support\}$;

First, the algorithm initializes the itemsets in FCC_1 to the items in the database which does not require any database pass. Then in iteration k of the main for loop of the algorithm, the closure and support of each itemset in FCC_k are initialized. The algorithm then finds candidate frequent closed itemsets of size k, FCC_k, in line 7. Frequent closed itemsets, FC_k, are found in the next step where the minimum support threshold, minSupport, is used to prune out the infrequent itemsets. Finally, in line 14 the algorithm generate candidates FCI of size $k+1$, FCC_{k+1}.

Closure of an itemset $X \subseteq \mathcal{I}$, is by definition the smallest closed itemset containing X which is equal to the intersection of all frequent itemsets containing X. Therefore, closure(X) is found using the following formula

$$closure(X) = \bigcap_{t \in T}\{\beta(t) \mid X \subseteq \beta(t)\}$$

which is incorporated in the following algorithm. This algorithm requires one database pass to find closure of all elements in FCC_k.

Algorithm 2 *Generate-Closures(FCC_k)*
1) forall transactions $t \in T$ do begin
2) $\beta_t = \{X \in FCC_k \mid X \subseteq \beta(t)\}$; //all itemsets that are contained in $\beta(t)$
3)   forall itemsets $X \in \beta(t)$ do begin
4)   if (X.closure = $\emptyset$) then
5)   X.closure $\leftarrow \beta(t)$;
6)   else
7)   X.closure $\leftarrow$ X.closure $\cap \beta(t)$;
8)   end if
9)   X.Support++;
10)   end forall
11) end forall
12) Return $\bigcup\{X \in FCC_k \mid X.closure \neq \emptyset\}$;

The algorithm *Generate-Candidates(FC_k)* that finds candidate closed itemsets of size $k+1$ uses the result of the following property [11].

Property 4. Let X be an itemset of length k and $\mathcal{X} = \{X_1, X_2, \cdots, X_m\}$ be a set of $(k-1)$-subsets of X where $\bigcup_{X_i \in \mathcal{X}} = X$. If $\exists X_i \in \mathcal{X}$ such that $X \subseteq closure(X_i)$, then $closure(X) = closure(X_i)$.

This property means that the itemset X results in redundant computations of frequent closed itemsets because $closure(X)$ which is equal to $closure(X_i)$ is already generated. Therefore, X can be removed from FCC_{k+1}. The *Generate-Candidates(FC_k)* is similar to the *AprioriGen* algorithm that was first given in [2] except for the second pruning step at line 6.

Algorithm 3 *Generate-Candidates(FC_k)*
1) insert into $FCC_{(k+1)}$.itemset
2)   select X.itemset$_1$, X.itemset$_2$, $\cdots$, X.itemset$_k$, Y.itemset$_k$
3)   from FC_k.itemset X, FC_k.itemset Y
4)   where X.itemset$_1$=Y.itemset$_1$ $\wedge$ X.itemset$_2$=Y.itemset$_2$ $\wedge \cdots \wedge$
        X.itemset$_{k-1}$=Y.itemset$_{k-1}$ $\wedge$ X.itemset$_k$ <Y.itemset$_k$;
     //prune all supersets of infrequent itemsets
5)   delete all itemsets $X \in FCC_{(k+1)}$.itemset where some k-subset of X is
     not in FC_k;
     //prune all itemsets with closures already generated
6)   delete all itemsets $X \in FCC_{(k+1)}$.itemset where the closure of some
     k-subset of X contains X;
7) Return $\bigcup \{X \in FCC_{(k+1)}\}$;

The *Close-FCI* algorithm requires one database pass in each iteration. That pass is needed in the *Generate-Closure* algorithm.

Example 3. Applying the Close-FCI on the database represented in Table 1 for minimum support $s = 3/5 = 0.6$ and minimum confidence $c = 0.75$, the following frequent closed itemsets are found:

$$FC = \{ABCDE, BCDE, ABE, BE\}$$

with support 3/5, 4/5, 4/5, and 5/5, respectively. On the other hand, the Apriori algorithm produced 31 different frequent itemsets (the nonempty subsets of $ABCDE$) for the same values of s and c.

4.2 Generating Representative Association Rules

In this section, we present an algorithm for generating representative association rules which we call *Generate-RAR*. *Generate-RAR* takes as input the set of all FCI and produces the set of all RAR. *Generate-RAR* is a modification of the *FastGenAllRepresentatives* given in [8] and it uses Properties 1-3 from Section 2. It also uses the result that for an itemset X, $sup(X) = sup(closure(X))$ which was mentioned in Section 4.1. The *Generate-RAR* algorithm is given below.

Algorithm 4 *Generate-RAR(all frequent closed itemsets FC)*
1) $c \leftarrow minConfidence$; //user specified value for minimum confidence
2) $RAR \leftarrow \emptyset$; //initialize set of RAR
3) $k \leftarrow 0$
     //split frequent closed itemsets according to size
4)   forall $X \in FC$ do begin
5) $FC_{|X|} \leftarrow FC_{|X|} \cup \{X\}$;
6)   if $(k < |X|)$ then $k \leftarrow |X|$;
7)   end forall
8)   for $(i \leftarrow k; i > 1; i - -)$ do
9)   forall $Z \in FC_i$ do begin
10) $maxSup = max(\{sup(Z') \mid Z \subset Z' \in FC\} \cup \{0\})$;
11)   if $(Z.support \neq maxSup)$ then begin //see Property 3
12) $A_1 = \{\{Z[1]\}, \{Z[2]\}, \cdots, \{Z[i]\}\}$; //create 1-antecedents
                 //start loop2
13)   for $(j = 1; (A_j \neq \emptyset)$ and $(j < i); j + +)$ do begin
14)   forall $X \in A_j$ do begin
15) $Y \leftarrow$ smallest closed itemset containing X;
16)   X.support = Y.support;
                         //check if $X \Rightarrow Z \backslash X$ is a representative association rule
17)
18)   if $(Z.support/X.support \geq c$ and
                              $maxSup/X.support < c)$ then
19) $RAR \leftarrow RAR \cup \{X \Rightarrow Z \backslash X\}$;
20) $A_j = A_j \setminus \{X\}$;
21)   end if
22)   end forall
23) $A_{j+1} \leftarrow AprioriGen(A_j)$;
24)   end for //end loop2
25)   end if
26)   end forall
27) Return RAR;

First, RAR is initialized to be empty. Then, each frequent closed itemset X of length i is added to FC_i and the length of the maximal closed itemsets k is found. Line 8 controls the generation of representative association rules. First, the largest rules (of size k) are generated and added to RAR. Next, representative association rules of size $(k-1)$ are generated and added to RAR and so on. Finally, representative association rules of size 2 are generated and added to RAR. The generation of association rules of size i is controlled in lines 9 through 26 as follows:

Let Z be a frequent closed itemset of size i. First, $maxSup$ is found in line 10. If there is no superset of Z, $maxSup$ will be assigned the value zero. If $maxSup$ has the same value as Z.support, then according to Property 3, no representative association rule will be generated from Z. Otherwise, the process of generating representative association rules from Z starts. First, the set A_1 is assigned all possible 1-itemset antecedents of Z. The loop in lines 13 through

24 controls the generation of representative association rules with antecedents of length j. All possible antecedents $X \in A_j$ are considered. Support of X (which is equal to support(closure(X))) is found. $X \to Z \backslash X$ is a valid representative association rule if its confidence is greater then or equal to c and if it satisfies the second condition of property 2.i in which case X will be removed from A_j. After all representative association rules with antecedents of length j are generated from Z, A_j may not be empty. The *AprioriGen* function is called with argument A_j to find antecedents of length $j+1$. The *AprioriGen* function is given in [2] and it consists of lines 1 through 5 of Algorithm 3. Condition ii of Property 2 is satisfied by making sure that no itemset in A_{j+1} is a superset of an antecedent of an already generated representative association rule which is taking care of in line 20.

Example 4. Running the Generate-RAR *algorithm using as input the frequent closed itemsets produced in the previous example, the following representative association rules are generated:*
$RAR = \{A \to BCDE, C \to ABDE, D \to ABCE, B \to CDE,$
$E \to BCD, B \to AE, E \to AB\}$.
They are the same rules generated from the FastGenAllRepresentatives *in [8].*

5 Conclusion

This paper presents a new approach for generating representative association rules using frequent closed itemsets which constitute only a subset of the set of all frequent itemsets. Our approach results in a big reduction in the input size and thus faster algorithms for generating representative association rules. We also presented a new algorithm, called *Generate-RAR* for generating representative association rules.

Traditional association mining algorithms first generate all frequent itemsets and then use frequent itemsets to generate all the association rules. For future work, we will investigate if FCI can be used directly to generate all the association rules. We will also investigate if our approach for finding RAR using FCI can result in faster generation of all the association rules than the traditional algorithms.

References

1. R. Agrawal, T. Imielinski, and A. Swami, "Mining Association Rules Between Sets of Items in Large Databases," Proceedings of the ACM SIGMOD Conference on Management of Data, Washington, D.C. 207-216, 1993.
2. R. Agrawal and R. Srikant, "Fast Algorithms for Mining Association Rules," Proceedings of the 20th VLDB Conference, Santiago, Chile, 478-499, 1994.
3. S. Brin, R. Motwani, J. Ullman, and S. Tsur, "Dynamic Itemset Counting and Implication Rules for Market Basket Data," Proceedings of the ACM SIGMOD international conference on Management of Data, 255-264, 1997.

4. R. Bayardo, R. Agrawal, and D. Gunopupulos, "Constraint-based rule mining in large, dense databases," in ICDE-99, 1999.
5. B. Ganter and R. Wille, "Formal Concept Analysis: Mathematical Foundations," (Springer, Berlin, 1999).
6. R. Godin and R. Missaoui, "An Incremental Concept Formation for Learning from Databases," Theoretical Computer Science, 133, 387-419, 1994.
7. M. Kryszkiewicz, "Representative Association Rules," in Proc. PAKDD '98, Lecture Notes in Artificial Intelligence, vol. 1394, (Springer-Verlag 1998), 198-209.
8. M. Kryszkiewicz, "Fast Discovery of Representative Association Rules," in Proc. RSCTC '98, Lecture Notes in Artificial Intelligence, vol. 1424, (Springer-Verlag 1998), 214-221.
9. Ng. R. T. Lakshmanan, and L. Han, "Exploratory Mining and Pruning Optimization of Constrained Association Rules," in SIGMOD-98, 1998.
10. B. Liu, W. Hsu, and Y. Ma, "Pruning and Summarizing the Discovered Associations," in ACM SIGKDD'99, 1999.
11. N. Pasquier, Y. Bastide, R. Taouil, and L Lakhal, "Efficient Mining of Association Rules Using Closed Itemset Lattices," Information Systems 24 (1999) 25-46.
12. A. Savasere, E. Omiencinsky, and S. Navathe. "An Efficient Algorithm for Mining Mining Association Rules in Large Databases," Proceedings of the 21st VLDB Conference, Zurich, Switzerland, 1995.
13. R. Srikant, Q. Vu, and R. Agrawal, "Mining Association Rules with Item Constraints," KDD-97, pp. 67-73, 1997.
14. H. Toivonen, "Sampling Large Databases for Association Rules," Proceedings of the 22nd VLDB Conference, Bombay, India, 134-145, 1996.
15. R. Wille 1982, "Restructuring Lattice Theory: an Approach Based on Hierarchies of Concepts," in: Ivan Rivali, ed., Ordered sets, 445-470, 1982.

Legitimate Approach to Association Rules under Incompleteness

Marzena Kryszkiewicz and Henryk Rybinski

Institute of Computer Science, Warsaw University of Technology,
Nowowiejska 15/19, 00-665 Warsaw, Poland
mkr@ii.pw.edu.pl, hrb@ii.pw.edu.pl

Abstract. A notion of legitimate definitions of support and confidence under incompleteness is defined. Properties of generic legitimate definitions of support and confidence are investigated. We show that in the case of incompleteness legitimate association rules can be derived from legitimate representative rules by the cover operator. It is proved that the minimum condition maximum consequence association rules under incompleteness constitute a subset of representative rules of the same type. Algorithms for generating association rules under incompleteness are offered.

1 Introduction

The problem of association rules discovery was introduced in [1] for sales transaction database. The problem of data incompleteness did not occur for a transaction database, however, it is often unavoidable in relational databases. Missing data may result from errors, measurement failures, changes in the database schema etc.

Incompleteness of the data in the database introduces a confusion of how to treat requests for a given user-specified support and confidence of rules. We cannot evaluate exact values of support and confidence. Instead, we can evaluate optimistic and pessimistic support and confidence of a rule. For a marginal incompleteness we could expect that the difference between optimistic and pessimistic parameters is not essential. If however the incompleteness grows, the gap between optimistic and pessimistic parameters will also grow. We could therefore foresee that the user is also interested in "expected" support and confidence.

Data incompleteness in the context of association rules was addressed in [5]. It was offered in [5] how to compute pessimistic and optimistic estimations of support and confidence of an association rule. In [6] a set of properties that characterize a legitimate approach to incompleteness has been proposed. An example of a legitimate probabilistic approach was presented in [6] as well as examples of popular approaches ignoring missing values that turn out not to be legitimate.

In this paper, we investigate properties of generic legitimate definitions of support and confidence in the case of incomplete databases. We will show that in case of incompleteness legitimate association rules can be derived from legitimate representative rules by the cover operator [3]. We also prove that the minimum condition maximum consequence association rules of any type constitute a subset of representative rules of the same type. Finally, we offer the algorithms generating association rules from incomplete databases.

2 Association Rules in Complete Relational Databases

Let us consider a table $D = (O, AT)$, where O is a non-empty finite set of *tuples* and AT is a non-empty finite set of *attributes*, such that $a: O \rightarrow V_a$ for any $a \in AT$, where V_a denotes the domain of a. Any *attribute-value pair* (a,v), where $a \in AT$ and $v \in V_a$ will be called an *item*. A set of items will be called *itemset*. *Support* of an itemset X is denoted by $sup(X)$ and defined as the number (or the percentage) of tuples in D that contain X.

An *association rule* is an expression of the form: $X \Rightarrow Y$, where X and Y are items and $X \cap Y = \emptyset$. Support of a rule $X \Rightarrow Y$ is denoted by $sup(X \Rightarrow Y)$ and is defined as $sup(X \cup Y)$. *Confidence* of the rule $X \Rightarrow Y$ is denoted by $conf(X \Rightarrow Y)$ and defined as $sup(X \cup Y) / sup(X)$. Usually, one is interested in discovering rules that have support greater than a specified minimum support s and confidence not less than a user specified minimum confidence c. Such set of rules will be denoted by $AR(s,c)$, i.e. $AR(s,c) = \{r|\ sup(r) > s \land conf(r) \geq c\})$. If s and c are understood we will write briefly AR.

Usually ARs are too numerous for practical use. One of the most popular methods of restricting the number of rules is to generate only those with the minimum condition part. Those rules are in particular useful in classification procedures. It was shown in [9] that for any rule with minimum antecedent one can reduce its consequent without any loss of support and confidence; instead, it may even lead to the rule with better parameters. This observation justifies generating the rules with minimum antecedents and maximum consequents [10]. A set of *minimum condition maximum consequence association rules* wrt. support s and confidence c ($MMR(s,c)$) is defined as follows:

$$MMR(s,c) = \{r: (X \Rightarrow Y) \in AR(s,c) |\ \neg \exists r': (X' \Rightarrow Y') \in AR(s,c),\ r' \neq r \land X' \subseteq X \land Y' \supseteq Y\}.$$

Recently another way of reducing the size of the set of generated rules was proposed in [3]. Let us recall the notion of *representative association rules* and *cover* operator. Informally speaking, a set of all *representative association rules* is a least set of rules that covers all association rules by means of a *cover operator*. The *cover C* of the rule $X \Rightarrow Y$, $Y \neq \emptyset$, is defined as follows:

$$C(X \Rightarrow Y) = \{X \cup Z \Rightarrow V|\ Z, V \subseteq Y \land Z \cap V = \emptyset \land V \neq \emptyset\}.$$

Each rule in $C(X \Rightarrow Y)$ consists of a subset of items occurring in the rule $X \Rightarrow Y$. The antecedent of any rule r covered by $X \Rightarrow Y$ contains X and perhaps some items in Y, whereas r's consequent is a non-empty subset of the remaining items in Y. For the cover C the following property holds:

Property 1. Let $r: (X \Rightarrow Y)$ and $r': (X' \Rightarrow Y')$ be association rules. Then:

$r' \in C(r)$ iff $X' \cup Y' \subseteq X \cup Y \land X' \supseteq X$ iff $X' \cup Y' \subseteq X \cup Y \land X' \supseteq X \land Y' \subseteq Y$.

As proved in [3], for an association rule r having support s and confidence c, each rule in the cover $C(r)$ belongs to $AR(s,c)$. Hence, if $r \in AR(s,c)$, then every rule in $C(r)$ also belongs to $AR(s,c)$. This property can be applied for looking for a base of rules covering all others. In the sequel, such a minimal base of rules will be called a set of *representative association rules* and will be denoted by RR. Formally, a set of *representative association rules* wrt. support s and confidence c is defined as follows:

$$RR(s,c) = \{r \in AR(s,c)|\ \neg \exists r' \in AR(s,c),\ r' \neq r \land r \in C(r')\}.$$

Each rule in *RR* is called a *representative association rule*. From the definition of *RR* it results that no representative association rule belongs to the cover of another association rule and $AR(s,c) = \bigcup_{r \in RR(s,c)} C(r)$. It was proved in [4] that $MMR \subseteq RR$ and *MMR* can be derived from *RR*.

3 Support and Confidence in Incomplete Databases

Computation of a real support of itemsets and real confidence of association rules is not feasible in the case of databases with missing attribute values. However, one can always calculate possible least and greatest values of support and confidence, as well as, try to predict their "expected" values. In order to express the properties of data incompleteness we will apply the following notions:

- Missing values will be denoted by "*",
- The *maximal set of tuples that certainly match an itemset X* is denoted by $n(X)$ and is defined as follows: $n(X) = \{t \in D | \forall (a,v) \in X: a(t) = v\}$,
- By $m(X)$ we denote the *maximal set of tuples that possibly match the itemset X* in D, i.e. $m(X) = \{t \in D | \forall (a,v) \in X: a(t) \in \{v,*\}\}$,
- The set-theoretical difference $m(X) \setminus n(X)$ is denoted by $d(X)$,
- By $n(-X)$ we denote the *maximal set of tuples that certainly do not match the itemset X* in D, i.e. $n(-X) = D \setminus m(X)$,
- By $m(-X)$ we denote the *maximal set of tuples that possibly do not match the itemset X* in D, i.e. $m(-X) = D \setminus n(X)$.

Example 1. Given the incomplete database D presented in Fig. 1, Fig. 2 illustrates the notions of certainly and possibly matching an itemset.

Id	X1	X2
1	*	c
2	a	*
3	a	c
4	a	c
5	*	d
6	b	*
7	b	*
8	b	*

Itemset X	{(X1,a),(X2,c)}
n(X)	{3,4}
m(X)	{1,2,3,4}
d(X)	{1,2}
n(-X)	{5,6,7,8}
m(-X)	{1,2,5,6,7,8}

Fig. 1. Example incomplete database   **Fig. 2.** Tuples matching itemset X

Let least possible support of an itemset *X* be denoted by *pSup(X)* (pessimistic case) and greatest possible support be denoted by *oSup(X)* (optimistic case). Clearly,

$$pSup(X) = |n(X)| \text{ and } oSup(X) = |m(X)|.$$

Let $X \subset Y$. It is easy to observe that $pSup(X) \geq pSup(Y)$ and $oSup(X) \geq oSup(Y)$.

Let $pConf(X \Rightarrow Y)$ and $oConf(X \Rightarrow Y)$ denote *least possible confidence* and *greatest possible confidence* of $X \Rightarrow Y$, respectively. These values can be computed according the following equations (see [5] for proof):

$$pConf(X \Rightarrow Y) = |n(X) \cap n(Y)| / [|n(X) \cap n(Y)| + |m(X) \cap m(-Y)|], \quad (1)$$

$$oConf(X \Rightarrow Y) = |m(X) \cap m(Y)| / [|m(X) \cap m(Y)| + |n(X) \cap n(-Y)|]. \quad (2)$$

The differences between optimistic and pessimistic estimations for rules can be high. It would be desirable to have a method of predicting support and confidence close to real (though unknown) values. There were proposed in the literature several definitions of expected support and expected confidence. One can argue which definition is better or when it should be applied. Whatever is the definition of expected support and confidence in an incomplete database, we do not accept it if anyone of the postulates below is not satisfied:

Postulates. Let X and Y be itemsets, $iSup(X)$ denote support under incompleteness, and $iConf(X)$ denote confidence under incompleteness, $A \subseteq AT$, and $Instances(A)$ denote the set of all possible tuples over the set of attributes A.

$$iSup(X) \in [pSup(X), oSup(X)], \qquad \textbf{(P1)}$$

$$iSup(X) \geq iSup(Y) \text{ for } X \subset Y, \qquad \textbf{(P2)}$$

$$iConf(X \Rightarrow Y) = iSup(X \cup Y) / iSup(X), \qquad \textbf{(P3)}$$

$$iConf(X \Rightarrow Y) \in [pConf(X \Rightarrow Y), oConf(X \Rightarrow Y)], \qquad \textbf{(P4)}$$

$$\Sigma_{X \in Instances(A)}, iSup(X) = 1 \text{ for any } A \subseteq AT. \qquad \textbf{(P5)}$$

The definitions of *support under incompleteness* and *confidence under incompleteness* that satisfy all postulates P1-P5 will be called *legitimate*. Below we will show an example of legitimate approach to incompleteness (see [6] for proof).

Example 2. Let $\mu: AT \times V \rightarrow [0,1]$ denote a frequency with which value $v \in V_a$ occurs for the attribute $a \in AT$ in D, defined as $\mu(a,v) = |n(a,v)| / |D - d(a,v)|$. Based on the notion of μ_t^a, the probability $probSup_t$ of supporting an item (a,v) by the tuple $t \in D$ is defined:

$$probSup_t(a,v) = \begin{cases} 1 & \text{if } a(t) = v \\ \mu_t^a(v) & \text{if } a(t) = * \\ 0 & \text{otherwise}. \end{cases}$$

The probability ($probSup_t$) of supporting an itemset $X = \{(a_1, v_1), \ldots, (a_k, v_k)\}$ by a tuple $t \in D$ is defined as follows:

$$probSup_t(X) = probSup_t(a_1, v_1) * \ldots * probSup_t(a_k, v_k).$$

Probable support ($probSup$) of an itemset X in the database D is defined below:

$$probSup(X) = [\Sigma_{t \in D}, probSup_t(X)].$$

Probable confidence ($probConf$) of a rule $X \Rightarrow Y$ is defined in usual way:

$$probConf(X \Rightarrow Y) = probSup(X \cup Y) / probSup(X). \qquad \bullet$$

Incompleteness of the data introduces a confusion of how to treat requests for a given user-specified support and confidence of rules. One can imagine that a user is interested in the rules whose pessimistic support and expected confidence are above requested thresholds, or in the rules whose pessimistic support and pessimistic confidence are above the thresholds. Other variations of user requirements are also likely. To this end we proposed in [7] a generic definition of types of association rules:

$$AR_{\alpha\beta}(s,c) = \{r \mid \alpha Sup(r) > s \wedge \beta Conf(r) \geq c\},$$

where α/β could be substituted by either p (pessimistic), or o (optimistic) support / confidence. For instance, $AR_{po}(s,c) = \{r|\ pSup(r) > s \land oConf(r) \geq c\}$.

In the sequel, we investigate properties of generic legitimate definitions of support and confidence under incompleteness, therefore, we presume that α/β can be also substituted by i (legitimate support / confidence under incompleteness). Obviously, the most natural combinations of requests for support and confidence of rules are: $AR_{pp}(s,c)$, $AR_{oo}(s,c)$, $AR_{ii}(s,c)$. Please note that for the complete database all these definitions are equivalent.

Analogously to RR for a complete database, we introduced the notion of representative rules for incomplete databases in [7]. Here we apply the same generic notation as we did for $AR_{\alpha\beta}$ above. A set of *representative association rules* wrt. minimum support s and minimum confidence c is denoted by $RR_{\alpha\beta}(s,c)$ and defined as follows:

$$RR_{\alpha\beta}(s,c) = \{r \in AR_{\alpha\beta}(s,c)|\ \neg \exists r' \in AR_{\alpha\beta}(s,c),\ r' \neq r \land r \in C(r')\}.$$

Property 2 [7]. Let r, r' be rules and $r' \in C(r)$. Then:
a) $pSup(r') \geq pSup(r)$ and $oSup(r') \geq oSup(r)$,
b) $pConf(r') \geq pConf(r)$ and $oConf(r') \geq oConf(r)$.

By analogy we introduce here the same generic $\alpha\beta$ notation for MMR as we did for AR and RR. A set of *minimum condition maximum consequence association rules* wrt. support s and confidence c will be denoted by $MMR_{\alpha\beta}(s,c)$ and defined as follows:

$$MMR_{\alpha\beta}(s,c)=\{r:(X{\Rightarrow}Y) \in AR_{\alpha\beta}(s,c)|\neg\exists r':(X'{\Rightarrow}Y') \in AR_{\alpha\beta}(s,c),\ r' \neq r \land X' \subseteq X \land Y' \supseteq Y\}.$$

In the next sections we will investigate relationships among legitimate $AR_{\alpha\beta}$, $RR_{\alpha\beta}$ and $MMR_{\alpha\beta}$, in the case of an incomplete database.

4 Legitimate Representative Rules

In this section we will show that legitimate $AR_{\alpha\beta}(s,c)$ can be derived syntactically from legitimate $RR_{\alpha\beta}(s,c)$. Let us start with examining properties of rules related by the cover operator.

Property 3. Let r, r' be some rules and $r' \in C(r)$. Then:
a) $iSup(r') \geq iSup(r)$,
b) $iConf(r') \geq iConf(r)$.

Proof: Let $r: X{\Rightarrow}Y$, $r': X'{\Rightarrow}Y'$ and $r' \in C(r)$.
Ad. a) Follows immediately from Property 1 and Postulate P2.
Ad. b) $iConf(X'{\Rightarrow}Y') = iSup(X' \cup Y') / iSup(X')$ and $iConf(X{\Rightarrow}Y) = iSup(X \cup Y) / iSup(X)$. It follows from Property 1 and Postulate P2 that $iSup(X' \cup Y') \geq iSup(X \cup Y)$ and $iSup(X') \leq iSup(X)$. Hence, $iConf(X'{\Rightarrow}Y') \geq iConf(X{\Rightarrow}Y)$. ∎

Properties 2-3 allow us to conclude with the following property:

Property 4. Let r, r' be rules and $r' \in C(r)$. If $r \in AR_{\alpha\beta}(s,c)$, then $r' \in AR_{\alpha\beta}(s,c)$.

By definition of $RR_{\alpha\beta}$ and from Property 4 one can infer that all association rules of any $\alpha\beta$ type can be derived by means of cover operator from representative rules of the same type, which is stated by Property 5.

Property 5. $\qquad AR_{\alpha\beta}(s,c) = \bigcup_{r \in RR_{\alpha\beta}(s,c)} C(r).$

5 Legitimate Minimum Condition Maximum Consequence Rules

In Section 5, we investigate how incompleteness influences the relationship between RR and MMR. In particular, we prove, that the minimum condition maximum consequence rules of $\alpha\beta$ type constitute a subset of representative rules of the same type.

Property 6. $\qquad MMR_{\alpha\beta}(s,c) \subseteq RR_{\alpha\beta}(s,c).$

Proof: We will write shortly $RR_{\alpha\beta}$, and $MMR_{\alpha\beta}$ instead of $RR_{\alpha\beta}(s,c)$, and $MMR_{\alpha\beta}(s,c)$, respectively. Property 1 allows us to express $RR_{\alpha\beta}$ as follows:
$RR_{\alpha\beta} = \{r: (X \Rightarrow Y) \in AR_{\alpha\beta} | \neg \exists r': (X' \Rightarrow Y') \in AR_{\alpha\beta}, r' \neq r \land X \cup Y \subseteq X' \cup Y' \land X' \subseteq X \land Y' \supseteq Y\}.$
On the other hand, $MMR_{\alpha\beta}$ is defined as follows:
$MMR_{\alpha\beta} = \{r: (X \Rightarrow Y) \in AR_{\alpha\beta} | \neg \exists r': (X' \Rightarrow Y') \in AR_{\alpha\beta}, r' \neq r \land X' \subseteq X \land Y' \supseteq Y\}.$

Using the two formulae above one can easily observe that each rule in $MMR_{\alpha\beta}$ belongs also to $RR_{\alpha\beta}$, but not necessarily vice versa. •

Next, we prove that the minimum condition maximum consequence rules of any $\alpha\beta$ type can be extracted from representative rules of the same type.

Property 7.

$MMR_{\alpha\beta}(s,c) = \{r:(X \Rightarrow Y) \in RR_{\alpha\beta}(s,c) | \neg \exists r':(X' \Rightarrow Y') \in RR_{\alpha\beta}(s,c), r' \neq r \land X' \subseteq X \land Y' \supseteq Y\}.$

Proof: We will write shortly $RR_{\alpha\beta}$, and $MMR_{\alpha\beta}$ instead of $RR_{\alpha\beta}(s,c)$, and $MMR_{\alpha\beta}(s,c)$, respectively. By Property 6, $MMR_{\alpha\beta}$ are contained in $RR_{\alpha\beta}$. So, $MMR_{\alpha\beta} = \{r: (X \Rightarrow Y) \in RR_{\alpha\beta} | \neg \exists r': (X' \Rightarrow Y') \in AR_{\alpha\beta}, r' \neq r \land X' \subseteq X \land Y' \supseteq Y\}$. Now, we have only to prove that for any $r: (X \Rightarrow Y) \in AR_{\alpha\beta}$, the expression (3) is equivalent to (4).

$$\exists r': (X' \Rightarrow Y') \in AR_{\alpha\beta}, r' \neq r \land X' \subseteq X \land Y' \supseteq Y \qquad (3)$$

$$\exists r'': (X'' \Rightarrow Y'') \in RR_{\alpha\beta}, r'' \neq r \land X'' \subseteq X \land Y'' \supseteq Y. \qquad (4)$$

Let $r': (X' \Rightarrow Y') \in AR_{\alpha\beta}$ and $r' \neq r$ and $X' \subseteq X$ and $Y' \supseteq Y$. Each association rule belongs to the cover of some representative rule, so there is some $r'': (X'' \Rightarrow Y'')$ in $RR_{\alpha\beta}$, such that $r' \in C(r'')$. Hence, $X'' \subseteq X' \subseteq X$ and $Y'' \supseteq Y' \supseteq Y$ and thus, (3) implies (4). The inverse implication is trivial (any representative rule is association one). •

6 The Algorithms

In this section, we will present the algorithms of generating AR_{pp}, AR_{oo} and AR_{ii}. Other $AR_{\alpha\beta}$ combinations can be computed similarly. The problem of generating association rules is usually decomposed into two subproblems:

1. Generate all itemsets whose support exceeds the minimum support *minSup*. The itemsets of this property are called *frequent* (*large*).
2. Generate association rules from frequent itemsets. Let X be a frequent itemset and $\emptyset \neq Y \subset X$. Then any rule $X \setminus Y \Rightarrow Y$ holds if $(sup(X) / sup(X \setminus Y)) \geq minConf$.

Step (1) of our algorithms is based on the algorithm in [8], in that we use the lists of transaction identifiers. Step (2) is performed with the *ap-genrules* algorithm [2], except for computing confidence, which we show below. Further on, we will use the following auxiliary notions: *k-itemset* is a set of size k, a *regular itemset* is an itemset without missing values.

6.1 Generation of Frequent Itemsets under Incompleteness

First, let us remind briefly the main idea of the *gen_large_itemsets* algorithm (see Fig. 3) computing frequent itemsets [8]. Then, we will offer modifications of this algorithm allowing to compute frequent itemsets for given threshold of supports: pessimistic, optimistic, and under incompleteness as defined in Example 2. The following notation is used in the *gen_large_itemsets* function:

- F_k - set of frequent k-itemsets;
- $f[1] \bullet f[2] \bullet ... \bullet f[k]$ – k-itemset consisting from items $f[1], f[2], ... , f[k]$;
- *tIdList* – a list of transaction identifiers;

Associated with each itemset is a *support* field to store the support for this itemset and *tIdList* to store identifiers of transactions containing the itemset.

```
function gen_large_itemsets(D: database);
1)   compute the family F₁ of frequent 1-itemsets and their tIdLists;
2)   for (k = 2; Fk ≠ ∅; k++) do {
3)      forall f₁ ∈ Fk-1 do {
4)         forall f₂ ∈ Fk-1 do {
5)            if f₁[1]=f₂[1] ∧...∧ f₁[k-2]=f₂[k-2] ∧ f₁[k-1]<f₂[k-1] then {
6)               c = f₁[1] • f₁[3] • ... • f₁[k-1] • f₂[k-1];
7)               if c has k subsets in Fk-1 then {
8)                  c.tIdList = f₁.tIdList ∩ f₂.tIdList;
9)                  c.support = |c.tIdList|;
10)                 if c.support > minSup then Fk = Fk ∪ {c}; }}}}
11)  return ∪k Fk;
```

Fig. 3. Function *gen_large_itemsets*

The *gen_large_itemsets* function reads the database only once in order to create lists of transaction identifiers for each item occurring in the database. F_1 is assigned those 1-itemsets that have support not less than *minSup*. Next, each k-th iteration, $k \geq 2$, generates candidate k-itemsets from the pairs of frequent $(k-1)$-itemsets (see [8] for details). To avoid unnecessary computations of *tIdLists*, the candidate k-itemsets that do not have some subset in frequent $(k-1)$-itemsets are pruned. The *tIdList* for each remaining candidate c is computed as the intersection of *tIdLists* of $(k-1)$-itemsets that were used for constructing c. Its length determines the support for c. The k-itemsets with support greater than *minSup* are included into F_k.

Now, we will present necessary modification of *gen_large_itemsets* for computing frequent itemsets in the case when the database is incomplete.

The Generate-P-FrequentItemsets Function

It computes all frequent itemsets X such that $pSup(X) > minSup$. It differs from *gen_large_itemsets* only in line 1 that initializes 1-itemsets. The appropriate code corresponding to line 1 of *gen_large_itemsets* is as follows:

```
compute frequent regular 1-itemsets F₁ and their tIdLists, where tIdList for
each c in F₁ is the list of identifiers of transactions that certainly contain c.
```

The Generate-O-FrequentItemsets Function

It computes all frequent itemsets X such that $oSup(X) > minSup$. It differs from *gen_large_itemsets* only in line 1 that initializes 1-itemsets. The appropriate code corresponding to line 1 of *gen_large_itemsets* is as follows:

```
compute frequent regular 1-itemsets F₁ and their tIdLists, where tIdList for
each c in F₁ is the list of identifiers of transactions that possibly contain c.
```

The Generate-I-FrequentItemsets Function

It computes all frequent itemsets X such that $probSup(X) > minSup$. The function uses a modified *tIdList* structure:

Let c be a candidate k-itemset. The elements of *c.tIdList* will be pairs (*tId*, *iVec*) where, *tId* is the transaction identifier and *iVec* is a Boolean vector of size k that for each item $c[j] \in c$, $j=1, ..., k$, indicates if the transaction identified by *tId* contains the item certainly. If so, *iVec[j]* is assigned 1, otherwise it is equal 0. In addition, 1-itemsets have a component μ that stores information on frequencies of the items.

The function differs from *gen_large_itemsets* in line 1 (that initializes 1-itemsets) and in lines 8-9 (that compute *tIdList* and *support*, respectively). The appropriate code corresponding to line 1 of *gen_large_itemsets* is as follows:

```
compute frequent regular 1-itemsets F₁ as well as their frequency μ and tIdLists
keeping information on transactions that possibly contain c;
```

The code corresponding to lines 8-9 should be as follows:

```
forall t₁ ∈ f₁.tIdList do
    forall t₂ ∈ f₂.tIdList do
        if t₁.tId = t₂.tId then
            add (t₁.tId, t₁.iVec • t₂.iVec[k-1]) to c.tIdList;
c.support = 0;
forall t ∈ c.tIdList do {
    probSubₜ = 0;
    for j=1 to k do {
        probSubₜ = probSubₜ * max(iVec[j], μ(c[j])); };
    c.support = c.support + probSubₜ; };
```

The expression $max(iVec[j], \mu(c[j]))$ returns 1 if the j-th item of c is contained by the respective transaction certainly, otherwise it returns the frequency of this item.

6.2 Computing Confidence of Association Rules under Incompleteness

The problem of finding AR_{pp}, AR_{oo} and AR_{ii} consists in computing *pConf*, *oConf* and *probConf* for candidate rules, respectively.

Computing AR_{pp}

Rules AR_{pp} are computed from frequent itemsets F generated by *Generate-P-FrequentItemsets*. Let $Z \in F_k$, $k \geq 2$, and $X \cup Y = Z$. Then $X \Rightarrow Y \in AR_{pp}$ if $pConf(X \Rightarrow Y) \geq minConf$. In order to compute $pConf(X \Rightarrow Y)$, we have to know support of both $n(X) \cap n(Y)$ and $m(X) \cap m(-Y)$ (see Eq. (1)). Clearly, $|n(X) \cap n(Y)| = |n(X \cup Y)| = pSup(Z)$, which was computed when looking for frequent itemsets. Still, $|m(X) \cap m(-Y)|$ (equal to $|m(X) \setminus n(Y)|$) must be computed. Assume, the computation of AR_{pp} was preceded by generating *nTIdLists* and *mTIdLists* for each $f \in F_1$. These lists consist of identifiers of transactions containing f certainly and possibly, respectively. The *nTIdList* and *mTIdList* of X and Y, can be computed by intersecting *nTIdLists* and *mTIdLists* of all items in these itemsets, respectively. Then, $pConf(X \Rightarrow Y)$ can be computed as follows:

```
X.mTIdList = X[1].mTIdList ∩ X[2].mTIdList ∩ ... ∩ X[|X|].mTIdList;
Y.nTIdList = Y[1].nTIdList ∩ Y[2].nTIdList ∩ ... ∩ Y[|Y|].nTIdList;
pConf = sup(Z) / [sup(Z) + |X.mTIdList \ Y.nTIdList|];
```

Computing AR_{oo}

Rules AR_{oo} are computed from frequent itemsets F generated by *Generate-O-FrequentItemsets*. Let $Z \in F$ and $X \cup Y = Z$. Then $X \Rightarrow Y \in AR_{oo}$ if $oConf(X \Rightarrow Y) \geq minConf$. By analogy to *pConf*, $oConf(X \Rightarrow Y)$ will be computed as follows (see Eq. (2)):

```
X.nTIdList = X[1].nTIdList ∩ X[2].nTIdList ∩ ... ∩ X[|X|].nTIdList;
Y.mTIdList = Y[1].mTIdList ∩ Y[2].mTIdList ∩ ... ∩ Y[|Y|].mTIdList;
oConf = sup(Z) / [sup(Z) + |X.nTIdList \ Y.mTIdList|];
```

Computing AR_{ii}

Rules AR_{ii} are computed from frequent itemsets F generated by *Generate-I-FrequentItemsets*. Let $Z \in F$ and $X \cup Y = Z$. Then $X \Rightarrow Y \in AR_{ii}$ if $probConf(X \Rightarrow Y) \geq minConf$. Fortunately, *probConf* can be computed as usual *conf* of rules.

7 Conclusions

We investigated the notion of legitimate definition of support and confidence under incompleteness. It was shown that for incomplete datasets legitimate $AR_{\alpha\beta}$ can be derived from legitimate $RR_{\alpha\beta}$ by the cover operator. We also proved that $MMR_{\alpha\beta}$ under incompleteness constitute a subset of $RR_{\alpha\beta}$ of the same type. Algorithms for generating $AR_{\alpha\beta}$ under incompleteness were offered.

References

1. Agrawal, R., Imielinski, T., Swami, A.: Mining Associations Rules between Sets of Items in Large Databases. In: Proc. of the ACM SIGMOD Conference on Management of Data. Washington, D.C. (1993) 207-216
2. Agrawal, R., Mannila, H., Srikant, R., Toivonen, H., Verkamo A.I.: Fast Discovery of Association Rules. In: Fayyad, U.M., Piatetsky-Shapiro, G., Smyth, P., Uthurusamy, R. (eds.): Advances in Knowledge Discovery and Data Mining. AAAI, CA (1996) 307-328
3. Kryszkiewicz, M.: Representative Association Rules. In: Proc. of PAKDD '98. Melbourne, Australia. LNAI 1394. Springer-Verlag (1998) 198-209
4. Kryszkiewicz M.: Representative Association Rules and Minimum Condition Maximum Consequence Association Rules. In: Proc. of PKDD '98. Nantes, France. LNAI 1510. Springer-Verlag (1998) 361-369
5. Kryszkiewicz, M.: Association Rules in Incomplete Databases: In: Proc. of PAKDD '99. Beijing, China. LNAI 1574. Springer-Verlag (1999) 84-93
6. Kryszkiewicz, M.: Probabilistic Approach to Association Rules in Incomplete Databases, Proc. of WAIM '2000. Shanghai, Chiny. LNCS 1846. Springer-Verlag (2000), 133-138
7. Kryszkiewicz M., Rybi•ski H.: Incomplete Database Issues for Representative Association Rules. In: Proc. ISMIS '99. Warsaw, Poland. LNAI 1609. Springer (1999) 583-5916
8. Savasere, A, Omiecinski, E., Navathe, S.: An Efficient Algorithm for Mining Association Rules in Large Databases. In: Proc. of the 21st VLDB Conference. Zurich, Swizerland (1995) 432-444
9. Toivonen H., Klemettinen M., Ronkainen P., Hätönen K., and H. Mannila H.: Pruning and grouping discovered association rules. In: MLnet Workshop on Statistics, Machine Learning, and Discovery in Databases. Heraklion, Crete, Greece (1995) 47-52
10. Washio, T., Matsuura, H., Motoda, H.: Mining Association Rules for Estimation and Prediction. In: Proc. of PAKDD '98. Melbourne, Australia. LNAI 1394. Springer-Verlag (1998) 417-419

A Simple and Tractable Extension of Situation Calculus to Epistemic Logic

Robert Demolombe and Maria del Pilar Pozos Parra *

ONERA Toulouse, France

1 Introduction

The frame problem and the representation of knowledge change have deserved a lot of works. In particular, at the Cognitive Robotics Group, at Toronto, several researchers in the last ten years have produced quite interesting papers in a uniform logical framework based on Situation Calulus [Rei91, SL93, LR94, LL98]. In [Rei91] Reiter has proposed a simple solution to the frame problem. Scherl and Levesque in [SL93] have defined an extension to Epistemic Logic to represent knowledge dynamics in contexts where some actions may produce knowledge, like, for instance, sensing actions for a robot. This approach has been extended by Lakemeyer and Levesque in [LL98] to modal operators of the kind "I know and only know". Also, they have given a formal semantics and axiomatics, and they proved soundness and completeness of the axiomatics.

These extensions to Epistemic Logic offer a large expressive power. Indeed, there is no restriction on formulas in the scope of modal operators. However, they have lost the simplicity of the solution to the frame problem initially proposed in [Rei91], and the possibility to find a tractable implementation of these extensions is far to be obvious. As far as we know, at the present time there is no such implementation.

In this paper a simple extension to Epistemic Logic of Reiter's initial solution is presented that could easily be implemented. In exchange we have to accept strong restrictions on the expressive power of the epistemic part of the logical framework. However, we believe that for a large class of applications these restrictions are not real limitations. In the following intuitive ideas of the proposed solution are presented with a simple example. Then, we give the general logical framework, and, finally a comparison is made with the solutions that we have mentioned before.

2 The frame problem in the context of extended situation calculus: an example

Situation Calculus [McC68, Rei99] is a sort of classical first order logic where predicates may have an argument (the last argument) of a particular sort, which

---

* ONERA/CERT, 2 Avenue E. Belin B.P. 4025, 31055 Toulouse Cedex, France. e-mail: {demolomb,pozos}@cert.fr.

is called a "situation"; these predicates are called "fluents". This argument is intended to represent the sequence of actions which have been performed from the initial state to the current state. A situation is syntactically represented by a term of the form $do(a, s)$ where a denotes an action, and s denotes a situation. The initial situation is denoted by $S0$.

For instance, $position(x, s)$ represents the fact that a given object is at the position x in the situation s. Action variables and situation variables can be quantified. For instance, $\neg \exists s(position(2, s))$ represents the fact that in no situation a given object is at the position 2. Action quantification is an essential feature in the solution to the frame problem proposed by Reiter. Indeed, the fact that, for example, there is no other possibility to change the position of an object than to perform the action $move$ can be represented by: $\forall s \forall a \forall x (position(x, s) \land \neg position(x, do(a, s)) \rightarrow a = move)$. To intuitively present how the solution to the frame problem can be extended to epistemic logic, we use the following scenario.

Let's consider a simple robot that can move forward (action adv) or backward (action rev) along a railtrack. Performance of actions adv or rev changes his position of one distance unit. There may be obstacles on the railtrack, like branches of trees that have fallen. Suppose the robot is moving during the night and there is a pilot in the robot. The pilot can recognise obstacles, provided he has switched on a spotlight (action $obs.obstacle$), and the obstacle is not beyond the visibility distance d. The spotlight is not always on because it consumes battery ressources, which are limited. When the robot moves he computes his new position, and this position is indicated on a screen which can be seen by the pilot (action $inf.position(x)$). The pilot performs the action $inf.position(x)$ before the action $obs.obstacle$ in order to know his position and to determine the position of visible obstacles, if there are. The pilot can inform the robot about the existence of an obstacle at x (action $inf.obstacle(x)$), and the robot stops if he knows that there is an obstacle in a short distance sd.

We see that the description of this scenario involves evolution of the world and evolution of what the pilot and the robot believe [2]. We first show how the frame problem can be solved if we only consider evolution of the world.

For each fluent, two axioms define the positive effects or the negative effects of the actions. For instance, for the fluent $position(x, s)$, the effect of performing the action adv (respectively rev) when the robot is at the position $x-1$ (respectively $x+1$) in the situation s, is that it is at the position x in the situation $do(a, s)$ [3]:

(1) $(a = adv \land position(x - 1, s) \lor a = rev \land position(x + 1, s)) \rightarrow position(x, do(a, s))$

The negative effect axiom expresses that if the robot is at the position x in the situation s and he performs either the action adv or the action rev, then in the situation $do(a, s)$ he is no more at the position x:

---

[2] We have no room here to give a complete formal description of this scenario. Also, some assumptions are not perfectly realistic, but we mainly want to show how such scenarios can be formalised.

[3] All the variables are implicitly universally quantified.

(2) $(a = adv \lor a = rev) \land position(x, s) \rightarrow \neg position(x, do(a, s))$

One of the most important features to solve the frame problem in the approach presented in [Rei99] is the "causal completeness assumption". This assumption expresses that the positive effect axioms and the negative effect axioms "characterize **all** the conditions underwhich action a can cause the fluent *position* to become true (respectively false) in the successor situation". If, in addition to (1) and (2), we accept this assumption, then we have (see axiom (G2) for the general form):

(3) $position(x, do(a, s)) \leftrightarrow [a = adv \land position(x-1, s) \lor a = rev \land position(x+1, s)] \lor position(x, s) \land \neg[(a = adv \lor a = rev) \land position(x, s)]$

This axiom defines the objective representation of the evolution of the world. If we want to define the subjective representation of the evolution of the world, we can extend the language with epistemic modal operators. For that purpose, we introduce modal operators like B_r, such that $B_r \phi$ is intended to mean that the robot r believes that ϕ holds in the present situation. To represent, in a similar approach, the evolution of what the robot believes, we have to consider four effect axioms for each fluent. For example, for the fluent $position(x, s)$, there are four distinct possible attitudes of the robot which are formally represented by: $B_r position(x, s)$, $\neg B_r position(x, s)$, $B_r \neg position(x, s)$ and $\neg B_r \neg position(x, s)$. The corresponding axioms (4), (5), (6) and (7) are given below.

The effect of performing action adv (respectively rev) when the robot believes that he is at the position $x - 1$ (respectively $x + 1$) in the situation s is that he believes that he is at the position x in the situation $do(a, s)$:

(4) $(a = adv \land B_r position(x - 1, s) \lor a = rev \land B_r position(x + 1, s)) \rightarrow B_r position(x, do(a, s))$

The effect of performing either action adv or rev when the robot believes that he is at the position x in the situation s is that he does not believe that he is at the position x in the situation $do(a, s)$:

(5) $(a = adv \lor a = rev) \land B_r position(x, s) \rightarrow \neg B_r position(x, do(a, s))$

We have two similar axioms to define the attitude of the robot with respect to the fact that he believes that he is not at the position x in the situation $do(a, s)$:

(6) $(a = adv \lor a = rev) \land B_r position(x, s) \rightarrow B_r \neg position(x, do(a, s))$

(7) $(a = adv \land B_r position(x - 1, s) \lor a = rev \land B_r position(x + 1, s)) \rightarrow \neg B_r \neg position(x, do(a, s))$

If we extend the causal completeness assumptions to the robot's beliefs, we get, after some simplifications, the two axioms (8) and (9) (see axioms (G3) and (G4) for the general form):

(8) $B_r position(x, do(a, s)) \leftrightarrow [a = adv \land B_r position(x - 1, s) \lor a = rev \land B_r position(x+1, s)] \lor B_r position(x, s) \land \neg[(a = adv \lor a = rev) \land B_r position(x, s)]$

(9) $B_r \neg position(x, do(a, s)) \leftrightarrow [(a = adv \lor a = rev) \land B_r position(x, s)] \lor B_r \neg position(x, s) \land \neg[a = adv \land B_r position(x - 1, s) \lor a = rev \land B_r position(x + 1, s)]$

Notice that in the definition of these axioms we have implicitly assumed that if the robot performs either the action adv or the action rev, he believes that he has performed these actions. However, if some action is performed by the pilot, like the action $obs.obstacle$, the robot is not necessarily informed about this fact.

It is interesting to see with this example how the pilot's beliefs and the robot's beliefs about the fluent $obstacle$ may evolve in two different way. We have the following effect axioms (10), (11), (12) and (13) for this fluent.

If the pilot has switched on the spot light, and there is an obstacle at some position x which is visible by the pilot, then the pilot believes that there is an obstacle at x [4]:

(10) $a = obs.obstacle \land obstacle(x, s) \land position(y, s) \land y \leq x \leq y + d \rightarrow B_p obstacle(x, do(a, s))$

If the pilot has switched on the spot light, and there is no obstacle at some position x which is visible by the pilot, then the pilot does not believe that there is an obstacle at x:

(11) $a = obs.obstacle \land \neg obstacle(x, s) \land position(y, s) \land y \leq x \leq y + d \rightarrow \neg B_p obstacle(x, do(a, s))$

We have two similar effect axioms for $\neg obstacle(x, do(a, s))$.

(12) $a = obs.obstacle \land \neg obstacle(x, s) \land position(y, s) \land y \leq x \leq y + d \rightarrow B_p \neg obstacle(x, do(a, s))$

(13) $a = obs.obstacle \land obstacle(x, s) \land position(y, s) \land y \leq x \leq y + d \rightarrow \neg B_p \neg obstacle(x, do(a, s))$

Then, from the causal completion assumption we have the axioms (14) and (15).

(14) $B_p obstacle(x, do(a, s)) \leftrightarrow [a = obs.obstacle \land obstacle(x, s) \land position(y, s) \land y \leq x \leq y + d] \lor B_p obstacle(x, s) \land \neg[a = obs.obstacle \land \neg obstacle(x, s) \land position(y, s) \land y \leq x \leq y + d]$

(15) $B_p \neg obstacle(x, do(a, s)) \leftrightarrow [a = obs.obstacle \land \neg obstacle(x, s) \land position(y, s) \land y \leq x \leq y + d] \lor B_p \neg obstacle(x, s) \land \neg[a = obs.obstacle \land obstacle(x, s) \land position(y, s) \land y \leq x \leq y + d]$

If the only way for the robot to be informed about the fact that there is an obstacle at x is to perform the action $inf.obstacle(x)$, then we have the axioms (16) and (17) below.

(16) $B_r obstacle(x, do(a, s)) \leftrightarrow a = inf.obstacle(x) \lor B_r obstacle(x, s)$

(17) $B_r \neg obstacle(x, do(a, s)) \leftrightarrow B_r \neg obstacle(x, s) \land \neg(a = inf.obstacle(x))$

Let's assume that in the initial situation $S0$ the pilot and the robot both ignore whether there are obstacles in any places. This is formally represented by: $\neg \exists x B_r obstacle(x, S0)$, $\neg \exists x B_r \neg obstacle(x, S0)$, $\neg \exists x B_p obstacle(x, S0)$ and $\neg \exists x B_p \neg obstacle(x, S0)$. If in the situation $S0$ there is an obstacle at the position 3, the pilot and the robot have wrong beliefs. If the distance d is equal to 10, after performance of the action $a_1 = obs.obstacle$, the pilot believes that there is an obstacle at the position 3, while the robot ignores that there this an obstacle at the position 3, i.e. we have: $B_p obstacle(3, do(a_1, S0))$ and $\neg B_r obstacle(3, do(a_1, S0))$. Finally, if after action a_1 the pilot performs the action $a_2 = inf.obstacle(3)$,

---

[4] As a matter of simplification, it is assumed here that the pilot only looks at obstacles that are foreward.

the robot and the pilot have the same beliefs about this obstacle. We have: $B_p obstacle(3, do(a_2, do(a_1, S0)))$ and $B_r obstacle(3, do(a_2, do(a_1, S0)))$.

In fact these actions can be performed only if some preconditions are satisfied. These preconditions are expressed with a particular predicate $Poss$ (see [Rei99]). The formula $Poss(a, s)$ means that in the situation s it is possible to perform the action a. For example, a precondition to perform the action adv is that the robot does not believe that there is an obstacle in a short distance sd, and there is no obstacle in front of him.

(18) $Poss(adv, s) \leftrightarrow \neg \exists x \exists y (B_r position(x, s) \wedge B_r obstacle(y, s) \wedge y - x \leq sd) \wedge \neg \exists x \exists y (position(x, s) \wedge obstacle(y, s) \wedge y = x + 1)$

3 General framework

Now we present the general framework of the extended Situation Calculus. Let L be a first order language with equality with the constant symbol $S0$, the function symbol do, and the predicate symbol $Poss$. Let L_M be an extension of language L with modal operators denoted by $B_1, \ldots, B_i, \ldots$, where modal operators can only occur in modal literals. Modal literals are of the form $B_i l$, where l is a literal of L, and l is **not formed with equality** predicate. Let's consider the theory T which contains the following axioms.

Action precondition axioms.
For each action a there is in T an axiom of the form [5]:
(G1) $Poss(a, s) \leftrightarrow \pi_a(s)$
where π_a is a formula in L_M.

Successor state axioms.
For each fluent F there is in T an axiom of the form:
(G2) $F(do(a, s)) \leftrightarrow \Gamma_F^+(a, s) \vee F(s) \wedge \neg \Gamma_F^-(a, s)$
where Γ_F^+ and Γ_F^- are formulas in L.

Successor belief state axioms.
For each modal operator B_i and each fluent F [6], there are in T two axioms of the form:
(G3) $B_i(F(do(a, s))) \leftrightarrow \Gamma_{i_1, F}^+(a, s) \vee B_i(F(s)) \wedge \neg \Gamma_{i_1, F}^-(a, s)$
(G4) $B_i(\neg F(do(a, s))) \leftrightarrow \Gamma_{i_2, F}^+(a, s) \vee B_i(\neg F(s)) \wedge \neg \Gamma_{i_2, F}^-(a, s)$
where $\Gamma_{i_1, F}^+$, $\Gamma_{i_1, F}^-$, $\Gamma_{i_2, F}^+$, and $\Gamma_{i_2, F}^-$ are formulas in L_M.

We also have in T unique name axioms for actions and for situations, and we assume that modal operators obey the (KD) logic (see [Che88]).

---

[5] As a matter of simplification the arguments of function symbols are not explicited, and, for fluents, the only argument which is explicited is the situation. For instance, we could have $a(x_1)$ and $F(x_1, x_2, s)$. Also, it is assumed that all the free variables are universally quantified.

[6] To avoid to have equality in the scope of modal operators, we assume that fluent functions are expressed via fluent predicates, i.e. $y = f(x, s)$ is expressed by $F(y, x, s)$.

Moreover, it is assumed that for each fluent F we have [7]:
(H1) $\vdash T \to \forall\neg(\Gamma_F^+ \wedge \Gamma_F^-)$
(H2) $\vdash T \to \forall\neg(\Gamma_{i_1,F}^+ \wedge \Gamma_{i_1,F}^-)$
(H3) $\vdash T \to \forall\neg(\Gamma_{i_2,F}^+ \wedge \Gamma_{i_2,F}^-)$
(H4) $\vdash T \to \forall\neg(\Gamma_{i_1,F}^+ \wedge \Gamma_{i_2,F}^+)$
(H5) $\vdash T \to \forall(B_i(F(s)) \wedge \Gamma_{i_2,F}^+ \to \Gamma_{i_1,F}^-)$
(H6) $\vdash T \to \forall(B_i(\neg F(s)) \wedge \Gamma_{i_1,F}^+ \to \Gamma_{i_2,F}^-)$

The three assumptions (H4), (H5) and (H6) guarantee that if agents' beliefs are consistent in the intial state, they are consistent in all the successor states.

It can easily be shown that, if we have (H1), in the context of the theory T, successor state axioms like (G2) are equivalent to the conjunction of properties (A1), (A2), (A3), and (A4).

(A1) $\Gamma_F^+(a,s) \to F(do(a,s))$
(A2) $\Gamma_F^-(a,s) \to \neg F(do(a,s))$
(A3) $\neg\Gamma_F^-(a,s) \to [F(s) \to F(do(a,s))]$
(A4) $\neg\Gamma_F^+(a,s) \to [\neg F(s) \to \neg F(do(a,s))]$

In a similar way we have shown that, if we have (H2) and (H3), in the context of the theory T, successor belief state axioms of the form (G3) (resp. (G4)) are equivalent to the conjunction of properties (B1), (B2), (B3) and (B4) (resp. (C1), (C2), (C3) and (C4)).

(B1) $\Gamma_{i_1,F}^+(a,s) \to B_i(F(do(a,s)))$
(B2) $\Gamma_{i_1,F}^-(a,s) \to \neg B_i(F(do(a,s)))$
(B3) $\neg\Gamma_{i_1,F}^-(a,s) \to [B_i(F(s)) \to B_i(F(do(a,s)))]$
(B4) $\neg\Gamma_{i_1,F}^+(a,s) \to [\neg B_i(F(s)) \to \neg B_i(F(do(a,s)))]$
(C1) $\Gamma_{i_2,F}^+(a,s) \to B_i(\neg F(do(a,s)))$
(C2) $\Gamma_{i_2,F}^-(a,s) \to \neg B_i(\neg F(do(a,s)))$
(C3) $\neg\Gamma_{i_2,F}^-(a,s) \to [B_i(\neg F(s)) \to B_i(\neg F(do(a,s)))]$
(C4) $\neg\Gamma_{i_2,F}^+(a,s) \to [\neg B_i(\neg F(s)) \to \neg B_i(\neg F(do(a,s)))]$

Properties (B3) and (C3) show that positive beliefs remain unchanged after performance of an action as long as $\neg\Gamma_{i_1,F}^-(a,s)$ and $\neg\Gamma_{i_2,F}^-(a,s)$ holds. Properties (B4) and (C4) show that negative beliefs remain unchanged after performance of an action as long as $\neg\Gamma_{i_1,F}^+(a,s)$ and $\neg\Gamma_{i_2,F}^+(a,s)$ holds.

Definition 1. Regression operator. We define a regression operator R_T from formulas in L_M to formulas in L_M.

1. When W is a non fluent atom, including equality atoms, or when W is a fluent atom whose situation argument is the constant $S0$, $R_T[W] = W$.
2. When W is an atom formed with fluent F of the form $F(\mathbf{t}, do(\alpha, \sigma))$ whose successor state axiom in T is [8] $\forall a \forall s \forall \mathbf{x}[F(\mathbf{x}, do(a,s)) \leftrightarrow \Phi_F]$ then:

---

[7] Here, we use the symbol $\forall$ to denote the universal closure of all the free variables in the scope of $\forall$.
[8] We use the notation $\mathbf{x}$ for the tuple of variables $x_1, \ldots, x_n$, and $\forall\mathbf{x}$ for $\forall x_1 \ldots \forall x_n$; $\Phi_F.\{\mathbf{x}/\mathbf{t}, a/\alpha, s/\sigma\}$ denotes the result of the application of the substitution $\{\mathbf{x}/\mathbf{t}, a/\alpha, s/\sigma\}$ to formula Φ_F.

$R_T[F(\mathbf{t}, do(\alpha, \sigma))] = R_T[\Phi_F.\{\mathbf{x}/\mathbf{t}, a/\alpha, s/\sigma\}]$

3. When W is an atom of the form $Poss(\alpha(\mathbf{t}), \sigma)$, whose action precondition axiom is $\forall s \forall \mathbf{x} Poss(\alpha(\mathbf{x}), \sigma) \leftrightarrow \Pi_\alpha(\mathbf{x}, s)$ then:
$R_T[Poss(\alpha(\mathbf{t}), s)] = R_T[\Pi_\alpha(\mathbf{x}, s).\{\mathbf{x}/\mathbf{t}, s/\sigma\}]$

4. When W is a modal literal of the form $B_i(F(\mathbf{t}, do(\alpha, \sigma)))$ or $B_i(\neg F(\mathbf{t}, do(\alpha, \sigma)))$ whose successor belief state axioms in T are $\forall a \forall s \forall \mathbf{x}[B_i(F(\mathbf{x}, do(a, s))) \leftrightarrow \Phi_{i_1, F}]$ and $\forall a \forall s \forall \mathbf{x}[B_i(\neg F(\mathbf{x}, do(a, s))) \leftrightarrow \Phi_{i_2, F}]$ then:
$R_T[B_i(F(\mathbf{t}, do(\alpha, \sigma)))] = R_T[\Phi_{i_1, F}.\{\mathbf{x}/\mathbf{t}, a/\alpha, s/\sigma\}]$ and
$R_T[B_i(\neg F(\mathbf{t}, do(\alpha, \sigma)))] = R_T[\Phi_{i_2, F}.\{\mathbf{x}/\mathbf{t}, a/\alpha, s/\sigma\}]$

5. When W is a formula in L_M [9], $R_T[\neg W] = \neg R_T[W]$ and $R_T[\exists x W] = \exists x R_T[W]$.

6. When W_1 and W_2 are formulas in L_M, $R_T[W_1 \vee W_2] = R_T[W_1] \vee R_T[W_2]$.

Theorem 2. *Let T_0 be a set of closed sentences in L_M, without $Poss$ predicate, and whose situation argument in fluents is $S0$. Let T_{ss} be the set of precondition axioms and of successor axioms for the fluents of a given application. Let T_u be the set of unique name axioms. We use notations $T = T_u \cup T_{ss} \cup T_0$ and $T' = T_u \cup T_0$. Let $R_T^*(\phi)$ be the result of repeated applications of R_T until the result is unchanged. Let s_{gr} be a ground situation term.*
We have $\vdash T \rightarrow W(s_{gr})$ *iff* $\vdash T' \rightarrow R_T^*[W(s_{gr})]$

For the proof we can use the same technique as Scherl and Levesque in [SL93], but the proof is much more simple because we do not have explicit accessibility relation to represent modal operators (see next section).

This theorem shows that to prove W in situation s_{gr} comes to prove $R_T^*[W]$ in situation $S0$, droping axioms of the kind (G1), (G2), (G3) and (G4).

Theorem 2 can be used for different purposes. The most important of them, as mentioned by Reiter in [Rei99], is to check whether a given sequence of actions is executable, in the sense that after performing any of these actions, the preconditions to perform the next action are satisfied. Another one, is to check whether some property holds after performance of a given sequence of actions. These two features are essential for plan generation.

We also have the following theorem.

Theorem 3. *Let A be a formula of the form F, $B_i F$ or $B_i \neg F$, where F is an atom formed with a fluent predicate. Let T be a theory such that for every successor axiom of the form: $A(\mathbf{x}, do(a, s)) \leftrightarrow \Gamma_A^+(\mathbf{x}, a, s) \vee A(\mathbf{x}, s) \wedge \neg \Gamma_A^-(\mathbf{x}, a, s)$, there is no other variable that occurs in Γ_A^+ or Γ_A^- than the variables in $\mathbf{x}$, or a or s. Let $\phi(s)$ be a formula in L_M such that the only variable that occurs in ϕ is s.*
If for every ground formula $A(S0)$ we have either $\vdash T \rightarrow A(S0)$ or $\vdash T \rightarrow \neg A(S0)$, then, for every ground situation term s_{gr}, which is a successor situation of $S0$, we have either $\vdash T \rightarrow \phi(s_{gr})$ or $\vdash T \rightarrow \neg \phi(s_{gr})$.

---
[9] The definition of R_T for universal quantifier $\forall$, conjunction $\wedge$, implication $\rightarrow$ and equivalence $\leftrightarrow$, is directly obtained from the usual definitions of these quantifier and logical connectives in function of $\exists$, $\neg$ and $\vee$.

The proof is by induction on the complexity of the formula ϕ in $S0$, and by induction on the depth of the term s_{gr}. Theorem 3 intuitively says that if we have a complete description of what the agents believe in $S0$, then we have a complete description of their beliefs in every successor situation.

4 Related works

In [SL93] Scherl and Levesque have defined an extension of Situation Calculus to Epistemic Logic for a unique modal operator, but without any restriction about formulas that are in the scope of the modal operator.

In their approach, the first idea is to define the modal operator $Knows$ in terms of an accessibility relation K which is explicitly represented in the axiomatics. Formally, they have: $Knows(\phi, s) \stackrel{def}{=} \forall s'(K(s', s) \to \phi(s'))$. The second idea is to define knowledge change by defining accessibility relation change. Moreover, two kinds of actions are distinguished: knowledge-producing actions, denoted by $\alpha_1, \ldots, \alpha_n$, and non-knowledge-producing actions. Each action α_i informs the agent in the situation $do(\alpha_i, s)$ about the fact that some formula p_i is true or false in the situation s. It is assumed that the action α_i does not change the state of the world. From a technical point of view, after the performance of action α_i, relation K selects, in the situation $do(\alpha_i, s)$, those situations where p_i has the same truth value as it has in the situation s. For instance, if p_i is true in s, then situations s', which are accessible from s and where p_i is false, are no more accessible from $do(\alpha_i, s)$. Notice that if p_i is false in all the situations which are accessible from s, there is no situation accessible from $do(\alpha_i, s)$. That means that the agent believes any formula.

This problem disappear in the logical framework presented by Shapiro et al. in [SPL00], where a plausibility degree $pl(s)$ is assigned to all the situations. From the accessibility relation $B(s', s)$, an accessibility relation $B_{max}(s', s)$ between s and the most plausible situations can be defined by $B_{max}(s', s) \stackrel{def}{=} B(s', s) \land \forall s''(B(s'', s) \to pl(s') \leq pl(s''))$. Then, the fact that an agent believes ϕ in s is defined as $Bel(\phi, s) \stackrel{def}{=} \forall s'(B_{max}(s', s) \to \phi(s'))$. Here, an agent can consistently believe ϕ in $do(a, s)$, while he believed $\neg \phi$ in s, provided there exists at least one most plausible situation related to $do(a, s)$ where ϕ holds.

For a non-knowledge-producing action a, it is assumed that knowledge changes in the same way as the world does. That is, if a situation s' is accessible from s, the situation $do(a, s')$ is accessible from $do(a, s)$ as well. In formal terms, the evolution of relation K is defined by the following axiom [10]:
$Poss(a, s) \to$
$[K(s'', do(a, s)) \leftrightarrow \exists s' K(s', s) \land s'' = do(a, s') \land Poss(a, s') \land$
$((\neg(a = \alpha_1) \land \ldots \land \neg(a = \alpha_n)) \lor$
$a = \alpha_1 \land (p_1(s) \leftrightarrow p_1(s')) \lor$
$\ldots$
$a = \alpha_n \land (p_n(s) \leftrightarrow p_n(s')))]$

---
[10] In fact, condition $Poss(a, s')$ is not present in [SL93], it was added in [LL98].

This successor axiom does not explicitly define which formulas are true or false in $do(a, s')$. From the examples presented in their paper we understand that the truth value of formulas in situations like s'' is defined by the successor state axioms of the type (G2). That implicitly means that: i) whenever some action has been performed the agent knows that this action has been performed, ii) the agents knows the effects of all the actions, i.e he knows all the successor state axioms, and iii) when the agent get information through a knowledge-producing action, this information is always true information, in the sense that this information is true in the situation s where he is.

How this formalisation could be extended to the context of multi-agents? The fact i) cannot be accepted in general. We can accept that an agent knows that an action has been performed when it has been performed by himself, but not necessarily when it has been performed by another agent. This problem could be solved by defining as many accessibility relations K_i as there are distinct agents, and by distinguishing for each agent those actions $\beta_1, \ldots, \beta_m$ which are performed by the agent i. For an action a which is neither of the sort β_j nor α_k, the fact that knowledge does not change can be represented by the fact that accessible situations from $do(a, s)$ are the same as accessible situations from s. That could lead to successor axioms for each relation K_i of the form:

$Poss(a, s) \rightarrow$
$[K_i(s'', do(a, s)) \leftrightarrow$
$(K_i(s'', s) \wedge \neg(a = \alpha_1) \wedge \ldots \wedge \neg(a = \alpha_n) \wedge \neg(a = \beta_1) \wedge \ldots \wedge \neg(a = \beta_m)) \vee$
$(\exists s' K_i(s', s) \wedge s'' = do(a, s') \wedge Poss(a, s') \wedge$
$(a = \beta_1 \vee \ldots \vee a = \beta_m \vee$
$a = \alpha_1 \wedge (p_1(s) \leftrightarrow p_1(s')) \vee$
$\ldots$
$a = \alpha_n \wedge (p_n(s) \leftrightarrow p_n(s')))]$

However, even with this extension there are still the problems related to ii) and iii). For ii), the problems is that in real situations agents may have wrong beliefs about the the evolution of the world. For instance, an agent may believe that droping a fragile object make it broken, while another agent may believe that the object is not necessarily broken, depending on his weight or on other particular circumstances. This raises the question of how to represent in this framework different evolutions of the world in the context of different agents beliefs? May be a possible answer is to have different successor state axioms, for the same fluent, to represent the "true" evolution of the world, and to represent the evolution of the world in the context of each agent's beliefs. That is, more or less, the idea we have proposed in this paper with the axioms of the type (G3) and (G4).

For iii), the problem is that there are applications where agents may receive information from different sensors, or from other agents, some of them are not necessarily reliable and may communicate wrong information. Here again, we believe that axioms like (G3) and (G4) are a possible solution, because they allow us to represent communication actions whose consequences are to generate wrong agents beliefs.

5 Conclusion

In conclusion, we have presented a general framework to solve the frame problem in the context of a limited extension of Situation Calculus to Epistemic logic. Even if for this solution strong restrictions are imposed on the language L_M, we can express non trivial properties like: $\forall s \forall x (B_r position(x, s) \rightarrow position(x, s))$, which means that in every situation the robot has true beliefs about his position, or $\forall s \forall x (B_r obstacle(x, s) \rightarrow B_p obstacle(x, s))$, which means that the robot's beliefs about obstacles are a subset of the pilot's beliefs about obstacles. Also, since in the (KD) logics we have $B_i(l \wedge l') \leftrightarrow B_i l \wedge B_i l'$, it would be a trivial extension to L_M to allow conjunction of literals in the scope of modal operators. Finally, the regression operator R_T allows us to check whether these kinds of properties can be derived from T_0. The implementation of a modal theorem prover for the restricted language L_M should not be a big issue. We are currently working on this aspect.

References

[Che88] B. F. Chellas. *Modal Logic: An introduction*. Cambridge University Press, 1988.

[LL98] G. Lakemeyer and H. Levesque. AOL: a logic of acting, sensing, knowing and only knowing. In *Proc. of the 6th Int. Conf. on Principles of Knowledge Representation and Reasoning*. 1998.

[LR94] F. Lin and R. Reiter. State constraints revisited. *Journal of Logic and Computation*, 4:655–678, 1994.

[McC68] J. McCarthy. Programs with Common Sense. In M. Minski, editor, *Semantic Information Processing*. The MIT press, 1968.

[Rei91] R. Reiter. The frame problem in the situation calculus: a simple solution (sometimes) and a completeness result for goal regression. In V. Lifschitz, editor, *Artificial Intelligence and Mathematical Theory of Computation: Papers in Honor of John McCarthy*, pages 359–380. Academic Press, 1991.

[Rei99] R. Reiter. Knowledge in Action: Logical Foundations for Describing and Implementing Dynamical Systems. Technical report, University of Toronto, 1999.

[SL93] R. Scherl and H. Levesque. The Frame Problem and Knowledge Producing Actions. In *Proc. of the National Conference of Artificial Intelligence*. AAAI Press, 1993.

[SPL00] S. Shapiro and M. Pagnucco and Y. Lespérance and H. Levesque. Iterated Belief Change in the Situation Calculus. In *Proc. of the 7th Conference on Principle on Knowledge Representation and Reasoning (KR2000)*. Morgan Kaufman Publishers, 2000.

Rule Based Abduction

Sai Kiran Lakkaraju and Yan Zhang

School of Computing and Information Technology
University of Western Sydney, Nepean
Kingswood, NSW 2747, Australia
E-mail: {slakkara,yan}@cit.nepean.uws.edu.au

Abstract. This paper introduces a procedural approach to perform rule based abduction in a knowledge base. In this context a knowledge base is realised as a normal abductive logic program, and an observation can be either a literal or a rule. A SLDNF resolution based proof procedure is employed to achieve this rule based abduction. It is shown that using this algorithm, one can always find a minimal explanation for the observation if there exists such an explanation.
Key words: Abduction, knowledge representation, nonmonotonic reasoning.

1 Introduction

Abduction plays a vital role in commonsense reasoning, knowledge representation, and database update. Basically, if the set of abducibles is constructed not only with the facts/rules from the knowledge base but also from the beliefs of other agents regarding the observations, then there is always a chance of making the knowledge base more consistent and flexible. In nonmonotonic reasoning a knowledge base always changes when ever a new observation is provided. In such a situation there are three possible effects on the existing knowledge base.

1. The observation is already deducible from the knowledge base, i.e. the observation can be explained with the help of the existing knowledge.
2. A part of the knowledge base with the new observation is able to derive the other part of the knowledge base. if so the size of the existing knowledge base can be reduced.
3. Some facts/rules must be added to the existing knowledge base to explain the observation.

An observation can be a *fact*, *rule* or a *program* . By finding the explanation to the observation and by adding it to the knowledge base we eventually change the knowledge base from an old state to a new state.
In this paper we mainly concentrate on such situations where observations are rules. When an observation is a rule, we propose that the body part of the rule must be explained first to form a new knowledge base which intern explains the head of the rule. Then the union of the explanations gives the complete explanation for the observation (the rule).

We will provide a SLDNF resolution approach, to update the knowledge base when an observation is either a literal or a rule. The paper is organized as follows. In the next section we introduce basic definitions and concepts about abductive logic programs. In section 3 we outline the basic approach for our rule based abduction through different examples. In section 4, based on the basic idea presented in section 3, we formalize the procedures of abduction. Finally, in section 5 we discuss related work and conclude the paper.

2 Definitions and Concepts

We first briefly explain the SLDNF proof procedure that we will use throughout this paper. The *linear resolution with selection function* (or SL-resolution) is a restricted form of linear resolution. The main restriction is effected by a selection function which chooses from each clause one single literal to be resolved upon in that clause. SL-resolution operates on chains rather than clauses and hence strictly is not a form of resolution. It does, however employ ideas of unification and resolution. SLD resolution (SL resolution for definite clauses) is described as follows. Let P be a definite program and G be a goal. An unrestricted SLD derivation of $\mathcal{P} \cup \mathcal{G}$ consists of a sequence $G_0 = G, G_1, \cdots$ of goals, a sequence $C_1, C_2, \cdots$ of variants of clauses in program $\mathcal{P}$ (called the input clauses of the derivation), and a sequence $\emptyset_{(1)}, \emptyset_{(2)}, \cdots$ of substitutions. Each non-empty goal G_i contains one atom, which is selected atom of G_i. The clause G_{i+1} is said to be *derived* from G_i and C_i with substitutions $\emptyset_i$ and is carried out as follows. Suppose G_i is $\leftarrow A_1, \cdots, A_k, \cdots, A_m$ and A_k is the selected atom. Let $C_i = A \leftarrow B_1, \cdots, B_n$ by any clause in P such that A and A_k are unifiable with any unifier $\emptyset$. Then G_{i+1} is $\leftarrow (A_1, \cdots, A_{k-1}, B_1, \cdots, B_n, A_{k+1}, \cdots, A_m)\emptyset$ and $\emptyset_{i+1}$ is $\emptyset$. An unrestricted SLD-refutation is a derivation ending at an empty clause. SLDNF resolution is essentially SLD-resolution scheme augmented by *negation as failure* inference rule. The completeness and soundness are discussed by Lloyd [4].

A *rule* is of form

$$A_0 \leftarrow A_1, \cdots, A_m, notB_{m+1}, \cdots, notB_n, \qquad (1)$$

where $A_0, \cdots, A_m, B_{m+1}, \cdots, B_n$ are atoms of language $\mathcal{L}$. A *normal logic program* is a finite set of rules of form (1). A rule of form $\leftarrow A_1, \cdots, A_m, notA_{m+1}, \cdots, notA_n$ is called a *normal goal*. In a logic program P, a rule without body $A \leftarrow$ is called a *fact*. A term,atom,literal , rule or program is *ground* if no variable occurs in it. It should be also noted that any free variable in a rule is assumed to be universally quantified. If $\mathcal{P}$ does not contain any constant symbols, we will assume one in $\mathcal{P}$.

Now we define a *normal abductive logic program* to be a pair $< \mathcal{P}, \mathcal{A} >$, where $\mathcal{P}$ is a normal Logic Program and $\mathcal{A}$ is the set of *Abducibles*. An abducibles is a rule. Let $< \mathcal{P}, \mathcal{A} >$ be a normal abductive logic program and $\mathcal{G}$ is a goal (observation). Note that $\mathcal{G}$ can be a literal or a rule. We first consider the case that $\mathcal{G}$ is a literal. If $\mathcal{G}$ is a ground atom (we also call $\mathcal{G}$ is a positive observation), then proper hypothesis is introduced to or removed form the current knowledge base to explain the observation by showing that $\mathcal{G}$ can be derived from the resulting knowledge base. If $\mathcal{G}$ is a negative ground atom (we also call $\mathcal{G}$ is a negative observation), then proper hypothesis is introduced to or removed form the current knowledge base to explain the observation by showing that the corresponding ground atom of $\mathcal{G}$ *cannot* be derived from the resulting knowledge base.

Definition 1. *A pair* $(\mathcal{E}, \mathcal{F})$ *is an* explanation *of a positive observation (or negative observation,resp.)* $\mathcal{G}$ *with respect to an abductive logic program* $< \mathcal{P}, \mathcal{A} >$ *if*

1. $(\mathcal{P} \cup \mathcal{E}) - \mathcal{F} \vdash \mathcal{G}$, *(or* $(\mathcal{P} \cup \mathcal{E}) - \mathcal{F} \nvdash \overline{\mathcal{G}}^1$, *resp.);*
2. $(\mathcal{P} \cup \mathcal{E}) - \mathcal{F}$ *is consistent; and*
3. $\mathcal{E} \subseteq \mathcal{A}$ *and* $\mathcal{F} \subseteq \mathcal{A} \cap \mathcal{P}$.

An explanation $(\mathcal{E}, \mathcal{F})$ *of an observation* $\mathcal{G}$ *is* minimal *if for any explanation* $(\mathcal{E}', \mathcal{F}')$ *of* $\mathcal{G}$, $\mathcal{E}' \subseteq \mathcal{E}$ *and* $\mathcal{F}' \subseteq \mathcal{F}$ *imply* $\mathcal{E}' = \mathcal{E}$ *and* $\mathcal{F}' = \mathcal{F}$.

3 The Approach

In this section we describe the basic idea of our approach of abductive reasoning by illustrating several examples.

Example 1. Let $< \mathcal{P}, \mathcal{A} >$ be an abductive program such that

$\mathcal{P}$:
$Bird(tweety) \leftarrow,$
$Bird(opus) \leftarrow,$
$Broken\text{-}Wing(tweety) \leftarrow,$
$Ab(x) \leftarrow Broken\text{-}Wing(x),$
$Flies(x) \leftarrow Bird(x), not Ab(x),$
$\mathcal{A}$:
$Broken\text{-}Wing(x) \leftarrow.$

The explanation for a positive observation $\mathcal{G} = Flies(tweety)$ can be derived from the SLDNF tree showed in Figure 1. That is, the observation $\mathcal{G} = Flies(tweety)$ has an explanation $(\mathcal{E}, \mathcal{F}) = (\emptyset, \{Broken\text{-}Wing(tweety) \leftarrow\})$.

On the other hand, an explanation $(\mathcal{E}, \mathcal{F})$ for the negative observation $\mathcal{G} = notFlies(opus)$ is $(\{Broken\text{-}Wing(opus) \leftarrow\}, \emptyset)$ and can be derived from the SLDNF tree showed in Figure 2. Note that to explain the fact $notFlies(opus)$ we have to add the fact $Broken\text{-}wing(opus) \leftarrow$ to the logic program $\mathcal{P}$ ∎

Now we consider the case that the observation is a rule. An *observation rule* is a rule of the form

$$A_0 \leftarrow A_1, \cdots, A_m, notB_{m+1}, \cdots, notB_n, \qquad (2)$$

---
[1] Here we use $\overline{\mathcal{G}}$ to denote the complementary literal of $\mathcal{G}$. For instance, if $\mathcal{G}$ is $notF$, then $\overline{\mathcal{G}}$ is F.

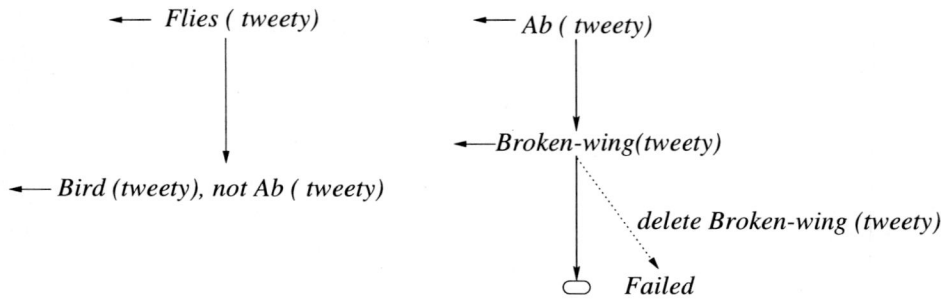

Fig. 1. The SLDNF tree for $\mathcal{P} \cup \{\leftarrow Flies(tweety)\}$.

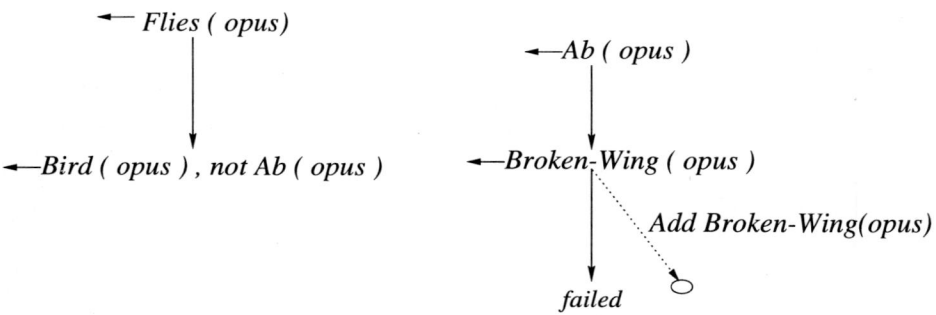

Fig. 2. The SLDNF tree for $\mathcal{P} \cup \{\leftarrow Flies(opus)\}$.

or of the form

$$notA_0 \leftarrow A_1, \cdots, A_m, notB_{m+1}, \cdots, notB_n. \tag{3}$$

The former is called a *positive observation rule*, while the latter is called a *negative observation rule*.

Definition 2. *Given an abductive program* $< \mathcal{P}, \mathcal{A} >$, *a pair* $(\mathcal{E}, \mathcal{F})$ *is an explanation of the observation rule* $\mathcal{G}$ *if there exist* $\mathcal{E}_1, \cdots, \mathcal{E}_n$ *and* $\mathcal{F}_1, \cdots, \mathcal{F}_n$ *such that*

1. $(\mathcal{P} \cup \mathcal{E}_1) - \mathcal{F}_1 \vdash A_1, \cdots, (\mathcal{P}_m \cup \mathcal{E}_m) - \mathcal{F}_m \vdash A_m, (\mathcal{P}_{m+1} \cup \mathcal{E}_{m+1}) - \mathcal{F}_{m+1} \not\vdash B_{m+1}, \cdots, (\mathcal{P} \cup \mathcal{E}_n) - \mathcal{F}_n \not\vdash B_n$;
2. $(\mathcal{P} \cup \mathcal{E}) - \mathcal{F} \vdash A_0$ (or $\mathcal{P} \cup \mathcal{E}) - \mathcal{F} \not\vdash A_0$ if the head of $\mathcal{G}$ is $notA_0$), where $\mathcal{E}_1 \cup \cdots \cup \mathcal{E}_n \subseteq \mathcal{E}$, and $\mathcal{F}_1 \cup \cdots \cup \mathcal{F}_n \subseteq \mathcal{F}$;
3. $\mathcal{E} \cap \mathcal{F} = \emptyset$ and $(\mathcal{P} \cup \mathcal{E}) - \mathcal{F}$ is consistent;
4. $\mathcal{E} \subseteq \mathcal{A}$ and $\mathcal{F} \subseteq \mathcal{A} \cap \mathcal{P}$.

$(\mathcal{E}, \mathcal{F})$ *is a* minimal *explanation for* $\mathcal{G}$ *as there does not exist another explanation* $(\mathcal{E}', \mathcal{F}')$ *of* $\mathcal{G}$ *such that* $\mathcal{E}' \subset \mathcal{E}$ *and* $\mathcal{F}' \subset \mathcal{F}$.

Now let us see how we can use the SLDNF resolution proof to achieve abductive reasoning in which an observation is a rule.

Example 2. Consider an abductive program $<\mathcal{P}, \mathcal{A}>$ where

$\mathcal{P}$:
$Sci(Kiran) \leftarrow,$
$Math(Kiran) \leftarrow,$
$Crazy(x) \leftarrow Sci(x), notCs(x),$
$Happy(x) \leftarrow Crazy(x), Math(x),$
$Crazy(x) \leftarrow Phi(x).$
$\mathcal{A}$:
$Happy(x) \leftarrow Crazy(x), Math(x),$
$Phi(Kiran) \leftarrow.$

Consider the observation $\mathcal{G}$:

$notHappy(kiran) \leftarrow Crazy(kiran), notCs(kiran).$

According to our definition, we need to find $\mathcal{E}_1, \mathcal{E}_2$ and $\mathcal{F}_1, \mathcal{F}_2$ such that $(\mathcal{P} \cup \mathcal{E}_1) - \mathcal{F}_1 \vdash Crazy(Kiran)$, $(\mathcal{P} \cup \mathcal{E}_2) - \mathcal{F}_2 \not\vdash Cs(Kiran)$, and $(\mathcal{P} \cup \mathcal{E}_1 \cup \mathcal{E}_2) - (\mathcal{F}_1 \cup \mathcal{F}_2) \not\vdash Happy(Kiran)$.

Note any finitely failed SLDNF tree of $\mathcal{P} \cup \{\leftarrow Q\}$ implies that $notQ$ is derived from $\mathcal{P}$. The detailed discussion is referred to in [2]. Now we have the following revised SLDNF trees for deriving $Crazy(Kiran)$ and $notCs(Kiran)$ respectively.

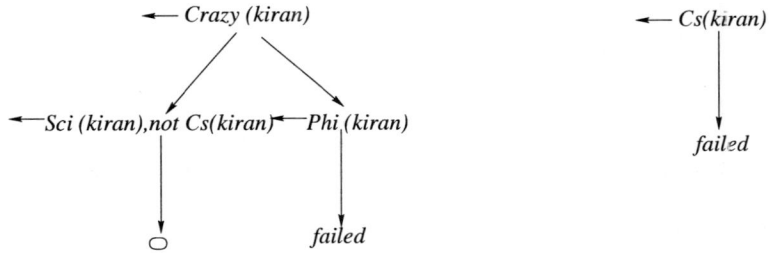

Fig. 3. The SLDNF tree for $\mathcal{P} \cup \{\leftarrow Crazy(Kiran)\}$.

According to our previous discussion, from Figure 3 we can see that a minimal explanation for $Crazy(Kiran)$ is $(\emptyset, \emptyset)$. Since $P \not\vdash CS(Kiran)$, the minimal explanation for $notCs(Kiran)$ is $(\emptyset, \emptyset)$.

From Figure 4 we can see that to achieve $(\mathcal{P} \cup \mathcal{E}) - \mathcal{F} \not\vdash Happy(Kiran)$, there are four possible explanations:

1. $(\mathcal{E}, \mathcal{F}) = (\emptyset, \{Happy(Kiran) \leftarrow Crazy(Kiran), Math(Kiran)\}),$
2. $(\mathcal{E}, \mathcal{F}) = (\emptyset, \{Sci(Kiran) \leftarrow\}),$
3. $(\mathcal{E}, \mathcal{F}) = (\emptyset, \{Math(kiran) \leftarrow\}),$
4. $(\mathcal{E}, \mathcal{F}) = (\{Cs(Kiran) \leftarrow\}, \emptyset).$

However, explanations 2 and 3 are not satisfying our definition $\mathcal{F} \subseteq \mathcal{A} \cap \mathcal{P}$, explanation 4 is not satisfying our definition that $\mathcal{E} \subseteq \mathcal{A}$. Also, as explanations for both $Crazy(Kiran)$ and $notCs(kiran)$ is $(\emptyset, \emptyset)$, we can conclude that explanation 1 is the only and final minimal explanation for $\mathcal{G}$ ∎

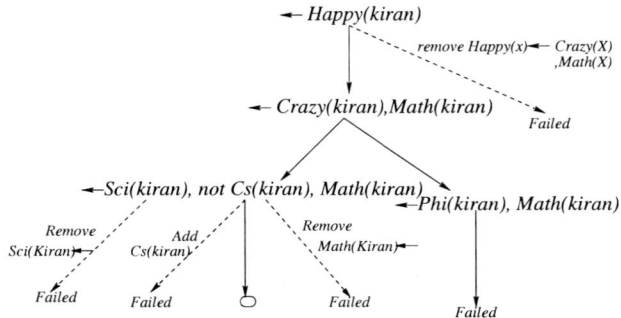

Fig. 4. The SLDNF tree for $(\mathcal{P} \cup \{Phi(kiran) \leftarrow\}) \cup \{\leftarrow Happy(Kiran)\}$.

4 Formal Descriptions

Based on the ideas presented previously, in this section we describe the formal procedures for our rule based abduction. As it has been described in section 3, our approach is based on the SLDNF resolution. In the SLDNF resolution proof, the negation is proved from a logic program by finite failure. However, it is well known that it is possible that a SLDNF tree may include infinite branch. In this case, no result can be proved from the SLDNF resolution proof. For instance, given a logic program $\mathcal{P} = \{A \leftarrow notB, B \leftarrow notA\}$, no finite SLDNF tree exists for $\mathcal{P} \cup \{\leftarrow A\}$. To avoid this problem, in this context, we assume that during the abduction process, each SLDNF tree is *finite* in the sense that each branch in the SLDNF tree is finite.

To simplify our following description, we also introduce some useful notions about SLDNF trees. Observing a SLDNF tree described previously, e.g. figure Fig.3 on page 6, if we use $n_0, \cdots, n_i, \cdots$ to denote all nodes in a SLDNF tree, a SLDNF tree can be actually presented by a set of *branches* $n_0 \rightarrow n_1 \rightarrow \cdots$, $\cdots, n_0 \rightarrow n_k \rightarrow \cdots$. Each branch in the SLDNF tree starts from the root node n_0. Consider the SLDNF tree in Fig.3, it is quite clear that it can be described as two branches $n_0 \rightarrow n_1 \rightarrow n_2 \rightarrow \Diamond$ and $n_0 \rightarrow n_3 \rightarrow$ **Failed**, where nodes n_0, n_1, n_2, n_3 are $\leftarrow Crazy(Kiran), \leftarrow Sci(Kiran), notCs(Kiran), \leftarrow notCs(Kiran)$ and $\leftarrow Phi(Kiran)$ respectively. Node $\Diamond$ indicates the end of a successful branch, and **Failed** indicates the end of a finitely failed branch. We also call a segment of a branch starting from the root node a *path* in a SLDNF tree.

Algorithm 1. Main$(<\mathcal{P},\mathcal{A}>, \mathcal{G})$
Function: Explain the observation $\mathcal{G}$ in logic program $<\mathcal{P},\mathcal{A}>$
Input: A logic program (knowledge base) $<\mathcal{P},\mathcal{A}>$ and an observation $\mathcal{G}$, where the body of $\mathcal{G}$ consists of $P_1, \cdots, P_m, notP_{m+1}, \cdots, notP_n$, Q is the head of the observation
Output: $<\mathcal{E},\mathcal{F}>, \mathcal{P}'$
Begin
    $\mathcal{E} = \emptyset$; $\mathcal{F} = \emptyset$;
    For each i $(1 \leq i \leq m)$ Do **Insert Hypothesis** ($<\mathcal{P},\mathcal{A}>,P_i$);

For each i $(m+1 \leq i \leq n)$ Do **Delete Hypothesis** ($<\mathcal{P},\mathcal{A}>,P_i$);
If the observation is a positive observation
    Then **Insert Hypothesis**($<\mathcal{P},\mathcal{A}>,Q$) ;
If the observation is a negative observation
    Then **Delete Hypothesis** ($<\mathcal{P},\mathcal{A}>,Q$) ;
Print ($\mathcal{P}'$, $<\mathcal{E},\mathcal{F}>$);
End.

The function of algorithm **Main** ($<\mathcal{P},\mathcal{A}>,\mathcal{G}$) is to split the observation, which is a rule, into a set of literals, deriving the explanation for each literal and updating the program with the explanations starting from the body part of the observation. Once the body part is explained and program is updated, the updated program is used. It should be noted that we do not restrict an observation to be ground. For a non-ground observation, as we mentioned earlier, we need to consider all its ground forms which are obtained by substituting variables in the observation with elements of program universe $\mathcal{U}(\mathcal{P})$. This consideration will be adopted in our following algorithms.

Algorithm 2. Insert Hypothesis($<\mathcal{P},\mathcal{A}>,\mathcal{G}$)
Function: Explain the observation $\mathcal{G}$ WRT $<\mathcal{P},\mathcal{A}>$
Input: A logic program $<\mathcal{P},\mathcal{A}>$ and an observation $\mathcal{G}$
Output: An explanation $<\mathcal{E},\mathcal{F}>$ such that ($\mathcal{P} \bigcup \mathcal{E}$) - $\mathcal{F} \models \mathcal{G}$
Begin
    Let $\mathcal{P}' = \mathcal{P}$ and $Path = \emptyset$;
    Loop
        T = a finite SLDNF tree for $\mathcal{P}' \cup \{\leftarrow \mathcal{G}\}$;
        If there is a successful branch
        $n_0 \to n_1 \to \cdots \to n_i \to \diamond$ in T and $n_i \in A$
        then $\mathcal{E}_n = n_i$, $\mathcal{E} = \mathcal{E} \cup \mathcal{E}_n$, $\mathcal{P}' = \mathcal{P}' \cup \mathcal{E}$;
        Return ($<\mathcal{E},\mathcal{F}>,\mathcal{P}'$).
        Select a finitely failed branch B in T such that $Path \subseteq B$:
        $n_0 \to \cdots \to n_i \to$ `Failed`;
        (1) If n_i has form $\leftarrow P_{i1}, [not]P_{i2}, \cdots$
            then If $P_i \subseteq \mathcal{A}$ then $\mathcal{E}_n = P_{i1}$, $\mathcal{E} = \mathcal{E} \cup \mathcal{E}_n$, $\mathcal{P}' = (\mathcal{P}' \cup \mathcal{E})$;
            Return ($<\mathcal{E},\mathcal{F}>,\mathcal{P}'$);
        (2) If n_i has form $\leftarrow notP_{i1}, [not]P_{i2}, \cdots$,
            $\mathcal{P}'$ =**DeleteHypothesis**($\mathcal{P}', notP_{i1} \leftarrow$);
    EndLoop
End.

The detailed explanation of the above algorithm is referred to our full paper due to the space limit of this paper. Here it is sufficient to highlight some key ideas embedded in this algorithm.

The algorithm **Main** splits the observation into a set of *literals* and passes them to the algorithm **Insert hypothesis**. If there is a successful branch in the SLDNF tree, then the process stops. Otherwise, we need to add or remove some rules from $\mathcal{P}$ so that the goal $\leftarrow Q$ can be achieved. It is worth noting that if a sub goal $\leftarrow notP$ fails in some branch of the SLDNF tree, we need to call algorithm **DeleteHypothesis**($\mathcal{P}', notP \leftarrow$), which will be described below, in order to achieve the sub goal $\leftarrow notP$. Finally, after change $\mathcal{P}$ to $\mathcal{P}'$ such that $\mathcal{P}' \vdash \mathcal{G}$. First, the body of the observation is explained using this algorithm by SLDNF tree for $\mathcal{P} \cup \{\leftarrow \mathcal{G}\}$ where $\mathcal{G}$ is an atom from the body of the observation. Once the body of the observation is explained completely and a new program $\mathcal{P}' = (\mathcal{P} \cup \mathcal{E}) - \mathcal{F}$ is formed to explain the head of the observation, thereafter from the SLDNF tree for $\mathcal{P}' \cup \{\leftarrow \mathcal{G}\}$ where $\mathcal{G}$ is the head of the observation we will get another set of explanation which combined gives the explanation for the observation.

Algorithm 3. Delete Hypothesis($< \mathcal{P}, \mathcal{A} >, \mathcal{G}$)
Function: Explain the observation $\mathcal{G}$ WRT $< \mathcal{P}, \mathcal{A} >$
Input: A logic program $< \mathcal{P}, \mathcal{A} >$ and an observation $\mathcal{G}$
Output: An explanation $< \mathcal{E}, \mathcal{F} >$ such that ($\mathcal{P} \bigcup \mathcal{E}$) - $\mathcal{F} \nvdash \mathcal{G}$)
Begin
    T = a finite SLDNF tree for $\mathcal{P}' \cup \{\leftarrow \mathcal{G}\}$;
    Loop
        Let T = a finite SLDNF tree for $\mathcal{P}' \cup \{\leftarrow \mathcal{G}\}$;
        For each successful branch B in T,
            (1) If there exist two nodes n_i and n_{i+1} in B such that $n_i \rightarrow n_{i+1}$, where n_i and n_{i+1} have forms $\leftarrow P_{i1}, [not]P_{i2}, \cdots$ and $\leftarrow [not]P_{i2}, \cdots$ respectively, then $\mathcal{F}_n = P_{i1}$;
            If $\mathcal{F}_n \subseteq \mathcal{A} \cap \mathcal{P}$ Then $\mathcal{F} = \mathcal{F} \cup \mathcal{F}_n, \mathcal{P}' = \mathcal{P}' - \mathcal{F}$;
            Return ($\mathcal{P}'$, $< \mathcal{E}, \mathcal{F} >$);
            (2) If there exist two nodes n_i and n_{i+1} in B such that $n_i \rightarrow n_{i+1}$, where n_i and n_{i+1} have forms $\leftarrow notP_{i1}, [not]P_{i2}, \cdots$ and $\leftarrow [not]P_{i2}, \cdots$ respectively, then $\mathcal{E}_n = P_{i1}$;
            if $\mathcal{E}_n \subseteq \mathcal{A}$ then $\mathcal{E} = \mathcal{E} \cup \mathcal{E}_n, \mathcal{P}' = \mathcal{P}' \cup \mathcal{E}$;
            Return ($\mathcal{P}'$, $< \mathcal{E}, \mathcal{F} >$);

            (3) If there exist two nodes n_i and n_{i+1} in B such that $n_i \rightarrow n_{i+1}$, where n_i and n_{i+1} have forms $\leftarrow P_{i1}, [not]P_{i2}, \cdots$ and $\leftarrow [not]P_{i11}, \cdots, [not]P_{i1q}, [not]P_{i2}, \cdots$ respectively in which a rule r: $P_{i1} \leftarrow [not]P_{i11}, \cdots, [not]P_{i1q}$ is used to derive node n_{i+1}, then $\mathcal{F}_n = r$;
            If $\mathcal{F}_n \subseteq \mathcal{A} \cap \mathcal{P}$ Then $\mathcal{F} = \mathcal{F} \cup \mathcal{F}_n, \mathcal{P}' = \mathcal{P}' - \mathcal{F}$;
            Return ($\mathcal{P}'$, $< \mathcal{E}, \mathcal{F} >$) ;
        T = a finite SLDNF tree for $\mathcal{P}'\{\leftarrow \mathcal{G}\}$;
    EndLoop
End.

Theorem 1. *Given an abductive logic program $< \mathcal{P}, \mathcal{A} >$ and an observation rule $\mathcal{G}$. If there exists an explanation for $\mathcal{G}$, then the algorithm* **Main**($< \mathcal{P}, \mathcal{A} >, \mathcal{G}$) *will always return an explanation* $(\mathcal{E}, \mathcal{F})$ *for $\mathcal{G}$, and this explanation is minimal.*

This theorem ensures that our procedure only finds the minimal explanation for an observation with respect to an abductive logic program if there exists some explanations for this observation.

5 Concluding Remarks

In this paper, we propose a SLDNF proof approach for abductive reasoning. Differently from previous approaches, our approach allow the observation to be a rule. Since within our framework, an abductive reasoning is translated into a SLDNF proof, our approach can be easily implemented based on the revision of traditional SLDNF proof procedure. Finally, we should mention that the system implementation based on the framework proposed in this paper is being undertaken currently.

References

1. A.C.Kakas, F. Toni and R.A. Kowalski, Abductive logic programming, *Journal of Logic and Computation* (1993) PP 719-770.
2. A.C. Kakas and P. Mancarella, Database updates through abduction. In *Proceedings of the 16th VLDB Conference*, pp 650-661, 1990.
3. S. Lakkaraju and Y. Zhang, A Procedural Approach for Rule Based Update In *Proceedings of AI-99*.
4. J.W. Lloyd, *Foundations of Logic Programming*, 2nd edition. Springer-Verlag, 1987.
5. C. Sakama and K. Inoue, Updating extended logic programs through abduction. In *Proceedings of LPNMR-99*, pp 147-161, 1999.
6. Y. Zhang and N.Y. Foo, Updating logic programs. In *Proceedings of the 13th Europen Conference on Artificial Intelligence (ECAI'98)*, pp 403-407. John Wiley & Sons, Inc., 1998.

An Efficient Proof Method for Non-clausal Reasoning

E. Altamirano, G. Escalada-Imaz

Artificial Intelligence Research Institute (IIIA)
Spanish Scientific Research Council (CSIC)
Campus UAB, s/n
08193 Bellaterra, Barcelona (Spain)
{ealtamir,gonzalo}@iiia.csic.es

Abstract Enlarging the class of tractable SAT problems is a relevant topic because of the repercussions in both practical applications and theoretical grounds. In this paper, it is proved that some non-clausal Horn-like formulas can be solved in linear time. In addition to its theoretical importance, this result has a special practical interest because Knowledge Based Systems could benefit of it due to the Horn-like structure of the formulas. In order to prove such linearity a correct refutational calculi is first provided and second, a linear algorithm is described.

1 Introduction

To represent knowledge and to reason with non-clausal formulas is a matter of high importance in Artificial Intelligence and more generally in Computer Science. However, most of the existing methods have been designed for clausal reasoning. Thus, the two main methods to solve problems in non-clausal formulas transform the original non-clausal formula to a clausal formula. Both transformations have severe drawbacks because either the number of propositions of the transformed formula increases exponentially as a consequence of the ∨/∧ distribution or a certain number of artificial literals are introduced in the transformed formula losing the logical equivalence relation which could be invalid for certain applications.

In this paper, we identify Negation Normal Form (NNF) formulas $F_1 \wedge F_2 \wedge \ldots \wedge F_n$ having a Horn-like structure. Each F_i is a disjunction of two optional terms, i.e. $F_i = NNF_i^- \vee C_i^+$: the first one is a NNF formula with only negated propositions, noted NNF^-, and the second one is a conjunction of non-negated atomic propositions, noted C^+.

The kind of formulas we deal with in this paper can arise from an original non-clausal representation of the problem. They are compact representations of Horn formulas given that they require less symbols than Horn formulas to codify identical problems; this reduction can be in an exponential rate.

This paper is structured as follows. In the next section we briefly review the research about non-clausal tractable reasoning and related issues. After having done the formal definition of the mentioned formulas, we define a sound and refutational complete logical calculi and finally, we detail an algorithm in pseudo-code with suitable data structures to resolve SAT-problems expressed in this kind of specified formulas. This algorithm is showed to be sound, refutational complete and it runs in strictly linear time.

2 Related Work

Several methods have been developed to infer with non-clausal formulas. This is the case of Matings [2], Matrix Connection [3], Dissolution [9], and TAS [7]. However, no studies relative to NNF tractability employing one of these methods have been carried out.

As far as we know, the first published results concerning non-clausal tractability comes from [4-6] where a strictly linear forward chaining algorithm to test for the satisfiability of certain NNF formulas subclass is detailed. Such a class embeds the Horn case as a particular case. In [8] a linear backward algorithm is given for the same NNF subclass of formulas. Following this research line, very recently a preliminary version of this article has been presented in [1].

New results concerning NNF tractability are reported in [11] where a method called Restricted Fact Propagation is presented which is a quadratic, incomplete non-clausal inference procedure. More recently, in [12, 13] a significant advance has been accomplished. The author defines a class of formulas by extending Horn formulas to the field of the NNF formulas. Such extension relies on the concept of polarity. A method to make inferences and potentially to detect refutational formulas is designed. In [12], a SLD-resolution variant with the property of being refutationally complete is showed but its computational complexity is not studied. In [13] a method for propositional NNF Horn-like formulas is described and it is stated that the method is sound, incomplete and linear.

However, concerning the last issue, no algorithm is specified, indeed the steps of the method are described as different propagations of some truth values in a sparse tree. Then, although it seems that the number of inferences of the proposed method is linear, it is not proved the resulting complexity (w.r.t. the number of computer instructions) of a linear number of truth value propagations on the employed sparse trees.

3 Our formulas

Firstly, we introduce our class of non normal formulas.

Definition 1. *A literal L is either an atomic proposition $p \in \mathbf{P}$, noted L^+ or its negation $\neg p$, noted L^-.*

Notation From now on, D stands for a disjunction of literals $(L_1 \vee \ldots \vee L_k)$ and C denotes a conjunction of literals $(L_1 \wedge \ldots \wedge L_k)$. D^+ and C^+ (D^- and C^-) include only positive (negative) literals.

Definition 2. *A CNF formula is a conjunction of disjunctions of literals ($D_1 \wedge \ldots \wedge D_m$) and DNF is a disjunction of conjunction of literals ($C_1 \vee \ldots \vee C_m$). Also, CNF^+ and DNF^+ (resp. CNF^- and DNF^-) includes only positive (resp. negative) literals.*

Definition 3. *A clause C^F is a disjunction of three optional terms $C^F = DNF^- \vee CNF^- \vee C^+$. Clauses with only the $DNF^- \vee CNF^-$ (resp. C^+) term are said negative (resp. positive) clauses. We denote the empty clause by $\square$. A formula F is a finite conjunction of clauses C^F. We note $F_\square$ any formula containing the empty clause $\square$.*

Example 1. An example of the kind of the defined formulas is $F = \{(p_1), (p_3), (p_6), (((\neg p_1 \wedge \neg p_2) \vee \neg p_3) \vee ((\neg p_4 \vee \neg p_5) \wedge \neg p_6) \vee (p_7 \wedge p_8)), (\neg p_8)\}$. For the non unitary clause we have $DNF^- = ((\neg p_1 \wedge \neg p_2) \vee \neg p_3)$, $CNF^- = ((\neg p_4 \vee \neg p_5) \wedge \neg p_6)$. This non unitary clause is equivalent to eight clauses, for instance two of them are: $(\neg p_1 \vee \neg p_3 \vee \neg p_4 \vee \neg p_5 \vee p_7)$ and $(\neg p_2 \vee \neg p_3 \vee \neg p_6 \vee p_8)$.

Definition 4. *An interpretation I assigns to each formula F one value in the set $\{0, 1\}$ and it satisfies:*

- *A literal p ($\neg p$) iff $I(p) = 1$ ($I(p) = 0$).*
- *A disjunction $D = L_1 \vee \ldots \vee L_k$, iff $I(L_i) = 1$, for at least one L_i.*
- *A conjunction $C = L_1 \wedge \ldots \wedge L_k$, iff $I(L_i) = 1$ for every L_i.*
- *A clause $C^F = DNF^- \vee CNF^- \vee C^+$, iff $I(DNF^-) = 1$ or $I(CNF^-) = 1$ or $I(C^+) = 1$.*
- *A formula F if I satisfies all clauses C^F of the formula.*

An interpretation I is a model *of a formula F if satisfies the formula. We say that F is* satisfiable *if it has at least one model, otherwise, it is* unsatisfiable.

Definition 5. Variable Truth Assignment (VTA) *If $(p) \in F$ then VTA derives a new formula F' resulting of removing from F the unit clause (p), the conjunctions $(\neg p \wedge \neg p_1 \wedge \ldots \wedge \neg p_k)$, $(\neg p \wedge D_2^- \wedge \ldots \wedge D_m^-)$, and all the occurrences of $\neg p$.*

Definition 6. And-Elimination (AE) *Inference rule AE derives from a positive conjunction clause $(p_1 \wedge \ldots \wedge p_i \wedge \ldots \wedge p_n)$, the unit clauses $(p_1), \ldots, (p_i), \ldots, (p_n)$.*

Definition 7. Clause Proof *A refutation of a formula F, is a finite succession of formulas $< F_1, F_2, \ldots, F_n >$ such that $F_1 = F$, $F_n = F_\square$ and for each $1 \leq i \leq n - 1$, either $F_{i+1} = VTA(F_i)$ or $F_{i+1} = AE(F_i)$.*

Example 2. A proof of the unsatisfiability of F in the previous example is:

$\{(p_1), (p_3), (p_6), (((\neg p_1 \wedge \neg p_2) \vee \neg p_3) \vee ((\neg p_4 \vee \neg p_5) \wedge \neg p_6) \vee (p_7 \wedge p_8)), (\neg p_8)\}$
$\vdash_{VTA} \{(p_3), (p_6), (\neg p_3 \vee ((\neg p_4 \vee \neg p_5) \wedge \neg p_6) \vee (p_7 \wedge p_8)), (\neg p_8)\}$
$\vdash_{VTA} \{(p_6), (((\neg p_4 \vee \neg p_5) \wedge \neg p_6) \vee (p_7 \wedge p_8)), (\neg p_8)\}$
$\vdash_{VTA} \{(p_7 \wedge p_8), (\neg p_8)\}$
$\vdash_{AE} \{(p_7), (p_8), (\neg p_8)\}$
$\vdash_{VTA} \{(p_7), \square\} = F_\square$

Theorem 1. Soundness $F \vdash_{(VTA+AE)} F' \Rightarrow F \models F'$.

The proofs of the soundness of each rule are trivial and the proof of the theorem follows straightforwardly from those proofs.

Theorem 2. Refutational Completeness *Let F be an unsatisfiable formula; then* $F \vdash_{(VTA+AE)} F_\square$.

The proof is by induction on the length of F, i.e. the number of occurrences of literals in F. The following theorem extends completeness to atomic clauses.

Theorem 3. Literal Completeness $F \models (L) \Rightarrow F \vdash_{(VTA+AE)} (L)$.

Comparing VTA/AE proofs with known automated deduction approaches as for instance Analytic Tableau, Path Dissolution, Matrix Connection, etc. is beyond the scope of this article. For a discussion about complexity of Tableau proofs and resolution approaches the reader can see [10, 14].

4 Algorithm Description

The principle of the algorithm is the application of the VTA and AE inference rules until one of the following facts arise: (1) the empty clause is derived, or (2) no more new unit clauses can be derived. In case (1) the original formula is detected to be unsatisfiable meanwhile in case (2) the formula results to be satisfiable.

Brief description of the algorithm We describe roughly the steps of the algorithm. Initially, it tests whether F has positive unit clauses. In the negative case, it returns "satisfiable" because all the clauses have at least one negative literal. So, assume that $(p) \in F$. Thus F is satisfiable iff $F.\{p \leftarrow 1\}$ is satisfiable. In other words F is satisfiable iff the formula F' resulting of removing from F the unit clause (p), the conjunctions $(\neg \mathbf{p} \wedge \neg p_1 \wedge \ldots \wedge \neg p_k)$ and $(\neg \mathbf{p} \wedge D_1^- \wedge \ldots \wedge D_m^-)$ and all the remaining occurrences of the literal $\neg p$, is satisfiable. Thus, whenever $(p) \in F$ the algorithm removes from F the mentioned elements, i.e. it applies the VTA inference rule. This operation is performed by the algorithm for each positive unit clause in F. Now, observe that some clauses in the initial formula can become positives ($C^F = C^+$) because of the removals of some negative parts. Also due to these removals, a pure negative clause ($C^F = DNF^- \vee CNF^-$) can become empty. Thus, at this point, three algorithmic states can arise:
1) An empty clause is produced. The algorithm ends by determining that the original formula is unsatisfiable; 2) No positive clauses have emerged. The algorithm ends by determining that the formula is satisfiable; and 3) A positive clause is produced. Then, the algorithm applies the And-Elimination rule and adds new unit clauses to the formula. Thus, a new iteration of the described operations above are carried out with these new unit clauses.

We begin the description of the algorithm by a very simple recursive version in order to help the reader to understand it: each inference rule is implemented by one procedure. Thus, procedures VTA and AE must perform the following operations according to the definition of their corresponding inference rules:

(VTA F p): It applies the VTA rule returning the formula F' resulting of removing from F the clause (p) and the the conjunctions $(\neg p \wedge \neg p_1 \wedge \ldots \wedge \neg p_k)$, $(\neg p \wedge D_1^- \wedge \ldots \wedge D_m^-)$ and all the remaining occurrences of $\neg p$.

(AE F C^+): It applies the And-Elimination rule returning the formula F' resulting of removing C^+ from F and adding the unit clauses (p) for each conjunct p in C^+.

We note F^+ the set of positive clauses in F which can be empty. The main procedure is configured by the simple next code.

VTA-AE-Propagation(F)
    If $F^+ = \{\}$ then return(sat)
    If $\square \in F$ then return(unsat)
    If $(p) \in F$ then return(VTA-AE-Propagation (VTA F p))
    If $C^+ \in F$ then return(VTA-AE-Propagation (AE F C^+))

Theorem 4. *This algorithm returns sat iff the input formula F is satisfiable.*

Theorem 5. *The maximal number of recursions is at most in $O(size(F))$.*

The previous algorithm is correct but not very efficient. Its efficiency is similar to that of the methods proposed in [11–13]. Although the number of recursions is bounded by O(n), the complexity of each line is clearly not constant and so, the algorithm's complexity measured in computer instructions number is not linear. One can check that searching for the clauses including some occurrence of $\neg p$ without a suitable data structure has $O(size(F))$ computational cost. Hence, the real complexity of the algorithm in number of computer instructions is at least in $O(n^2)$.

Optimal algorithm description. Next, we proceed introducing suitable data structures to bound the worst-case complexity and we shall do it progressively in order to facilitate the understanding of the whole algorithm since its complete definition in pseudo-code contains many details. Thus first we shall discuss the VTA operations relative to the CNF^- term, second to the DNF^- and finally to both together. So, the structure of (VTA F p) is as follows:

(VTA F p)
    (VTA-CNF p)
    (VTA-DNF p)
    (VTA-DNF-CNF p)

The function in the last line potentially returns conjunctions C_i^+ if both negative terms of their respective clauses CNF_i^- and DNF_i^- have been falsified. Now let us focus in each one of these three parts of VTA.

CNF^- processing. A $CNF^- = D_1^- \wedge \ldots \wedge D_n^-$ is falsified when all the propositions in the negative literals of at least one disjunction D_i^- in the CNF^- are derived. According to the VTA inference for each $(p) \in F$ the negatives occurrences $\neg p$ must be removed. To perform this step, we should search the $\neg p$ occurrences and remove them from the CNF term. But this search has an O(n) overhead. To render this cost constant we use a counter $Neg.Counter(D_i^-)$ associated to each disjunction $D_i^- \in CNF^-$. Thus the physical removal is substituted by the decrement operation. Each decrement, done in O(1) time, is equivalent to one removal. In this way, the counter associated with a disjunction indicates the number of literals whose proposition has not been deduced yet. Whenever a counter is set to zero a flag $flag(C^F, CNF)$ is turned on and this will be used in the conjunct processing of the CNF^- and DNF^- terms.

Therefore, to handle the CNF^- term we require two data structures: $Neg.D(p)$: Set of pointers to couples (D^-, C^F) such that $\neg p \in D^- \in C^F$; and $Neg.Counter(D^-)$: Counter of negated propositions $\neg p$ in the disjunction D^- such that p has not been deduced yet.

Notation. Henceforth, [X] denotes a pointer to the object X. For example $[C^F]$ is a pointer to the clause C^F.

The algorithmic application of VTA over the CNF term is as follows:

VTA-CNF (p)
    Remove (p) from F
    for $\forall([D^-], [C^F]) \in Neg.D(p)$ **do**:
        Decrement $Neg.Counter(D^-)$
        if $Neg.Counter(D^-) = 0$ **then** $flag(C^F, CNF) \leftarrow 1$

DNF^- processing. The processing of the DNF by the VTA when a clause (p) is deduced is based also in the adequate handling of counters. However, these counters operate differently that in the CNF case. Indeed for each DNF a unique counter is associated. Thus each decrement must correspond to one and only one conjunction $C^- \in DNF$. So, the counter is set to zero when at least one proposition p for each conjunction is deduced and each of these propositions falsifies each conjunction $C^- \in DNF^-$, because $\neg p \in C^-$.

But, notice that further deduced propositions whose complements belong to the same negative conjunction must not provoke decrements of the counter. Indeed, only one decrement for each conjunction can be enabled. Otherwise, the counter could be set to 0 without having falsified the whole DNF^-. Indeed that could happen if $DNF^- = C_1^- \vee C_2^- \vee \ldots \vee C_i^- \vee \ldots \vee C_k^-$ and the propositions deduced are not distributed in all the k negative conjunctions.

To ensure that the counter is decremented only once per each negative conjunction, we require a flag $First(C_i^-)$ which indicates whether any proposition whose negation is in C_i^- has been already derived. Thus, the meaning of the aforementioned flag is $First(C_i^-) = 0$ (false) if no proposition in C_i^- has been deduced. Once the first proposition is deduced $First(C_i^-)$ is set to 1 (true).

Similarly, to the CNF case, if the DNF part of a clause is falsified then a flag $flag(C^F, DNF)$ is turned on. With the two new data structures the algorithm corresponding to the execution of the VTA over the DNF term is the following

VTA-DNF (p)
    for $\forall([C^-], [C^F]) \in Neg.C(p)$ do:
        if $First(C^-) = 0$ then do:
            Decrement $Neg.Counter(C^F)$
            $First(C^-) \leftarrow 1$;
            if $Neg.Counter(C^F) = 0$ then $flag(C^F, DNF) \leftarrow 1$

CNF^- and DNF^- conjunct processing.
Both algorithmic operations above have been done independently, now we require to joint the effect of the algorithms by considering the state of both flags $flag(C^F, CNF)$ and $flag(C^F, DNF)$. Thus, if both flags are turned on means that the disjunction $CNF^- \vee DNF^-$ of C^F has been falsified and then the conjunction C^+ of C^F is deduced since $C^F = CNF^- \vee DNF^- \vee C^+$. If C^+ is the empty conjunction, the initial formula is unsatisfiable, otherwise the procedure AE will be launched. Thus the algorithmic steps are:

VTA-DNF-CNF (F p)
    for $\forall [C^F] \in Neg.C(p) \cup Neg.D(p)$ do:
        if $flag(C^F, CNF) = flag(C^F, DNF) = 1$ then:
            if $\exists C^+ = \square \in C^f$ then return (unsat)
            else $C^+ = (p_1 \wedge \ldots \wedge p_n)$

AE algorithm. We will now describe the application of the $AE(C^+)$ procedure, i.e. the And-Elimination inference rule. We observe first that a same proposition p can be deduced in more than one conjunction C^+, and then the counters could be decremented more than once. To disallow these multiple decrements, we use a boolean variable as follows: $Val(p) = 1$ iff p has already been derived from F. So, the truth propagation of variable p is allowed only when the flag $Val(p)$ is set to 0, and once the propagation has been performed, the flag is set to 1 disallowing further propagations. Also, a list C^+ of non-negated propositions in C^F is required. Thus, the procedure AE becomes:

AE (C^+)
    for $\forall p \in C^+$ do:
        if $Val(p) = 0$ then do:
            Add (p) to F
            $Val(p) \leftarrow 1$

Complete algorithm. Finally, we present the definitive version of the whole algorithm. By lack of space we omit the procedure to initialise all data structures employed by the algorithm. To improve the search of unitary clauses (p) and conjunctions (C^+) these are placed, when they are deduced, into two respective stacks stack.p and stack.C with the purpose of avoiding searching them inside the formula. Thus, the complete algorithm is given below.

while stack.p $\neq$ {} do:
    $p \leftarrow pull(stack.p)$
;**VTA-CNF (p)**
    for $\forall [D^-, C^F] \in Neg.D(p)$ do:
        Decrement $Neg.Counter(D^-)$
        if $Neg.Counter(D^-) = 0$ then $flag(C^F, CNF) \leftarrow 1$
;**VTA-DNF (p)**
    for $\forall [C^-, C^F] \in Neg.C(p)$ do:
        if $First(C^-) = 0$ then do:
            Decrement $Neg.Counter(C^F)$;
            $First(C^-) \leftarrow 1$;
            if $Neg.Counter(C^F) = 0$ then $flag(C^F, DNF) \leftarrow 1$
;**VTA-DNF-CNF (p)**
    for $\forall (C^F) \in Neg.D(p) \cup Neg.C(p)$ do:
        if $flag(C^F, CNF) = flag(C^F, DNF) = 1$ then:
            if $\exists C^+ = \square \in C^f$ then return (unsat)
            Else push(C^+, stack.C)
;**AE (C^+)**
    while stack.C $\neq$ {} do:
        $C^+ \leftarrow$ pull(stack.C)
        for $\forall p \in C^+$ do:
            if $Val(p) = 0$ then do:
                push(p , stack.p)
                $Val(p) \leftarrow 1$
return (sat)

Theorem 6. Correctness. *The algorithm described in the above lines returns "unsat" iff F is unsatisfiable.*

The proof follows from the correctness of the Logical Calculi given that the algorithm is an implementation step by step of the inference rules. Each one of the operations performed by the algorithm has its counterpart in the pure inference process as it has been described.

Theorem 7. *The complexity of this algorithm is strictly in $O(size(F))$.*

The proof of this theorem basically resides in the fact that each proposition is introduced at most once in stack.p and each conjunction C^+ at most once in stack.C. Hence, the total cost of the "for" loops is bounded by the number of occurrences of literals in the initial formula.

5 Conclusions

We have defined a new class of non-clausal formulas having a Horn-like shape. This class includes Horn formulas as a particular case. Secondly, we have presented a calculus and showed its soundness and completeness. And, finally, we have designed a strictly linear algorithm to solve the SAT problem in this class. Our formulas are of relevant interest in many applications as for instance those based on Rule Based Systems where they can benefit of the use of a richer language than Horn classical formalisms. The proposed formulas represent logically equivalent pure Horn problems but with exponentially less symbols. Hence, as the described algorithm runs in linear time, the gain of time can be of an exponential order with respect to the known linear algorithms running on the Horn formulas.

Acknowledgements This research was developed under a bilateral collaboration between the IIIA-CSIC and the CINVESTAV-CONACyT and was partially financed by the Universidad Autónoma de Guerrero (Mexico).

References

1. E. Altamirano and G. Escalada-Imaz. Two efficient algorithms for factorized Horn theories (in Spanish). In *V Conferencia de Ingeniería Eléctrica, CIE'1999*, Mexico, D.F., september 1999.
2. P.B. Andrews. Theorem proving via general matings. *Journal of Association Computing Machinery*, 28, 1981.
3. W. Bibel. *Automated theorem proving*. Fiedr, Vieweg and Sohn, 1982.
4. G. Escalada-Imaz. Linear forward inferences engines for a class of rule based systems (in French). Technical Report LAAS-89172, Laboratoire D'Automatique et Analyse des Systemes, Toulouse, France, 1989.
5. G. Escalada-Imaz. *Optimisation d'algorithmes d'inference monotone en logique des propositions et du premier ordre*. PhD thesis, Université Paul Sabatier, Toulouse, France, 1989.
6. G. Escalada-Imaz and A.M. Martínez-Enríquez. Forward chaining inference engines of optimal complexity for several classes of rule based systems (in Spanish). *Informática y Automática*, 27(3):23–30, 1994.
7. G.Aguilera, I.P. de Guzman, and M. Ojeda. Increasing the efficiency of automated theorem proving. *Journal of Applied Non-classical Logics*, 5(1):9–29, 1995.
8. M. Ghallab and G. Escalada-Imaz. A linear control algorithm for a class of rule-based systems. *Journal of Logic Programming*, (11):117–132, 1991.
9. N.V. Murray and E. Rosenthal. Dissolution: making paths vanish. *Journal of the ACM*, 3:504–535, 1993.
10. N.V. Murray and E. Rosenthal. On the computational intractability of analytic tableau methods. *Bulletin of the IGPL*, 2(2):205–228, September 1994.
11. R. Roy-Chowdhury-Dalal. Model theoretic semantics and tractable algorithm for CNF-BCP. In *Proc. of the AAAI-97*, pages 227–232, 1997.
12. Z. Stachniak. Non-clausal reasoning with propositional definite theories. In *International Conferences on Artificial Intelligence and Symbolic Computation*, volume 1476 of *Lecture Notes in Computer Science*, pages 296–307. Springer Verlag, 1998.
13. Z. Stachniak. Polarity guided tractable reasoning. In *International American Association on Artificial Intelligence, AAAI-99*, pages 751–758, 1999.
14. A. Urquhart. The complexity of propositional proofs. *The Bulletin of Symbolic Logic*, 1(4):425–467, 1995.

An Intelligent System Dealing with Complex Nuanced Information within a Statistical Context

D. Pacholczyk, F. Dupin de Saint Cyr

LERIA, University of Angers, 2, Bd Lavoisier, F-49045, Angers CEDEX 01, FRANCE

Abstract. The main object of this paper is to propose an intelligent system dealing with affirmative or negative information. We do not refer to a logical negation but to a linguistic one. Moreover, not only atomic but also complex nuances can be denied. Among the intended meanings of a linguistic negation, the choice is made by using the strength of the user negation and a preference principle which takes into account the answer simplicity.

1 Introduction

In this paper, we present a general model dealing with nuanced information expressed in affirmative or negative forms as they may appear in knowledge bases including, rules like "If the patient is not vaccinated, the inflammation due to the test is not moderate or high" and facts like, "the inflammation due to the test is not small". In our work, we do not refer to classical logic negation but to a new kind of negation called *linguistic negation*. The representation of *affirmative information* can be made by using the model proposed in [1] which can deal with nuanced information within a fuzzy context [11], and the representation of *negative information* can be based on the methodology proposed in [6–8]. Previous models [6], [8] have been improved in [7] in such a way that a user can now deny a combination of nuanced properties called here a *complex nuance*. The *first object of this paper* has been to improve previous negation operators in such a way that all *negation forms* F_t of complex nuances U come from the mechanism, denoted "All ... except ...", which defines $\text{Neg}_t(x, U)$, the *reference frame* of the linguistic negation of "x is U", given F_t. The strength ρ with which the user denies "x is U" defines $\text{Neg}_t^\rho (x, U)$ a set containing V the affirmative translations of the denied nuance U. In many cases, the user wishes only one affirmative translation, so a choice strategy extracts one intended meaning belonging to $\text{Neg}_t^\rho (x, U)$. In [6,7], the choice is made among the solution leading to the greatest membership degree $\mu_V(x)$. When the linguistic negation is restricted to one nuanced property, statistical data have been exploited to make this uncertain choice [3]. *The second object of this paper* has been to enrich this choice strategy in order to deal with denied *complex nuances*. This is made by taking into account *statistical data* on the discourse universe, such as the *frequency of use* of a *fuzzy property* associated with a concept, and of a *nuance* applied to a property. We also suggest taking into account *linguistic statistics* about the use of negation in natural language. In order to illustrate this model, we consider an example of *reasoning on medical facts and rules in a medical diagnosis field*. The statistical information are calculated by exploiting index cards written by a doctor after his consultations. From the following three index cards (**ICi**) and five rules (**Rj**), we wish to deduce a diagnosis.

IC1: The temperature is not very low, the eardrum color is really very red, fat eating is low, the tension is perfectly (i.e. "exactly") normal

IC2: The temperature is not very high, the eardrum color is normal, fat eating is high, and the tension is normal.

IC 3: The temperature is rather little high, the eardrum color is normal, fat eating is very low or really moderate, the tension is normal, and the inflammation due to the monotest is not small.

R1: If the temperature is high and the eardrum color is very red, the disease is an otitis.

R2: If the temperature is not normal, the patient is ill.

R3: If fat eating is not moderate, the cholesterol risk is not low.

R4: If the cholesterol risk is high, a diet with no fat is recommended.

R5: If the patient is not vaccinated, the inflammation due to the test is not moderate or high.

2 Nuanced Expression Probability

We suppose that the discourse universe is characterized by a finite number of *concepts* C_i. A set of *basic properties* P_{ik} is associated with each C_i, whose *description domain* is denoted as D_i. For example, the concept "temperature" can be characterised by the basic properties "low", "normal" and "high" ; the concept "eardrum colour" by the basic properties "normal" and "red" ; the concept "cholesterol risk" by the properties "not-existent", "low", "moderate", "high" and "tremendous", the concept "feat eating" by the properties "zero", "low", "moderate" and "high", and the concept "inflammation" by the properties "not-existent", "small", "moderate" and "high". Moreover, *linguistic modifiers* allow us to express *nuanced knowledge* like "the temperature is *rather* normal". In this paper, we use the methodology proposed in [1] to cope with affirmative information like "x is $f_\alpha m_\beta P_{ik}$" and with negative information like "x is not $f_\alpha m_\beta P_{ik}$" (where f_α and m_β are *linguistic modifiers*). In the following, $f_\alpha m_\beta$ is called the *nuance* applied to P_{ik}. The modifiers f_α and m_β are taken from two ordered sets of *fuzzy modifiers*. The first one groups *translation modifiers* which operate both a translation and a precision variation on the basic property: in this paper, we use $M_9 = \{m_\beta\}_{\beta \in [1..9]}$ = {*extremely little, very very little, very little, rather little, moderately, rather, very, very very, extremely*} (Figure 1). The second set contains *precision modifiers* which make it possible to modify the precision of the properties. Here, we use $F_6 = \{f_\alpha\}_{\alpha \in [1..6]}$ = {*vaguely, neighboring, more or less, moderately, really, exactly*} (Figure 2). These two sets are totally ordered by the relation: $m_\alpha < m_\beta$ (resp. $f_\alpha < f_\beta$) $\Leftrightarrow \alpha < \beta$. The modifier *"moderately"* is equivalent to the empty word.

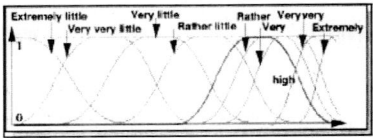

Fig. 1. Translation Modifiers

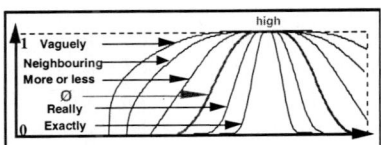

Fig. 2. Precision Modifiers

From the set of index cards owned by the doctor, we can extract statistical knowledge about the use of nuance expressions in his language. Note that these statistics are computed only on *affirmative* expressions contained on these index cards. After index cards analysis, *the percentages of use of every basic property*, when the doctor speaks about a given concept, are available.

Example: *For the concept of temperature, the statistical probability that he uses the term "low" is calculated by counting the occurrences of the term "low" (possibly with a nuance) with regard to the total number of expressions concerning temperature. For instance: 5% of positive expressions concerning "temperature" use the term "low", 80% the term "normal" and 15% the term "high".*

In the same way, we can obtain statistics about the use of every possible nuance for every basic property. In practice, we only need statistics about the use of *relevant* nuances: a combination of modifiers $f_\alpha m_\beta$ is *relevant* if it is used in English. Out of 54 possible nuances, finally only 20 are relevant. The statistics concerning the use of these 20 relevant combinations are computed in the same way as for basic properties.

Example: *For the property "low" associated with the concept of "temperature", the statistical probability that the doctor uses the nuance "really very" is computed by counting the occurrences of "really very low" with regard to the total number of expressions containing "low" and concerning temperature. For instance: 5% of positive expressions containing "low" use the nuance "really very", 20% the empty nuance, 1% "very little", 8% "very"...*

Proposition 1. *For any expression* $e =$ *"x is* $f_\alpha m_\beta P_{ik}$*", the statistical probability that the doctor uses this expression in an index card when he speaks about the concept* C_i *is: if* $Pr(bp(e) = P_{ik}) > 0$ *then* $Pr(e) = Pr(bp(e) = P_{ik}) \times Pr(n(e) = f_\alpha m_\beta | bp(e) = P_{ik})$ *else we set* $Pr(e) = 0$*, where* $n(e)$ *and* $bp(e)$ *are respectively the nuance and the basic property of* e.

Example: *The statistical probability that "very low" appears in affirmative expressions concerning a patient "temperature" is equal to* $0.05 \times 0.08 = 0.004$.

Definition 1. *Given* $G = \{A_h\}_{h=1,...p}$ *a subset of* N_i *(the set of all nuanced properties associated with the concept* C_i*) any finite combination of nuanced properties of G based upon the operators* $\wedge$ *and* $\vee$*, denoted as* $U(A_1,..,A_h,..,A_p)$*, is said to be a* **complex nuance induced from G***. The* **set of all complex nuances induced from** G *is denoted* $\mathbf{G^*}$.
If no confusion is possible, then U *stands for* $U(A_1, \ldots, A_h, \ldots, A_p)$.
More formally, if we suppose that any complex nuance is defined in the **prefixed form***, then* $\mathbf{G^*}$ *can be recursively defined as follows:*
1 - $\forall A_h \in G$*,* $A_h \in G^*$*, 2 -* $\forall U \in G^*$*,* $\forall V \in G^*$*,* $\gamma UV \in G^*$*, with* $\gamma \in \{\wedge, \vee\}$*, 3 - Any element of* G^* *results from a finite number of uses of rules 1 and 2.*

Definition 2 (First preference principle). *Let* $U(A_1,..,A_p)$ *and* $V(B_1,..,B_q)$ *be to complex nuances where every* A_i *(resp.* B_j*) are distinct:* U *is preferred to* V *if and only if* $\prod_{j=1..p} Pr(A_j) > \prod_{j=1..q} Pr(B_j)$.

Example: *The expression "x is low" is preferred to "x is low or really very low" since* $Pr($ *"x is low"* $) > Pr($ *"x is low"* $) \times Pr($ *"x is really very low"* $)$*,* $0.1 > 0.1 \times 0.0025 = 0.00025$.

3 Linguistic Negation Translations

Taking into account several linguistic results [5, 4, 2], Pacholczyk showed in [6] that when the user says *"x is not U"*:(1) he *rejects* the sentence *"x is U"*, and (2) he *possibly refers* either to another object y (i. e., "y is U") or to another property V (i. e., "x is V") distinct from U but which concerns the same concept C_i. The linguistic negation concept used here is defined, as in [9], by a one-to-many mapping from E into $\mathfrak{P}(\mathfrak{E})$ (E parts), called here a *multi-set function*. In this paper we propose a *new definition of the concept of linguistic negation* which can be viewed as a *generalization of the one* proposed in [8]. Let M be the set of every possible modifier combinations (including irrelevant combinations) ($M = \{f_\alpha\}_{\alpha \in [1..6]} \times \{m_\beta\}_{\beta \in [1..9]}$). Given a concept C_i, let $B_i = \{P_{ik}\}_{k \in \mathbb{N}}$ be the set of basic properties associated with C_i having as definition domain D_i, N_{ik} be the set of all nuanced properties associated with property P_{ik}, and N_i be the set of all nuanced properties associated with the concept C_i. Given $G \subset N_i$, G^* denotes the set of all complex nuances induced from G. Then, for any concept C_i, we define a linguistic negation operator as a *parameterized function* Neg_t:

Definition 3. *For any concept C_i, let U and V be two complex nuances induced from N_i^*, such that:* $U = U(n_1 P_{ik_1}, \ldots, n_j P_{ik_j}, \ldots, n_p P_{ik_p})$ *and* $V = V(m_1 P_{il_1}, \ldots, m_j P_{il_j}, \ldots, m_q P_{il_q})$. *and let* $L = \{k_1, \ldots, k_j, \ldots, k_p, l_1, \ldots, l_j, \ldots, l_q\}$ *A* **linguistic negation operator** *is a function* $Neg_t : D_i \times N_i^* \to \mathfrak{P}(\mathfrak{D}_i) \times \mathfrak{P}(\mathfrak{N}_i^*)$ *defined as follows, knowing that* $n \in M, \gamma \in \{\wedge, \vee\}$ *and* $t \in \{0, 1, 2, 3, 4, 5\}$:
- $Neg_0(x,u) = (\emptyset, \emptyset)$; $Neg_1(x, u) = (D_i \backslash \{x\}, \{u\})$; $Neg_2(x, u) = (\{x\}, N_i^* \backslash \{u\})$
- $Neg_3(x, nP_{ik}) = (\{x\}, N_{ik}^* \backslash \{nP_{ik}\})$
and $Neg_3(x, \gamma UV) = (\{x\}, (\cup_{h \in L} N_{ih})^* \backslash \{\gamma UV\})$
- $Neg_4(x, nP_{ik}) = (\{x\}, N_i^* \backslash N_{ik}^*)$; $Neg_4(x, \gamma UV) = (\{x\}, N_i^* \backslash (\cup_{h \in L} N_{ih})^*)$
- $Neg_5(x, nP_{ik}) = (\{x\}, N_{im}^*)$ where $P_{im} \neq P_{ik}$ is a precise property in B_i
and $Neg_5(x, \gamma UV) = (\{x\}, (\cup_{m \in G} N_{im})^*)$ where G is a set of index such that $L \cap G = \emptyset$ and $\{P_{im} \in N_i \mid m \in G\} \subset B_i$.

It is possible to associate a **standard form F_t** for "x is not U" with *each scope of the negation operator*. When a speaker says *"x is not U"*, he means that:
- **F_0**: *For this x, "x is U " is rejected and there is not any corresponding affirmative expression.* Saying "the disease is not an otitis" may not admit any affirmative translation, the only thing that the doctor knows about the disease is that it is not an otitis.
- **F_1**: *Another object of the same domain satisfies the same nuanced property U.* "x is not U" means "not(x) is U". The doctor can note that "Mary is not ill" because he knows that "it is John and Jack who are ill".
- **F_2**: *The same x satisfies another complex nuance of N_i^*.* For instance, the doctor can say that "The temperature is not very high" because he hesitates between "the temperature is normal" and "the temperature is little high".
- **F_3**: *The same x satisfies another complex nuance of N_i^* induced from the same basic properties*. The doctor can say "the temperature is not low" because he thinks that "the temperature is really low".
- **F_4**: *The same x satisfies a complex nuance of N_i^* which is not induced from the same basic properties.* The doctor can say "cholesterol risk is not low" because he thinks that "cholesterol risk is at least medium".
- **F_5**: *The same x satisfies a complex nuance of N_i^* induced from other **precise** basic properties.* The doctor can say that "the patient is not very big" since he thinks that "he is rather thin".
- **F_6**: *The same x satisfies a complex nuance induced from new affirmative basic properties.* In this case, "x is not U" means that the new property "not-U" is associated to the same concept as U. "The patient is not seriously ill" may introduce a new basic property "not-seriously-ill".

Note that, in this paper, we are not concerned with the forms F_1 or F_6.

Remark 1. Let $t \in \{2, ..., 5\}$, "x is not U" means that there exists a complex nuance V such that we have "x is V". This V is defined as follows: $V \in F$ such that $Neg_t(x, U) = (\{x\}, F)$ with $F \in \mathfrak{P}(\mathfrak{N}_i^*)$. In other words, **"x is V" is an affirmative translation of "x is not U" in form F_t**.
If no confusion is possible, in the following, we simply write $V \in Neg_t(x, U)$.

Proposition 2. $Neg_5(x, u) \subset Neg_4(x, u) \subset Neg_2(x, u)$; $Neg_3(x, u) \subset Neg_2(x, u)$.

Assumption : Statistics on the most used negation forms are available for any nuanced expression. These statistics come from a linguistic analysis of English negation.

For instance, "Temperature is not normal" usually corresponds to a nuance of "low" or "normal" or "high" and refers to the form F_2 in 70% of cases. On the other hand, "Temperature is not very low" is related in 60% of cases to a nuance of "high" (F_5). "Fat eating is not moderate" usually means that it is at least "high" (F_5). "Cholesterol risk is not low" means usually that it can be "moderate", "high" or "tremendous" (F_4). "Temperature is not very high" usually corresponds to another nuance of "high" (F_3).

4 Linguistic Negation Strength

Let us notice that the subset $Neg_t(x, U)$ systematically excludes U but it must also exclude complex nuances which are close to U. So, a complex nuance V may be chosen as a negation of an expression $u = $"$x$ is U" if V is a complex nuance associated with the same concept as U and if for a given x: if $\mu_U(x)$ (resp. $\mu_V(x)$) is close to 1 then $\mu_V(x)$ (resp. $\mu_U(x)$) is close to 0.

Definition 4. *Let $0.5 \leq \rho \leq 1$. Let us suppose that C_i is a concept, F_t a standard form ($t \in [2..5]$), Neg_t a linguistic negation operator and $U(A_1, , A_h, , A_p)$ (or U) is a* **complex nuance induced from** N_i. *The* **linguistic negation** *of U* **applied to** x, *given ρ and the* **standard form** F_t, *denoted as* **$Neg_t^\rho(x, U)$**, *is a set of complex nuances induced from N_i. More precisely, $V(B_1, , B_m, , B_q)$ (or V) being a complex nuance induced from N_i, then $V \in Neg_t^\rho(x, U)$ iff:*
L0: $\forall B_m, \exists A_h$ such that: L00: $B_m \in Neg_t(x, A_h)$, L01 : $\forall z, \mu_{A_h}(z) \geq \rho \Longrightarrow \mu_{B_m}(z) \leq 1 - \rho$, L02 : $\forall z, \mu_{B_m}(z) \geq \rho \Longrightarrow \mu_{A_h}(z) \leq 1 - \rho$.
L1: $\forall z, \{\mu_U(z) \geq \rho \Longrightarrow \mu_V(z) \leq 1-\rho\}$, and **L2:** $\forall z, \{\mu_V(z) \geq \rho \Longrightarrow \mu_U(z) \leq 1-\rho\}$.
Any $V \in Neg_t^\rho(x, U)$ is said to be a **linguistic negation** *of U* **applied to** x, *given ρ and the* **standard form** F_t.

Definition 5. *Let U and V be associated with the same concept, $Vo_x(U, V) = min\{\mu_U(x) \rightarrow_L \mu_V(x), \mu_V(x) \rightarrow_L \mu_U(x)\}$ where $\rightarrow_L$ is Lukasiewicz's implication. Let the neighborhood between U and V be: $Vo(U, V) = min_x Vo_x(U, V)$.*

Proposition 3. *If $V \in Neg_i^\rho(x, U)$ then $Vo(V, U) \leq 2(1 - \rho)$*

Definition 6. *Knowing that $0.5 \leq \rho \leq 1$, the value $2\rho - 1$ defines the rejection strength associated with a property $V \in Neg_i^\rho(x, U)$.*

Proposition 4. $Neg_i^1(x, U) \subset Neg_i^{0.9}(x, U) \subset ... \subset Neg_i^{0.5}(x, U)$.

Example: *Using proposition 3, Vo("normal", "rather little high")=0.2, then $\rho \leq 0.9$, i.e., "rather little high" belongs at most to the subsets of $Neg_i^{0.9}(x,$ "normal"). For Q = "extremely high", Vo("normal",Q) = 0 means that $\rho \leq 1$, i.e. "extremely high" belongs to the subsets of $Neg_i^1(x,$ "normal").*

Definition 7 (Second Preference Principle). *Given a standard form F_t and an expression "x is U", let $\rho(V)$ be the greatest ρ such that $V \in Neg_t^\rho(x, U)$. V is **preferred to** W for the negation of U if $\rho(V) \geq \rho(W)$.*

Example: *For the negation of "normal" in form F_5, "extremely high" is preferred to "rather little high". Since $\rho($"extremely high"$) = 1$ meanwhile $\rho($"rather little high"$) = 0.9$.*

The following properties result from previous definition of linguistic negation.

Proposition 5. *(1): Given $U \in N_i^*$ and $V \in N_i^*$: $V \in Neg_t^\rho(x, U)$ if $U \in Neg_t^\rho(x, V)$. (2): A complex nuance $V \in Neg_t^\rho(x, U)$ if: (2a): $U = A \wedge B$ with $A \in N_i^*$ and $B \in N_i^*$, $V = P \vee Q$ with $P \in Neg_t^\rho(x, A)$, $Q \in Neg_t^\rho(x, B)$, (2b): $U = A \vee B$ with $A \in N_i^*$ and $B \in N_i^*$, $V = P \wedge Q$ with $P \in Neg_t^\rho(x, A)$, $Q \in Neg_t^\rho(x, B)$.*

Proposition 6 (Contraposition Laws). *Let us suppose that the implication $\rightarrow$ and its associated T-norm T satisfy the properties: $u \rightarrow v = 1$ iff $u \leq v$; $T(u \rightarrow v, v \rightarrow w) \leq u \rightarrow w$ (weak transitivity law); $u \rightarrow v = \neg v \rightarrow \neg u$ (contraposition law). Then, the extended linguistic negation possesses the following properties:*
(i) If there exists $P \in Neg_t^\rho(x, A)$ and $Q \in Neg_t^\rho(y, B)$ such that $Q \subset \neg B$ and $\neg A \subset P$, then the rule if "x is A" then "y is B" implies the rule if "y is not B" then "x is not A".
(ii) If there exists $P \in Neg_t^\rho(x, A)$ and $Q \in Neg_t^\rho(y, B)$ such that $P \subset \neg A$ and $\neg B \subset Q$, then the rule if "y is not B" then "x is not A" implies the rule if "x is A" then "y is B".
(iii) If there exists $P \in Neg_t^\rho(x, A)$ and $Q \in Neg_t^\rho(y, B)$ such that $\neg A = P$ and $\neg B = Q$, then the rules if "y is not B" then "x is not A", and if "x is A" then "y is B" are equivalent.

Example 1. Let us suppose that:
P=($\vee$rather low $\vee$ very low extremely low)$\in Neg_2^\rho$(x, $\vee$(medium high)),
($\vee$ rather low $\vee$ very low extremely low) $\subset \neg$($\vee$medium high),
Q=$\neg$vaccinated. In the rule R5 :"If the patient is not vaccinated, the inflammation due to the test is not medium or not high", the conclusion is translated into "the inflammation is rather or very or extremely low". By using previous proposition, this rule implies the rule "If the inflammation is medium or high then the patient is vaccinated".

5 Choice Strategy

We extend the *strategy* proposed in [3] to the negation of complex nuances:
1. First, select the negation form F_t according to linguistic statistics concerning the complex nuance U (assumption presented in 3)

2. Compute the set $Neg_t(x, U)$.
3. Select in $Neg_t(x, U)$, the complex nuances V such that $\rho(V)$ is maximum.
4. Among these complex nuances, use the first preference principle in order to make a choice.
5. Discriminate ex-aequo complex nuances by taking into account their complexity degrees. This last point is justified by Sperber and Wilson's simplicity principle [10].

6 The Model in Action

First of all, we are going to translate all negative complex nuances appearing in rules and facts. They concern the expressions: "temperature is not normal", "fat eating is not moderate", "risk of cholesterol is not weak", "temperature is not very low", "temperature is not very high", "the inflammation is not small", "the patient is not vaccineted" and "the inflammation due to mono-test is not moderate or high". Let us decompose the affirmative translation of the assertion "Temperature is not normal": according to linguistic statistics on the different forms of linguistic negation, this negation has the standard form F_2. So, let us calculate Neg_2(x, "normal"). This set contains all nuances of all properties associated with the concept "temperature" (except the precise nuance expression "∅ normal"): Neg_2(x, "normal") $= N^*_{\text{temperature}} \setminus \{\text{"normal"}\}$. Then we must select the properties having a maximal rejection strength. The following properties have a strength equal to 1: properties containing "high" or "low" without nuance, or with translation nuances greater than "rather", or with precision nuances greater than "more or less", or with combinations "really very", "really very very", "really extremely", and also the three properties "extremely little normal", "very very little normal" and "very little normal". Computing the statistical probability of all these nuanced properties, it appears that the best expression is "high". "Temperature is not very low" is translated into "Temperature is high". "Temperature is not very high" by the form F_2 leads to all nuances of "high", which finally gives "Temperature is little high". "Inflammation is not small" is translated into "Inflammation is medium or high" "Fat eating is not moderate", by the form F_3, leads to the nuances of "zero" "low" and "high", and finally is translated into "Fat eating is high". "The cholesterol risk is not low" is translated into "The cholesterol risk is high". The rule R5 is translated as shown in example 1. Finally, we obtain:

IC1: The temperature is high, the eardrum color is really very red, fat eating is low, the tension is exactly normal.
IC2: The temperature is little high, the eardrum color is normal, fat eating is high, and the tension is normal.
IC3: The temperature is rather little high, the eardrum color is normal, fat eating is very low or really moderate, the tension is normal, and the inflammation due to the mono-test is medium or high.
R1: If the temperature is high and the eardrum color is very red, the disease is an otitis.

R2: If the temperature is *high*, the patient is ill.
R3: If fat eating is *high*, the cholesterol risk is *high*.
R4: If the cholesterol risk is high, a diet with no fat is recommended.
R5: If the inflammation due to the monotest is *medium or high* then the patient is *vaccinated*.

From index card 1, rule R1 can be fired, since "really very red" implies "very red", so we can infer that "The disease is an otitis". Rule R2 leads us to deduce that "The patient is ill". The other two rules can not be fired. From index card 2, rule R3 gives that "the cholesterol risk is high", then rule R4 leads us to recommend "a diet with no fat". By using rule R5, we can conclude that the third patient is vaccinated.

7 Conclusion

We have defined a general model of linguistic negation of assertions like "x is not U" in the fuzzy context. This approach to negation, in accordance with linguistic analysis pursues preceding works. The strategy of choice uses the rejection strength: one selects the complex nuances which belong to the strongest negations of the initial complex nuance then statistical information on the language and on the customs of the speaker are used.

References

1. Desmontils E., Pacholczyk D.: Towards a linguistic processing of properties in declarative modelling. Int. Journal of CADCAM and Comp. Graphics 12:4, 351-371, 1997
2. Ducrot O., Schaeffer J.-M. : Nouveau dictionnaire encyclopédique des sciences du langage. Seuil Paris, 1995
3. Dupin de Saint-Cyr F., Pacholczyk D.: De la Portée et de la Force d'un Opérateur de Négation à l'Interprétation de la Négation Linguistique d'Informations Nuancées, Proc. of RFIA00, Paris, II, 421-430, 2000
4. Horn L.R.: A Natural History of Negation. The Univ. of Chicago Press 1989
5. Muller C.: La négation en français. Publications romanes et françaises Genève 1991
6. Pacholczyk D.: A New Approach to the Intended Meaning of Negative Information. Proc. of ECAI-98, Brighton, UK, August 1998, Pub. by J. Wiley & Sons, 114-118, 1998
7. Pacholczyk D.: An Extension of a Linguistic Negation Model allowing us to Deny Nuanced Property Combinations. Proc. of ECSQARU 99, LNIA 1638, 316-327, 1999
8. Pacholczyk D., Levrat B.: Coping with Linguistically Denied Nuanced Properties: A Matter of Fuzziness and Scope, in Proc. IEEE ISIS/CIRA/ISAS Joint Conference, Gaitherburg MD, USA, (1998) 753-758
9. Torra V.: Negation Functions Based Semantics for Ordered Linguistic Labels. Int. Jour. of Intelligent Systems 11, 975-988, 1996
10. D. Sperber, D. Wilson : La pertinence. Communication et cognition. Paris, Les éditions de Minuit (1989)
11. Zadeh L.A.: Fuzzy Sets. Information and Control 8, 338-353, 1965

On the Complexity of Optimal Multisplitting

Tapio Elomaa[1] and Juho Rousu[2]

[1] Department of Computer Science, P. O. Box 26 (Teollisuuskatu 23),
FIN-00014 University of Helsinki, Finland, `elomaa@cs.helsinki.fi`
[2] VTT Biotechnology, P. O. Box 1500 (Tietotie 2),
FIN-02044 VTT, Finland, `juho.rousu@vtt.fi`

Abstract. Dynamic programming has been studied extensively, e.g., in computational geometry and string matching. It has recently found a new application in the optimal multisplitting of numerical attribute value domains. We reflect the results obtained earlier to this problem and study whether they help to shed a new light on the inherent complexity of this time-critical subtask of machine learning and data mining programs.

The concept of monotonicity has come up in earlier research. It helps to explain the different asymptotic time requirements of optimal multisplitting with respect to different attribute evaluation functions. As case studies we examine Training Set Error and Average Class Entropy functions. The former has a linear-time optimization algorithm, while the latter—like most well-known attribute evaluation functions—takes a quadratic time to optimize. It is shown that neither of them fulfills the strict monotonicity condition, but computing optimal Training Set Error values can be decomposed into monotone subproblems.

1 Introduction

Consider the *optimal multisplitting problem* faced by classifier learning algorithms in processing numerical attributes. Given a sample S containing b indivisible subsets and an evaluation function F to rank partition candidates, find the F-optimal partition of S with at most k intervals. We denote a partition with k intervals by $\uplus_{i=1}^{k} R_i$. The examples are instances of m classes.

Numerical attribute domain partitioning is a time-critical subtask in machine learning and data mining algorithms. Recently there have been many attempts to enhance the efficiency of this task [2,3,5,6,7,8,9,14]. Many commonly-used functions conform to *cumulativity*: the score of a partition is a weighted sum of its interval scores [9,5]. This property lets us apply dynamic programming to combine the solution efficiently from optimal partitions of subsequences. The time complexity of the algorithm is only quadratic in b.

The inherent complexity of the multisplitting task is uncharted territory. However, mathematically similar problems have been encountered in computational geometry and string matching [11]. This paper reflects that work to the multisplitting framework. It turns out that the optimal multisplitting algorithms solve as a subproblem an instance of the *column minima problem* [11], for which lower bound results are already known. This problem takes $\Omega(b^2)$ time if the only knowledge that we have about the function is that it is cumulative. If the function is *monotone*, it can be optimized in $\Omega(b \log b)$ time and, further, a so-called *totally monotone* function is optimizable in linear time. Commonly-used attribute evaluation functions do not fulfill these properties. In particular, we show that two functions, *Average Class Entropy (ACE)* [8,13] and *Training Set Error (TSE)* [2,3,9], are not monotone.

It is known that many evaluation functions—including *ACE* and *TSE*—fulfill Jensen's inequality [4]. Its consequence is convexity of the function over the data set from which it follows that each partitioning of the sample leads to a better partition score. We study whether Jensen's inequality alleviates the inherent complexity of optimal multipartitioning.

TSE is the only commonly-used evaluation function that is known to be optimizable in linear time. Even though the function itself is not monotone, its optimization algorithms can combine the result from optimal values for (totally) monotone subproblems in a single scan through the data [2,3,9]. Similar decomposition of more complex functions does not seem possible.

Section 2 describes the optimal multisplitting problem. Then we examine, in Section 3, the monotonicity formulations that have emerged in string matching and computational geometry and translate Jensen's inequality to the same vocabulary. It corresponds to a weak form of monotonicity. In Sections 4 and 5 we study the functions *ACE* and *TSE*, which have different known optimization requirements. Section 6 discusses the observations presented in this paper and gives the concluding remarks of this study.

2 The Optimal Multisplitting Problem

Assume that the sample $S = \bigcup_{i=1}^{b} S_i$ with b indivisible subsets has been given. In a partition $\biguplus_{i=1}^{k} R_i$ of S each interval is composed of consecutive sample subsets, $R_i = \bigcup_{\ell=h}^{j} S_\ell$. The cumulative attribute evaluation function scoring partition candidates is defined as

$$p\left(\biguplus_{i=1}^{k} R_i\right) = \sum_{i=1}^{k} w(R_i),$$

where $w(R) = |R| f(R)$ is the score given to interval R by the "impurity" function f.

The score of an interval consisting of subsets $S_h, \ldots, S_j$ is denoted by $w(h, j) = w(\bigcup_{\ell=h}^{j} S_\ell)$. Furthermore, $p(k, j)$ denotes the minimum value of p on k-partitions of $S_1, \ldots, S_j$.

The dynamic programming algorithm for optimal multisplitting [5,9,14] calculates the recurrence

$$p(k, j) = \begin{cases} \min_{k \leq h < j}\{p(k-1, h) + w(h+1, j)\} & \text{if } k \leq j \\ \infty & \text{otherwise} \end{cases} \quad (1)$$

and outputs $p(k,b)$ as the answer. In other words, the best partition of subsets $S_1, \ldots, S_j$ into k intervals consists of some non-empty final interval $R_k = \bigcup_{\ell=h+1}^{j} S_\ell$ together with the optimal $(k-1)$-partition of the sample prefix composed of subsets $S_1, \ldots, S_h$, where $h \geq k-1$.

In recurrence 1 the computation of best k-partitions of a set depends on the scores of the best $(k-1)$-partitions of the proper prefixes of the set. Thus, when calculating values on row k, the algorithm consults the values on row $k-1$.

The time complexity of the dynamic programming computation of recurrence 1 is $O(kb^2)$, excluding the work needed to compute $w(h,j)$ for each $1 \leq h \leq j \leq b$. There are $\Theta(b^2)$ such terms; the computation of each value requires scanning the class frequency distribution in $\Theta(m)$ time. Thus, the total complexity for obtaining the $w(h,j)$ values is $\Theta(mb^2)$. We cannot avoid computing all $w(h,j)$ values without knowing more about the evaluation function than just its cumulativity.

3 The Column Minima Problem and Optimal Multiplitting

The calculation of one row of the matrix p is an instance of the *one-dimensional dynamic programming problem* introduced in string matching context by Galil and Park [10].

Definition 1. *Given a real-valued function $w(h,j)$ for integers $0 \leq h \leq j \leq b$ and $C[0]$, the* one-dimensional dynamic programming problem *is to compute*

$$C[j] = \min_{0 \leq h < j} \{D[h] + w(h,j)\} \text{ for } 1 \leq j \leq b, \tag{2}$$

where $D[h]$ is computed from $C[h]$ in constant time.

From the above formula we obtain recurrence 1 by replacing $C[j]$ with $p(k,j)$, $D[h]$ with $p(k-1,h)$, and $w(h,j)$ with $w(h+1,j)$.

The one-dimensional dynamic programming problem, in turn, is equivalent to finding the *column minima* of a $b \times b$ upper triangular matrix C, defined by

$$C[h,j] = p(k-1,h) + w(h+1,j). \tag{3}$$

This problem is well-studied and lower bounds for its time complexity are known. Most importantly, it has been shown that if no further information on the function w is available, solving the column minima problem takes time $\Omega(b^2)$ [11]. This agrees with time complexity $O(b^2)$ of computing one row of the optimal multisplitting matrix p [5]. Hence, any algorithm that needs the scores of $(k-1)$-splits when computing the optimal k-split of a sample must take $\Theta(kb^2)$ time if no extra information about the function being optimized is available.

Aggarwal et al. [1] studied the closely related problem of finding the *row maxima* of rectangular matrices. They considered two kinds of monotonicity in the matrices. We translate the properties to the column minima problem. Finding the column minima of a matrix A is equivalent to finding the row maxima of its negated transpose, $-A^T$.

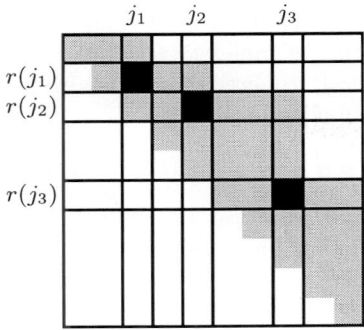

Fig. 1. Illustration of the effect of monotonicity: the black squares denote the locations of column minima. Only the gray area of the upper triangular matrix needs to be examined to discover the column minima of the matrix.

Definition 2. *Let $r(j)$ be the smallest row index for the minimum value in the column j in a matrix. The matrix is* monotone, *if $1 \leq j_1 < j_2 \leq b$ implies that $r(j_1) \leq r(j_2)$. If every submatrix of A is monotone, then A is* totally monotone.

An equivalent definition of total monotonicity was given by Galil and Park [10]. Given a pair of row indices, $h_1 < h_2$, and a pair of column indices, $j_1 < j_2$, in a totally monotone matrix C the following holds.

$$C[h_1, j_1] > C[h_2, j_1] \Longrightarrow C[h_1, j_2] > C[h_2, j_2]. \qquad (4)$$

In the whole matrix both monotonicity and total monotonicity imply that the minimum row indices are nondecreasing; $r(1) \leq r(2) \leq \cdots \leq r(k)$.

Aggarwal et al. [1] showed that the time complexity of the column minima problem is reduced to $\Theta(b \log b)$ when the matrix is monotone and introduced an $O(b)$ algorithm for the totally monotone case. In their work the matrix is a square, while the one-dimensional dynamic programming matrix is upper triangular. This does not affect the asymptotic time complexity, because a $b \times b$ upper triangular matrix contains a square submatrix of size $\lceil b/2 \rceil \times \lceil b/2 \rceil$. It is easy to construct a matrix where the respective column minima fit inside the square submatrix. Since the column minima problem needs to be solved also for this submatrix, the same $\Theta(b \log b)$ bound applies to upper triangular matrices.

In searching for the minimum element of column j_2, $j_1 < j_2$, in a monotone matrix, we never have to look rows with lesser index than $r(j_1)$. Thus, the area that needs to be covered in the matrix decreases as we go towards higher column indices (see Fig. 1). The matrix manipulation never needs to back-up to the rows with a smaller index than the least-recently found column minimum has.

These strong monotonicity conditions require the optimized function to be simple. Unfortunately, no attribute evaluation function is known to satisfy them. Hence, they do not seem to help in bringing the time complexity of optimal multisplitting down. Subsequently we show explicitly that two commonly-used attribute evaluation functions are not monotone. On the other hand, we are able to explain *TSE*'s linear-time complexity $O(kmb)$ with the help of monotonicity.

The most useful general property that *is known to hold* for many evaluation functions is Jensen's inequality [4]:

$$w(h, i) + w(i + 1, j) \leq w(h, j), \text{ for any } h \leq i \leq j. \quad (5)$$

Intuitively, this inequality states that we obtain a better score for a partition of the sample by dividing the final interval into two rather than keeping it together as one. Taking this further, any splitting of the data will make the partition score better.[1]

Inequality 5 is equivalent with the following *weak monotonicity* condition:

$$\forall i < j : C[h, i] \geq C[i, i] \implies C[h, j] \geq C[i, j]. \quad (6)$$

Weak monotonicity has the following interpretation. If a $(k-2)$-split of a prefix of the data has a better score than a $(k-1)$-split of the same prefix, the optimal multisplit cannot contain the $(k-1)$-split as its part. Elomaa and Rousu [6] use property 6 to dynamically prune the search space. In conjunction of the *ACE* function, the search space could be reduced in half, which indicates that the the expected case behavior of the search algorithm differs considerably from the above outlined worst case.

This property is not as helpful as the stronger monotonicity conditions. In the worst case, the items $C[i, i]$ are all column maxima and the location of the column minima cannot be constrained at all. Thus, the asymptotic behavior of the multisplitting algorithms cannot be improved below $\Omega(b^2)$ with the help of Jensen's inequality. It relates partitions of different arity, while the monotonicity conditions only talk about partitions with the same arity. In the following, we show by two counterexamples that Jensen's inequality and monotonicity are not equivalent.

4   Average Class Entropy

The Average Class Entropy of a partition $\biguplus_i R_i$ of a sample S is

$$ACE\left(\biguplus_i R_i\right) = \sum_i \frac{|R_i|}{|S|} H(R_i) = \frac{1}{|S|} \sum_i |R_i| H(R_i),$$

where H is the *entropy* function,

$$H(R) = -\sum_{j=1}^{m} P(C_j, R) \log P(C_j, R),$$

in which m denotes the number of classes and $P(C, R)$ stands for the proportion of the examples in R that belong to the class C. *ACE* is a concave function by the concavity of the entropy function [4]. Thus, it fulfills Jensen's inequality.

---

[1] This is the reason why we need to bound the partition arities in practical situations.

ACE is a component function in many well-known evaluation functions used, e.g., in decision tree learning; *Information Gain* and *Gain Ratio* [13] as well as *Normalized Distance* measure [12] are examples of evaluation functions that build upon ACE.

4.1 Average Class Entropy Is Not Monotone

We show by a simple example that ACE does not satisfy monotonicity.

Let the sample $S = \bigcup_{i=1}^{5} S_i$ consist of five indivisible subsets with the following class distributions for the two classes:

$$S_1 = \langle 0, 3 \rangle, S_2 = \langle 3, 0 \rangle, S_3 = \langle 17, 5 \rangle, S_4 = \langle 4, 5 \rangle, S_5 = \langle 4, 2 \rangle.$$

It is not possible to split the first indivisible subset into two partition intervals. Hence, $C[1,4] = C[1,5] = \infty$. The optimal ACE scores of the other 3-partitions are:

$$C[2,4] = ACE(S_1 \uplus S_2 \uplus (S_3 \cup S_4)) \approx 26.328$$
$$C[3,4] = ACE(S_1 \uplus (S_2 \cup S_3) \uplus S_4) \approx 25.031$$

$$C[2,5] = ACE(S_1 \uplus S_2 \uplus (S_3 \cup S_4 \cup S_5)) \approx 31.847$$
$$C[3,5] = ACE(S_1 \uplus (S_2 \cup S_3) \uplus (S_4 \cup S_5)) \approx 31.963$$
$$C[4,5] = ACE(S_1 \uplus (S_2 \cup S_3 \cup S_4) \uplus S_5) \approx 35.225$$

The values $C[3,4]$ and $C[2,5]$ are the minima for the columns 4 and 5, respectively. Their row indices, $r(4) = 3$ and $r(5) = 2$, violate the monotonicity condition, which requires non-decreasing row indices for the column minima.

5 Training Set Error

The *majority* class of a set R, denoted by $\text{maj}_C(R)$, is its most frequently occurring class. The number of *disagreeing* instances is given by

$$\delta(R) = |\{r \in R \mid \text{val}_C(r) \neq \text{maj}_C(R)\}|.$$

Training Set Error is the number of training instances falsely classified in the partition. For a partition $\biguplus R_i$ of S it is defined as

$$TSE\left(\biguplus_i R_i\right) = \sum_i \delta(R_i).$$

Also TSE is concave and, thus, fulfills Jensen's inequality.

The linear-time optimization of TSE is best understood from the relatively simple algorithm for it. Auer's [2] $O(kmb)$ time algorithm for optimal TSE

Table 1. Auer's [2] algorithm for optimal *TSE* partitions.

```
Partition optimalTSE(k, S, b, m)
/* Partition sample S into at most k intervals so that the training set
   error is minimized. There are b possible cut points and m classes. */

// DATA STRUCTURES: P[h,j] is the optimal partition of the processed
// data into h intervals with the last one labeled by j. E[h,j] is the
// corresponding number of disagreements. E' stores intermediate results.
Partition [k][m] P; int [k][m] E; int [m] E';

for ( h= 1 to k )                            // initialize
    for ( j= 1 to m ) {P[h,j]= ⟨S₁,j⟩; E[h,j]= 0;}
for ( i= 1 to b )                            // go through segments
{ for ( j= 1 to m ) {E'[j]= E[1,j]; E[1,j]+= δⱼ(Sᵢ);}
  for ( h= 2 to k )
  { j*= arg minⱼE'[j]; E*= E[j*];
    for ( j= 1 to m )
    { E'[j]= E[h,j];
      if ( E* < E[h,j] ) {P[h,j]= P[h-1,j*] ⊎ ⟨Sᵢ,j⟩; E[h,j]= E* + δⱼ(Sᵢ);}
      else {P[h,j]= P[h,j]; E[h,j]+= δⱼ(Sᵢ);}}}}
j*= arg minⱼE[k,j];
return P[k,j*];
```

multipartitioning (Table 1) assumes that the sample has been sorted into ascending order by the value of the numerical attribute under consideration. It maintains for each class j the optimal partitions of the prefixes of the sorted data into $1, \ldots, k$ intervals such that the last interval is labeled by j. In the algorithm $\delta_j(S)$ denotes the number of disagreements with class j in the subset S; $\delta_j(S) = |\{s \in S \mid \text{val}_C(s) \neq j\}|$.

In processing each new indivisible example segment a simple update of partitions suffices. We only need to check whether the new segment labeled by class j obtains less total disagreements when it is combined together with the previously known best partition of the data into k intervals with the last one labeled by j or together with the previously known best partition of the data into $k-1$ intervals with the last one labeled by an other class than j.

This is a consequence of the monotonicity condition. When scanning through the data the disagreement with respect to each class grows monotonically. Auer's algorithm can be seen to compute m (totally) monotone optimization problems in parallel.

5.1 Training Set Error Is Not Monotone

We show that *TSE* does not satisfy monotonicity, even though optimizing it can be decomposed into m monotone subproblems.

Again, let the sample $S = \bigcup_{i=1}^{5} S_i$ consist of five indivisible subsets with the following class distributions:

$$S_1 = \langle 2,1 \rangle, S_2 = \langle 0,1 \rangle, S_3 = \langle 6,2 \rangle, S_4 = \langle 1,3 \rangle, S_5 = \langle 4,2 \rangle.$$

Like above, $C[1,4] = C[1,5] = \infty$. The optimal *TSE* scores of the other 3-partitions are:

$$C[2,4] = TSE(S_1 \uplus S_2 \uplus (S_3 \cup S_4)) = TSE(\langle 2,1\rangle, \langle 0,1\rangle, \langle 7,5\rangle) = 6$$
$$C[3,4] = TSE(S_1 \uplus (S_2 \cup S_3) \uplus S_4) = TSE(\langle 2,1\rangle, \langle 6,3\rangle, \langle 1,3\rangle) = 5$$

$$C[2,5] = TSE(S_1 \uplus S_2 \uplus (S_3 \cup S_4 \cup S_5)) = TSE(\langle 2,1\rangle, \langle 0,1\rangle, \langle 11,7\rangle) = 8$$
$$C[3,5] = TSE(S_1 \uplus (S_2 \cup S_3) \uplus (S_4 \cup S_5)) = TSE(\langle 2,1\rangle, \langle 6,3\rangle, \langle 5,5\rangle) = 9$$
$$C[4,5] = TSE(S_1 \uplus (S_2 \cup S_3 \cup S_4) \uplus S_5) = TSE(\langle 2,1\rangle, \langle 7,6\rangle, \langle 4,2\rangle) = 9$$

The row indices, $r(4) = 3$ and $r(5) = 2$, of the column minima are increasing, contrary to the monotonicity condition.

5.2 Remarks

TSE differs from other functions that satisfy Jensen inequality by its optimization efficiency that is below the inherent complexity of dynamic programming algorithms based on recurrence 1. Nevertheless, *TSE* itself is not a monotone function and the optimization algorithm may have to back-up to recover the optimal partition. This can also be seen from the algorithm: an extra, one-dimensional table E' is needed to store intermediate results in the dynamic programming. The number of back-up points, though, is constant; there is only a single location per class to go back to.

A single left-to-right scan over the ordered data suffices to reveal the optimal *TSE* value, because for a fixed class and arity finding the optimal *TSE* value is a totally monotone problem. Thus, the lower bound for the restricted case is $\Omega(b)$. Combining this for all k arities and m classes gives total time requirement of $\Omega(kmb)$, which is the asymptotic time requirement of Auer's algorithm.

6 Discussion

The results in this article suggest that it may be hard to improve the asymptotic time complexity of the optimal multisplitting algorithm. If the algorithm relies on the best $(k-1)$-partitions to be computed for all prefixes when computing the k-partitions, $\Omega(b^2)$ bound seems inevitable for each row, which implies total complexity bound $\Omega(kb^2)$ for the whole problem. Hence, if more efficient algorithms are desired, the computation should be arranged in some other way than row by row or the rows should be made sparse by utilizing special properties of the evaluation function.

The strongest general property that is known to hold for many evaluation functions is Jensen's inequality, which guarantees that a partition cannot have a worse score than its subpartitions. However, this alone does not reduce the asymptotic time complexity. Stronger properties than Jensen's inequality—monotonicity or total monotonicity—lead to better time complexities. However, as shown by the counterexamples, neither of the properties hold for the commonly-used *ACE* and *TSE* functions.

The concept of monotone function is potentially relevant for the attribute evaluation functions used in machine learning algorithms. Designing monotone or even totally monotone functions would allow efficient optimization of these functions. Recent advances in solving the optimal multisplitting problem seem to bring the solution closer to the problem's inherent complexity bound.

Acknowledgements. We thank Esko Ukkonen for bringing into our attention the work done in the field of string matching.

References

1. Aggarwal, A., Klawe, M.M., Moran, S., Shor, P., Wilber, R.: Geometric Applications of a Matrix Searching Algorithm. Algorithmica **2** (1987) 195–208
2. Auer, P.: Optimal Splits of Single Attributes. Unpublished manuscript, Institute for Theoretical Computer Science, Graz University of Technology (1997)
3. Birkendorf, A.: On Fast and Simple Algorithms for Finding Maximal Subarrays and Applications in Learning Theory. In: Ben-David, S. (ed.): Computational Learning Theory. Lecture Notes in Artificial Intelligence, Vol. 1208, Springer-Verlag, Berlin Heidelberg New York (1997) 198–209
4. Cover, T.M., Thomas, J.A.: Elements of Information Theory. John Wiley and Sons, New York (1991)
5. Elomaa, T., Rousu, J.: General and Efficient Multisplitting of Numerical Attributes. Mach. Learn. **36** (1999) 201–244
6. Elomaa, T., Rousu, J.: Speeding Up the Search for Optimal Partitions. In: Żytkow, J., Rauch, J. (eds.): Principles of Data Mining and Knowledge Discovery. Lecture Notes in Artificial Intelligence, Vol. 1704, Springer-Verlag, Berlin Heidelberg New York (1999) 89–97
7. Elomaa, T., Rousu, J.: Generalizing Boundary Points. In: Proceedings of the Seventeenth National Conference on Artificial Intelligence. AAAI Press, Menlo Park, CA (2000) to appear
8. Fayyad, U.M., Irani, K.B.: On the Handling of Continuous-Valued Attributes in Decision Tree Generation. Mach. Learn. **8** (1992) 87–102
9. Fulton, T., Kasif, S., Salzberg, S.: Efficient Algorithms for Finding Multi-Way Splits for Decision Trees. In: Prieditis, A., Russell, S. (eds.): Machine Learning: Proceedings of the Twelfth International Conference. Morgan Kaufmann, San Francisco, CA (1995) 244–251
10. Galil, Z., Park, K.: A Linear-Time Algorithm for Concave One-Dimensional Dynamic Programming. Inf. Process. Lett. **33** (1990) 309–311
11. Galil, Z., Park, K.: Dynamic Programming with Convexity, Concavity and Sparsity. Theor. Comput. Sci. **92** (1992) 49–76

12. López de Màntaras, R.: A Distance-Based Attribute Selection Measure for Decision Tree Induction. Mach. Learn. **6** (1991) 81–92
13. Quinlan, J.R.: Induction of Decision Trees. Mach. Learn. **1** (1986) 81–106
14. Zighed, D.A., Rakotomalala, R., Feschet, F.: Optimal Multiple Intervals Discretization of Continuous Attributes for Supervised Learning. In: Heckerman, D. et al. (eds.), Proceedings of the Third International Conference on Knowledge Discovery and Data Mining. AAAI Press, Menlo Park, CA (1997) 295–298

Parametric Algorithms for Mining Share Frequent Itemsets

Brock Barber and Howard J. Hamilton

Department of Computer Science, University of Regina, Regina, SK., Canada S4S 0A2

Abstract. Itemset share, the fraction of some numerical total contributed by items when they occur in itemsets, has been proposed as a measure of the importance of itemsets in association rule mining. The IAB and CAC algorithms [4], are able to find share frequent itemsets that have infrequent subsets. These algorithms perform well but do not always find all possible share frequent itemsets. In this paper, we describe the incorporation of a threshold factor into these algorithms. The threshold factor can be used to increase the number of frequent itemsets found at a cost of an increase in the number of infrequent itemsets examined. The modified algorithms are tested on a large commercial database. Their behavior is examined using principles of classifier evaluation from machine learning.

1 Introduction

A data mining problem receiving considerable attention is the discovery of association rules from market basket data. The problem was first introduced in the context of bar code data analysis [1]. The goal of bar code data analysis is to identify buying patterns by examining itemsets, groups of items purchased together in transactions. From any itemset, an association rule can be derived which, given the purchase of a subset of the items in an itemset, predicts the probability of the purchase of the remaining items. The problem of discovering association rules from transaction data can be decomposed into two subtasks [1]: (1) find all itemsets meeting a minimum frequency requirement, and (2) generate association rules from the frequent itemsets. The second step is relatively easy compared to the first [14]. The focus of this paper is the first task, the extraction of frequent itemsets from transaction data. While the problem and our methods are general, we present the problem in terms of the retail sales domain, because it provides easily accessible intuitions for explaining problems, concepts and solutions.

In general, examination of all possible combinations of products and services offered by a retail organization is impractical, so methods are needed to focus effort on itemsets considered important to an organization. Itemset share, the fraction of some numerical value, such as total quantity of items sold or total profit, contributed by items when they occur in an itemset, has been proposed as a measure of itemset importance [5]. Unlike support [1], itemset share can be applied to non-binary numerical data that are typically associated with items in a transaction, allowing for a more in sightful analysis of the impact of itemsets in terms of stock, cost or profit. In practice, itemset ranking by support and share can be significantly different [5].

The support measure is downward closed since all subsets of a frequent itemset, are themselves frequent [3]. This property has permitted the development of efficient algorithms that traverse only a portion of the itemset lattice, yet find all possible frequent itemsets, e.g. [3, 9, 14]. However, since share can work with non-binary numerical values, the share of an itemset can be higher than the share of its subsets. Thus, if the frequency requirement is based on the total share of the itemset, frequent itemsets might contain infrequent subsets. Algorithms that do not rely on the property of downward closure have been proposed to extract this class of frequent itemset from transaction data [4]. The CAC and IAB algorithms perform well when applied to a large commercial database, finding most of the frequent itemsets while counting few infrequent itemsets. In this paper, we introduce parametric versions of these algorithms. A parameter called the threshold factor is used to increase (decrease) the ability of the algorithms to find frequent itemsets at the cost of increasing (decreasing) the number of infrequent itemsets examined. Algorithm behavior is evaluated using methods developed for classification system evaluation in machine learning.

2 The Share Measure and Share Frequent Itemsets

We summarize itemset methodology formally as follows [2]. Let $I = \{I_1, I_2, ..., I_m\}$ be a set of literals, called *items*. Let $D = \{T_1, T_2, ..., T_n\}$ be a set of n transactions, where for each transaction $T \in D$, $T \subseteq I$. A set of items $X \subseteq I$ is called an *itemset*. A transaction T contains an itemset X if $X \subseteq T$. Each itemset X is associated with a set of transactions $T_X = \{T \in D \mid T \supseteq X\}$, the set of transactions containing itemset X.

A *measure attribute* (MA) is a numerical attribute associated with each item in each transaction, such as quantity sold [4]. The *transaction measure value* of item I_p in transaction T_q, $tmv(I_p, T_q)$, is the value of a measure attribute associated with I_p in T_q. The *global measure value* of item I_p, $MV(I_p)$, is the sum of the transaction measure values of I_p in every transaction in which I_p appears, given as

$$MV(I_p) = \sum_{T_q \in T_{I_p}} tmv(I_p, T_q). \tag{1}$$

The *total measure value* (MV) is the sum of the global measure values for all items in I in every transaction in D, given as

$$MV = \sum_{p=1}^{m} MV(I_p). \tag{2}$$

We use x_i to denote the ith item of an itemset X. The *item local measure value* of item x_i in itemset X, $lmv(x_i, X)$, is the sum of the transaction measure values of the item x_i in all transactions containing X, given by

$$lmv(x_i, X) = \sum_{T_q \in T_X} tmv(x_i, T_q). \tag{3}$$

The *itemset local measure value* of itemset X, $lmv(X)$, is the sum of the local measure values of each of the k items in X in all transactions containing X, given by

$$lmv(X) = \sum_{1}^{k} lmv(x_i, X). \quad (4)$$

The *item share* of an item x_i in itemset X, $SH(x_i, X)$, is the ratio of the local measure value of x_i in X to the total measure value, as given by

$$SH(x_i, X) = lmv(x_i, X)/MV. \quad (5)$$

The *itemset share* of itemset X, $SH(X)$, is the ratio of the local measure value of X to the total measure value, as calculated by

$$SH(X) = lmv(X)/MV. \quad (6)$$

We illustrate share using the sample transaction database in Table 1. The TID column gives the transaction identifier values. Beneath each item name are values indicating quantity of item sold, the measure attribute. The table values are transaction measure values. For example, $tmv(D,T1)$ is 14. Global measure values for the items are indicated in the last row. The total measure value MV is 100. Table 2 provides local measure values, item shares and itemset shares for the itemset ACD and its subsets. For itemset AD, $lmv(AD) = lmv(A,AD) + lmv(D,AD) = 6+25 = 31$ and $SH(AD) = lmv(AD)/MV = 31/100 = 0.31$. Support (sup) is shown for comparison.

Table 1: Sample Transaction Database

TID	Item A	Item B	Item C	Item D	
T1	1	0	1	14	
T2	0	0	6	0	
T3	1	0	2	4	
T4	0	0	4	0	
T5	0	0	3	1	
T6	0	0	1	13	
T7	0	0	8	0	
T8	4	0	0	7	
T9	0	1	1	10	
T10	0	0	0	18	MV
$MV(I_p)$	6	1	26	67	100

Table 2: Sample Database Summary

X	sup	Itemset X		Item A		Item C		Item D	
		lmv	SH	lmv	SH	lmv	SH	lmv	SH
A	0.3	6	0.06	6	0.06	-	-	-	-
C	0.8	26	0.26	-	-	26	0.26	-	-
D	0.7	67	0.67	-	-	-	-	67	0.67
AC	0.2	5	0.05	2	0.02	3	0.03	-	-
AD	0.3	31	0.31	6	0.06	-	-	25	0.25
CD	0.5	50	0.50	-	-	8	0.08	42	0.42
ACD	0.2	23	0.23	2	0.02	3	0.03	18	0.18

To find frequent itemsets with infrequent subsets, we employ the following definition of share frequency. An itemset X is *share frequent*, or simply *frequent*, if $SH(X) \geq minshare$, a user defined minimum share value. This definition of frequency is not downward closed. A property P is *downward closed* with respect to the lattice of all itemsets if, for each itemset with the property P, all of its subsets also have the property P [12]. However, the share of an itemset may increase or decrease as the itemset is extended by the adding an item. Adding an item x_i to a k-itemset X to create a new $(k+1)$-itemset Y, adds a restriction to the measure values of the items in X. The measure values associated with the items in X contribute to the local measure value of Y, only when they occur with the new item x_i. Their contribution towards the local measure value of Y must be less than or equal to their contribution to the local measure value of X. However, the local measure value of x_i is added to the local measure value of Y, which may be less than, equal to, or greater than the local measure value of X. If share frequency is measured against the share of the itemset, an itemset with share above *minshare*, may have component itemsets with share below *minshare*.

3 Description of Algorithms

The Combine All Counted (CAC) and Item Add-Back (IAB) algorithms extract share-frequent itemsets from transaction data, including those with infrequent subsets [4]. Parametric versions of the algorithms, which we refer to as parametric CAC (PCAC) and parametric IAB (PIAB), are introduced here. The major modification is the addition of a threshold factor.

The first pass through the data collects information about all 1-itemsets in the data. Summary information is compiled, including MV and TC_T, the total number of transactions. C_k is the set of candidate itemsets for the kth pass. C_2 is generated using information about the 1-itemsets and information about the candidate 2-itemsets is collected in pass 2. The process of building C_k using itemsets in C_{k-1} continues until no candidate itemsets are added to C_k. After the kth pass, the local measure value and transaction count is available for each counted k-itemset.

In the IAB algorithm, each item with a non-zero transaction count is given the chance to contribute in the generation of candidate itemsets for every pass. In the kth pass, candidate itemsets are generated by adding to each itemset in C_{k-1}, any item found in the first pass that is not contained in the itemset. In the absence of pruning, this would produce an exhaustive algorithm. To prevent this, three types of pruning are used. First, *zero pruning* removes any itemset $X_i \in C_{k-1}$ for which $TC_{Xi} = 0$. Second, *share infrequency pruning* removes any itemset $X_i \in C_{k-1}$ whose actual share $SH(X_i) < minshare$. Third, *predictive pruning* uses a heuristic to calculate the predicted share of a potential candidate itemset X_{pc}, $PSH(X_{pc})$, and prunes any X_{pc} where $PSH(X_{pc}) < minshare$. $PSH(X_{pc})$ is based on the actual share of components of X_{pc} [4].

We modify predictive pruning to create the PIAB algorithm. The intuition is that itemsets having nearly enough share should not be pruned since their supersets may have enough. We define the *threshold factor* (*TF*) as a parameter, ranging from 0.0 to 1.0, that is applied to the share threshold prior to comparison of a predicted share value. Potential candidate itemsets are added to C_k only if $PSH(X_{pc}) \geq TF*minshare$. For $TF = 1.0$, the parametric versions of the algorithms behave identically to the non-parametric versions. The threshold factor is similar to the *relaxation factor*, a parameter that has been applied to the support measure as a means of adjusting algorithm accuracy [10].

The processes of candidate itemset generation and itemset pruning are encapsulated in a procedure **GenerateCandidateItemsets**. The generation of the next potential candidate itemset X_{pc}, is represented by an iterator procedure **GenerateNextItemset**. The first call to the procedure returns the first generated itemset, repeated calls cycle through all possible generated itemsets and when no more itemsets can be generated, the procedure returns false. The value of $PSH(X_{pc})$ is returned by the function **GetPredictedShare**. For PIAB, the procedure **GenerateCandidateItemsets** is written as:

GenerateCandidateItemsets (C_{k-1})

```
1    foreach X_i ∈ C_{k-1}
2        if TC_{Xi} = 0 or SH(X_i) < minshare then
3            remove X_i from C_{k-1}
4    while X_pc := GenerateNextItemset() do
5        PSH(X_pc) := 0, SubsetCount := 0
6        foreach x_i ∈ X_pc
7            foreach x_j ∈ X_pc where i ≠ j
8                add x_j to X_s
9            if X_s ∉ C_{k-1} then
10               continue
11           PSH(X_pc) := PSH(X_pc) + GetPredictedShare(x_i,X_s)
12           SubsetCount := SubsetCount +1
13       if PSH(X_pc)/SubsetCount ≥ TF*minshare then
14           Add X_pc to C_k
```

In the CAC algorithm, each counted itemset is given a chance to contribute to the generation of a larger frequent itemset. Itemsets are generated by combining itemsets in C_{k-1} which differ only in their last item. Again, in the absence of pruning, the algorithm is exhaustive. Zero pruning and predictive pruning are done as with IAB. Infrequency pruning is done, but after the generation of candidate itemsets. In addition, *subset pruning* prunes any generated itemset with a k-1 subset not found in C_{k-1}. To create the PCAC algorithm, the threshold factor is incorporated. For PCAC, the procedure **GenerateCandidateItemsets** is written:

GenerateCandidateItemsets (C_{k-1})

```
1    foreach X_i ∈ C_{k-1}
2        if TC_{Xi} = 0 then
3            remove X_i from C_{k-1}
4    while X_pc := GenerateNextItemset() do
5        PSH(X_pc) := 0, SubsetCount := 0
```

```
6         foreach $x_i \in X_{pc}$
7             foreach $x_j \in X_{pc}$ where $i \neq j$
8                 add $x_j$ to $X_s$
9             if $X_s \notin C_{k-1}$ then
10                break;
11            $PSH(X_{pc}) := PSH(X_{pc}) + GetPredictedShare(x_i, X_s)$
12            $SubsetCount := SubsetCount + 1$;
13        if $SubsetCount = k$ and $PSH(X_{pc})/SubsetCount \geq TF*minshare$ then
14            Add $X_{pc}$ to $C_k$
```

We illustrate the effect of the threshold factor by comparing the behavior of the CAC and PCAC algorithms. Figure 1 gives the itemset lattice for the sample data set in Table 1. Each node in the lattice is labeled with the itemset name. Below the itemset name are the total measure value of the itemset in all transactions and the number of transactions in which the itemset appears, separated by a forward slash. The total measure value *MV* is equal to 100 and, assuming *minshare* is 0.20, any item with a measure value greater than or equal to 20 is share frequent. Frequent itemsets are shaded in Figure 1. A threshold factor of 0.60 is assumed.

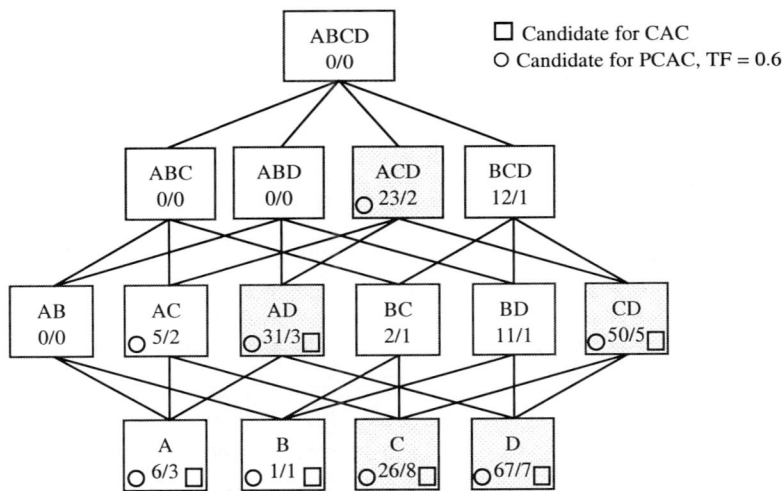

Figure 1: Itemset Lattice

CAC: The first pass identifies 1-itemsets C and D as frequent itemsets. All counted 1-itemsets are used to generate the candidate itemsets for pass 2. However, before any 2-itemset is added to C_2, we prune based on predicted share value [4]. Consider itemset AC. $SH(A) = 0.06$, $TC_A = 3$ and $SH(C) = 0.26$, $TC_C = 8$. Since $TC_A < TC_C$, $PSH(AC) = SH(A) + SH(C)*(TC_A/TC_T) = 0.06+0.26*(3/10)=0.14$ which is less than *minshare*, so AC is pruned. Only AD and CD meet the minimum share requirement so they are the only 2-itemsets counted in pass 2. The algorithm terminates after pass 2 because AD and CD cannot be used to generate a 3-itemset with all subsets existing in C_{k-1}. The frequent 3-itemset ACD is missed because the subset AC was not counted.

PCAC: PCAC performs in the same way as CAC until predictive pruning, where PSH(AC) = 0.14 is now compared to $TF*minshare$ = 0.12. Since the less stringent requirement is met, AC is also counted in pass 2. After the second pass we find that AC is infrequent, but before discarding it, we use it to generate candidate itemsets. It is combined with itemset AD to generate ACD. Since we counted AC, AD and CD in pass 2, ACD is not subset pruned. PSH(ACD) = 0.26 which is greater than $TF*minshare$, so ACD is counted in the third pass. The algorithm terminates after the third pass since no 4-itemsets can be created from a single 3-itemset. We counted one additional itemset and it was frequent. By setting a less stringent criteria for deciding which itemsets to count, PCAC may find frequent itemsets that CAC missed. However, the increased effectiveness comes at the cost of counting more infrequent itemsets.

4 Evaluation Criteria

An algorithm for extracting frequent itemsets from a data set can be thought of as a method for classifying itemsets into two classes, frequent itemsets (positive instances) and infrequent itemsets (negative instances). A confusion matrix [7] gives information about actual and predicted classifications done by a classification system. Performance of such systems is commonly evaluated using terms defined on the matrix data. These terms are listed in Table 3. Here, a is the number of correct predictions that an itemset is infrequent, b is the number of incorrect predictions that an itemset will be frequent, c is the number of incorrect of predictions that an itemset will be infrequent, and d is the number of correct predictions that an itemset will be frequent.

Table 3: Classifier Evaluation Terms

Term	Proportion of	Equation
Accuracy (AC)	total number of predictions that were correct	$(a+d)/(a+b+c+d)$
True Positive Rate (TP)	positive cases correctly identified	$d/(c+d)$
False Positive Rate (FP)	negatives cases incorrectly classified as positive	$b/(a+b)$
True Negative Rate (TN)	negatives cases classified correctly	$a/(a+b)$
False Negative Rate (FN)	positives cases incorrectly classified as negative	$c/(c+d)$
Precision (P)	predicted positive cases that were correct	$d/(b+d)$

Most itemsets in an itemset lattice are infrequent or do not occur at all. Accuracy may not be an adequate performance measure when the number of negative cases is much greater than the number of positive cases [8].

We use ROC graphs to examine the general behavior of the PIAB and PCAC algorithms with variation of the threshold factor. An ROC graph is a plot with the true positive rate (TP) on the Y axis and the false positive rate (FP) on the X axis [13]. Point (0,1) is the perfect classifier: classifying all cases correctly. Point (0,0) represents a classifier that predicts all cases to be negative, while the point (1,1) corresponds to a classifier that predicts every case to be positive. Point (1,0) is the classifier that is incorrect for all classifications. The CAC and IAB algorithms produces a single ROC point, a (*FP*,*TP*) pair for a particular data set. In the PCAC and PIAB algorithms, each *TF* value produces a (*FP*,*TP*) pair for the data set. A series of such pairs can be used to plot an ROC curve.

It has been suggested that the area beneath an ROC curve can be used as a measure of accuracy in many applications [13]. In [11], it is argued that using classification accuracy to compare classifiers is not adequate unless cost and class distributions are completely unknown and a single classifier must be chosen to handle any situation. They propose evaluation of classifiers using a ROC graph and imprecise cost and class distribution information.

Itemset algorithms must have sufficient generality that they can be applied to any transaction database and so, we choose not to include assumptions regarding cost and class distributions in our analysis. ROC graphs provide a visual tool for examining the tradeoff between the ability of a classifier to correctly identify positive cases and the number of negative cases incorrectly classified. We use ROC graphs to examine algorithm behavior, without making claims about the relative accuracy of the parametric algorithms.

5 Experimental Results

All experiments were carried out using a 450 Mhz Pentium II PC with 384 MB of RAM. Test data were drawn from an eight million record customer database provided to us by a commercial partner. Transaction information is contained in a three million record table representing the purchase of 2200 unique items by over 500,000 customers. Five discovery tasks were performed on different subsets of the data. Baseline information was obtained using the ZSP algorithm [4], which is guaranteed to find all frequent itemsets but requires a large amount of space and time. By knowing the true answers, we are able to evaluate the behavior and performance of our parametric algorithms. A share threshold of 0.02 was used, since this produced tasks small enough to run the ZSP algorithm on but not too small to be interesting. Information about the tasks is contained in Table 4. Tasks T1, T2, T3 and T4 were run for threshold factors 1.0, 0.9, 0.8, 0.7, 0.6, 0.5, 0.4, 0.3, 0.2, 0.1 and 0.0. T5 was tested for threshold factors 1.0, 0.9, 0.8, 0.7, 0.6, 0.5, 0.1 and 0.0.

Table 4: Task Metrics

Task Identifier	T1	T2	T3	T4	T5
Transaction Count	599	2058	2374	5092	14257
Record Count	1844	4373	7051	24725	43700

Figure 2 shows the ROC graphs for PCAC and PIAB for four of the five tasks. Only the upper portions of the graphs are shown because this includes all ROC points. The ROC graph for T3 is not shown because both algorithms found all frequent itemsets for all values of *TF*. Decreasing the threshold factor simply moves successive ROC points further to the right on the top axis of the ROC graph.

At *TF* equal to 1.0, the PCAC and PIAB algorithms are equivalent to CAC and IAB. As *TF* is decreased towards 0.0, the share frequency criterion becomes less stringent and more of the itemsets that are generated may be selected for counting. The general trend is for the ROC points to move upward and to the right as the threshold factor is decreased. However, the true positive rate tends to move up in a series of steps. This is best illustrated in Figure 2(a) where the initial ROC points for the PIAB algorithm in T1 move to the right but not upward. The true positive rate for the PIAB algorithm remains the same until the threshold factor equals 0.4. Similar behavior is displayed for the PCAC algorithm, although it seems more prevalent in the IAB algorithm. The step behavior of the true positive rate occurs because a decrease in threshold factor will have no effect until certain discrete events occur. Additional frequent itemsets must be generated, predicted to be frequent, not subset pruned for the PCAC algorithm and counted. The false positive rate always increased with a

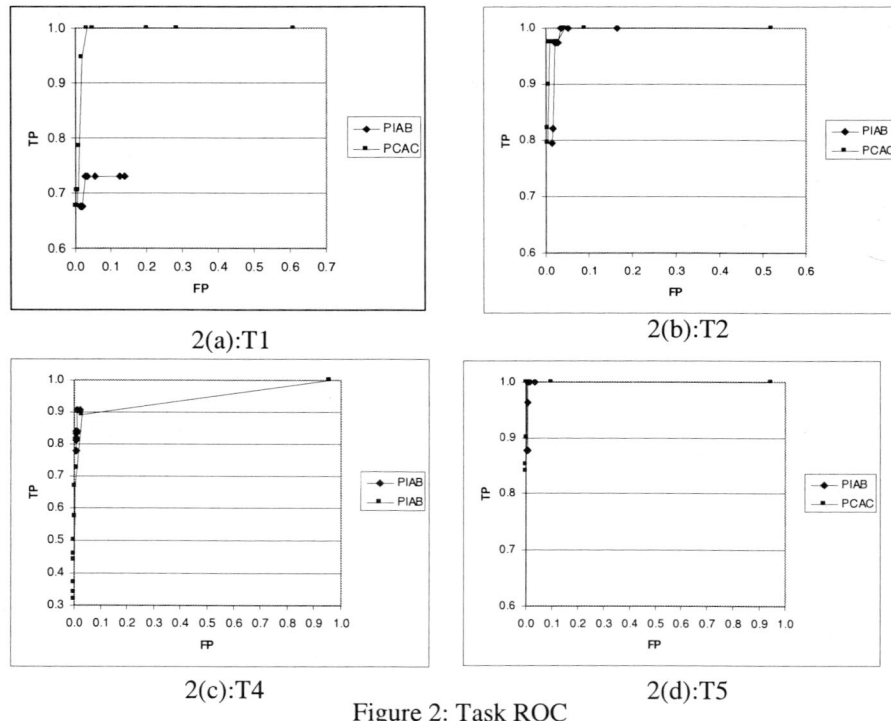

Figure 2: Task ROC

decrease in the threshold factor, although in some cases, the increase was small.

In the PCAC algorithm, all counted itemsets are used to generate potential candidate itemsets. As the threshold factor is decreased, the number of counted itemsets increases and thus, more itemsets are available for itemset generation and fewer itemsets are subset pruned. For $TF = 0$, the PCAC accepts any itemset that is generated, provided that none of its subsets have been zero pruned, so the CAC algorithm becomes equivalent to the ZSP algorithm, with $TP_{PCAC} = TP_{ZSP} = 1$ and $FP_{PCAC} = FP_{ZSP}$. In other words, if the threshold factor is decreased far enough, the algorithm is guaranteed to find all frequent itemsets. On the other hand, the PIAB algorithm does not guarantee this. The upper limit of TP equals 1 and this limit was reached in T2 and

T5. However, in T1 and T4, TP_{PIAB} is less than 1 for a threshold factor of 0.0. Itemsets are generated by adding single items to frequent itemsets from the previous pass. If the set of frequent itemsets for the kth pass does not change with a decrease in the threshold factor, then the foundation of the generated itemsets remains the same. Once all itemsets that can be generated by adding single items to frequent itemsets are counted, the true and false positive rates are at their maximum. Thus, there is a trade-off between algorithms. The PCAC algorithm will find all frequent itemsets, provided the threshold factor is low enough, but it may count many infrequent itemsets. The PIAB algorithm counts fewer infrequent itemsets at lower threshold factors but may not find some frequent itemsets, regardless of the threshold factor.

The degree of variation between the ROC curves for different tasks run against the same transaction database indicates that we may not be able to provide a single ROC curve that is applicable to all tasks or transaction databases. Further tests against different transaction databases will be required to confirm or refute this. However, the general shape of the ROC curve seems consistent enough to provide the user with insight into the general behavior that can be expected as the threshold factor is varied.

6 Conclusions

We described the incorporation of a threshold factor into algorithms for discovering share frequent itemsets, including those that contain infrequent subsets. The threshold factor can be used to increase (decrease) the effectiveness of the algorithms at a cost of an increase (decrease) in the number of infrequent itemsets examined. The CAC and IAB are useful heuristic algorithms for finding share frequent itemsets. This study indicates that a threshold factor can be successfully incorporated into these algorithms to increase the number of frequent itemsets found.

References

1. Agrawal A., Imielinksi T., Swami A. Mining Association Rules between Sets of Items in Large Databases. *Proceedings of the ACM SIGMOD International Conference on the Management of Data*. Washington D.C., (1993) 207-216.
2. Agrawal A., Mannila H., Srikant R., Toivonen H., Verkamo A.I. Fast Discovery of Association Rules. Fayyad U.M., Piatetsky-Shapiro G., Smyth P. and Uthurusamy R. Eds., *Advances in Knowledge Discovery and Data Mining*, Menlo Park, California, (1996) 307-328.
3. Agrawal A., Srikant R. Fast Algorithms for Mining Association Rules. *Proceedings of the 20th International Conference on Very Large Databases*. Santiago, Chile, (1994) 487-499.
4. Barber B., Hamilton H.J. Algorithms for Mining Share Frequent Itemsets Containing Infrequent Subsets. *4th European Conference on Principles and Practices of Knowledge Discovery in Databases*, Lyon, France (2000).
5. Carter C.L., Hamilton H.J., Cercone N. Share Based Measures for Itemsets. *Proceedings of the First European Conference on the Principles of Data Mining and Knowledge Discovery*. Trondheim, Norway, (1997) 14-24.
6. Hilderman R.J., Carter C., Hamilton H.J., Cercone N. Mining Association Rules from Market Basket Data using Share Measures and Characterized Itemsets. *International Journal of Artificial Intelligence Tools*. 7(2): (1998) 189-220.
7. Kohavi R., Provost F. Glossary of Terms. Machine Learning, 30(2): (1998) 271-274.
8. Kubat M., Holte R.C., Matwan S. Machine Learning for the Detection of Oil Spills in Satellite Radar Images. *Machine Learning*, 30(1): (1998) 195-215.

9. Mannila H., Toivonen H., Verkamo A.I. Efficient Algorithms for Discovering Association Rules. *Proceedings of the 1994 AAAI Workshop on Knowledge Discovery in Databases.* Seattle, Washington, (1994) 144-155.
10. Park J.S., Yu P.S., Chen M. Mining Association Rules with Adjustable Accuracy. *Proceedings of the Sixth International Conference on Information and Knowledge Management.* (1997) 151-160.
11. Provost F., Fawcett T. Analysis and Visualization of Classifier Performance: Comparison under Imprecise Class and Cost Distribution. *Proceedings of the Third International Conference on Knowledge Discovery and Data Mining.* (1997) 43-48.
12. Silverstein C., Brin S., Motwani R. Beyond Market Baskets: Generalizing Association Rules to Dependence Rules. *Data Mining and Knowledge Discovery,* 2(1): (1998) 39-68.
13. Swets J.A. 1988. Measuring the Accuracy of Diagnostic Systems. *Science,* 240, (1988) 1285-1293.
14. Zaki M.J., Parthasarathy, M., Ogihara M., Li W. New Algorithms for Fast Discovery of Association Rules. *Proceedings of the Third International Conference on Knowledge Discovery & Data Mining.* Newport Beach, California, (1997) 283-286.

Discovery of Clinical Knowledge in Hospital Information Systems: Two Case Studies

Shusaku Tsumoto

Department of Medicine Informatics, Shimane Medical University, School of Medicine, 89-1 Enya-cho Izumo City, Shimane 693-8501 Japan
E-mail: tsumoto@computer.org

Abstract. Since early 1980's, the rapid growth of hospital information systems stores the large amount of laboratory examinations as databases. Thus, it is highly expected that knowledge discovery and data mining(KDD) methods will find interesting patterns from databases as reuse of stored data and be important for medical research and practice because human beings cannot deal with such a huge amount of data. However, there are still few empirical approaches which discuss the whole data mining process from the viewpoint of medical data.In this paper, KDD process from a hospital information system is presented by using two medical datasets. This empirical study show that preprocessing and data projection are the most time-consuming processes, in which very few data mining researches have not dicussed yet and that application of rule induction methods is much easier than preprocessing.

1 Introduction

Medical practice and research has been changed by rapid growth of life science, including biochemistry and immunology(Levinson, 1996). The mechanism of a disease can be explained as a biochemical process or cell disorder and the diagnostic accuracy of medical experts is increasing due to the development of laboratory examinations. However, it is also true that data analysis is very indispensable to generating a hypothesis. For instance, discovery of HIV infection and Hepatitis type C were inspired by analysis of clinical courses unexpected by experts on immunology and hepatology, respectively (Fauci et al, 1997). Although the life science has been rapidly advanced, mechanisms of many diseases are still unknown: especially, neurological diseases were very difficult to analyze because their prevalence is very low (Adams and Victor, 1993). Even the mechanism of diseases with high prevalence, such as cancer, is partially known to medical experts. In this sense, medical research always need a good hypothesis, which is one of the most important motivations to data mining and knowledge discovery for medical people.

Also, another aspects interest medical researchers in data mining. Since early 1980's, the rapid growth of hospital information systems (HIS) stores the large amount of laboratory examinations as databases (Van Bemmel, and Musen, 1997). For example, in a university hospital, where more than 1000 patients visit from Monday to Friday, a database system stores more than 1 GB numerical data of laboratory examinations for each year. Furthermore, storage of medical image and other types of data are discussed in medical informatics as research topics on electronic patient records and all the medical data will be stored in hospital information systems within the 21th century. Thus, it is highly expected that data mining methods will find interesting patterns from databases as reuse of stored data and be important for medical research and practice because human beings cannot deal with such a huge amount of data.

In this paper, knowledge discovery and data mining (KDD) process (Fayyad, et. al, 1996) for two medical datasets extracted from a hospital information system is presented. This empirical study show that preprocessing and data proejction are the most time-consuming processes, in which very few data mining researches have not dicussed yet and that application of rule induction methods is much easier than preprocessing.

2 Data Selection

In this paper, we use the following two datasetsfor data mining, which are extracted from two different hospital information systems. One is bacterial test data, which consists of 101,343 records, 254 attributes. This data includes past and present history, physical and laboratory examinations, diagnosis, therapy, a type of infectious disease, detected bacteria, and sensitivities for antibiotics. The other one is a dataset on the side effect of steroid, which consists of 31,119 records, 287 basic attributes. This data includes past and present history, physical and laboratory examinations, diagnosis, therapy, and the type of side effects. The characteristic of the second dataset is that it is a temporal database: although it includes 287 basic attributes, 213 attributes of which have more than 100 temporal records. These datasets are obtained through the first to third steps of KDD process: data selection, data cleaning and data reduction.

In the first step of KDD process, these databases are extracted from two different hospital information systems by simple queries. Table 1 gives results for data selection: the second column shows the total size of HISs when data were selected (December, 1998). The third column presents the size of data selected from HIS. Finally, the fourth column gives the computational time required for data selection. Since each HIS is implemented on different computers, it may be difficult to compare each computational time, but those values suggests that time is not dependent on the selected data, but on the total HIS size.

Table 1. Data Selection Results

Datasets	HIS size	Target Data	Time Required
Bacterlal Test	1,275,242(52GB)	361,932(14GB)	2.3 Days
Side-Effect	2,631,549(100GB)	135,749(6GB)	7.3 Days

3 Data Cleaning as Preprocessing

After data selection, data cleaning is required since the data obtained from the first step are not clean, including data records not suitable for data analysis. Even though these records are selected by matching the query condition, they do not include enough amount of information. In the case of bacterial tests, some patients have information about bacterial tests, but they made very few laboratory examinations. On the other hand, in the case of side effects, some patients have a record on steroid therapy, but they made two or three examinations for each year since the status of allergic diseases was stable and the patients do not suffer from any side-effects. These data may be removed if the queries in the first step are refined. However, it should be pointed out that the refinement of queries is not so a easy task because we have many types of insuitable data, which is difficult to predict and will occur not only due to the factors of patients but also due to the factors of medical doctors. For example, some patients may not come to the outpatient clinic when they recover from allergic disorders. Some doctors may tell the patients not to come to the clinic so often. In some cases, some patients are found to suffer from very a severe disease and they should be admitted to a university hospital. In other cases, patients may move from a university hospital to a municipal hospital. Many factors not included in a database, mainly social factors, will be a cause for degrading the cleaness of data.

Thus, it is much easier to clean the data not by using complex queries but by using simple statistics. Since each domain has attributes indispensabe to evaluating the status of patients, the number of attributes used to describe each record is a good index for removing not-clean records. So, we define the two-fold cleaning steps: first we select the records which have no missing values in the pre-defined indispensable attributes. Then, we calculate how many attributes in the remaining attributes are used to describe the records in the first step. If the number of attributes used for a case is not sufficient, then this case will removed. For those steps, the indispensable attributes and the threshold for the second selection are given a priori by domain experts. In the case of bacterial tests, 254 attributes are very important. Furthermore, 27 attributes are indispensable to describe each case. Thus, in the data cleaning step, first select the records which have missing values in the 27 attributes. Then, calculate how many attributes have missing values in remaining 227 attributes. If 75% of them are missing , then remove these records. In the case of steroid side-effects, the same strategy is applied. 74 non-temporal attributes are indispensable and 217 temporal attributes are used for second selection.

Table 2. Data Cleaning Results

Datasets	Data	Cleaned Data	Time Required
Bacterlal Test	361,932(14GB)	101,343(3GB)	5.2 Days
Side-Effect	135,749(6GB)	31,119(1,5GB)	2.5 Days

Table 2 summarizes results for data cleaning. The second column shows the number of records selected in the data selection. The third column shows the size of data after the data cleaning process is applied. The fourth column gives the computational time required for the whole steps.

4 Data Reduction and Projection as Preprocessing

4.1 Data Projection for Bacterial Test Database

From the viewpoint of table processing, data cleaning can be viewed as cleaning steps in the direction of records, that is, in the direction of row. On the other hand, data reduction can be viewed as cleaning steps in the direction of attributes, that is, in the direction of column. Although the data cleaning process is a time-consuming process, it takes much more time to reduce and project data in clinical databases due to the characteristics of biological science, including medicine. The tradition of classification in biology tends to have a large scale of classification systems. For example, let us consider the classfication of bacteria. If we look at the classification tree of bacteria, more than millions of bacteria are classified neatly in one classification tree. However, the number of bacteria on which we want to focus in medicine is very few, compared with this total classification. But, even the number of bacteria used in bacterial tests are too many, compared with the total number of classes used in data mining techniques. In the case of bacterial test database, 1194 kinds of bacteria and other bioorganism are used as a target class. If conventional classification or statistical methods are applied to this data, most of induced rules or patterns may be useless because these methods tend to extract knowledge for differentiation between 1194 classes. These tendencies are unexpected for medical experts who want to have more generalized results.

Thus, generalization of such over-classified attributes is required to discover rules which are easy for medical experts to interpret. For bacterial test databases, a simple concept hierarchy was used for generalization of values. All the bioorganism are classified into 1124 bacteria and 70 non-bacteria bioorganism. Then, 1124 bacteria are classified into aerobic (973) and anaerobic (151). ¿From the levels below, which is not shown in this figure, conventional bacterial classification sytem is used for construction of tree, into which totally five levels of hierarchy is implemented. For each data mining task, we set up which hierarchical level is suitable for data analysis. Although data projection for this dataset is parallel with data mining process, we show how much time is needed for each data projection in Table 3.

The first column denotes the level of concept hierarchy used for generalization. The second one shows the number of total values for compairson. Ther third column gives the number of total values for generalization.The fourth column gives computational time for generalization. Finally, the fifth column gives time for construction of a hierarchical tree by interaction between data analyzers and domain experts. Especially, since a construction of hierarchical tree can be viewed as a knowledge acquisition process, it cannot be automated. In addition to

Table 3. Generalization of Bacteria

Projection Level	Generalized Values	Time Computation	Time Construction
2. (Bacteria)	2	24 hours	1.0 Days
3.	5	25 hours	1.4 Days
4	52	26 hours	3.0 Days
5	175	27 hours	7.0 Days
6	1194	0	-

this hierarchical tree, the following hiearchical tre is also needed for generalization of this dataset: a classification tree for chronic diseases which each patient suffers from. This process also takes 7 days to complete the tree structure with domain experts.

4.2 Data Reduction for Steroid Side-Effect Database

Characteristics of Medical Temporal Databases. Since incorporating temporal aspects into databases is still an ongoing research issue in database area (Abiteboul, et. al., 1995), temporal data are stored as a table in hospital information systems (H.I.S.) with time stamps. The characteristics of medical temporal data are as follows(Tsumoto 1999): (1)The Number of Attributes are too many, (2) Irregularity of Temporal Intervals, and (3)Missing Values.

Data Reduction using Moving Average Method. The way to how to deal with medical temporal databases is discussed in (Tsumoto, 1999). Tsumoto introduces extended moving average method, which automatically set up the scale of the temporal interval. For example, if the scale factor is set to be 2, then the temporal interval for moving average is calculated from 2, 4, 8, 16, and so on. Each temporal interval is called window. In general, let s and yi denote a scale factor and a value for laboratory test. Then moving average for y is defined as:

$$\hat{y}_w = \sum_{j=1}^{w} \frac{y_j}{w}.$$

where n denotes an integer which is set up for temporal interval. Thus, sn gives the size of window. One of the disadvantages of moving average method is that it cannot deal with categorical attributes. To solve this problem, we will classify categorical attributes into three types, whose information should be given by users. The first type is constant, which will not change during the follow-up period. The second type is ranking, which is used to rank the status of a patient. The third type is variable, which will change temporally, but ranking is not useful. For the first type, extended moving average method will not be applied. For the second one, integer will be assigned to each rank and extended moving average method for continuous attributes is applied. On the other hand, for the third one, the temporal behavior of attributes is transformed into statistics by using frequencies.

For further discussion on data reduction of temporal data, the readers may refer to (Tsumoto, 1999).

Table 4. Computational Time for Data Summarization

Window Size	Computational Time
$2^7 (= 128)$	12.0 hours
$2^8 (= 256)$	8.0 hours
∞	7.0 hours

Results of Data Reduction for Steroid Side Effects. Steroid side-effects is known as long-term side-effects, usually observed when a patient takes steroid for more than several years. Thus, to capture long-term effects, the window size is set to $2^7 (= 128)$ and $2^8 (= 256)$. It is true that a significant amount of temporal information is lost by using data reduction, but we should remember that the first objective of data mining is to find simple useful and unexpected patterns from data. As discussed in 4.2.1, medical temporal data suffer from many types of irregularities. Table 4 shows the computational time required for data summerization. It is notable that this table shows the trade-off relationship between the window-size and computational time: if the window-size is smaller, the computational time grows much larger.

4.3 Total Time Required for Data Reduction

In summary, Table 5 gives the total time required for data reduction and projection for each data set, including knowledge acquisition process. The second column gives the type of preprocessing. The third column shows total time required for each process. Finally, the fourth column shows the time required for acquisition of knowledge from domain experts.

Table 5. Total Time Required for Data Reduction

Dataset	Preprocessing	Total Time	Time for Acquisition
Bacterial Test	Projection	15.25 Days	7.0 Days
Side-Effect	Summarization	2.3 Days	0

This table suggests the generalization of values in attributes should be a time-consuming process, especially when domain knowledge is given.

5 Rule Induction as Data Mining

After the third step, rule induction based on rough set model (Pawlak, 1991) was applied to two medical datasets. Tsumoto (1998, 2000) extends rough-set-based rule induction methods into probabilistic domain.

In this section, we skip this part due to the limitation of space. For further discussion, the readers may refer to (Tsumoto, 1998; Polkowski and Skowron, 1998). The algorithm introduced in (Tsumoto 1999) was implemented on the Sun Spaarc station and was applied to the above two medical databases, the information of which is summarized in Table 6. For rule induction, the thresholds for accuracy and coverage are set to 0.5 and 0.5, respectively.

For bacterial test databases, six attributes for which domain experts want to find simple pattern are assigned to decision attributes. For side-effect databases, one attribute (side-effect) is assigned to a decision attribute. Table 7 summarizes the results of data mining. The second and third columns give information about data. The fourth column shows the number of rules induced by rule induction methods. Finally, the fifth column presents the computational time required. It is notable that the computational time is rather small, compared with computational time.

Table 6. Summary of Data Mining

Data	Size	Attributes	Rules	Computational Time
Bacterial Test	101,343 (3GB)	254	24,335	60 hours (2.5 Days)
Side-Effects	31,119 (1.5GB)	287	14,715	18 hours (0.75 Days)

6 Interpretation of Induced Rules

After the data mining step, we obtain many rules to be interpreted by medical experts. Even if the amount of information is very small compared with the original databases, it still takes about one week to evaluate all the induced rules.

6.1 Induced Rules of Bacterial Tests

Of 24,335 rules, only 114 rules are unexpected or interesting to medical experts. From these discovered results, nine rules are shown below.

1. β-lactamase(+) $\rightarrow$ Bacteria_Detection (+)
2. β-lactamase(3+) $\rightarrow$ Bacteria_Detection (+)

These two results are interesting from the viewpoints of history of bacterial infection. Since penicillin has been introduced as antibiotics, many bacteria have acquired to generate enzymes that decompose penicillin, called β-lactamase. The above two results show that such penicillin-resistant bacteria can be more easily found than penicillin-sensitive ones.

3. Pneumonia → Bacteria_Detection (-)
  4. Fever (BT>39) → Bacteria_Detection(-)
  5. Malignant Tumor → Bacteria_Detection (-)

These three results are unexpected by medical experts. As for the third rule, it is well known that bacterial infection is the main cause for pneumonia. However even if pneumonia comes from the bacterial infection, it would be difficult to detect bacteria. Concerning the fourth rule, high fever suggests that the degree of infection is high. However, it would be difficult to detect bacteria even if the degree is high. Finally, in the case when a patient suffers from malignant tumor, he/she may suffer from a severe infection due to immunological insufficiency. However, it may be difficult to detect bacteria for this case.

6.2 Induced Results of Steroid Side-Effects

Of 14,715 rules, only 106 rules are unexpected or interesting to medical experts. From these discovered results, four rules are shown below For simplicity, these rules are given in the summarized form, though they are originally represented as the conjunction of temporal attributes.

  1. [Steroid>3.0years] & [Headache(+)>0.5] → Glaucoma

This rule shows that headache is an important sign for glaucoma due to steroid side-effect.

  2.. [Steroid>2.5years] & [Blurred Vision(+)>0.75] → Cataracta

This result shows that if a patient takes steroid for more than two years, the side effects may be frequently observed. Those unexpected/interesting rules are also feedbacked to the university hospital, which donates a dataset to us, and the staff in the hospital is evaluating them.

7 Discussion: Time Required for KDD Process

After the data interpretation phase, about one percent of induced rules are found to be interesting or unexpected to medical experts. In this section, the total KDD process is reviewed with respect to computational time.

Table 7 shows the total time required for KDD process. Each column shows two data sets for each process. Each row includes computational time required for each process. Totally, it takes about one month and three weeks to complete the whole KDD process for bacterial test database and side-effect database, respectively. It is notable that more than 60% of the process is devoted to the three preprocessing processes: data selection, cleaning and reduction. Especially, as for the bacterial test dataset, 22.75 days (79.5%) are used for three processes. It is because domain knowledge should be acquired for generalization of data, as discussed in Section 3. On the other hand, only 4 to 8 percent of total time is spent for data mining process. In the case of bacterial test database, only 2.5 days (8%) is used for rule induction. Therefore, these empirical results suggest that the main KDD processes should be preprocessing rather than discovery of patterns from data.

Another important point is that data interpretation is also time-consuming process, compared with data mining process because it needs interpretation by domain experts. In summary, if we want to make KDD process faster, then we should consider the automation of processes which needs interaction between computers and domain experts. Especially if domain knowledge is easily incorporated into the program, the computational time for the third step may be significantly improved. Therefore, more intensive research on automation of data preprocessing is required for future research.

Table 7. Total Time Required for KDD process

KDD Process	Bacterial Test	Side-Effect
Data Selection	2.3	7.3
Data Cleaning	5.2	2.5
Data Reduction	15.25	2.3
Data Mining	2.5	0.75
Data Interpretation	7.0	7.0
Total Time	32.25	19.85

References

1. Adams, R.D. and Victor, M. 1993. Principles of Neurology, 5th edition, McGraw-Hill, NY.
2. Abiteboul, S., Hull, R., and Vianu, V. 1995. Foundations of Databases, Addison-Wesley, NY.
3. Fauci, A.S., Braunwald, E., Isselbacher, K.J. and Martin, J.B. (eds.) 1997. Harrison's Principles of Internal Medicine (14th Ed), McGraw Hill, NY.
4. Fayyad, U., Piatetsky-Shapiro, G. and Smyth, P. 1996. The KDD Process for Extracting Useful Knowledge from Volumes of data. CACM, 39: 27-34.
5. Levinson, W.E. and Jawetz, E. 1996. Medical Microbiology & Immunology : Examination and Board Review (4th Ed), Appleton & Lange.
6. Holt, J.G. (Ed.) Bergey's Manual of Systematic Bacteriology (Vol 1) Lippincott, Williams & Wilkins.
7. Pawlak, Z. 1991. Rough Sets, Kluwer Academic Publishers, Dordrecht.
8. Polkowski, L. and Skowron, A. 1998. Rough Sets in Knowledge Discovery Vol.1 and 2, Physica-Verlag, Heidelberg.
9. Tsumoto, S. 1998. Automated extraction of medical expert system rules from clinical databases based on rough set theory, Information Sciences, 112, 67-84.
10. Tsumoto, S. 1999. Rule Discovery in Large Time-Series Medical Databases. In: Proc. 3rd European Conference on Principles of Knowledge Discovery and Data Mining (PKDD), LNAI 1704, Springer Verlag, 23-31.
11. Van Bemmel,J. and Musen, M. A.1997. Handbook of Medical Informatics, Springer-Verlag, NY.

Foundations and Discovery of Operational Definitions

Jan M. Żytkow and Zbigniew W. Raś

Computer Science Dept. University of North Carolina, Charlotte, N.C. 28223
e-mail: zytkow@uncc.edu & ras@uncc.edu
also Institute of Computer Science, Polish Academy of Sciences

Abstract. Empirical equations and rules are important classes of regularities that can be discovered in databases. We concentrate on their role as definitions of attribute values. Such definitions can be used in many ways in a single database and for transfer of knowledge between databases. We analyze *quests* for definitions of an attribute in a given database. A quest triggers a discovery mechanism that specializes in searching recursively a system of databases and returns a set of partial definitions. We introduce the notion of shared operational semantics founded on an equation-based and rule-based system of partial definitions. It gives necessary foundations for designing local query answering systems in a distributed knowledge system (DKS).

1 Shared Semantics for Distributed Autonomous DBs

In many fields, such as medical, manufacturing, banking, military and educational, similar databases are kept at many sites. Each database stores information about local events and uses attributes suitable for locally collected information, but since the local situations are similar, the majority of attributes are compatible among databases. Yet, an attribute may be missing in one database, while it occurs in many others. For instance, different military units may apply the same battery of personality tests, but some of these tests may be not used in one unit or another.

Missing attributes lead to problems. A recruiter new at a given unit may query a local database S_1 to find candidates who match a desired description, only to realize that one component a_1 of that description is missing in S_1 so that the query cannot be answered. The same query would work in other databases but the recruiter is interested in identifying suitable candidates in S_1.

1.1 System Architecture

Operational semantics introduced in [15] provides definitions of missing attributes through search for definitions in many databases. Figure 1 shows the architecture of a distributed knowledge system. Discovery Layer for each database is initially formed from rules and equations extracted from that database. They define some of the attributes by other attributes in the same database and are discovered by an automated process. They are used for knowledge exchange between databases and jointly form an integrated semantics for a distributed knowledge system which defines the meaning of queries. Query answering system QAS uses definitions extracted at other databases and/or available in the local discovery layer to answer queries which otherwise would not be reachable.

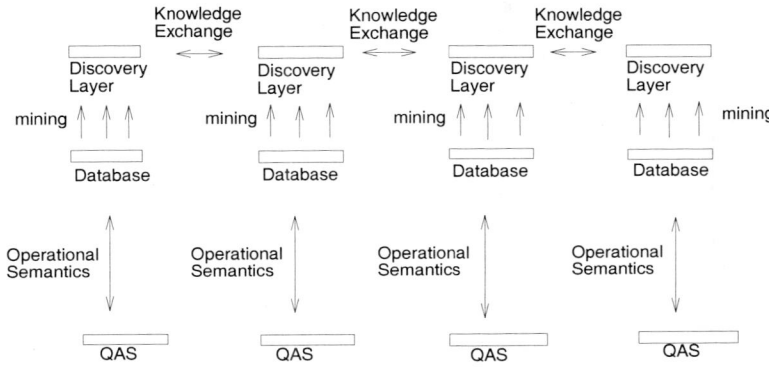

Fig. 1. Distributed Knowledge System

1.2 Links to Previous Research

QAS is a natural knowledge-discovery-based extension of the query answering system for a system of databases, presented in [11], [12], [13]. In these papers rules discovered in one database define values of missing attributes in other databases. The search for rules can use many strategies which find rules describing decision attributes in terms of classification attributes. It has been used in conjunction with such systems like $LERS$ (developed by J. Grzymala-Busse [3]) or $AQ15$ (developed by R. Michalski and his collaborators [8]).

The task of integrating established database systems can be complicated not only by the differences between the sets of attributes but also by differences in structure and semantics of data. The notion of an intermediate model, proposed by Wiederhold [7], is very useful in dealing with such a problem, because it describes the database content at a relatively high abstract level, sufficient for a homogeneous representation of all databases. In our paper a discovery layer can be seen as an application of the ideas of an intermediate model for a distributed DB system.

1.3 Operational Definition

Definitions that are used to compute attribute values of objects are often called operational definitions. They are common in science, where values of each attribute are determined in many ways, depending on different applications. Operational semantics has been introduced by Bridgman [1] and developed by Carnap [2] and many others, including semantics of coherent sets of operational definitions developed by Zytkow [16] and applied in robotic experiments [18].

Operational semantics can be applied to databases. Many computational mechanisms can be used to define values of an attribute. We call them operational definitions because each is a mechanism by which the values of a defined attribute can be computed. Many are partial definitions, as they apply to subsets of records that match the "if" part of a definition. In 1989-1990, Ras et al. [6], [14] introduced a mechanism which first seeks and then applies as definitions rules in the form "If Boolean-expression(x) then a(x)=w" which are partial definitions of attribute a applicable to all objects x that satisfy Boolean-expression(x).

The 49er system can find knowledge in many forms, including equations, that can be used to define one attribute by other attributes in a relational table. We conducted experiment, using this mechanism in addition to rule-based definitions [4]. The growing interest in KDD will make the discovery of operational definitions increasingly popular. Recently, Prodromidis & Stolfo [10] argued that attribute definitions are a useful target for discovery in databases.

1.4 Shared Semantics in Action: Query Answering

Many query-answering situations can benefit from the following generic scenario. A query q is issued at database S_1, but it is "unreachable" in S_1 because it uses an attribute a which is missing in S_1. A request for a definition of a is issued to other sites in the distributed autonomous database system. The request specifies attributes $a_1, ..., a_n$ available at S_1. When attribute a and a subset $\{a_{i_1}, ..., a_{i_k}\}$ of $\{a_1, ..., a_n\}$ are available in another database S_2, a discovery mechanism is invoked to search for operational definitions at S_2, by which values of a can be computed from values of some of $a_{i_1}, ..., a_{i_k}$. If discovered, such a definition is returned to the discovery layer over S_1 and used to compute the unknown values of a that occur in query q.

The same mechanism can apply if attribute a is available at S_1, but some values of a are missing. In that case, the discovery mechanism can be applied at S_1, if the number of the available values of a is sufficiently large.

1.5 Other Applications

Functional dependencies in the form of equations are a succinct, convenient form of knowledge, useful in many ways. The equation $a = f(a_1, a_2, ..., a_m)$ can be directly used to predict values $a(x)$ of a for object x by substituting the values of $a_1(x), a_2(x), ..., a_m(x)$ if all are available. If some are not directly available, they may be predicted by other operational definitions.

When we suspect that some values of a may be wrong, an equation imported from another database may be used to verify them. An equation acquired at the same database may be used, too. For instance, patterns discovered in clean data can be applied to distinguish wrong values in the raw data.

Equations that are generated at different sites can be used to cross-check the consistency of knowledge and data coming from different databases. If the values of a that are computed by two independent equations are approximately equal, this confirms consistency of both definitions.

All equations by which values of a can be computed expand the understanding of a. Attribute understanding is often initially inadequate when we receive a new dataset for the purpose of data mining. We may know the domain of values of a, but we do not understand a's detailed meaning, so it is difficult to apply background knowledge and the knowledge discovered about a. In such cases, an equation that relates a poorly understood attribute a with attributes of known meaning, explains some of the meaning of a.

2 Recursive Search for Equations

Let us present in algorithmic details a recursive discovery mechanism that supports global query answering. When an attribute a is needed but unreachable in database S_1, a request for a definition of a is issued to other sites in the distributed database system. The request specifies the attributes $a_1, ..., a_n$ available in S_1, because only those attributes can be included in a definition useful at S_1.

In this section we present a recursive algorithm that searches for equations and we analyze an application os this algorithm. But that algorithm can be used to search for rules. In section 3 we will present an example of recursive search for rules.

When the attribute a and a subset $\{a_{i_1}, ..., a_{i_k}\}$ of $\{a_1, ..., a_n\}$ are available in another database S_2, 49er's discovery mechanism is invoked to search S_2 for equations by which values of a can be computed from values of some of $a_{i_1}, ..., a_{i_k}$. If discovered, such equations are returned to S_1 and can be used in numerous ways.

In [15] we considered a computational mechanism that searches at each database individually for equations suitable in a role of definitions of a. But there are numerous situations when this mechanism must be expanded and applied recursively.

2.1 Non-overlapping Attribute Sets

First, there may be no database which contains a and any of $\{a_1, ..., a_n\}$. This can be illustrated with the following example of simple relation schemas, one relation per database:

$S_1(a_1, a_2, ..., a_n)$; definition of attribute a is sought
$S_2(a, b_1, ..., b_k)$
$S_3(b_1, a_2, a_3)$

Suppose that an equation $a = f(b_1)$ has been discovered in S_2. It cannot be used in S_1, because b_1 is unavailable. But S_3 includes b_1 and some of $\{a_1, ..., a_n\}$. An equation $b_1 = f_1(a_2, a_3)$ may be discovered that defines b_1 in terms a_2 and a_3. That equation can be substituted into $a = f(b_1)$ leading to equation $a = f(f_1(a_2, a_3))$ that can be applied in S_1.

2.2 Search for a Sufficient Fit

Second, there may be a database S_4 that includes a and some of $\{a_1, ..., a_n\}$. But no equation that defines a through any of $\{a_1, ..., a_n\}$ has a fit sufficient to play the role of a definition. In this situation, the search for a definition can be expanded. Perhaps an equation is discovered that has a sufficient fit to play the role of a definition, but in addition to some of $\{a_1, ..., a_n\}$ it uses b_1, unavailable in S_1. We already discussed the steps appropriate for this situation.

2.3 Empirical Contents in a Set of Definitions

There is a more systematic reason why the search for equations should continue, even if it has been successful. Equations that are used to compute missing values are empirical generalizations. Although they may be reliable, we cannot trust them unconditionally, and it is a good practice to seek their further verification, especially if they are applied to the expanded range of values of a. The verification may come from additional knowledge that can be used as alternative definitions. Ras et al. [12], [13] used rules coming from various sites and verified their consistency.

Multiple equations give a chance for cross-verification, as their predictions can be compared. Each consistent prediction provides extra justification for the system of definitions, while each inconsistency calls for further empirical analysis of data and definition improvements.

2.4 Recursive Discovery Algorithm

The following algorithm can be used to search recursively for an attribute definition:

Algorithm: Find definitions of attribute a that are applicable in DB
```
    Let A be the set of attributes in DB
    For each database X
        Let A_X be the set of attributes in both A and X
        If a is available in X and A_X is not empty, then
            seek all definitions of a in A_X; push them on def(a,A)
    If def(a,A) is non-empty, then HALT
    For each X
        If a is available in X, then
            seek all definitions of a in X; push them on def(a,X)
            For each definition DEF in def(a,X)
                For each attribute b in DEF that is missing in DB
                    Find definitions of attribute b that are applicable in DB
                If all attributes in DEF are defined by attributes in A
                    then add DEF to list of definitions of a
```

2.5 An Example of Recursive Search for a Definition

Consider the following database schemas

$S_1(a_1, a_2, a_3)$,
$S_2(a, a_1, b)$,
$S_3(a, b, a_3)$,
$S_4(a_2, b)$

also illustrated in Figure 2. Discovery layer is assigned to each of these four databases and contains definitions of attributes and/or attribute values extracted from them. Definition of attribute a is sought.

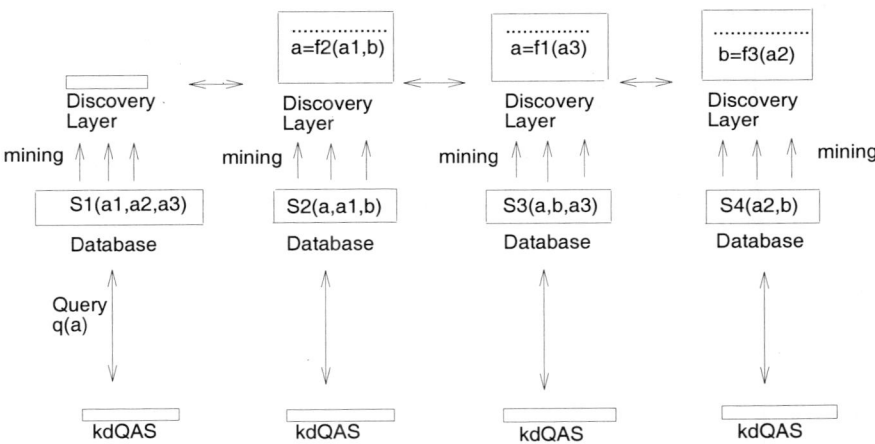

Fig. 2. Search for equations in support of query answering; an example

The recursive search for a definition follows these steps:
 1. S_1 sends a request: define a, use a_1, a_2, a_3
 2. S_2 and S_3 try to answer.
 3a. Situation 1: definition found in S_3: $a = f_1(a_3)$
 3b. Situation 2: no definition found, so the search is expanded to additional attributes: define a, use a_1, a_2, a_3 and any other parameters available.
 4. S_2 tries to answer.
 5a. Situation 3: definition not found so the search halts.
 5b. Situation 4: definition found in S_2: $a = f_2(a_1, b)$
 6. A new quest is issued: define b, use a_1, a_2, a_3
 7. S_3 and S_4 try to answer.
 8a. Situation 5: definition not found so the search halts.
 8b. Situation 6: definition found in S_4: $b = f_3(a_2)$
 9. Equation in 8b is substituted into equation in 5b: $a = f_2(a_1, f_3(a_2))$
 10. The search halts.

3 Query Answering System Based on Reducts

In this section we recall the notion of a reduct and show how it can be used to improve the query answering process in distributed autonomous database systems ($DADS$). We assume that information stored in all databases is consistent.

Let us assume that $S = S(A)$ is a database schema and $S(X, A)$ represents its view. Each $a \in A$ is interpreted here as a function $a : X \longrightarrow Dom(a)$, where $Dom(a)$ is a domain of a. For simplicity reason we assume that $Dom(a), Dom(b)$ are disjoint for any $a, b \in A$ such that $a \neq b$.

Let $B \subset A$. We say that $x, y \in X$ are indiscernible by B in S, denoted $[x \approx_B y]$, if $(\forall a \in B)[a(x) = a(y)]$.

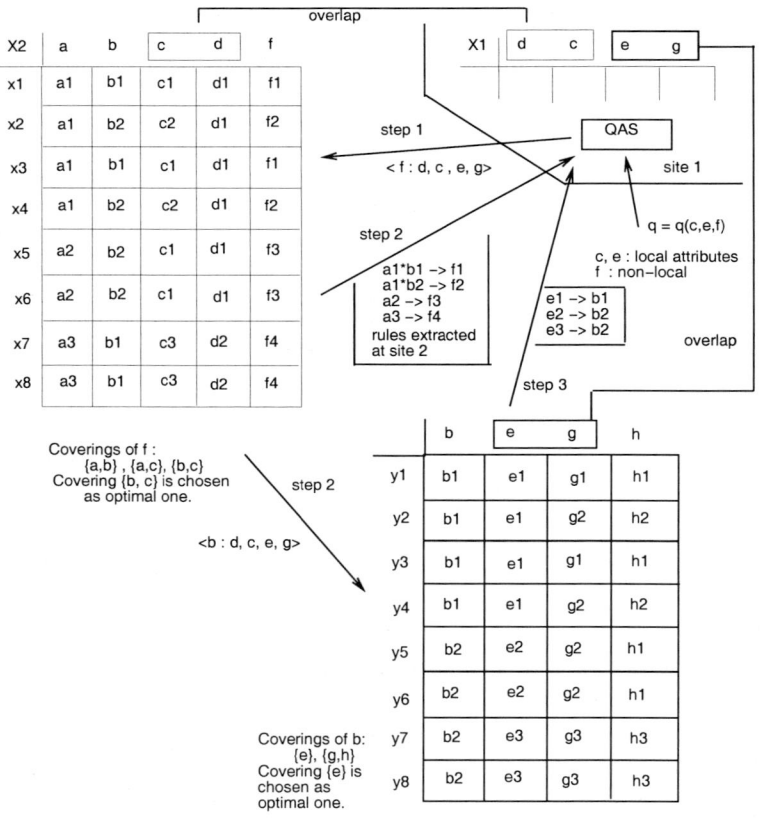

Fig. 3. Process of resolving a query by QAS in $DADS$

Now, assume that both B_1, B_2 are subsets of A. We say that B_1 depends on B_2 if $\approx_{B_2} \subset \approx_{B_1}$. Also, we say that B_2 is a reduct of B_1 (B_1-reduct) if B_1 depends on B_2 and B_2 is minimal. If B is a singleton set ($B = \{f\}$) then instead of B-reduct we say f-reduct.

Example. Assume the following scenario:

- $S_1 = (X_1, \{c, d, e, g\})$, $S_2 = (X_2, \{a, b, c, d, f\})$, $S_3 = (X_3, \{b, e, g, h\})$ are views of database schemas S_1, S_2, S_3, respectively.
- User submits a query $q = q(c, e, f)$ to the query answering system QAS associated with database S_1,
- Databases S_1, S_2, S_3 form a distributed autonomous database system $DADS$.

Attribute f is non-local for a database S_1 so the query answering system associated with S_1 has to contact other sites of $DAKS$ requesting a definition of f in terms of $\{d, c, e, g\}$. Such a request is denoted by $< f : d, c, e, g >$. Assume that the database S_2 is contacted. The definition of f, extracted from S_2, involves only attributes $\{d, c, e, g\} \cap \{a, b, c, d, f\} = \{c, d\}$. There are three f-reducts (coverings of f) in S_2. They are: $\{a, b\}, \{a, c\}, \{b, c\}$. The optimal f-reduct is the one which has minimal number of elements outside $\{c, d\}$; in our case $\{a, c\}$ and $\{b, c\}$. Let us assume that $\{b, c\}$ is chosen as an optimal f-reduct in S_2.
Then, the definition of f in terms of attributes $\{b, c\}$ may be extracted from S_2 and the query answering system of S_2 will contact other sites of $DADS$ requesting a definition of b (which is non-local for S_1) in terms of attributes $\{d, c, e, g\}$. If a definition of b is found, then it is sent to QAS of S_1. Figure 4 shows the process of resolving query q in the example above.

4 Conclusion

The discovery layer at each site is formed from partial definitions either extracted from that site or imported from other sites. All these partial definitions (if consistent) define a local operational semantics and the meaning of queries seen by the local query answering system. Discovery processes update discovery layers associated with all databases in $DADS$. They do it in real-time whenever a local query cannot be answered with the help of local operational semantics. As a result of operational definitions discovery, local semantics is augmented with a relevant selection of definitions found in $DADS$.

References

1. Bridgman, P.W., *The Logic of Modern Physics*, The Macmillan Company, 1927
2. Carnap, R., "Testability and Meaning", in *Philosophy of Science, 3*, 1936

3. Grzymala-Busse, J., "LERS - A system for learning from examples based on rough sets", in *Intelligent Decision Support, Handbook of Applications and Advances of the Rough Sets Theory*, Slowinski, R. (ed), Kluwer Academic Publishers, 1992, 3-18
4. Klopotek, M., Michalewicz, M., Michalewicz, Z., Ras, Z., Wierzchon, S., Zytkow, J., "Discovering knowledge in distributed databases", in *Proc. of 6th International Workshop on Intelligent Information Systems*, 1997, 128-138
5. Kryszkiewicz, M., Rybinski, H., "Reducing information systems with uncertain attributes", in *ISMIS'96 Proceedings*, LNCS/LNAI, Springer, Vol. 1079, 1996, 285-294
6. Maitan, J., Ras, Z., Zemankova, M., "Query handling and learning in a distributed intelligent system", in *Methodologies for Intelligent Systems, IV*, (Ed. Z.W. Ras), North Holland, 1989, 118-127
7. Maluf, D., Wiederhold, G., "Abstraction of representation for interoperation", in *Proceedings of Tenth International Symposium on Methodologies for Intelligent Systems*, LNCS/LNAI, Springer-Verlag, No. 1325, 1997, 441-455
8. Michalski, R.S., Mozetic, I., Hong, J. & Lavrac, N. "The multipurpose incremental learning system AQ15 and its testing application to three medical domains." in *Proceedings of the Fifth National Conference on Artificial Intelligence*, Morgan Kaufmann, 1986, 1041-1045
9. Pawlak, Z., "Rough classification", in *International Journal of Man-Machine Studies*, Vol. 20, 1984, 469-483
10. Prodromidis, A.L. & Stolfo, S., "Mining databases with different schemas: Integrating incompatible classifiers", in *Proceedings of The Fourth Intern. Conf. onn Knowledge Discovery and Data Mining*, AAAI Press, 1998, 314-318
11. Ras, Z., "Query answering in DAKS based on reducts", in *Proceedings of FQAS'2000*, Advances in Soft Computing, Physica-Verlag, 2000, will appear
12. Ras, Z., "Resolving queries through cooperation in multi-agent systems", in *Rough Sets and Data Mining* (Eds. T.Y. Lin, N. Cercone), Kluwer Academic Publishers, 1997, 239-258
13. Ras, Z., Joshi, S., "Query approximate answering system for an incomplete DKBS", in *Fundamenta Informaticae Journal*, IOS Press, Vol. 30, No. 3/4, 1997, 313-324
14. Ras, Z., Zemankova, M, "Intelligent query processing in distributed information systems", in *Intelligent Systems: State of the Art and Future Directions*, Z.W. Ras, M. Zemankova (Eds), Ellis Horwood Series in Artificial Intelligence, London, England, November, 1990, 357-370
15. Ras, Z., Żytkow, J.,"Discovery of equations to augment the shared operational semantics in distributed autonomous BD System", in *PAKDD'99 Proceedings*, LNCS/LNAI, No. 1574, Springer-Verlag, 1999, 453-463
16. Żytkow, J., "An interpretation of a concept in science by a set of operational procedures", in *Polish Essays in the Philosophy of the Natural Sciences*, Krajewski W. ed. Boston Studies in the Philosophy of Science, Vol.68, Reidel 1982, p.169-185.
17. Żytkow, J. & Zembowicz, R., "Database exploration in search of regularities", in *Journal of Intelligent Information Systems*, No. 2, 39-81
18. Żytkow, J.M., Zhu, J., and Zembowicz R., "Operational definition refinement: a discovery process", in *Proceedings of the Tenth National Conference on Artificial Intelligence*, The AAAI Press, 1992, p.76-81.

A Multi-agent Based Architecture for Distributed KDD Process

Chunnian Liu[1], Ning Zhong[2], and Setsuo Ohsuga[3]

[1] School of Computer Science, Beijing Polytechnic University, China
[2] Dept. of Information Eng., Maebashi Institute of Technology, Japan
[3] Dept. of Information and Computer Science, Waseda University, Japan

Abstract. Focuses in KDD research are being extended from individual techniques to KDD process, while KDD systems have been rapidly evolving from the early stand-alone ones to the current large, Distributed KDD (DKDD) systems. In this paper, we concentrate on the architectural aspect of Distributed KDD systems from the perspective of CSCW (treating DKDD as a special case of CSCW). After summarizing the requirements needed to support Distributed KDD, we describe a Client/Server architecture for DKDD that is "traditional" for CSCW in general. Then we propose a Multi-Agent (MAS) based architecture for DKDD. Compared with the traditional Client/Server architecture, the MAS based architecture is better in terms of simplicity and flexibility, and particularly useful in modeling and providing support to cooperative activities (communication, negotiation, coordination and collaboration).

1 Introduction

Data Mining or *Knowledge Discovery in Databases (KDD)* means discovering new, useful knowledge (called *models* in this paper) from vast amount of data accumulated in an organization's databases. *KDD process* is the set of activities needed to transform raw datasets into usable models. In real-world applications, KDD process is an integrated part of a whole business process, with activities such as data sampling, pre-processing, mining (model building), model analysis, visualization and integration into the business process. Now it is well-recognized that real-world KDD process can be very complex, similar in many aspects to Software Development Process [Hum89].

KDD is essentially a demand-driven field. Although early work in KDD inevitably concentrated on individual mining techniques, what really important is the KDD systems combining various KDD techniques and their successful applications to real-world databases. KDD systems have rapidly evolved [Ce99]. While the first generation of KDD systems was stand-alone mining applications over files, the second generation has been integrated with data management, and the third (current) generation is characterized by distribution of data and computation over enterprises' Intranets or across the global Internet. We will call this kind of KDD systems *Distributed KDD Systems - DKDD*. Further development of DKDD systems includes dynamically adding new computational resources to a network, and the mobility of code (for example, mining components move to DBMS sites and execute within the databases). Thus, we can view the Distributed KDD as a special case of *CSCW (Computer-Supported Cooperative Work)* that is a multidisciplinary research area focusing on effective methods of sharing information and coordinating activities [Gru94].

In recent years, along with developing new KDD techniques, we pay increasing attention to the process and architecture aspects of KDD systems. In order to increase both autonomy and versatility of KDD systems, we proposed a Global Learning Scheme (GLS) [?, ZLO97], as a framework for organizing complex KDD process. GLS has two levels: the meta-level of process and the object level of process. On the meta-level, it provides mechanisms and facilities for modeling, planning, scheduling, controlling and management of KDD process; On the object level, various mining components (methods and algorithms) are grouped according to the different stages of data mining. Within this framework, we have been investigating the planning meta process in depth [ZLKO97, LZO97]. In particular, we propose to handle iterations in KDD process by integrating planning and execution [LZO98], and to deal with KDD process changes by incremental replanning [ZLKO98].

In this paper, we concentrate on the architectural aspect of Distributed KDD (DKDD) systems from the perspective of CSCW (treating DKDD as a special case of CSCW). First we summarize the requirements needed to support Distributed KDD. Then we describe a Client/Server architecture for DKDD that is "traditional" for CSCW in general and based on the current status of Internet technology. Then we propose a Multi-Agent (MAS) based architecture for DKDD, which is adapted from the generic MAS architecture for CSCW [MMP98, WCL99]. Compared with the traditional architecture, the agent-based architecture is better in terms of simplicity and flexibility, and particularly useful in modeling and providing support to cooperative activities (communication, negotiation, coordination and collaboration). The proposed architecture can model the information flow as well as the process flow in DKDD. Finally in the Conclusion section, implementation issues are also briefly discussed.

2   Requirements for Distributed KDD Architecture

Distributed KDD (such as Enterprise Distributed Data Mining [Ce99]) faces unique challenges and needs architectural support to cope with them. The requirements for architectural support to Distributed KDD can be summarized as follows.

- *Multiple roles*: Unlike simple, stand-alone, prototype KDD work, real-world KDD process involves multiple human roles. We can identify at least three types of them: the analysts (for KDD task planning and result analysis), the knowledge engineers (executing the mining tasks), and the end-users (people managing and optimizing the business process within that the KDD process occurs). Multiple people may access to the data and the analytical results (the models), so the KDD system must provide multiple access points.

(NB: in this paper, the "user" of KDD systems mainly refers to analysts or knowledge engineers).
- *Mining on data of huge size*: Gigabytes or even terabytes of data have been accumulated in large organizations. Mining on such large scale of data has the following implications on KDD architecture:
  - We need large computational power (high-performance servers) for mining tasks, and visualization tools for data analysis and model analysis.
  - The mining operation should be run close to the databases, because it is not practical to move the vast data between the sites of individual analysts. This requirement can be supported either by mobile mining components traveling to the database sites and executing there, or by setting up high-performance servers close to databases.
  - The user should be allowed to browse and sample data during planning and editing his/her mining tasks.
- *Mining on diverse and distributed data sources*: Because various types of data are accumulated on many sites in a large organization. A user may need to access to multiple datasets. So the KDD system must support distributed mining and combining partial results into a meaningful total.
- *KDD process planning*: There are several stages in KDD process (the three major stages are : pre-processing, model-building and model analysis and refinement). For each stage, there is a large number of available KDD techniques and algorithms. Some of them may be out-of-date soon while new ones come continuously. So, good combination of KDD techniques and easy integration with new techniques are very desirable, and this demands careful planning of the KDD tasks. Note that from different kinds of data resources, different KDD techniques are needed, so the planning involves browsing and sampling data.
- *Interactions among KDD roles*: Because the KDD process is iterative through the cycle of data-selection, pre-processing, model building, and model analysis and refinement, high degree of interactions among analysts, knowledge engineers and the end-users is needed.
- *Flexibility*: Wide range of configuration options is needed to fulfill different needs of large organizations, so that the applications can be scaled from a few client workstations to high-performance server machines.
- *Open-ended-ness* for future extension.
- *Conceptual and architectural simplicity* is important in designing such a complex system to ensure/enhance its correctness, flexibility and openness, etc.

On the implementation level, the rapid development of Internet and related technologies such as *software component technology* and various Java/CORBA packages do provide solutions to Distributed KDD. But on the design level, we need conceptual and architectural clarity and simplicity for complex systems like Distributed KDD systems. And this is the focus for the remaining part of this paper.

3 A Client/Server Architecture for Distributed KDD

We regard Distributed KDD (DKDD) as a special case of Computer-Supported Cooperative Work (CSCW), trying to apply the generic Client/Server architecture of CSCW [LC98, WCL99] to DKDD. We also investigate the architectures of some existing Distributed KDD systems such as [Ce99] (though mainly on the implementation level). As a result, we can describe here the "traditional" Client/Server architecture for Distributed KDD systems as Figure 1.

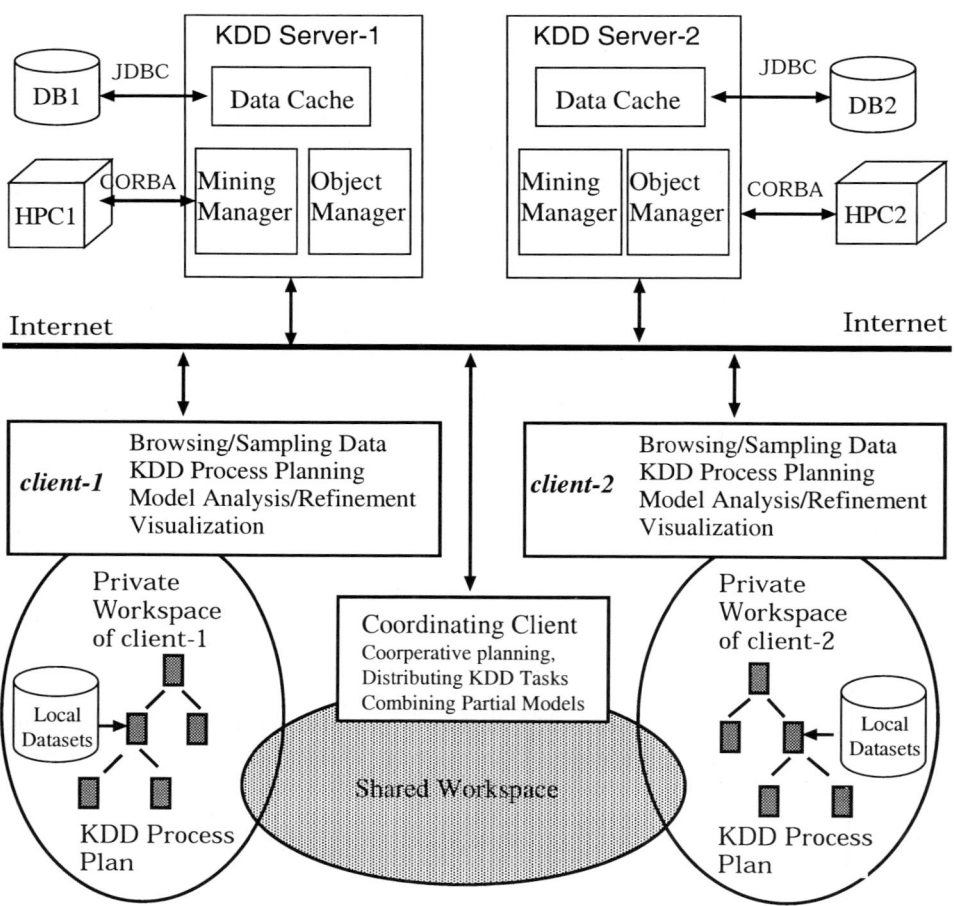

Fig. 1. A Client/Server Architecture of Distributed KDD Systems

On the server side, there could be multiple *KDD Server* sites, as well as multiple *DBMS* sites and *HPC* (High-Performance Computation) sites, distributed globally (that is, they may be allocated across the whole Internet).

Obviously, a DBMS site provides the traditional database service. On the other hand, a HPC site provides parallel mining algorithms executed on a dedicated separate parallel machine. The algorithms include traditional ones for classification, clustering, and association rule discovery, as well as new techniques such as Inductive Logic Programming (ILP). Full range of statistical service is also available. As we mentioned before, a HPC site should be set up close to the Database site from where the data is mined.

Each KDD server consists of three main components:

1. *Object Manager*: Manages persistent objects such as data mining tasks and results (models) for various users, authenticates users and controls users' access to the persistent objects.
2. *Mining Manager*: Provides the interface with a HPC site via CORBA, etc. This component controls data conversion, data transfer, and parameter passing to the mining algorithms.
3. *Data Cache*: Provides the interface with a Database via JDBC. The loaded and cached data are used for browsing the Database from the client.

On the client side, there could be multiple *KDD Client* sites, also across the whole Internet. KDD client sites communicate with each other and with KDD Servers. Connections between KDD client sites and DBMS/HPC sites are indirect via KDD Server sites. A KDD client has the following components (the first three interact with KDD Servers, the others are for internal work):

– *Object Browser*: Viewing the user's objects held in the *Object Manager* of KDD Server.
– *KDD Task Invoker*: Submitting the planned and edited KDD tasks via the *Mining Manager* of KDD Server, which in turn asks the service of HPC site. Meanwhile some simple KDD tasks using only local datasets can also be executed locally on the client site (see items below).
– *Data Browser*: Providing the function of browsing remote databases, by access to the *Data Cache* of KDD Server.
– *KDD Process Planner*: Tries to make a plan for the KDD process. A plan is a partially ordered network of mining activities needed for pre-processing, model-building, model analysis and refinement. Remote data browsing and sampling are needed for the planner to select relevant data, to choose appropriate mining techniques, and to organize them into a KDD process plan.
– *Visualization Tool*: For large-scale mining tasks, visualization is necessary for data analysis as well as model analysis.
– *Work Space*: The place where the client works. A group of KDD clients may have shared work spaces.

There is a special, coordinating KDD client (for the manager of the data mining group in an organization, for example) with the following functionalities:

– *Cooperative Planning*: Partitioning the overall KDD tasks to a group of knowledge engineers (KDD clients).

- *Scheduling*: Distributing KDD tasks planned and submitted by other KDD clients to appropriate KDD servers. The selection of KDD servers is based on some resource allocation policy (for example, to execute the mining operations in a KDD server that contains the data).
- *Synthesis*: Combining partial results from individual clients into a meaningful total.

4 Multi-Agent Based Architecture for Distributed KDD

The previous section shows how complex a DKDD environment could be. Similar situations exist in other Distributed Artificial Intelligence (DAI) systems [ONJ96], and in Cooperative Software Engineering (CSE) [WCL99]. They are all examples of Computer-Supported Cooperative Work (CSCW) [Gru94, LC98]. Nowadays it is widely agreed that the Multi-Agent Systems (MAS) [MMP98] are a better way to model these decentralized, distributed, open-ended systems and environments. A MAS is a loosely-coupled networks of problem solvers (agents) that work together to solve a given problem.

In this section we propose a MAS-based architecture for Distributed KDD, which is inspired by our MAS-based architecture for CSE [WCL99]. In such a Multi-Agent architecture, we have the following main components:

- *Agents*:
  An agent is a piece of software created by and acting on behalf of the user (or some other agent). It is set up to achieve a modest goal, with the characteristics of autonomy, interaction, reactivity to environment, as well as pro-activeness. We have the following different types of agents in the DKDD architecture: *Assistant agents* assisting KDD people in various work, such as browsing and sampling data, planning KDD process, analyzing/refining models, etc. *Interacting agents* helping participants in their cooperative work such as communication, negotiation, coordination, and mediation. *Mobile agents* as mining software that can move to DBMS sites and execute within the databases (in order to reduce data transfer). And *System agents* for the administration of the multi-agent architecture such as to register and manage large numbers of components; to monitor events in and status of the WorkSpaces as well as the AgentMeetingPlaces (see below), and to collect relevant measurement according predefined metrics.
- *AgentMeetingPlaces (AMPs)*:
  AgentMeetingPlaces are where agents meet and interact with each other (for communication, negotiation, coordination and collaboration). AMPs are built on the underlying communication mechanisms, but must provide agents with more intelligent means to facilitate their interaction. First of all, AMPs should provide agent communication languages such as KQML [Fe97], defining communication types and the common syntax of the messages transmitted. AMPs must also provide a set of information models, by which a recipient can understand what the message means. For negotiation agents,

the purpose is to reach an agreement. The progress of negotiation depends mainly on the negotiation *strategies* employed by the negotiating agents, but MeetingPlaces should provide mechanisms to minimize communication overheads, to help the negotiating agents in minimizing their computation efforts.

- *WorkSpaces (WSs)*:
  Just as a software engineer needs a WorkSpace (WS) to perform software development tasks, a knowledge engineer needs a WS to perform mining tasks. A WS is primarily a container (often realized as a storage area in a file system, in a web-directory, or in a sub-database) for relevant data in a suitable format, together with the processing tools. The relevant data include sampled source data as well as KDD process data (the KDD process plan, etc). The processing tools include KDD algorithms for experiments on the sampled data, KDD process planner, and visualization tools. In a large organization, there may be a KDD division involving a number of analysts and knowledge engineers. The KDD people work in groups. Each group may have a shared WorkSpace while each knowledge engineer having his/her private WS. That is, the contents of WorkSpaces may overlapped with data shared by several people.

- *Repositories*:
  The architecture contains various Repositories (global, local or distributed). The most fundamental one is the databases to mine and the model repositories to store the results of data mining. There could be other important repositories. For example, in a large organization with frequent KDD activities, it is useful to maintain an *Experience Base (EB)*. The idea is that for each (previous) mining project, its project/operative profile is recorded in EB. All new projects are initiated with a specific own project/operative profile. The project manager uses it as the criteria to search EB for those previously completed projects that are similar to the current project. The best-matching candidate is then selected as the *baseline* project. With the baseline project as a guidance, the KDD process planning (hence the quality) of the new project could be improved in various ways.

- *High-Performance Computers (HPCs)*:
  A HPC in this context means a parallel machine dedicated to mining tasks. This is necessary when the size of data to be mined is so big that it becomes unrealistic to perform the real data mining tasks in the WorkSpaces.

Within our architecture, the main components are interconnected and interoperated as follows:

1. *Set up WorkSpaces according to people grouping*: In large KDD process, we can perceive various groups of people working as a team. Each KDD people has his/her private WorkSpace, and the group has a share WorkSpace. There are controlling WSs where the managing, controlling and scheduling people (or their agents) work. They are responsible for the overall cooperative planning for the current KDD project (while other KDD people plan their own

part of KDD tasks), for accessing to global databases, and for distributing real mining tasks to HPCs according to some resource allocation policy.
2. *Create agents*: Assistant agents are created by people to help them work; Interacting agents are created for communication purpose; Mobile agents are created and sent to perform data mining tasks within databases sites; and system agents are created by default to manage various components in the architecture. Note that agents creation is a process of instantiation of the corresponding agent classes. That is, agents are created from templates, typically by supplying few parameters.
3. Communication between agent groups is via AgentMeetingPlaces (AMPs). Some system agents are created by default to manage the AMP (creation, deletion, and bookkeeping).
4. Repositories can be global, local to one people or a group of people, or distributed. For the databases and model repositories, we mention that: (1). Local databases and model repositories can be accessed from associated WorkSpaces; (2). Global ones can be accessed only from controlling WSs where the controlling and scheduling people (or their agents) work. (3). In the architecture, HPC components are optional, and set close to the relevant global database components. (4). Mobile agents can travel to global databases and execute there.
5. Within a WorkSpace, any existing KDD process models are allowed. For example, the planning and replanning techniques described in [ZLKO97, LZO97] can be applied to them.

Figure 2 shows some key issues in the architecture about the interconnection and interoperation among the five components. To prevent the figure from being too complicated, we do not show every details of the architecture described above. The features shown in the figure include: WorkSpaces and their local KDD process and local databases; The controlling WS and its access to the global database; Mobile agents traveling to the global database; The close neighborhood of the global database and the HPC; And interacting agents communicating via an AgentMeetingPlace. When the normal WS completes a KDD plan and tries to perform real mining tasks on the global database, it communicates with the controlling WS by interacting agents, and the latter schedules the tasks, accesses to the global database, caches relevant data, and invokes the nearby HPC to perform the mining tasks. The results will be passed to the normal WS also via interacting agents. The normal WS can also send a mobile agent to the global database and let it execute there.

Compared with the "traditional" Client/Server architecture, we can mention the main advantages of MAS-based architecture, when applied to Distributed KDD, as follows:

- *Decentralization*: being able to break down a complex system into a set of decentralized, cooperative subsystems. Here we may have distributed databases and unbounded numbers of agents, WorkSpaces and AgentMeetingPlaces.

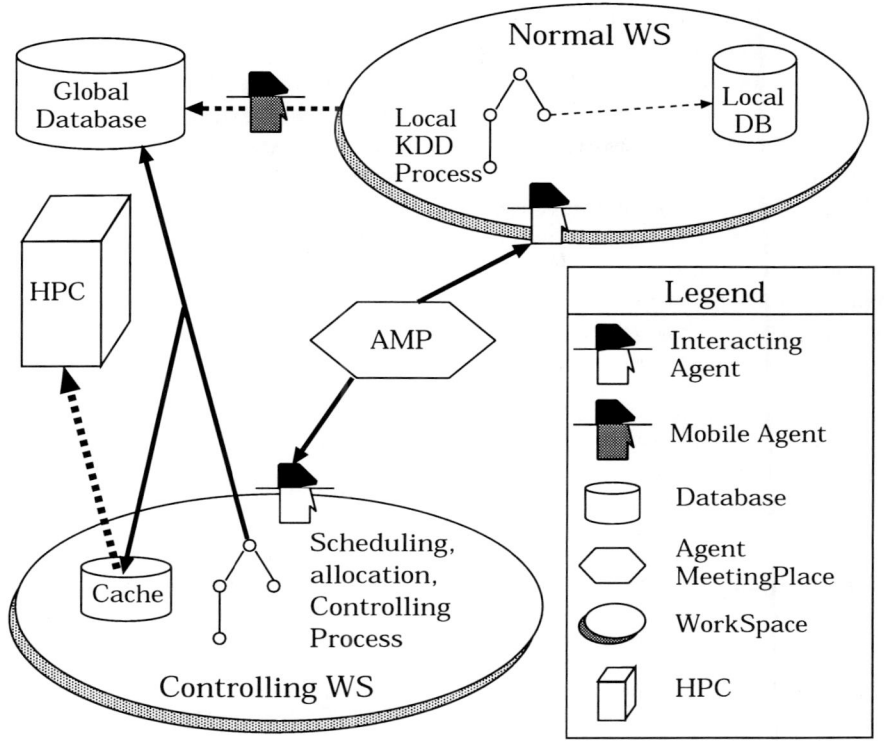

Fig. 2. Multi-Agent Architecture for Distributed KDD

- *Reuse of previous components/subsystems*: That is, building a new and possibly larger system by interconnection and interoperation of existing (sub)systems, even though they are highly heterogeneous.
- *Cooperative Work Support*: being able to better model and support the spectrum of interactions (communication, negotiation, coordination and collaboration) in cooperative work.
- *Flexibility*: being able to cope with the characteristic features of a distributed environment such as DKDD, namely incomplete specification, evolution, and open-endedness.
- *Simplicity*: being able to offer conceptual clarity and simplicity in modeling and design.

5 Conclusions

This paper investigated the architectural aspects of Distributed KDD systems, viewing DKDD as a special case of CSCW. We listed the the requirements needed to support Distributed KDD, described a Client/Server architecture for DKDD that is "traditional" for CSCW in general, and proposed a Multi-Agent (MAS) based architecture for DKDD. Compared with the traditional Client/Server architecture, the MAS based architecture is better in terms of simplicity and flexibility, and particularly useful in modeling and providing support to cooperative activities (communication, negotiation, coordination and collaboration).

In terms of implementation, both architectures rely on Internet related technologies such as *software component technology* and various Java/CORBA packages. For the MAS based architecture, we also need communicating agents and mobile agents. On this part, the rapid development of agent technology provides many options for implementation, such as KQML [Fe97] for agent communication, Aglets [LO98] for mobile agents, etc. Prototyping work based on these techniques is under way.

Acknowledgment

C. Liu's work was supported in part by the Natural Science Foundation of China (NSFC), Beijing Municipal Natural Science Foundation (BMNSF), and the 863 High-Tech Program of China.

References

[Ce99] J. Chattratichat et al. An Architecture for Distributed Enterprise Data Mining. *Proc. 7th International Conference on High Performance Computing and Networking (HPCN'99)*, 1999.

[Fe97] T. Finin et al. KQML as an Agent Communication Languege. J.M. Bradshaw (ed.) *Software Agents*, MIT Press, 1997.

[Gru94] J. Grudin. Computer-Supported Cooperative Work: History and Focus. *IEEE Computer*, 27(5):19–26, 1994.

[Hum89] W. S. Humphrey. *Managing the Software Process*. Addison-Wesley, 1989.

[LC98] C. Liu and R. Conradi. Process View of CSCW. *Proc. of ISFST98*, 1998.

[LO98] D.B. Lange and M. Oshima. *Programming and Developing Java Mobile Agents with Aglets*. Addison-Wesley, 1998.

[LZO97] C. Liu, N. Zhong, and S. Ohsuga. Planning and Replanning of KDD Process. *Proc. IASTED International Conference: AI and Soft Computing*, pages 339–342, 1997.

[LZO98] C. Liu, N. Zhong, and S. Ohsuga. Handling KDD Process Iteration by Integration of Planning and Controlling. *Proc. SMC'98: 1998 IEEE International Conference on System, Man, and Cybernetics*, pages 411–416, 1998.

[MMP98] M. Divitini M. Matskin and S. Petersen. An Architecture for Multi-Agent Support in a Distributed Information Technology Application. *Proc. International Workshop on Intelligent Agents in Information and Process Management*, pages 47–58, 1998.

[ONJ96] G.M.P. O'Hare and eds. N. Jennings. *Foundations of Distributed Artificial Intelligence*. John Wiley & Sons, Inc., 1996.

[WCL99] A.I. Wang, R. Conradi, and C. Liu. A multi-Agent Architecture for Cooperative Software Engineering. *Proc. of SEKE'99, the 11th International Conference on Software Engineering and Knowledge Engineering*, pages 162–169, 1999.

[ZLKO97] N. Zhong, C. Liu, Y. Kakemoto, and S. Ohsuga. KDD Process Planning. *Proc. Third International Conference on Knowledge Discovery and Data Mining (KDD-97)*, AAAI Press, pages 291–294, 1997.

[ZLKO98] N. Zhong, C. Liu, Y. Kakemoto, and S. Ohsuga. Handling KDD Process Changes by Incremental Replanning. J. Zytkow and M. Quafafou (eds.) *Principles of Data Mining and Knowledge Discovery*, LNAI 1510, Springer-Verlag, pages 111–120, 1998.

[ZLO97] N. Zhong, C. Liu, and S. Ohsuga. A Way of Increasing both Autonomy and Versatility of a KDD System. Z.W. Ras and A. Skowron (eds.) *Foundations of Intelligent Systems*, LNAI 1325, Springer-Verlag, pages 94–105, 1997.

Towards a Software Architecture for Case-Based Reasoning Systems

Enric Plaza and Josep-Lluís Arcos

IIIA - Artificial Intelligence Research Institute
CSIC - Spanish Council for Scientific Research
Campus UAB, 08193 Bellaterra, Catalonia, Spain.
Email: {enric,arcos}@iiia.csic.es
WWW: http://www.iiia.csic.es

Abstract. We present a software architecture model of adaptation in CBR. A software architecture is defined by its components and their connectors. We present a software architecture for CBR systems based on three components (a task description, a domain model, and adaptors) connected by a type of connectors called bridges. Adaptors are basic inference components that perform specific transformations to cases. Two kinds of adaptors are introduced: domain adaptors (*d-adaptors*) and case-based adaptors (*c-adaptors*). Adaptors are applied to a given problem, performing search until a sequence of adaptor instantiations is found such that a solution is achieved. Thus, in the **ABC** architecture adaptation is viewed as a search process on the space of adaptors. The **ABC** components have been used in the SaxEx application, a CBR system for generating expressive musical phrases.

1 Introduction

The goal of software architectures is learning from system developing experience in order to provide the abstract recurring patterns for improving further system development. As such, software architectures contribution is mainly methodological in providing a way to specify systems. In this paper we present a software architecture for adaptation in CBR—called **ABC** for "Adaptors and Bridges as Connectors"—based on the notion of connectors and inspired on object-oriented and component-based methodologies.

The three main elements of the **ABC** software architecture are (i) a task description—characterizing the goal that a CBR system pursues; (ii) a domain model—characterizing the ontology and properties of the knowledge content; and (iii) a library of adaptors—performing transformations to case-specific models. The connector linking these three elements are called bridges.

---

* This research has been supported by the Project IST-1999-19005 **IBROW** *An Intelligent Brokering Service for Knowledge-Component Reuse on the World-Wide Web*, and the CICYT Project **SMASH**: *Systems of Multiagents for Medical Services in Hospitals*.

ABC follows the "problem solving as modeling" view, i.e. solving a problem consists of building a model specific to the problem that satisfies the task requirements; we call it the *case-specific model*. In this view, a knowledge system uses a domain model to enlarge the input model until a complete and correct case-specific model is built—where "complete and correct" are with respect to the requirements of the task.

We have considered two kinds of adaptors: domain adaptors (*d-adaptors*) and case-based adaptors (*c-adaptors*). *D-adaptors* use some domain-specific knowledge to transform the case-specific model (in a way specified by the adaptor's competence). *C-adaptors* also transform the case-specific model but use domain knowledge that includes precedent cases retrieved from case memory.

Adaptors are applied to the case-specific model, performing search until a sequence of adaptor instantiations is found such that transforms the initial case-specific model into a correct case-specific model that satisfies the task goals. Thus, adaptation is viewed as a search process on the space of adaptors. Since new adaptors can be applied to the first adapted object, several search strategies (such as depth-first, breadth-first, and beam search) are possible.

We are applying the ABC theory in the SaxEx application[1], a complex real-world case-based reasoning system for generating expressive performances of melodies based on examples of human performances that are represented as structured cases.

In general terms, a *software architecture* describes the (i) components, (ii) connectors, and (iii) a configuration of how the components should be connected [7]. We can consider CBR systems as a specific variant of knowledge systems that furthermore use experiential knowledge [3]. Because of this we have taken UPML, a software architecture being developed for reuse of knowledge systems, and we are developing a variant adequate for CBR systems. The Unified Problem Solving Method Development Language UPML is currently in development by the IBROW consortium, and the first version is currently released [6]. Although UPML can still have future modifications we expect them to be minor and maintain stable the core ideas.

The organization of this paper is as follows. In Section 2 we present the ABC architecture. Section 3 describes how two different families of adaptors (c-adaptors and d-adaptors) are incorporated in the Noos language. Finally, in Section 4 we present the conclusions and discuss related work.

2 The ABC architecture

The three main elements of the ABC software architecture are (i) a task description, (ii) a domain model, and (iii) a library of adaptors. Figure 1 shows these three elements connected with a special kind of connector called *bridge*. In addition, the problem to be solved is called input in the figure and for simplicity we will include the case base into the domain model element. More specifically, we will consider that each solved problem is a model per se, and we will call it

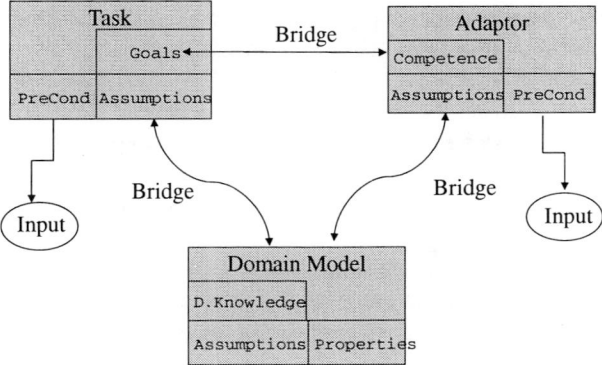

Fig. 1. The ABC software architecture consists of three elements: a task description, a domain model, and a library of adaptors. These three elements are connected by connectors called bridges. The problem to be solved is called input in this picture.

case-specific model—in other words, it is the model of an episode of solving that problem [3].

These three elements are taken from UPML where the main goal is the reuse of Problem Solving Methods (PSMs); since our goal is the *reuse of cases* we propose the specific architectural variation where adaptors play the role of PSMs. This transformation makes sense since PSMs are the components that perform the inferences for a knowledge system to build the *case-specific model* of the problem (i.e. the "solution" to the problem). In our approach the final case-specific model is build by the adaptors that transform the *case-specific model* imported from the case(s) retrieved from the case-base.

Tasks, domain models, and adaptors are conceptually distinct entities, although in practice CBR systems use an implicit description of the task and domain knowledge is tightly integrated with the CBR engine. From a methodological stance, however, it is better to consider they separate and possibly coming from distinct sources.

In a similar manner, specification of tasks is also being studied [11] to provide a vocabulary capable of describing tasks across a range of different domains of application, i.e. independent of the domain-specific vocabulary. For instance, a task description of diagnosis [6] is specified in terms of findings and hypothesis. A *bridge* is then needed to connect in a meaningful way task descriptions and domain models. For instance, in a medical diagnosis domain the bridge from the task description to the domain model maps findings and hypothesis to the terms manifestation and cause respectively.

Therefore the methodological approach we are endorsing takes two main aspects of knowledge modeling techniques: explicit representation and conceptual separation of tasks, domain knowledge, and adaptors. From UPML software architecture [6] we adopt the bridge connectors among ABC architecture components but we change the main elements of the architecture for CBR systems. In the rest of this section we make explicit those components, while in later sections we show a particular adaptation engine developed following this methodology.

2.1 The ABC components

Tasks A task provides a way to characterize what a CBR system is intended to achieve.

Task Description consist of a *task name* and
1. pragmatics (author, explanation, URL, last change date)
2. ontology (the vocabulary)
3. specification
   - goals (expressions characterizing the output case-models)
   - preconditions (expressions characterizing valid input case-models)
   - assumptions (expressions characterizing requirements on domain knowledge)

The main elements for characterizing a task are goals, preconditions and assumptions. These elements are described in some logical language, the option of which is open to the designer. Preconditions state constraints to be satisfied by the problems to be solved (input case models). Goals specify properties to be satisfied by the solved problem, i.e. by the output case-specific model. Finally assumptions determine assumptions made by the task description upon the content of the domain model.

Domain Models Domain models are specified using a specific vocabulary (domain ontology) and is characterized by properties, assumptions, and domain knowledge.

Domain Model Description consist of a *domain model name* and
1. pragmatics (author, explanation, URL, last change date)
2. ontology (the vocabulary)
3. specification
   - properties (meta-expressions characterizing domain knowledge)
   - domain knowledge (expressions describing knowledge)
   - assumptions (expressions characterizing assumptions on domain model)

Properties and assumptions both are used to characterize the knowledge content of a KB. These characteristics can be directly inferred from the domain knowledge or can be derived from requirements introduced by other components of the specification. While properties deal with characteristics of the knowledge content assumptions deal with external requirements like the environment of the system. As before, properties, assumptions, and domain knowledge are expressed in a specific formal language of choice.

Task-domain bridge The *td-bridge* is a connector that translates (refines) the task specification to a particular domain specified in domain-model. This bridge may add assumptions (on domain knowledge) to ensure that the translation result is valid. The only formal requirement is the union of both task and domain specifications is logically consistent.

Adaptors An adaptor is a special kind of connector between case-specific models—i.e. between "models of cases".

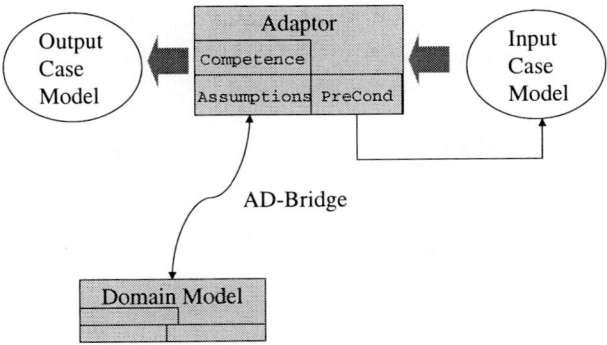

Fig. 2. The adaptor is a connector between "models of cases", here called case-specific models.

Adaptor Description consist of an *adaptor name* and
1. pragmatics (author, explanation, URL, last change date)
2. ontology (the vocabulary)
3. specification
    - preconditions (expressions characterizing valid input case-models)
    - assumptions (expressions characterizing domain knowledge needed by the adaptor to succeed)
    - competence (expressions characterizing the output case-models)

The preconditions of an adaptor specify the requirements to be satisfied by the input case-specific model for the adaptor's result be a valid one. The competence is a description of the transformation resulting from the application of the adaptor. Finally, the assumptions express the kind of domain knowledge the adaptor requires in order to be able to function. These assumptions may enlarge the requirements on domain knowledge already specified by the task. Since the case base is considered as a specific type of domain knowledge, case-based adaptation is considered to be realized by adaptors that use the experiential knowledge of the case base. Case-based adaptors are later discussed on § 3.

Task-Adaptor bridge The *ta-bridge* works like the tb-bridge above but now is connecting the task goals with the adaptors competence. Since the task goals specify the conditions for a problem to be correctly and completely solved the problem solving process is finished when an adaptor with a corresponding competence is available.

There are different ways to realize the ta-bridge depending on the strategy used to implement adaptors. A common strategy is designing a component library of adaptors. Moreover, depending on the complexity of the application domain the designers may implement one-shot adaptors—i.e. adaptors with a competence that directly fulfills the task goals. In more complex situations, the "total" adaptor need to be constructed from the elementary components in the adaptor library to fit the needs of each particular problem. This is the implementation we have chosen for the SaxEx system[9]. In this setting, adaptation is then a search problem over the space of adaptors whose goal is finding a combination of adaptor instantiations such that the final competence satisfies, via the bridge, the task goals.

From the software architecture stance what is formally required to establish a ta-bridge is only that the adaptor competence logically implies the task goals. The ABC architecture does not establish control constraints on the implementation nor distinguishes the situations where the adaptors already exist or have to be constructed from elementary adaptor components (and whether this construction is automated or performed by hand).

Domain-Adaptor bridge The *da-bridge* connects the assumptions upon domain knowledge specified by the adaptor with the domain model, in a similar way to how td-bridge maps task assumptions to the domain model. Some requirements of PSM upon domain models have already been established by connecting method with task and task with a domain. Now we only need to map the knowledge requirements that are exclusively for the method.

2.2 ABC and CBR systems design

The very idea of software architectures is learning from system developing experience in order to provide the abstract recurring patterns for improving further system development. As such, software architectures contribution in mainly methodological in providing a way to specify systems. If we consider existing CBR systems and the ABC architecture we can observe that ABC is making explicit issues that CBR system developers already know but treat implicitly when developing new systems and that they are not explicit either on the actual CBR system. Let's take the task of a CBR system, for instance. The specific task a CBR system has always to be specified, albeit informally, in the system design phase. The ABC approach considers this a specification of the task but also provides a specific way to relate that specification to each other component of the architecture: preconditions relate to the input problem, assumptions relates with the availability of knowledge, and goals relate to the search process performed by the CBR system.

Furthermore, let us consider domain knowledge. Some CBR systems use cases (and similarity) as the unique source of knowledge available to solve problems—e.g. instance-based learning approaches. However, a great number of CBR systems use domain knowledge, for different purposes and in different ways, in addition to cases. Commonly, this domain knowledge is not described as such, but it is described by explaining the implementation of the CBR system. In other words, what is described is the representation used to encode it (rules, constraints) and the role it plays in the system implementation (mainly concerning control issues). It is our personal opinion that a clarification of the role of domain knowledge in CBR systems is needed to improve the understanding of CBR and the development of CBR systems.

As a result of focusing on the adaptation process, ABC suggests that retrieval (and similarity assessment) is also a type of domain knowledge. In our approach, solving a problem is constructing a case-specific model of the "input problem"—as was established in the knowledge-level description of CBR [3]. A software architecture is a much refined level of description, so solving a problem in ABC involves building a case-specific model that satisfies the task description goals. Domain knowledge is used to perform the inference necessary to build this model[1]. The ABC architecture does not deal with control aspects of the implementation, thus the order in which domain-specific inference and case retrieval are performed is unspecified.

3 Implementing Adaptors

We will consider two kinds of adaptors: domain adaptors (*d-adaptors*) and case-based adaptors (*c-adaptors*). "Transformational adaptation" is realized by d-adaptors, i.e. by adaptors that use some domain-specific knowledge to transform the case-specific model (in a way specified by the adaptor's **competence**). Moreover, that domain knowledge is the one explicitly required by the adaptor's **assumptions**.

"Derivational replay" is realized by c-adaptors. Case-based adaptors also transform the case-specific model but use some domain knowledge that includes a precedent case retrieved from case memory[2]. In the simplest scenario there is only on retrieved case, but in CBR systems where parts of cases are also cases each part can be adapted in a case-based way by c-adaptors. Derivational replay in planning is one example and the SaxEx system [9] is another example — adaptation in the SaxEx system combines d-adaptors and c-adaptors.

The main issue to go from a specification like ABC to an actual implementation is deciding how is 1) the representation of components and bridges, and 2) the control scheme. We are implementing adaptors in Noos, a representation language designed for supporting knowledge modeling approaches to problem solving and learning [2] in which different CBR systems have been built, including SaxEx. In Noos cases are represented as feature terms [8], a formalism for representing structured cases in which any subpart of a case (feature term) is also a term—and thus is also a case. Inference is provided by problem solving methods (PSMs) that use domain knowledge to build models (or parts of models). A problem is solved when a case-specific model is completed, and then it is retained in the case base. Retrieval is performed by specialized PSMs, retrieval methods, that use domain knowledge or heuristic principles to search the case base. Concerning the control scheme, Noos inference is on demand, i.e. follows a lazy evaluation strategy. The chain of control is thus backwards: retrieval methods determine the features of a case that they need, thus forcing the evaluation of

---

[1] Some CBR papers distinguish between primary and derived feature cases. Primary features are those appearing on the "input case" and derived features are inferred by the system from primary features. In our approach, inference uses domain knowledge (including cases) to build a model of the problem.

[2] Recall that, for the ABC architecture, the base of cases is also part and parcel of the domain model.

the PSMs that infer those features needed that were not part of the input problem model. Moreover, c-adaptors use retrieval methods so the retrieval process is in fact directed by the adaptation strategy.

The main ABC elements incorporated in Noos are i) an explicit description of a task, ii) adaptors, iii) and ta-bridges. Since the rest of the ABC elements is obviated, some parts of this elements need not be represented explicitly: the reason being that Noos will not be reasoning about them. Thus, a task holds only goals and preconditions, while adaptors holds only competence and preconditions. Assumptions are not present since we are not representing td-bridges nor da-bridges. The contents of these slots (goals, competence, preconditions) are expressed by feature terms. Satisfaction is represented as feature term subsumption ($\sqsubseteq$), thus a case-specific model C satisfies an adaptor preconditions A_P^i when $A_P^i \sqsubseteq C$ (A_P^i subsumes C).

The overall adaptation process is realized following an "Adaptation as Search" strategy. The initial state is the case-specific model of the problem; this begins with the information given as input, but the domain PSMs can enlarge this model performing inference as needed. The goal state is a complete and correct case-specific model C_F that satisfies the task goals T_G. The ta-bridge provides a translation from the task description vocabulary to the domain vocabulary used in adaptors and case specific models. Thus, the task goals expressed in domain vocabulary are obtained applying the bridge B_{TA} to the task goals $B_{TA}(T_G)$ and therefore a solution is defined as a case-specific model C_F such that $B_{TA}(T_G) \sqsubseteq C_F$.

Adaptors are applied to the case-specific model, performing search until a sequence of adaptor instantiations is found such that transforms the initial case-specific model into C_F. A classical means ends analysis technique is used with the adaptors, where *preconditions* establish if the adaptor is applicable to a particular case-specific model, and *competence* establishes the goals or subgoals achievable by instantiating the adaptor. Since Noos provides automatic backtracking, selection of adaptors and adaptor instantiation following several search strategies —such as depth-first, breadth-first, and beam search— can be easily implemented for a particular CBR system. An interesting issue left for future work is performing a case-based search of adaptor selection and instantiation: since adaptors are feature terms, they are stored in memory by Noos and they are thus amenable to be retrieved. This case-based adaptation process would be able to use both c-adaptors and d-adaptors, unifying "transformational" and "generative" adaptation in a case-based reuse of cases.

4 Discussion and Related work

A conceptual framework for describing CBR systems is Richter's knowledge containers [10]. An approach towards a formal model of transformational adaptation based on the knowledge containers framework is presented in [4]. The purpose of Bergmann and Wilke's paper is to characterize when properties such as soundness and completeness can be formally proven to hold in transformational adaptation. Interestingly, their approach centered on adaptation also seems to downplay the importance of retrieval (and similarity) in CBR systems, in a similar way as the ABC architecture conceives of retrieval as a part of domain knowledge. In our approach it is up to the designer of a CBR system to decide whether completeness is required or possible. Moreover, the designer may decide to use a logical language for specifying a ABC architecture and then formally prove that certain formal properties hold. For an approach of using UPML with automated reuse see [5]. It is an interesting question whether the knowledge containers framework could be refined to provide a software architecture with the containers as components—in which case appropriate connectors should be defined.

References

1. Josep Lluís Arcos, Ramon López de Mántaras, and Xavier Serra. Saxex : a case-based reasoning system for generating expressive musical performances. *Journal of New Music Research*, 27 (3):194–210, 1998.
2. Josep Lluís Arcos and Enric Plaza. Inference and reflection in the object-centered representation language Noos. *Journal of Future Generation Computer Systems*, 12:173–188, 1996.
3. E. Armengol and E. Plaza. A knowledge level model of case-based learning. In S. Wess, K.D. Althoff, and M. Richter, editors, *Topics in Case-Based Reasoning*, number 837 in Lecture Notes in Artificial Intelligence, pages 53–64. Springer-Verlag, 1993.
4. R. Bergmann and W. Wilke. Towards a new formal model of transformational adaptation in case-based reasoning. In *European Conference on Artificial Intelligence (ECAI'98)*, 1998.
5. D. Fensel and V. R. Benjamins. Key issues for automated problem-solving methods reuse. In *Proceedings of the 13th European Conference on Artificial Intelligence (ECAI-98)*, pages 63–67, 1998.
6. D. Fensel, V. R. Benjamins, M. Gaspari S. Decker, R. Groenboom, W. Grosso, M. Musen, E. Motta, E. Plaza, G. Schreiber, R. Studer, and B. Wielinga. The component model of upml in a nutshell. In *Proceedings of the International Workshop on Knowledge Acquisition KAW'98*, 1998.
7. D. Garland and D. Perry (Eds.). Special issue on software architectures. *IEEE Transactions on Software Engineering*, 1995.
8. Enric Plaza. Cases as terms: A feature term approach to the structured representation of cases. In M. Veloso and A. Aamodt, editors, *Case-Based Reasoning, ICCBR-95*, number 1010 in Lecture Notes in Artificial Intelligence, pages 265–276. Springer-Verlag, 1995.
9. Enric Plaza and Josep L. Arcos. The ABC of adaptation: Towards a software architecture for adaptation-centered CBR systems. Technical Report 99-21, IIIA-CSIC, 1999. Available online at http://www.iiia.csic.es/Projects/cbr/ABC/abc.html.
10. M. M. Richter. The knowledge contained in similarity measures, 1995. Invited talk to ICCBR-95. Available at http://wwwagr.informatik.uni-kl.de/ lsa/CBR/.
11. K. Seta, M. Ikeda, T. Shima, O. Kakusho, and R. Mizoguchi. Clepe: a task ontology based conceptual level programming environment. *Trans. of IEICE*, (9), 1999.

Knowledge Representation in Planning: A PDDL to OCL_h Translation

R. M. Simpson, T. L. McCluskey, D.Liu, D.E.Kitchin

Department of Computing Science University of Huddersfield, UK
r.m.simpson@hud.ac.uk t.l.mccluskey@hud.ac.uk
d.liu@hud.ac.uk d.e.kitchin@hud.ac.uk

Abstract. Recent successful applications of AI planning technology have highlighted the knowledge engineering of planning domain models as an important research area. We describe an implemented translation algorithm between two languages used in planning representation: PDDL, a language used for communication of example domains between research groups, and OCL_h, a language developed specifically for planning domain modelling. The algorithm is being used as part of OCL_h's tool support to import models expressed in PDDL to OCL_h's environment. Here we outline the translation algorithm, and discuss the issues that it uncovers. Although the tool performs reasonably well when its output is measured against hand-crafted OCL_h, it results in only partially specified models. Analyis of the translation results shows that this is because many natural assumptions about domains are not captured in the PDDL encodings.

1 Introduction

Despite many years of research into AI Planning, knowledge engineering for applications of planning technology is in its infancy. Recent successful AI planning applications [10, 14, 1] have nonetheless highlighted the problems facing knowledge engineering in planning. Questions include how to choose appropriate planner technology for a given application, and how to encode knowledge into domain models for use with planning algorithms. The engineering of knowledge-based planners has resulted in a set of workshops and initiatives, including those in references [2, 12]. An accepted syntax for exchange of domain encodings is PDDL, a *planning domain definition language*, and many established planners can be obtained via the internet with a set of domains encoded in this syntax. PDDL emerged from the need to construct a common language for the biannual AIPS competitions (for details of PDDL and domain examples consult reference [3]). Language conventions such as PDDL help the research community to some extent in the problems of exchanging research information, and in the independent validation of research results.

Domain definition languages such as PDDL, however, are not designed with the same criteria in mind as a *domain modelling language*. The latter would be associated with a domain building methodology, be structured to allow the expeditious capture of knowledge, and have the benefit of a tools environment for knowledge engineering. OCL_h stems from a family of fairly simple planning-oriented domain modelling languages deriving from the work in reference [8]. The benefit in using OCL_h is seen as twofold: to improve the planning knowledge acquisition and validation process; and to improve and clarify the plan generation process in planning systems. A range of planners have been implemented for use with OCL_h and the language is being used as a prototype for a collaborative UK project to create a knowledge engineering platform for planning [13]. OCL_h is structured to allow the capture of object and state-centred knowledge, as well as action-centred knowledge, [1] and it is encased in a tools environment.

In this paper we discuss the issues raised in the construction of one of the tools in OCL_h's environment: a translator, to help import models written in PDDL into the OCL_h environment. The translation is feasible because PDDL and OCL_h share similar underlying assumptions about worlds - they are assumed closed, actions are deterministic and instantaneous. We outline the translation process, and the results in applying it to example domain models. The tool's output is used to both make comparisons with hand-crafted models, and to identify omissions and insecurities in the PDDL encodings. In effect, it is not possible for the tool to derive secure OCL encodings as the notion of "valid state" is present neither implicitly or explicitly in the PDDL encodings. This has serious consequences for the communication and maintenance of domain descriptions within this medium, as the set of states in which an action can be executed wll generally contain many states which are not sensible. For example, while it is understood that domain writers encode all the positive preconditions that must be true before an action is executed, the language does not encourage the recording of propositions that must not be true for the execution to make sense.

2 The Planning Domain Definition Language

PDDL was established by the AIPS-98 Competition Committee to enable competitors to have a common language for defining domains, and to aid the development of a set of problems written in PDDL on which the different planners could be tested [3]. PDDL has been incrementally extended to include a wide range of syntactic features, although most planners cited in the literature utilise only basic features. Planners can be restricted to a PDDL subset by declaring those language features required when the domain is defined. Here we only mention those features relevant to the paper.

PDDL's basic level of representation is the literal, and a model's central element is a set of operator schemae representing generalised domain actions (very much in the style of 'classical planning' literature with its roots in STRIPS [4]). Each operator is defined with a precondition and effect, where the semantics are interpreted under the STRIPS assumptions. Below are two examples of simple PDDL operator definitions which use typed parameters. They belong to an encoding of an example domain called the Tyre World which was taken from the distribution examples associated with reference [3]. A planner using the Tyre World should be able to output sequences of ground operators to solve goals involving changing a flat tyre. We will use this domain as a "running example".

---

[1] traditionally, models of planning domains were equated with a set of action specifications, and were therefore only 'action-centred'

```
(:action loosen
   :parameters (?n - nut ?h - hub)
   :precondition (and (have wrench)(tight ?n ?h)(on-ground ?h))
   :effect (and (loose ?n ?h) (not (tight?n ?h))))
(:action fetch
   :parameters (?x - (either tool wheel) ?c - container)
   :precondition (and  (in ?x ?c) (open ?c))
   :effect (and (have ?x) (not (in ?x ?c))))
```

The *loosen* operator models the action of undoing (but not removing) the nuts that fasten a wheel onto a hub. The *fetch* operator models the action of removing a tool or a wheel from a container in which it was stored (such as a car's trunk).

Problems for a planner to solve are posed in PDDL as an initial state (a set of ground literals) and a goal condition. Although the current PDDL version includes many other features (hierarchically-defined operators, domain axioms, safety constraints, quantification over parameter domains etc) the majority of domain encodings and test-sets available via the internet input the simple form of PDDL similar to that described above.

3 The Object-Centred Language OCL_h

OCL_h was designed to be a kind of 'lifted' STRIPS-language, aimed to keep the generality of classical planning but to incorporate a model-building method and be structured to help the validation and operationalisation of domain models [12, 6, 8]. An OCL_h world is populated with dynamic/static objects grouped into *sorts*[2]. Each dynamic object exists in one of a well defined set of states (called 'substates'), where these substates are characterised by predicates. On this view the application of an operator will result in some of the objects in the domain moving from one substate to another. In addition to describing the actions in the problem domain, OCL_h provides information on the objects, their sort hierarchy and the permissible states that the objects may be in. Relations and propositions are not fully independent entities – rather they now belong to collections that can be manipulated as a whole. So instead of dealing with literals planning algorithms reason with objects. Similarly to a typed PDDL specification, the objects and the sorts they belong to are predefined, as is the sort of each argument of each predicate in the OCL_h model.

An object description in a planning world is specified by a tuple (s, i, ss), where s is a sort identifier, i is an object identifier of sort s, and ss is its *substate*. A substate is a set of predicates which all *describe* i. For example, $(nut, nut0, [loose(nut0, hub1)])$ is an object description meaning that $nut0$ of sort nut is loosely done up on $hub1$. Or again, $(container, trunk1, [closed(trunk1), locked(trunk1)])$ is an object description meaning that container $trunk1$ is closed and locked. Only a restricted set of predicates are allowed to describe an object and appear in its substate. Substates operate under a *closed world* assumption local to this restricted set - thus in the last example, the predicate $open(trunk1)$ is false because (a) it is used to describe objects of this

---

[2] we use the name 'sorts' rather than 'object classes' to emphasis that OCL_h is an abstract object-centred modelling language - in contrast to an OO implementation language

sort (b) it does not appear in the substate. The domain modeller defines the predicates used to describe objects, and the form of each substate, using **substate class definitions**. The predicate expressions in such definitions are constructed to form a complete, disjoint covering of the space of substates for objects of each sort. When fully ground, an expression from a substate class definition forms a legal substate. For example, the substate class definitions for the sorts container, nut and hub are[3]:

```
substate_classes(container,C,
   [[closed(C)],[open(C)],[closed(C),locked(C)]])
substate_classes(nut,N,[[loose(N,H)],[tight(N,H)],[off_hub(N)]])
substate_classes(hub,H,[[on_ground(H),fastened(H)],
   [jacked_up(H,J),fastened(H)],[unfastened(H),jacked_up(H,J)],
   [free(H),jacked_up(H,J),unfastened(H)]])
```

The first example means that objects of sort container can be either closed, not open and not locked, *or* open, not closed and not locked, *or* closed, locked and not open (*or* here is exclusive). Thus, in OCL_h *negation* is implicit: if it is the case that $\neg open(trunk1)$, then this means that $trunk1$ must be in one of two substates, its object description being $(container, trunk1, [closed(trunk1)])$ or $(container, trunk1, [closed(trunk1), locked(trunk1)])$.

A domain model is built up in OCL_h by creating the operator set at the same time as creating the substate class definitions. We define an **object expression** to be a tuple (s, i, se) such that the expression part se is a generalisation of one or more substates (se is normally a set of predicates containing variables). An **object transition** to be an expression of the form $(s, i, se \Rightarrow ssc)$ where i is an object identifier or a variable of sort s, (s, i, se) forms a valid object expression, and ssc is taken from one of the substate class definitions. Thus when ssc is ground it will always be a valid substate. An action in a domain is represented by an **operator schema** with an identifier, a prevail condition, and a list of transitions. Each expression in the prevail condition must be true before execution of the operator, and will remain true throughout operator execution.

Two OCL_h operators hand-crafted to (loosely) correspond to the PDDL operators above are as follows:

```
operator(loosen(N,H,W),
     [ (wrench,W,[have(W)]), (hub,H,[on_ground(H),fastened(H)]) ],
     [(nut,N,[tight(N,H)]=>[loose(N,H)])] )
operator(fetch(T,C),
     [(container,C,[open(C)])],
     [(tool_or_wheel,T,[in(T,C)]=>[have(T)])])
```

As with PDDL, OCL_h has many other features such as conditional operators, hierarchical operators, atomic and general invariants, but due to lack of space we refer the reader to the literature for these details.

---

[3] whereas in PDDL we write a variable as an identifier beginning with '?', in OCL_h variables are identifiers with leading capitals

4 Lifting PDDL To OCL_h

The general framework : We base the translation on two main assumptions: (1) the input to the translator will be any model written in the subset of PDDL that includes STRIPS-like operators with literals having typed arguments. This translator will be adequate for our purposes as the test sets in use in the AIPS competitions and those available in resource web sites are generally no more expressive. Where the model is written without typed arguments, it can be augmented by hand or a tool such as TIM [5] can be used to provide the typing. (2) the translation should keep, as far as possible, the names and structure of the input model. This leads us to the following general framework for translation:

PDDL parameter type $name$ $\Rightarrow$ OCL_h sort $name$
PDDL predicate $\Rightarrow$ OCL_h predicate
PDDL operator $name$ $\Rightarrow$ OCL_h operator $name$

The first association preserves the correspondence of PDDL primitive types and OCL_h sorts but does not guarantee conformity to the OCL_h requirements for a sort hierarchy as PDDL's sort hierarchy is not required to form a tree structure. This problem can however be ignored for the STRIPS-like domains we are interested in here as non-primitive types are just a device to allow single operators to describe transformations to objects of diverse types. The second association raises problems concerning the issue of grouping predicates to form substate class definitions as discussed below. Once this is done, re-writing the PDDL operators by extracting the *object transitions* and the *prevail clauses* from the raw STRIPS operator is relatively straightforward.

Inducing Substate Classes : Steps in the OCL_h method that are used to derive substate class definitions are as follows:
1. Identify the sorts that are dynamic and those that are static
2. For each dynamic sort, identify those predicates that are to be included in defining its substate classes
3. For each dynamic sort, define its substate classes

For step 1, a sufficient condition for a sort s to be dynamic is that PDDL type s is described by a property which can be changed by a PDDL operator. Those types that have no changeable properties, but are referred to within a changing relation may or may not be mapped to a dynamic sort – this choice will become clear after our discussion of step 2. In step 2, the problem alluded to above arises in that given a predicate $p(s1, s2, .., sn)$, what subset of the OCL_h sorts $s1, s2, .., sn$ should it be associated with, to describe that sorts' substates classes? In the method associated with OCL_h it is proposed that normally each predicate describe a single sort (although if the sort were not primitive the predicate would be used in distinct primitive sorts). To illustrate this problem consider the PDDL predicate *in*, with two arguments of type *tool* and *container* respectively. Both types are mapped over to OCL_h dynamic sorts, and the question arises: should the predicate "in" be used to describe the state of an object of sort *tool*, the state of an object of sort *container*, or both? Though from a logical point of view there is no more reason to say that the predicate *in* characterises a *tool* than there is to say it characterises a *container* there are strong pragmatic reasons to classify the predicate as belonging to only one of the objects referenced. If we allow predicates to describe all its sorts' states then there is a clear redundancy in our representation, in that we record the same information twice. More serious than this, allowing a relational predicate to characterise all referenced sorts introduces the *frame problem* in a particularly acute manner. Recall that the right hand sides of OCL_h transitions must fully characterise the resulting substate of the dynamic object participating in the transition, without default persistence but with a closed world assumption local to the predicates describing that sort. Then to record the possible substates of the *container* we would have to consider the possible combinations of the container being open, closed and locked along with all possible combinations of objects such as the tools and wheels being either in or not in the container; this would lead to a proliferation of object transitions and operators. The discussion above shows that it is not practicable to let a predicate be used in the substate descriptions of objects in every one of its argument sorts. Our solution to this frame problem is to try to follow the intuition in building an OCL_h model manually: let the algorithm choose *one* single sort. This distinguished sort is said to *own* the predicate. Though from a logical point of view this may seem arbitrary it coincides with intuition in the sense that we would not naturally think of the action of opening the trunk as having a different result depending on the trunk's contents. In English an action verb is typically thought of as characterising the subject of the sentence rather than the object. In this spirit we say the the predicate *in(wrench,trunk)* describes the state of the *wrench* but not that of the *trunk*. Given that we will only allow a predicate to characterise a single sort the choice of sort could be made in a number of competing ways. We could try to allocate predicates to sorts in a way to try and minimise the frame problem or to minimise the number of sorts that change state in the actions concerned, or we could simply allocate them to the first mentioned object in each predicate. Up to now our experiments have shown the third strategy gives satisfactory results when the auto-generated OCL_h model has been compared to a hand crafted version. Returning to step 1, this analysis determines the split of static and dynamic sorts: if some dynamic predicate has the property that its first argument can contain object identifiers of sort s, then s is a dynamic sort; otherwise, s will not be described by any dynamic predicates and hence will be static.

Dealing with Negation : Negation in OCL_h is not represented explicitly, because of the local closed world assumption used in substates. It may be the case, however, that a negative form (or opposite) is required. We deal with this by potentially creating for each predicate a negative form, identified by prepending the predicate with *not_*. Though we start with the availability of all such possible negations, not all are used in the final translation. This would be the case in the Tyre-World if the container/trunk is described as either *open* or *closed* in which case the negative forms *not_open* and *not_closed* are not needed.

5 The Translation

Overall Results : The associations described above have been implemented in a PDDL-OCL translation tool. The translator has been tested on four typed PDDL worlds: Tyre-World, Ferry, Gripper and Fridge. Details of the algorithm and results, which have been omitted here due to lack of space are given in [9].

The translation results are encouraging in that they are close to those produced by hand translation from the same PDDL source. To indicate the nature of the results, of the thirteen actions in the Tyre World two of the actions contained anomalies flagged up by the translation. Eight of the translated actions contained unnecessary, though correct, negations on their right hand sides and two actions had incomplete object transitions. In the Ferry world the translation was capable of running by a planner by the addition of a missing object state to the problem specification.

Anomalies : The translation forms a basis for hand completion but also has the power to flag up potential problems and insecurities with the PDDL domain specification. The most interesting of the anomalies uncovered in the Tyre World domain occurs with the *jack-down* action which is translated as follows:

```
(:action jack-down
    :parameters (?h - hub)
    :precondition (not (on-ground ?h))
    :effect (and (on-ground ?h) (have jack)))
operator(jack_down(H),
    [],
    [(hub,H,[not_on_ground(H)]=>[on_ground(H)]),
     (tool,jack,[]=>[have(jack)])] )
```

The transition for the jack indicates that it may be in any state prior to being possessed as a result of jacking down the wheel. This is not adequate as the mechanic only possesses the jack after execution of the action because it was used to jack-up the wheel in the first place. The PDDL formulation works (operationally) because in the version of the domain used there is no alternative way of getting the wheel off the ground (although we might have an alternative jack-up action, such as use a block and tackle).

A second anomaly which arises with the encoding of the *jack_down* action is that we treat [on_ground(H)] as a complete substate of the *hub*. From the auto-generated substate class definition for the *hub* we see that either the predicate *fastened(H)* or *unfastened(H)* and either the predicate *free(H)* or *not_free(H)* must also apply to the the hub and this raises the following question: should it not be the case that we should make it a precondition of the action that the hub has the wheel on and fastened to it prior to jacking down? In effect, in OCL_h terms the transition should be

```
(hub,H,[not_on_ground(H),fastened(H),not_free(H) ]=>
       [on_ground(H),fastened(H),not_free(H)]),
```

This example raises a more general problem with the PDDL representation of planning domains. As noted OCL_h requires the right hand side of an object transition to completely characterise the resulting state of the object. In the PDDL representation the state of an object is not fully determined by the application of an operator as some of the object's properties my simply persist without being referenced by the operator from an earlier state. This makes it impossible to determine which states of objects are legal in the domain. In general if we have n predicates characterising an object sort, excluding their negations, there are 2^n possible substates for objects of that sort. Accordingly for example for the *hub* described by the predicates *(on_ground(H), fastened(H), free(H)* there are eight possible such substates. From the PDDL operators some of these states may be reachable and others not but we cannot definitively exclude any from the substate class definition as all may be candidate start states for an imaginable problem. But this is inadequate as some of these states may not be possible such as *on_ground(H)* ∧ *fastened(H)* ∧ *free(H)* i.e. hub on the ground with the nuts fastened up but no wheel on. Operationally this may not occur if the hub starts off in a sensible state but we should not be relying on such a procedural definition of an object's states to determine what is possible.

Implicit agents : A problem with agents of actions being implicit in PDDL domain specifications arises when translating to OCL_h. The problem is amply illustrated by the following translation of the *move* rule from the Gripper domain where we have a robot that can move from a named location to another named location.

```
operator(move(TO, FROM),
  [],
  [ (room,TO,[]=>[at_robby(TO)]),
    (room,FROM,[at_robby(FROM)]=>[not_at_robby(FROM)])] )
```

The rooms *TO* and *FROM* are being classed as dynamic objects subject to change, but we would more naturally want to say that it is the robot that has changed. In OCL_h parlance locations should be treated as static. To solve this problem we need to recode the *at_robby* predicate and introduce an agent i.e. the robot. *at(Agent,Location)*. If the agent has, as in this case been implicitly encoded into the predicate then there will only be one such agent and we can effectively introduce a new constant *agent0* and a new type *Agent*.

6 Discussion

The basic strategies of re-casting domain knowledge from a predicate base into an object-centred base are not new and have been discussed in the literature for some period (e.g. see [11]). OCL_h contrasts with previous domain modelling languages for planning such as SIPE-2 and O-Plan [16, 15] in its simplicity and clarity. On the other hand, OCL_h is less sophisticated than these system (for a comparison of O-Plan's TF and OCL_h see reference [7]).

Fox and Long in reference [5] show that the limitation of requiring arguments to be typed in the PDDL specification is not fundamental to the translation. They demonstrate that type information can be extracted from a set of PDDL operator schemae only. Fox and Long's TIM uses the operator schemae to analyse the domain and produce types such that objects belonging to them are identical up to naming. It therefore appears to produce a type structure more appropriate to OCL_h. Our future work will involve merging the translation algorithm with the TIM engine to create a more powerful translation tool.

It has been acknowledged since the modern inception of AI that the representation of knowledge has a critical bearing on the performance of a problem solver. In planning especially, there have been relatively few insights or research projects in this area - instead the planning literature has tended to concentrate on the efficiency issues of planners, or the adequacy of expression of their domain model languages. We see our ongoing work on the translation from PDDL to OCL_h as promoting the debate on the relative merits of planning domain encodings, and, in time, the matching up of appropriate planner technology to application domain. Working with a domain modelling language such as OCL_h gives opportunities for higher level domain validation with rich tool support that eases domain modelling. Our translator from PDDL to OCL_h gives us access to a rich source of research examples written in PDDL, although it shows up the lack of "knowledge content" in these encodings. Also, it has highlighted those issues in representation such as use of negation, completeness and security of models, and construction of object hierarchies that are fundamental to the creation of a planning domain model.

References

1. A. Tate (editor). *Advanced Planning Technology: Technological Achievements of the ARPA/Rome Laboratory Planning Initiative.* IOS Press, 1996.
2. Benjamins, Nunes de Barros, Shahar, Tate and Valente (eds). *Workshop on Knowledge Engineering and Acquisition for Planning: Bridging Theory and Practice.* Proceedings of AIPS, 1998.
3. AIPS-98 Planning Competition Committee. PDDL - The Planning Domain Definition Language. Technical Report CVC TR-98-003/DCS TR-1165, Yale Center for Computational Vision and Control, 1998.
4. R. E. Fikes and N. J. Nilsson. STRIPS: A New Approach to the Application of Theorem Proving to Problem Solving. *Artificial Intelligence*, 2, 1971.
5. M. Fox and D. Long. The Automatic Inference of State Invariants in TIM. *JAIR vol. 9*, pages 367–421, 1997.
6. D. Liu and T.L.McCluskey. The OCL Language Manual, Version 1.2. Technical report, Department of Computing Science, University of Huddersfield, 2000.
7. T. L. McCluskey, P. Jarvis, and D. E. Kitchin. OCL_h: a sound and supportive planning domain modelling language. Technical report, Department of Computer Science, The University of Huddersfield, 1999.
8. T. L. McCluskey and J. M. Porteous. Engineering and Compiling Planning Domain Models to Promote Validity and Efficiency. *Artificial Intelligence*, 95:1–65, 1997.
9. T. L. McCluskey and R.M.Simpson. Adequacy of Planning Domain Descriptions. *Technical Report, The University of Huddersfield*, 2000.
10. B. Pell N. Muscettola, P. P. Nayak and B. C. Williams. Remote Agent: To Boldly Go Where No AI System Has Gone Before. *Artificial Intelligence*, 103(1-2):5–48, 1998.
11. N. J. Nilsson. *Principles of Artificial Intelligence.* Springer-Verlag, 1982.
12. PLANET. *First Workshop of the PLANET Knowledge Acquistion Technical Coordination Unit.* http://helios.hud.ac.uk/planet, 1999.
13. Planform. *An Open Environment for Building Planners.* http://helios.hud.ac.uk/planform.
14. S. Chien (editor). *Proceedings, 1st NASA Workshop on Planning and Scheduling in Space Applications.* NASA, Oxnard CA, 1997.
15. A. Tate, B. Drabble, and J. Dalton. O-Plan: a Knowledged-Based Planner and its Application to Logistics. AIAI, University of Edinburgh, 1996.
16. D. Wilkins. Using the SIPE-2 Planning System: A Manual for SIPE-2, Version 5.0. SRI International, Artificial Intelligence Center, 1999.

A Method and Language for Constructing Multiagent Systems

Hiroyuki Yamauchi[1] and Setsuo Ohsuga[2]

[1] Research Center for Advanced Science and Technology, University of Tokyo
4-6-1 Komaba, Meguro-ku, Tokyo 153-8904, JAPAN
yama@ai.rcast.u-tokyo.ac.jp
[2] Faculty of Software and Information Science, Iwate Prefectural University
152-52 Sugo Takizawa, Iwate 020-0172, JAPAN
ohsuga@soft.iwate-pu.ac.jp

Abstract. This paper describes a method to construct multiagent systems. The method proposed here is explored accounting for the Gibson's ecological view of information, i.e. affordance. We apply the idea of affordance not only to the reactive models of agents but also to the deliberative models of the agents. By this approach, we can avoid the frame problems that emerge from the dynamic environment including the agent's mental world. We describe a basic representation scheme for agent modules which reflects the Gibson's view of information resources. As a system description language, we use knowledge processing language KAUS (knowledge acquisition and utilization system) based on first order logic and axiomatic set theory. We consider as an example the multi-strata modeling scheme for developing human-computer interactive problem solving systems that are essentially multiagent systems.

1 Introduction

The more the information systems become complex and large-scaled, the more it becomes important for the systems that these have the property of autonomy for information gathering, processing and management. The agent technology developed so far [1,2,3] has been much applied to design and implement such autonomous systems (autonomous agent systems). We can see the typical applications of the agent technology in intelligent robots [1], enterprise models [10] and information assistants in the Web environment [3].

The common problem to be solved for constructing agent systems is how to describe agent structures (organizations), its functionality, properties and the control structures in general. We have to consider especially the autonomy of the system, that is, the capabilities of *learning from environments* through the interaction and *self decision-making* for attaining his/her goals. In the multiagent environment, the *coordination*, *cooperation* and *communication* with each individual agent should be considered. We believe that it is inevitable for us to apply knowledge processing technology to solve these problems.

The other points to be considered about agent systems are the ability of *perception* of information resources by the agents. If the environment with which the agents interact is very complex, large, uncertain and dynamical, it is very hard and even impossible for them to possess beforehand all information about it. This means that the agent should have a mechanism to find automatically information necessary in the environment. The more importance is how to solve the *frame problem* in artificial intelligence.

In this paper, we show that the Gibson's ecological view of information resources denoted by *affordance* [7] is valuable to solve these problems described above. In Chapter 2, we briefly summarize the notion of affordance, the frame problem in artificial intelligence [6] and a link between them. In Chapter 3, we describe a basic representation scheme for agent modules which reflects the Gibson's view of information resources. In Chapter 4, we describe a language which is suitable for implementing agent systems, specifically, KAUS language [11,12] developed by us. In Chapter 5, we consider as an example the multi-strata modeling scheme for developing human-computer interactive problem solving systems. Finally, we give concluding remarks in Chapter 6.

2 A Link between the Frame Problem in AI and Affordance in Psychology

The frame problem is the problem of describing, computationally, what properties persist and what properties change as actions are performed. In the literature [6], the two frame problems are discussed. One is the mathematical frame problem and another is the commonsense frame problem. The former is concerned with the intractability of time and space for computing frame axioms. The latter is concerned with the intrinsic difficulty of axiomatization and conceptualization of the significant portion of the real world, the world which is complex, large, uncertain and dynamical, and ill-defined by nature. The qualification problem in such circumstances has to be solved.

On the other hand, the theory of affordance established by ecological psychologist James J. Gibson (1904-1979) gives us a new point of view on perception of objects in the real world. According to Gibson, the correct context of our perception is defined by the *interaction* between us and the real world. It differs from the notion in traditional cognitive psychology in the sense that perception and action are not treated as separate processes in affordance theory. Organisms move in the world, finding the available information in it and move again using the information found. It is assumed that information available emerges from what is maintained in the real world. An organism's activities are said to be directly linked to what the objects in the real world afford (for example, a chair affords sitting).

To summarize, the frame problem means that it is fundamentally difficult for the practical AI systems to provide all information necessary for problem solving in advance and in the exhaustive way. It also denotes that the centralized information management is intractable for complex and large-scaled problem domains. On the other hand, the notion of affordance shows us that we can configure AI systems in such a manner that they maintain information resources in a distributed objects in the problem world and they accept on-demand information processing through the interaction with the objects. The concurrent objects [13], agent-oriented programming [14] and more broadly saying distributed artificial intelligence (DAI) and multiagent systems (MAS) [15] involve the potential use of the notion of affordance. Brook's subsumption architecture [1, 2] in robotics also involves the potential use of the notion of affordance to avoid the frame problem in AI. However, the Brook's architecture is only concerned with a *reactive* level of actions. As a result, the potential use of affordance in the *deliberative* level is not considered. In fact, the Gibson's original ecological view of information also relates only to the reactive level of perception. It denies all the mental states, cognitive maps and inference in the organism's brain (similarly the Brook's architecture). The higher level of intelligence is not at all referred to therein. We claim this restrictive consideration of affordance is not enough to construct truly intelligent systems. In general settings, intelligent systems should have all the reactive, cognitive and deliberative performances. The seamless applications of affordance to all the reactive, cognitive and deliberative performances of the agents should be considered. Our goal in this paper is thus motivated to take away this restrictive consideration in developing multiagent systems.

3 Basic Representation Scheme for Individual Agents

As described in the previous chapter, every constituent of the environment, namely organisms and artifacts, can be regarded as ecological information mediators and processors. From the ecological viewpoint of information resources, not only organisms but also artifacts can be regarded as agents having certain affordances. Consequently the basic representation scheme for agents described in this chapter can be applied both to the representation of organic agents and to that of artificial agents.

3.1 Skeleton Structure of an Agent Functional Module

An agent functional module(M) is defined with a set of variables for holding values of the specified attributes and a set of primitive/compound methods (procedures) that achieve the specified goal using the input and the current state of the defined variables. We define the skeleton structure of an agent functional module M as follows (see Fig.1).

(1) The name of the module.
(2) The layer or type of the module.
(3) The state of the module.
(4) The affordance.
(5) A set of methods for perception, action, and communication.
(6) The local working memory.

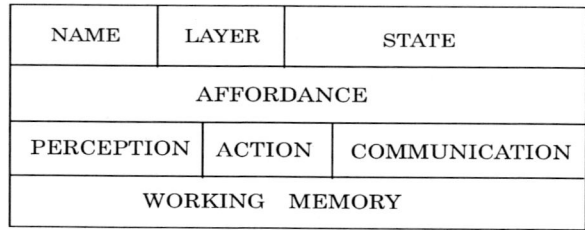

Fig. 1. Skeleton structure of an agent functional module

The name of the module is a unique identifier that can discriminates itself from other agent modules. The state of the agent module holds both the current state and the past state (history) in the predefined time period. The *history* describes the time series of the results of perceptions and actions executed by this module. The *layer* or *type* of the agent module is used for classifying this module (see the next section). The layers considered here are those of reactive modules which perform reactive goals, mental modules which perform local reflective thinking, and social modules which perform cooperative tasks. A rational agent is made up from these three layers [1].

Perception, action and *communication* are component modules that characterize the agent functionality. The *affordance* describes the set of attributes and their values the agent module can afford to the other agent modules. This set may be either invariable or variable. The variable set means that some attributes are added or deleted by the interaction with the environment. Some attributes may denote *roles* of the module. We want to stress here that the perception module is a *context dependent searcher* which searches for the necessary information from the affordances that are maintained by the other agent modules. In this sense, the perception module is *active* input function but *not passive* input function. [1]

An *action* module describes an output function that changes the internal state of the agent module and affects the external state by executing an action in the environment. A *communication* module describes the protocol for message passing and the intention of acceptance/rejection of messages to and from the other agent modules. The communication module is triggered in the *if-needed* mode. The *local working memory* is a memory used only within its agent module. It is a temporary memory used at the activation time of this agent module.

To conclude, the frame problems described in the previous chapters can be avoided by making cooperative use of perception and communication modules. That is, it is needless for each individual agent to know all things about his/her

---

[1] In constraint logic programming, active constraints and passive constraints are considered. The author believe that the Gibson's view of perception corresponds to the active constraint solving in constraint logic programming.

environment but only required to interact with it and perceive (search for) the necessary information which is afforded by the other agent modules. The result of this interaction affects only the state of the agents concerned with the interaction and does not affect the states of the other agents.

3.2 Organizing Modules

The skeleton structure of an agent module defined in Sec.3.1 consists of three functional modules; perception, action and communication. A specific reactive agent is described by the set of specialized reactive modules. A cognitive agent is described by the set of specialized reactive modules and cognitive modules. A social agent is described by the set of reactive modules, cognitive modules and social modules. A rational agent is either a cognitive agent or a social agent. These various agents are organized to construct a multiagent system. Fig.2 shows

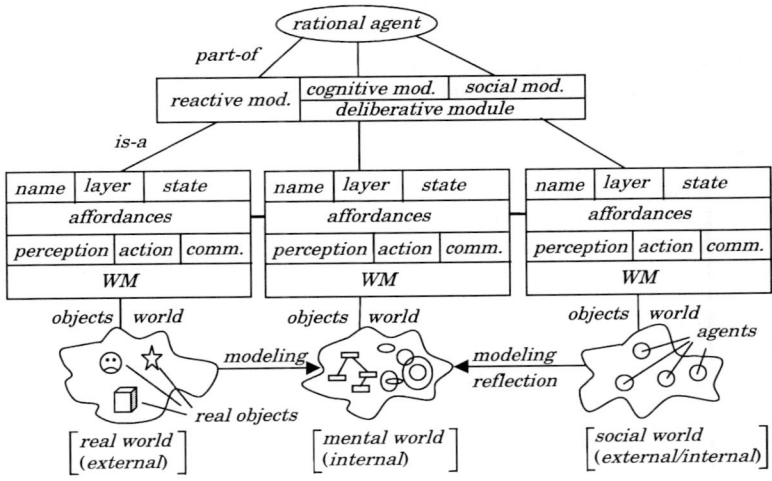

Fig. 2. Representation of Rational Agent

that a social rational agent is composed of the bottom layer (reactive modules), the middle layer (cognitive module) and the top layer (social module). The activities of each layer are defined over the object world which is directly linked to it and also over the adjacent layer or layers shown in the figure. The perception modules in each layer are context dependent searchers which search for the necessary information from the affordances maintained in the object world which is directly linked to it. For example, the object world for the reactive layer is the real world which is the external world for the rational agent. The perception modules in the reactive layer perceive the real objects in the real world, and the agent makes actions in the world. The object world for the cognitive layer is the mental world of the rational agent. The perception modules in the cognitive layer perceive the necessary information from the affordances maintained in the mental world, and according to the results of perception modules, the agent changes its mental state and makes intentions, plans and decisions for actions performed in the environment. The maintenance of the mental world would be also performed by analyzing the current state of the mental world.

Agent modules and auxiliary submodules described above are organized in the three abstraction hierarchies, i.e., aggregation, generalization and association. First, we can define an *aggregate* of modules in such a way that it exhibits a collective functionality of the aggregate. A social behavior in a multiagent system is a typical example. Second, the *generalization* (classification) of modules whatever they are primitive or compound is also considered taking account of their functionality, i.e., perception modules, action modules, etc. Third, we can define an *association* of aggregates in such a way that it exhibits a collective *group* functionality of the association. A typical example is seen in coordination and cooperation with different task groups of agents. A selection and scheduling problem for different plans, i.e., selecting a set of appropriate goals and then scheduling the selected goals is also another example. The figure 3 summarizes these three abstraction hierarchies. As seen in the next chapter, KAUS language facilitates commands for describing these abstraction hierarchies in the coherent way.

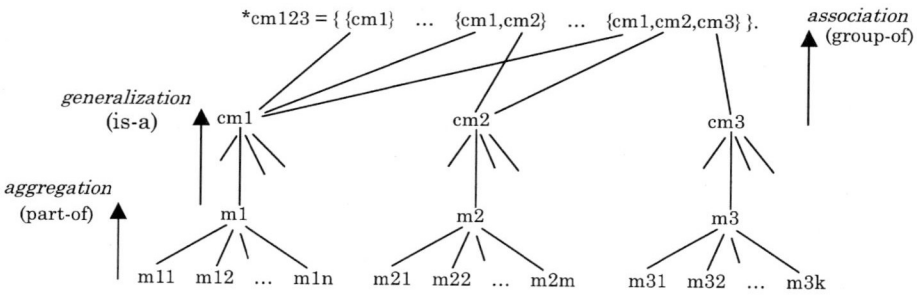

Fig. 3. Organizing modules

We note here that in the practical implementation we cannot explicitly describe all members of an association because the cardinality of the association set is exponential (2^u). Because of this, in practical applications the members of an association would be given generatively. In fact, for example, a candidate set of plans is to be generated by the planning module if required.

3.3 Execution Control of Modules

As for the execution of agent modules and the associated submodules, sequential, parallel (concurrent), synchronous and asynchronous executions are considered.

The autonomy of the control is emerged from each agent's interaction with the environment. To put it in another way, the affordances described in the agent modules make it easy to produce such a control dynamics: One agent module perceives the other agent module's affordances and then reacts or makes a deliberative decision for the next action, and vice versa. The frame problems would not occur in such autonomous control systems.

4 KAUS Language as a System Description Language

We have developed KAUS (Knowledge Acquisition and Utilization System) language from our motivation that a language based on sets and logic would be advantageous for modelling intelligent systems. For example, in intelligent design systems based on the multiagent technology, we are required to describe both structures and functionality of designed objects and the participated agents.

Using KAUS we can describe object structures in terms of set hierarchies (*isa*, *part-of* and *group-of* hierarchies mentioned in 3.2). Agents can be also organized in the set hierarchy. On the other hand, we describe their functionality and relations in terms of the extended predicate logic in which primitive procedural functions are incorporated. Regarding sets as types of variables, we can write *typed* logic programs in KAUS for problem solving. The following shows the syntax of KAUS language[15].

```
CLAUSE         ::=  LITERAL | PREFIX LITERAL | AND-OR_FORMULA | PREFIX AND-OR_FORMULA
LITERAL        ::=  ATOMIC_FORMULA | ~ATOMIC_FORMULA
PREFIX         ::=  [QUANTIFIER VARIABLE C_FLAG / TYPE] |
                    [QUANTIFIER VARIABLE C_FLAG / TYPE] PREFIX
AND-OR_FORMULA ::=  (CONNECTIVE BODY BODY ...)
ATOMIC_FORMULA ::=  (PREDICATE_NAME ARGS)
BODY           ::=  LITERAL | AND-OR_FORMULA
TYPE           ::=  BASE_SET | POWERSET | COMPONENT_SET | NAMELESS_SET | VARIABLE
ARGS           ::=  TERM | TERM ARGS | TERM, ARGS
PREDICATE_NAME ::=  NTA_NAME | PTA_NAME
NTA_NAME       ::=  BASE_SET
PTA_NAME       ::=  $NAME
QUANTIFIER     ::=  A | E
C_FLAG         ::=  ? | # | ?#
CONNECTIVE     ::=  | | &
```

A clause is used both for asserting a rule or a fact in the rule base and for describing a goal to be resolved by the inference system. An assertion clause ends with the period (.) and a query clause ends with the question mark (?). For example, a rational agent shown in Fig. 2 is represented as follows.

```
!ins_e rational_agent agent1;        |* define agent1 is-a rational_agent. *|
!ins_e agent1:reactive_mod rmod1;    |* agent1 has a reactive_mod rmod1.   *|
!ins_e agent1:cognitive_mod cmod1;   |* agent1 has a cognitive_mod cmod1.  *|
!ins_e agent1:social_mod smod1;      |* agent1 has a social_mod smod1.     *|
.....
[A Agents/*rational_agent][A Agent/Agents][A SocialActivity/socialActivity]
(act Agent SocialActivity).  |* all members of rational agents perform a social
activity. *|
.....
```

Note that in the above example expressions, *rational_agent* denotes the powerset of rational agents. Hence the variable *Agents* is declared as a variable of a *set* type.

The KAUS inference system is composed of those modules shown in Fig. 4. Among these, the *elimination of tautology*, *type check* and *world operation* characterize the KAUS inference system. To our knowledge, the standard Prolog and its extended versions do not have these inference facilities.

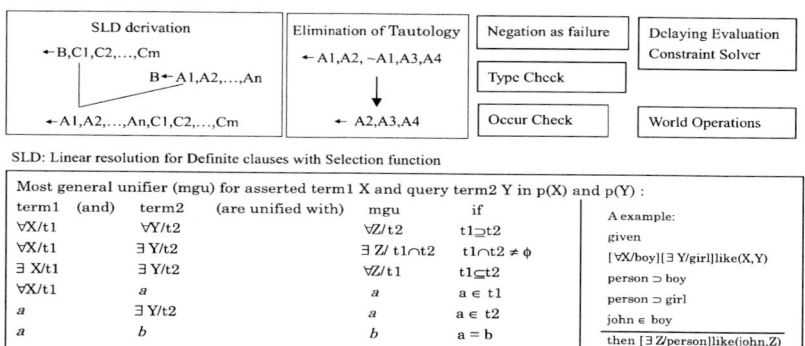

Fig. 4. Inference modules of KAUS

In the next chapter we issue the problem for implementing intelligent agent systems using the multi-strata modelling scheme [8,9,10].

5 Representation of Multi-strata Models by KAUS

The multi-strata model is a new modelling scheme for developing human-computer interactive problem solving systems. For example, business models in enterprises and many kinds of design project models for developing new products are instances of the multi-strata model. Fig.5 shows the skeleton structure of the multi-strata model and its KAUS description. In the figure, $S1$, $S2$ and $S3$ are called *subjects* each of whom undertakes a subtask for problem solving. A subject may be either a human or computer. For example, $S3$'s task is to make the object model (denoted by $S3 : model$) of the object $S3 : obj$. $S3 : obj$ is explicated by the lower level subject $S2$. The subject $S2$ in turn undertakes his own task specified by $S2 : obj$, and makes the model of it. This process continues until the lowest level of the subject's task is clarified.

We can describe the functionality of each agent (subject) in the multi-strata model using the *!define_agent* command, which implements the agent module given in Section 3.1. For example, the subject $S2$ is described like followings:

```
!ins_e rational_agent s2;   |* s2 is-a rational agent. *|
!ins_e human s2;            |* s2 is-a human. *|
!define_agent s2 {
    |* attribute declarations *|
```

```
         (privateData name("Jack Jones"), age(32), sex(male)).
         .........
        |* body of s2 *|
        (s2 type:designer state([<0,idle>]),
             afford([design(s1:Requirement Model),eval(Model,Status)]),
             perception(getValue(s1,Obj)),
             action(putValue(Obj,s1:Requirement)),
             communication(inform(Status,Agent))
        ).
        |* local rules maintained by s2 *|
        .........
};
```

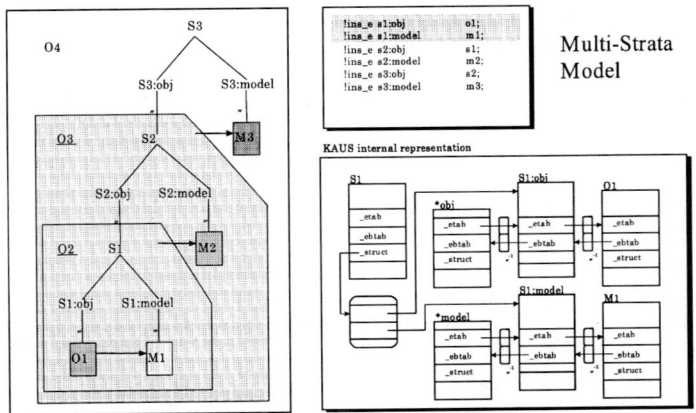

Fig. 5. Multi-strata model(skeleton structure)

6 Concluding Remarks

We have described a representation scheme for agent modules which is available as a skeleton for real application modules in multiagent systems. The point in this paper is that we have applied affordance to such higher levels of intelligent activities as cognition and deliberative thinking processes. We have shown that it is possible to avoid the frame problems existing in the three layers by embedding the notion of affordances into each agent module which is allocated in the reactive, cognitive and social layers respectively.

Another point is that we have showed KAUS language is suitable for implementing layered systems. However, we have not clarify the details of the control structures of the multiagent systems which are built up from the described modules. This subject should be solved in future.

Acknowledgement. The authors would like to thank the anonymous referees for their useful comments on this work.

References

1. Jorg P. Muller: The Design of Intelligent Agents - A Layered Approach, Springer, 1996.
2. Jacques Ferber: Multi-Agent Systems, An Introduction to Distributed Artificial Intelligence, Addison-Wesley, 1999.
3. Feffrey M. Bradshaw (edited): Software Agents, The MIT Press, 1997.
4. Donald A. Norman: Some Observations on Mental Models, Readings in Human-Computer Interaction, A Multidisciplinary Approach, (eds.) Ronald M. Baecker, and William A.S. Buxton, Morgan Kaufmann, pp.241-244, 1987.
5. M.T. Turvey, Kevin Shockley and Claudia Carello: Affordance, Proper function, and the physical basis of peceived heaviness, Cognition 73 B17-B26, Elsevier Science, 1999.
6. Frank M. Brown (edited): Proceedings of 1987 Workshop, 'The Frame Problem in Artificial Intelligence', Morgan Kaufmann, 1987.
7. Gibson, J.J., The ecological approach to visual perception, Boston, MA: Houghton Mifflin, 1979.
8. Setsuo Ohsuga: Multi-Strata Model and Its Applications - Particularly to Automatic Programming, International Conference on Industrial & Engineering Application of Artificial Intelligence and Expert Systems'96, 1996.
9. Setsuo Ohsuga: Toward Trully Intelligent Information Systems - From Expert Systems to Automatic Programming, Knowledge-Based Systems, vol.10, No.3, Elsevier, 1998.
10. Setsuo Ohsuga: A Medelling Scheme for New Information Systems - An Application to Enterprise Modelling and Program Specification, Proc. IEEE, Conference on Systems, Man and Cybernetics, 1999.
11. H. Yamauchi and S. Ohsuga : Modelling Objects by Extensions and Intensions, in Information Modelling and Knowledge Bases III, S. Ohsuga, et.al. (eds.), pp.160-173, IOS Press, 1992.
12. Hiroyuki Yamauchi: KAUS User's Manual v.6.502 (PDF file), RCAST, Univ. of Tokyo, 1999.
13. Mario Tokoro: The Society of Objects, Readings in Agents, Morgan Kaufmann, 1998.
14. Yoav Shoham: Agent-oriented programming, Readings in Agent, 1998.
15. H. Friedrich, M. Kaiser, O. Rogalla, and R. Dillmann: Learning and Communication in Multi-Agent Systems, in Distributed Artificial Intelligence Meets Machine Learning, Gerhard Weis (edited), 1997.

A Formalism for Building Causal Polytree Structures Using Data Distributions

M. Ouerd[1], B. J. Oommen[2], and S. Matwin[1]

[1]School of Information Technology and Engineering, University of Ottawa,
Ottawa, Canada K1S 5B6
{ouerd, matwin}@site.uottawa.ca
[2]School of Computer Science, Carleton University, Ottawa Canada K1S 5B6
oommen@scs.carleton.ca

Abstract. In this paper we have considered the problem of approximating an underlying distribution by one derived from a dependence polytree. This paper proposes a formal and systematic algorithm, which traverses the undirected tree obtained by the Chow method [2], and by using the independence tests it successfully orients the polytree. Our algorithm uses an application of the Depth First Search (DFS) strategy to multiple causal basins. The algorithm has been formally proven and rigorously tested for synthetic and real-life data.

1. Introduction and Background

Over the last decade Bayesian learning principles have received a fair amount of attention. Although they are elegant, they usually involve summations or integrals along all possible instantiations of the parameters and along all possible models. In the case of learning of Bayesian networks (which is distinct from Bayesian Learning itself) this can be perceived as a discrete optimization problem [5]. Precise solutions of this can be obtained by using search if we assume that there are only a few relevant models. This has proven to be the method of choice in many real-life applications [1].

Many of the Bayesian models, which are studied, are intractable. The challenge is to find general-purpose, tractable approximation algorithms for reasoning with these elegant and expressive stochastic models. For example, if we are to use Bayesian learning to improve performance of distributed database applications where there can be millions of transactions every day, we will need an efficient technique to build a model of the use of the database. The belief network that underlies the Bayesian learning is at the heart of the approach. The connection between Bayesian learning and belief networks is that one can *use* Bayesian techniques to induce a belief network referred to as a Bayesian Belief Network (BBN). Often, due to the lack of domain knowledge and in the interest of simplicity, it is assumed that the underlying structure is in a particularly simple form, representing reciprocal independence of variables involved. This results in a simple variant of Bayesian learning called Naïve Bayes.

The learning benefits if a more comprehensive and causal model of interaction between the variables is available. Such a model, represented as a Bayesian network, plays the role of a restricted hypothesis bias [9]. The method allows us to obtain the approximating probability distribution P(X) by a well-defined and easily computable density function $P_a(X)$. Indeed, it is impractical to store all estimates of the joint function P(X) for all possible values of the vector X. Our goal is to build a probabilistic network from the distribution of the data, which adequately represents it. Once constructed, such a network can provide insight into probabilistic dependencies that exist between the variables.

In order to measure the "goodness" of the approximation, an information theoretic measure can be specified in terms of the Kullback-Leibler [8] cross-entropy metric to compare joint probability distributions. *Chow and Liu* [2] used this measure to approximate discrete distributions by collecting the entire first and second order marginals. They derived a relationship between the measure of closeness between the probabilities and the measure of independence between all the pairs of the variables. A Maximum Weight-Spanning Tree (MWST) called the "Chow Tree" was built using the information measure between the variables forming the nodes of the tree. An alternative method of obtaining such a tree using the χ^2 metric was later proposed by *Valiveti and Oommen* [17]. A subsequent work due to *Rebane and Pearl* [15] used the Chow Tree as the starting point of an algorithm which builds a polytree (singly connected network) from a probability distribution. This algorithm orients the Chow tree by assuming the availability of independence tests on various multiple parent nodes.

The works of *Rebane and Pearl* [15] are commendable. Although, as we shall see, they did not answer all the questions regarding polytrees, their inference and their characterization, in our opinion their work was pioneering (i.e., with regard to polytree representations) and represented a quantum jump since the work done on trees in the late 1960's. In our opinion, their most fundamental contribution was to discover and utilize the edge "orienting" principle [18] referred to later.

Numerous authors have built on the foundation of the work of Rebane and Pearl. Noteworthy are the results of Srinivas *et al.*, [16] who worked with independence, and the recent results of Dasgupta [3] which explicitly specifies the complexity of the underlying problem. Friedman has also worked in the area and has modified the traditional EM algorithm to devise the "Structural" EM algorithm [4] to learn BBNs, and also demonstrated how one can learn Bayesian Networks from massive data sets using the "Sparse Candidate Algorithm" [16].

Much of the current work has made substantial progress on learning the structure of multiply connected networks and dynamic Bayesian networks, and this has even been achieved in the presence of hidden variables and for real data sets for which perfect independence tests are not realistic.

This paper deals with the problem of automatically building a belief network in terms of a directed polytree with the assumption that the observations have been presented to the system in terms of joint probability distributions. Thus we assume that the joint dependence relations represented by these observations is available. *Pearl* [14] discussed this process using a two-phase dependence learning scheme. Our aim is to find causal polytree structures that fit the data presented in terms of joint probability distributions. The question of inferring the polytree structure from the data (as opposed to the data distribution) is the study of a subsequent paper presently being compiled and is described in [11].

The reader must observe that polytrees represent much richer dependency models than undirected trees, because their joint probability density functions are products of higher order distributions. Consequently, the problem itself can be shown to be a much harder problem than that of finding the best tree [14]. First of all, the algorithm is not guaranteed to find a polytree structure if the underlying distribution is degenerate and not of a polytree type distribution i.e., if the distributions do not fit into a polytree representation. Secondly, the algorithm relies on the repeated use of the independence tests that determines categorically whether two random variables X_i and X_j are statistically independent. As shown in [12], even if the random variables are statistically independent, the experimental evaluation may never yield conclusive independence decisions.

1.1 Problem Statement and Outline of Solution

If we consider the polytree construction algorithm as given by [15] we observe that the order of traversing the tree so as to orient it is unspecified. The implementation strategy of determining the order of dependence tests is unanswered and left to the reader. In this paper we shall formally develop an algorithm which answers these questions. First of all, the algorithm determines the network structure or the tree using the MWST algorithm described earlier, and subsequently orients the tree by beginning with the assumption that we have marginal independence between *at least* two parents of any node. Thus if, there is no independence of any two parents of a node the algorithm will terminate by informing the user that the underlying tree structure cannot be oriented to yield a polytree.

The problem of orienting the tree is solved in two steps. The first step identifies all the independencies. In fact, every two nodes X_i and X_j are independent if the following equality is satisfied: $P(X_i, X_j) = P(X_i) * P(X_j)$.

Although, as mentioned above, this equality is not always satisfied with sample data, in this paper, we assume that such independence inferences are available. We also assume that we are provided with this information whenever it is requested. The second step is as follows: after inferring all the statistical independence between the pairs of variables, we use the following *Orienting Principle (T)* due to Pearl et al. [14] to completely orient the tree.

<u>Orienting Principle (T):</u>
>For every unoriented triplet of variables X, Y and Z ordered as: X—Z—Y, we test for the independence of X and Y. If X and Y are independent then X is a parent of Z and Y is a parent of Z. For any triplet X, Y and Z such that: X→ Z — Y, we test if X and Y are independent, and if this is so Y is parent of Z otherwise Y is a child of Z.

The details of why this principle works is omitted here but explained in greater detail in [10] and [13]. Utilizing all this information we shall show that the polytree can be efficiently computed if the underlying tree structure is systematically traversed.

2. A Depth First Search Algorithm for Building Polytrees

Our algorithm for inducing the polytree is an application of the Depth-First Search (DFS) algorithm to causal components of the undirected tree. Let $T = (V, E)$ be a connected, undirected tree where V is the set of vertices, and E the set of edges.

A vertex Z is said to be an *articulation point* between vertices X and Y if we have independence between X and Y. As defined by Pearl [14] and used by all the researchers since, a Causal Basin starts with a multi-parent cluster (a child node and all of its direct parents) and continues in the direction of causal flow to include all of the child's descendants and all of the direct parents of those descendants.

An example of this is given in the Figure 1.

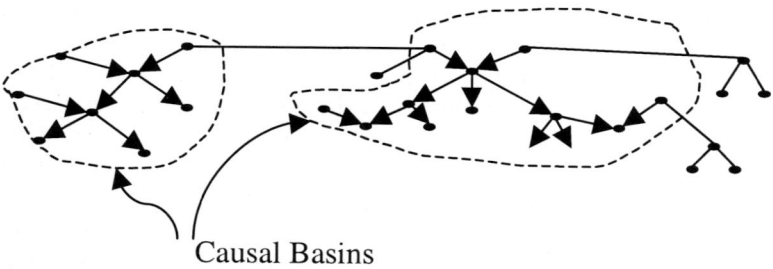

Fig. 1. A Causal Basin as defined by Pearl [14].

2.1 Problems with Pearl's Algorithm

Although the above definition is consistent, there are some unanswered questions which arise from the work of Rebane and Pearl [15]. In fact, although they specify a formal algorithm to compute the causal basins, they leave the following questions unanswered:
1. The question of what is meant by the outermost layer is not clear since it "depends on the tree" and its representation.
2. The question of how the traversal is done is not completely defined.
3. The algorithm introduces ambiguity regarding the edges that are already traversed.
4. The notion of causal basins depends on the starting point.

The last of these issues can be seen from the following figure in which the Chow tree of this polytree is taken from [14].

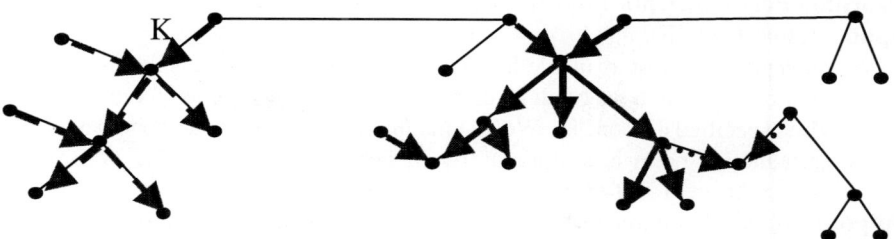

Fig. 2. Three Causal Basins starting respectively at nodes H, K and C.

Observe that the Chow tree of Fig. 1 and Fig. 2 is the same. In Fig. 1, the first articulation point (starting point) is node C and the second articulation point is node K. Having this order for the choice of the starting points we detect two causal basins as given in Fig. 1. In Fig. 2, we use node H as the starting point, node C as the second and node K as the third to be able to complete the same orientation of the Chow tree as in Fig. 1 using three causal basins instead of two. From this it is easy to see that the starting point determines the individual causal basins.

2.2 Motivation for a DFS Strategy

Consider the process of visiting the vertices of an undirected tree in the following manner. We select and "visit" a starting vertex Z which is one of the articulation points in T and in particular, an articulation point between two nodes X and Y. First of all we orient the edges (X, Z) and (Y, Z) as pointing to node Z following the orienting principle since we have independence between them. Then we select any edge (Z, W) incident upon Z. We check for independence between nodes X and W to determine the orientation of edge (Z, W). We observe two possible scenarios: If there is no independence between X and W then edge (Z, W) is pointing to node W. We then visit node W and begin to search for a new edge starting at vertex W. After completing the search through all causal paths beginning at W, the search returns either to Z, the vertex from which W was first reached, to search through all nodes in the adjacency list of Z, or to another non-visited articulation point. If there is independence between X and W the edge (Z, W) is pointing to node Z, and the search returns either to Z, the vertex from which W was first reached, to search through all the nodes in the adjacency list of Z, or to another non-visited articulation point.

The process of selecting unexplored edges incident on Z is continued until this list is exhausted. This is formalized in the algorithm Polytree-Depth-First-Search.

The input to the algorithm is mainly the set of nodes, and for every node X_i we specify its "adjacency" list which consists of a list of nodes X_j such that arc X_i —X_j exists in the tree structure of the underlying tree. Also provided are the independence tests between nodes whenever required. The algorithm is formally given below.

Algorithm Polytree-Depth-First-Search
Input: A tree T=(V, E).
  Independence test available for every pair of nodes when it is required.
  For every node, a list of all its direct neighbours specified as a connected
  List specified as ConList. We assume that the test for a node being an
  articulation point is a straightforward operation. Also, a node W is in the
  causal basin of X if there is a path from X to W.
Output: A directed polytree if the orientation exists. It returns T, the undirected tree
  if any orientation is not possible.
Method
Begin
  For (all X in V) **Do**
    Visited [X] = false   /* Visited is an array holding the nodes */
  EndFor
  For (all X in V) **Do**
    /* always start with an articulation point */
    If (!Visited[X] and X is an articulation point) **Then**
      Call Processing (X)
    EndIf
  EndFor
End Algorithm Polytree-Depth-First-Search

Procedure Processing (X)
Begin
  /* node X is not a leaf */
  If ((Visited[X] = false) and (ConList(X)) > 1)) **Then**
    /* orient the adjacent edges */
    Call IndepOrient(X)
    Visited[X] = true
    /* traverse the adjacency list of X */
    For (all W in the ConList of X and W is in causal basin of X) **Do**
      /* Processing is recursive because of Depth-First Search */
      Call Processing(W)
    EndFor
  EndIf
End Algorithm Processing

The above algorithm uses a DFS strategy. It is easy to devise an analogous algorithm, which uses a Breadth-First Search strategy, or any systematic search scheme

Procedure IndepOrient (X)
Begin
   For (every distinct N_1 and N_2 in ConList(X)) **Do**
      If Indep(N_1, N_2) = True **Then**
          print arcs from (N_1 to X) and (N_2 to X)
      EndIf
   EndFor
   For (every distinct N_1 and N_2 in ConList(X)) **Do**
      If (arc from N_1 to X exists and edge from N_2 to X is unoriented) **Then**
          print arc from X to N_2
      Else If (arc from N_2 to X exists and edge from N_1 to X is unoriented) **Then**
          print arc from X to N_1
      EndIf
   EndFor
End Algorithm IndepOrient

2.3 Analytic Properties of the Algorithm

The formal proof that the above algorithm works follows the arguments of the DFS traversal of a graph.

Theorem 1:
The algorithm Polytree-Depth-First-Search correctly computes the polytree given the skeleton tree structure and the underlying independence relationships.
Proof: The proof is done inductively and found in [10]. ♦♦♦

We shall now state results regarding the complexity of the above algorithm. This is done by considering the number of independence tests that need to be performed to be able to orient the tree.

Theorem 2:
For a tree in which every node has up to k adjacent nodes, at depth d in the tree, the total number of independence tests which have to be done is: $k * (k-1)^{d-1} * \binom{k}{2}$.

Proof: The proof is not too involved, but follows from an induction on the size of the DFS tree. It is omitted here, but found in the unabridged paper [10]. ♦♦♦

From a straightforward examination the overall burden of the computation can be obtained by observing that for each node we have to do pairwise independence tests between its neighbors. Theorem 2 explains what happens at every level of the tree.

3. Experimental Results

In order to check the algorithm developed in this paper, we have done numerous experiments. We assumed that we were to learn an underlying polytree, which is unknown to the algorithm. Also, we assumed that independence tests, which are consistent with the polytree orientation, were available whenever they were needed by the algorithm. The polytree learning algorithm applicable for discrete data (called *Discrete-Polytree-Depth-First-Search*) was invoked using the skeleton and the sequence of independence tests.

The entire project involved the testing of the algorithm for numerous trees and polytrees. The details of the results are omitted here but included in [10], [13]. The descriptions in [10], [13] include specific polytrees, and clearly demonstrate how the structure is learnt as the independence tests are provided. The test cases also show examples in which the polytrees have one or multiple causal basins. In every case, the polytree was exactly inferred.

Our experimental results consistently demonstrate that the algorithm successfully orients the polytree. However, if the independence information for any pair of nodes is not provided to the algorithm, the algorithm will terminate without orienting the tree. The advantages of our algorithm are numerous; first of all it is computationally efficient since it uses a DFS scheme. It also extends itself to both discrete and continuous variables, and provides a very efficient way of traversing the tree. Finally, we mention that the scheme also handles multi-feature variables. It has been used quite successfully in a real-life application where the problem is to improve performance in systems using repeated queries which access distributed databases [13]. The scheme has also been used for the ALARM data [7].

4. Conclusion

In this paper we have considered the problem of approximating an underlying distribution by one derived from a dependence polytree. The skeletal form of the polytree is known to be the MWST of a complete graph, with $I_t(X_i, X_j)$, the information theoretic metric, as the edge weight between the pair of nodes X_i and X_j. Once the tree is derived, *Rebane and Pearl*, in [15], proposed to use an independence test to determine if a variable has multiple parents. They dictated that every two node neighbors X and Y of a node Z must be tested for marginal independence to decide if Z has parents X and Y.

This paper proposes a formal and systematic algorithm to traverse the tree obtained by the Chow method. It uses an application of the DFS strategy to multiple causal basins. Experimental results clearly demonstrate that when the required independence tests are available to the algorithm the orientation of the polytree is completed, and always correct. The algorithm has also been used in two real-life applications [13] involving distributed databases and the ALARM data.

References

1. Cooper, G.F. and Herskovits, E.H., (1992). A Bayesian Method for the Induction of Probabilistic Networks from Data. *Machine Learning*, Vol. 9: 309-347.
2. Chow, C.K. and Liu, C.N., (1968). Approximating Discrete Probability Distributions with Dependence Trees". *IEEE Transactions on Information Theory*, Vol. 14: 462-467.
3. Dasputa, S. (1999). Learning Polytrees. *Proceedings of the Workshop on Uncertainty in Artificial Intelligence*. Available from http://www2.sis.pitt.edu/~dsl/UAI/uai99.html.
4. Friedman, N. (1998). The Bayesian Structural EM Algorithm. *Uncertainty in Artificial Intelligence*.
5. Friedman, N., Nachman, I., and Pe'er, D. (1999). Learning Bayesian Network Structure from Massive Datasets: The "Sparse Candidate" Algorithm. *Proceedings of the Workshop on Uncertainty in Artificial Intelligence*. Available from http://www2.sis.pitt.edu/~dsl/UAI/uai99.html.
6. Geiger, D., Paz, A., and Pearl, J. (1990). Learning Causal Trees from Dependence Information. *Proceedings of AAAI*, Boston, MA: MIT Press, pp. 770-776.
7. Herskovits, E. H., (1991) *Computer-Based Probabilistic-Network Construction*, Doctoral Dissertation, Medical Information Sciences, Stanford University, CA.
8. Kullback, S., and Leibler, R.A. (1951). On Information and Sufficiency. *Annals Mathematics Statistics*, Vol. 22: 79 – 86.
9. Mitchell, Tom M. (1997). *Machine Learning*. McGraw-Hill.
10. Oommen, B.J., Matwin, S., Ouerd, M., (2000). A Formalism for Building Causal Polytree structures using Data Distributions. Unabridged version of this paper.
11. Ouerd, M., Oommen, B.J., and Matwin, S. (2000). Generation of Random Vectors for Underlying DAG Structures given First Order Marginals. In preparation.
12. Ouerd, M., Oommen, B.J., and Matwin, S. (2000). Inferring Polytree Dependencies from Sampled Data. In preparation.
13. Ouerd, M., (2000). *Building Probabilistic Networks and Its Application to Distributed Databases*. Doctoral Dissertation, SITE, University of Ottawa, Ottawa, Canada.
14. Pearl, J. (1988). *Probabilistic Reasoning in Intelligent Systems: Networks of Plausible Inference*. Morgan Kaufmann, San Mateo, CA.
15. Rebane, G. and Pearl, J. (1987). "The Recovery of Causal Polytrees from Statistical Data". *Proceedings of the Workshop on Uncertainty in Artificial Intelligence*. Seattle, Washington, 222- 228.
16. Srinivas, S., Russell, S., and Agogino, A. (1989). Automated Construction of Sparse Bayesian Networks from Unstructured Probabilistic Models and Domain Information. *Proceedings of the Workshop on Uncertainty in Artificial Intelligence*. Available from http://www2.sis.pitt.edu/~dsl/UAI/uai89.html.
17. Valiveti, R.S. and Oommen, B.J. (1992). On Using the Chi-Squared Metric for Determining Stochastic Dependence. *Pattern Recognition*. Vol. 25, No. 11: 1389-1400.
18. Verma, T., and Pearl, J. (1991). An Algorithm for Deciding if a Set of Observed Independencies Has a causal Explanation. *Proceedings of the Workshop on Uncertainty in Artificial Intelligence*. Available from http://www2.sis.pitt.edu/~dsl/UAI/uai.html.

Abstraction in Cartographic Generalization

Sébastien Mustière[1], Lorenza Saitta[2], and Jean-Daniel Zucker[3]

[1]IGN – COGIT / LIP6
2-4, avenue Pasteur, 94165 St-Mandé Cedex - FRANCE
sebastien.mustiere@ign.fr

[2]Univ. Del Piemonte Orientale
Dip di Scienze e Tecnologie Avanzate
Corso Borsalino 54, 15100 Alessandria, ITALY
saitta@di.unito.it

[3]Univ. Paris VI – CNRS – LIP6
4, place Jussieu, 75252 Paris - FRANCE
jean-daniel.zucker@lip6.fr

Abstract. This article shows that cartographic generalization is best viewed as representing (formulating, renaming knowledge) and abstracting (simplifying a given representation). The general process of creating map is described so as to show how it fits into an abstraction framework developed in artificial intelligence to emphasize the difference between abstraction and representation. The utility of the framework lies in its efficiency to automate knowledge acquisition for the cartographic generalization as a combined acquisition of knowledge for abstraction and knowledge for changing a representation.

1 Introduction

In this paper we address the problem of automating cartographic generalization. This automation is needed for several reasons: first to decrease cost and time necessary to produce maps, then to allow geography experts who are not necessary cartography specialists to create their own maps with a good quality, and finally to facilitate the crucial need of multi-level analysis of geographic data.

The lack of efficient generalization tools in GIS is due to the fact that generalization is a difficult task: it is guided by a lot of geographic and cartographic knowledge. An approach to face this need for automation is to build expert systems that have proved to be efficient in numerous fields where knowledge require to be introduced. Many authors emphasize that the main problem for the use of expert systems is the «**knowledge acquisition bottleneck**».

Moreover, the analysis of first-generation expert systems 0 stress the need to differentiate, separate, and structure the different types of knowledge in second-generation expert systems. We present in this article a description of the knowledge used in cartographic generalization well fitted to its acquisition. We analyze generalization along two dimensions: **knowledge abstraction** and **knowledge representation**, as proposed by 0. This distinction is necessary to differentiate, and so acquire, the different knowledge types involved in generalization

2 Differentiating Representation and Abstraction

Representing knowledge is one of the main research topic in Artificial Intelligence since its birth. The AI community has come out in the past fifty years with a large variety of languages that are each more or less adapted to represent different field of humans knowledge that require to be represented and processed 0. Although a large amount of human expertise can be formulated as a set of specific procedures or inferences in one given language or paradigm, the cartographic generalization process clearly requires several knowledge representation languages to capture the different types of knowledge manipulated, ranging from the raw data from the world to their final representation as a usable map.

Saitta and Zucker have recently proposed a model of abstraction (hereafter called the KRA model), supporting reasoning in a wide context 0. They distinguish two fundamental processes, namely the process of changing the language of representation and the process of abstracting the language of representation.

The KRA model originates from the observation that the conceptualization of a domain involves four different levels. Underlying any source of experience there is the world (W), where *concrete* objects reside. However, the world is not really known, because we only have a mediated access to it, through our perception. Then, what is important for an observer is not the world *in se,* but the perception P(W) that s/he has of it. At this level the percepts «exist» only for the observer and only during their being perceived. Their reality consists in the «physical» stimuli produced on the observer. In order to let these stimuli become available over time, for retrieval and further reasoning, they must be first of all memorized and organized into a *structure* S. This structure is an *extensional* representation of the perceived world, in which stimuli related one to another are stored together into tables. The set of these tables constitutes a relational database, on which relational algebra operators can be applied. Finally, in order to symbolically describe the perceived world, and to communicate with other agents, a *language* L is needed. L allows the perceived world to be described *intensionally*. Finally, a theory T might be needed to reason about the world. The theory may also contain general knowledge, which does not belong to the specific domain, and allows inferences to be drawn. At the theory level we operate through inference rules. Let us define R = < P(W), S, L, T > as a *Reasoning Context.* The relationships among the four considered levels are represented in Figure 1.

There is an infinity of ways in which the world can be perceived by an intelligent agent, according to the observation tools, the goal of the observation, the agent's cultural background, and so on. This variability is captured by the diversity of the world perceptions P(W). It is at this layer that the type and amount of information the agent will memorize, speak about, and reason about later is established. The less detailed the perception, the more abstract. Sometimes the agent has control over the perception in such a way to collect exactly the information it needs to achieve its goals. Sometimes, the agent can not control the perception, so that it may receive much more information than it currently needs, or maybe it wants to perform several tasks, each one requiring different pieces of information, which, on the other hand, are easy to collect together. The preceding considerations suggest that it would be very useful to have methods to actually or virtually transform a perception into a more abstract one. The following definition of abstraction tries to capture this process.

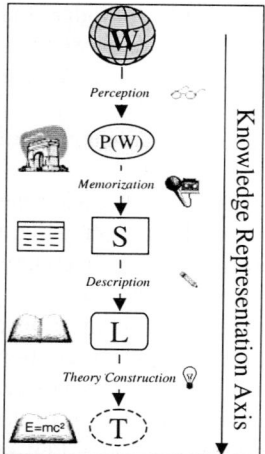

Figure 1- *Levels of knowledge representation*

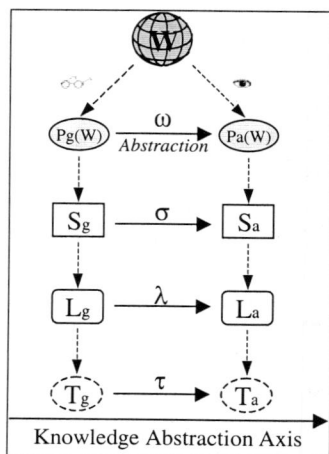

Figure 2 - *Changing levels of details*

Definition – Given a world W, let $R_g = (P_g(W), S_g, L_g, T_g)$ and $R_a = (P_a(W), S_a, L_a, T_a)$ be two reasoning contexts, which we label as *ground* and *abstract*. An *Abstraction* is a functional mapping $A : P_g(W) \to P_a(W)$ between a perception $P_g(W)$ and a *simpler* perception $P_a(W)$ of the same world W.

Some comments are needed about this definition. In 0 a formal definition of "simpler" in terms of relative information gain has been given. Obviously, the process of abstracting a perception can be iterated, leading to several levels of abstraction. If no perception can be identified as a preeminent one, then any level can be selected as "ground", being the notion of "simpler" only a relative one. Another important point is that abstraction should be a reversible process; in fact, **to abstract does not mean to delete information,** but only to *hide* information, in such a way that the opposite process (concretion) becomes possible, as well. Finally, according to this definition, the abstraction process starts at the perception level, but propagates toward the layers of Figure 1. However, the abstraction relations between the structures, the languages and the theories are determined by the relations defined on the perceptions.

In Figure 2, the view on abstraction presented in this paper is synthetically described. The symbols ω, σ, λ and τ denote abstraction operators working between entities of the same layer.

The perceptual stimuli of the perceived world are classified according to categories that are proved useful to organize information (including ontologies, objects, attributes, functions, and relations). Within the KRA model framework 0, a set of fundamental *abstraction operators* have been defined. These operators are defined at the perceived world P(W) level. The proposed set of fundamental operators $\Omega = \{\omega_{hide}, \omega_{ind}, \omega_{set}, \omega_{ta}, \omega_{eq\text{-}val}, \omega_{red\text{-}arg}, \omega_{prop}\}$ is not exhaustive. In particular contexts (such as cartography), it can be either reduced or augmented with domain-specific abstraction types.

The operator ω_{hide} is a fundamental abstraction operator that consists in hiding any kind of knowledge be it an object, an attribute of an object or a relation between objects. For example an isolated street may be hidden. The operator ω_{ind} consists in making several objects indistinguishable. For example only one of several close isolated trees will be considered as a typical representative of them. The operator ω_{set} consists in grouping several objects that are considered not to be distinguished. For example the grouping of a set of trees into an object "forest". The operator ω_{ta} consists in grouping a set of *different* objects to form a new *compound object*. For example grouping several streets and buildings to form a town. The operator $\omega_{eq\text{-}val}$ specifies what subset of the attribute or function values can be merged, because they are considered indistinguishable. For example, two objects with close altitudes will be considered at the same altitude. Finally, the operators ω_{red_arg} specifies a relation and a subset of its arguments, which must be dropped from the relation, obtaining thus a relation with reduced arity. For example the argument "type of crossing" may be hidden in the relation between two roads. From these operators defined at the perception level, the operators between the structures, languages and theories are deduced.

3 Cartography in the KRA Model

The KRA model exhibits several key properties for cartography. It allows to distinguish the process of representation (change of language) from the process of abstraction (change of level of detail). These two processes are usually very much entangled in cartography. This distinction provides the basis for automating knowledge acquisition in cartography, as a combined acquisition of knowledge for abstraction and knowledge for changing representation, as we will explain in the following section.

The topographic map production process closely parallels the KRA model, because it can be analyzed according to the two dimensions, representation and abstraction. Let us first consider the scheme of Figure 1 applied to cartography. The first step of cartography is to collect data from the geographic world, or part of it (W). This is usually done through aerial photographs or satellite images. These data are the perceived world P(W). Objects contained in these photographs are located and labeled to create a geographic database (GDB). This GDB is the set of geographic data organized in a Structure (S). Then, this GDB is displayed by means of cartographic symbols applied to objects stored in it. This is the creation of a map, an iconic language (L). Finally, maps are created for tasks like space analysis, search for itineraries, town planning, or geographic theory construction. The theory T contains all the background facts and laws allowing to reason about geographic configuration, and may be different for different tasks.

Cartography is not just knowledge representation. All the steps of map creation do not involve only a knowledge representation, but also involve a knowledge abstraction. In particular, Map creation (which contains the generalization process) is both a knowledge representation process, when objects are symbolized, and a knowledge abstraction process, when objects relevant to the theory construction are identified. So described, map creation is represented as a diagonal process in Fig. 3.

Generalization Process in the KRA Model

Knowledge abstraction in generalization is the identification of abstracted geographic objects relevant to the theory construction that will be done from the map. "Objects" have to be taken here in a very wide sense: they may represent any basic geographic objects (like a house, a road…) or any set of basic objects having a geographical meaning (the set of streets of town, a street and the buildings along it…).

In our model, the abstraction process is to go from a detailed description of a geographic object, describing each part of the object, to a more abstract description of the object, describing only properties of the object relevant to the map users needs. For example, an abstraction is to go from a complete description of a set of streets in a town to the description «this is a streets network».

As we explained in the KRA model presentation, abstractions at the structure level (i.e. on objects of the geographic database) shall only be considered as consequences abstractions at the geographic world perception level.

Knowledge representation in generalization is the process of symbolizing abstracted objects. For example, this representation process is to determine which symbolized subset of streets is the best suited in order to well represent «a street network». This choice is guided by the necessity to well represent the abstracted object and restricted by the drawing possibilities (we can not represent all the symbolized streets because they will overlap themselves on the paper).

Difference but not independence. It is important to notice that knowledge abstraction and representation can not be performed independently one from the other, nor that when abstraction has been done the "ground" GDB is no more necessary. For example, in a street network the drawn streets are a subset of the "ground" streets. The abstracted object «street network» helped us to change our view of the world, but the representation process needs to look again at the ground

GDB to represent actual objects of the world. In this way we imitate the human perception, which continuously change the level of abstraction to well analyze space. These inter-links between abstraction and representation explain why, manually, these two steps have always been performed in one time by the cartographer. Anyway, it has been shown that efficient expert systems need to clearly separate the different types of knowledge involved in human processes 0. We so believe that this distinction between abstraction and representation is necessary for the creation of an automated process of cartographic generalization.

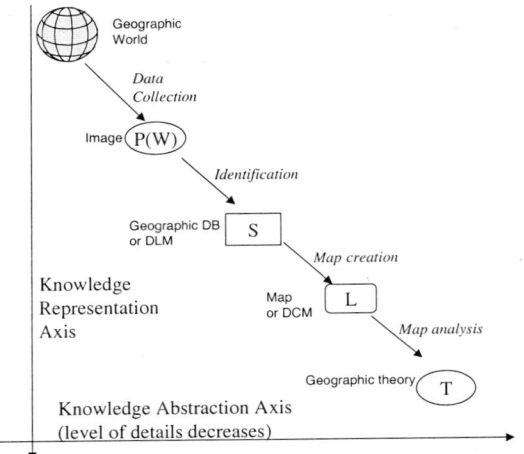

Figure 3 - *Representation and abstraction in cartography*

Our model is related to Brassel and Weibel's famous model of cartographic generalization 0, in the sense that it differentiates the drawing process from other processes. But the main differences comes from the knowledge abstraction step which, in our model, is far from being a simple «statistical generalization». More, we think that this step is the most complex step of cartographic generalization, even if the representation process is still far from being well mastered in the field of automatic cartographic generalization.

For these reasons, our model is closely related to Nyerges' view of cartographic generalization 0 which splits generalization in two phases: *«Geographical information abstraction mainly concerns managing geographical meaning in databases, and map generalization mainly concerns structuring map presentations»*.

4 Structuring Knowledge for Its Acquisition

The distinction enhanced between abstraction and representation is necessary for efficient knowledge acquisition in cartographic generalization. Because of space limitation we do not detail this section. We only quote the different types of knowledge involved in these two processes.

One the one hand, the knowledge abstraction process is seen as the identification of geographic objects that are relevant to the user needs (and need to be drawn). It manipulates 1/ geographic knowledge (e.g. town model) which could be acquired through geographers interview 2/ perception knowledge (Gestalt) which has been a lot studied in psychology and cartography 3/ space analysis knowledge, which should be acquired in computers through space analysis tools (e.g. Delaunay triangulation).

One the other hand, representation knowledge manipulates
1. graphic knowledge to define when a map is legible, which is well known by cartographers
2. drawing knowledge to define how to represent an abstracted object and which could be acquired through cartographers interviews or manually drawn maps analysis
3. algorithm knowledge which could be acquired through cartographers drawing analysis by machine learning techniques.

5 Conclusion

One of the key problems limiting the automation of cartographic knowledge acquisition lies in the heterogeneity of the knowledge that is involved throughout the process of creating maps from geographic databases. In this article, we have adapted an Artificial Intelligence model to distinguish two fundamental transformations used in the field of cartography, namely *abstraction* and *representation* of knowledge. The first contribution of this work is therefore to propose a classification of cartographic and geographic knowledge along this dimension in order to facilitate its acquisition.

References

1. Brassel K., Weibel R. (1988). A review and conceptual framework of automated map generalization. *International Journal Geographical Information Systems, 2,* 229-244.
2. Clancey W. (1983). The Epistemology of a rule-based expert system- A framework for Explanation. *Artificial Intelligence, 20.*
3. Ginsberg, (1997). *Essentials of Artificial Intelligence*, Morgan Kaufmann.
4. Nyerges (1991) *Representing Geographic Meaning in Map Generalization,* Longman Scientific & Technical, pp 59-85.
5. Saitta, L. and J.-D. Zucker (1998). Semantic Abstraction for ConceptRepresentation and Learning. *Symposium on Abstraction, Reformulation and Approximation* (SARA-98), Asilomar Conference Center, Pacific Grove, California.

Author Index

Aha, David W.   358
Ali, Abdelbaset   444
Alsaffar, Ali H.   435, 453
Altamirano, E.   534
An, Aijun   119
Arcos, Josep-Lluís   601
Armani, B.   68

Badia, Antonio   321
Barber, Brock   562
Barruffi, R.   228
Bassett, Jeffrey K.   157
Bertino, E.   68
Boisson, Cécile   463
Botta, Marco   31
Boussouf, M.   139
Breslow, Leonard A.   358
Butz, C.J.   247

Catania, B.   68
Cercone, Nick   119
Chen, Jianhua   331
Chu, Wesley W.   400
Croft, W. Bruce   1
Cruz, Alfredo   186
Cundell, Diana R.   349

De Jong, Kenneth A.   157
Demolombe, Robert   515
Deogun, Jitender S.   435, 453, 495
Desai, Bipin C.   444
Dupin de Saint Cyr, F.   543

El-Kwae, Essam A.   97
Elomaa, Tapio   552
Emmerman, Philip J.   12
Escalada-Imaz, G.   534
Esposito, Floriana   40, 109

Fanizzi, N.   109
Ferilli, S.   109
Fernandes, Chris   389
Fu, Ada W.-c.   59

Giordana, Attilio   31
Gordon, Diana F.   166

de Groeve, Michelle   426
Grzymala-Busse, Jerzy W.   148

Hacid, Mohand-Saïd   77
Haddad, Sami S.   444
Hafez, Alaaeldin   130
Hamilton, Howard J.   266, 562
Han, L.   283
Henschen, Lawrence   389
Higashiyama, Michiko   237
Hoffmann, Jörg   216

Johnson, C.A.   377

Karimi, Kamran   266
Kashmirian, Jennifer   205
Kawasaki, Shuhei   176
Kim, Minkoo   453
Kitchin, D.E.   610
Kostoff, Ronald N.   86
Kramer, Stefan   426
Kryszkiewicz, Marzena   505
Kwong, Renfrew W.-w.   59

Lakkaraju, Sai K.   525
Laradi, D.   68
Lanza, Antonietta   40
Lin, Tsau Y.   486
Lin, Weiqiang   49
Lisi, Francesca A.   40
Liu, Chunnian   591
Liu, D.   610
Liu, Jiming   474
Louie, Eric   486
Lowden, Barry G.T.   256
Lu, James J.   301

Malerba, Donato   40
Maluf, David A.   368, 474
Marin, B.   68
Matwin, Stan   274, 629
McCluskey, T.L.   610
Michalski, Ryszard   21
Milano, M.   228
Morin, Johanne   274
Movva, Uma Y.   12

Mukaidono, Masao 311
Muñoz-Ávila, Hector 358
Murray, Neil V. 301
Mustière, Sébastien 638

Ninomiya, Tomoko 311
Núñez, Arturo 196

Ohsuga, Setsuo 176, 237, 591, 619
Oommen, B.J. 629
Orgun, Mehmet A. 49
Ouerd, M. 629
Oussalah, Mourad 340

Pacholczyk, D. 543
Padgham, Lin 205
Park, Sanghyun 400
Peters, J.F. 283
Pfahringer, Bernhard 426
del Pilar Pozos Parra, Maria 515
Plaza, Enric 601
Puuronen, Seppo 417

Quafafou, M. 139
Ouerd, M. 629

Raghavan, Vijay V. 453
Ramanna, S. 283
Raś, Zbigniew W. 582
Riff, María-Cristina 196
Robinson, Jerome 256
Rosenthal, Erik 301
Rousu, Juho 552
Rybinski, Henryk 505

Saitta, Lorenza 31, 409, 638

Sanders, Robyn 349
Saquer, Jamil 495
Sebag, Michele 31
Semeraro, G. 109
Sever, Hayri 435
Shah, Pankaj 148
Shahmehri, Nahid 463
Silibovsky, Randy S. 349
Simpson, R.M. 610
Skowron, A. 283
Spears, William M. 166
Suraj, Z. 283
Sztandera, Les M. 349

Tamzalit, Dalila 340
Tanaka, Katsuaki 237
Tang, Jian 59
Torroni, P. 228
Toumani, Farouk 77
Tsumoto, Shusako 573
Tsymbal, Alexey 417

Weber, Rosina 358
Widmer, Gerhard 426
Wieczorkowska, Alicja 292
Wiederhold, Gio 368
Wong, S.K.M. 247

Yamauchi, Hiroyuki 619

Zarri, G.P. 68
Zhang, Yan 525
Zhong, Ning 591

Zucker, Jean-Daniel 409, 638
Żytkow, Jan M. 582

Lecture Notes in Artificial Intelligence (LNAI)

Vol. 1791: D. Fensel, Problem-Solving Methods. XII, 153 pages. 2000.

Vol. 1792: E. Lamma, P. Mello (Eds.), AI*IA 99: Advances in Artificial Intelligence. Proceedings, 1999. XI, 392 pages. 2000.

Vol. 1793: O. Cairo, L.E. Sucar, F.J. Cantu (Eds.), MICAI 2000: Advances in Artificial Intelligence. Proceedings, 2000. XIV, 750 pages. 2000.

Vol. 1794: H. Kirchner, C. Ringeissen (Eds.), Frontiers of Combining Systems. Proceedings, 2000. X, 291 pages. 2000.

Vol. 1804: B. Azvine, N. Azarmi, D.D. Nauck (Eds.), Intelligent Systems and Soft Computing. XVII, 359 pages. 2000.

Vol. 1805: T. Terano, H. Liu, A.L.P. Chen (Eds.), Knowledge Discovery and Data Mining. Proceedings, 2000. XIV, 460 pages. 2000.

Vol. 1809: S. Biundo, M. Fox (Eds.), Recent Advances in AI Planning. Proceedings, 1999. VIII, 373 pages. 2000.

Vol. 1810: R. López de Mántaras, E. Plaza (Eds.), Machine Learning: ECML 2000. Proceedings, 2000. XII, 460 pages. 2000.

Vol. 1813: P.L. Lanzi, W. Stolzmann, S.W. Wilson (Eds.), Learning Classifier Systems. X, 349 pages. 2000.

Vol. 1821: R. Loganantharaj, G. Palm, M. Ali (Eds.), Intelligent Problem Solving. Proceedings, 2000. XVII, 751 pages. 2000.

Vol. 1822: H.H. Hamilton, Advances in Artificial Intelligence. Proceedings, 2000. XII, 450 pages. 2000.

Vol. 1831: D. McAllester (Ed.), Automated Deduction – CADE-17. Proceedings, 2000. XIII, 519 pages. 2000.

Vol. 1834: J.-C. Heudin (Ed.), Virtual Worlds. Proceedings, 2000. XI, 314 pages. 2000.

Vol. 1835: D. N. Christodoulakis (Ed.), Natural Language Processing – NLP 2000. Proceedings, 2000. XII, 438 pages. 2000.

Vol. 1836: B. Masand, M. Spiliopoulou (Eds.), Web Usage Analysis and User Profiling. Proceedings, 2000. V, 183 pages. 2000.

Vol. 1847: R. Dyckhoff (Ed.), Automated Reasoning with Analytic Tableaux and Related Methods. Proceedings, 2000. X, 441 pages. 2000.

Vol. 1849: C. Freksa, W. Brauer, C. Habel, K.F. Wender (Eds.), Spatial Cognition II. XI, 420 pages. 2000.

Vol. 1856: M. Veloso, E. Pagello, H. Kitano (Eds.), RoboCup-99: Robot Soccer World Cup III. XIV, 802 pages. 2000.

Vol. 1860: M. Klusch, L. Kerschberg (Eds.), Cooperative Information Agents IV. Proceedings, 2000. XI, 285 pages. 2000.

Vol. 1861: J. Lloyd, V. Dahl, U. Furbach, M. Kerber, K.-K. Lau, C. Palamidessi, L. Moniz Pereira, Y. Sagiv, P.J. Stuckey (Eds.), Computational Logic – CL 2000. Proceedings, 2000. XIX, 1379 pages.

Vol. 1864: B. Y. Choueiry, T. Walsh (Eds.), Abstraction, Reformulation, and Approximation. Proceedings, 2000. XI, 333 pages. 2000.

Vol. 1865: K.R. Apt, A.C. Kakas, E. Monfroy, F. Rossi (Eds.), New Trends Constraints. Proceedings, 1999. X, 339 pages. 2000.

Vol. 1866: J. Cussens, A. Frisch (Eds.), Inductive Logic Programming. Proceedings, 2000. X, 265 pages. 2000.

Vol. 1867: B. Ganter, G.W. Mineau (Eds.), Conceptual Structures: Logical, Linguistic, and Computational Issues. Proceedings, 2000. XI, 569 pages. 2000.

Vol. 1881: C. Zhang, V.-W. Soo (Eds.), Design and Applications of Intelligent Agents. Proceedings, 2000. X, 183 pages. 2000.

Vol. 1886: R. Mizoguchi, J. Slaney (Eds.), PRICAI 2000: Topics in Artificial Intelligence. Proceedings, 2000. XX, 835 pages. 2000.

Vol. 1898: E. Blanzieri, L. Portinale (Eds.), Advances in Case-Based Reasoning. Proceedings, 2000. XII, 530 pages. 2000.

Vol. 1889: M. Anderson, P. Cheng, V. Haarslev (Eds.), Theory and Application of Diagrams. Proceedings, 2000. XII, 504 pages. 2000.

Vol. 1891: A.L. Oliveira (Ed.), Grammatical Inference: Algorithms and Applications. Proceedings, 2000. VIII, 313 pages. 2000.

Vol. 1902: P. Sojka, I. Kopeček, K. Pala (Eds.), Text, Speech and Dialogue. Proceedings, 2000. XIII, 463 pages. 2000.

Vol. 1904: S.A. Cerri, D. Dochev (Eds.), Artificial Intelligence: Methodology, Systems, and Applications. Proceedings, 2000. XII, 366 pages. 2000.

Vol. 1910: D.A. Zighed, J. Komorowski, J. Żytkow (Eds.), Principles of Data Mining and Knowledge Discovery. Proceedings, 2000. XV, 701 pages. 2000.

Vol. 1919: M. Ojeda-Aciego, I.P. de Guzman, G. Brewka, L. Moniz Pereira (Eds.), Logics in Artificial Intelligence. Proceedings, 2000. XI, 407 pages. 2000.

Vol. 1925: J. Cussens, S. Džeroski (Eds.), Learning Language in Logic. X, 301 pages 2000.

Vol. 1932: Z.W. Raś, S. Ohsuga (Eds.), Foundations of Intelligent Systems. Proceedings, 2000. XII, 646 pages.

Vol. 1934: J.S. White (Ed.), Envisioning Machuine Translation in the Information Future. Proceedings, 2000. XV, 254 pages. 2000.

Vol. 1937: R. Dieng, O. Corby (Eds.), Knowledge Engineering and Knowledge Management. Proceedings, 2000. XIII, 457 pages. 2000.

Lecture Notes in Computer Science

Vol. 1899: H.-H. Nagel, F.J. Perales López (Eds.), Articulated Motion and Deformable Objects. Proceedings, 2000. X, 183 pages. 2000.

Vol. 1900: A. Bode, T. Ludwig, W. Karl, R. Wismüller (Eds.), Euro-Par 2000 Parallel Processing. Proceedings, 2000. XXXV, 1368 pages. 2000.

Vol. 1901: O. Etzion, P. Scheuermann (Eds.), Cooperative Information Systems. Proceedings, 2000. XI, 336 pages. 2000.

Vol. 1902: P. Sojka, I. Kopeček, K. Pala (Eds.), Text, Speech and Dialogue. Proceedings, 2000. XIII, 463 pages. 2000. (Subseries LNAI).

Vol. 1903: S. Reich, K.M. Anderson (Eds.), Open Hypermedia Systems and Structural Computing. Proceedings, 2000. VIII, 187 pages. 2000.

Vol. 1904: S.A. Cerri, D. Dochev (Eds.), Artificial Intelligence: Methodology, Systems, and Applications. Proceedings, 2000. XII, 366 pages. 2000. (Subseries LNAI).

Vol. 1905: H. Scholten, M.J. van Sinderen (Eds.), Interactive Distributed Multimedia Systems and Telecommunication Services. Proceedings, 2000. XI, 306 pages. 2000.

Vol. 1906: A. Porto, G.-C. Roman (Eds.), Coordination Languages and Models. Proceedings, 2000. IX, 353 pages. 2000.

Vol. 1907: H. Debar, L. Mé, S.F. Wu (Eds.), Recent Advances in Intrusion Detection. Proceedings, 2000. X, 227 pages. 2000.

Vol. 1908: J. Dongarra, P. Kacsuk, N. Podhorszki (Eds.), Recent Advances in Parallel Virtual Machine and Message Passing Interface. Proceedings, 2000. XV, 364 pages. 2000.

Vol. 1910: D.A. Zighed, J. Komorowski, J. Żytkow (Eds.), Principles of Data Mining and Knowledge Discovery. Proceedings, 2000. XV, 701 pages. 2000. (Subseries LNAI).

Vol. 1912: Y. Gurevich, P.W. Kutter, M. Odersky, L. Thiele (Eds.), Abstract State Machines. Proceedings, 2000. X, 381 pages. 2000.

Vol. 1913: K. Jansen, S. Khuller (Eds.), Approximation Algorithms for Combinatorial Optimization. Proceedings, 2000. IX, 275 pages. 2000.

Vol. 1914: M. Herlihy (Ed.), Distributed Computing. Proceedings, 2000. VIII, 389 pages. 2000.

Vol. 1917: M. Schoenauer, K. Deb, G. Rudolph, X. Yao, E. Lutton, J.J. Merelo, H.-P. Schwefel (Eds.), Parallel Problem Solving from Nature – PPSN VI. Proceedings, 2000. XXI, 914 pages. 2000.

Vol. 1918: D. Soudris, P. Pirsch, E. Barke (Eds.), Integrated Circuit Design. Proceedings, 2000. XII, 338 pages. 2000.

Vol. 1919: M. Ojeda-Aciego, I.P. de Guzman, G. Brewka, L. Moniz Pereira (Eds.), Logics in Artificial Intelligence. Proceedings, 2000. XI, 407 pages. 2000. (Subseries LNAI).

Vol. 1920: A.H.F. Laender, S.W. Liddle, V.C. Storey (Eds.), Conceptual Modeling – ER 2000. Proceedings, 2000. XV, 588 pages. 2000.

Vol. 1921: S.W. Liddle, H.C. Mayr, B. Thalheim (Eds.), Conceptual Modeling for E-Business and the Web. Proceedings, 2000. X, 179 pages. 2000.

Vol. 1922: J. Crowcroft, J. Roberts, M.I. Smirnov (Eds.), Quality of Future Internet Services. Proceedings, 2000. XI, 368 pages. 2000.

Vol. 1923: J. Borbinha, T. Baker (Eds.), Research and Advanced Technology for Digital Libraries. Proceedings, 2000. XVII, 513 pages. 2000.

Vol. 1924: W. Taha (Ed.), Semantics, Applications, and Implementation of Program Generation. Proceedings, 2000. VIII, 231 pages. 2000.

Vol. 1925: J. Cussens, S. Džeroski (Eds.), Learning Language in Logic. X, 301 pages 2000. (Subseries LNAI).

Vol. 1926: M. Joseph (Ed.), Formal Techniques in Real-Time and Fault-Tolerant Systems. Proceedings, 2000. X, 305 pages. 2000.

Vol. 1927: P. Thomas, H.W. Gellersen, (Eds.), Handheld and Ubiquitous Computing. Proceedings, 2000. X, 249 pages. 2000.

Vol. 1931: E. Horlait (Ed.), Mobile Agents for Telecommunication Applications. Proceedings, 2000. IX, 271 pages. 2000.

Vol. 1766: M. Jazayeri, R.G.K. Loos, D.R. Musser (Eds.), Generic Programming. Proceedings, 1998. X, 269 pages. 2000.

Vol. 1791: D. Fensel, Problem-Solving Methods. XII, 153 pages. 2000. (Subseries LNAI).

Vol. 1932: Z.W. Raś, S. Ohsuga (Eds.), Foundations of Intelligent Systems. Proceedings, 2000. XII, 646 pages. (Subseries LNAI).

Vol. 1933: R.W. Brause, E. Hanisch (Eds.), Medical Data Analysis. Proceedings, 2000. XI, 316 pages. 2000.

Vol. 1934: J.S. White (Ed.), Envisioning Machine Translation in the Information Future. Proceedings, 2000. XV, 254 pages. 2000. (Subseries LNAI).

Vol. 1937: R. Dieng, O. Corby (Eds.), Knowledge Engineering and Knowledge Management. Proceedings, 2000. XIII, 457 pages. 2000. (Subseries LNAI).

Vol. 1938: S. Rao, K.I. Sletta (Eds.), Next Generation Networks. Proceedings, 2000. XI, 392 pages. 2000.

Vol. 1939: A. Evans, S. Kent (Eds.), «UML» – The Unified Modeling Language. Proceedings, 2000. XIV, 572 pages. 2000.